Simulation of Industrial Processes for Control Engineers

Simulation of Industrial Processes for Control Engineers

Philip Thomas
Honorary Visiting Professor,
Department of Electrical, Electronic and Information Engineering, City University, London

OXFORD AUCKLAND BOSTON JOHANNESBURG MELBOURNE NEW DELHI

Butterworth-Heinemann
Linacre House, Jordan Hill, Oxford OX2 8DP
225 Wildwood Avenue, Woburn, MA 01801-2041
A division of Reed Educational and Professional Publishing Ltd

A member of the Reed Elsevier plc group

First published 1999

British Library Cataloguing in Publication Data
A catalogue record for this book is available from the British Library

Library of Congress Cataloguing in Publication Data
A catalogue record for this book is available from the Library of Congress

ISBN 0 7506 4161 4

Typeset in 10/12pt Times by Laser Words, Madras, India
Printed in Great Britain by The Bath Press, Bath.

Contents

Foreword

by Prof. Dr.-Ing. Dr. h.c. mult. Paul M. Frank, Gerhard-Mercator-Universität, Duisburg, Germany

Mathematical modelling and simulation are of fundamental importance in automatic control. They form the backbone of the analytical design methodology for open-loop and closed-loop control systems. They represent the first step that a control engineer has to take when he has the task of designing a control system for a given plant. Not only is the analytical model an essential part of the design method, it is also indispensable in the analysis of the resulting control concept. On the one hand, it is needed for the analysis of stability and robustness of the control system, on the other hand it is used for the (nowadays exclusively digital) computer simulation of the plant in order to perform an online check of the resulting electronic controller within the closed-loop control systems.

Besides this, mathematical modelling and simulation play an increasing role in computer-aided approaches for control systems design and optimization. Due to the present tremendous progress in computer technology, analytical optimization techniques are being more and more replaced by systematic trial and error methods and evolutionary algorithms using digital simulations of the processes. There is a clear trend at the moment towards such computer-assisted approaches. This implies that mathematical modelling and simulation as a pre-condition will gain increasing importance. This is especially true for the field of automation and optimization in the chemical and process industries, because here it is common for the plants and their models to be rather complex and non-linear, so that analytical design and optimization techniques fail or at least are extremely cumbersome. Maybe it is no exaggeration to anticipate that in the future the mathematical model will belong within the technical specification of any dynamic device used in a technical plant.

The work of Professor Thomas is a highly important contribution to the attainment of these objectives in the field of process engineering. On the solid grounds of his long practical experience and expertise in the design of process control systems, he uses the systematic approach to modelling and simulation of dynamical systems in the process industries, ranging from the detailed understanding of the physical processes occurring on the plant to the codification of this understanding into a consistant and complete set of descriptive equations. With thoroughness and lucidity, the text explains how to simulate the dynamic behaviour of the major unit processes found in the chemical, oil, gas and power industries. Determined attempts have been made to derive the descriptive equations from the balance equations – the first principles – in a clear, step by step, systematic manner, with every stage of the argument included. Thus, the book contributes to both the simulation of industrial plants by control engineers and a deep understanding of the quantitative relationships that govern the physical processes. Reflecting his exceptionally broad expertise in a wide variety of areas in applied control theory, systems theory and engineering, Professor Thomas's treatment of modelling and simulation of industrial processes casts much light on the underlying theory and enables him to extend it in many important directions.

The present volume is concerned, in the main, with the fundamental concepts of dynamic simulation – including thermodynamics and balance equations – and their application to the great variety of processes and their components in the process industries. This provides indeed a good grounding for all those wishing to apply dynamic simulations for industrial process plant control. It serves for both undergraduate engineering students in electrical, mechanical and chemical engineering specializing in process control, starting from their second year, and for postgraduate control engineering students. However, it may also be considered as a very valuable reference book and practical help to control and chemical engineers already working in industry. The great variety of subsystems and technical devices occurring in plants of chemical and process industry are tackled in full detail and can be used directly to setup digital computer programms. Therefore, the book can be highly recommended to practical control engineers in this field.

Professor Thomas's treatise is clearly a very important and comprehensive accomplishment. It deepens the understanding of the dynamic behaviour of technical plants and their components and stimulates a more extensive application of modelling and simulation in the field of the process industries.

Notation

The wide range of subjects covered by the book causes occasional problems with duplication of symbols. Use has been made of generally recognized notation wherever possible, and normally the meaning of each symbol is clear enough in its context. However, a particular difficulty arises in any process engineering text from conflicting demands for the use of the letter v: both specific volume and velocity have strong claims. It has been decided in this book to use v to denote specific volume, and to assign to velocity the symbol, c, on the basis that c has an association with speed for most scientists and engineers, albeit the speed of light. SI units are assumed.

Symbols

Symbol	*Meaning*	*Units*
a	stoichiometric coefficient	
a_{ij}	stoichiometric coefficient of the ith component in the jth reaction	
a_j	jth nominally constant parameter	
$\hat{a}_j$	value of jth nominally constant parameter expected in advance	
A	cross-sectional area	m^2
A	constant used in Antoine equation for vapour pressure	
A_i	mass of chemical	kmol
A_{nm}	constants used in Margules correlation for distillation	
A_t	throat area of nozzle; effective throat area of valve at a given valve opening	m^2
A_T	effective throat area of valve at fully open	m^2
$\{A\}$	6.023×10^{26} molecules of chemical A, = one kilogram-mole's worth	
$[A]$	concentration of chemical A	$kmol/m^3$
$\mathbf{a}$	vector of constant parameters	
$\mathbf{a}_{opt}$	vector of optimally chosen constant parameters	
$\mathbf{A}$	$n \times n$ state matrix for a linear system	
$\mathbf{A}^{(i)}$	matrix associated with distillation plate i	
b	constant	
b	stoichiometric coefficient	
b	half of velocity deadband, half of backlash width	
b_0	adjustment coefficient used with pipeflow function	
B	constant used in Antoine equation for vapour pressure	
B_{ij}	boiloff rate of component j from the liquid in plate i	kmol/s
$\mathbf{b}^{(i)}$	vector of boiloff rates on plate i	
$\mathbf{B}$	$n \times l$ input matrix for a linear system	
c	signal produced by controller	
c	velocity	m/s
c	stoichiometric coefficient	
$c_{a,j}$	gain of filter for white noise for parameter j	
c_B	average linear speed of turbine blade	m/s
c_c	critical velocity = speed of sound at local conditions	m/s
c_{ij}	gain of transfer function, g_{ij}	
$c_{\max}$	maximum value of controller output signal	
$c_{\min}$	minimum value of controller output signal	
c_n	neutron speed	m/s
c_p	specific heat at constant pressure	J/(kg K)
c_{ri}	velocity of incoming gas relative to turbine blade	m/s
c_{ro}	velocity of outgoing gas relative to turbine blade	m/s
c_{son}	speed of sound in the fluid	m/s
c_v	specific heat at constant volume	J/(kg K)
$\mathbf{c}^{(i)}$	vector associated with distillation plate i	
C	conductance	m^2
C	constant used in Antoine equation for vapour pressure	
C_1	$= C_g/C_v$, ratio of valve gas flow conductance to liquid flow conductance at a given valve opening	
C_1^*	$= C_g^*/C_v^*$, ratio of gas sizing coefficient to liquid sizing coefficient, both at a given valve opening	$[(scf/US\ gall).(min/h)/(psi)^{1/2}]$

Symbol	Definition	Units
C_{1t}	$= C_g/C_{vt}$, ratio of valve gas flow conductance to liquid flow conductance for the valve as far as the throat only. Both conductances at a given valve opening	
C^*_{1t}	$= C^*_g/C^*_{vt}$, ratio of gas sizing coefficient to liquid sizing coefficient, for the valve as far as the throat only, both at a given valve opening	[(scf/US gall). (min/h)/ (psi)$^{1/2}$]
C_d	discharge coefficient	
C_{fgh}	valve friction coefficient for gas at high-pressure ratios	
C_{FGH}	valve friction coefficient for gas at high-pressure ratios at fully open	
C_{fl}	valve friction coefficient for liquid flow	
C_g	$= yC_G$, gas flow conductance at valve opening, y	m^2
C^*_g	gas sizing coefficient at a given valve opening	scf/h/psi
C_G	gas flow conductance for fully open valve	m^2
C_i	concentration of precursor group i	nuclei/ m^3
C_I	value of C_I when the valve is fully open: $C_I = C_G/C_V$	
C_L	line conductance	m^2
C_T	total conductance of line plus valves and fittings	m^2
C_v	$= yC_V$, liquid flow conductance at valve opening, y	m^2
C^*_v	liquid sizing coefficient at a given valve opening, equal to the valve capacity for water at 60°F	[US gall/ min/ (psi)$^{1/2}$]
C_V	liquid flow conductance for fully open valve	m^2
C'_V	constant of proportionality for fully open valve, assuming that the differential pressure and specific volume are constant	kg/s
C_{vel}	ratio of measured velocity downstream of nozzle to the velocity that would have occurred if the expansion had been isentropic	
C_{vt}	liquid flow conductance at a given valve opening for the valve as far as the throat only	m^2
C_{VT}	valve conductance to the valve throat at fully open	m^2
C^*_{vt}	liquid sizing coefficient at a given valve opening, for the valve as far as the throat only	[US gall/ min/ (psi)$^{1/2}$]
d	constant defined by local text	
d_j	weighting fraction	
D	derivative term in controller output signal	
D	diameter	m
D	constant used in Riedel equation for vapour pressure	
D_i	specific enthalpy drop across the ith stage of a turbine under isentropic conditions	J/kg
D_j	average partial heat of solution of component j	J/kmol
D_v	valve size	m
dF	work done against friction in the small element by unit mass of the working fluid	J/kg
dq	heat flux into the small element per unit mass flow = heat input per unit mass of the working fluid	J/kg
dw	useful power abstracted from the small element per unit mass flow = useful work done by unit mass of the working fluid	J/kg
e	error, =difference between measured variable and setpoint	
e_m	error term after modification by limiting	
E	energy	J
E	expression involved in estimating the pressure ratio across the valve that will lead to choked gas flow	
E	activation energy for reaction	J/kmol
E	sum of the squared flow errors	kg^2/s^2
E_i	total vapour flow from distillation plate i to plate $i+1$	kmol/s
$\mathbf{e}$	vector of differences between model and plant measured transients	
$\mathbf{e}_{\min}$	vector of differences between model and plant measured transients with the optimal set of constant parameters	
f	Fanning friction factor	
f	function	
f_b	multiplying factor to account for the additional metal contained in the baffles, assumed to be at the same temperature as the heat exchanger shell	

Symbol	Meaning	Units
f_{comb}	combination function, combining f_{hpr} and f_{lpr}	
f_F	function derived from Fisher Universal Gas Sizing Equation	
f_{flow}	generalized mass-flow function	kg/s
f_{hpr}	high-pressure-ratio function	
f_{lpa}	long-pipe approximation flow function	
f_{lpr}	low-pressure-ratio function	
f_{LG}	liquid–gas function, used to approximate gas flow through a valve by analogy with the liquid flow case	
f_{noz}	nozzle flow function	
f_{NV}	nozzle–valve function used to model gas flow through the valve by analogy with nozzle flow	
f_{NVA}	approximating function for f_{NV}	
f_{pipe}	pipeflow function	
f_{P1}	function relating head to volume flow at design speed for a centrifugal pump	m
f_{P2}	function relating pump power demand to volume flow at design speed for a centrifugal pump	W
f_{P3}	efficiency function, dependent on volume flow and speed for a centrifugal pump	
f_{shock}	shock correction factor for blade efficiency	
F	frictional loss per unit mass of the working fluid along whole length of the pipe	J/kg
F	force	N
F	mass flow in kilogram-mole units	kmol/s
F	fission rate of the reactor per m^3 of fuel	
F_{Li}	liquid feed flow to plate i	kmol/s
$\mathbf{f}$	n-dimensional vector function of the state, $\mathbf{x}$, and forcing variables, $\mathbf{u}$	
g	acceleration due to gravity	m/s^2
g	function	
g_{corr}	neutron thermalization correction factor	
$g_{ij}(s)$	elemement of transfer function matrix, $\mathbf{G}(s)$	
G	mass velocity, $= W/A$	$kg/(m^2\ s)$
G	specific gravity with respect to water at 60°F	
G_a	specific gravity of gas with respect to air at same temperature	
G_j	constant used in converting activity coefficient for component j to a different temperature range	K
$\mathbf{g}$	vector function dependent on the vector $\mathbf{z}$	
$\mathbf{G(s)}$	transfer function matrix	
h	specific enthalpy	J/kg
h	sum of weighted squared deviations	
H	pump head	m
H	Lagrange function	
H_p	polytropic head	J/kg
H_s	isentropic head	J/kg
$\mathbf{h}$	vector function	
i	$\sqrt{-1}$	
I	integral term in controller output signal	
I_A	adjusted value of integral term	
I_D	desaturated integral term	
j	general integer index	
J	moment of inertia	kgm^2
$\mathbf{J}$	Jacobian state matrix	
$\mathbf{J_a}$	Jacobian matrix for parameter variations in companion model	
$\mathbf{J_B}$	Jacobian input matrix	
$\mathbf{J_x}$	Jacobian state matrix for companion model	
k	controller gain	
k	general constant, meaning dependent on local text	
k	forward velocity constant	
k	multiplication factor for the nuclear reactor	
k'	backward velocity constant	
k_∞	frequency factor for the reaction	
k_d	delayed neutron component of multiplication factor	
k_{di}	component of multiplication factor associated with delayed neutron group i	
k_p	prompt neutron component of multiplication factor	
k_α	$= \Delta a_j^{(i)}/(\alpha_{jd,1} + \alpha_{jd,2})$	
K	vapour pressure function	Pa
K	energy loss in velocity heads	
K	heat transfer coefficient	$W/(m^2\ K)$
K_b	energy loss in velocity heads due to bend or fitting	
K_c	cavitation coefficient for a rotary valve at a given valve opening	
K_c	effective thermal conductivity of catalyst bed	W/m

Symbol	Description	Units
K_C	cavitation coefficient for a rotary valve at fully open	
K_{con}	energy loss in velocity heads due to contraction at the inlet	
K_f	energy loss in velocity heads due to pipe friction	
K_m	pressure recovery coefficient for liquid flow through valve at a given opening	
K_M	pressure recovery coefficient for liquid flow through valve at fully open	
K_T	total energy loss in velocity heads	
K_v	energy loss in velocity heads due to valve at a given opening	
K_{V0}	energy loss in velocity heads at the fully open valve when the valve size matches the pipe diameter	
l	dimension of vector of forcing functions	
l	level	m
l	average neutron lifetime	s
L	length of component or pipe	m
L_{eff}	effective pipelength	m
L_i	total liquid flow from distillation plate i to plate i-1	kmol/s
m	mass	kg
m	polytropic exponent for frictionally resisted adiabatic expansion	
m_0	polytropic exponent for frictionally resisted adiabatic expansion over the convergent part of the nozzle	
m_{0c}	polytropic exponent for frictionally resisted adiabatic expansion over the convergent part of the nozzle when the flow is critical	
m_{0D}	polytropic exponent for frictionally resisted adiabatic expansion over the convergent part of the nozzle for the design flow	
m_b	coefficient used in calculating pipe-flow coefficient, b_0, at different pressure ratios	
M	mass in kilogram-moles	kmol
M_A	mass of chemical A	kmol
M_B	mass of chemical B	kmol
M_C	mass of chemical C	kmol
M_{Li}	total liquid mass in distillation plate i	kmol
M_{Lij}	mass of component j in the liquid phase in distillation plate i	kmol
M_R	kilogram-moles of reaction	kmol rxn
M_{Vi}	total vapour mass in distillation plate i	kmol
M_{Vij}	mass of component j in the vapour phase in distillation plate i	kmol
$\mathbf{M}$	Mach number = ratio of velocity to the local sound velocity	
n	general index	
n	dimension of state vector	
n	polytropic index of gas expansion	
n	concentration of neutrons	neutrons/m^3
n_d	concentration of delayed neutrons	neutrons/m^3
n_{di}	concentration of delayed neutrons in group i	neutrons/m^3
n_G	number of neutrons in one of the M groups	neutrons/m^3
n_p	concentration of prompt neutrons	neutrons/m^3
N	number of cells	
N	rotational speed in revolutions per second	r/s
N_{AK}	number of molecules in a kilogram-mole, $= 6.023 \times 10^{26}$ (Avogadro's number $\times$ 1000)	
N_f	concentration of fissile nuclei	nuclei/m^3
N_F	number of degrees of freedom for a gas	
N_{RE}	Reynolds number	
p	pressure	Pa
p_{cp}	pressure at the critical point for the fluid (point of indefinite transition between liquid and vapour)	Pa
p_{ij}	partial pressure of component j in distillation plate i	Pa
p_t	throat pressure for the nozzle or valve	Pa
p_{tcav}	throat pressure at cavitation	Pa
p_{tvap}	vapour pressure at the valve throat temperature	Pa
p_{vap}	vapour pressure	Pa
P	power	W
P	proportional term in controller output signal	
P_D	power demanded by the pump	W

Symbol	Meaning	Unit
P_F	power expended against friction	W
P_{IMP}	power expended by the impeller	W
P_m	modified proportional term	
P_P	pumping power, i.e. useful power spent in raising the pressure of the fluid	W
P_S	power supplied to the pump	W
q	quality of steam	
Q	heat	J
Q	volumetric flow rate	m^3/s
Q^*	volume flow in US gallons per min	US gall/min
Q_{cfh}	volume flow in cubic feet per hour	ft^3/h
Q_{crit}	critical or choked flow of gas through the valve	standard ft^3/h
Q_{scfh}	equivalent volume flow in standard cubic feet per hour	standard ft^3/h
r	ratio of pressures at stations '1' and '2'	
r_c	critical pressure ratio for a gas	
r_j	reaction rate density, referred to the volume of the packed bed	kmol rxn/(m^3 s)
$r_{\lim}$	fraction of pressure ratio down to which an ordered expansion can occur	
r_{pvc}	ratio of the pressure at valve inlet to the pressure at the critical point for the fluid	
r_{vap}	ratio of valve throat pressure at a given opening to the vapour pressure of the fluid at the valve-inlet temperature	
R	universal gas constant, value = 8314	J/(kmolK)
R	remainder term, equal to the adjusted integral term less the integral term	
R_1	exponentially lagged version of the remainder term	
R_B	ratio of blade speed to incoming gas speed	
R_i	rate of radioactive decay of precursor group i	nuclei/(m^3 s)
R_v	valve rangeability, = ratio of maximum to minimum valve opening	
R_w	characteristic gas constant, $= R/w$	J/(kg K)
s	specific entropy	J/(kg K)
S	stiffness	
S	entropy	J/K
S_i	total sidestream flow extracted from distillation plate i	kmol/s
t	time	s
$t_{1/2,i}$	half-life of delayed neutron precursor group i	s
T	time constant	s
T	temperature	K
T_d	derivative action time	s
T_i	integral action time or reset time	s
u	specific internal energy	J/kg
U	$= QN_0/N$, ratio of flow to normalized speed	m^3/s
U_i	total internal energy of the contents of distillation plate i	J
$\mathbf{u}$	l-dimensional vector of forcing variables	
v	specific volume	m^3/kg
V	volume	m^3
V_{scf}	volume at standard conditions (pressure = 14.7 psia, T = 520°R) of an arbitrary mass of gas that has volume V at arbitrary conditions p_1, T_1	standard cubic feet
w	molecular weight	
w_p	polytropic specific work	J/kg
w_s	isentropic specific work	J/kg
W	mass flow	kg/s
W_c	critical flow for a gas	kg/s
W_{cav}	cavitating flow for liquid through a rotary valve	kg/s
W_{choke}	choking flow of liquid through a valve	kg/s
x	(fractional) valve travel, fully shut = 0, fully open = 1	
x	distance	m
x_{ij}	mole fraction of component j in the liquid phase in distillation plate i	
x_o	steam dryness fraction at the start of the expansion	
$\mathbf{x}$	n-dimensional vector of system states	
$\mathbf{X}$	n-dimensional vector of system states driven with variations in the nominally constant parameters	
y	(fractional) valve opening, fully shut = 0, fully open = 1	
y_{ij}	mole fraction of component j in the vapour phase in distillation plate i	
Y	expansion factor	
$\mathbf{y}$	k-dimensional vector of model outputs	

Symbol	Meaning	Units
$\mathbf{Y}$	k-dimensional vector of model outputs when the model is driven with variations in the nominally constant parameters	
z	height relative to datum	m
z_{Lij}	mole fraction of component j in the liquid feed to distillation plate i	
Z	compressibility factor, dependent on temperature and pressure. $Z = 1$ for an ideal gas	
$\mathbf{z}$	vector of unknowns defined by nonlinear, simultaneous equations $\mathbf{g}(\mathbf{z}) = \mathbf{0}$	
$\mathbf{z}$	vector of plant transient measurements	
α	power to which concentration of chemical A is raised in forward reaction	
α'	power to which concentration of chemical A is raised in backward reaction	
α_1	angle of turbine nozzle, measured relative to the direction of turbine wheel motion	degrees
α_1	composite term for net heat input to boiling vessel	W
α_2	angle of gas stream leaving turbine stage, measured relative to the direction of turbine wheel motion	degrees
α_2	composite term for total heat capacity of contents of boiling vessel	J/K
$\alpha_n^{(i)}$, $\alpha_n^{(ij)}$ ($n = 1$ to 7)	combinations of variables for distillation plate i, sometimes making particular reference to component j.	
α_{jd}	steady deviation from optimal value of nominally constant parameter j that causes the mean squared error to double	
α_{js}	steady deviation from optimal value of nominally constant parameter j	
$\boldsymbol{\alpha}$	vector of variations to constants, $\mathbf{a}$	
β	constant used in polynomial	
β	delayed neutron fraction	
β	power to which concentration of chemical B is raised in forward reaction	
β'	power to which concentration of chemical B is raised in backward reaction	
β_1	blade inlet angle	degrees
β_2	blade outlet angle	degrees
β_i	delayed neutron fraction for group i	
β_{in}	angle of approach to turbine blade of the incoming gas jet	degrees
γ	ratio of the specific heats, c_p/c_v, = index for isentropic expansion for a gas	
γ	power to which concentration of chemical C is raised in forward reaction	
γ'	power to which concentration of chemical C is raised in backward reaction	
γ_{ij}	activity of component j on distillation plate i	
δ	small increment of quantity following	
$(\delta a_j)^2$	contributory variance of parameter j	
δC_{fi}	production of nuclei of delayed neutron precursor group i due to absorption of neutrons in a fission event	nuclei
δM_R	increase in the kilogram-moles of reaction	kmol rxn
Δ	denoting incremental quantity of variable following	
$\Delta a_j^{(i)}$	standard deviation required by parameter j acting on its own to match measured variable i	
Δc_w	change of speed in the direction of turbine wheel motion	m/s
Δh_N	actual change in specific enthalpy through the nozzle	J/kg
Δh_{Ns}	change in specific enthalpy through the nozzle for an isentropic expansion	J/kg
Δp	differential pressure	Pa
Δt	integration time-step	s
ΔH_j	enthalpy of reaction j	J/kmol rxn
ΔU_j	internal energy of reaction j	J/kmol rxn
ε	a measure of the average height of the excrescences on the pipe surface	m
ε	reactor elongation factor	
ε/D	relative roughness of the pipe surface	
ζ_{ij}	damping factor in transfer function $g_{ij}(s)$, associated with output i and parameter j	
η	efficiency	
η_B	blade efficiency	

η_{Ba}	blade efficiency when there is no entry loss	
η_{BN}	nozzle efficiency for the expansion taking place in the moving blades of a reaction stage	
η_c	distillation column efficiency	
η_N	nozzle efficiency	
η_P	pump efficiency	
η_S	stage efficiency	
θ	angle	rad
θ_i	height of the liquid on tray i	m
θ_m	measured value of plant variable, θ_p	
θ_p	plant variable	
θ_s	setpoint for plant variable, θ_p	
θ_w	height of the weir on distillation tray i	m
κ	bulk modulus of elasticity of the fluid	Pa
λ	eigenvalue	s^{-1}
λ	molar fraction	
μ	dynamic viscosity	Pa s
μ	constant used in polynomial	
μ	mass fraction	
ξ	nuclear power density averaged over the core	W/m^3
π	constant used in pressure ratio polynomial	
ρ	degree of reaction in a turbine stage	
ρ	reactivity	dollars, niles
σ_f	effective cross-sectional area for fission of each fissile nucleus	m^2
$\sigma^2_{\alpha,j}$	variance of the nominally constant parameter, j	
σ^2_θ	variance to be associated with predicted variable, θ	
$\sigma^2_{\psi,i}$	variance of the companion model output, i	
τ	time constant	s
τ	frictional shear stress	N/m^2
τ_j	standard deviation expected in advance for parameter j	
τ_{stroke}	valve stroking time	s
ϕ	heat flux per unit length	W/m
$\phi(T)$	'phi', $= \int_0^T (c_p/T)\,dT$, the temperature-dependent component of specific entropy	J/(kg K)
Φ	heat flux	W
Φ	white noise intensity	
$\boldsymbol{\phi}$	state difference vector: $\mathbf{X} - \mathbf{x}$	
χ_{ave}	reactor flux averaged over the complete core	neutrons/ $(m^2\,s)$
$\chi^{(i)}$	vector of state deviations associated with state subvector $\mathbf{x}^{(i)}$	
ψ	useful power extracted per unit length	W/m
$\boldsymbol{\psi}$	vector of outputs of companion model	
ω	rotational speed in radians per second	rad/s
$\omega_{a,j}$	break frequency defining frequency content of the variation of parameter j	rad/s
ω_{ij}	undamped natural frequency of tranfer function g_{ij} relating the variance of output i to nominally constant parameter, j	rad/s

Additional subscripts and superscripts

0	at time zero
0	at datum position
0	over convergent part of nozzle
0	at design conditions
0	model matching
0c	over convergent part of nozzle in critical conditions
0D	over convergent part of nozzle in design conditions
0,1,2, 3...	enumerative identifiers
1	at upstream station or inlet
2	at downstream station or outlet
a	station identifier
a	air
at	atmospheric
ave	average
b	due to bends and fittings
b	station identifier
B	'boiloff' or evaporation; condensation when flow is negative
B	blade
c	critical or choked
c	of catalyst bed pellets
c	of the controller
cc	from fuel-pin cladding to coolant
$clad$	of the fuel-pin cladding
con	contraction
$cool$	of the coolant
$crit$	critical
cs	critical and isentropic
C	relating to the distillate side of the distillation column condenser
d	demanded

d	downcomer
d	normally downstream
di	downcomer inlet
do	downcomer outlet
D	at design conditions
e	evaporator
eff	effective
f	friction
f	feed
f	of the fuel
fc	between fuel and cladding
fuel	fuel
F	friction
g	gas
G	denotes fully open valve, for gas flow
G	gas
i	general index
i	for isothermal expansion
in	inlet
j	general index
k	general index
liq	liquid
L	liquid
L	over the whole length
m	metalwork
nuc	nuclear
N	nozzle
opt	optimal
out	outlet
over	overall
p	at constant pressure
p	for a polytropic expansion
P	needed in practice
P	pump
pins	of the fuel pins
prior	prior
r	riser
r	due to the reaction
recalc	recalculated
rev	reversibly
ri	relative at the inlet
ro	relative at the outlet
s	setpoint
s	shell-side
s	under conditions of constant entropy
sa	under conditions of constant entropy from mid-stage to stage outlet
si	shell inside
stroke	associated with the stroke of the valve
sw	shell wall
sws	shell wall to shell-side fluid
sys	system
S	stage
t	throat
t	total
t	tube-side
ti	tube, inside
to	tube, outside
trans	transferred
tw	tube wall
twt	from tube wall to tube-side fluid
T	total
T	in the stagnation state
T	theoretical
tc	throat, critical
tot	total
tray	associated with the distillation tray
up	normally upstream
v	associated with a valve; associated with liquid flow through valve
v	at constant volume
vap	vapour
V	vapour
V	denotes fully open valve, for liquid flow
vc	vena contracta
w	wall
w	water
w	in the direction of turbine wheel motion
z	due to height difference
$\wedge$	specified per kilogram-mole
*	in US units

1 Introduction

Much of control engineering literature has concentrated on the problem of controlling a plant when a mathematical model of that plant is at hand, at which time a large number of effective techniques become available to help design the control system. Unfortunately for the control engineer working in the process industries, the assumption that mathematical models exist for his plants is seriously flawed in practice. Coming to a plant for the first time, the best the control engineer can realistically expect is steady-state models for a subset of the key plant items, perhaps supplemented by steady-state, plant-performance data if the plant has begun operating.

The predominantly steady-state nature of most available models arises from their origin as tools for the design engineers. The design engineer will be concerned almost exclusively with producing a flowsheet for a single operating point. Very properly, he will wish to optimize the performance of the plant at that point, first through choosing the right structure for the plant and then by specifying the right equipment, including the right sizes of pipe, of pumps, of chemical reactor and so on. This will be a complicated, iterative process and, to simplify it, the designer will normally assign to the control engineer the equally difficult job of ensuring that the plant as designed will remain at the operating point that has been chosen. A result of this division of labour is that the designer's mathematical model will be constructed under the assumption that solution is necessary only at the design point, so that a steady-state model suffices.

While it is desirable for the control engineer to make an early input to the design process, it is nevertheless often the case that the major items of plant equipment will have been chosen by the time the control engineer appears on the scene. Even though such a procedure may make good control more difficult (as, for example, when vessels sized for steady-state performance are too small to give ideal buffering against disturbances), the practice has the beneficial effect of reducing the 'problem space' for the dynamic simulation: the sizes and characteristics of the major plant vessels and machinery will often be fixed at the time of writing the program. However, unlike the flowsheet package used by the design engineer, the control engineer's simulation model must calculate conditions not just once at the designed-for steady-state, but over and over again as the plant's conditions change with time in response to disturbances and interactions with connected plant. Further, the design engineer's model may well neglect conditions a long way from the design point, under the implicit but overly optimistic assumption that such conditions will not be met in practice. Experience shows, however, that plants are often operated a very long way from their design points, either temporarily because of an unexpected plant upset, or at the direction of plant management, who may wish to maximize production despite part of the plant being down for maintenance. The control scheme will be expected to cope with these eventualities, and so must the control engineer's simulation model.

It may be seen from the above that the modelling and simulation task facing the control engineer is significantly different from that facing the design engineer. Some, noting the significant effort implicit in the design model when finalized for flowsheet conditions, have argued that this steady-state model can be 'dynamicized' so as to transform it into a dynamic simulation model capable of calculating transient behaviour. But such a strategy represents an attempt to move from the particular to the general, since the most general statement of the plant's physics, chemistry and engineering will be dynamic, and the steady state is just one special case. The proper starting point for the dynamic simulation model lies with the time-dependent laws for the conservation of mass, energy and momentum. It is by applying these fundamental physical principles to the unit processes making up the plant that the modeller may construct an elegant dynamic simulation that will be computationally efficient.

The current availability of a number of effective continuous simulation languages means that the control engineer has excellent tools at his disposal to set down his mathematical description into a form that will produce a time-marching simulation. Some simulation languages offer a number of advanced features in addition, such as linearization about one or more chosen operating points to produce the canonical control matrices, A, B, C and D, and numerical evaluation of the frequency responses for stability assessment and control system design. But the riches available from the present generation of continuous simulation languages should not deceive the reader into thinking that the control engineer's job has been thereby rendered nugatory. Far from it. These features will be of use only after the mathematical model has been derived. The major task facing the control engineer working in the process industries is the detailed understanding of the physical processes occurring on the plant and the codification of this understanding into a consistent and complete set of descriptive equations.

This is the background against which the book has been written. The text sets out to explain how to simulate the dynamic behaviour of the major unit processes found in the chemical, oil-and-gas and power industries. A determined attempt has been made to derive the descriptive equations from first principles in a clear, step-by-step manner, with every stage of the argument included. The book is designed allow the control engineer to simulate his industrial plant and understand quantitatively how it works.

The two chapters following introduce the subject. Chapter 2 covers the fundamental principles of dynamic simulation, including the nature of a solution in principle, model complexity, lumped and distributed systems, the problem of stiffness and ways to overcome it. Chapter 3 provides the thermodynamic background required for process simulation and derives the conservation equations for mass and energy applied to lumped systems, including the equation for the conservation of energy for a rotating component such as a turbine. The chapter goes on to apply the conservation equations for mass, energy and momentum to the important case of one-dimensional fluid flow through a pipe.

Chapters 4 through to 10 are devoted to deriving and explaining the equations for calculating the flow of fluid between plant components. Such flow may usually be assumed to be in an evolving steady state because the time constants associated with establishing flow are usually much smaller than those of the other plant components being simulated. (Situations where this assumption is untenable are covered in Chapter 19, which deals with the transient behaviour of long pipelines.) Chapter 4 deals with steady-state, incompressible flow, deriving the necessary relationships from the steady-state energy equation. The chapter introduces the Fanning friction factor, as well as pressure drops associated with bends and at pipe entry and exit. Finally an equation is presented to calculate mass flow from the pipe inlet conditions and outlet pressure, applicable to liquids and also to gases and vapours where the total pressure drop is less than about 5%.

Moving on to compressible flow, it is first of all necessary to explain the physics of flow through an ideal, frictionless nozzle. Chapter 5 shows how the behaviour of such a nozzle may be derived from the differential form of the equation for energy conservation under a variety of constraint conditions: constant specific volume, isothermal, isentropic and polytropic. The conditions for sonic flow are introduced, and the various flow formulae are compared. Chapter 6 uses the results of the previous chapter in deriving the equations for frictionally resisted, steady-state, compressible flow through a pipe under adiabatic conditions, physically the most likely case on a process plant. Full allowance is made for choked flow. The resulting equations are implicit and nonlinear, but a simple solution scheme is given, iterating on the single variable of the pressure just downstream of the effective nozzle at the pipe's entrance. A number of methods are presented to replace the implicit set of compressible flow equations with simpler, explicit equations without significant loss of accuracy. Full details of the explicit approximating functions are given in Appendix 2 for four values of the specific-heat ratio, corresponding to the cases of dry, saturated steam, superheated steam, diatomic gas and monatomic gas.

Chapter 7 describes liquid flow through a control valve, including flashing and cavitation effects. The effect of partial valve openings is covered, as well as the various forms of valve characteristic: equal percentage, butterfly, linear and quick-opening. The control valve on the plant will be preceded and succeeded by finite line conductances, and it is necessary to allow for these in calculating the effect of the control valve on flow. The situation is complicated for liquid flow by the possibilities of choking and cavitation within the valve. Chapter 8 presents an explicit procedure for calculating liquid flow from the pipe's upstream and downstream pressures.

Chapter 9 describes a model for gas flow through a control valve based on nozzle concepts, including sonic effects. The long-established Fisher Universal Gas Sizing Equation is also explained, with a detailed derivation given in Appendix 3 and a comparison with the nozzle-based model given in Appendix 4. Chapter 10 presents three methods for calculating the flow of gas through a line containing a control valve, making full allowance for potential sonic flow both in the valve and at pipe outlet. The first two methods are dependent on the satisfaction of a convergence criterion and so require an indefinite number of iterations, but the third, more approximate method allows the number of iterations to be fixed at a low number in advance.

Chapter 11 considers the accumulation of liquids and gases in process vessels, both when the temperature is constant and when it varies as a result of heat exchange. The usefulness of kilogram-mole units (kmol) in modelling gas mixtures is explained. Chapter 12 treats the more complex case of liquid and vapour mixtures in vapour–liquid equilibrium. The new Method of Referred Derivatives is employed to generate explicit solutions for the behaviour both of boiling vessels, such as are used in steam plant and refrigeration systems, and for the more complex system comprising a multicomponent distillation column. The latter set of equations allows for the use of activity coefficients, and it is proposed that the Margules correlation will give sufficient accuracy for control engineering purposes. Chapter 13 explains the principles underlying chemical reactions, generalizing these to the case of several concurrent reactions with large numbers of reagents and products. The principles of time-dependent mass and energy balance are then extended to the case of chemical reaction so that the transient behaviour can be calculated. Finally the chapter explains in detail how to simulate both a gas reaction taking place inside a reaction vessel and a liquid reaction inside a continuously stirred, tank reactor.

The next four chapters are devoted to process machines, starting with turbines. An accurate model of a turbine requires consideration of the inefficiency introduced by frictional losses in its nozzles. Chapter 14 builds on the introduction to nozzles given in Chapter 5 to allow for the effect of friction. The chapter also introduces the concept of stagnation properties of thermodynamic variables to account for the non-negligible velocities found at the nozzle inlet in a real turbine. The problem of accounting for conditions a long way from the design point is often neglected by the design engineer, but, as noted previously, can be one of great significance to the control engineer, whose control schemes will be expected to cope with potentially major deviations from the nominal operating point. New results are therefore presented on explicit methods for calculating the efficiencies of both convergent-only and convergent–divergent nozzles over the full pressure range, not just at the design point. Details of comparisons with experimental data are given in Appendix 6. Chapter 15 continues the consideration of off-design conditions, and presents new, explicit methods of calculating the efficiency of impulse and reaction blading in a turbine over the full range possible for the ratio of blade speed to gas/steam speed. The chapter goes on to list the sequence of steps necessary to calculate the power of the turbine. Chapter 16 presents a number of simplifications that can be made without degrading significantly the accuracy of the turbine-power calculation, including neglecting the effect of interstage velocities, utilizing the concept of a stage efficiency calculated as a function of the nozzle and blade efficiencies, and, when simulating a steam turbine, using simple analytic functions to approximate steam table data.

Chapter 17 describes the modelling of turbo pumps and compressors. Dimensional analysis is applied to the pump in order to derive the affinity laws from first principles. The energy equation is used to derive the differential equation describing the dynamics of pump speed, and a method of calculating the flow of liquid being pumped through a pipe is given, which can be made fully explicit if the head versus flow characteristic is approximated by a polynomial of third order or lower. The chapter goes on to explain the foundations for the two methods used to calculate the performance of a rotary compressor: the first, often used in the USA, is based on polytropic head characteristics, while the second, often used by European manufacturers, is based on the pressure ratio characteristics. Methods of modelling the flows and pressures associated with a general multistage compressor are given using each of the two performance models.

The principles for modelling flow networks with rapidly settling flow are laid out in Chapter 18, which covers both liquid and gas flow networks. The chapter begins by setting down explicit equations for combining simple parallel and series conductances and then moves on to consider more complex networks where a direct explicit solution is not available. Two methods of solution are presented. The first is iterative, based on the Newton–Raphson method. The basis of the method is explained, as are the difficulties caused to the method by the points of inflexion that are inherent in the flow equations near the point of flow reversal. The chapter explains how the flow equations may be modified with little loss of accuracy to speed up the solution. The second technique presented is based on the Method of Referred Derivatives, which converts the set of implicit, nonlinear, simultaneous equations into an equivalent set of linear equations which may be solved for the time-derivatives of the original variables, either explicitly or by Gauss elimination. Finally, the chapter shows a way of modelling liquid networks containing nodes of significant volume whose temperatures may vary.

The next two chapters deal with distributed systems. Chapter 19 considers the situation of a long pipeline, when the establishment of flow takes an appreciable time. The equations governing the dynamics of long liquid and gas pipelines are derived from first principles, based on the conservation of mass and momentum. The Method of Characteristics is explained, including how to interface it to practical boundary conditions such as pumps, in-line valves and pipe junction headers. The application of finite differences is also considered, and a practical scheme based on central differencing is outlined, together with recommendations for the spatial and temporal step-lengths. Chapter 20 derives the equations for a typical, shell-and-tube heat exchanger from the mass balance and energy balance equations for both liquids and gases. A solution sequence using finite differences is presented to calculate the dynamic performance of a counter-current heat exchanger. The chapter goes on to derive the equations governing the behaviour of a catalyst bed reactor operating on gaseous reagents. Chemical kinetics equations from Chapter 13 are combined with the equations for conservation of mass and energy in order to produce a fully dynamic model. A solution scheme based on finite differences is given.

Nuclear reactors produce nearly a fifth of the world's electricity, and so must now be accounted a common unit process in the power generation industry. Chapter 21 explains the process of nuclear fission and emphasizes the importance of delayed neutrons in both thermal and fast reactors. Neutron kinetics equations are derived from first principles based on a point model. The chapter explains the process of heat transfer to the reactor coolant, and how reactor temperature effects feed back to the neutron kinetics through the reactivity temperature coefficients.

Chapter 22 provides equations for typical process controllers and control valve dynamics. The controllers considered are the proportional controller, the proportional plus integral (PI) controller and the proportional plus integral plus derivative (PID) controller. Integral desaturation is an important feature of PI controllers, and mathematical models are produced for three different types in industrial use. The control valve is almost always the final actuator in process plan. A simple model for the transient response of the control valve is given, which makes allowance for limitations on the maximum velocity of movement. In addition, backlash and velocity deadband methods are presented to model the nonlinear effect of static friction on the valve.

The last two chapters are concerned with ensuring that the final simulation model is fit for the purpose intended. Chapter 23 deals with linearization, which provides a valuable, diverse technique for checking that the main simulation model has been programmed correctly. This is most important in the real industrial world, where the control engineer may be modelling a particular plant or plant area for the first time. The concept of linearization is relatively easy to set down, but the difficulties inherent in linearizing the equations for a complex plant should not be underestimated. Accordingly extensive examples are given, based on actual plant experience. The last chapter, Chapter 24, deals with model validation: the testing of the model, preferably as a whole, but at least in part, against empirical data. The earliest control engineering models tended to be simplified, analytic linearizations of system behaviour about an operating point, used more or less exclusively for the selection of control parameters. Not too much was expected from the dynamic model, and so the requirement for rigorous model validation, as opposed to intuitive feel, was small. Nowadays, however, the advent of massive computing power at a low cost means that more and more is expected of simulation models, beginning with control parameter selection, but moving on to trip system evaluation and safety studies on the one hand and process optimization on the other. Hence the increased importance of formal model validation. Chapter 24 describes the basis of the formal validation technique known as Model Distortion. The chapter concludes the book by explaining how the technique may be applied to real empirical data to produce a quantitative validation of the simulation model.

The text makes a feature of setting down, where appropriate, the sequence in which the modelling equations may be solved. Detailed worked examples are also provided throughout the text.

Given that literally thousands of equations are presented in total, it is appropriate to comment on the way in which the algebraic arguments have been built up. It should be observed first of all that every equation represents an enormous compression over the natural language that would have been needed to express the same idea. Despite this, I suspect that I am not alone in having noticed and indeed suffered from the custom of a good many mathematical authors whose habit is to skip lines of equations in their enthusiasm to develop an idea. Excusing themselves with such comments as 'Clearly...', or 'It is obvious that...', they proceed to omit several vital steps in the argument, forcing the reader to devote several tens of minutes chasing them down before he can get back on track, if at all. No doubt there have been many authors for whom the omitted steps were indeed obvious (at the time of writing, at least), but perhaps there have also been those who, feeling that the steps left out should have been obvious, have hesitated to provide further explanation for fear of hinting at a less than sure intuitive grasp on their own part. I myself have made no attempt to save space by omitting equations, but, on the contrary, have tried my best to put in every step. My feeling is that it is difficult enough to convey mathematical ideas without including unofficial 'exercises for the student' as deliberate pitfalls along the way! Besides, I want to be able to understand the book myself when I refer to it in future years. But inevitably there will be places where I shall have failed, and have left out a stepping stone, or worse, more than one, for which I can only crave the indulgence of the reader.

The material contained in the book is based on many years' experience of modelling and simulation in the chemical and power industries. It is intended to provide a good grounding for those wishing to program dynamic simulations for industrial process plant. It is judged to be appropriate for undergraduate engineering students (electrical, mechanical or chemical) specializing in process control in their second year or later, and for post-graduate control engineering students. It aims also to be of practical help to control and chemical engineers already working in industry. The level is suitable for control engineering simulations for industrial process plant and simulations aimed at evaluating different plant operational strategies, as well as the programming of real-time plant analysers and operator-training simulators.

2 Fundamental concepts of dynamic simulation

2.1 Introduction

This chapter introduces the basic ideas of dynamic simulation by considering a very simple unit on a process plant and showing how a mathematical model of its dynamic behaviour may be built up. This model is used to illustrate the general simulation problem, and conditions are given for when the simulation problem may be considered solved in principle. The chapter goes on to show how it is possible to produce different but equally valid models of the same plant using different state variables, and how extending the range of physical phenomena considered leads to an increase in the complexity and order of the model. The implications of modelling distributed systems are considered, and ways of introducing partial differential equations into the simulation are discussed. The problems of stiffness are reviewed and illustrated by reference to the simple unit process model. A number of different ways are then presented whereby stiff systems may be simulated without using excessive computing time.

2.2 Building up a model of a simple process-plant unit: tank liquid level

Figure 2.1 shows a tank taking in two inlet flows and giving out a single outflow. Such an arrangement might form part of an effluent conditioning system at the back end of a chemical plant, for example. The inlet flows are modulated by valves 1 and 2, while the outlet flow is modulated by valve 3. A level controller receives a measurement of level from a level transducer, compares this with its setpoint, and then sends out a control signal to adjust the travel of valve 3. The function of the level controller is to maintain the liquid level at or near the setpoint despite any deliberate changes or random fluctuations in the inlet flows.

Let us set down a set of governing equations, starting with the mass balance: the rate of change of mass in the tank *equals* the mass inflow *minus* the mass outflow or in mathematical symbols:

$$\frac{dm}{dt} = W_1 + W_2 - W_3 \tag{2.1}$$

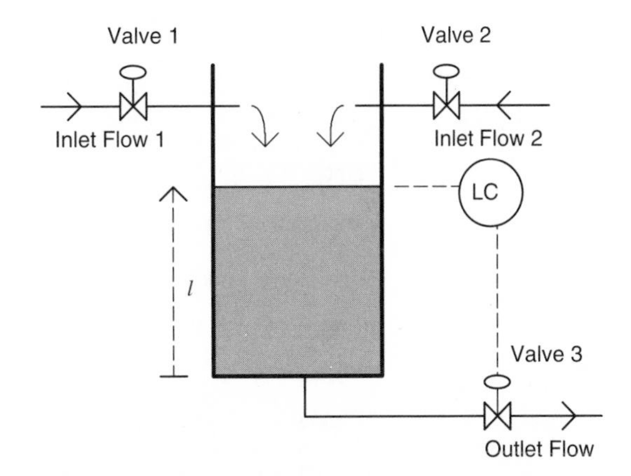

Figure 2.1 Tank liquid level.

where m is the mass of liquid in the tank (in kg), W_1 and W_2 are the inlet flows and W_3 is the outlet flow (all in kg/s). We need now to derive expressions for the flows cited in equation (2.1).

The outlet mass flow, W_3, will depend on the pressure difference across the valve, Δp (Pa), on the specific volume of the liquid in the tank, v (m^3/kg), and on the fractional valve opening of valve 3, y_3, defined as the ratio of the valve's existing flow area to its flow area when fully open. Using a general expression for flow through a valve that will be derived later in the book, W_3 may be written as:

$$W_3 = C_{V3} y_3 \sqrt{\frac{\Delta p}{v}} \tag{2.2}$$

Here C_{V3} is the valve's conductance at fully open (m^2) (see Chapter 7 for a full discussion of the flow through control valves).

For this model, we will assume for simplicity that changes in the differential pressures across inlet valves 1 and 2 are insignificant and that the specific volume of neither inlet stream varies. Then the mass flows W_1 and W_2 will depend solely on the fractional valve openings, y_1 and y_2:

$$W_1 = C'_{V1} y_1 \tag{2.3}$$

$$W_2 = C'_{V2} y_2 \tag{2.4}$$

where C'_{V1} and C'_{V2} are constants. If the pressures above the liquid in the tank and at the outflow are atmospheric, the differential pressure will result solely from the level of the liquid:

$$\Delta p = \frac{lg}{v} \tag{2.5}$$

where g is the acceleration due to the Earth's gravity (9.81 m/s^2). Each of the fractional valve openings, y_i, referred to above will depend on both the position of the valve actuator stem, known as 'valve travel', and the valve's flow-area vs. travel characteristic. Let us assume that for the valves on our particular plant, the two inlet valves are linear, while the designer has chosen a square-law characteristic for the outlet valve:

$$y_1 = x_1 \tag{2.6}$$

$$y_2 = x_2 \tag{2.7}$$

$$y_3 = x_3^2 \tag{2.8}$$

where x_i are the fractional valve travels for each of the valves.

Each valve will be driven by a valve-positioner, which is a servomechanism designed to drive the valve travel, x, to its demanded travel, x_d. This valve positioner will take a certain time to move the valve, and we will use the simplest possible model of the dynamics of the valve plus positioner, namely a first-order exponential lag:

$$\frac{dx_1}{dt} = \frac{x_{d1} - x_1}{\tau_1} \tag{2.9}$$

$$\frac{dx_2}{dt} = \frac{x_{d2} - x_2}{\tau_2} \tag{2.10}$$

$$\frac{dx_3}{dt} = \frac{x_{d3} - x_3}{\tau_3} \tag{2.11}$$

Here τ_i is the time constant associated with the ith valve positioner, typically of the order of a few seconds.

Let us assume that inlet valves, 1 and 2, are in manual-control mode, and thus may be moved by operator action on the plant. Demanded valve position may thus be modelled by an imposed 'forcing function'. A typical forcing function suitable for testing a control system would be a step increase followed later by a step decrease to the original value.

The travel of valve 3 is governed by the action of the level controller. For simplicity, we will suppose that the level controller has a purely proportional action so that the demanded valve travel, x_{d3}, is given by

$$x_{d3} = k(l - l_s) \tag{2.12}$$

where l is the measured level (m), l_s is the level setpoint (m) and k is the level control gain (m^{-1}). Again, the level setpoint will be made a forcing function defined externally to the model.

The level is found from the cross-sectional area, A, of the tank, and the specific volume, v, of the liquid and, of course, the mass of liquid contained in the tank:

$$l = \frac{mv}{A} \tag{2.13}$$

We shall assume, for simplicity, that there is negligible variation in the specific volume of the liquid in the tank, v. (Such an assumption could be reasonable in practice if one or more of the following conditions applied: (i) if the specific volumes of the two inlet streams had similar values, (ii) if the variations in the two inlet flows about their mean values were small, or (iii) if one inlet stream was very much smaller than the other.)

The thirteen equations derived above contain algebraic expressions for flows, level, differential pressure, fractional valve openings and demanded valve travel, as well as expressions for the rate of change of liquid mass and for the rate of change of valve travel for each valve. Given a knowledge of the constants contained in our equations, we can calculate all these algebraic expressions at any instant in time, *once we know the present values of the liquid mass in the tank and of the three valve travels*. These last four variables are vital indicators of the condition of the system, and are called the 'state variables' or, more colloquially, the 'states' of the system. What prevents the flow of the calculation being circular is that we may integrate numerically the state derivatives with respect to time from any given starting values for the state variables to find their values at any later time. At time t_0, the liquid mass and the valve travels will be at their initial conditions, assumed known:

$$\begin{aligned} m &= m_0 \\ x_1 &= x_{1,0} \\ x_2 &= x_{2,0} \\ x_3 &= x_{3,0} \end{aligned} \tag{2.14}$$

Thereafter the liquid mass at any subsequent time is found by integrating equation (2.1):

$$m = m_0 + \int_{t_0}^{t} \left(\frac{dm}{dt}\right) dt \tag{2.15}$$

and the valve travels are determined by integration of equations (2.9) to (2.11):

$$x_1 = x_{1,0} + \int_{t_0}^{t} \left(\frac{dx_1}{dt}\right) dt \tag{2.16}$$

$$x_2 = x_{2,0} + \int_{t_0}^{t} \left(\frac{dx_2}{dt}\right) dt \tag{2.17}$$

$$x_3 = x_{3,0} + \int_{t_0}^{t} \left(\frac{dx_3}{dt}\right) dt \tag{2.18}$$

We have now derived a model for the tank liquid level system, and by programming these equations into a simulation language on a digital computer, we can examine the behaviour of the system over time. In a typical use of such a model, we would examine the response of liquid level to a range of forcing functions imposed on inlet valve demanded travels and on the setpoint for liquid level. We would then adjust the gain of the level controller to give good control over the range of liquid levels expected in plant operation.

We shall now use the mathematical model just derived to illustrate some general features of dynamic simulation.

2.3 The general form of the simulation problem

The variables used in the model of the tank liquid level system above may be characterized as in the Table 2.1.

Table 2.1 Categorizing the variables in the liquid level model

Constants	$C'_{V1}, C'_{V2}, C_{V3}, k, v, g, \tau_1, \tau_2, \tau_3, A$
Initial conditions	$m_0, x_{1,0}, x_{2,0}, x_{3,0}$
Forcing variables	x_{d1}, x_{d2}, l_s
Algebraics	$W_1, W_2, W_3, \Delta p, y_1, y_2, y_3, x_{d3}, l$
Derivatives	$\frac{dm}{dt}, \frac{dx_1}{dt}, \frac{dx_2}{dt}, \frac{dx_3}{dt}$
State variables	m, x_1, x_2, x_3

The most important variables in the system are the state variables, since it is their evolving behaviour in time that is the basis of the dynamic response of the system. The importance of their role may be brought out further by rearranging the equations in Section 2.1 to eliminate all the algebraic equations and leave just the four state equations, integration of which enables us to trace the response of the system.

Substituting into equation (2.1) for each of the mass flows, W, from equations (2.2) to (2.4), and further substituting for the dependencies contained in equations (2.5) to (2.8) and in equation (2.13) gives:

$$\frac{dm}{dt} = C'_{V1}x_1 + C'_{V2}x_2 - C_{V3}x_3^2\sqrt{\frac{g}{vA}}\sqrt{m} \tag{2.19}$$

while substituting into equation (2.11) from equations (2.12) and (2.13) gives

$$\frac{dx_3}{dt} = -\frac{kl_s}{\tau_3} - \frac{x_3}{\tau_3} + \frac{kv}{A\tau_3}m \tag{2.20}$$

Hence, using equations (2.9), (2.10), (2.19) and (2.20), we may write down the equations describing the dynamics of the liquid tank system as:

$$\begin{aligned}
\frac{dx_1}{dt} &= \frac{x_{d1}}{\tau_1} - \frac{x_1}{\tau_1} \\
\frac{dx_2}{dt} &= \frac{x_{d2}}{\tau_2} - \frac{x_2}{\tau_2} \\
\frac{dx_3}{dt} &= -\frac{kl_s}{\tau_3} - \frac{x_3}{\tau_3} + \frac{kv}{A\tau_3}m \\
\frac{dm}{dt} &= C'_{V1}x_1 + C'_{V2}x_2 - C_{V3}x_3^2\sqrt{\frac{g}{vA}}\sqrt{m}
\end{aligned} \tag{2.21}$$

Thus in order to solve for the essential dynamic behaviour of the liquid tank system, it is sufficient to integrate just these four equations (2.21) in the derivatives of the four states x_1, x_2, x_3, m, from the initial conditions $x_{1,0}$, $x_{2,0}$, $x_{3,0}$, m_0.

Equations (2.21) have been written in the order and manner above to bring out the dynamic interdependence of the states that will normally emerge as a feature of models of typical industrial processes. While the derivative of one state may depend only on the current value of that state, as in the case of the valve travels, x_1 and x_2, others will depend not only on their own state but also on a number of others. This latter situation arises above in the cases of control valve travel, x_3, and the liquid mass in the tank, m. The dependence may be linear in some cases, but in any normal process model, there will be a large number of nonlinear dependencies, as exhibited above by the derivative for tank liquid mass, which is dependent on a term multiplying the square of one state by the square-root of another. This is an important point to grasp for those more accustomed to thinking of linear, multivariable control systems: such systems are idealizations only of a nonlinear world.

Equation (2.21) also shows how state behaviour depends on the forcing variables, in this case the externally determined setpoint for liquid level, l_s, and the demanded valve travels for inlet valve 1, x_{d1}, and inlet valve 2, x_{d2}.

We may write down the basic form for a soluble simulation problem as:

$$\begin{aligned}
\frac{d\mathbf{x}(t)}{dt} &= \mathbf{f}(\mathbf{x}(t), \mathbf{u}(t), t) \\
\mathbf{x}(t) &= \mathbf{x}(t_0) + \int_{t_0}^{t} \mathbf{f}(\mathbf{x}(t), \mathbf{u}(t), t)\, dt
\end{aligned} \tag{2.22}$$

where

x is an n-dimensional vector of system states, whose values are known at time $t = t_0$, **u** is an l-dimensional vector of forcing variables, **f** is a vector function that depends on the states, **x**, on the forcing variables, **u**, and (sometimes) directly on time itself, t. (The direct dependence on time can allow for the change in parameters over time in a known manner, such as the ageing of catalyst in a catalyst bed. It would normally be possible to include an extra state in the model to account for the gradual change in such a parameter, but there may be times when it is easier to insert a direct, algebraic dependence on time.) The differential of the vector, **x**, with respect to time is defined as the vector of the differentials of the components of **x**.

The fundamental point to be noted is that we may regard a simulation problem as solved in principle as soon as

(i) we have a consistent set of initial conditions for all the state variables, and
(ii) we are able to equate the time differential of each state variable to a defined expression involving some or all of the state variables, some or all of the inputs and time.

For example, in the case of the liquid-level system, the vector of states, **x**, is 4-dimensional and given by:

$$\mathbf{x} = \begin{bmatrix} x_1 \\ x_2 \\ x_3 \\ m \end{bmatrix} \tag{2.23}$$

The vector of forcing variables, **u**, is 3-dimensional and given by:

$$\mathbf{u} = \begin{bmatrix} x_{d1} \\ x_{d2} \\ l_s \end{bmatrix} \tag{2.24}$$

and the vector function, **f**, consists of 4 rows and is given by:

$$\mathbf{f}(\mathbf{x}, \mathbf{u}) = \begin{bmatrix} f_1(\mathbf{x}, \mathbf{u}) \\ f_2(\mathbf{x}, \mathbf{u}) \\ f_3(\mathbf{x}, \mathbf{u}) \\ f_4(\mathbf{x}, \mathbf{u}) \end{bmatrix} \tag{2.25}$$

where the functions f_i are defined by the right-hand sides of equations (2.21). In this case, **f** has no explicit dependence on time.

The derivative of the state vector in this case is given by:

$$\frac{d\mathbf{x}}{dt} = \begin{bmatrix} \frac{dx_1}{dt} \\ \frac{dx_2}{dt} \\ \frac{dx_3}{dt} \\ \frac{dm}{dt} \end{bmatrix} \tag{2.26}$$

If the model of the system to be simulated can be reduced to the form of equations (2.22), then a time-marching, numerical solution becomes possible by repeated application of, for instance, the first-order Euler integration formula:

$$\mathbf{x}(t + \Delta t) = \mathbf{x}(t) + \Delta t\, \mathbf{f}(\mathbf{x}(t), \mathbf{u}(t), t) \tag{2.27}$$

There are a number of proprietary simulation packages available, and many will offer a number of more complex integration algorithms. Nevertheless the first-order Euler method can prove a very robust and efficient algorithm for many simulation problems, especially those with a large number of discontinuities. But whatever the integration routine, the principle is the same: establish the starting condition of the system, i.e. the initial values of the system's states, then integrate forward in a time-marching manner to determine their subsequent behaviour, using the algebraic equations to link together the effects of changes in state values on different parts of the system.

Very often the simulation program in a commercial package is divided up for ease of reference and modification, as well as computational efficiency into sections similar to the categories of Table 2.1:

a section for constants that will be input or evaluated only once;
a section for initial conditions, again evaluated only once;
a section where the algebraic equations needed for derivative evaluation are calculated;
a section where the numerical integration is performed;
an output section, where the output form is specified, e.g. graphs for some variables, numerical output for others.

2.4 The state vector

Once programmed, the dynamic simulation will be used to understand the various processes going on inside a complex plant and to make usable predictions of the behaviour that will result from any changes or disturbances that may occur on the real plant, represented on the simulation by forcing functions or alterations to the chosen starting conditions. A basic first step is to characterize the condition of the plant at any given instant in time, and it is the state vector that, taken in conjunction with its associated mathematical model, allows us to do this. The state vector is an ordered collection of all the state variables. For a typical chemical plant, the state vector will consist of a number of temperatures, pressures, levels and valve positions, and the total number of state variables will be the 'dimension' or 'order' of the plant. For those who normally associate dimensions with directions in geometrical space, it might seem strange to describe a process plant as twenty-dimensional, and one might imagine that such a plant would be horrendously complicated. In fact, as industrial process plants go, such a plant would be of only moderate complexity.

In view of the fundamental importance of the state vector to the way in which we look at the plant, it might be supposed that only one set of state variables could emerge from a valid mathematical description of the plant, and that the composition of the state vector would have to be unique. In fact, this is not so. It will normally be possible to choose several different ways of describing a process plant, and each description will lead to a different set of variables making up the state vector, and a different associated mathematical model.

To demonstrate this, let us consider our example, the tank liquid level system of Figure 2.1.

Trivially, we should get a different set of numbers if we measured our fourth state, mass, in tonnes rather than kilograms. Slightly less trivially, we should get a different set of numbers if we chose the fourth state to be not mass in kilograms, but level in metres. Using level as opposed to mass changes the magnitude and units of the numbers comprising the state, but does not alter the completeness of the description. In this case, level and mass are simply, indeed linearly, related by equation (2.13), repeated below:

$$l = \frac{mv}{A} \tag{2.13}$$

But the relationship between state variables arising from different mathematical descriptions of the same process does not have to be linear. Let us assume that we wish to recast our equations in terms of valve openings rather than valve travels. This is a simple business for the linear valves 1 and 2, where fractional valve openings are identical with fractional valve travels (equations (2.6) and (2.7)). But the outlet valve has a square-law characteristic:

$$y_3 = x_3^2 \tag{2.8}$$

Clearly, our state vector should contain the same information if we substituted valve opening, y_3, instead of valve travel, x_3, but what is the precise effect of the change?

A formal differentiation of (2.8) with respect to time gives:

$$\frac{dy_3}{dt} = 2x_3\frac{dx_3}{dt} = 2\sqrt{y_3}\frac{dx_3}{dt} \tag{2.28}$$

To recast our model so that level and valve travels are the new states, we substitute from equations (2.6), (2.7), (2.8), (2.13) and (2.28) into equation set (2.21), to achieve a new mathematical description of the system dynamics:

$$\begin{aligned}
\frac{dy_1}{dt} &= \frac{x_{d1}}{\tau_1} - \frac{y_1}{\tau_1}\\
\frac{dy_2}{dt} &= \frac{x_{d2}}{\tau_2} - \frac{y_2}{\tau_2}\\
\frac{dy_3}{dt} &= -\frac{2kl_s}{\tau_3}\sqrt{y_3} - \frac{2}{\tau_3}y_3 + \frac{2k}{\tau_3}\sqrt{y_3}l\\
\frac{dl}{dt} &= \frac{C'_{V1}v}{A}y_1 + \frac{C'_{V2}v}{A}y_2 - \frac{C_{V3}\sqrt{g}}{A}y_3\sqrt{l}
\end{aligned} \tag{2.29}$$

Once again we have four states, but this time the state vector is not given by (2.23) but by:

$$\mathbf{x} = \begin{bmatrix} y_1 \\ y_2 \\ y_3 \\ l \end{bmatrix} \tag{2.30}$$

Here the states are all different from those of the previous formulation, but it is clear that the description is equally valid, and the new states have equally sensible, physical meanings.

2.5 Model complexity

We changed from one set of state variables to another in Section 2.4 and, although the meanings and values of the state variables changed, the number of state variables remained the same. Intuitively, this is not surprising, since we had introduced no new physical phenomena into our modelling, and the two descriptions of the plant were based on different manipulations of the same descriptive equations. The fact that different mathematical descriptions based on the same set of modelled phenomena give rise to the same number of state variables leads us to look on the dimension of our model as a measure of its complexity.

It will be appreciated that our description of the plant is, in reality, only an approximation covering as few features as we can get away with, while still capturing the essential behaviour of the plant. For instance, in the example above of the tank liquid level, no mention was made of liquid temperature, entailing an implicit assumption that temperature variations would be small over the period of interest. If it had been necessary to allow for temperature effects, perhaps because of fear of excessive evaporation or because of environmental temperature limits set for a waste water stream, then liquid temperature would have had to be included as an additional state variable, and the dimension or order of the plant as we modelled it would go up from 4 to 5. If we had needed to make an allowance for the temperature of the metal in the tank, then the additional state variable would have pushed the order up to 6. Of course, the plant itself would not have changed, merely our perception of how it worked.

The question of when the model is adequate is a deep one, and treated at greater length in Chapter 24 on model validation. At this stage, it is worth nothing that the control engineer will normally have a purpose in mind for his model, usually designing and checking for stability and control. In a large plant, he should first identify the subsystems that have only a low degree of interaction with each other and can, to a first approximation, be regarded as independent. He should then devise a separate mathematical model leading to a separate simulation of each of the important subsystems, including only the physical phenomena that are in his best judgment likely to cause significant effects. When the study concerns uprating the control of an existing plant, he should take every opportunity to test his model against data coming from that plant to test its validity. If a model fails in such a test against real data, it will need to be modified so that it can pass the test, usually by introducing additional physical phenomena, and raising the model's dimension.

The situation when designing a new plant is more difficult, since it is easy to be lulled into a false sense of security, assuming that the output from the model is correct because there is nothing around to contradict it. But, in practice, the new plant is likely to be similar in many respects to forerunning plants, and the modeller should in the first instance take the opportunity of testing a modified version of his model against an existing plant, applying the rules just set out. It is difficult to conceive that the new plant is really totally novel (or else how on earth did the designers manage to persuade the company board to invest their money in a plant with absolutely no track record?), but if such is indeed the situation then there will be no previous plant data against which to validate the model. In this case the best that the modeller can do is perform a sensitivity study for the parameters about which he feels most concern, and use the differences in resulting predictions as error bounds. There must always be a higher level of scepticism about the predictions from such an unvalidated model.

2.6 Distributed systems: partial differential equations

The assumption implicit in the discussion so far is that the system to be modelled consists of lumped-parameter elements and thus may be described adequately using ordinary differential equations in time. This will be true for a large number of process plant systems to a high degree of accuracy. But there are plant components that fit uneasily within this characterization, since they are inherently distributed in nature. It may be possible to model their responses using a simple, lumped-parameter approach if they are relatively unimportant items in a larger system, but sometimes the degree of error introduced will be unacceptable for the system under study. Accurate modelling requires that they be described by partial differential equations in time and space. Examples are very long pipelines, heat exchangers and catalyst beds, and detailed models are derived for these components in Chapters 19 and 20.

The distributed parameter component can be introduced into the larger system being simulated in one of two ways: either it can be introduced as an integral part through finite differencing in the distance dimension (or dimensions), or else it can be kept as a separate computational entity that communicates with the main simulation only at specified communication intervals.

To illustrate these concepts, let us take the example of a heat exchanger, where the temperature of the fluid within the tube will vary continuously throughout the length of the heat exchanger. The describing equations will have the form:

$$\frac{\partial T}{\partial t} + c\frac{\partial T}{\partial x} = k_1(T_w - T) \tag{2.31}$$

$$\frac{\partial T_w}{\partial t} = k_2(T - T_w) + k_3(T_s - T_w) \tag{2.32}$$

where

T is the temperature of the fluid inside the tube (°C),
T_w is the temperature of the tube wall (°C),
c is the velocity of the fluid inside the tube (m/s),
T_s is the temperature of the shell-side fluid (°C),
k_1, k_2, k_3 are all heat transfer constants (s^{-1}).

In many cases there would be a partial differential equation similar to (2.31) for the shell-side fluid also. An exception occurs when the shell-side fluid consists of condensing steam, when the shell-side fluid temperature can be characterized by a single value and described by an ordinary differential equation. For simplicity we will consider here this last case.

We may divide the heat exchanger along the length of its tube as shown in Figure 2.2 below so that we may apply a finite-difference approximation to the equations.

We use the finite difference approximation for the temperature gradient along the heat exchanger:

$$\frac{\partial T}{\partial x} \approx \frac{\Delta T}{\Delta x} \tag{2.33}$$

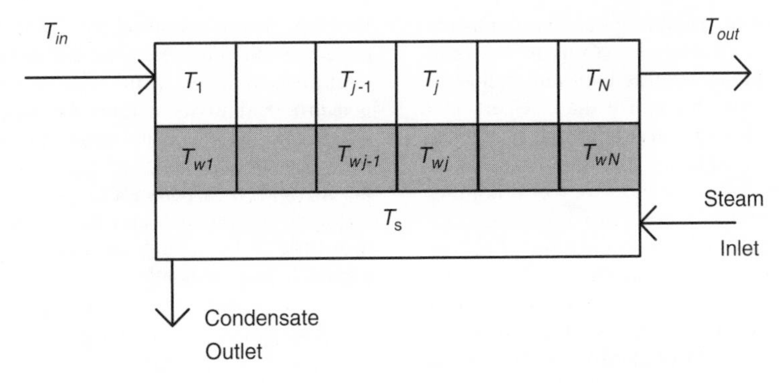

Figure 2.2 Schematic of a heat exchanger divided into N cells.

Setting

$$\Delta x = \frac{L}{N} \tag{2.34}$$

where L is the length of the heat exchanger tube (m) and N is the number of cells, and putting

$$\Delta T = T_j - T_{j-1} \tag{2.35}$$

allows us to recast equations (2.31) and (2.32) as:

$$\frac{dT_1}{dt} = k_1(T_{w1} - T_1) - c\frac{(T_1 - T_{in})}{\Delta x} \tag{2.36}$$

$$\frac{dT_j}{dt} = k_1(T_{wj} - T_j) - c\frac{(T_j - T_{j-1})}{\Delta x} \quad \text{for } j = 2 \text{ to } N \tag{2.37}$$

$$\frac{dT_{wj}}{dt} = k_2(T_j - T_{wj}) + k_3(T_s - T_{wj}) \quad \text{for } j = 1 \text{ to } N \tag{2.38}$$

The formulation of (2.36) and (2.37) shows how the rate of change of the tube-fluid temperature in a given cell will vary with time, dependent on two opposing driving forces:

(1) the heat passing from the hot tube wall to warm the tube-side fluid, and
(2) the cooling effect of the tube fluid flowing from the cooler previous cell at velocity, c.

Equation (2.38) shows how the corresponding section of the tube wall is warmed by the steam on the shell-side, but cooled by the tube-side fluid.

Once a value for Δx has been fixed by choosing the number of cells, N, the equations (2.36), (2.37) and (2.38) are in the canonical form of (2.22). The cell temperatures for the tube-side fluid and for the tube wall may be added to the state vector of the overall system simulation, as indicated by equation (2.39):

$$\mathbf{x} = \begin{bmatrix} \vdots \\ \vdots \\ T_1 \\ \vdots \\ T_j \\ \vdots \\ T_N \\ T_{wl} \\ \vdots \\ T_{wj} \\ \vdots \\ T_{wN} \\ \vdots \\ \vdots \end{bmatrix} \tag{2.39}$$

A point to be noted is that the selection of the number of cells and hence the cell length, Δx, cannot be totally free in any finite difference scheme. The Courant condition suggests that the time integration should not attempt to calculate beyond the spatial domain of influence by using a temperature at a distance beyond the range of influence determined by the characteristic velocity of temperature propagation. Hence

$$c\Delta t \leq \Delta x \tag{2.40}$$

In practice, Δx will be normally be set in advance by the modeller by his choice of the number of cells, while the integration routine may well seek to vary the integration timestep. The resulting restriction on the integration time interval is:

$$\Delta t \leq \frac{\Delta x}{c} \tag{2.41}$$

But the modeller may choose to use the method of characteristics in preference to a finite difference scheme, since this method is generally accepted to be the most accurate way of dealing with hyperbolic partial differential equations such as the heat exchanger equation (2.31). Here we recognize that the left-hand side of that equation is the total differential of temperature with respect to time:

$$\frac{\partial T}{\partial t} + c\frac{\partial T}{\partial x} = \frac{\partial T}{\partial t} + \frac{\partial T}{\partial x}\frac{dx}{dt} = \frac{dT}{dt} \tag{2.42}$$

This differential holds along the characteristic defined by the velocity, c. In effect, we may calculate the temperature of a packet of fluid moving with the stream through the heat exchanger. But to use this method, we need to fix the time interval absolutely as

$$\Delta t = \frac{\Delta x}{c} \tag{2.43}$$

which is a stricter condition than (2.41). The modeller sets Δx by his choice of N, but the velocity, c, may need to vary over a large range as the simulation progresses. Accommodating such variations has some slightly awkward (although soluble) implications for when we use the method of characteristics to simulate the heat exchanger on its own.

But having the timestep determined completely by just one plant component out of a very much larger simulation can lead to 'the tail wagging the dog' and may bring unacceptable consequences for the simulation of the plant as a whole, such as instability or an excessive time taken to run the simulation. An alternative way of dealing with the problem is to run an independent sub-simulation of the distributed-parameter component, and cause results to be exchanged between the main simulation and the sub-simulation only at specified communication intervals. Running two (or more) concurrent simulations that communicate at specified time intervals is a perfectly acceptable way of working. It has the practical advantage, too, that the program for the subsystem involving the solution of partial differential equations can be developed and debugged quite separately from the overall simulation. Since any program involving the solution of partial differential equations is likely to be fairly complicated, this can be a significant benefit.

2.7 The problem of stiffness

It is obviously beneficial for the simulation program to run as fast as possible, both in terms of cost and in terms of convenience for the control engineer who has to interact with it. But a major problem arising with process plants is the wide variety of time constants inherent in them. If the integration timestep for the system as a whole must be kept short to cope with the shortest time constant, then clearly the overall integration speed will be low. This is a real concern, and gives rise to the notion of 'stiffness', which may be quantified using the concept of 'system time constants', discussed below. A system is said to be stiff when it possesses time constants of widely different magnitudes, and the stiffness, S, of a model is measured by taking the ratio of the largest to the smallest time constant:

$$S = \frac{\tau_{max}}{\tau_{min}} \tag{2.44}$$

All models of realistic physical systems will possess a range of time constants, and hence a degree of stiffness. A system will not be seen as stiff if $S < 10$, but a system with $S > 100$ will certainly be regarded as stiff. The boundary between what is and what is not a stiff system lies somewhere in between, perhaps with $S = \sim 30$.

The time constant is a linear concept, which derives from the solution of a linear differential equation such as that used to model valve 1 in Section 2.2:

$$\frac{dx_1}{dt} = \frac{x_{d1} - x_1}{\tau_1} \tag{2.9}$$

The solution to this equation from a starting condition of $x_1 = 0$ at $t = 0$ is:

$$x_1 = x_{d1}\left(1 - e^{-(t/\tau_1)}\right) \tag{2.45}$$

The time constant, τ_1, determines how rapidly x_1 approaches x_{d1}. For example, after a step input in x_{d1}, the difference between x_1 and x_{d1} will be less than 5% after a time period of 3 time constants has elapsed.

Generalizing to the multivariable case, a time-invariant, linear dynamic system may be defined by the vector differential equations:

$$\frac{d\mathbf{x}}{dt} = \mathbf{Ax} + \mathbf{Bu} \tag{2.46}$$

where:
A is an $n \times n$ matrix with constant elements, and
B is an $n \times l$ matrix of constants.

The system time constants are taken as the negative reciprocals of the real parts of the non-zero eigenvalues of the matrix **A**. These determine the time responses of various parts of the system in an analogous way to τ_1 in the example above.

While the time constant is strictly a linear concept, the basic idea can be transferred to nonlinear systems by linearizing about an operating point. Now the time 'constants' will not be constant at all, but will depend, at any instant of time, on the values of the states and, in some cases, the plant inputs. Nevertheless, for reasons of custom and familiarity, we continue to use the term 'time constant' in the context of nonlinear systems, but with the above caveat in the back of our minds.

The linearized equivalent for the system defined by equation set (2.22) is

$$\frac{d\tilde{\mathbf{x}}}{dt} = \mathbf{J}\tilde{\mathbf{x}} + \mathbf{J_B}\tilde{\mathbf{u}} \tag{2.47}$$

where:

$\tilde{\mathbf{x}}$ is an $n \times 1$ vector of state deviations from an operating point defined by the state vector, $\mathbf{x}$, and the input vector, $\mathbf{u}$,

$\tilde{\mathbf{u}}$ is an $l \times 1$ vector of input deviations from the input vector, $\mathbf{u}$,

$\mathbf{J}$ is the $n \times n$ Jacobian matrix, defined as:

$$\mathbf{J} = \frac{\partial \mathbf{f}}{\partial \mathbf{x}} = \begin{bmatrix} \frac{\partial f_1}{\partial x_1} & \frac{\partial f_1}{\partial x_2} & \cdots & \frac{\partial f_1}{\partial x_n} \\ \frac{\partial f_2}{\partial x_1} & \frac{\partial f_2}{\partial x_2} & \cdots & \frac{\partial f_2}{\partial x_n} \\ \vdots & \vdots & \cdots & \vdots \\ \frac{\partial f_n}{\partial x_1} & \frac{\partial f_n}{\partial x_2} & \cdots & \frac{\partial f_n}{\partial x_n} \end{bmatrix} \tag{2.48}$$

while the $n \times l$ input-Jacobian matrix, $\mathbf{J_B}$, is defined as:

$$\mathbf{J_B} = \frac{\partial \mathbf{f}}{\partial \mathbf{u}} = \begin{bmatrix} \frac{\partial f_1}{\partial u_1} & \frac{\partial f_1}{\partial u_2} & \cdots & \frac{\partial f_1}{\partial u_l} \\ \frac{\partial f_2}{\partial u_1} & \frac{\partial f_2}{\partial u_2} & \cdots & \frac{\partial f_2}{\partial u_l} \\ \vdots & \vdots & \cdots & \vdots \\ \frac{\partial f_n}{\partial u_1} & \frac{\partial f_n}{\partial u_2} & \cdots & \frac{\partial f_n}{\partial u_l} \end{bmatrix} \tag{2.49}$$

We evaluate the Jacobian matrices at a particular operating condition, defined by its states and the system inputs. It is important to emphasize that the linearized equation (2.47) and the Jacobian matrices it contains are valid only near that operating point.

For the multivariable, nonlinear system, the time constants are the negative reciprocals of the non-zero eigenvalues of the Jacobian matrix, $\mathbf{J}$, which are the roots of the equation:

$$|\mathbf{J} - \lambda \mathbf{I}| = 0 \tag{2.50}$$

To put some flesh on these theoretical bones, let us consider again the tank liquid level system. Linearization of the equation set (2.21) allows us to set down the Jacobian matrix in terms of the states and the system constants as:

$$\mathbf{J} = \begin{bmatrix} -\frac{1}{\tau_1} & 0 & 0 & 0 \\ 0 & -\frac{1}{\tau_2} & 0 & 0 \\ 0 & 0 & -\frac{1}{\tau_3} & \frac{kv}{A\tau_3} \\ C'_{V1} & C'_{V2} & -2C_{V3}\sqrt{\frac{g}{vA}} \times x_3\sqrt{m} & -\frac{1}{2}C_{V3}\sqrt{\frac{g}{vA}} \times x_3^2 \frac{1}{\sqrt{m}} \end{bmatrix} \tag{2.51}$$

This may be simplified using equation (2.2), repeated below:

$$W_3 = C_{V3} y_3 \sqrt{\frac{\Delta p}{v}} \tag{2.2}$$

to the form:

$$\mathbf{J} = \begin{bmatrix} -\frac{1}{\tau_1} & 0 & 0 & 0 \\ 0 & -\frac{1}{\tau_2} & 0 & 0 \\ 0 & 0 & -\frac{1}{\tau_3} & \frac{kv}{A\tau_3} \\ C'_{V1} & C'_{V2} & -\frac{2W_3}{x_3} & -\frac{W_3}{2m} \end{bmatrix} \tag{2.52}$$

We find the eigenvalues by setting the determinant to zero:

$$|\mathbf{J} - \lambda \mathbf{I}| = \begin{vmatrix} -\frac{1}{\tau_1} - \lambda & 0 & 0 & 0 \\ 0 & -\frac{1}{\tau_2} - \lambda & 0 & 0 \\ 0 & 0 & -\frac{1}{\tau_3} - \lambda & \frac{kv}{A\tau_3} \\ C'_{V1} & C'_{V2} & -\frac{2W_3}{x_3} & -\frac{W_3}{2m} - \lambda \end{vmatrix} = 0 \tag{2.53}$$

or:

$$\left(\frac{1}{\tau_1} + \lambda\right)\left(\frac{1}{\tau_2} + \lambda\right)\left[\left(\frac{1}{\tau_3} + \lambda\right) \times \left(\frac{W_3}{2m} + \lambda\right) + \frac{2kvW_3}{A\tau_3 x_3}\right] = 0 \tag{2.54}$$

The solution to this quartic equation can be seen by inspection to be the two roots:

$$\lambda_1 = -\frac{1}{\tau_1}$$
$$\lambda_2 = -\frac{1}{\tau_2} \tag{2.55}$$

and the roots of the quadratic contained in the square brackets, given by:

$$\lambda_3 = \frac{a+b}{2} + \frac{1}{2}\sqrt{(a-b)^2 + 4de}$$
$$\lambda_4 = \frac{a+b}{2} - \frac{1}{2}\sqrt{(a-b)^2 + 4de} \tag{2.56}$$

where

$$a = -\frac{1}{\tau_3}$$
$$b = -\frac{W_3}{2m}$$
$$d = \frac{kv}{A\tau_3} \tag{2.57}$$
$$e = -\frac{2W_3}{x_3}$$

To evaluate these eigenvalues, we need data at an operating point, such as the physically feasible data-set is given in Table 2.2. It will now be shown how easy it is for stiffness to creep into a simulation.

Using these data, we calculate that the eigenvalues are:

$$\lambda_1 = -0.3333$$
$$\lambda_2 = -0.1666$$
$$\lambda_3 = -0.0894 \tag{2.58}$$
$$\lambda_4 = -0.0112$$

so that the corresponding system time constants in seconds are:

$$\tau_{sys\,1} = 3$$
$$\tau_{sys\,2} = 6$$
$$\tau_{sys\,3} = 11.19 \tag{2.59}$$
$$\tau_{sys\,4} = 88.99$$

We may see that the time constants are of significantly different value and that the system has a stiffness ratio of nearly 30. In fact, the stiffness ratio depends strongly on the gain that is selected for the level controller. When the gain, k, is reduced to 0.5, the time constants change to:

$$\tau_{sys\,1} = 3$$
$$\tau_{sys\,2} = 6$$
$$\tau_{sys\,3} = 10.53 \tag{2.60}$$
$$\tau_{sys\,4} = 178.26$$

so that the stiffness ratio is nearly 60. A stiffness ratio in excess of 500 can result if the controller gain, k, is reduced to a very low value.

Table 2.2 Operating point data for the tank liquid level system

Constants	C'_{V1} (kg/s) 10	C'_{V2} (kg/s) 10	C_{V3} (m^2) 0.004517	k (m^{-1}) 1.0	v (m^3/kg) 0.001
	g (m/s^2) 9.8	A (m^2) 3	τ_1 (s) 3	τ_2 (s) 6	τ_3 (s) 10
Forcing variables	x_{d1} (dimensionless) 0.5	x_{d2} (dimensionless) 0.5	l_s (m) 2		
State variables	x_1 (dimensionless) 0.5	x_2 (dimensionless) 0.5	x_3 (m) 0.707	m (kg) 8121	
Algebraics	W_1 (kg/s) 5	W_2 (kg/s) 5	W_3 (kg/s) 10	Δp (Pa) 26529	l (m) 2.707
	y_1 (dimensionless) 0.5	y_2 (dimensionless) 0.5	y_3 (dimensionless) 0.5	x_{d3} (dimensionless) 0.707	

It is thus apparent from the simple but quite feasible example of the tank liquid level system that stiffness can easily become a significant feature of the simulation of a process plant. While stiffness in such a small simulation as this will not cause a major computational burden, stiffness in a larger process plant system will result in a very significant slowing of the integration, and special measures need to be taken to counter its influence.

2.8 Tackling stiffness in process simulations: the properties of a stiff integration algorithm

Explicit integration algorithms have the advantage that all calculations proceed from known data and the integration progresses in an entirely straightforward, time-marching manner. Unfortunately, for the simplest of these, the Euler integration algorithm of equation (2.27), numerical instability will occur if the timestep is greater than twice the smallest time constant, so that the we must constrain the timestep to:

$$\Delta t \leq 2\tau_{\min} \tag{2.61}$$

This constraint will hold throughout the calculation, so that the speed of the simulation is limited by the shortest time constant, even though the rapid dynamics of the associated part of the model will come very quickly to have little effect on the solution. It might be hoped that this constraint could be eased by choosing a more complex, but still explicit, integration algorithm. But this is not the case: the condition for a fourth-order algorithm such as Runge–Kutta is little better at:

$$\Delta t \leq 3\tau_{\min} \tag{2.62}$$

These restrictions cause us to consider implicit algorithms as an alternative.

Here a finding of Dahlquist's gives useful guidance. Dahlquist defines an integration algorithm as having the highly desirable property of 'A-stability' if it is stable for *all* step lengths when applied to the linear differential equation describing an unconditionally stable physical system

$$\frac{d\mathbf{x}}{dt} = -\lambda \mathbf{x} \tag{2.63}$$

with λ strictly positive. Only implicit algorithms of order two or below are A-stable.

To illustrate the difference in stability properties between explicit and implicit integration algorithms, consider again the equation used to describe valve dynamics in Section 2.2. Dropping the subscripts from equation (2.9) for clarity and generality, and setting the demanded valve travel, x_d, to zero, indicating a demand for closure, we have:

$$\frac{dx}{dt} = -\frac{x}{\tau} \tag{2.64}$$

which is in the form of equation (2.63). The explicit, forward Euler approximation yields the equation

$$x_{k+1} = x_k - \Delta t \frac{x_k}{\tau} = \left(1 - \frac{\Delta t}{\tau}\right) x_k \tag{2.65}$$

where $x_k = x(t_0 + k\Delta t)$. Stability requires

$$\left| \frac{x_{k+1}}{x_k} \right| \leq 1 \qquad \text{for all } k \tag{2.66}$$

which leads to the condition stated in the first paragraph of this section, namely

$$\Delta t \leq 2\tau \tag{2.67}$$

However, the implicit, backward Euler approximation gives

$$x_{k+1} = x_k - \Delta t \frac{x_{k+1}}{\tau} \tag{2.68}$$

or

$$x_{k+1} = \frac{x_k}{1 + \dfrac{\Delta t}{\tau}} \tag{2.69}$$

From inspection, this is stable for all positive values of the timestep, Δt, in-line with Dahlquist.

But while the implicit integration algorithm above can be rearranged easily into an explicit form for the single-variable, linear case, the same cannot be said for the multivariable, nonlinear cases that we will normally be dealing with in process modelling. If we examine the general simulation case given by equation set (2.22), then applying the implicit, backward Euler algorithm produces the set of equations:

$$\mathbf{x}_{k+1} - \Delta t\, \mathbf{f}_{k+1}(\mathbf{x}_{k+1}, \mathbf{u}_{k+1}) - \mathbf{x}_k = \mathbf{0} \tag{2.70}$$

which represents a system of n nonlinear simultaneous equations in the n unknowns of the vector $\mathbf{x}$ at the $(k+1)$th timestep. We will need to solve a similar set of simultaneous equations at each timestep. Thus in order to get the boon of an algorithm with much better stiffness properties, allowing us to take much bigger timesteps, we have had to pay for it by involving ourselves in significantly more computation at each timestep.

We will now illustrate the way that equation (2.70) could be solved as part of a stiff integration package. The solution relies partly on using the Newton–Raphson technique for solving nonlinear simultaneous equations, the principles of which will now be explained. We may describe a system of n nonlinear, simultaneous equations in the n unknowns of the vector $\mathbf{z}$ by the vector equation:

$$\mathbf{g}(\mathbf{z}) = \mathbf{0} \tag{2.71}$$

Applying a truncated Taylor's formula in the vicinity of the jth estimate of the roots, $\mathbf{z}^{(j)}$, gives

$$\mathbf{g}(\mathbf{z}^{(j+1)}) = \mathbf{g}(\mathbf{z}^{(j)}) + \frac{\partial \mathbf{g}(\mathbf{z}^{(j)})}{\partial \mathbf{z}} \left[\mathbf{z}^{(j+1)} - \mathbf{z}^{(j)}\right] \tag{2.72}$$

To obtain the $(j+1)$th estimate of the roots, namely $\mathbf{z}^{(j+1)}$, we note that these roots should ideally satisfy equation (2.71) precisely:

$$\mathbf{g}(\mathbf{z}^{(j+1)}) = \mathbf{0} \tag{2.73}$$

Substituting from (2.73) into (2.72) allows us to write down our next estimate for $\mathbf{z}$, $\mathbf{z}^{(j+1)}$, as:

$$\mathbf{z}^{(j+1)} = \mathbf{z}^{(j)} - \left[\frac{\partial \mathbf{g}(\mathbf{z}^{(j)})}{\partial \mathbf{z}}\right]^{-1} \mathbf{g}(\mathbf{z}^{(j)}) \tag{2.74}$$

This formula is used to converge iteratively on the true roots of the equation set.

To use this result to solve equation (2.70) for the values of the states at the $(k+1)$th timestep, $\mathbf{x}_{k+1}$, we put

$$\mathbf{g}(\mathbf{x}_{k+1}) = \mathbf{x}_{k+1} - \Delta t\, \mathbf{f}_{k+1}(\mathbf{x}_{k+1}, \mathbf{u}_{k+1}) - \mathbf{x}_k = \mathbf{0} \tag{2.75}$$

and note that

$$\frac{\partial \mathbf{g}}{\partial \mathbf{x}_{k+1}} = \mathbf{I} - \Delta t \frac{\partial \mathbf{f}_{k+1}}{\partial \mathbf{x}_{k+1}} \tag{2.76}$$

Since $\partial \mathbf{f}_{k+1}/\partial \mathbf{x}_{k+1}$ is simply the Jacobian, $\mathbf{J}$, as defined in equation (2.48) at the time instant, $t_0 + (k+1)\Delta t$, we may rewrite (2.76) as:

$$\frac{\partial \mathbf{g}}{\partial \mathbf{x}_{k+1}} = \mathbf{I} - \Delta t\, \mathbf{J}_{k+1} \tag{2.77}$$

Hence we may generate the $(j+1)$th estimate for $\mathbf{x}_{k+1}$ from the jth such estimate by:

$$\mathbf{x}_{k+1}^{(j+1)} = \mathbf{x}_{k+1}^{(j)} - [\mathbf{I} - \Delta t\, \mathbf{J}_{k+1}^{(j)}]^{-1} \times [\mathbf{x}_{k+1}^{(j)} - \Delta t\, \mathbf{f}_{k+1}^{(j)}(\mathbf{x}_{k+1}^{(j)}, \mathbf{u}_{k+1}) - \mathbf{x}_k] \tag{2.78}$$

A starting value for $\mathbf{x}_{k+1}$ will be guessed (and once the integrations have begun a good value of this vector will be its value at the last timestep, $\mathbf{x}_k$) and equation (2.78) can be evaluated repeatedly until some criterion of convergence is satisfied.

Note that $\mathbf{u}_{k+1}$ and $\mathbf{x}_k$ on the right-hand side are invariant throughout all the iterations at each timestep. The Jacobian is evaluated numerically, and strictly this should be done at each iteration. But this is a time-consuming procedure, and in practice the Jacobian is not usually calculated at each iteration, and not even at every timestep. Further time is saved by using sparse matrix techniques to take advantage of the fact that the Jacobian usually possesses many zero elements (cf. equation (2.52) for example). Sparse matrix techniques are similarly used in solving equation (2.78) once the Jacobian has been found. Finally, the integration routine will seek to lengthen the timestep to the maximum extent consistent with a defined accuracy criterion, to take advantage of the strong stability properties of the implicit method.

As a result of including an implicit, 'stiff' integration algorithm such as the first-order algorithm described in outline above, a simulation package may speed up markedly the execution speed of the simulation as a whole. Several different integration routines, some designed for stiff systems and some not, may well be provided in the simulation package, and it will be for the control engineer to decide which he wishes to use. Often this will be through a process of trial and error, with speed, stability and accuracy as the objectives.

2.9 Tackling stiffness in process simulations by modifications to the model

The modeller will not normally wish to tamper with the stiff integration algorithms provided with his modelling package – it would almost always be counterproductive for him to repeat programming carried out by the package designer and already tested to a high degree. Nevertheless, the modeller's physical grasp of the problem can allow him to reduce the stiffness of the equations finally presented for numerical integration. Considering equation (2.44), repeated below

$$S = \frac{\tau_{\max}}{\tau_{\min}} \tag{2.44}$$

it is clear that stiffness may be reduced by increasing the value of $\tau_{\min}$. This may be done in two ways:

(i) the modeller may decide to increase artificially the minimum time constant so that it is closer to the other time constants dominating what he considers to be the essential behaviour of the model, or
(ii) by assuming that the fastest part of the model responds instantaneously, he may decrease the minimum time constant to zero and thus take that time constant out of consideration.

It may seem paradoxical but it is nevertheless true that these two opposite courses of action have essentially the same effect on the stiffness ratio, S. In the first case, the effect is to raise the value of the smallest time constant towards the next smallest:

$$\tau_{\min} \Rightarrow \tau_{\min-1} \qquad (2.79)$$

where $\tau_{\min-1}$ is the second smallest time constant, so that

$$S \Rightarrow \frac{\tau_{\max}}{\tau_{\min-1}} \qquad (2.80)$$

In the second case, the smallest time constant disappears from consideration, so that the new stiffness ratio is given by the similar formula:

$$S = \frac{\tau_{\max}}{\tau_{\min-1}} \qquad (2.81)$$

To gain an understanding of the physical significance of these two courses of action, let us refer once more to the tank liquid level system of Section 2.2, working near the operating point set out in Table 2.2. The dominant time constant is that associated with liquid transit time, namely 88.99 s. If our principal concern is merely level control, it will make little difference to the result of the simulation if we increase the values of the time constants associated with valves 1 and 2 to 10 s, say. But by doing so, we reduce the stiffness ratio from 29 to 9, a useful gain. Equally, it will make little difference if we artificially reduce the time constants for valves 1 and 2 to zero. The differential equations of (2.9) and (2.10) are then replaced by the simple algebraics:

$$x_1 = x_{d1} \qquad (2.82)$$

and

$$x_2 = x_{d2} \qquad (2.83)$$

The stiffness ratio is reduced to 8, and the simplification has brought the additional bonus of reducing the number of states from 4 to 2.

It goes without saying that care is always needed in applying either method above. The modeller needs to keep in mind at all times the ultimate purpose of his simulation, and he must be particularly careful if the purpose of the simulation should change, when he will need to go back and check whether his artificial manipulation of the time constants is still valid.

2.10 Solving nonlinear simultaneous equations in a process model: iterative method

The procedure of replacing one or more differential equations by an algebraic equation is, of course, universal in modelling. The assumption being made is that the dynamics of certain parts of the process are so fast that they reach a steady state almost instaneously. Such a component may be regarded as continuously in a steady state that evolves as different conditions are encountered at the component's boundaries. For example, we did not attempt to model the settling of the electronic currents in the level controller when we modelled the tank liquid level system: we knew that this process would occur as near instantaneously as would make no difference for our purposes. Unfortunately there are cases where the perfectly reasonable assumption of zero time constants leads to an implicit set of nonlinear, simultaneous equations. A case in point of importance to process modelling is fluid flow in a network. The establishment of liquid flow or of high-pressure gas or steam flow in a network of pipes will normally be very rapid compared with the more gradual changes in levels and pressures induced in connected vessels. Hence it is usually valid to assume that the flow network is continuously in a steady state. But the resulting algebraic equations cannot normally be solved simply, since they are nonlinear simultaneous equations.

Consider the system of Figure 2.3, which represents a flow network with six nodes. Liquid flows from an upstream accumulator, at pressure p_1, to three downstream accumulators, at pressures p_3, p_5 and p_6. The flow passes through a pipeline network with line conductances C_{12}, C_{23}, C_{24}, C_{45} and C_{46}. Let us assume that the network forms part of a larger model,

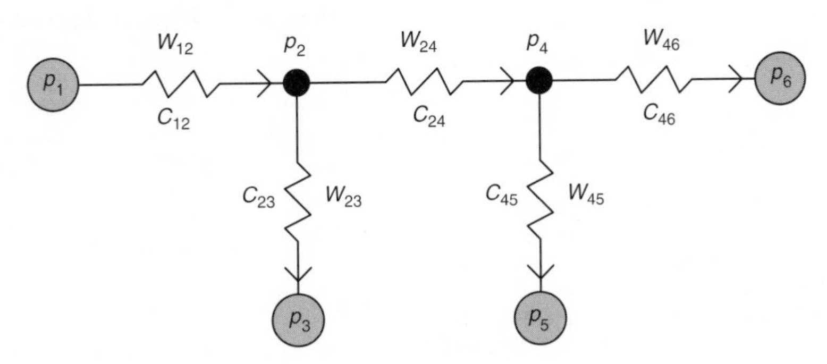

Figure 2.3 Schematic of liquid flow network, discharging to three accumulators.

the interface to which is provided by the pressures p_3 and p_5, assumed to be state variables. Further, let us assume for purposes of illustration that the pressures, p_1 and p_6, are externally determined, so that they should be regarded as inputs. Because flow establishes itself so quickly, pressures p_2 and p_4 will be modelled as algebraic variables.

Assuming that the liquid stays at the same temperature, the specific volume, v, will be constant throughout the network at its inlet value: $v = v_1$, so we may write the flow equations as

$$W_{12} = C_{12}\sqrt{\frac{p_1 - p_2}{v}} \tag{2.84}$$

$$W_{23} = C_{23}\sqrt{\frac{p_2 - p_3}{v}} \tag{2.85}$$

$$W_{24} = C_{24}\sqrt{\frac{p_2 - p_4}{v}} \tag{2.86}$$

$$W_{45} = C_{45}\sqrt{\frac{p_4 - p_5}{v}} \tag{2.87}$$

$$W_{46} = C_{46}\sqrt{\frac{p_4 - p_6}{v}} \tag{2.88}$$

$$W_{12} - W_{23} - W_{24} = 0 \tag{2.89}$$

$$W_{24} - W_{45} - W_{46} = 0 \tag{2.90}$$

Given that the boundary pressures p_1, p_3, p_5 and p_6 are either input variables or state variables (or explicitly derivable from the model's states), we have in equations (2.84) to (2.90) a set of seven nonlinear simultaneous equations in the seven unknowns: W_{12}, W_{23}, W_{24}, W_{45}, W_{46}, p_2 and p_4. We can in this case reduce the order of the problem easily by substituting for the flows into equations (2.89) and (2.90) to give

$$C_{12}\sqrt{\frac{p_1 - p_2}{v}} - C_{23}\sqrt{\frac{p_2 - p_3}{v}} - C_{24}\sqrt{\frac{p_2 - p_4}{v}} = 0 \tag{2.91}$$

$$C_{24}\sqrt{\frac{p_2 - p_4}{v}} - C_{45}\sqrt{\frac{p_4 - p_5}{v}} - C_{46}\sqrt{\frac{p_4 - p_6}{v}} = 0 \tag{2.92}$$

But we are still left with two nonlinear simultaneous equations in the two pressure unknowns p_2 and p_4.

It is a characteristic of pumped liquid systems and of the steam or gas flow networks with turbines and compressors that no explicit solution is generally available. It is clear from the form of equations (2.91) to (2.92) that no explicit solution can be expected even for the simple flow network above. Instead, an iterative method is needed in order to achieve a solution.

A common method of solution is the Newton–Raphson method, already described in connection with a stiff integration algorithm in Section 2.7, equations (2.71) to (2.74). The equations above are in the form

$$\mathbf{g}(\mathbf{z}(t), \mathbf{x}(t), \mathbf{u}(t)) = \mathbf{0} \tag{2.93}$$

with, in this particular case,

$$\mathbf{z}(t) = \begin{bmatrix} p_2(t) \\ p_4(t) \end{bmatrix} \quad \mathbf{x}(t) = \begin{bmatrix} p_3(t) \\ p_5(t) \end{bmatrix} \quad \mathbf{u}(t) = \begin{bmatrix} p_1(t) \\ p_6(t) \end{bmatrix} \tag{2.94}$$

In general, the vector of model states, $\mathbf{x}$, and the vector of inputs, $\mathbf{u}$, will be held constant throughout each set of iterations.

At each time, t, successively better estimates for $\mathbf{z}$ may be calculated from

$$\mathbf{z}^{(j+1)} = \mathbf{z}^{(j)} - \left[\frac{\partial \mathbf{g}(\mathbf{z}^{(j)}, \mathbf{x}(t), \mathbf{u}(t))}{\partial \mathbf{z}}\right]^{-1} \times \mathbf{g}(\mathbf{z}^{(j)}, \mathbf{x}(t), \mathbf{u}(t)) \tag{2.95}$$

A good subroutine for solving nonlinear simultaneous equations may be provided within the overall simulation package, or it may be necessary for the modeller to introduce such a routine himself. Commercial software is available if not already provided within the simulation package. Further detail on iterative methods for solving implicit equations is given in Chapter 18, Section 18.5, which includes a discussion on how to speed up convergence in flow networks.

2.11 Solving nonlinear simultaneous equations in a process model: the Method of Referred Derivatives

An alternative method of solving nonlinear simultaneous equations within a simulation is based on the properties of equation (2.93). Since the vector function, $\mathbf{g}$, is constant (at zero) throughout all time, it follows that its time differential is also zero at all times:

$$\frac{d}{dt}\mathbf{g}(\mathbf{z}(t), \mathbf{x}(t), \mathbf{u}(t)) = 0 \tag{2.96}$$

We may expand equation (2.96) to give

$$\frac{\partial \mathbf{g}}{\partial \mathbf{z}}\frac{d\mathbf{z}}{dt} + \frac{\partial \mathbf{g}}{\partial \mathbf{x}}\frac{d\mathbf{x}}{dt} + \frac{\partial \mathbf{g}}{\partial \mathbf{u}}\frac{d\mathbf{u}}{dt} = 0 \tag{2.97}$$

or

$$\frac{\partial \mathbf{g}}{\partial \mathbf{z}}\frac{d\mathbf{z}}{dt} = -\frac{\partial \mathbf{g}}{\partial \mathbf{x}}\frac{d\mathbf{x}}{dt} - \frac{\partial \mathbf{g}}{\partial \mathbf{u}}\frac{d\mathbf{u}}{dt} \quad (2.98)$$

where, assuming there are k equations in k unknowns, $\partial \mathbf{g}/\partial \mathbf{z}$ is the $k \times k$ matrix:

$$\frac{\partial \mathbf{g}}{\partial \mathbf{z}} = \begin{bmatrix} \frac{\partial g_1}{\partial z_1} & \frac{\partial g_1}{\partial z_2} & \cdots & \frac{\partial g_1}{\partial z_k} \\ \frac{\partial g_2}{\partial z_1} & \frac{\partial g_2}{\partial z_2} & \cdots & \frac{\partial g_2}{\partial z_k} \\ \vdots & \vdots & \cdots & \vdots \\ \frac{\partial g_k}{\partial z_1} & \frac{\partial g_k}{\partial z_2} & \cdots & \frac{\partial g_k}{\partial z_k} \end{bmatrix} \quad (2.99)$$

$d\mathbf{z}/dt$ is the $k \times 1$ vector given by:

$$\frac{d\mathbf{z}}{dt} = \begin{bmatrix} \frac{dz_1}{dt} \\ \frac{dz_2}{dt} \\ \vdots \\ \frac{dz_k}{dt} \end{bmatrix} \quad (2.100)$$

$\partial \mathbf{g}/\partial \mathbf{x}$ is the $k \times n$ matrix:

$$\frac{\partial \mathbf{g}}{\partial \mathbf{x}} = \begin{bmatrix} \frac{\partial g_1}{\partial x_1} & \frac{\partial g_1}{\partial x_2} & \cdots & \frac{\partial g_1}{\partial x_n} \\ \frac{\partial g_2}{\partial x_1} & \frac{\partial g_2}{\partial x_2} & \cdots & \frac{\partial g_2}{\partial x_n} \\ \vdots & \vdots & \cdots & \vdots \\ \frac{\partial g_k}{\partial x_1} & \frac{\partial g_k}{\partial x_2} & \cdots & \frac{\partial g_k}{\partial x_n} \end{bmatrix} \quad (2.101)$$

$d\mathbf{x}/dt$ is the $n \times 1$ vector given by:

$$\frac{d\mathbf{x}}{dt} = \begin{bmatrix} \frac{dx_1}{dt} \\ \frac{dx_2}{dt} \\ \vdots \\ \frac{dx_n}{dt} \end{bmatrix} \quad (2.102)$$

$\partial \mathbf{g}/\partial \mathbf{u}$ is the $k \times l$ matrix:

$$\frac{\partial \mathbf{g}}{\partial \mathbf{u}} = \begin{bmatrix} \frac{\partial g_1}{\partial u_1} & \frac{\partial g_1}{\partial u_2} & \cdots & \frac{\partial g_1}{\partial u_l} \\ \frac{\partial g_2}{\partial u_1} & \frac{\partial g_2}{\partial u_2} & \cdots & \frac{\partial g_2}{\partial u_l} \\ \vdots & \vdots & \cdots & \vdots \\ \frac{\partial g_k}{\partial u_1} & \frac{\partial g_k}{\partial u_2} & \cdots & \frac{\partial g_k}{\partial u_l} \end{bmatrix} \quad (2.103)$$

and $d\mathbf{u}/dt$ is the $l \times 1$ vector given by:

$$\frac{d\mathbf{u}}{dt} = \begin{bmatrix} \frac{du_1}{dt} \\ \frac{du_2}{dt} \\ \vdots \\ \frac{du_l}{dt} \end{bmatrix} \quad (2.104)$$

Equation (2.98) represents a set of k **linear** simultaneous equations in the k unknowns, dz_i/dt, which are calculated by reference to the derivatives of the state variables and the derivatives of the inputs – hence the name Method of Referred Derivatives.

The derivatives of the state variables are immediately available, since they are calculated in the normal course of the simulation. The derivatives of the input vector, $\mathbf{u}$, may be calculated to any required degree of accuracy off-line to the simulation. The only restriction on $\mathbf{u}$ is that its differentiation must not lead to a discontinuity, so that, for instance, a step change must be replaced by a steep ramp function (likely to be physically more realistic in any case).

Using the Method of Referred Derivatives, it is possible to integrate the vector $d\mathbf{z}/dt$ in the same way as the vector $d\mathbf{x}/dt$. Thus this method replaces the need to solve a set of nonlinear, simultaneous equations at each timestep by the simpler requirement of solving a set of linear, simultaneous equations, followed by integration of the resultant time-differentials from a feasible initial condition, $\mathbf{z}(0)$.

The initial condition may be determined by an iterative solution of equation (2.93) just once at the beginning of the simulation, or, indeed, by a prior, off-line calculation. Alternative techniques based on integrating an artificial 'prior transient' are given in Chapter 18, Sections 18.7 to 18.9, where a more detailed worked example is given.

As an example, taken the flow network of Figure 2.3, described by equations (2.91) and (2.92). Differentiating the two equations with respect to time gives:

$$\frac{W_{12}}{2(p_1 - p_2)}\left(\frac{dp_1}{dt} - \frac{dp_2}{dt}\right) - \frac{W_{23}}{2(p_2 - p_3)}\left(\frac{dp_2}{dt} - \frac{dp_3}{dt}\right) - \frac{W_{24}}{2(p_2 - p_4)}\left(\frac{dp_2}{dt} - \frac{dp_4}{dt}\right) = 0 \quad (2.105)$$

$$\frac{W_{24}}{2(p_2-p_4)}\left(\frac{dp_2}{dt}-\frac{dp_4}{dt}\right) - \frac{W_{45}}{2(p_4-p_5)}\left(\frac{dp_4}{dt}-\frac{dp_5}{dt}\right) - \frac{W_{46}}{2(p_4-p_6)}\left(\frac{dp_4}{dt}-\frac{dp_6}{dt}\right)=0 \quad (2.106)$$

Rearranging into the form of equation (2.98) yields:

$$\begin{bmatrix} \frac{W_{12}}{p_1-p_2}+\frac{W_{23}}{p_2-p_3}+\frac{W_{24}}{p_2-p_4} & -\frac{W_{24}}{p_2-p_4} \\ -\frac{W_{24}}{p_2-p_4} & \frac{W_{24}}{p_2-p_4}+\frac{W_{45}}{p_4-p_5}+\frac{W_{46}}{p_4-p_6} \end{bmatrix}\begin{bmatrix} \frac{dp_2}{dt} \\ \frac{dp_4}{dt} \end{bmatrix} = \begin{bmatrix} \frac{W_{23}}{p_2-p_3} & 0 \\ 0 & \frac{W_{45}}{p_4-p_5} \end{bmatrix}\begin{bmatrix} \frac{dp_3}{dt} \\ \frac{dp_5}{dt} \end{bmatrix} + \begin{bmatrix} \frac{W_{12}}{p_1-p_2} & 0 \\ 0 & \frac{W_{46}}{p_4-p_6} \end{bmatrix}\begin{bmatrix} \frac{dp_1}{dt} \\ \frac{dp_6}{dt} \end{bmatrix} \quad (2.107)$$

Given the fact that equation (2.107) is linear, an explicit solution is possible and, indeed would be efficient in this case of only two unknowns. Explicit solutions cease to be the most efficient method of solution when the number of unknowns exceeds three, when Gaussian elimination becomes an attractive direct method.

Solving equation (2.107) allows integration from the initial conditions ($p_2(0)$, $p_4(0)$):

$$\begin{aligned} p_2(t) &= p_2(0) + \int_0^t \frac{dp_2}{dt}dt \\ p_4(t) &= p_4(0) + \int_0^t \frac{dp_4}{dt}dt \end{aligned} \quad (2.108)$$

2.12 Bibliography

Franks, R.G.E. (1972). *Modelling and Simulation in Chemical Engineering*, John Wiley and Sons.

Hempel, O. (1961). On the dynamics of steam-liquid heat exchangers. *Trans. ASME*, June.

Roache, P.J. (1972, revised 1982). *Computational Fluid Dynamics*, Hermosa Publishing.

Smith, G.D. (1965, revised 1974). *Numerical Solution of Partial Differential Equations*, Oxford University Press.

Thomas, P.J. (1997). The Method of Referred Derivatives: a new technique for solving implicit equations in dynamic simulation, *Trans. Inst.M.C.*, **19**, 13–21.

Watson, H.D.D. and Gourlay, A.R. (1976). Implicit integration for CSMP III and the problem of stiffness, *Simulation*, February, 57–61.

3 Thermodynamics and the conservation equations

3.1 Introduction

Every process is subject to the laws of thermodynamics and to the conservation laws for mass and momentum, and we can expect every dynamic simulation of an industrial process to need to invoke one or more of these laws. The interpretation of these laws as they apply to different types of processes leads to different forms for the describing equations. This chapter will begin by reviewing the thermodynamic relations needed for process simulation, and it will go on to derive the conservation equations necessary for modelling the major components found in industrial processes. Finally, the different equations arising from lumped-parameter and distributed-parameter systems containing fluids will be brought out.

3.2 Thermodynamic variables

The thermodynamic state of unit mass of a homogeneous fluid is definite when fixed values are assigned to any two of the following three variables: pressure, p, temperature, T, and specific volume, v. These variables will be connected by an equation of state of the form:

$$f(p, T, v) = 0 \tag{3.1}$$

In particular, it is useful to emphasize that the thermodynamic state is defined completely if the pressure and the temperature of the fluid under consideration are known.

For a gas or a vapour, we may express the equation of state with good accuracy by

$$pv = Z\frac{R}{w}T = ZR_wT \tag{3.2}$$

where

p is the pressure (Pa),
v is the specific volume (m^3/kg),
$Z = Z(p, T)$ is the compressibility factor, where for an ideal gas $Z = 1$,
R is the universal gas constant = 8314 J/(kmol K),
T is the absolute temperature (K),
w is the molecular weight of the gas, and
$R_w = R/w$ is the characteristic gas constant, which applies only to the gas or gas mixture in question (J/kgK).

The compressibility factor, Z, is unity for an ideal gas. Real gases show deviations from the ideal, especially when exposed to a large range of pressures and temperatures. However, we will often wish to calculate gas behaviour over a reasonably restricted range of pressures and temperatures, in which case it is often possible to assign a constant (non-unity) value to Z. Many of the useful results applicable to an ideal gas then carry over to the real gas. We shall call the gas 'near-ideal' when it may be characterized over its operating range by equation (3.2) with Z = constant $\neq 1.0$; we shall use the term 'semi-ideal' when a good characterization requires $Z = Z(T)$.

Pressure, temperature and specific volume have a claim to be regarded as the most basic of the thermodynamic variables because of the ease with which they can be sensed and measured, and hence their familiarity to the practising physicist or engineer. However, there are a number of other thermodynamic variables necessary for the simulation of industrial processes that will be considered here.

The first of these is specific internal energy, u (J/kg). This is the energy possessed by the fluid due to the random motion of its molecules and to their internal potential and vibrational energies. Specific internal energy is strongly dependent on temperature, completely so for an ideal gas, although there may in practice be a small pressure dependency also.

The second additional thermodynamic variable to be considered is entropy, S (J/K). Entropy is a non-obvious variable that was introduced by Clausius in 1854 in connection with his work on the Second Law of Thermodynamics. He considered a reversible cycle converting heat into work, where the heat, Q (J), is supplied and subsequently rejected over a continuous range of temperature, T (K). He deduced that the heat supplied and rejected over the complete cycle and the temperature of the working fluid at the time of the heat transfer obeyed the equation:

$$\oint \frac{dQ}{T}\bigg|_{rev} = 0 \tag{3.3}$$

As a result, he was led to define a new term, S, through the differential:

$$dS = \frac{dQ}{T}\bigg|_{rev} \tag{3.4}$$

He christened this new term 'entropy'.

It should be noted that equation (3.3) applies to a reversible expansion, but does not depend on a particular outward nor return path in thermodynamic space. Entropy is thus a function purely of the state and not the path. The term dS is therefore a perfect differential, and we may integrate (3.4) between thermodynamic states 1 and 2 to give:

$$\int_1^2 dS = S_2 - S_1 = \int_1^2 \left.\frac{dQ}{T}\right|_{rev} \tag{3.5}$$

It follows that the change in entropy is the same for any reversible transition between the same thermodynamic states. Further, it is the change in entropy that is important, and so an arbitrary, convenient reference state is selected to which zero entropy is assigned, e.g. 0°C and 1 bar.

Specific entropy, s (J/(kg K)), is found by dividing the entropy of the working fluid by its mass, and is clearly also a thermodynamic variable dependent solely on the thermodynamic state, like pressure and temperature.

It is possible, as a general procedure, to form a new thermodynamic variable dependent solely on the thermodynamic state by combining any two or more of the thermodynamic variables above. An example of which we will make extensive use is specific enthalpy. Specific enthalpy, h (J/kg), is formed by amalgamating specific internal energy with two basic thermodynamic variables, pressure and specific volume:

$$h = u + pv \tag{3.6}$$

This particular grouping arises naturally in the equations for the conservation of energy, as will be shown later.

As already noted, all the thermodynamic variables introduced above are dependent purely on the thermodynamic state of the fluid under consideration. Since the thermodynamic state of the fluid may be completely defined by its pressure and temperature, it follows that we may regard all other thermodynamic variables as functions of pressure and temperature alone. Hence we may write:

$$\begin{aligned} v &= v(p, T) \\ u &= u(p, T) \\ s &= s(p, T) \\ h &= h(p, T) \end{aligned} \tag{3.7}$$

and, further, we may write the total differentials as:

$$\begin{aligned} dv &= \frac{\partial v}{\partial p} dp + \frac{\partial v}{\partial T} dT \\ du &= \frac{\partial u}{\partial p} dp + \frac{\partial u}{\partial T} dT \\ s &= \frac{\partial s}{\partial p} dp + \frac{\partial s}{\partial T} dT \\ h &= \frac{\partial h}{\partial p} dp + \frac{\partial h}{\partial T} dT \end{aligned} \tag{3.8}$$

where the partial differentials, $\partial v/\partial p$, $\partial v/\partial T$, $\partial u/\partial p \dots$ are themselves functions of p and T:

$$\frac{\partial v}{\partial p} = \frac{\partial v}{\partial p}(p, T), \; \frac{\partial v}{\partial T} = \frac{\partial v}{\partial T}(p, T),$$

$$\frac{\partial u}{\partial p} = \frac{\partial u}{\partial p}(p, T), \dots$$

3.3 Specific heats of gases

There are a number of relationships concerning the specific heats of gases that are of significant use to the process modeller. The specific heat, c, is defined as the amount of heat that must be supplied to raise the temperature of unit mass of a substance by one degree:

$$c = \frac{dq}{dT} \tag{3.9}$$

where dq (J/kg) is the (small) amount of heat that causes the temperature of unit mass of the substance to rise by the small amount dT (K). The specific heat, c, will have the units J/(kg K). Different amounts of heat will be necessary to raise the temperature of the gas, depending on whether and to what extent the gas is allowed to expand during the heating process, and thus there are any number of possible specific heats. Two specific heats are particularly important: the specific heat arising when the heating takes place with the gas volume kept constant, c_v, and the specific heat arising when the gas is kept at a constant pressure during the heating process, c_p. These are known as the principal specific heats.

It is possible to express the principal specific heats in terms of the thermodynamic variables we have introduced previously. The first law of thermodynamics may be written

$$dq = du + p\,dv \tag{3.10}$$

where the term $p\,dv$ represents the work done by the gas in expanding. But if the volume is held constant during the heating process, then there will be no expansion, and so all the heat will appear as a change in internal energy:

$$dq|_v = du|_v \tag{3.11}$$

Thus the specific heat at constant volume is found by substituting from (3.11) into (3.9) to give

$$c_v = \left.\frac{du}{dT}\right|_v \tag{3.12}$$

In fact, specific internal energy is dependent solely on temperature for an ideal gas, and so the constant-volume subscript, v, may be dropped:

$$c_v = \frac{du}{dT} \tag{3.13}$$

For the case when the pressure is kept constant during the heating process, we begin by noting the definition of specific enthalpy, namely:

$$h = u + pv \tag{3.6}$$

The incremental change in specific enthalpy for a constant-pressure heating will be

$$dh|_p = d(u + pv)|_p = (du + p\,dv)|_p \tag{3.14}$$

But applying equation (3.10) to the case of constant-pressure heating, we have

$$dq|_p = (du + p\,dv)|_p \tag{3.15}$$

Hence $dq|_p = dh|_p$, so that, from equation (3.9):

$$C_p = \left.\frac{dh}{dT}\right|_p \tag{3.16}$$

But we may demonstrate that specific enthalpy is a function of temperature only for a near-ideal or semi-ideal gas by substituting for the term pv from the equation of state (3.2) into equation (3.6):

$$h = u + ZR_wT \tag{3.17}$$

Since specific internal energy is a function of temperature alone, it follows from equation (3.18) that specific enthalpy will be a function solely of temperature provided that Z and R_w are either constant or functions of temperature only – conditions that are met for a near-ideal and semi-ideal gas. Hence equation (3.16) may be replaced by the simpler:

$$C_p = \frac{dh}{dT} \tag{3.18}$$

It may be noted in passing that the specific heat for a liquid or a solid cannot be determined at constant volume because each will expand when it is heated. For all practical purposes, the specific heat for a liquid or a solid has a single value, namely the specific heat at constant pressure. The symbol c_p is retained, and equation (3.18) applies.

3.3.1 Relationships between the principal specific heats for a near-ideal gas

Differentiating equation (3.17) with respect to temperature for a near-ideal gas when Z is constant gives:

$$\frac{dh}{dT} = \frac{du}{dT} + ZR_w \tag{3.19}$$

or, using the definitions of c_v and c_p given in equations (3.12) and (3.16),

$$c_p = c_v + ZR_w \tag{3.20}$$

We can proceed further, since the kinetic theory of gases gives us the result for a near-ideal gas that:

$$c_v = \frac{N_F}{2} ZR_w \tag{3.21}$$

where N_F is the number of degrees of freedom for the gas in question, given by:

$$\begin{aligned} N_F &= 3 \text{ for a monatomic gas} \\ &= 5 \text{ for a diatomic gas} \\ &= 6 \text{ for a polyatomic gas} \end{aligned} \tag{3.22}$$

Hence, by (3.20),

$$c_p = \left(1 + \frac{N_F}{2}\right) ZR_w \tag{3.23}$$

The ratio, γ, of the principal specific heats of a gas is of importance in expansion and compression processes, since it may be shown that a reversible, adiabatic (isentropic) expansion or compression will obey the law:

$$pv^\gamma = \text{constant} \tag{3.24}$$

where

$$\gamma = \frac{c_p}{c_v} \tag{3.25}$$

or, using (3.21) and (3.23),

$$\gamma = 1 + \frac{2}{N_F} \tag{3.26}$$

Equation (3.26) predicts $\gamma = 1.67$ for a monatomic gas such as helium or argon, $\gamma = 1.4$ for a diatomic gas such as oxygen, and $\gamma = 1.33$ for a polyatomic gas such as superheated steam. All these figures are in good general agreement with the values measured for real gases, although $\gamma = 1.3$ is normally a more realistic value for superheated steam.

3.4 Conservation of mass in a bounded volume

The principle of the conservation of mass will be invoked in just about every process simulation. Often we will need to consider the behaviour of tanks and vessels receiving one or more inflows and supplying one or more outflows. The fluid may be gas in a vessel, liquid in a tank or vapour in a vessel containing both liquid and vapour. Note that in the first case the volume is fixed by the confines of the tank, but when two phases are present in the same vessel, the volumes of each phase will change as the liquid boundary rises or falls.

Figure 3.1 depicts a bounded volume, where one or more of the boundaries is free to move. The principle of conservation of mass requires that

the rate of increase of mass
= the mass inflow – the mass outflow

or, applied to the system of Figure 3.1 and expressed as a differential equation,

$$\frac{dm}{dt} = W_1 + W_2 + W_3 - W_4 - W_5 \qquad (3.27)$$

where m is the mass in the bounded volume, and W_i are the mass flows.

In the general case,

$$\frac{dm}{dt} = \sum W_{in} - \sum W_{out} \qquad (3.28)$$

3.5 Conservation of energy in a fixed volume

Let us consider the fixed volume shaded in Figure 3.2. Conservation of energy requires that:

> the change of energy in the fixed volume *equals* the heat input *minus* the work output *plus* the work done on the fixed volume by the incoming fluid *minus* the work done by the outgoing fluid *plus* the energy brought into the fixed volume by the incoming fluid *minus* the energy leaving the fixed volume with the outgoing fluid.

The energy contained in the inlet and outlet streams will exist in three forms: internal energy, mu, kinetic energy, $\frac{1}{2}mc^2$, and potential energy relative to a given datum, mgz, where m is the mass under consideration, c is its velocity and z is its height above the datum.

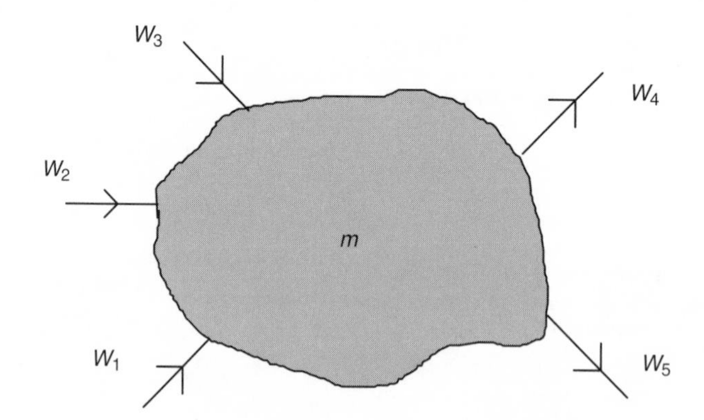

Figure 3.1 Bounded volume with mass inflows and outflows.

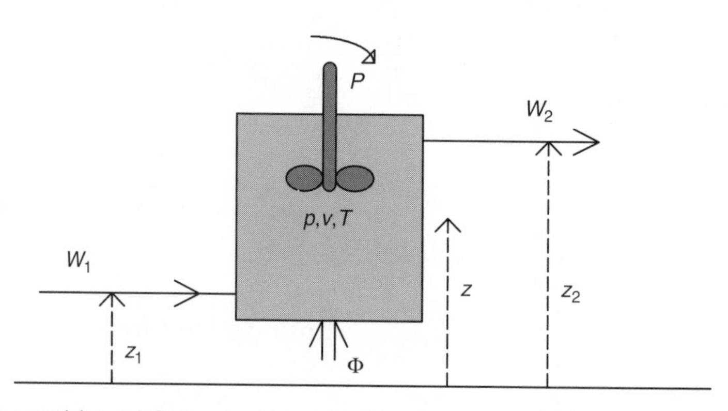

Figure 3.2 Fixed volume receiving an inflow and outflow, subject to a heat input and doing work.

Applying the principle of the conservation of energy to the fixed volume shown in Figure 3.2 during a time interval δt gives:

$$\delta E = \Phi\,\delta t - P\,\delta t + W_1\delta t\left(u_1 + \frac{1}{2}c_1^2 + gz_1\right) - W_2\delta t\left(u_2 + \frac{1}{2}c_2^2 + gz_2\right) + p_1 W_1\delta t\,v_1 - p_2 W_2\delta t\,v_2 \qquad (3.29)$$

where:

E is the energy contained in the fixed volume (J),
Φ is the heat flux into the fixed volume (W),
P is the mechanical power abstracted from the fixed volume (W),
W_1, W_2 are the inlet and outlet mass flows (kg/s),
u_1, u_2 are the inlet and outlet specific internal energies (J/kg),
c_1, c_2 are the inlet and outlet flow velocities (m/s),
z_1, z_2 are the heights above the datum of the inlet and outlet flows (m),
p_1, p_2 are the pressures of the inlet flow and the outlet flow (Pa).
Note that $W_1\,\delta t\,v_1 = \delta V_1$, the volume of fluid introduced in time δt, and similarly $W_2\,\delta t\,v_2 = \delta V_2$, the volume of fluid leaving in time δt.

Dividing equation (3.29) by δt and letting $\delta t \to 0$ allows us to write the differential equation:

$$\frac{dE}{dt} = \Phi - P + W_1\left(u_1 + \frac{1}{2}c_1^2 + gz_1\right) - W_2\left(u_2 + \frac{1}{2}c_2^2 + gz_2\right) + W_1 p_1 v_1 - W_2 p_2 v_2 \qquad (3.30)$$

We will now make the assumption that the contents are well mixed so that we may characterize each of the variables temperature, specific internal energy and specific volume by a single, bulk value that holds throughout the volume. In particular, the values at the outlet are the same as the bulk values:

$$T_2 = T \qquad u_2 = u \qquad v_2 = v \qquad (3.31)$$

Further, we will assume that there is no frictional loss due to flow from the inlet to the outlet. As a result, the pressure at the outlet and pressure at the mid-point differ only by the head difference:

$$p_2 = p + \frac{g}{v}(z - z_2) \qquad (3.32)$$

where z is the height above the datum of the centre of gravity of the fluid, taken as the position where the pressure is equal to its bulk value, p. Hence

$$p_2 v_2 = p_2 v = pv + g(z - z_2) \qquad (3.33)$$

Using equation (3.33) and noting also the definition of specific enthalpy as $h = u + pv$, we may rewrite (3.30) as:

$$\frac{dE}{dt} = \Phi - P + W_1\left(h_1 + \frac{1}{2}c_1^2 + gz_1\right) - W_2\left(h + \frac{1}{2}c_2^2 + gz\right) \qquad (3.34)$$

Now the energy, E, contained in the fixed volume is the sum of the fluid's internal, kinetic and potential energies:

$$E = m\left(u + \frac{1}{2}c^2 + gz\right) \qquad (3.35)$$

where m is the mass of the fluid in the fixed volume and c is its bulk velocity. But it is shown in Appendix 1 that we may neglect the kinetic energy and potential energy terms in the fixed volume in the normal process-plant and so may simplify equation (3.35) to

$$E = mu \qquad (3.36)$$

so that the time-differential of the energy in the fixed volume is

$$\frac{dE}{dt} = u\frac{dm}{dt} + m\frac{du}{dt} \qquad (3.37)$$

Hence we may set down the full form of the equation for the conservation of energy in a fixed volume as:

$$u\frac{dm}{dt} + m\frac{du}{dt} = \Phi - P + W_1\left(h_1 + \frac{1}{2}c_1^2 + gz_1\right) - W_2\left(h + \frac{1}{2}c_2^2 + gz\right) \qquad (3.38)$$

Applying the principle of the conservation of mass to this system gives:

$$\frac{dm}{dt} = W_1 - W_2 \qquad (3.39)$$

Equations (3.38) and (3.39) are two simultaneous equations in the two unknowns dm/dt and du/dt. Their convenient form allows us to reframe (3.38) as:

$$m\frac{du}{dt} = \Phi - P + W_1\left(h_1 + \frac{1}{2}c_1^2 + gz_1 - u\right) - W_2\left(h + \frac{1}{2}c_2^2 + gz - u\right) \qquad (3.40)$$

In many cases the remarks made in Appendix 1 about the relatively negligible values of kinetic energy and potential energy in the fixed volume will apply equally to the incoming and outgoing flows, so that it will then be possible to neglect these terms on the right–hand side of equation (3.40), leading to the simpler form:

$$m\frac{du}{dt} = \Phi - P + W_1(h_1 - u) - W_2(h - u) \quad (3.41)$$

Now

$$h - u = pv \quad (3.42)$$

Thus equation (3.41) becomes

$$m\frac{du}{dt} = \Phi - P + W_1(h_1 - u) - W_2 pv \quad (3.43)$$

where p is the average, 'bulk' pressure in the fixed volume. Equation (3.43) may now be integrated numerically with respect to time to solve for the specific internal energy, u.

We will normally wish to see the effect on the temperature of the enclosed volume. It is almost always possible to assume that specific internal energy is a function of temperature alone, specifically in the following cases:

(i) where the fluid in the fixed volume is a liquid, since pressure has only a slight effect on the specific internal energy;
(ii) when the fluid in the fixed volume is an ideal or near-ideal gas, when the specific internal energy is a function only of temperature, so that du/dT is a constant;
(iii) when the fluid in the fixed volume is a real gas with temperature more than about twice its critical value and pressure up to about five times its critical value; the overwhelming majority of gases as used in industrial processes come into either this category or category (ii) above;
(iv) when the fluid in the fixed volume is in vapour–liquid equilibrium because of boiling or condensation. In this condition pressure is a function of temperature, so any dependence on pressure is automatically a dependence on temperature.

We may thus replace du/dt by $(du/dT) \times (dT/dt)$ in equation (3.43), thus transforming it into a differential equation in temperature:

$$\frac{dT}{dt} = \frac{\Phi - P + W_1(h_1 - u) - W_2 pv}{m\dfrac{du}{dT}} \quad (3.44)$$

3.6 Effect of volume change on the equation for the conservation of energy

We assumed in Section 3.5 that the vessel had a fixed geometry, so that no power was spent in bulk expansion. In the absence of any other power output, we could set $P = 0$ in equation (3.43). But if the bounded volume had one or more free surfaces (e.g. a inside a piston chamber, or above a liquid in a vessel with a gas over-blanket), then we would need to take account of the work done against the imposed pressure. Let us take the case where the top surface in Figure 3.2 moves up a small amount in the time interval δt, so that the volume of the fluid increases by an amount δV (m^3). Assuming that the pressure above this surface is p_t (Pa), the work done is given by:

$$P\,\delta t = p_t\,\delta V \quad (3.45)$$

Dividing by δt and then letting $\delta t \to 0$ allows us to evaluate the power, P, as

$$P = p_t\frac{dV}{dt} \quad (3.46)$$

The pressure, p_t, differs from the bulk pressure of the fluid, p, only by the head difference, which may often be neglected. In this case,

$$P = p\frac{dV}{dt} \quad (3.47)$$

3.7 Conservation of energy equation for a rotating component

A number of rotating components are in common use in process plants, e.g. turbines, compressors, pumps, centrifuges and stirring paddles, and it is important to understand how such pieces of equipment are affected by the conservation of energy. For a rotating component, the principle of conservation of energy states that

> the rate of change of rotational energy *equals* the power in *minus* the useful power out and *minus* the power lost in friction.

The rotational energy is given by

$$\begin{aligned} E &= \tfrac{1}{2}J\omega^2 \\ &= 2\pi^2 J N^2 \end{aligned} \quad (3.48)$$

where:
J is the moment of inertia (kg m^2),
ω is the rotational speed in radians per second,
N is the rotational speed in revolutions per second.

A formal differentiation of (3.48) with respect to time gives:

$$\frac{dE}{dt} = 4\pi^2 JN\frac{dN}{dt} + 2\pi^2 N^2\frac{dJ}{dt} \tag{3.49}$$

The term in dJ/dt has been retained, since it is needed routinely in, for example, the modelling of solids–liquid separation centrifuges and it may be needed to model other rotating components in fault situations where change of geometry is possible.

The conservation of energy equation is therefore:

$$4\pi^2 JN\frac{dN}{dt} + 2\pi^2 N^2\frac{dJ}{dt} = P_{in} - P_{out} - P_F \tag{3.50}$$

where

P_{in} is the input power (W),
P_{out} is the useful output power (W),
P_F is the power expended against friction (W).

Power lost to friction is likely to have a component proportional to N^2 for motion against a lubricated surface and a component proportional to N^3 for motion through fluids.

3.8 Conservation of mass in a pipe

So far we have considered the conservation equations applied to a systems where temperature, pressure, specific volume, etc. could each be characterized by a single bulk variable, valid throughout the system. But in a long pipe or tube carrying a flowing fluid, such quantities will vary continuously along its length, making a long pipe an inherently distributed system. We need to take account of this variation in space as well as time in order to model the behaviour of such diverse systems as once-through boilers in power stations, where the tubes are tens of metres long, and oil and gas pipelines that may be tens of kilometres long.

The distributed nature of long pipes has implications for how the conservation principles apply. We will begin by developing the equation for the conservation of mass.

We assume that the fluid is homogeneous, that its flow is purely parallel to the axis of the pipe and of uniform velocity across the pipe, i.e. one-dimensional. Consider the element of pipe shown in Figure 3.3.

All variables are functions of both time and distance. Hence at any time, t, the values at the reference distance x are:

$$\text{specific volume} = v$$

$$\text{velocity} = c$$

where c will have units of m/s. At the same time instant, the values at $x + \delta x$ are:

$$\text{specific volume} = v + \delta v = v + \frac{\partial v}{\partial x}\delta x$$

$$\text{velocity} = c + \delta c = c + \frac{\partial c}{\partial x}\delta x$$

By geometry, the mass flow into the pipe element is:

$$W_{in} = A\frac{c}{v} \tag{3.51}$$

where A is the cross-sectional area of the pipe (m^2), while the flow out may be written formally as:

$$W_{out} = A\frac{c}{v} + \frac{\partial}{\partial x}\left(A\frac{c}{v}\right)\delta x \tag{3.52}$$

The mass of the pipe element is the volume divided by the specific volume:

$$m = \frac{A\,\delta x}{v} \tag{3.53}$$

so that the rate of change of mass in the pipe element is:

$$\frac{dm}{dt} = \frac{d}{dt}\left(\frac{A\,\delta x}{v}\right) \tag{3.54}$$

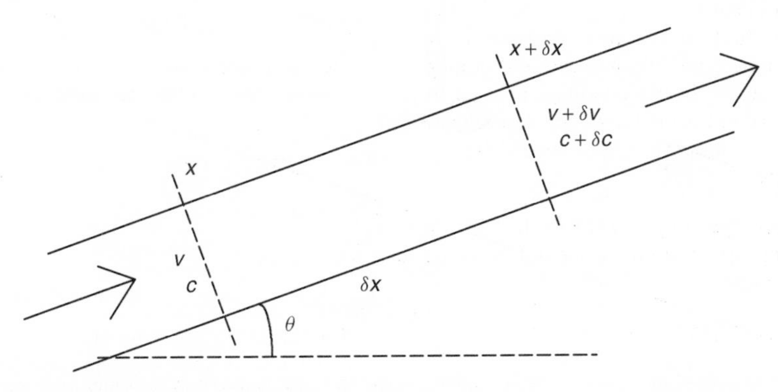

Figure 3.3 Element of pipe at an angle θ to the horizontal, showing specific volume and velocity variations.

We may now invoke the principle of the conservation of mass as applied to the pipe element, as given by equation (3.28):

$$\frac{dm}{dt} = \sum W_{in} - \sum W_{out} \tag{3.28}$$

Substituting from equations (3.51), (3.52) and (3.53) gives:

$$\frac{\partial}{\partial t}\left(\frac{A\,\delta x}{v}\right) = -\frac{\partial}{\partial x}\left(A\frac{c}{v}\right)\delta x \tag{3.55}$$

It will be noted that a partial derivative is needed in order to specify the time differential at any given point in space along the pipe. Since δx does not vary with time, it may be cancelled from the above equation to give the general form:

$$\frac{\partial}{\partial t}\left(\frac{A}{v}\right) = -\frac{\partial}{\partial x}\left(A\frac{c}{v}\right) \tag{3.56}$$

In many cases the cross-sectional area of the part of the pipe under consideration will not vary with either time or distance, so that equation (3.54) can be written:

$$\frac{\partial}{\partial t}\left(\frac{1}{v}\right) = -\frac{\partial}{\partial x}\left(\frac{c}{v}\right) \tag{3.57}$$

or

$$\frac{\partial v}{\partial t} = v^2\frac{\partial}{\partial x}\left(\frac{c}{v}\right) \tag{3.58}$$

Since the only time differential is specific volume, v, we may, in theory, apply finite differences in space to the partial differential in distance and solve by numerical integration for the later behaviour of v, provided we are first given

(i) the starting distribution of the specific volume $v(x, t)$ for all x at $t = 0$, and
(ii) the behaviour of velocity $c(x, t)$ for all x and all t under consideration

The first requirement is likely to be met relatively easily, but the second is unlikely to be known in advance. Hence equation (3.58) is unlikely to fulfil the modeller's needs on its own.

3.9 Conservation of energy in a pipe

We will consider the element of pipe redrawn in Figure 3.4. The pipe element is subject to a heat flux per unit length of ϕ watts per metre, and we have allowed for the possibility of the flow contributing useful work of ψ watts per metre.

To derive the equation for the conservation of energy, we proceed in the same way as we derived the conservation of mass equation, regarding all variables as functions of both time and distance. Hence at any time, t, the values at the reference distance x are:

$$\text{pressure} = p$$

$$\text{specific volume} = v$$

$$\text{velocity} = c$$

$$\text{specific internal energy} = u$$

$$\text{height} = z$$

while at the same time instant, the values at $x + \delta x$ are:

$$\text{pressure} = p + \delta p = p + \frac{\partial p}{\partial x}\delta x$$

$$\text{specific volume} = v + \delta v = v + \frac{\partial v}{\partial x}\delta x$$

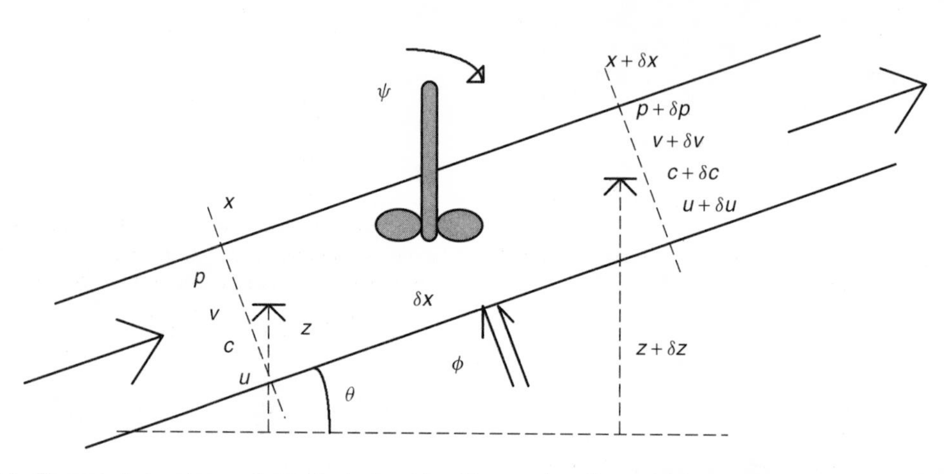

Figure 3.4 Element of pipe at an angle θ to the horizontal, subject to a heat flux and doing useful work: variations in pressure, specific volume, velocity and specific internal energy.

$$\text{velocity} = c + \delta c = c + \frac{\partial c}{\partial x}\,\delta x$$

$$\text{specific internal energy} = u + \delta u = u + \frac{\partial u}{\partial x}\,\delta x$$

$$\text{height} = z + \delta z = z + \frac{\partial z}{\partial x}\,\delta x$$

$$= z + \delta x \sin\theta$$

The principle of conservation of energy is the same as stated at the beginning of Section 3.4, namely:

> the change of energy in the pipe element *equals* the heat input *minus* the work output *plus* the energy brought into the pipe element by the incoming fluid *minus* the energy leaving the pipe element with the outgoing fluid *plus* the work done on the pipe element by the incoming fluid *minus* the work done by the outgoing fluid

The principle applied for a time interval δt gives:

$$\begin{aligned}\delta E &= \phi\,\delta x\,\delta t - \psi\,\delta x\,\delta t + \frac{Ac}{v}\,\delta t\left(u + \frac{1}{2}c^2 + gz\right)\\ &\quad - \left[\frac{Ac}{v}\,\delta t\left(u + \frac{1}{2}c^2 + gz\right)\right.\\ &\quad \left. + \frac{\partial}{\partial x}\left(\frac{Ac}{v}\,\delta t\left(u + \frac{1}{2}c^2 + gz\right)\right)\delta x\right]\\ &\quad + pAc\,\delta t - \left[pAc\,\delta t + \frac{\partial}{\partial x}(pAc\,\delta t)\,\delta x\right] \qquad (3.59)\end{aligned}$$

Cancelling terms gives

$$\begin{aligned}\delta E &= \phi\,\delta x\,\delta t - \psi\,\delta x\,\delta t\\ &\quad - \frac{\partial}{\partial x}\left(\frac{Ac}{v}\left(u + \frac{1}{2}c^2 + gz\right)\right)\delta x\,\delta t\\ &\quad - \frac{\partial}{\partial x}(pAc)\,\delta x\,\delta t \qquad (3.60)\end{aligned}$$

Noting first that $pAc = (Ac/v)pv$, and then using the identity $h = u + pv$ allows us to re-express (3.60) as:

$$\begin{aligned}\delta E &= \phi\,\delta x\,\delta t - \psi\,\delta x\,\delta t\\ &\quad - \frac{\partial}{\partial x}\left(\frac{Ac}{v}\left(h + \frac{1}{2}c^2 + gz\right)\right)\delta x\,\delta t \qquad (3.61)\end{aligned}$$

Dividing by δt and letting $\delta t \to 0$ gives the partial differential equation:

$$\begin{aligned}\frac{\partial E}{\partial t} &= \phi\,\delta x - \psi\,\delta x\\ &\quad - \frac{\partial}{\partial x}\left(\frac{Ac}{v}\left(h + \frac{1}{2}c^2 + gz\right)\right)\delta x \qquad (3.62)\end{aligned}$$

Now the energy in the pipe element, E, is given by the mass *times* the total of the specific internal energy plus the kinetic energy per unit mass plus the potential energy per unit mass, or, in mathematical symbols:

$$E = \frac{A\,\delta x}{v}\left(u + \frac{1}{2}c^2 + gz\right) \qquad (3.63)$$

where z is the height of the centre of gravity of the element above the entrance to the pipe, which is taken as the datum. We may express this height as

$$z = x\sin\theta \qquad (3.64)$$

where x is the distance measured from the start of the pipe. For a rigidly connected pipe, the element will be fixed in space and so z will not change with time.

Hence the time differential is given by:

$$\begin{aligned}\frac{\partial E}{\partial t} &= \left(u + \frac{1}{2}c^2 + gz\right)\frac{\partial}{\partial t}\left(\frac{A}{v}\right)\delta x\\ &\quad + \frac{A}{v}\frac{\partial}{\partial t}\left(u + \frac{1}{2}c^2\right)\delta x \qquad (3.65)\end{aligned}$$

Substituting from equation (3.65) into equation (3.62) and dividing throughout by δx yields the required equation for the conservation of energy:

$$\begin{aligned}&\left(u + \frac{1}{2}c^2 + gz\right)\frac{\partial}{\partial t}\left(\frac{A}{v}\right) + \frac{A}{v}\frac{\partial}{\partial t}\left(u + \frac{1}{2}c^2\right)\\ &\quad = \phi - \psi - \frac{\partial}{\partial x}\left(\frac{Ac}{v}\left(h + \frac{1}{2}c^2 + gz\right)\right)\end{aligned} \qquad (3.66)$$

We have from a differentiation of equation (3.64):

$$\frac{\partial z}{\partial x} = \sin\theta \qquad (3.67)$$

and thus we may rewrite (3.66) as:

$$\begin{aligned}&\left(u + \frac{1}{2}c^2 + gx\sin\theta\right)\frac{\partial}{\partial t}\left(\frac{A}{v}\right) + \frac{A}{v}\frac{\partial}{\partial t}\left(u + \frac{1}{2}c^2\right)\\ &\quad = \phi - \psi - \frac{Ac}{v}\frac{\partial}{\partial x}\left(h + \frac{1}{2}c^2\right)\\ &\qquad - \left(h + \frac{1}{2}c^2 + gx\sin\theta\right)\frac{\partial}{\partial x}\left(\frac{Ac}{v}\right)\\ &\qquad - \frac{Ac}{v}g\sin\theta \qquad (3.68)\end{aligned}$$

For simple pipe-flow, no work is done, so $\psi = 0$, and the cross-sectional area, A, is constant, so that the conservation of energy equation may be rewritten:

$$\begin{aligned}&\left(u + \frac{1}{2}c^2 + gx\sin\theta\right)\frac{\partial}{\partial t}\left(\frac{1}{v}\right) + \frac{1}{v}\frac{\partial}{\partial t}\left(u + \frac{1}{2}c^2\right)\\ &\quad = \frac{\phi}{A} - \frac{c}{v}\frac{\partial}{\partial x}\left(h + \frac{1}{2}c^2\right) - \left(h + \frac{1}{2}c^2 + gx\sin\theta\right)\\ &\qquad \times \frac{\partial}{\partial x}\left(\frac{c}{v}\right) - \frac{c}{v}g\sin\theta \qquad (3.69)\end{aligned}$$

Assuming as in the previous section that finite differences can be applied to the right-hand side of the equation, we have effectively a single equation in the three unknowns: $\partial v/\partial t$, $\partial u/\partial t$ and $\partial c/\partial t$. We may add the mass conservation equation (3.55) in $\partial v/\partial t$ to give us two equations in three unknowns, but this still leaves us one equation short. For a full dynamic simulation, we need to consider momentum as well.

3.10 Conservation of momentum in a pipe

Figure 3.3 has now been redrawn showing the forces acting on the fluid element in the pipe (Figure 3.5).

The conservation of momentum requires that:

> the change in momentum in the pipe element *equals* the momentum brought in by the incoming fluid *minus* the momentum leaving with the outgoing fluid *plus* the impulse of the force on the fluid due to pressure at the inlet *minus* the impulse of the force on the fluid from the pressure at the outlet *minus* the impulse of the frictional force acting against the flow *minus* the impulse of the gravitational force acting against the flow.

Expressed mathematically for a period of length δt:

$$\delta\left(\frac{A\,\delta x}{v}c\right) = \frac{Ac}{v}\,\delta t\,c - \left[\frac{Ac}{v}\,\delta t\,c + \frac{\partial}{\partial x}\left(\frac{Ac}{v}\,\delta t\,c\right)\delta x\right] + pA\,\delta t - \left[pA\,\delta t + \frac{\partial}{\partial x}(pA\,\delta t)\,\delta x\right] - \tau\pi D\,\delta x\,\delta t - \frac{A\,\delta x}{v}g\sin\theta\,\delta t \tag{3.70}$$

where:

D is the diameter of the pipe (m), and

τ is the frictional shear stress (N/m^2), given by the equation:

$$\tau = \frac{1}{2}\frac{c|c|}{v}f \tag{3.71}$$

where f is the dimensionless Fanning friction factor; the modulus is used to ensure that the frictional force always acts to oppose the direction of flow.

We now substitute (3.71) into (3.70), cancel out terms, divide by $\delta x\,\delta t$, then let $\delta t \to 0$ to give:

$$\frac{\partial}{\partial t}\left(\frac{Ac}{v}\right) = -\frac{\partial}{\partial x}\left(\frac{Ac^2}{v}\right) - \frac{\partial}{\partial x}(pA) - \frac{\pi Dc|c|}{2v}f - \frac{Ag}{v}\sin\theta \tag{3.72}$$

For the common case where $A = \pi D^2/4 = \text{constant}$, we may rewrite (3.72) as:

$$\frac{\partial}{\partial t}\left(\frac{c}{v}\right) = -\frac{\partial}{\partial x}\left(p + \frac{c^2}{v}\right) - \frac{2c|c|}{vD}f - \frac{g}{v}\sin\theta \tag{3.73}$$

Assuming once more that the right-hand side of this equation can be subjected to a finite-difference scheme, equation (3.73) provides a further equation in $\partial c/\partial t$ and $\partial v/\partial t$, so that in equations (3.58) and (3.69) and (3.73) we have three equations in the three unknowns $\partial v/\partial t$, $\partial u/\partial t$ and $\partial c/\partial t$. But we should note that the right-hand side of (3.73) contains not only c and v, but also pressure p. Also, we should note that in practice, we will want to know also the temperature T as a function of time and space. Since v and u are independent thermodynamic functions, it is possible to express a third, such as pressure or temperature,

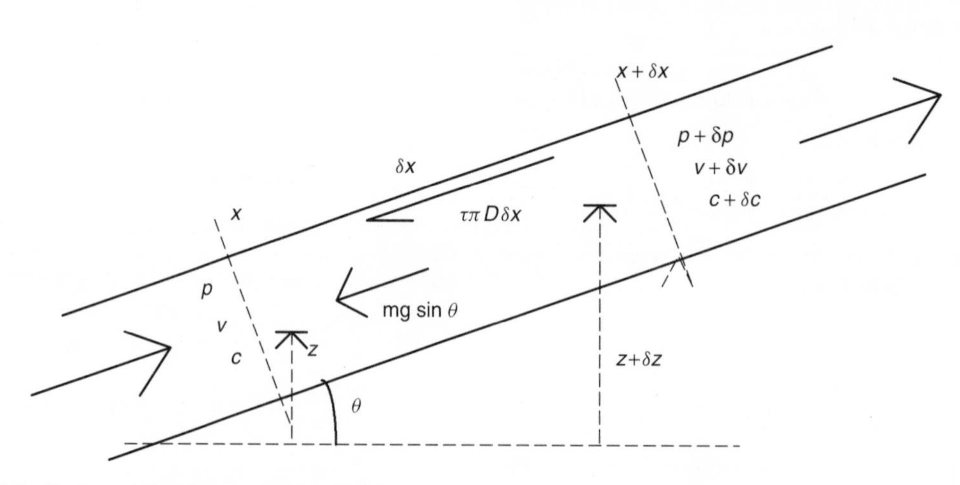

Figure 3.5 Element of pipe showing the forces acting.

as a function of them. So that we should be able to substitute

$$p = p(u, v) \tag{3.74}$$

in equation (3.73) and proceed with the solution for the three unknowns $\partial v/\partial t$, $\partial u/\partial t$ and $\partial c/\partial t$. In practice, the thermodynamic properties are not usually tabulated in terms of u and v, and an alternative procedure is to use the expressions foreshadowed in equation set (3.8), so that we substitute into (3.58), (3.69) and (3.73):

$$\frac{\partial v}{\partial t} = \frac{\partial v}{\partial p}\frac{\partial p}{\partial t} + \frac{\partial v}{\partial T}\frac{\partial T}{\partial t} \tag{3.75}$$

$$\frac{\partial u}{\partial t} = \frac{\partial u}{\partial p}\frac{\partial p}{\partial t} + \frac{\partial u}{\partial T}\frac{\partial T}{\partial t} \tag{3.76}$$

At this stage, we have three equations in the three unknowns: $\partial p/\partial t$, $\partial T/\partial t$ and $\partial c/\partial t$, and we use the relations:

$$\begin{aligned} u &= u(p, T) \\ v &= v(p, T) \end{aligned} \tag{3.77}$$

to calculate the terms needed on the right-hand side of the equations so that the solution can proceed.

To summarize: the three conservation equations (3.58), (3.69) and (3.73), taken together with the thermodynamic relations of (3.77) allow a solution to proceed by time integration at discrete points in space along the pipe. The conceptual solution has the following characteristics: given an initial distribution of pressure, temperature and velocity along the pipe, $p(x, 0)$, $T(x, 0)$, $c(x, 0)$ for $0 \le x \le L$, where L is the length of the pipe, all future distributions $p(x, t)$, $T(x, t)$, $c(x, t)$ may be calculated (at least approximately) by numerical integration with respect to time at a number of discrete points along the pipe, with, in general, the accuracy being greater if a larger number of points is chosen.

In practice, modellers normally seek to simplify the equations when applying them to particular situations. It is usual when considering pipeline dynamics, for example, to ignore changes in energy and hence temperature, and solve only the two conservation equations for mass and momentum (see Chapter 18). On the other hand, the energy equation dominates, understandably, in the simulation of heat exchangers (see Chapter 19). Chapter 19 also deals with the further complexities involved when the flow is accompanied by chemical reaction, as in gas flow through a catalyst bed.

3.11 Bibliography

BNES (1980). *Boiler Dynamics and Control in Nuclear Power Stations 2, Proceedings of the Second International Conference* held in Bournemouth, 23–25 October 1979 (British Nuclear Energy Society, 1980), particularly the following papers:

Fincham, A.E. and Goldwater, M.H. (1979). Simulation models for gas transmission networks, *Trans. Inst. Measurement and Control*, **1**, 3–12.

Lockey, J. (1966). *The Thermodynamics of Fluids*, Heinemann Educational Books Ltd, London.

Partington J.R. (1913). *Thermodynamics*, Constable and Co. Ltd, London (4th revised edition 1950).

BESBET II – A computer code for the simulation of severe transients in LMFBR boilers, R.K. Thomasson.

Numerical comparison between a reference and simplified two-phase flow models as applied to steam generator dynamics, J.-F. Dupont, G. Sarlos, D.M. Le Febve and P. Suter.

NUMEL – A computer aided design suite for the assessment of the steady state, static-dynamic stability and transient responses of once-through steam generators, P. Lightfoot, R.T. Deam, C.H. Green and J. Rea.

4 Steady-state incompressible flow

4.1 Introduction

The flow equations developed in Chapter 3, Sections 3.8, 3.9 and 3.10 cover the full, dynamic case, but very often flow will reach a stable, steady-state value very quickly, certainly much quicker than equilibrium can establish itself in the other parts of the plant being simulated. In such instances, following the precepts of Chapter 2, Section 2.9, it will be sensible to neglect the small time constants associated with flow settling, and assume that the flow reaches its steady-state value instantaneously. One method of proceeding would be to set the time-differentials to zero in the equations developed for conservation of mass, energy and momentum, and integrate numerically with respect to distance, with boundary values of upstream and downstream pressure defined. But very often this procedure will be overelaborate. For the vast majority of cases we want a reasonably accurate characterization of flow in terms of the known upstream parameters (pressure, specific volume) and the downstream pressure. This chapter begins with a development of general equations describing both compressible and incompressible flow. It will go on to introduce simplifications applicable for incompressible flow, defined as flow where the change in specific volume may be considered small. This categorization applies not only to the flow of liquids, but also to the flow of gases and vapours when the pressure drop is sufficiently small – of the order of 5%.

4.2 The energy equation for general steady-state flow

To develop the appropriate equations, we begin with the energy equation for a small element of pipe of length δx from Chapter 3, Section 3.9:

$$\frac{\partial E}{\partial t} = \phi\,\delta x - \psi\,\delta x - \frac{\partial}{\partial x}\left(\frac{Ac}{v}\left(h + \frac{1}{2}c^2 + gz\right)\right)\delta x \tag{3.62}$$

where

E is the energy contained in the element (W),

ϕ is the rate of energy input per unit length, (W/m),

ψ is the power abstracted per unit length, (W/m), and

$(Ac/v) = W$ is the mass flow rate through the pipe element, (kg/s).

The energy contained in the element will be unchanged with time in the steady state:

$$\frac{\partial E}{\partial t} = 0 \tag{4.1}$$

Further, the partial differential on the right-hand side of equation (3.62) may be expanded as

$$\begin{aligned}
&\frac{\partial}{\partial x}\left(\frac{Ac}{v}\left(h + \frac{1}{2}c^2 + gz\right)\right)\delta x \\
&= \frac{d}{dx}\left(W\left(h + \frac{1}{2}c^2 + gz\right)\right)\delta x \\
&= W\frac{d}{dx}\left(h + \frac{1}{2}c^2 + gz\right)\delta x \\
&= W\delta\left(h + \frac{1}{2}c^2 + gz\right)
\end{aligned} \tag{4.2}$$

since in the steady state, the only variation occurs with distance, and the flow rate, W, will be independent of distance. Substituting back into equation (3.62) gives:

$$\frac{\phi\,\delta x}{W} - \frac{\psi\,\delta x}{W} - \delta\left(h + \frac{1}{2}c^2 + gz\right) = 0 \tag{4.3}$$

The term $\phi\,\delta x/W$ is the heat flux into the small element per unit mass flow, or, alternatively, the heat input per unit mass of the passing fluid, in units of Joules per kilogram. As $\delta x \to 0$, we may name this dq:

$$dq = \frac{\phi\,\delta x}{W}, \quad \delta x \to 0 \tag{4.4}$$

The term $\psi\,\delta x/W$ is the useful power abstracted from the small element per unit mass flow, or, alternatively, the useful work abstracted per unit mass of the passing fluid, in units of Joules per kilogram. As $\delta x \to 0$, we may name this dw:

$$dw = \frac{\psi\,\delta x}{W}, \quad \delta x \to 0 \tag{4.5}$$

Thus equation (4.3) as $\delta x \to 0$ may be rewritten:

$$dq - dw - d\left(h + \tfrac{1}{2}c^2 + gz\right) = 0 \tag{4.6}$$

or:

$$dq - dw - dh - c\,dc - g\,dz = 0 \tag{4.7}$$

This equation explains how the steady-state values of specific enthalpy and speed of the fluid emerging from an infinitesimal pipe element will be affected by the heat input per unit mass the fluid receives as it passes through the element, the work per unit mass the fluid does, and the rise in height it experiences as it traverses the element.

The quantity dq is normally taken as the specific external heating to the differential element, and the quantity dw as the specific useful work abstracted. However, the relationship defined by equation (4.7) applies equally when the specific heat input and the specific work done are the total specific heat input and the total specific work, where the latter include frictional heating and work, as we shall now show.

In a frictionally resisted flow, the total work done will be the sum of the useful work, dw, and the work done against friction, dF, per unit mass of fluid, in units of Joules per kilogram:

$$dw_t = dw + dF \tag{4.8}$$

But the frictional work will reappear as heat, so the total heating per unit mass is now the sum of the externally imposed heating and the frictional heating:

$$dq_t = dq + dF \tag{4.9}$$

Substituting from equation (4.8) and (4.9) back into equation (4.7) gives

$$\begin{aligned} 0 &= dq_t - dF - (dw_t - dF) - dh \\ &\quad - c\,dc - g\,dz \qquad (4.10) \\ &= dq_t - dw_t - dh - c\,dc - g\,dz \end{aligned}$$

Notice that equations (4.7) and (4.10) have the same form, except that dq and dw have been replaced by dq_t and dw_t in the latter. We will now develop equation (4.10) for the case of pipe flow.

In pipe-flow, there will be no useful work done, i.e.

$$dw = 0 \tag{4.11}$$

In addition, the First Law of Thermodynamics states that the change in internal energy and the work done in expansion are driven by the total heat input according to the equation:

$$dq_t = du + p\,dv \tag{4.12}$$

Further, we may formally differentiate $h = u + pv$ to give:

$$dh = du + p\,dv + v\,dp \tag{4.13}$$

Substituting the conditions of (4.11), (4.12) and (4.13) into (4.10) gives:

$$\begin{aligned} &du + p\,dv - dF - (du + p\,dv + v\,dp) \\ &\quad - c\,dc - g\,dz = 0 \qquad (4.14) \end{aligned}$$

or

$$v\,dp + c\,dc + g\,dz + dF = 0 \tag{4.15}$$

Equation (4.15) is the general equation describing frictionally resisted flow in a pipe element. We may rewrite it in terms of rates of change with distance along the pipe by dividing by dx:

$$v\frac{dp}{dx} + c\frac{dc}{dx} + g\frac{dz}{dx} + \frac{dF}{dx} = 0 \tag{4.16}$$

4.3 Incompressible flow

The velocity of fluid in a pipe is given by

$$c = v\frac{W}{A} \tag{4.17}$$

In the steady state, the mass flow will be the same all along the pipe, and we may also expect the cross-sectional area to be constant along its length. Accordingly, we may differentiate (4.17) with respect to x to give:

$$\frac{dc}{dx} = \frac{W}{A}\frac{dv}{dx} \tag{4.18}$$

Changes in specific volume can result only from changes in pressure and/or temperature. For an essentially incompressible fluid such as a liquid, the change in pressure along the pipe will cause only a minor change in the specific volume. Further, on the basis that pipes much hotter or colder than the surrounding environment are normally lagged on process plants, the changes in temperature along a pipe are likely to be small, and the resulting changes in specific volume insignificant. Accordingly, we may expect that the rate of change of specific volume will be small: $dv/dx \approx 0$. It follows from equation (4.18) that $dc/dx \approx 0$. Hence we may simplify equation (4.16) to:

$$\frac{dp}{dx} = -\frac{g}{v}\frac{dz}{dx} - \frac{1}{v}\frac{dF}{dx} \tag{4.19}$$

The frictional loss of energy, dF/dx, is given by the Fanning formula in its differential form:

$$\frac{dF}{dx} = \frac{4f}{D}\frac{c^2}{2} \tag{4.20}$$

where f is the dimensionless Fanning friction factor. [Note: some authors work with an alternative definition of friction factor, λ, where $\lambda = 4f$ for reasons that will be apparent from inspecting equation (4.20). This text will stick to the Fanning friction factor, f, throughout.]

Using the fact that $c = v(W/A)$, we may rewrite equation (4.20) as

$$\frac{dF}{dx} = \frac{2f}{D}\frac{W^2}{A^2}v^2 \tag{4.21}$$

We may substitute from equation (4.21) into (4.19) and then integrate between the limits corresponding to the beginning and end of the pipeline:

$$\int_{p_1}^{p_2} dp = -g \int_0^L \frac{1}{v}\frac{dz}{dx}\,dx - \frac{2}{D}\frac{W^2}{A^2}\int_0^L vf\,dx \tag{4.22}$$

to produce the result:

$$\Delta p = \Delta p_z + \Delta p_f \tag{4.23}$$

where

$\Delta p \; = p_1 - p_2$ is the total pressure drop (Pa),

$\Delta p_z = g \int_0^L \frac{1}{v}\frac{dz}{dx}\,dx$ is the pressure drop due to height difference,

$\Delta p_f = \frac{2}{D}\frac{W^2}{A^2}\int_0^L vf\,dx$ is the pressure drop due to friction, and

L is the length of the pipe (m).

It will be noted that specific volume has been retained within the integrals above to allow for the possibility of some variation in specific volume with distance.

4.4 Magnitude of the Fanning friction factor, *f*

Considering laminar flow first, the Fanning friction factor introduced in equation (4.20) depends only on the Reynolds number, N_{RE}:

$$f = \frac{16}{N_{RE}} \tag{4.24}$$

where N_{RE} is dimensionless and given by

$$N_{RE} = \frac{cD}{v\mu} \tag{4.25}$$

in which μ is the dynamic viscosity (Pa.s). Since $W = Ac/v$, the Reynolds number may also be written:

$$N_{RE} = \frac{WD}{A}\frac{1}{\mu} \tag{4.26}$$

Equation (4.24) allows the Fanning friction factor to be found during laminar flow for both incompressible fluids and compressible fluids, i.e. liquids and gases.

Whether flow is laminar or turbulent depends on the size of the Reynolds number. Flow is laminar when the Reynolds number is below 2000, and it will be essentially turbulent when the Reynolds number is in excess of about 4000, but between Reynolds numbers of about 2000 and about 4000 there is a region where the flow may be either laminar or turbulent, or a mixture of the two flow regimes. Turbulent flow is by far the most frequently encountered condition on a process plant. For turbulent flow, the friction factor, f, depends on two parameters: (i) the Reynolds number and (ii) the relative roughness of the pipe, ε/D, where ε is a measure of the average height of the excrescences on the pipe surface. The Colebrook–White equation, based partly on theory and partly on extensive experiment, gives an implicit relationship for the friction factor:

$$\frac{1}{\sqrt{f}} + 4\log_{10}\left(\frac{1}{3.7}\frac{\varepsilon}{D} + \frac{1.256}{N_{RE}\sqrt{f}}\right) = 0 \tag{4.27}$$

Once more, this expression applies equally to liquids and gases.

Equation (4.24) allows the Fanning friction factor to be plotted against Reynolds number in the laminar region, while equation (4.27) allows the friction factor to be plotted in the turbulent region with ε/D as parameter. A diagram containing the two plots is known as a Moody diagram, see Figure 4.1.

Since commercial pipe is normally relatively smooth, with an ε/D of between 0.002 and 0.0002, it follows approximately Blasius's equation for perfectly smooth pipe, with $\varepsilon = 0$, up to a Reynolds number of 100 000:

$$f = \frac{0.0791}{N_{RE}^{0.25}} \tag{4.28}$$

For flow at a Reynolds number in excess of this value, the friction factor is constant with a value for commercial pipe given approximately by

$$f \approx \frac{0.0791}{(100\,000)^{0.25}} \approx 0.0045 \tag{4.29}$$

Knowledge of the true ε/D value will allow a more exact value to be calculated from the implicit relationship of equation (4.27). It should be noted from Figure 4.1, however, that the friction factor will vary by no more than a factor of two across all commercial grades of pipe, provided the flow regime is fully turbulent (corresponding to Reynolds numbers greater than 10 000).

Let us consider the case of a pipe having the same diameter along its length and carrying a given steady flow of liquid, W. It is clear from equation (4.26) that the Reynolds number will depend only on the dynamic viscosity. Dynamic viscosity, μ, is a function essentially only of temperature for liquids, so provided the temperature remains approximately constant along the pipe-length, as one would expect with a lagged pipe, the Reynolds number will stay constant along the length of the pipe. Accordingly a single value of the Fanning friction factor will be valid along the length of a pipe carrying any given liquid flow.

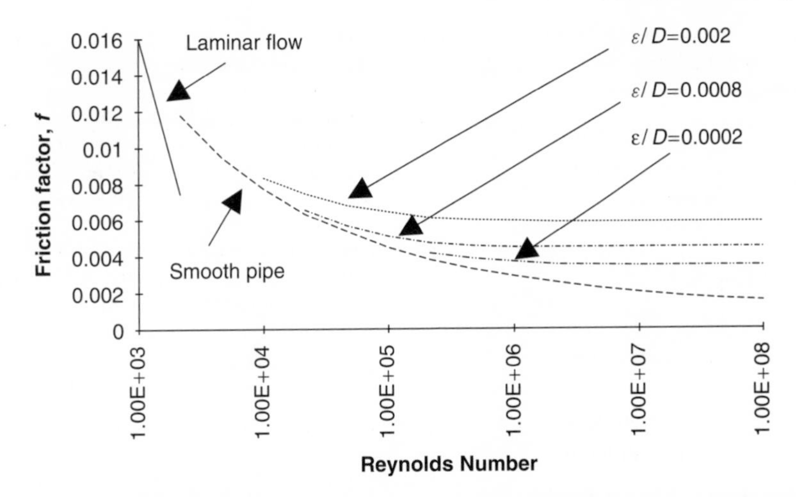

Figure 4.1 Moody diagram: Fanning friction factor, f, vs. Reynolds number for the range of commercial pipe relative roughnesses.

Gas flow in lagged process pipes will be essentially adiabatic, and some drop in temperature may occur. The drop in temperature will lead to a decrease in viscosity, but the decrease in viscosity will usually be rather small. For example, a drop in temperature of methane from 200°C to 100°C causes the viscosity to drop by 25%, and from equation (4.25), this will lead to an increase of Reynolds number of a similar percentage. This increase in Reynolds number will have no effect on friction factor for Reynolds numbers above 100 000 because of the flatness of the Moody curve in this region. We may evaluate the effect for Reynolds numbers below 100 000 by differentiating equation (4.28) to give:

$$\frac{df}{f} = -0.25\frac{dN_{RE}}{N_{RE}} \tag{4.30}$$

It follows from equation (4.30) that even the very large, 100°C drop in temperature discussed will cause only about a 6% change in friction factor in the more sensitive case of $N_{RE} < 100\,000$. We may conclude that, for the purposes of simulation, a single value of the Fanning friction factor will be valid along the length of a pipe carrying any given gas flow. This value of f should be calculated from equation (4.27) or from a Moody diagram using a Reynolds number ideally evaluated at mid-pipe conditions.

In simulation studies, we shall, of course, be interested in calculating flow as a function of pressure drop for a variety of conditions. For constant geometry, the Reynolds number will vary in proportion to variations in the flow rate, W. If the Reynolds number is at all times above 100 000 then there will be no consequential change in friction factor because of the flatness of the Moody curve in this region. Moreover, equation (4.30) tells us that even when the Reynolds number is below 100 000, the change in the friction factor in percentage terms will be only a quarter of the percentage change in Reynolds number and hence only a quarter of the percentage change in mass flow rate. As a result, we shall normally make the approximation that the friction factor is constant across the complete flow range.

4.5 Frictionally resisted, incompressible flow through a real pipe

Equation (4.23) applies to a theoretical pipe, but for a real pipeline additional allowances need to be made for pipe entry and exit losses, bends and fittings. Consider a fluid flowing in a typical process plant pipeline as shown in Figure 4.2.

The pressure drop through a pipe is the sum of the hydrostatic pressure due to the increase in level and the frictional pressure drop due to the flow through the pipe, to the flow past fittings such as bends and valves and to the entry and exit losses:

$$p_1 - p_4 = \Delta p_z + \Delta p_f + \Delta p_b + \Delta p_{in} + \Delta p_{out} \tag{4.31}$$

where

Δp_z is the pressure difference due to the difference in levels at the entrance to and exit from the pipe (Pa),
Δp_f is the pressure drop due to friction (Pa),
Δp_b is the pressure drop caused by bends and fittings (Pa),
$\Delta p_{in} = p_1 - p_2$ is the pressure drop caused by an untapered inlet (Pa),
$\Delta p_{out} = p_3 - p_4$ is the pressure drop caused by a residual velocity at the outlet (Pa).

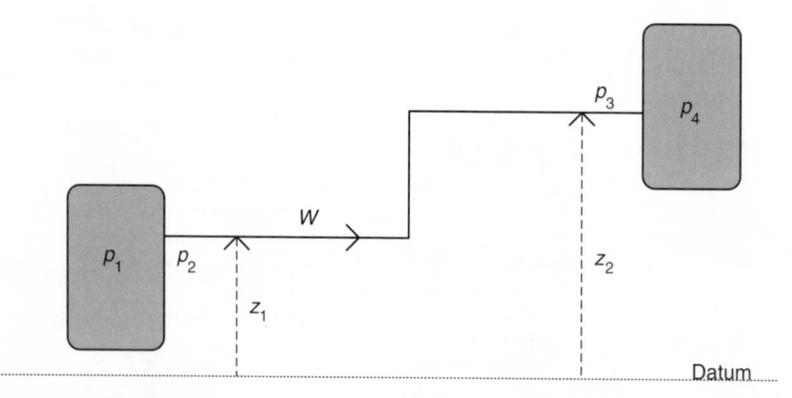

Figure 4.2 Flow through a typical pipeline.

4.6 Pressure drop due to level difference

As noted above, the pressure drop due to level difference is given by:

$$\Delta p_z = \int_0^L \frac{g}{v}\frac{dz}{dx}dx = g\int_{z_1}^{z_2} \frac{1}{v}dz \tag{4.32}$$

where L is the length of the pipe. Let us define an average specific volume, v_z, by:

$$\frac{1}{v_z} = \frac{1}{z_2 - z_1}\int_{z_1}^{z_2} \frac{1}{v}dz \tag{4.33}$$

or

$$\frac{z_2 - z_1}{v_z} = \int_{z_1}^{z_2} \frac{1}{v}dz \tag{4.34}$$

Substituting into (4.32) gives:

$$\Delta p_z = \frac{g(z_2 - z_1)}{v_z} \tag{4.35}$$

Strictly, we should evaluate v_z by equation (4.33). In the case of the diagram of Figure 4.2, this integration would take place over the region between the 90° bends. In most cases, however, little will be lost by taking an average of the specific volume over the whole length of the pipe:

$$v_z \approx v_{ave} \tag{4.36}$$

For cases where there is no heat exchange, the average specific volume will be the same as that at the inlet, so

$$v_z \approx v_1 \tag{4.37}$$

4.7 Frictional pressure drop

As noted in Section 4.3, the frictional pressure drop is given by the Fanning equation:

$$\Delta p_f = \frac{2}{D}\frac{W^2}{A^2}\int_0^L v f\, dx \tag{4.38}$$

where f is the Fanning friction factor (dimensionless).

We will define the average value of the specific volume, v_{ave}, along the pipe-length by

$$v_{ave} = \frac{1}{L}\int_0^L v\, dx \tag{4.39}$$

We may substitute from equation (4.39) into equation (4.38), and, since from Section 4.4 we may assume that the friction factor, f, is constant along the pipe, this integration yields

$$\Delta p_f = 2f\frac{L}{D}\frac{v_{ave}}{A^2}W^2 \tag{4.40}$$

This may be rewritten in terms of the inlet specific volume as:

$$\Delta p_f = 2f\frac{L}{D}\frac{v_1}{A^2}\frac{v_{ave}}{v_1}W^2 \tag{4.41}$$

Equation (4.41) may be rearranged to give mass flow in terms of frictional pressure drop:

$$W = C_f\sqrt{\frac{\Delta p_f}{v_1}} \tag{4.42}$$

where the frictional conductance, C_f (m^2), is a constant for the pipeline given by:

$$C_f = A\sqrt{\frac{1}{2f}\frac{D}{L}\frac{v_1}{v_{ave}}} \tag{4.43}$$

We may note that $v_{ave} \approx v_1$ for liquid flow without heat exchange.

4.8 Pressure drop due to bends and fittings

The frictional effects of bends and fittings are often expressed in terms of a quantity called the 'velocity head drop'. To introduce this concept, we begin by observing that the specific energy lost to friction, F(J/kg), for incompressible flow follows from an integration of the Fanning equation (4.20) with friction factor and velocity constant over the length of pipe:

$$F = \int_0^F dF = \int_0^L \frac{4f}{D}\frac{c^2}{2}\,dx = 4f\frac{L}{D}\frac{c^2}{2} \qquad (4.44)$$

This can be rewritten

$$F = K\frac{c^2}{2} \qquad (4.45)$$

where the dimensionless number, K, is the number of velocity heads lost to friction, and is given by:

$$K = \frac{F}{\frac{1}{2}c^2} = 4f\frac{L}{D} \qquad (4.46)$$

It may be seen that K is the ratio of the energy lost to friction to the kinetic energy of the fluid flowing through the pipe.

The change in pressure due to friction follows from integration of equation (4.19) when $dz/dx = 0$:

$$\int_0^{-\Delta p} \frac{dp}{dx}\,dx = -\int_0^F \frac{1}{v}\frac{dF}{dx}\,dx \qquad (4.47)$$

where Δp is the upstream pressure minus the downstream pressure. Assuming the specific volume is independent of distance, the integration yields:

$$\Delta p = \frac{F}{v} = K\frac{1}{v}\frac{c^2}{2} \qquad (4.48)$$

Estimates of losses in velocity heads for bends and a selection of fittings is given in Table 4.1. See Chapter 7, Section 7.8 for a discussion of and data on the velocity-head losses of globe and butterfly valves.

Table 4.1 Frictional losses in bends and fittings

Type of bend or fitting	*Frictional loss, K, in velocity heads*
Standard 45° bend	0.35
Standard 90° bend	0.75
180° bend (close return)	2.2
Diaphragm valve – fully open	2.3
Gate valve – fully open	0.17

For the ith such bend or fitting, the pressure drop is given by:

$$\Delta p_{bi} = K_{bi}\frac{1}{v_{bi}}\frac{c^2}{2} \qquad (4.49)$$

where the subscript, bi, connotes the ith bend or fitting, and, in particular, K_{bi} is the number of velocity heads lost there. Using $W = Ac/v$, equation (4.49) may be reframed in terms of mass flow as:

$$\Delta p_{bi} = \frac{K_{bi}v_{bi}}{2A^2}W^2 \qquad (4.50)$$

Equation (4.50) may in turn be rearranged to give mass flow in terms of pressure drop:

$$W = C_{bi}\sqrt{\frac{\Delta p_{bi}}{v_1}} \qquad (4.51)$$

where the bend conductance, C_{bi} (m^2), is given by:

$$C_{bi} = A\sqrt{\frac{2}{K_{bi}}\frac{v_1}{v_{bi}}} \qquad (4.52)$$

in which we may normally take $v_{bi} \approx v_1$ for liquid flow without heat exchange. Equation (4.52), derived for the ith bend or fitting expresses the general relationship between any velocity-head drop, K, and its equivalent flow conductance, C.

The total pressure drop due to all bends and fittings is found by summation of equation (4.50):

$$\Delta p_b = \sum_i \frac{K_{bi}v_{bi}}{2A^2}W^2 \qquad (4.53)$$

4.9 Pressure drop at pipe outlet

In Figure 4.2, consider the conditions at the point where the pipe joins the downstream vessel, which vessel may, in general, be taken to represent also a pipe with a larger diameter. The upstream pipe has an internal diameter D, while the downstream vessel or larger pipe has a greater (effective) internal diameter, D_4, as in Figure 4.3 below. Station '3' marks the junction where the pipe meets the larger pipe/vessel, while station '4' denotes a point downstream of the junction, where eddying has ceased.

If there were no losses present, one would expect that the velocity would decrease as the flow entered the larger vessel/pipe, with the reduction in kinetic energy leading to a pressure increase. But the situation is changed by the frictional losses caused by turbulence eddies at the vessel inlet. The extent of any increase in pressure may be investigated using

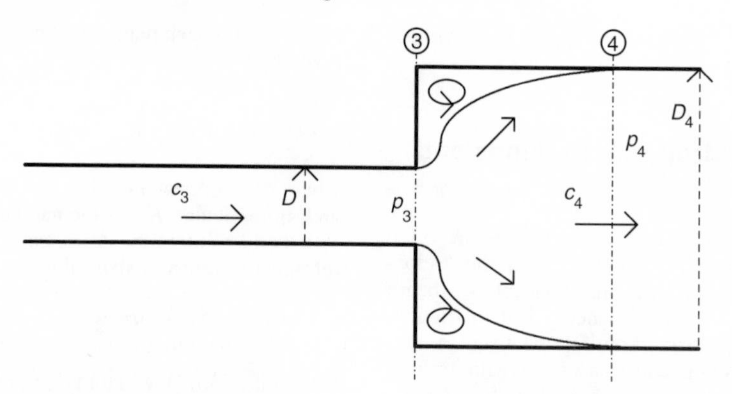

Figure 4.3 Pressure drop at the outlet from the pipe.

the energy equation, given in differential form by equation (4.15), repeated below:

$$v\,dp + c\,dc + g\,dz + dF = 0 \tag{4.15}$$

Integrating this between stations '3' and '4', where the specific volume remains constant at v but the pipe cross-sectional area changes, we achieve:

$$p_4v - p_3v + \frac{c_4^2}{2} - \frac{c_3^2}{2} + gz_4 - gz_3 + F_{3-4} = 0 \tag{4.54}$$

or

$$p_3v + \frac{c_3^2}{2} + gz_3 = p_4v + \frac{c_4^2}{2} + gz_4 + F_{3-4} \tag{4.55}$$

where the term F_{3-4} is the energy loss per unit mass flow due to frictional effects between stations '3' and '4' (in units of J/kg). In the absence of a height change, and if we could assume that frictional effects were negligible, equation (4.55) would give:

$$\Delta p_{out} = p_3 - p_4 = \frac{1}{2v}(c_4^2 - c_3^2) \tag{4.56}$$

For the important case of discharge into a large vessel, the velocity $c_4 \approx 0$, and so (4.56) becomes

$$\Delta p_{out} = -\frac{1}{2v}c_3^2 \tag{4.57}$$

indicating the pressure rise discussed above. However, all this supposes that there is no frictional loss, and we may see in Figure 4.3 that an abrupt outlet into a pipe or vessel of greater diameter causes a region of stagnant fluid to be set up away from the main flow. This region will be subject to eddies that drain energy from the main flow, so that the assumption of zero frictional losses, F_{3-4}, is not valid.

Allowing for a non-zero value of F_{3-4}, equation (4.55) may be rearranged to give:

$$F_{3-4} = p_3v - p_4v + \frac{c_3^2}{2} - \frac{c_4^2}{2} \tag{4.58}$$

To evaluate the term, $p_3v - p_4v$, we apply the principle of the conservation of linear momentum to the section of flow between stations '3' and '4'. The net force in the direction of flow is equal to the rate of momentum leaving minus the rate of momentum entering the section. Assuming pressure equalization at p_3 along the whole flow front at station '3', we may write:

$$p_3A_4 - p_4A_4 = \frac{A_4c_4}{v}c_4 - \frac{Ac_3}{v}c_3 \tag{4.59}$$

where $A = \pi D^2/4$ is the cross-sectional area of the upstream pipe, while $A_4 = \pi D_4^2/4$ is the cross-sectional areas of the downstream pipe/vessel. Since conservation of mass dictates that in the steady state

$$\frac{A_4c_4}{v} = \frac{Ac_3}{v} \tag{4.60}$$

we may eliminate the area, A_4, in equation (4.59) to give

$$p_3v - p_4v = c_4^2 - c_3c_4 \tag{4.61}$$

Substituting from (4.61) into (4.58) gives:

$$F_{3-4} = c_4^2 - c_3c_4 + \frac{c_3^2}{2} - \frac{c_4^2}{2} = \frac{c_3^2 - 2c_3c_4 + c_4^2}{2}$$
$$= \frac{(c_3 - c_4)^2}{2} \tag{4.62}$$

From (4.60), we may write

$$c_4 = \frac{A}{A_4}c_3 = \frac{D^2}{D_4^2}c_3 \tag{4.63}$$

Hence (4.62) becomes:

$$F_{3-4} = \frac{1}{2}c_3^2\left(1 - \left(\frac{D}{D_4}\right)^2\right)^2 \tag{4.64}$$

In the important case of a pipe entering a large vessel, the ratio $D/D_4 \to 0$, so that

$$F_{3-4} = \tfrac{1}{2}c_3^2 \tag{4.65}$$

Substituting this value into equation (4.58), together with the equivalent assumption that the velocity in the vessel, $c_4 \approx 0$, gives

$$\begin{aligned} p_3 &= p_4 + \frac{1}{v}F_{3-4} - \frac{c_3^2}{2v} + \frac{c_4^2}{2v} \\ &= p_4 + \frac{c_3^2}{2v} - \frac{c_3^2}{2v} + 0 \\ &= p_4 \end{aligned} \tag{4.66}$$

Accordingly, the effect of frictional turbulence on entering the vessel has exactly cancelled out the depression in end-of-pipeline pressure below discharge-vessel pressure that would be expected in a frictionless expansion. We may therefore set

$$\Delta p_{out} = 0 \tag{4.67}$$

in equation (4.31).

4.10 Pressure drop at pipe inlet

Figure 4.4 shows the effect of an abrupt inlet to the pipeline. Notice the vena contracta formed just downstream of the inlet.

Three stations have been marked on Figure 4.4: the upstream station '1', the position of the vena contracta, 'vc', and the downstream station '2', where uniform pipe-flow has become established.

In the absence of height differences and if frictional losses are ignored, the application of the energy equation (4.55) to the system of Figure 4.4 gives, after adjustment of the station subscripts:

$$p_1 + \frac{c_1^2}{2v} = p_{vc} + \frac{c_{vc}^2}{2v} = p_2 + \frac{c_2^2}{2v} \tag{4.68}$$

Since geometry and the conservation of mass imply that

$$c_1 < c_2 < c_{vc} \tag{4.69}$$

it follows that the pressure drops to a low point at the vena contracta, station 'vc', as potential energy is converted into kinetic energy, and then makes a partial recovery at station '2' as some of the kinetic energy is converted back into potential energy.

The process of converting potential energy into kinetic energy at the vena contracta is known to be very efficient. However, the pressure recovery process is less so. Accordingly we make the assumption that all the frictional losses associated with pipe inlet occur over the pipe section 'vc' to '2'. As a result, we may now use the analysis of Section 4.9 directly in order to estimate the frictional loss. Changing subscripts ($3 \to vc$, $4 \to 2$), equation (4.62) gives this loss as:

$$F_{1-2} = F_{vc-2} = \frac{(c_{vc} - c_2)^2}{2} \tag{4.70}$$

Applying the conservation of mass for the steady flow between station 'vc' and station '2' gives

$$\frac{A_{vc}c_{vc}}{v} = \frac{Ac_2}{v} \tag{4.71}$$

which allows us to eliminate c_{vc} from equation (4.70):

$$\begin{aligned} F_{1-2} &= \frac{1}{2}c_2^2\left(\frac{A}{A_{vc}} - 1\right)^2 \\ &= \frac{1}{2}c_2^2\left(\frac{1}{C_{vc}} - 1\right)^2 \end{aligned} \tag{4.72}$$

where $C_{vc} = A_{vc}/A$ is the coefficient of contraction for the vena contracta, about 0.6 when the pipe is being supplied from a large vessel. In practice, the energy loss is normally taken as

$$F_{1-2} = K_{con}\frac{c_2^2}{2} \tag{4.73}$$

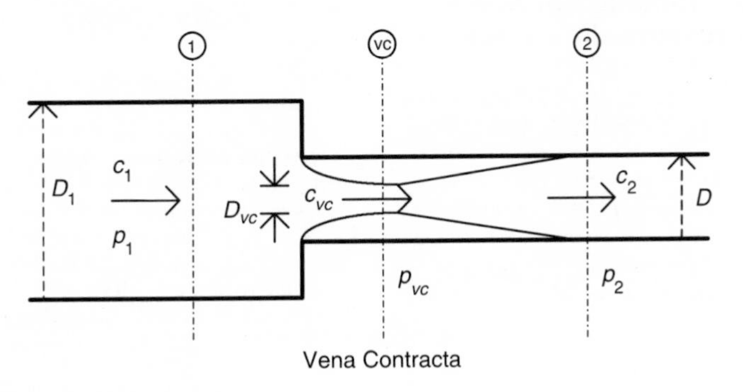

Figure 4.4 Pressure drop at the inlet to the pipe.

where K_{con} is the contraction coefficient for pipe entry, the size of which depends on the ratio of the square of the diameters of the pipes at the junction. Dividing by the specific volume gives the loss in pressure at pipe inlet as:

$$\Delta p_{in} = K_{con}\frac{c_2^2}{2v} \tag{4.74}$$

The contraction coefficient, K_{con}, which has units of velocity heads, is given in Table 4.2.

Table 4.2 Contraction coefficient for an abrupt entrance (velocity heads)

D^2/D_1^2	0	0.2	0.4	0.6	0.8	1.0
K_{con}	0.5	0.45	0.36	0.21	0.07	0

For an abrupt entrance from a large upstream vessel, $D_1^2/D_0^2 \approx 0$, so that a value of 0.5 is appropriate for K_{con}. When the entrance is tapered, as in a convergent nozzle, the loss is reduced very greatly. A value of 0.05 is then typical for K_{con}.

Using the notation of the pipe run shown in Figure 4.2, we have the following equations relating mass flow and inlet pressure drop:

$$\Delta p_{in} = K_{con}\frac{c_1^2}{2v_1} = \frac{K_{con}v_1}{2A^2}W^2 \tag{4.75}$$

or

$$W = C_{in}\sqrt{\frac{\Delta p_{in}}{v_1}} \tag{4.76}$$

where the flow conductance (in m^2) is given by:

$$C_{in} = A\sqrt{\frac{2}{K_{con}}} \tag{4.77}$$

4.11 Overall relationship between mass flow and pressure difference

We may now substitute into equation (4.31) the expressions derived for Δp_z, Δp_f, Δp_b, Δp_{in} and Δp_{out}:

$$p_1 - p_2 = \frac{g(z_2 - z_1)}{v_z} + 2f\frac{L}{D}\frac{v_{ave}}{A^2}W^2 + \sum_i \frac{K_{bi}v_{bi}}{2A^2}W^2 + \frac{K_{con}v_1}{2A^2}W^2 + 0 \tag{4.78}$$

or

$$p_1 - p_2 - \frac{g(z_2 - z_1)}{v_z} = \left(2f\frac{L}{D}\frac{v_{ave}}{v_1} + \sum_i \frac{K_{bi}}{2}\frac{v_{bi}}{v_1} + \frac{K_{con}}{2}\right)\frac{v_1}{A^2}W^2 \tag{4.79}$$

We may rearrange (4.79) to give an explicit equation for W:

$$W = \frac{A}{\sqrt{\left(2f\frac{L}{D}\frac{v_{ave}}{v_1} + \sum_i \frac{K_{bi}}{2}\frac{v_{bi}}{v_1} + \frac{K_{con}}{2}\right)}} \times \sqrt{\frac{p_1 - p_2 - \frac{g(z_2 - z_1)}{v_z}}{v_1}} \tag{4.80}$$

or

$$W = C_L\sqrt{\frac{p_1 - p_2 - \frac{g(z_2 - z_1)}{v_z}}{v_1}} \tag{4.81}$$

where the overall flow conductance, C_L, is given by

$$\frac{1}{C_L^2} = \frac{1}{C_f^2} + \sum_i \frac{1}{C_{bi}^2} + \frac{1}{C_{in}^2} \tag{4.82}$$

The flow conductances in equation (4.82) are those derived in the previous subsections. The overall line conductance, C_L, will be constant to the extent that the constituent flow conductances do not alter, and these may be regarded as invariant for incompressible flow at high Reynolds numbers.

Control valves in the line will clearly have a major effect, and ways of calculating this effect for both compressible and incompressible flow will be given in Chapter 7.

4.12 Bibliography

Glasstone, S. and Sesonke, A. (1981). *Nuclear Reactor Engineering*, 3rd edition, Van Nostrand Reinhold, New York.

Lapple, C.E. (1943). Isothermal and adiabatic flow of compressible fluids, *Trans. Am. Inst. Chem. Eng.*, **39**, 385–432.

Perry, R.H., Green, D.W. and Maloney, J.O. (1984). *Perry's Chemical Engineer's Handbook*, 6th edition, McGraw-Hill, New York. Chapter 5: Fluid and particle mechanics.

Streeter, V.L. and Wylie, E.B. (1983). *Fluid Mechanics, First SI Metric Edition*, McGraw-Hill, New York.

Webber, N.B. (1965). *Fluid Mechanics for Civil engineers*, Spon's Civil Engineering Series, London.

5 Flow through ideal nozzles

5.1 Introduction

The nozzle is a device or a phenomenon that reduces the area available to the flow and so causes the fluid's velocity to increase at the expense of its pressure. The flow area has a minimum value at the 'throat' of the nozzle; thereafter the area may either remain constant (in the case of a 'convergent nozzle') or increase (in the case of a 'convergent–divergent nozzle'). Nozzles may be deliberately and carefully engineered, or they may be an unintentional consequence of the geometry of flow. For example, the nozzle in a turbine is a precision-engineered device for producing a high level of kinetic energy in the working fluid, which energy is then converted into rotational mechanical energy in the turbine wheel. On the other hand, an unintentional nozzle is formed whenever a fluid flows from a large vessel into a pipeline, when a vena contracta is formed just downstream of the pipe inlet.

The widespread occurrence in process plant of nozzles, intentional and unintentional, means that it is necessary for the control engineer to have an understanding of their physical principles for modelling purposes. We will consider in this chapter the features of an ideal, frictionless nozzle. Results from this simplified model will be put to immediate use in Chapter 6 in order to account for the unavoidable nozzle formed at beginning of a gas pipe. Moreover, the idealized approach provides the groundwork for the more complex models of Chapter 14, where turbine nozzles with friction are considered in some detail.

5.2 Steady-state flow in a nozzle

The nozzle is a relatively short device, a few pipe diameters long at most, and the flow will establish itself so rapidly that the control engineer may safely content himself with a steady-state analysis. We may begin with the analysis of Section 4.2, which led to the differential form of the energy equation (4.15) for fluid flow where no useful work is done. This is repeated below:

$$v\,dp + c\,dc + g\,dz + dF = 0 \tag{4.15}$$

Because of the normally small length of the nozzle, it is reasonable to omit the effect of friction in this preliminary analysis of nozzles. (Frictional effects are customarily catered for in nozzle analysis by using the concept of nozzle efficiency.) Setting $dF = 0$ in equation (4.15) gives the differential for fluid speed as:

$$c\,dc = -v\,dp - g\,dz \tag{5.1}$$

This equation relates the change in the velocity of the fluid to the change in pressure across the fluid element and also to the change in height across the element. In practice, of course, we will be concerned with the pressure difference and the height difference at finite distances through the nozzle, including through the nozzle as a whole. We shall denote values at the entrance to the nozzle by the subscript '1' and at the selected downstream station by the subscript '2' (see Figure 5.1).

Figure 5.1 shows the general case of a convergent–divergent nozzle, with conditions at the throat being given the subscript 't'. (By contrast, a convergent nozzle will either end at the throat, or else will

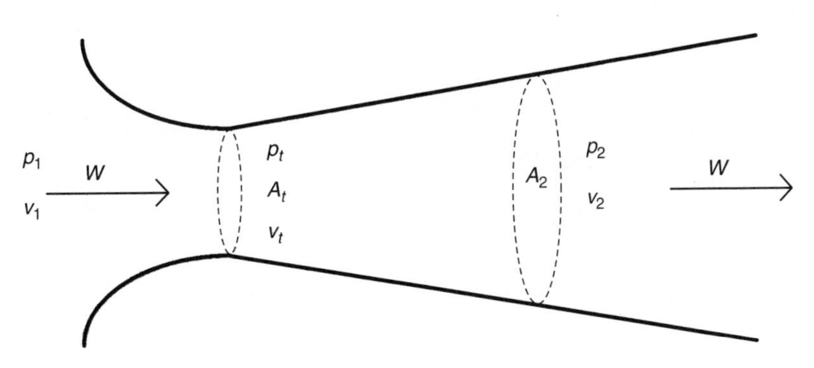

Figure 5.1 convergent–divergent nozzle.

have a section of parallel pipe replacing the divergent part of the nozzle shown in Figure 5.1. For such a convergent nozzle, the downstream area A_2 will be equal to the throat area, A_t.)

We may integrate equation (5.1) from the entrance to the downstream station:

$$\int_{c_1}^{c_2} c.dc = -\int_{p_1}^{p_2} v\,dp - g\int_{z_1}^{z_2} dz \qquad (5.2)$$

where

c_1 is the fluid velocity at the entrance to the nozzle (m/s),
c_2 is the fluid velocity at the downstream station (m/s),
p_1 is the pressure at the entrance to the nozzle (Pa),
p_2 is the pressure at the downstream station (Pa),
z_1 is the height of the entrance to the nozzle (m),
z_2 is the height of the downstream station (m).

This integration yields:

$$c_2^2 - c_1^2 = -2\int_{p_1}^{p_2} v\,dp - 2g(z_2 - z_1) \qquad (5.3)$$

To proceed with the integration of the second term, we need to specify how specific volume, v, varies along the nozzle with pressure. Three possible classes of behaviour are possible.

(i) The simplest behaviour to treat is where the reversible expansion occurs with a constant specific volume, i.e.

$$v = \text{constant} \qquad (5.4)$$

This will apply to a high degree of accuracy to the flow of liquid through a nozzle. It will be shown later that it serves reasonably well also as an approximation for gas flow at low pressure ratios.

(ii) A general, polytropic but reversible expansion for a gas or a vapour is characterized by the equation:

$$pv^n = \text{constant} \quad \text{for } 0 < n < \infty \qquad (5.5)$$

where n is the polytropic index of expansion. The value of n will depend on the ratio of specific heats, γ, of the gas being passed, and the degree to which the gas exchanges heat with its surroundings. [Note that this treatment does not account for friction, where irreversible heating occurs.] The most important value of polytropic index is $n = \gamma$, which corresponds to a reversible adiabatic or isentropic expansion. This implies that no heat is exchanged with the surroundings during the frictionless expansion, so that the gas will cool as it expands. An index $n > \gamma$ implies that heat is removed from the gas during expansion, while an index $n < \gamma$ implies that heat is supplied from the outside to the gas during the expansion. $n = 1$ implies that sufficient heat is supplied from the surroundings to keep the temperature constant, as will now be discussed under (iii).

(iii) Isothermal expansion for an ideal, near-ideal or semi-ideal gas or vapour is characterized by the equation:

$$pv = \frac{ZRT}{w} \qquad (3.2)$$

which yields for constant temperature, T:

$$pv = \text{constant} \qquad (5.6)$$

The limiting case for a polytropic expansion where $n = 1$ needs to be treated separately as the mathematics of the polytropic expansion lead to an indeterminate solution when n is precisely unity.

We shall now consider each of these cases in turn.

5.2.1 Steady-state flow through a nozzle with constant specific volume

The integration of equation (5.3) is very simple in this case:

$$\begin{aligned} c_2^2 - c_1^2 &= -2v\int_{p_1}^{p_2} dp - 2g(z_2 - z_1) \\ &= 2v_1(p_1 - p_2) + 2g(z_1 - z_2) \end{aligned} \qquad (5.7)$$

since $v = v_1 = v_2$.

If the nozzle is fed from a reservoir, we may set $c_1 = 0$ in equation (5.7) to give the outlet velocity as:

$$c_2 = \sqrt{2p_1v_1\left(1 - \frac{p_2}{p_1} + \frac{g(z_1 - z_2)}{p_1v_1}\right)} \qquad (5.8)$$

The mass flow is given by

$$W = \frac{A_2c_2}{v_2} = \frac{A_2c_2}{v_1} \qquad (5.9)$$

so that the equation for steady-state mass flow with constant specific volume is

$$W = A_2\sqrt{2\frac{p_1}{v_1}\left(1 - \frac{p_2}{p_1} + \frac{g(z_1 - z_2)}{p_1v_1}\right)} \qquad (5.10)$$

When $|g(z_1 - z_2)| \ll p_1v_1$, the equation simplifies to

$$W = A_2\sqrt{2\frac{p_1}{v_1}\left(1 - \frac{p_2}{p_1}\right)} \qquad (5.11)$$

5.2.2 Steady-state flow through a nozzle for a gas undergoing a polytropic expansion

The expansion defined by (5.5) may be re-expressed for convenience as:

$$pv^n = b^n \tag{5.12}$$

where b is a constant. Accordingly we have the following expression for the variation of specific volume with pressure during the polytropic expansion:

$$v = bp^{(-1/n)} \tag{5.13}$$

Substituting this into equation (5.3) gives:

$$c_2^2 - c_1^2 = -2b \int_{p_1}^{p_2} p^{-(1/n)} dp - 2g(z_2 - z_1) \tag{5.14}$$

We may integrate the first expression on the right-hand side of the equation to give:

$$c_2^2 - c_1^2 = -2b \left[\frac{p^{1-(1/n)}}{1 - \frac{1}{n}} \right]_{p_1}^{p_2} - 2g(z_2 - z_1) \tag{5.15}$$

or

$$c_2^2 - c_1^2 = -2b \frac{n}{n-1} \left(p_2^{1-(1/n)} - p_1^{1-(1/n)} \right) - 2g(z_2 - z_1) \tag{5.16}$$

From (5.13), the constant, b, is given by

$$b = vp^{1/n} \tag{5.17}$$

and, in particular, we may evaluate b from the inlet conditions, which will be known to us:

$$b = v_1 p_1^{1/n} \tag{5.18}$$

Substituting this expression into (5.10) gives:

$$\begin{aligned} c_2^2 - c_1^2 &= -2\frac{n}{n-1} p_1^{1/n} v_1 \left(p_2^{1-(1/n)} - p_1^{1-(1/n)} \right) - 2g(z_2 - z_1) \\ &= 2\frac{n}{n-1} v_1 \left(p_1 - p_2 \left(\frac{p_2}{p_1} \right)^{-(1/n)} \right) + 2g(z_1 - z_2) \\ &= 2\frac{n}{n-1} p_1 v_1 \left(1 - \left(\frac{p_2}{p_1} \right)^{1-(1/n)} \right) + 2g(z_1 - z_2) \end{aligned} \tag{5.19}$$

or

$$c_2^2 - c_1^2 = 2\frac{n}{n-1} p_1 v_1 \left(1 - \left(\frac{p_2}{p_1} \right)^{(n-1)/n} \right) + 2g(z_1 - z_2) \tag{5.20}$$

We may often simplify (5.20) by setting $c_1 = 0$, since the gas velocity in a vessel or header will have a very low or zero component in the direction of the nozzle. In addition, we may neglect the potential energy term on the basis of the enormous difference in magnitude of the terms on the right-hand side of equation (5.20). [Let us take the example of nitrogen passing from one vessel to another. Treating it as an ideal gas,

$$p_1 v_1 = \frac{RT_1}{w} \tag{5.21}$$

where gas constant $R = 8314$ J/(kg mole K) and molecular weight $w = 28$. At $T_1 = 293$ K, $p_1 v_1 = 87\,000$ J/kg, while at 493 K, $p_1 v_1 = 146\,386$ J/kg. Considering a reversible adiabatic expansion of a diatomic gas, $n = \gamma = 1.4$. Taking the pressure ratio, p_2/p_1 as 0.95, for example, allows us to evaluate the first term on the right-hand side, the pressure term, as:

$$\begin{aligned} 2\frac{n}{n-1} p_1 v_1 \left(1 - \left(\frac{p_2}{p_1} \right)^{(n-1)/n} \right) &= 8860 \text{ J/kg} \quad \text{for } T = 293 \text{ K} \\ &= 14\,908 \text{ J/kg} \quad \text{for } T = 493 \text{ K} \end{aligned}$$

Even if we assume a very large nozzle at 2 m long, pointing vertically, the potential energy term $2g(z_1 - z_2)$ is only 40 J/kg, which is insignificant compared with the values above.]

Neglecting the potential energy term and setting $c_1 = 0$ in equation (5.20) allows us to write down the velocity at the downstream station as:

$$c_2 = \sqrt{2\frac{n}{n-1} p_1 v_1 \left(1 - \left(\frac{p_2}{p_1} \right)^{(n-1)/n} \right)} \tag{5.22}$$

The mass flow rate will be given in terms of downstream parameters by:

$$W = \frac{A_2 c_2}{v_2} \tag{5.23}$$

where, from (5.13):

$$v_2 = v_1 \left(\frac{p_1}{p_2} \right)^{1/n} \tag{5.24}$$

Substituting from (5.23) and (5.24) into (5.22) gives the required expression for the polytropic, reversible expansion:

$$W = A_2\sqrt{2\frac{n}{n-1}\frac{p_1}{v_1}\left(\left(\frac{p_2}{p_1}\right)^{2/n} - \left(\frac{p_2}{p_1}\right)^{(n+1)/n}\right)} \tag{5.25}$$

This expression gives the flow in terms of a general downstream station. Most often we shall wish to work with the outlet pressure, in which case the corresponding area term is that at the nozzle outlet. We shall also use this expression to evaluate the mass flow in terms of the conditions at the throat, in which case

$$\begin{aligned} A_t &= A_2 \\ p_t &= p_2 \end{aligned} \tag{5.26}$$

It may be noted in passing that the throat pressure in a convergent–divergent nozzle will be greater than the outlet pressure only if the fluid is being accelerated downstream of the nozzle to supersonic speeds, which implies both careful design of the divergent section and care in matching the overall pressure ratio to that design. Otherwise the pressure will reduce as far as the throat, but will rise again in the divergent section of the nozzle as the fluid slows down. This aspect of nozzle behaviour is covered in more detail in Chapter 14.

5.2.3 Isentropic steady-state flow

A very important special case of the polytropic flow equation (5.25) is that describing a reversible, adiabatic expansion, where no heat is exchanged with the surroundings, i.e. an isentropic expansion. In this case, the ratio of specific heats, γ, is substituted for n in the mass-flow equation:

$$W = A_2\sqrt{2\frac{\gamma}{\gamma-1}\frac{p_1}{v_1}\left(\left(\frac{p_2}{p_1}\right)^{2/\gamma} - \left(\frac{p_2}{p_1}\right)^{(\gamma+1)/\gamma}\right)} \tag{5.27}$$

As was discussed in Section 3.3.1, the theoretical value of the index, γ, is 1.67 for a monatomic gas such as helium or argon, 1.4 for diatomic gases such as hydrogen, oxygen and nitrogen and 1.33 for a polyatomic gas such as carbon dioxide or superheated steam. The true indices will deviate somewhat from these values in practice: for instance the value of 1.3 is normally used as a better approximation for superheated steam.

5.2.4 Steady-state flow through a nozzle for a gas undergoing an isothermal expansion

In practice, the flow through a nozzle is likely to be much closer to adiabatic than isothermal. Nevertheless, the isothermal situation represents one limiting case ($n = 1$) of the reversible polytropic expansion and will be treated for the sake of completeness. Let us write equation (5.6) as:

$$pv = b \tag{5.28}$$

where b is a constant. Hence the specific volume is inversely proportional to pressure as pressure varies along the length of the nozzle:

$$v = \frac{b}{p} \tag{5.29}$$

Substituting this into equation (5.3) gives:

$$c_2^2 - c_1^2 = -2b\int_{p_1}^{p_2}\frac{dp}{p} - 2g(z_2 - z_1) \tag{5.30}$$

so that

$$\begin{aligned} c_2^2 - c_1^2 &= -2b[\ln p]_{p_1}^{p_2} - 2g(z_2 - z_1) \\ &= 2b\ln\left(\frac{p_1}{p_2}\right) + 2g(z_1 - z_2) \end{aligned} \tag{5.31}$$

We may evaluate b from the upstream conditions:

$$p_1 v_1 = b \tag{5.32}$$

and substitute into (5.31) to give

$$c_2^2 - c_1^2 = 2p_1 v_1 \ln\left(\frac{p_1}{p_2}\right) + 2g(z_1 - z_2) \tag{5.33}$$

As in the case of the general polytropic expansion, and for similar reasons, we may neglect the potential energy term. As before, we may also simplify this equation by setting the inlet velocity to zero for gas or vapour coming from a vessel or header, so that the velocity is given by:

$$c_2 = \sqrt{2p_1 v_1 \ln\left(\frac{p_1}{p_2}\right)} \tag{5.34}$$

Noting that the mass flow is given in terms of the downstream parameters by

$$W = \frac{A_2 c_2}{v_2} \tag{5.35}$$

and that, as a result of equation (5.28)

$$v_2 = \frac{p_1}{p_2}v_1 \tag{5.36}$$

we may combine equations (5.34), (5.35) and (5.36) to yield the mass flow, W:

$$W = A_2\sqrt{2\frac{p_1}{v_1}\left(\frac{p_2}{p_1}\right)^2 \ln\left(\frac{p_1}{p_2}\right)} \tag{5.37}$$

5.3 Maximum mass flow for a polytropic expansion

When a gas at constant inlet conditions of pressure, p_1, and specific volume, v_1, expands to a downstream pressure, p_2, at downstream area, A_2, flow will increase if the downstream pressure decreases until a maximum flow is reached. It follows from the form of equation (5.25) that maximum flow will occur at the maximum of the function, f, given by:

$$f\left(\frac{p_2}{p_1}\right) = \left(\frac{p_2}{p_1}\right)^{2/n} - \left(\frac{p_2}{p_1}\right)^{(n+1)/n} \tag{5.38}$$

From calculus, this will occur when the differential of f with respect to the pressure ratio, (p_2/p_1), is zero, i.e. when

$$\frac{2}{n}\left(\frac{p_2}{p_1}\right)^{(2-n)/n} - \frac{n+1}{n}\left(\frac{p_2}{p_1}\right)^{1/n} = \frac{n+1}{n}\left(\frac{p_2}{p_1}\right)^{1/n}\left(\frac{2}{n+1}\left(\frac{p_2}{p_1}\right)^{(1-n)/n} - 1\right) = 0 \tag{5.39}$$

Equation (5.39) is satisfied by the critical pressure ratio

$$\frac{p_2}{p_1} = \left(\frac{n+1}{2}\right)^{n/(1-n)} = \left(\frac{2}{n+1}\right)^{n/(n-1)} \tag{5.40}$$

Substituting from equation (5.40) into equation (5.25) gives the maximum flow as:

$$W_{\max} = A_2\sqrt{2\frac{n}{n-1}\frac{p_1}{v_1}\left(\left(\frac{2}{n+1}\right)^{2/(n-1)} - \left(\frac{2}{n+1}\right)^{(n+1)/(n-1)}\right)}$$

$$= A_2\sqrt{2\frac{n}{n-1}\frac{p_1}{v_1}\left(\frac{2}{n+1}\right)^{(n+1)/(n-1)}\left(\left(\frac{2}{n+1}\right)^{(1-n)/(n-1)} - 1\right)} \tag{5.41}$$

Since

$$\left(\frac{2}{n+1}\right)^{(1-n)/(n-1)} - 1 = \frac{n+1}{2} - 1 = \frac{n-1}{2} \tag{5.42}$$

it follows that

$$W_{\max} = A_2\sqrt{\frac{p_1}{v_1}n\left(\frac{2}{n+1}\right)^{(n+1)/(n-1)}} \tag{5.43}$$

Equation (5.40) has specified the value of downstream pressure, p_2, for maximum flow, but it has not specified the associated downstream area, A_2. In fact, solving equation (5.25) for the area, A_2, associated with downstream pressure, p_2, shows that the same condition of a maximum value of $f(p_2/p_1)$ causes A_2 to be a minimum for any given flow, including the maximum flow, $W_{\max}$. Hence we may conclude that the locations of the critical pressure ratio and the minimum area coincide. The nozzle's minimum area occurs at its throat, and accordingly it is the throat area that determines the maximum flow. We may thus write $A_t = A_2$ in (5.43):

$$W_{\max} = A_t\sqrt{\frac{p_1}{v_1}n\left(\frac{2}{n+1}\right)^{(n+1)/(n-1)}} \tag{5.44}$$

It is instructive to examine how the maximum flow for any given inlet conditions varies with polytropic index, n. For fixed nozzle geometry and constant upstream conditions, the behaviour of $W_{\max}$ will be determined by the behaviour of the function

$$g = n\left(\frac{2}{n+1}\right)^{(n+1)/(n-1)} \tag{5.45}$$

Differentiating with respect to n gives:

$$\frac{dg}{dn} = \left(\frac{2}{n+1}\right)^{(n+1)/(n-1)} \times\left(1 - \frac{2n}{(n-1)^2}\ln\left(\frac{2}{n+1}\right)\right) \tag{5.46}$$

which is positive for all $n > 1$. The reversible adiabatic has the largest value of polytropic index for a flow process without heat rejection, and thus the adiabatic mass flow will be the greatest that an uncooled nozzle can pass.

5.4 Sonic flow

5.4.1 Sonic flow for a polytropic expansion

It is shown in specialized texts on fluid dynamics that a convergent–divergent nozzle is needed to accelerate a gas from subsonic to supersonic conditions, since gas acceleration in the subsonic regime requires the flow area to diminish with speed, while gas acceleration from sonic to supersonic speeds requires the flow area to expand with speed. The subsonic, convergent part of the nozzle is linked to the supersonic, divergent part of the nozzle by a duct of constant flow area, known as the throat, which is kept very short in practice in order to avoid frictional losses. The throat is the only section of the nozzle in which sonic flow can occur, and it is impossible for the throat to support any speed greater than sonic. The above remarks apply to all polytropic expansions, although the isentropic expansion has the greatest practical significance.

It follows from the previous paragraph that a convergent-only nozzle cannot produce supersonic velocities. Such nozzles are most often used to produce subsonic outlet velocities, but it is possible for them to yield a sonic velocity at the outlet, since the outlet in a convergent-only nozzle is equivalent to the throat of a convergent–divergent nozzle, and we will often refer to it as the 'throat'.

For either type of nozzle, the speed of the gas in the nozzle throat will increase as the pressure ratio decreases, until the speed of sound sets a limit. The speed of sound in the throat can be shown to be a function of the local thermodynamic state:

$$c_c = \sqrt{\gamma p_c v_c} \tag{5.47}$$

where the subscript 'c' denotes the critical value at which the speed of sound is reached in the throat of the nozzle. Reducing the outlet pressure further once the throat pressure has fallen to p_c will not alter the mass flow of the gas, which is now known as 'choked'. To find the critical pressure ratio in a polytropic expansion that will cause this speed to be reached, we substitute $p_c = p_2$ into equation (5.22) and equate the speed to that given in equation (5.47):

$$\sqrt{2\frac{n}{n-1}p_1 v_1\left(1-\left(\frac{p_c}{p_1}\right)^{(n-1)/n}\right)} = \sqrt{\gamma p_c v_c} \tag{5.48}$$

or

$$\frac{2n}{\gamma(n-1)}\frac{p_1 v_1}{p_c v_c}\left(1-\left(\frac{p_c}{p_1}\right)^{(n-1)/n}\right) = 1 \tag{5.49}$$

Noting that as a consequence of equation (5.12),

$$\frac{v_1}{v_c} = \left(\frac{p_c}{p_1}\right)^{1/n} \tag{5.50}$$

we may substitute into (5.49) and find the critical pressure ratio as:

$$\frac{p_c}{p_1} = \left(\frac{2\left(\dfrac{n}{\gamma}\right)}{n-1+2\left(\dfrac{n}{\gamma}\right)}\right)^{n/(n-1)} \tag{5.51}$$

To give an idea of the pressure ratios involved, consider a diatomic gas such as nitrogen, where the specific-heat ratio, $\gamma = 1.4$. For an isentropic expansion, the polytropic index is equal to the specific-heat ratio, and substituting $n = \gamma = 1.4$ into equation (5.51) gives a critical pressure ratio $p_c/p_1 = 0.5283$. As the expansion approaches isothermal, so $n \to 1.0$. Equation (5.51) is indeterminate at $n = 1.0$, but nevertheless we can gain an insight into the behaviour by setting $n = 1.01$, when $p_c/p_1 = 0.4978$. (A more complete treatment of the theoretical case of sonic velocity in an isothermal expansion is given in Section 5.4.3.)

The expression from equation (5.51) can be back-substituted in (5.50) to give the critical ratio of specific volumes:

$$\frac{v_c}{v_1} = \left(\frac{2\left(\dfrac{n}{\gamma}\right)}{n-1+2\left(\dfrac{n}{\gamma}\right)}\right)^{1/(1-n)} \tag{5.52}$$

Combining the two expressions (5.51) and (5.52) with the equation for the sonic velocity, (5.47), gives the downstream speed of sound in terms of the upstream conditions for a polytropic expansion:

$$c_c = \sqrt{\frac{2n}{n-1+2\left(\dfrac{n}{\gamma}\right)}p_1 v_1} \tag{5.53}$$

from which it can be seen that the velocity at the exit of the nozzle depends only on the upstream conditions once $p_t \le p_c$. Note, however, that the speed of sound in the throat for the same gas inlet conditions will depend on the temperature attained at the throat and thus on the mode of expansion. For example, in the case of an ideal diatomic gas with $\gamma = 1.4$, an isothermal expansion ($n = 1.0$) will give rise to a sonic velocity that is 10% greater than the sonic velocity that occurs with an isentropic expansion ($n = 1.4$). (This may be explained in physical terms by noting that the speed of sound rises with temperature (substitute equation (3.2) into equation (5.47)), and the temperature in the throat will be higher during an isothermal expansion than an isentropic expansion).

The corresponding mass flow is found from

$$W_c = \frac{A_t c_c}{v_c} \tag{5.54}$$

so that

$$W_c = A_t\sqrt{\gamma\left(\frac{2\left(\dfrac{n}{\gamma}\right)}{n-1+2\left(\dfrac{n}{\gamma}\right)}\right)^{(n+1)/(n-1)}\frac{p_1}{v_1}} \tag{5.55}$$

and, of course, this mass flow depends only on the upstream conditions. When the expansion is isentropic, the mass flow at sonic conditions will be the same as the maximum mass flow, but if the expansion is subject to a degree of heating, then $n < \gamma$, and the mass flow at sonic conditions will be less than the maximum mass flow.

5.4.2 Sonic flow for an isentropic expansion

The most important case of sonic flow is that where the expansion is isentropic: no heat is exchanged with the surroundings and frictional effects may be neglected. For an ideal gas undergoing an isentropic expansion, $n = \gamma$. Substituting into (5.51) gives

$$\frac{p_c}{p_1} = \left(\frac{2}{\gamma+1}\right)^{\gamma/(\gamma-1)} \tag{5.56}$$

which is identical to equation (5.40) when $\gamma = n$, confirming that the maximum flow for an isentropic expansion occurs when the velocity at the throat goes sonic.

The ratio of critical specific volume to inlet specific volume follows from equation (5.52):

$$\frac{v_c}{v_1} = \left(\frac{2}{\gamma+1}\right)^{1/(1-\gamma)} \tag{5.57}$$

The sonic velocity follows from equation (5.53):

$$c_c = \sqrt{\frac{2}{\gamma+1}\gamma p_1 v_1} \tag{5.58}$$

and the corresponding mass flow is:

$$W_c = A_t\sqrt{\gamma\left(\frac{2}{\gamma+1}\right)^{(\gamma+1)/(\gamma-1)}\frac{p_1}{v_1}} \tag{5.59}$$

For the special case where the upstream temperature is constant, we may usefully progress equation (5.59) further by the use of the characteristic gas equation:

$$p_1 v_1 = ZR_w T_1 \tag{5.60}$$

Substituting for v_1 from (5.60) into (5.59) gives:

$$W_c = p_1 A_t\sqrt{\gamma\left(\frac{2}{\gamma+1}\right)^{(\gamma+1)/(\gamma-1)}\frac{1}{ZR_w T_1}} \tag{5.61}$$

If the upstream temperature is constant, and if the gas is either an ideal ($Z = 1$) or semi-ideal gas ($Z = Z(T)$), then the term

$$A_t\sqrt{\gamma\left(\frac{2}{\gamma+1}\right)^{(\gamma+1)/(\gamma-1)}\frac{1}{ZR_w T_1}}$$

is a constant, making the mass flow dependent only on the upstream pressure. This simplifying result, that $W \propto p_1$, is often convenient in simulating the interface of a boiler plant to a turbine, where sonic conditions are frequently designed for in the first stage.

5.4.3 Sonic flow during an isothermal expansion

Achieving sonic flow during an isothermal expansion is a highly improbable eventuality because the rapid heat transfer needed would take a great effort to contrive. Nevertheless, it is a theoretical possibility and has an interest as the limiting case of a reversible polytropic expansion. Accordingly a brief treatment will be provided here for completeness.

To estimate the critical pressure ratio, we equate the sonic velocity to the outlet velocity calculated via equation (5.34). Substituting $p_c = p_t = p_2$ gives:

$$\sqrt{2p_1 v_1 \ln\left(\frac{p_1}{p_c}\right)} = \sqrt{\gamma p_c v_c} \tag{5.62}$$

Since, by equation (5.28)

$$p_c v_c = p_1 v_1 \tag{5.63}$$

it follows that

$$\ln\left(\frac{p_1}{p_c}\right) = \frac{\gamma}{2} \tag{5.64}$$

so that the critical pressure ratio is given by:

$$\frac{p_c}{p_1} = \mathrm{e}^{-(\gamma/2)} \tag{5.65}$$

The value of the critical pressure ratio clearly depends on the value of γ: when $\gamma = 1.4$, then $p_c/p_1 = 0.4966$, while when the value of γ appropriate for steam is taken, namely $\gamma = 1.3$, then $p_c/p_1 = 0.5220$.

The critical velocity can be found in terms of the upstream parameters by substituting from (5.65) back into equation (5.34) or directly by substituting from (5.63) into equation (5.47):

$$c_c = \sqrt{\gamma p_1 v_1} \tag{5.66}$$

while the mass flow, W_c, is given by substituting the critical pressure ratio into equation (5.37), with $A_t = A_2$:

$$W_c = A_t\sqrt{\frac{p_1}{v_1}\gamma \mathrm{e}^{-\gamma}} \tag{5.67}$$

5.5 Comparison between flow formulae

Let us compare the flow formulae for the cases of flow of a liquid at constant specific volume, general polytropic flow, and isothermal gas flow. The formulae for unchoked flow are summarized below.

Flow of a liquid at constant specific volume:

$$W = A_2\sqrt{2\frac{p_1}{v_1}\left(1-\frac{p_2}{p_1}\right)} \tag{5.11}$$

General polytropic flow:

$$W = A_2\sqrt{2\frac{n}{n-1}\frac{p_1}{v_1}\left(\left(\frac{p_2}{p_1}\right)^{2/n}-\left(\frac{p_2}{p_1}\right)^{(n+1)/n}\right)} \tag{5.25}$$

Isothermal flow of a gas or vapour:

$$W = A_2\sqrt{2\frac{p_1}{v_1}\left(\frac{p_2}{p_1}\right)^2\ln\left(\frac{p_1}{p_2}\right)} \tag{5.37}$$

Clearly the factor $A_2\sqrt{p_1/v_1}$ is common to all three expressions, and we may examine the differences in predicted flow by considering the behaviour of the flow functions:

$$f_v(r) = \sqrt{2(1-r)} \tag{5.68}$$

$$f_p(r,n) = \sqrt{\frac{2n}{n-1}\left(r^{2/n}-r^{(n+1)/n}\right)} \tag{5.69}$$

$$f_i(r) = \sqrt{2r^2\ln\left(\frac{1}{r}\right)} \tag{5.70}$$

where
f_v is the flow function for the expansion at constant specific volume,
f_p is the flow function for the general polytropic expansion,
f_i is the flow function for the isothermal expansion,
r is the pressure ratio p_2/p_1

The numerical behaviour of these flow functions is shown in tabular and graphical forms below. Table 5.1 shows three instances for the general, polytropic case: (i) an isentropic expansion of an ideal, diatomic gas, with $n = 1.4$, (ii) an isentropic expansion for superheated steam, with $n = 1.3$, and (iii) a polytropic expansion very close to an isothermal expansion, with $n = 1.01$.

Table 5.1 Tabulation of values of the various flow functions

r	f_v	$f_p(1.4)$	$f_p(1.3)$	$f_p(1.01)$	f_i
0.6	0.8944	0.6769	0.6627	0.6088	0.6065
0.7	0.7745	0.6383	0.6290	0.5928	0.5912
0.8	0.6325	0.5607	0.5555	0.5353	0.5344
0.9	0.4472	0.4226	0.4207	0.4135	0.4131
0.95	0.3162	0.3076	0.3070	0.3044	0.3043
0.98	0.2000	0.1978	0.1977	0.1970	0.1970
0.99	0.1414	0.1407	0.1406	0.1404	0.1404
1	0	0	0	0	0

Figure 5.2 shows the shape of the flow functions for the three main cases: an expansion with constant specific volume, an isentropic expansion with $n = 1.4$, and an isothermal expansion.

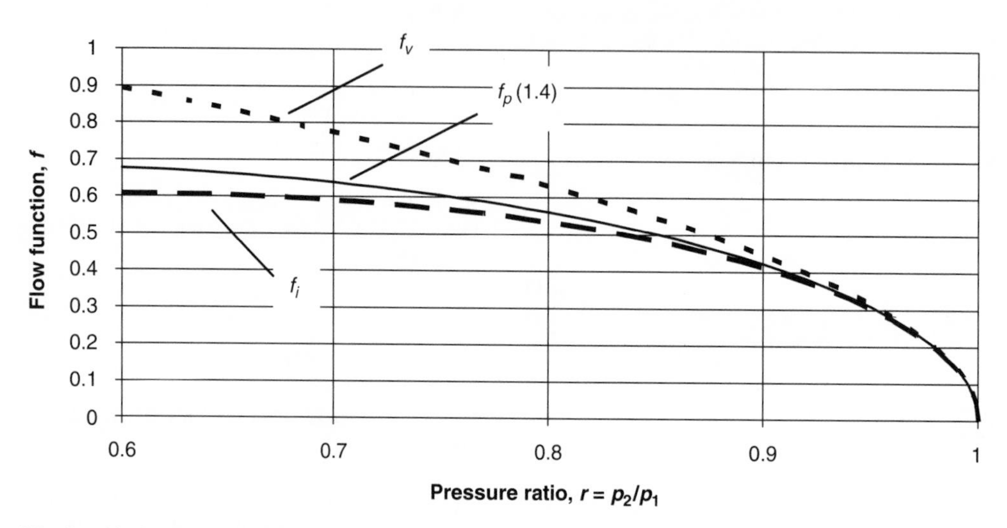

Figure 5.2 Graphical comparison of flow functions.

A number of points can be deduced from the graph and the table.

(i) All the expressions converge as the pressure ratio approaches unity, with the spread below 4% after $r = 0.95$, confirming that the simple form deduced by assuming constant specific volume is a reasonable approximation when the pressure ratio is small, even for gases and vapours.

(ii) The flow in a polytropic expansion approaches that of an isothermal expansion as $n \to 1.0$, as we should expect.

(iii) The flow functions for an isothermal expansion and an isentropic expansion have relatively similar values down to low pressure ratios, but the same is not true for the expansion at constant specific volume, where the flow function diverges significantly at pressure ratios below 0.9. The result demonstrates the need to allow for the change in specific volume for a gas when the pressure change is great.

5.6 Bibliography

Faires, V.M. (1957). *Thermodynamics*, 3rd edition, Macmillan, New York.

Kearton, W.J. (1922, 1958). *Steam Turbine Theory and Practice*, Pitman and Sons.

Lockey, J. (1966). *The Thermodynamics of Fluids*, Heinemann Educational Books Ltd, London.

Rogers, G.F.C. and Mayhew, Y.R. (1992). *Engineering Thermodynamics: Work and Heat Transfer*, 4th edition, Longmans.

Streeter, V.L. and Wylie, E.B. (1983). *Fluid Mechanics, First SI Metric Edition*, McGraw-Hill, New York.

6 Steady-state compressible flow

6.1 Introduction

We were able, in dealing with incompressible flow (Chapter 4), to simplify the general energy equation by assuming that the flow velocity was constant, while when dealing with ideal nozzle flow (Chapter 5), we were able to assume in the base case that no friction was present and that height differences were negligible. For compressible flow, however, the only simplification we can allow ourselves is the neglect of height differences, taking advantage of the fact that gases have a much greater specific volume than liquids, causing mass flow to be influenced much less by height differences when gas is being carried. But that aside, we must deal squarely with the general energy equation without further modification.

The first solution to the general energy equation for compressible flow in a horizontal pipe was provided by Lapple in an important paper in 1943. Believing that isothermal and adiabatic flow might occur in practical situations, Lapple analysed both cases. It is considered now, however, that the high heat flux required from the external environment to maintain an expansion isothermal is not representative of conditions on process plant, and so the adiabatic case is more physically realistic for the lagged pipes normally found. In fact, Lapple's analysis of the isothermal case contained an error concerning the sonic velocity. But while Lapple's use of the erroneous sonic velocity as a normalizing factor caused an anomaly in the graphs he presented on adiabatic flow, the fundamental equations he developed for the adiabatic case were correct. Accordingly Lapple's approach for analysing adiabatic, compressible flow will be followed in this chapter.

The full solution to the general energy equation will be seen to lead to an implicit set of equations for pipe-flow, requiring an iterative solution. However, it is possible to approximate the solution accurately with explicit formulations, and these will be presented at the end of the chapter.

6.2 General overview of compressible pipe-flow

Figure 6.1 shows the layout of the system being modelled. The compressible fluid is being carried from an upstream vessel at pressure p_1 to a downstream vessel at pressure p_4. The pressure just inside the entrance to the pipe is p_2, where in general $p_2 \neq p_1$. The pressure just inside the outlet of the pipe is p_3. For many cases, $p_3 = p_4$, but this does not hold for the case of choked flow, where $p_3 \geq p_4$.

The entrance to the pipe acts as a nozzle, accelerating fluid that is originally at rest in the upstream vessel from a velocity $c_1 = 0$ to a velocity c_2. The entrance to the pipe is shown as a well-rounded, convergent nozzle, assumed frictionless, with the pipe

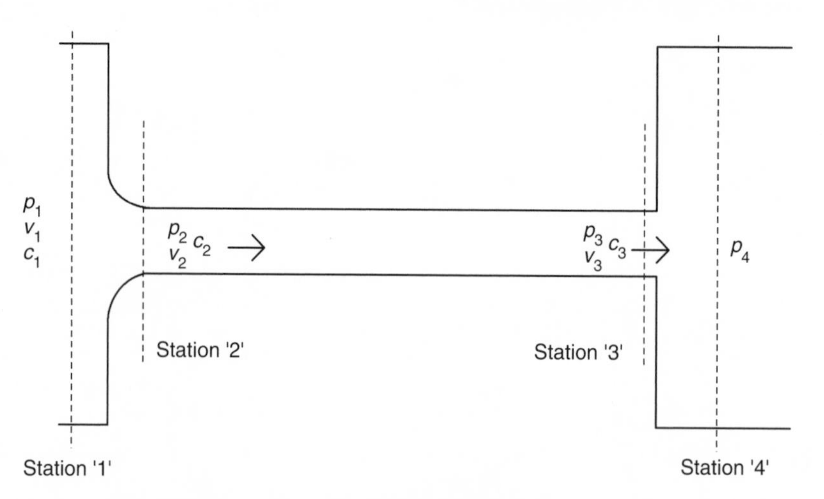

Figure 6.1 Schematic of a pipe carrying a compressible fluid between two vessels.

acting as the throat of the nozzle. (An allowance may be made for pipe systems where the pipe entrance is abrupt by assuming an additional velocity-head drop the same as that for a liquid, namely 0.5. See Chapter 4, Section 4.10.) If the pipe were frictionless, the thermodynamic properties of the fluid (e.g. pressure and temperature) would be maintained indefinitely as it passed down the length of the pipe. However, frictional losses cause the pressure to fall along the pipe, so that $p_3 < p_2$. The fluid will continue to accelerate along this pressure gradient, so that $c_3 > c_2$.

Chapter 5 presented a mathematical model for adiabatic flow through a frictionless nozzle, and we shall use some of that chapter's results. It will be helpful for our present purpose if we derive the Mach number at station '2', where the Mach number, **M**, is defined as the ratio of the velocity to the local speed of sound. Recalling equation (5.47), the Mach number at station '2' will be:

$$\mathbf{M}_2 = \frac{c_2}{\sqrt{\gamma p_2 v_2}} \tag{6.1}$$

Very little energy is lost to friction in the convergent part of any nozzle, and so we may write down the velocity at station '2' by using equation (5.22), setting $n = \gamma$ for frictionless, adiabatic flow:

$$c_2 = \sqrt{\frac{2}{\gamma - 1}\gamma p_1 v_1 \left(1 - \left(\frac{p_2}{p_1}\right)^{1-(1/\gamma)}\right)} \tag{6.2}$$

The isentropic expansion through the nozzle will, of course, obey the relation

$$p_1 v_1^\gamma = p_2 v_2^\gamma \tag{6.3}$$

so that

$$p_1 v_1 = p_2^{1/\gamma} v_2 p_1^{1-(1/\gamma)} \tag{6.4}$$

Substituting (6.4) into (6.2) gives:

$$c_2 = \sqrt{\frac{2}{\gamma - 1}\gamma p_2^{1/\gamma} v_2 p_1^{1-(1/\gamma)} \left(1 - \left(\frac{p_2}{p_1}\right)^{1-(1/\gamma)}\right)}$$

$$= \sqrt{\frac{2}{\gamma - 1}\gamma p_2^{1/\gamma} v_2 \left(p_1^{1-(1/\gamma)} - p_2^{1-(1/\gamma)}\right)}$$

$$= \sqrt{\frac{2}{\gamma - 1}\gamma p_2^{1/\gamma} p_2^{1-(1/\gamma)} v_2 \left(\left(\frac{p_1}{p_2}\right)^{1-(1/\gamma)} - 1\right)}$$

$$= \sqrt{\frac{2}{\gamma - 1}\gamma p_2 v_2 \left(\left(\frac{p_1}{p_2}\right)^{(\gamma-1)/\gamma} - 1\right)} \tag{6.5}$$

Combining equations (6.1) and (6.5) gives the Mach number at station '2' as:

$$\mathbf{M}_2 = \sqrt{\frac{2}{\gamma - 1}\left(\left(\frac{p_1}{p_2}\right)^{(\gamma-1)/\gamma} - 1\right)} \tag{6.6}$$

The specific volume at station '2' follows from equation (6.3):

$$v_2 = v_1 \left(\frac{p_1}{p_2}\right)^{1/\gamma} \tag{6.7}$$

It will be clear that the expressions on the right-hand sides of equations (6.6) and (6.7) both depend on the value of the pipe entrance pressure, p_2, which is at this stage unknown. To find p_2, we need to consider the flow through the rest of the pipe.

6.3 Frictionally resisted, adiabatic flow inside the pipe

6.3.1 Deriving a second equation for the Mach number at station '2'

We begin by recalling the general energy equation for frictionally resisted flow developed in Section 4.2 and repeated below:

$$v\,dp + c\,dc + g\,dz + dF = 0 \tag{4.15}$$

Formal differentiation of the expression pv gives

$$d(pv) = p\,dv + v\,dp \tag{6.8}$$

Also the energy loss due to friction may be found from equation (4.20) in terms of the increment of pipe-length, dx:

$$dF = \frac{2fc^2}{D}dx \tag{6.9}$$

where the Fanning friction factor, f, may be considered constant along the length of the pipe for the reasons given in Section 4.4.

When the pipe is horizontal,

$$dz = 0 \tag{6.10}$$

(In fact the term $g\,dz$ is usually small compared with dF, allowing the height term to be neglected in many cases; if necessary, an approximate allowance can be made in the final calculation by adding an additional effective pipe-length to the friction term.)

Substituting from equations (6.8), (6.9) and (6.10) into equation (4.15) gives the energy equation as:

$$d(pv) - (pv)\frac{dv}{v} + c\,dc + \frac{2fc^2}{D}dx = 0 \tag{6.11}$$

Now we need to find alternative expressions for $d(pv)$, (pv) and dv/v.

$d(pv)$

First we make use of the following thermodynamic relationships for a near-ideal gas, as discussed in Chapter 3, Section 3.3:

$$\frac{du}{dT} = c_v \tag{3.13}$$

$$c_p = c_v + ZR_w \tag{3.20}$$

where c_v is the specific heat at constant volume and c_p is the specific heat at constant pressure. The ratio of specific heats is given by

$$\gamma = \frac{c_p}{c_v} \tag{6.12}$$

While a formal differentiation of the characteristic gas equation, $pv = ZR_wT$, gives for a near-ideal gas:

$$d(pv) = ZR_w\,dT \tag{6.13}$$

Using (3.13), (3.20) and (6.12) in (6.13) gives the following progression:

$$\begin{aligned} d(pv) &= (c_p - c_v)\,dT = \frac{c_p - c_v}{c_v}\,du \\ &= \left(\frac{c_p}{c_v} - 1\right) du = (\gamma - 1)\,du \end{aligned} \tag{6.14}$$

so that

$$du = \frac{1}{\gamma - 1}\,d(pv) \tag{6.15}$$

The general energy equation applied to fluid may be written as in equation (4.7), repeated below:

$$dq - dw - dh - c\,dc - g\,dz = 0 \tag{4.7}$$

For an adiabatic expansion without work in a horizontal pipe, $dq = dw = dz = 0$. In addition, $dh = d(u + pv) = du + d(pv)$, so that (4.7) becomes:

$$du + d(pv) + c\,dc = 0 \tag{6.16}$$

Substituting from (6.15) into (6.16) produces the required relationship for $d(pv)$:

$$d(pv) = -\frac{\gamma - 1}{\gamma}\,c\,dc \tag{6.17}$$

pv

Equation (6.17) may be integrated to give the result:

$$pv + \frac{\gamma - 1}{2\gamma}c^2 = \text{constant} \tag{6.18}$$

It follows from equation (6.18) in particular that the fluid properties at a general point in the pipe are related to the properties at station '2' by the equation:

$$pv = p_2v_2 + \frac{\gamma - 1}{2\gamma}(c_2^2 - c^2) \tag{6.19}$$

which is the required expression for (pv).

dv/v

By continuity, the velocity at any location in the pipe is related to the mass flow, W, by:

$$c = v\frac{W}{A} \tag{6.20}$$

The mass flow is constant throughout the pipe in the steady state, and, assuming the pipe cross-sectional area remains constant, we may differentiate (6.20) to give

$$\frac{dc}{c} = \frac{dv}{v} \tag{6.21}$$

Substituting from equations (6.17), (6.19) and (6.21) into (6.11) and multiplying by γ/c^2 gives:

$$\begin{aligned} &(1 - \gamma)\frac{dc}{c} - \left(\gamma p_2v_2 + \frac{\gamma - 1}{2}(c_2^2 - c^2)\right)\frac{dc}{c^3} \\ &+ \gamma\frac{dc}{c} + \gamma\frac{2f}{D}\,dx = 0 \end{aligned} \tag{6.22}$$

Collecting terms gives the equation:

$$\begin{aligned} &(1 + \gamma)\frac{dc}{c} - (2\gamma p_2v_2 + (\gamma - 1)c_2^2)\frac{dc}{c^3} \\ &+ \gamma\frac{4f}{D}\,dx = 0 \end{aligned} \tag{6.23}$$

Equation (6.23) may now be integrated along the complete length of the pipe, between stations '2' and '3', i.e. from velocity c_2 to velocity c_3 and from $x = 0$ to $x = L$:

$$\begin{aligned} &(1 + \gamma)\ln\left(\frac{c_3}{c_2}\right) + \left(\frac{2\gamma p_2v_2 + (\gamma - 1)c_2^2}{2}\right) \\ &\times \left(\frac{1}{c_3^2} - \frac{1}{c_2^2}\right) + 4f\frac{L}{D}\gamma = 0 \end{aligned} \tag{6.24}$$

The middle term may be re-arranged as follows:

$$\begin{aligned} &\left(\frac{2\gamma p_2v_2 + (\gamma - 1)c_2^2}{2}\right)\left(\frac{1}{c_3^2} - \frac{1}{c_2^2}\right) \\ &= \left(\frac{2\gamma p_2v_2 + (\gamma - 1)c_2^2}{2c_2^2}\right)\left(\frac{c_2^2}{c_3^2} - 1\right) \end{aligned}$$

$$= \left(\frac{2 + (\gamma - 1)\dfrac{c_2^2}{\gamma p_2 v_2}}{2\dfrac{c_2^2}{\gamma p_2 v_2}} \right) \left(\frac{c_2^2}{c_3^2} - 1 \right)$$

$$= \left(\frac{2 + (\gamma - 1)\mathbf{M}_2^2}{2\mathbf{M}_2^2} \right) \left(\frac{c_2^2}{c_3^2} - 1 \right) \quad (6.25)$$

using the Mach number as defined in equation (6.1).

It follows from equation (6.20) that when flow and cross-sectional area are constant along the pipe, the ratio of velocities is the same as the ratio of specific volumes:

$$\frac{c_2}{c_3} = \frac{v_2}{v_3} \quad (6.26)$$

Substituting from equations (6.25) and (6.26) into (6.24) gives:

$$\left(\frac{2 + (\gamma - 1)\mathbf{M}_2^2}{2\gamma \mathbf{M}_2^2} \right) \left(1 - \frac{v_2^2}{v_3^2} \right) - \frac{\gamma + 1}{\gamma} \ln\left(\frac{v_3}{v_2} \right) - 4f\frac{L}{D} = 0 \quad (6.27)$$

Thus we now have two simultaneous equations for the Mach number at station '2'. The first, equation (6.6), depends on the flow through the entrance nozzle, in particular the ratio of pressures. The second, equation (6.27), depends on the subsequent flow through the pipe, and particularly on the ratio of specific volumes at stations '2' and '3'.

6.3.2 The ratio of specific volumes at pipe entrance and outlet, v_3/v_2

We may relate the fluid properties at station '3' to those at station '2' using equation (6.18):

$$p_3 v_3 + \frac{\gamma - 1}{2\gamma} c_3^2 = p_2 v_2 + \frac{\gamma - 1}{2\gamma} c_2^2 \quad (6.28)$$

Rearrangement of (6.28) gives:

$$\frac{p_3}{p_2}\left(\frac{v_3}{v_2} \right) = 1 + \frac{\gamma - 1}{2\gamma p_2 v_2} c_2^2 \left(1 - \frac{c_3^2}{c_2^2} \right) \quad (6.29)$$

Substituting for the Mach number from equation (6.1) and using the fact that the ratio of velocities is equal to the ratio of specific volumes (equation (6.26)) yields

$$\frac{p_3}{p_2}\left(\frac{v_3}{v_2} \right) = 1 + \frac{\gamma - 1}{2} \mathbf{M}_2^2 \left(1 - \left(\frac{v_3}{v_2} \right)^2 \right) \quad (6.30)$$

or

$$\frac{\gamma - 1}{2} \mathbf{M}_2^2 \left(\frac{v_3}{v_2} \right)^2 + \frac{p_3}{p_2}\left(\frac{v_3}{v_2} \right) - \left(1 + \frac{\gamma - 1}{2} \mathbf{M}_2^2 \right) = 0 \quad (6.31)$$

which is a quadratic in the ratio of specific volumes (v_3/v_2). Only the positive root is physically meaningful:

$$\frac{v_3}{v_2} = \frac{1}{(\gamma - 1)\mathbf{M}_2^2} \times \left[\sqrt{\left(\frac{p_3}{p_2} \right)^2 + 2(\gamma - 1)\mathbf{M}_2^2 + (\gamma - 1)^2 \mathbf{M}_2^4} - \left(\frac{p_3}{p_2} \right) \right] \quad (6.32)$$

Thus equation (6.32) expresses the ratio of specific volumes in terms of the ratio of pressures at the entrance and outlet of the pipe, as well as the Mach number at the entrance. We now need to find the pressure at the pipe outlet, p_3.

6.3.3 The pipe outlet pressure, p_3: the effect of choking

One might expect that the pipe outlet pressure would be the same as the pressure in the discharge vessel. This is indeed the case when the outlet velocity is subsonic. But when the pressure in the discharge vessel falls to a critical value, the outlet velocity reaches the speed of sound, which causes the pipe to choke in a way analogous to the choking of a nozzle. The outlet pressure, p_3, becomes independent of the pressure in the discharge vessel when the latter drops below this critical value.

From equation (6.17), the change in velocity along the pipe is given by:

$$dc = -\frac{\gamma}{(\gamma - 1)c} d(pv) \quad (6.33)$$

or, using the characteristic gas equation $pv = ZR_wT$,

$$dc = -\frac{\gamma Z R_w}{(\gamma - 1)c} dT \quad (6.34)$$

Since an adiabatic expansion must lead to cooling even in the case where friction is present, it follows from equation (6.34) that the velocity will increase continuously along the pipe, reaching its maximum at the outlet. We know from the theory of nozzles that supersonic conditions may be reached only when the section following the throat is divergent. Clearly this cannot be the case for the geometry of Figure 6.1, where the pipe is fed by a convergent-only nozzle and has a constant cross-sectional area downstream. The highest velocity that can be reached is therefore the sonic velocity, and this can occur only at the outlet. When sonic conditions occur at the outlet, the velocity there is given by:

$$c_{3c} = \sqrt{\gamma p_{3c} v_{3c}} \tag{6.35}$$

where the subscript 'c' indicates critical conditions, and the Mach number, $\mathbf{M}_3 = c_3/c_{3c}$, is unity. We may find the ratio of the pressure at station '2' and the pressure at station '3' at choking, p_{3c}/p_2, in the following manner.

We begin by deriving an equation equivalent to equation (6.27), except that it will refer to the Mach number at station '3' rather than station '2'. Considering equation (6.18), the fluid properties at station '3' may be related to those at a general point by

$$pv = p_3 v_3 + \frac{\gamma - 1}{2\gamma}(c_3^2 - c^2) \tag{6.36}$$

which is, of course, equation (6.19) with a change of subscript. Instead of substituting for pv from equation (6.19) into equation (6.11), we now choose to substitute from equation (6.36). Hence equation (6.22) is replaced by (6.37):

$$(1-\gamma)\frac{dc}{c} - \left(\gamma p_3 v_3 + \frac{\gamma - 1}{2}(c_3^2 - c^2)\right)\frac{dc}{c^3} + \gamma\frac{dc}{c} + \gamma\frac{2f}{D}\,dx = 0 \tag{6.37}$$

which yields, after rearrangement (cf. equation (6.23)):

$$(1+\gamma)\frac{dc}{c} - (2\gamma p_3 v_3 + (\gamma - 1)c_3^2)\frac{dc}{c^3} + \gamma\frac{4f}{D}\,dx = 0 \tag{6.38}$$

Equation (6.38) may now be integrated along the complete length of the pipe between stations '2' and '3', from $x = 0$ to $x = L$, and from velocity c_2 to velocity c_3, to give

$$(1+\gamma)\ln\left(\frac{c_3}{c_2}\right) + \left(\frac{2\gamma p_3 v_3 + (\gamma - 1)c_3^2}{2}\right) \times \left(\frac{1}{c_3^2} - \frac{1}{c_2^2}\right) + 4f\frac{L}{D}\gamma = 0 \tag{6.39}$$

Rearranging gives:

$$(1+\gamma)\ln\left(\frac{c_3}{c_2}\right) + \left(\frac{2 + (\gamma - 1)\dfrac{c_3^2}{\gamma p_3 v_3}}{2\dfrac{c_3^2}{\gamma p_3 v_3}}\right) \times \left(1 - \frac{c_3^2}{c_2^2}\right) + 4f\frac{L}{D}\gamma = 0 \tag{6.40}$$

Since

$$\mathbf{M}_3^2 = \frac{c_3^2}{\gamma p_3 v_3} \tag{6.41}$$

and the ratio of the specific volumes follows from continuity as the ratio of the velocities:

$$\frac{c_2}{c_3} = \frac{v_2}{v_3} \tag{6.26}$$

equation (6.40) becomes:

$$\left(\frac{2 + (\gamma - 1)\mathbf{M}_3^2}{2\gamma\mathbf{M}_3^2}\right)\left(1 - \left(\frac{v_3}{v_2}\right)^2\right) + \frac{\gamma + 1}{\gamma}\ln\left(\frac{v_3}{v_2}\right) + 4\frac{fL}{D} = 0 \tag{6.42}$$

At choking, $\mathbf{M}_3 = 1$, and so equation (6.42) becomes

$$\frac{\gamma + 1}{2\gamma}\left(1 - \left(\left.\frac{v_3}{v_2}\right|_{crit}\right)^2\right) + \frac{\gamma + 1}{\gamma}\ln\left(\left.\frac{v_3}{v_2}\right|_{crit}\right) + 4\frac{fL}{D} = 0 \tag{6.43}$$

This implicit equation may be solved for the single variable $(v_3/v_2|_{crit})$.

Meanwhile we may rearrange equation (6.28), not into equation (6.29) and hence equation (6.30), but into an alternative form in Mach 3 rather than Mach 2:

$$\frac{p_2}{p_3}\left(\frac{v_2}{v_3}\right) = 1 + \frac{\gamma - 1}{2\gamma p_3 v_3}c_3^2\left(1 - \frac{c_2^2}{c_3^2}\right) \tag{6.44}$$

Substituting specific volume ratio for velocity ratio (equation (6.26)) and using the definition of Mach 3 given in equation (6.41) allows us to transform equation (6.44) into

$$\frac{p_3}{p_2} = \frac{\left(\dfrac{v_2}{v_3}\right)}{1 + \dfrac{\gamma - 1}{2}\mathbf{M}_3^2\left(1 - \left(\dfrac{v_2}{v_3}\right)^2\right)} \tag{6.45}$$

Inserting the choking condition $\mathbf{M}_3 = 1$, gives the explicit equation for the critical pressure ratio:

$$\frac{p_{3c}}{p_2} = \frac{\left(\left.\dfrac{v_2}{v_3}\right|_{crit}\right)}{1 + \dfrac{\gamma - 1}{2}\left(1 - \left(\left.\dfrac{v_2}{v_3}\right|_{crit}\right)^2\right)} \tag{6.46}$$

Note that the critical pressure ratio, p_{3c}/p_2, depends only on the specific-heat ratio of the gas being passed, γ, and the friction term, $4fL/D$.

Equations (4.15), (6.9) and (6.10) may be combined into the form:

$$dp = -\frac{c}{v}\,dc - 2\frac{fc^2}{vD}\,dx \tag{6.47}$$

Equation (6.47) indicates that the increase in velocity between stations '2' and '3' in a horizontal pipe will cause the pressure to fall as a result of both the need to supply the additional kinetic energy associated with the higher velocity (term 1 on the right-hand side) and also the need to provide the energy required to overcome friction (term 2 on the right-hand side). The absolute value of each of terms 1 and 2 increases with velocity, c. Thus the pressure drop will reach its maximum when the terminal velocity (at station '3') reaches its highest possible value. Since the highest possible terminal velocity is the speed of sound, it follows that the pressure drop will be highest when the pressure ratio, p_3/p_2, is at its critical value, p_{3c}/p_2, given by equation (6.46) above. But the pressure drop between stations '2' and '3' may be written

$$\Delta p_{2-3} = p_2\left(1 - \frac{p_3}{p_2}\right) \tag{6.48}$$

from which it is clear that, for any given pressure, p_2, the maximum pressure drop, $\Delta p_{2-3}|_{\max}$, will occur when p_3/p_2 is at its minimum value, $p_3/p_2|_{\min}$. It follows that for any given pressure, p_2, the minimum pressure ratio and the critical pressure ratio must coincide:

$$\left.\frac{p_3}{p_2}\right|_{\min} = \frac{p_{3c}}{p_2} \tag{6.49}$$

so that

$$\frac{p_3}{p_2} \geq \frac{p_{3c}}{p_2} \tag{6.50}$$

Our knowledge of pipe choking indicates that the pressure, p_3, will be equal to the pressure in the receiving vessel, p_4, until the latter has fallen to below the critical pressure, but from then on, $p_3 = p_{3c}$. Bearing in mind equation (6.50), we may write the following conditional equation for pressure, p_3:

$$p_3 = \max\left(p_4,\, p_2\frac{p_{3c}}{p_2}\right) \tag{6.51}$$

6.4 Solution sequence for compressible flow through a pipe

The modeller will wish to calculate the flow in the pipe from a knowledge of the pressure and specific volume in the upstream vessel and the pressure in the downstream vessel. Thus he will need to solve the following set of four simultaneous equations:

$$\mathbf{M}_2 = \sqrt{\frac{2}{\gamma - 1}\left(\left(\frac{p_1}{p_2}\right)^{(\gamma-1)/\gamma} - 1\right)} \tag{6.6}$$

$$p_3 = \max\left(p_4,\, p_2\frac{p_{3c}}{p_2}\right) \tag{6.51}$$

$$\frac{v_3}{v_2} = \frac{1}{(\gamma - 1)\mathbf{M}_2^2} \times \left[\sqrt{\left(\frac{p_3}{p_2}\right)^2 + 2(\gamma - 1)\mathbf{M}_2^2 + (\gamma - 1)^2\mathbf{M}_2^4} - \left(\frac{p_3}{p_2}\right)\right] \tag{6.32}$$

$$\left(\frac{2 + (\gamma - 1)\mathbf{M}_2^2}{2\gamma\mathbf{M}_2^2}\right)\left(1 - \frac{v_2^2}{v_3^2}\right) - \frac{\gamma + 1}{\gamma}\ln\left(\frac{v_3}{v_2}\right) - 4f\frac{L}{D} = 0 \tag{6.27}$$

Given p_1, p_4, γ, f, L and D, the equations are dependent only on p_2, the pressure at station '2'. They are presented in the order above so as to indicate the following logical solution sequence.

(1) Guess p_2;
(2) work out first $\mathbf{M}_2$, then p_3 and hence p_3/p_2, and then v_3/v_2;
(3) ensure that equation (6.27) is satisfied by adjusting the value of p_2 iteratively.

When (6.27) is satisfied, it is then possible to calculate specific volume at station '2' from equation (6.7), repeated below:

$$v_2 = v_1\left(\frac{p_1}{p_2}\right)^{1/\gamma} \tag{6.7}$$

The velocity at station '2' will be given by

$$c_2 = \mathbf{M}_2\sqrt{\gamma p_2 v_2} \tag{6.52}$$

The mass flow through the pipe will be

$$W = A\frac{c_2}{v_2} \tag{6.53}$$

The specific volume at the outlet of the pipe follows from equation (6.32), and the velocity at the outlet follows from equation (6.26):

$$c_3 = \frac{v_3}{v_2}c_2 \tag{6.54}$$

The temperature just inside the pipe entrance will be given by

$$T_2 = \frac{p_2 v_2}{ZR_w} \tag{6.55}$$

while the temperature just inside the pipe outlet will be given by

$$T_3 = \frac{p_3 v_3}{Z R_w} \tag{6.56}$$

The temperature of the gas leaving the pipe will be much lower than that of the upstream vessel for cases where the pressure drop is large. However, the modeller should bear in mind that there will have been no change in the stagnation enthalpy, namely the enthalpy that the gas would possess if it had been brought to rest and its kinetic energy converted into enthalpy. This corresponds to the most likely circumstance in practice, since it is normal for essentially all the velocity of the gas to be dissipated and converted back into thermal energy in the receiving vessel. In this most common case, it is the stagnation enthalpy that will be needed in the energy balance for the gas in the receiving vessel.

6.5 Determination of the friction factor, *f*

The discussion of the friction factor given in Section 4.4 applies to compressible as well as incompressible flow. The friction factor will depend on the relative roughness and the Reynolds number, in-line with equation (4.27) and as illustrated in Figure 4.1. Since the Reynolds number depends on the velocity of the gas, it is strictly necessary to check the velocities coming from the calculation above in an iterative fashion in order to obtain the correct friction factor. However, Figure 4.1 shows that the friction factor is essentially independent of Reynolds number for commercial pipe once the Reynolds number has reached 100 000. Furthermore, it is shown in Section 4.4 that the variation in friction factor with flow is likely to be fairly small for all turbulent flows. Accordingly it will not normally be necessary to perform many iterations, if any, provided the region of operation is known approximately in advance. The control engineer will normally have the benefit of either flowsheet values or plant data to guide him here.

6.6 Determination of the effective length of the pipe

We begin by observing that the energy lost to friction is found by integrating the Fanning equation (4.20), repeated below:

$$\frac{dF}{dx} = \frac{4f}{D}\frac{c^2}{2} \tag{4.20}$$

If the velocity, c, may be assumed to be constant over the length, L, of the pipe, then the integration yields equation (4.44):

$$F = 4f\frac{L}{D}\frac{c^2}{2} \tag{4.44}$$

which can be rewritten as equation (4.45):

$$F = K\frac{c^2}{2} \tag{4.45}$$

where K is the dimensionless number of velocity heads lost to friction.

$$K = \frac{F}{\frac{1}{2}c^2} = 4f\frac{L}{D} \tag{4.46}$$

Expressing losses in terms of velocity heads as above obviously has its greatest utility when the velocity is truly constant throughout the pipe, i.e. in strictly incompressible flow. Hence the utility of Table 4.1 of Chapter 4, where the frictional losses for bends and fittings are presented for incompressible flow in terms of velocity heads. But it has been found in practice that the concept may be used to give an approximate characterization of the corresponding frictional losses in compressible flow also, even though the velocity will change along the pipe. Using the concept of frictional velocity heads embodied in equation (4.46) allows us to modify equation (6.27) to the form:

$$\left(\frac{2+(\gamma-1)\mathbf{M}_2^2}{2\gamma\mathbf{M}_2^2}\right)\left(1-\frac{v_2^2}{v_3^2}\right) - \frac{\gamma+1}{\gamma}\ln\left(\frac{v_3}{v_2}\right) - K_T = 0 \tag{6.57}$$

Here K_T represents the total number of velocity heads lost to friction, from pipe friction, entrance loss and losses due to bends and fittings:

$$K_T = 4f\frac{L}{D} + K_{con} + \sum_i K_{bi} \tag{6.58}$$

The values used for the additional losses are as given in Chapter 4.

Effectively an additional length of pipe has been added to the true length in order to find the effective pipe-length, L_{eff}:

$$L_{eff} = L + \frac{D}{4f}\left(K_{con} + \sum_i K_{bi}\right) \tag{6.59}$$

6.7 Sample calculation

The following is based on an example given in Perry's Chemical Engineering Handbook, which is in turn based on a calculation included in Lapple's original paper.

Calculate the mass flow of air from a large reservoir vessel at 1.101 MPa and 20°C through 10 m of horizontal, straight 2-in Schedule 40 steel pipe (inside diameter = 52.5 mm) and three standard elbows (i) to a large vessel at 0.8 MPa and (ii) to the atmosphere, assumed at 0.101 MPa. The pipe inlet is abrupt. Find the fluid properties at station '2' and station '3', as defined in Figure 6.1.

We will begin by assuming that the Reynolds number will be greater than 100 000, so that by equations (4.28) and (4.29), the Fanning friction factor, f, will be 0.0045. We shall check the calculated Reynolds number when the calculation of flow has been completed.

Using equation (6.58) and the results of Tables 4.1 and 4.2,

$$K_T = 4 \times 0.0045 \times \frac{10}{0.0525} + 0.5 + 3 \times 0.75$$

$$= 6.18 \text{ velocity heads (dimensionless)}$$

We may treat air over the range of pressures and temperatures expected as an ideal, diatomic gas. Hence $Z = 1$, and $\gamma = 1.4$.

Molecular weight, $w = 29$.

Hence $$R_w = \frac{8314}{29} = 286.69 \text{ J/(kg K)}$$

Pipe cross-sectional area, $$A, = \pi \frac{D^2}{4} = \pi \frac{(0.0525)^2}{4}$$

$$= 0.002165 \text{ m}^2$$

Iterative solutions of the equations given in Section 6.4 yield the following results:

Case (i)

	Pressure (Pa)	*Temperature (K)*	*Specific volume (m³/kg)*	*Velocity (m/s)*
Station '1'	1 101 000	293	0.07629	0.0
Station '2'	1 065 961	290.3	0.07808	73.54
Station '3'	800 000	288.3	0.10331	97.31
Station '4'	800 000			

Mass flow = 2.039 kg/s.

The average temperature along the pipe = 290.7 K, hence viscosity, $\mu = 1.88 \times 10^{-5}$ Pa.s.

Using equation (4.36),

$$N_{RE} = \frac{WD}{A\mu} = \frac{2.039 \times 0.0525}{0.002165 \times 1.88 \times 10^{-5}}$$

$$= 2.63 \times 10^6$$

Hence the condition that the Reynolds number be greater than 100 000 is easily satisfied, and we may regard our estimate of friction factor as reasonable.

Case (ii)

	Pressure (Pa)	*Temperature (K)*	*Specific volume (m³/kg)*	*Velocity (m/s)*
Station '1'	1 101 000	293	0.07629	0.0
Station '2'	1 041 415	288.4	0.07939	96.3
Station '3'	271 240	244.2	0.25807	313.0
Station '4'	101 000			

Mass flow = 2.626 kg/s.

The average temperature along the pipe = 268.6 K, hence viscosity, $\mu = 1.71 \times 10^{-5}$ Pa.s.

Using equation (4.36),

$$N_{RE} = \frac{WD}{A\mu} = \frac{2.626 \times 0.0525}{0.002165 \times 1.71 \times 10^{-5}}$$

$$= 3.72 \times 10^6$$

Hence the condition that the Reynolds number be greater than 100 000 is again easily satisfied, and we may regard our estimate of friction factor as reasonable.

Comparison of the two cases

Since the pressure in the receiving vessel is well above the critical pressure in case (i), the flow remains subsonic at all times. The velocity increases by about 1/3 as the air passes down the pipe, and cools as it does so. Nevertheless, the temperature fall is relatively small.

Choked flow occurs in the second case; the outlet flow becomes sonic and the pressure at station '3' is the critical pressure of 0.271 MPa rather than the pressure of 0.101 MPa that exists just below the outlet. Air velocity increases by a factor of about 3 over the length of the pipe, and the mass flow is greater than in case (i). However, the pressure at station '2' is only a few per cent lower than in case (i), and this is significant in restricting the increase in mass flow to only about 30% above the velocity observed in case (i).

6.8 Explicit calculation of compressible flow

The exact method for calculating compressible flow summarized in Section 6.4 and illustrated in Section 6.7 requires the iterative solution of a set of highly nonlinear simultaneous equations. However, it is shown in Appendix 2 that it is possible to use the results produced by the exact method as the basis for an approximate method of calculation that will give values of mass flow correct to better than 2%. This accuracy, which is of the same order as a well-maintained and calibrated orifice-plate flowmeter, for instance, will normally be quite sufficient for simulation purposes, and there is a major benefit to be had from eliminating the iterations otherwise needed at each timestep.

By choosing an appropriate correction factor,

$$b_0 = b_0\left(\frac{p_4}{p_1}, K_T, \gamma\right)$$

we may approximate the flow of a compressible fluid using a modified form of the incompressible flow equation:

$$W = b_0 C_L \sqrt{\frac{p_1 - p_3 - \dfrac{g(z_3 - z_1)}{v_{ave}}}{v_{ave}}} \tag{6.60}$$

where the pipe's conductance, C_L (m^2), is related to the frictional loss in velocity heads, K_T, by

$$C_L = A\sqrt{\frac{2}{K_T}} \tag{6.61}$$

the (approximate) average specific volume of the compressible fluid in the pipe is given by:

$$v_{ave} = v_1 \frac{\gamma + 1}{\gamma} \frac{1 - \dfrac{p_3}{p_1}}{1 - \left(\dfrac{p_3}{p_1}\right)^{(\gamma+1)/\gamma}} \tag{6.62}$$

and, referred to the inlet pressure, the pressure ratio just within the outlet of the pipe, p_3/p_1, is related to the pressure ratio in the discharge vessel, p_4/p_1, by:

$$\frac{p_3}{p_1} = \max\left(\frac{p_4}{p_1}, \frac{p_{3c}}{p_1}\right) \tag{6.63}$$

The critical pressure ratio, p_{3c}/p_1, will be a function of both the specific-heat ratio, γ, of the gas being passed and frictional loss in velocity heads, K_T.

Using equation (6.60) brings the additional, if minor, advantage over the exact method of allowing for the difference in height over the pipe. However, for a pipe where height differences are considered negligible, equations (6.60), (6.61) and (6.62) may be simplified to:

$$W = b_0 A \sqrt{\frac{p_1}{v_1}} f_{lpa} \tag{6.64}$$

Here f_{lpa} is the generalized, long-pipe approximation function:

$$f_{lpa} = \sqrt{\frac{1}{K_T}\frac{2\gamma}{\gamma + 1}\left(1 - \left(\frac{p_3}{p_1}\right)^{(\gamma+1)/\gamma}\right)} \tag{6.65}$$

A set of polynomial approximating functions are derived in Appendix 2 to be used with the explicit equations listed above. The critical pressure ratio just inside the pipe outlet is approximated by the polynomial:

$$\frac{p_{3c}}{p_1} = \pi_0 + \pi_1(\ln K_T) + \pi_2(\ln K_T)^2 + \pi_3(\ln K_T)^3 + \pi_4(\ln K_T)^4 \tag{6.66}$$

Similarly the two terms needed to calculate the correction factor, b_0, for a specified value of specific heat ratio, γ, take the form of a point value, $b_0(K_T, 0.95)$:

$$b_0(K_T, 0.95) = \beta_0 + \beta_1(\ln K_T) + \beta_2(\ln K_T)^2 + \beta_3(\ln K_T)^3 + \beta_4(\ln K_T)^4 \tag{6.67}$$

and a slope, $m_b(K_T)$:

$$m_b(K_T) = \mu_0 + \mu_1(\ln K_T) + \mu_2(\ln K_T)^2 + \mu_3(\ln K_T)^3 + \mu_4(\ln K_T)^4 + \mu_5(\ln K_T)^5 \tag{6.68}$$

The value of b_0 at any given pair of values $(K_T, p_3/p_1)$ is given by an equation that is linear in p_3/p_1:

$$b_0\left(K_T, \frac{p_4}{p_1}\right) = b_0(K_T, 0.95) + m_b(K_T)\left(\frac{p_3}{p_1} - 0.95\right) \tag{6.69}$$

Values of the coefficients, π_i, β_i and μ_i, are given in Appendix 2, Tables A2.5 to A2.8 for $\gamma = 1.67$, 1.4, 1.3 and 1.135. Each table is subdivided into sections (a) and (b) that give coefficients appropriate to short pipes ($K_T \leq 1$) and long pipes ($1 \leq K_T \leq 2048$), respectively.

6.9 Example using the long-pipe approximation

The example calculated in Section 6.7 will be recalculated using the explicit, long-pipe approximation. All the same physical data will be used.

Air is essentially a diatomic gas, with $\gamma = 1.4$, and so we may turn to Table A2.6 to find the appropriate polynomial coefficients. We have already calculated (Section 6.7) that $K_T = 6.18$, and so we select Table A2.6(b). Applying the polynomial equations (6.66), (6.67) and (6.68) using the coefficients from Table A2.6(b) gives the following results:

$$\frac{p_{3c}}{p_1} = 0.2466; \quad b_0(6.18, 0.95) = 0.9179; \quad m_b = 0.2106$$

Case (i): downstream pressure = 800 000 Pa

The pressure ratio, $p_4/p_1 = 800\,000/1\,101\,000 = 0.7266$ is greater than the critical pressure ratio, 0.2466, and so no choking will occur, and $p_3/p_1 = p_4/p_1$.

Hence from equation (6.69), the appropriate value of b_0 is

$$b_0(6.18, 0.7266)$$
$$= 0.9179 + 0.2106 \times (0.7266 - 0.95)$$
$$= 0.8724$$

The pipe is horizontal, and so we may use equation (6.65) to calculate $f_{lpa} = 0.2821$.

The flow follows from equation (6.64) as

$$W = 0.8709 \times 0.002165 \times \sqrt{\frac{1\,101\,000}{0.076294}} \times 0.2821 = 2.020\,\text{kg/s}$$

Case (ii): downstream pressure = 101 000 Pa

Now the pressure ratio, $p_4/p_1 = 101\,000/1\,101\,000 = 0.0917$ is less than the critical pressure ratio. Choking will occur, and $p_3/p_1 = p_{3c}/p_1 = 0.2466$.

Since the supply pressure is 1 101 000 Pa, it follows that the pressure at the pipe outlet is predicted to be 271 510 Pa.

From equation (6.69), the appropriate value of b_0 is

$$b_0(6.18, 0.2466)$$
$$= 0.9179 + 0.2106 \times (0.2466 - 0.95)$$
$$= 0.7698$$

Also, from equation (6.65), $f_{lpa} = 0.4143$.

The flow follows from equation (6.64) as

$$W = 0.7698 \times 0.002165 \times \sqrt{\frac{1\,101\,000}{0.076294}} \times 0.4143 = 2.623\,\text{kg/s}$$

These results are very close to those calculated in Section 6.7 using the full, iterative method, as shown below.

	Case (i)		*Case (ii)*	
	Flow (kg/s)	*Pipe outlet pressure (Pa)*	*Flow (kg/s)*	*Pipe outlet pressure (Pa)*
Full explicit model	2.039	800 000	2.626	271 240
Long-pipe approximation	2.020	800 000	2.623	271 510

6.10 Bibliography

Lapple, C.E. (1943). Isothermal and adiabatic flow of compressible fluids, *Trans. Am. Inst. Chem. Eng.*, **39**, 385–432.

Levenspiel, O. (1977). The discharge of gases from a reservoir through a pipe, *AIChE Journal*, **23**, 402–403.

Perry, R.H., Green, D.W. and Maloney, J.O. (1984). *Perry's Chemical Engineers' Handbook*, 6th edition, McGraw-Hill, New York. Chapter 5: Fluid and particle mechanics.

7 Control valve liquid flow

7.1 Introduction

Control of a process plant is effected almost exclusively through the use of control valves, making a good knowledge of their behaviour essential for the control engineer. Control valves may pass liquids or gases, and the flow behaviour of liquids is on the whole easier to understand than that of gases. This chapter introduces control valve flow by concentrating on how to calculate liquid flow.

There are, however, a number of concepts useful for understanding liquid flow that carry over directly to the understanding of gas flow. For example, the profile of pressure through the valve may be understood for both liquids and gases by appealing to the same thermodynamic equation; the valve characteristic will also be common to liquids and gases. It has been decided to present such common material in this chapter, and not to repeat it in the later chapter (Chapter 9) dealing with gas flow through the valve.

7.2 Types of control valve

The control valve adjusts the flow rate by changing the size and shape of the flow area offered to the fluid passing through it. There are two main categories of control valve used on process plant: globe valves and rotary valves. The flow area in a globe valve is dictated by the position of the valve plug relative to its seat, while the flow area in a rotary valve depends on the angular position of a disc or notched sphere within a tubular surround.

The position of the flow-restricting element in both categories of valve is governed by a piston actuator, either directly or through a gear mechanism. We will refer to the stroke of this piston as the *valve travel*. Valve travel is taken to be a normalized variable, with zero indicating that the valve opening is at its minimum, while unity implies a fully open valve. The *valve opening* is another normalized variable, being the ratio of the present flow to the flow at full-open under the same pressure and temperature conditions. In broad terms, valve opening is also the ratio of the flow area to the flow area at full-open. Valve opening and valve travel are related to each other via the valve characteristic curve.

Figure 7.1 shows the elements of a globe valve. Flow is regulated by moving a plug in and out of a fixed opening or port so as to restrict the flow area. (The valve type gets its name from the often globe-shaped cavity around the port region.) The globe valve has elements in common with the nozzle: a convergent inlet section, a divergent outlet section and a central region where the flow area is restricted: the throat in the nozzle, the port and cage in the valve. However, unlike the nozzles analysed in Chapter 5, the globe valve causes the flow to undergo two separate 90-degree changes of direction. Accordingly there is a considerable frictional pressure drop in the globe valve.

There are two classes of rotary valve: butterfly and rotating ball. A disc is rotated to vary the flow area

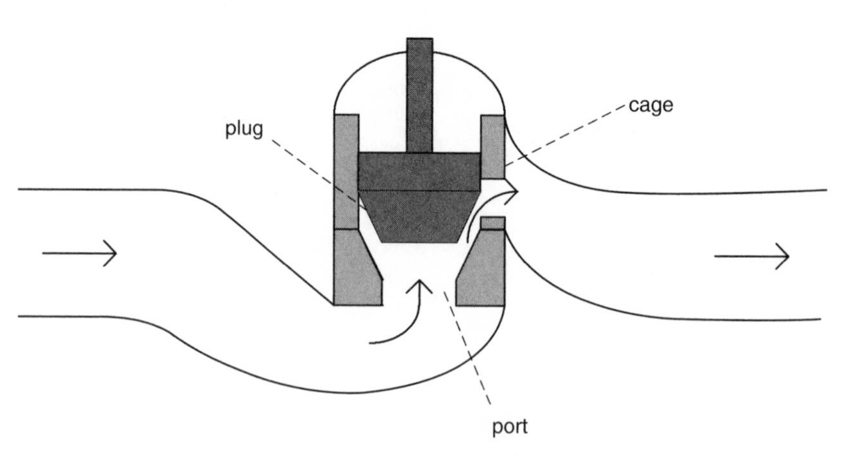

Figure 7.1 Schematic of a globe valve.

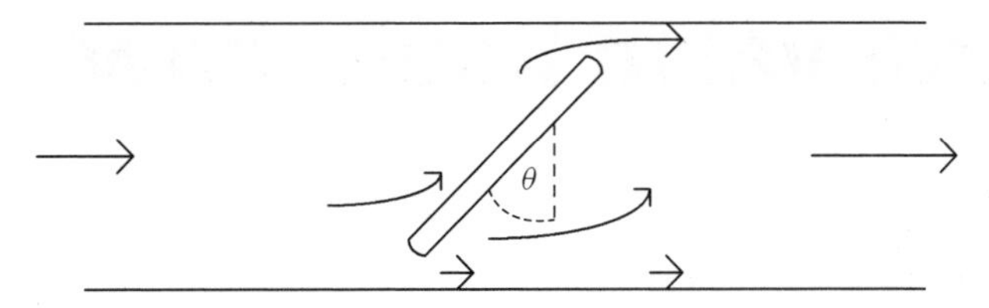

Figure 7.2 Schematic of butterfly valve showing angle of opening.

in the butterfly valve, see Figure 7.2. The ball valve extends the principle to the rotation of a complete ball, but one that has a flow passage cut into it, very often in the shape of a V-notch. Once again, the area available to the flow varies with the angle of rotation.

As shown in the figure, the flow through the butterfly valve has an upper and lower flow path. When in the fully open position, with $\theta = 90°$, the frictional resistance to flow is very small, and the pressure loss is low. But the full benefit of this low pressure drop appears only when the valve is used in a fully-on/fully-off mode, i.e. for isolation purposes. The excessively high torques that would be required to sustain the standard butterfly disc in steady and repeatable operation at large angles of opening mean that it is impractical to use the valve in control applications with openings in excess of 60°. From an inspection of Figure 7.2, it is therefore clear that when used in control applications, the entrance to the butterfly valve presents a convergent section to the upper part of the flow, while the exit from the valve presents a divergent section. In between there is a restriction. Thus, like the globe valve, the butterfly valve also possesses the three elements, namely convergent section, restriction and divergent section, that characterize the convergent–divergent nozzle as analyzed in Chapter 5, although the flow-field in a butterfly valve will be more complicated than in a pure nozzle.

The control valve's relationship to a nozzle is important in understanding and modelling control valve behaviour, particularly for gas flow.

7.3 Pressure distribution through the valve

Figure 7.3 represents the valve as a constriction imposed on the flow, and shows the inlet pressure, p_1, the throat pressure, p_t, and the outlet pressure, p_2, where the term, 'throat', has been imported from the study of nozzles to denote the position of minimum cross-section.

We may appeal to the energy equation for general, steady-state flow to gain an insight into the variation in pressure through the valve:

$$v\,dp + c\,dc + g\,dz + dF = 0 \tag{4.15}$$

Assuming for simplicity that the valve is horizontal, we may derive from (4.15) an expression for the change in pressure along the valve, dp:

$$dp = -\frac{dF}{v} - \frac{c\,dc}{v} \tag{7.1}$$

The change in pressure thus depends on two terms, which may be additive or in opposition. Since the energy lost to friction, dF, must always be positive, the effect of the first term will be to reduce pressure in the direction of flow. However, the sign of the second term depends on whether the flow velocity is increasing or decreasing.

We know from continuity that the velocity of a fluid with fixed specific volume (i.e. a liquid) must increase over the convergent section of the valve and decrease over the divergent part. Turning to gases, nozzle theory tells us that the velocity will increase over the convergent section of the valve. But the velocity will continue to increase over the divergent section only if the gas is being accelerated to supersonic speeds. This circumstance would require both a carefully designed divergent section and a carefully matched ratio of valve inlet and outlet pressures, and a conjunction of both these conditions is a very remote possibility on a process plant. Therefore we may be confident that the gas velocity will decrease over the divergent section of the control valve.

The argument above demonstrates that the velocity of both liquids and gases will increase as far as the valve's throat and decrease thereafter. Accordingly, the second

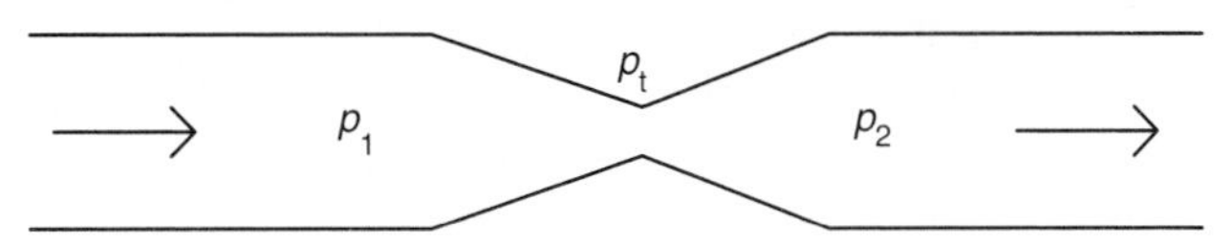

Figure 7.3 Schematic of valve showing the inlet pressure, the throat pressure and the outlet pressure.

term on the right-hand side of equation (7.1), namely $-c\,dc/v$, will accentuate the reduction in pressure up to the throat, but will tend to increase the pressure in the fluid after this point, since dc will now be negative. Whether the pressure then rises or falls and by how much will depend on the relative sizes of the opposing terms dF and $c\,dc$. The sizes of these terms will depend on whether the fluid being passed is a liquid or a gas, on the type of valve and on its opening.

It is found for practical valve designs that there is always a rise in pressure over the outlet section for liquids, so that the throat pressure, p_t, will be less than the outlet pressure, p_2:

$$p_t < p_2 \tag{7.2}$$

The pressure recovery at the outlet is more marked in rotary valves, where the less convoluted fluid path leads to a lower frictional loss, especially at large openings. For gases, there is usually a rise in pressure over the outlet section if the valve is of the rotary type, but the greater frictional dissipation in a globe valve usually causes a fall in pressure between the throat and valve outlet. The extent of the pressure recovery between valve throat and valve outlet will be considered in the next section for liquids.

7.4 Liquid flow through the valve

The mass flow (kg/s) of liquid through the valve has been found to be represented well by the simple expression:

$$W = C_v\sqrt{\frac{\Delta p}{v_1}} \tag{7.3}$$

where

v_1 is the inlet specific volume (m^3/kg),
C_v is the valve conductance (m^2),
and $\Delta p = p_1 - p_2$ is the pressure drop across the valve (Pa).

The valve conductance is a figure supplied for each valve by the manufacturer, often in US rather than SI units. In this case the parameter, C_v^*, is given, the valve capacity for water at 60°F in US gall/min/(psi)$^{1/2}$. It is shown in Appendix 3 that this may be related to the valve conductance, C_v, by the conversion equation:

$$C_v = 2.3837 \times 10^{-5} C_v^* \tag{7.4}$$

Equation (7.3) relates the flow to the pressure difference between the valve inlet and outlet, and it is reasonable to assume that a similar equation will relate the flow to the pressure difference between the valve inlet and the valve throat:

$$W = C_{vt}\sqrt{\frac{p_1 - p_t}{v_1}} \tag{7.5}$$

where C_{vt} is the valve conductance to the valve throat (m^2) and p_t is the pressure at the valve throat. The flow predicted by equation (7.5) should be the same as that predicted by equation (7.3), so that we may equate the right-hand sides of the two equations:

$$C_{vt}\sqrt{\frac{p_1 - p_t}{v_1}} = C_v\sqrt{\frac{p_1 - p_2}{v_1}} \tag{7.6}$$

Rearranging gives the throat pressure in terms of the valve upstream and downstream pressures as:

$$p_t = p_1 - \frac{(p_1 - p_2)}{\left(\dfrac{C_{vt}}{C_v}\right)^2} \tag{7.7}$$

Let us define a dimensionless friction coefficient for liquid flow, C_{fl}, by:

$$C_{fl} = \frac{C_{vt}}{C_v} \tag{7.8}$$

Hence equation (7.7) becomes

$$p_t = p_1 - \frac{(p_1 - p_2)}{C_{fl}^2} \tag{7.9}$$

Rearranging gives

$$p_t = p_2 + (C_{fl}^2 - 1)\frac{(p_1 - p_2)}{C_{fl}^2} \tag{7.10}$$

Since $p_1 > p_2$, it is clear that the throat pressure will be less than the valve outlet pressure as long as C_{fl} is less than unity. This condition will hold in the very important case, where some of the liquid turns (temporarily) into vapour as it passes through the throat of the valve, as will now be shown.

While manufacturers will always provide the value of the valve conductance, C_v, they do not give the value of the conductance to the valve throat, C_{vt}, so that it is not possible to calculate the friction coefficient, C_{fl}, directly from equation (7.8). However, equivalent information is normally made available for the special condition when the pressure in the valve throat is at the vapour pressure of the liquid being passed. This information is contained in the manufacturer-supplied cavitation coefficient, K_c, which is used in the following equation pair to define the pressure conditions at the onset of cavitation:

$$\begin{aligned} p_1 - p_2 &= K_c(p_1 - p_t) \\ p_t &= p_{tvap} \end{aligned} \tag{7.11}$$

Here p_{tvap} is the vapour pressure at the throat temperature. The first equation in the pair may be re-arranged to the form:

$$p_t = p_1 - \frac{(p_1 - p_2)}{K_c} \tag{7.12}$$

Comparing equation (7.12) with equation (7.7), it is clear that at cavitation

$$C_{fl}^2 = K_c \tag{7.13}$$

The value of K_c is found by experiment to be approximately independent of valve travel for globe valves. This is not surprising in the light of the interpretation of K_c as a friction parameter characterizing the valve downstream of the valve throat: this downstream geometry does not change with valve travel. There is not much variation in K_c-values from valve to valve for globe valves, and K_c for a globe valve will lie in the narrow range $0.7 \leq K_c \leq 0.9$.

The value of K_c at fully open for a rotary valve, K_C, is considerably lower, typically 0.25. Now, however, the location of the effective throat and the flow field downstream of it are both dependent on valve travel. This explains the dependence of K_c on valve travel, x, for rotary valves, where K_c has been found to obey approximately the following equation set:

$$\begin{aligned}\frac{K_c}{K_C} &= 2.308 && \text{for } 0 \leq x \leq 0.15 \\ &= 2.5385 - 1.5385x && \text{for } x > 0.15\end{aligned} \tag{7.14}$$

Good control will normally require the rotary valve to operate somewhere in mid-range. Substituting $x = 0.5$ in equation set (7.14) gives $K_c/K_C = 1.8$. For a rotary valve with $K_C = 0.26$ (say), this implies $K_c = 0.46$, substantially lower than for a globe valve.

It may be seen from the foregoing that the friction coefficient, C_{fl}, is less than unity at the point of liquid flashing for both globe and rotary values. It follows from equation (7.10) that the throat pressure will be lower than the outlet pressure at the onset of cavitation. In fact, inequality (7.2) contained in Section 7.3 will normally hold throughout the non-cavitating range also.

7.5 Cavitation and choking in liquid flow

The liquid passing through the valve will flash if the pressure in the throat of the valve falls to the vapour pressure corresponding to the liquid temperature in the throat. This has two effects. First it will induce cavitation as the pressure subsequently rises at the valve outlet, causing the bubbles of vapour to implode. If the bubbles are near to or in contact with a solid boundary when they collapse, the forces exerted by the liquid rushing into the cavities will create very high localized pressures that will cause pitting of the solid surface. This can cause significant damage to the plug and body of the valve. Secondly, the production of large volumes of vapour bubbles within the liquid stream can lead to choking in the valve. In this case the flow will be limited to its value just before choking. The modeller will wish to be able to predict both these phenomena.

7.5.1 Cavitation

Cavitation will start as soon as the throat pressure has reached the local vapour pressure. Theoretical considerations suggest that the throttling process as far as the throat will be isentropic, which will lead to a small decrease in liquid temperature. Hence the vapour pressure in the throat will be slightly less than the vapour pressure corresponding to the inlet temperature. This effect is small for a liquid, and the throat vapour pressure is unlikely to be reduced by more than 5% below the vapour pressure at the inlet to the valve at the liquid pressures normally found on process plant. For practical purposes, we may assume that the temperature of the liquid in the throat of the valve is the same as it was at the valve inlet. Thus the condition of cavitation onset may be written:

$$p_t = p_{vap}(T_1) \tag{7.15}$$

where $p_{vap}(T_1)$ is the vapour pressure at the inlet temperature, T_1.

7.5.2 Choking

It has been found in practice that choking does not always occur immediately bubbles form, but that the onset of choked flow may be delayed until the mean throat pressure has fallen to a fraction, r_{vap}, of the vapour pressure at the valve inlet:

$$p_t = r_{vap}\, p_{vap}(T_1) \tag{7.16}$$

where $p_{vap}(T_1)$ is the vapour pressure at temperature, T_1. The fraction, r_{vap}, has been found to correlate well with the ratio of the vapour pressure at valve inlet temperature to the pressure at the critical point of the substance being passed. The basis for this correlation is the fact that the specific volume of vapour gets closer and closer to that of the liquid being passed as the critical point is neared, coinciding at the critical point itself. Hence the relative volume of the vapour bubbles will fall as the vapour pressure gets closer to the critical pressure causing less obstruction. The fraction, r_{vap}, has been found to obey approximately the following equation:

$$\begin{aligned}r_{vap} &= 0.94 - 0.8625 r_{pvc} + 1.7257 r_{pvc}^2 \\ &\quad - 1.5617 r_{pvc}^3 + 0.4564 r_{pvc}^4\end{aligned} \tag{7.17}$$

where r_{pvc} is the ratio of the vapour pressure at valve inlet to the pressure at the critical point, p_{cp}:

$$r_{pvc} = \frac{p_{vap}(T_1)}{p_{cp}} \tag{7.18}$$

The pressure at the critical point for water, for example, is $p_{cp} = 22.12\,\text{MPa}$.

When choking occurs in the valve, the upstream and downstream pressures are related by a condition very similar to the equation pair (7.11) for cavitation, but now using equation (7.16) for the throat pressure, rather than (7.15), and replacing the cavitation coefficient, K_c, by a new choking coefficient, K_m:

$$\begin{aligned} p_1 - p_2 &= K_m(p_1 - p_t) \\ p_t &= r_{vap}\, p_{vap}(T_1) \end{aligned} \tag{7.19}$$

K_m is sometimes referred to as simply the 'pressure recovery coefficient'. The two coefficients are identical for globe valves: $K_m = K_c$, and often only K_m is quoted. The choking coefficient is somewhat larger than the cavitation coefficient for a rotary valve, with the full-open value, K_M, taking a value of typically 0.35, rather than the value 0.25 typical for K_C. The ratio K_m/K_M for a rotary valve follows a similar path to that of K_c/K_C, rising as the valve closes according to the approximate equation:

$$\begin{aligned} \frac{K_m}{K_M} &= 2.342 && \text{for } x \le 0.24 \\ &= 2.766 - 1.766x && \text{for } 0.24 < x \le 1.0 \end{aligned} \tag{7.20}$$

7.6 Relationship between valve capacity at part open and capacity at full open

It was noted in Section 7.1 that the 'valve opening', y, for a partly open valve is defined as the ratio of the current flow, W, to the flow, W_{max}, that the valve would pass when fully open between the same upstream and downstream pressures, i.e.

$$y = \frac{W}{W_{max}} \tag{7.21}$$

It is clear from equation (7.3) that the maximum flow for a given differential pressure will occur when the liquid flow conductance is at its maximum value, i.e. the value at fully open, $C_V = C_v|_{max}$. Accordingly

$$W_{max} = C_V \sqrt{\frac{\Delta p}{v_1}} \tag{7.22}$$

Substituting back into equation (7.21) gives the valve opening, y, as

$$y = \frac{W}{C_V \sqrt{\dfrac{\Delta p}{v_1}}} \tag{7.23}$$

Hence, by re-arrangement, the flow at a valve opening, y, is given by:

$$W = yC_V \sqrt{\frac{\Delta p}{v_1}} \tag{7.24}$$

Comparing equation (7.24) with equation (7.3), it is obvious that the liquid flow conductance at a valve opening, y, is related to the liquid flow conductance at fully open by

$$C_v = yC_V \tag{7.25}$$

7.7 The valve characteristic

The positions of the flow restrictors inside both globe and rotary valves depend on the action of a piston actuator, either directly or through a gear mechanism. As noted in Section 7.2, the stroke of this piston is called the 'valve travel'. We have already given this the symbol, x, with the convention that

$x = 0$ implies minimum valve opening

$x = 1$ implies maximum valve opening

The flow area will depend on the valve travel, with a characteristic dictated by the geometry of the plug and cage in a globe valve or by the detailed shape of the disc or notched ball in a rotary valve. Hence, in general, the valve opening is a function of the valve travel: $y = f(x)$. The geometrical basis for the valve characteristic means that the function, f, is applicable equally to liquid and gas flow.

The two most common globe valve characteristics are the linear valve and the equal percentage valve. The linear valve has the characteristic

$$y = x \tag{7.26}$$

The equal percentage valve has the characteristic behaviour that equal increments of valve travel give rise to equal percentage increases in valve opening. We may describe this behaviour mathematically by the differential equation:

$$\frac{dy}{dx} = k_1 y \tag{7.27}$$

where k_1 is a constant. This equation has the solution

$$y = \frac{1}{R_v} e^{k_1 x} \tag{7.28}$$

where R_v is a constant. Since the valve will be fully open when the valve is at full travel, we have $y = 1$ when $x = 1$. Substituting into (7.28) gives:

$$e^{k_1} = R_v \tag{7.29}$$

Taking logs gives

$$k_1 = \ln R_v \tag{7.30}$$

Substituting back into (7.28) allows us to write the relationship between valve opening, y, and valve travel, x, as:

$$y = \frac{1}{R_v} e^{k_1 x} = \frac{1}{R_v} (e^{\ln R_v})^x = \frac{R_v^x}{R_v} = R_v^{x-1} \qquad (7.31)$$

Since $y = 1/R_v$ when $x = 0$ and $y = 1$ when $x = 1$, it is clear that R_v is the ratio of the valve opening at maximum valve travel to the valve opening when the valve travel is at its minimum. Thus R_v is known as the 'rangeability' of the valve.

Relief valves require the flow to increase rapidly for a small change in valve travel, and such valves will have a 'quick-opening' characteristic, which may, depending on the manufacturer's detailed design, give 80%, say, of valve opening at 50% valve travel.

Geometrical analysis of a butterfly valve, neglecting the width of the disc, shows that the unobstructed area normal to flow is $A(1 - \cos\theta)$, where θ is the angle shown in Figure 7.2, and A is the flow area at full open. On this basis of area, we may write the relationship between valve opening and valve travel as:

$$y = 1 - \cos\left(\frac{\pi}{2}x\right) \qquad (7.32)$$

Equation (7.32) makes no attempt to account for the complex flow field that will be set up in a butterfly valve at low values of valve travel, and is approximate only.

Figure 7.4 displays a set of typical valve characteristics.

While the analytic derivations given above will be useful for many general studies, it should be remembered that the equations will not provide a precise match to the engineered characteristics. Manufacturers may provide test data on the performance of their valves at various valve travels, and these data are obviously to be preferred where available.

7.8 Velocity-head loss across the valve

The concept of a velocity head has its roots in the theory of incompressible flow travelling along a pipe of uniform cross-section, so that the velocity is constant. However, its use has been extended to the characterization of bends, fittings and valves, where the cross-section varies, and also to compressible, i.e. gas, flow, where there is an implicit assumption of an average specific volume.

The expression relating pressure drop, Δp (Pa), to the dimensionless velocity-head loss across the valve, K_v, may be written (cf. equation (4.48)) as

$$\Delta p = K_v \frac{1}{v_1} \frac{c^2}{2} \qquad (7.33)$$

where

v_1 is the specific volume (m^3/kg), *assumed to be constant throughout the pipe*, and

c is the velocity (m/s) of the fluid *in the main body of the pipe*.

A rearrangement of equation (7.3) for valve pressure drop and flow gives:

$$\Delta p = v_1 \frac{W^2}{C_v^2} \qquad (7.34)$$

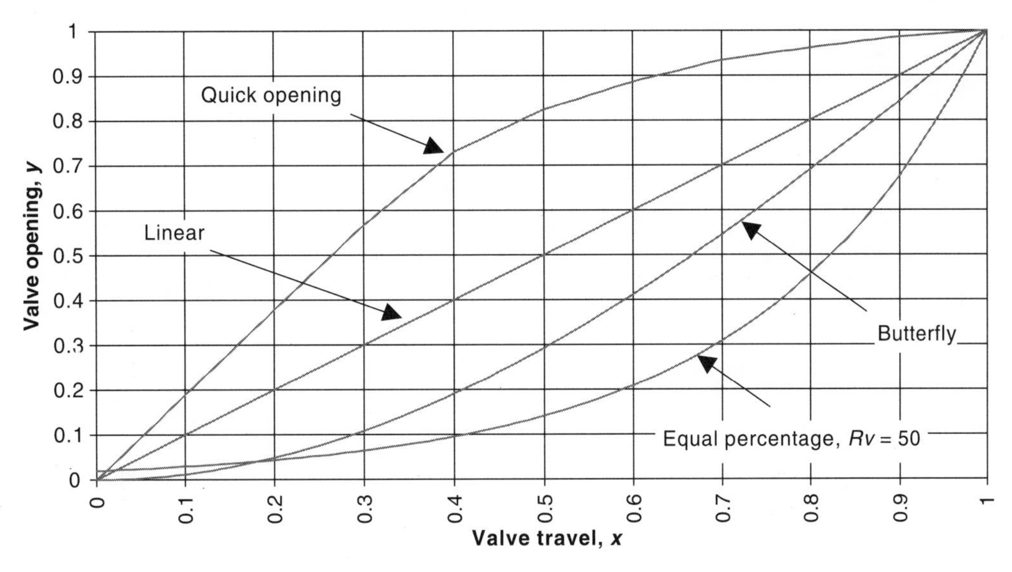

Figure 7.4 Typical valve characteristics.

Since the flow through the pipe is given by

$$W = \frac{Ac}{v_1} = \frac{\pi D^2}{4v_1}c$$

where D is the pipe diameter, equation (7.34) may be rearranged to give

$$\Delta p = \frac{\pi^2 D^4}{8C_v^2}\frac{1}{v_1}\frac{c^2}{2} \tag{7.35}$$

Hence by comparing (7.35) with (7.33), it is clear that the valve velocity-head drop, K_v, is given by

$$K_v = \frac{\pi^2 D^4}{8C_v^2} \tag{7.36}$$

where D is the pipe diameter in metres and C_v is the valve liquid flow conductance at a given valve opening, in square metres.

We may note that the pipe diameter and the valve diameter are by no means always the same: indeed the requirement for good control often necessitates a valve size only 50% of the line diameter. Accordingly it is sensible to include a factor to make explicit allowance for this. At the same time, we may relate the velocity-head drop to the velocity-head drop at full-open by using equation (7.25), namely $C_v = yC_V$. Hence equation (7.36) may be rewritten:

$$K_v = \frac{1}{y^2}\left(\frac{D}{D_v}\right)^4 \frac{\pi^2 D_v^4}{8C_V^2} \tag{7.37}$$

or

$$K_v = \frac{1}{y^2}\left(\frac{D}{D_v}\right)^4 K_{V0} \tag{7.38}$$

where

D_v is the valve size (m), approximately equal to the diameter of the valve throat at fully open, and

K_{V0} is the velocity head drop for the fully open valve when the pipe diameter is equal to the valve size:

$$K_{V0} = \frac{\pi^2 D_v^4}{8C_V^2} \tag{7.39}$$

We may expect the valve conductance to be roughly proportional to the effective throat area of the valve for any given category of valve. Further, since the valve size is approximately proportional to the throat area at full open, we may deduce that the relationship $C_V \propto D_v^2$ should apply, at least approximately. Applying this relationship to equation (7.39), we might expect that K_{V0} will be roughly constant, for a given category of valve. This is the basis for the figures that are sometimes quoted in engineering tables of velocity-head drop for valves, bends and fittings. There is a fair justification for this supposition in the case of a fully open butterfly valve, once it has reached the size where the thickness of the disc will have very little effect. Figure 7.5 shows K_{V0} versus valve size for 19 butterfly valves for which data is given in Fisher Controls (1977).

The K_{V0} value levels off in the range 0.30–0.45 once the valve size is greater than 150 mm.

The situation is not as clear-cut for the globe valve, as can be seen in Figure 7.6, which displays K_{V0} values for 44 globe valves for which data is available in Fisher Controls (1977). There is a tendency for the value of K_{V0} to increase with valve size, but there is a

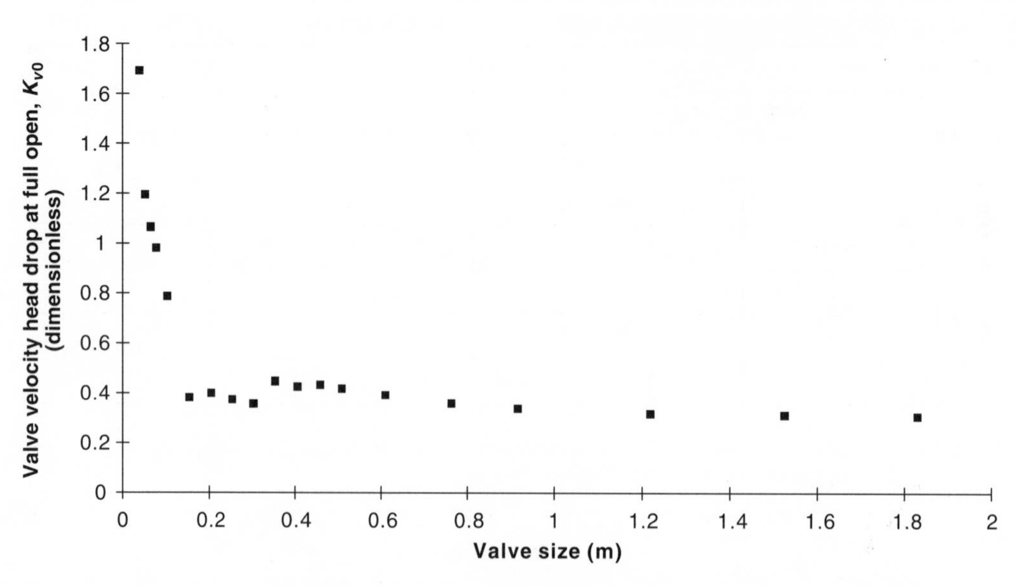

Figure 7.5 Velocity head drop across a fully open butterfly valve connected to a pipe with diameter equal to valve size.

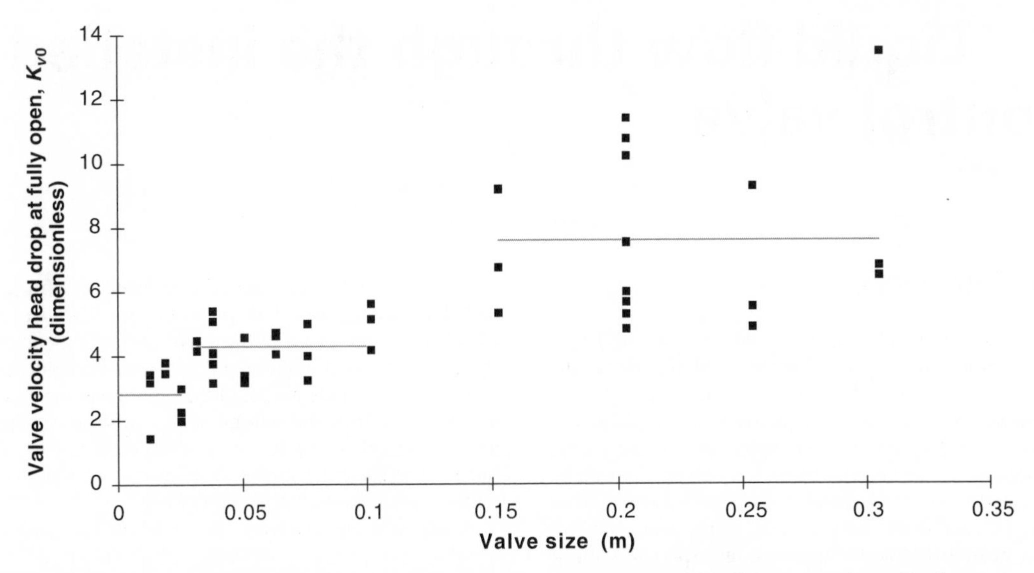

Figure 7.6 Velocity head drop across a fully open globe valve connected to a pipe with diameter equal to valve size.

considerable degree of scatter. In consequence, it is not very meaningful to quote a single figure to characterize all globe valves. But if better data is unavailable, one might use the following approximation for K_{V0} for a globe valve:

$$
\begin{aligned}
K_{V0} &\approx 2.8 \text{ for } 0 \leq D_v \leq 25\,\text{mm} \\
&\approx 4.3 \text{ for } 25\,\text{mm} \leq D_v \leq 100\,\text{mm} \qquad (7.40) \\
&\approx 7.6 \text{ for } D_v \geq 150\,\text{mm}
\end{aligned}
$$

These values have been marked in on Figure 7.6. Obviously a direct calculation using equation (7.39) is always to be preferred whenever the value of C_V is obtainable, whatever sort of valve is under consideration.

7.9 Bibliography

Fisher controls (1977). *Control Valve Handbook*, 2nd edition. Available directly from Fisher-Rosemount Ltd, Knight Road, Strood, Rochester, Kent, England, ME2 2EZ.

While, C. and Hutchison, S.J. (1979). Instrument models for process simulation, *Trans. Inst. MC*, **1**, 187–194.

8 Liquid flow through the installed control valve

8.1 Introduction

The modeller needs to know the pressure and specific volume immediately upstream of the valve and the pressure immediately downstream if he is to calculate control valve flow by the methods of Chapter 7. While these parameters may be known for the supply vessel and the receiving vessel, the valve may well be separated from each vessel by a significant length of pipe, perhaps containing several bends and other fittings. It follows that the pressure and specific volume immediately upstream of the valve will generally not be the same as in the supply vessel, nor will the pressure immediately downstream of the valve equal the pressure in the receiving vessel. The values for these variables must be calculated by combining the analysis of Chapter 7 for liquid flow through valves with that of Chapter 4 for liquid flow in pipes.

8.2 Liquid flow through an installed valve

Figure 8.1 shows a schematic of the general case of a valve installed in a pipeline, separated by significant conductances from both the supply vessel and the receiving vessel.

A valve of liquid conductance, C_v (m^2), is installed in a pipeline that runs between a supply vessel with pressure, p_1 (Pa), and specific volume, v_1 (m^3/kg), and a receiving vessel at pressure, p_4 (Pa). Let the conductance upstream of the valve (including entrance effects) be C_{L1} (m^2), and the downstream conductance be C_{L2} (m^2). Let the height of the pipeline at the supply vessel be z_1 (m), the height of the valve be z_{v1} (m), and the height of the line at discharge be z_4 (m). The pressure at the inlet to the valve is denoted p_{v1} (Pa), while that at the outlet is p_{v2} (Pa).

A pressure balance over the complete line yields:

$$p_1 = p_4 + \frac{g}{v_{ave}}(z_4 - z_{v1}) + \Delta p_{L2} + \Delta p_v + \frac{g}{v_{ave}}(z_{v1} - z_1) + \Delta p_{L1} \qquad (8.1)$$

where:

$$\begin{aligned} \Delta p_{L1} &= p_1 - p_2 \\ \Delta p_v &= p_{v1} - p_{v2} \\ \Delta p_{L2} &= p_{v2} - p_3 \\ p_3 &= p_4 + \frac{g}{v_{ave}}(z_4 - z_{v1}) \\ p_2 &= p_{v1} + \frac{g}{v_{ave}}(z_{v1} - z_1) \end{aligned} \qquad (8.2)$$

and where the incompressible nature of the liquid allows us to put $v_{ave} = v_1$. Using equation (4.42) for

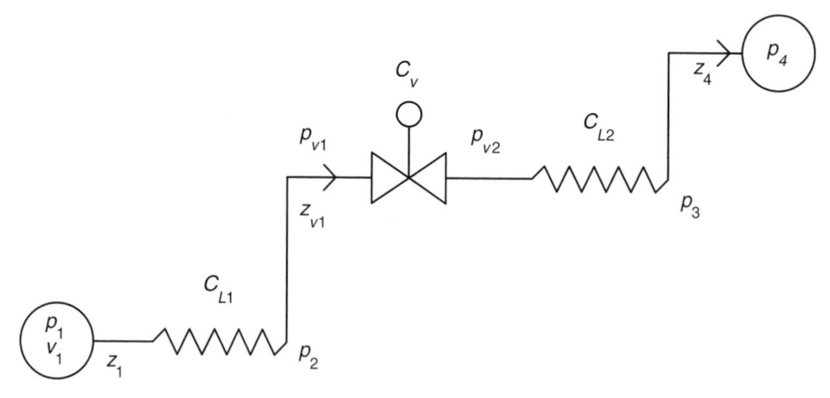

Figure 8.1 Control valve installed in a liquid pipeline.

the frictional, line pressure drops and equation (7.3) for the valve pressure drop and rearranging yields:

$$p_1 - p_4 + \frac{g}{v_{ave}}(z_1 - z_4) = v_{ave}\frac{W^2}{C_{L2}^2} + v_{ave}\frac{W^2}{C_v^2} + v_{ave}\frac{W^2}{C_{L1}^2} \tag{8.3}$$

Solving equation (8.3) for the flow, W, gives:

$$W = C_T\sqrt{\frac{p_1 - p_4 + \frac{g}{v_{ave}}(z_1 - z_4)}{v_{ave}}} \tag{8.4}$$

where C_T is the total conductance of line and valve (m^2), given by:

$$\frac{1}{C_T^2} = \frac{1}{C_{L1}^2} + \frac{1}{C_v^2} + \frac{1}{C_{L2}^2} \tag{8.5}$$

We may substitute $C_v = yC_V$ in order to find the total conductance in terms of the valve liquid conductance at full-open, C_V:

$$\frac{1}{C_T^2} = \frac{1}{C_{L1}^2} + \frac{1}{y^2C_V^2} + \frac{1}{C_{L2}^2} \tag{8.6}$$

8.3 Choking during liquid flow

Equation (8.4) will hold only as long as the flow is less than the choked flow, which is the maximum that the line will support for any given set of upstream conditions. When choking occurs, equation (7.19) will be valid, which, for the nomenclature of Figure 8.1, takes the form:

$$\begin{aligned} p_{v1} - p_{v2} &= K_m(p_{v1} - p_t) \\ p_t &= r_{vap}\,p_{vap}(T_{v1}) \end{aligned} \tag{8.7}$$

where the temperature at valve inlet, T_{v1}, will be the same as the temperature in the upstream vessel in the absence of heating:

$$T_{v1} = T_1 \tag{8.8}$$

Eliminating the throat pressure, p_t, and rearranging produces the following relationship between the valve upstream and downstream pressures:

$$p_{v2} = K_m r_{vap}\,p_{vap}(T_1) + (1 - K_m)p_{v1} \tag{8.9}$$

The flow will now be the choked flow, W_{choke}, which will obey the valve equation (7.3), so that

$$W_{choke} = C_v\sqrt{\frac{p_{v1} - p_{v2}}{v_{ave}}} \tag{8.10}$$

Substituting from equation (8.9) into equation (8.10) allows the elimination of valve downstream pressure, p_{v2}:

$$\begin{aligned} W_{choke} &= C_v\sqrt{\frac{p_{v1} - K_m r_{vap}\,p_{vap}(T_1) - (1 - K_m)p_{v1}}{v_{ave}}} \\ &= C_v\sqrt{\frac{K_m(p_{v1} - r_{vap}\,p_{vap}(T_1))}{v_{ave}}} \end{aligned} \tag{8.11}$$

This may be rearranged to yield an expression for the valve upstream pressure, p_{v1}:

$$p_{v1} = r_{vap}\,p_{vap}(T_1) + v_{ave}\frac{W_{choke}^2}{K_m C_v^2} \tag{8.12}$$

The flow must also satisfy the pressure-drop equation (4.42) for the pressure drop between the upstream vessel and station '2' shown in Figure 8.1. Hence:

$$W_{choke} = C_{L1}\sqrt{\frac{p_1 - p_2}{v_{ave}}} \tag{8.13}$$

or

$$p_1 = p_2 + v_{ave}\frac{W_{choke}^2}{C_{L1}^2} \tag{8.14}$$

Substituting for p_2 from (8.2) into equation (8.14) gives:

$$p_1 = p_{v1} + \frac{g}{v_{ave}}(z_{v1} - z_1) + v_{ave}\frac{W_{choke}^2}{C_{L1}^2} \tag{8.15}$$

Eliminating p_{v1} between equation (8.12) and (8.15) yields

$$p_1 = r_{vap}\,p_{vap}(T_1) + v_{ave}\frac{W_{choke}^2}{K_m C_v^2} + \frac{g}{v_{ave}}(z_{v1} - z_1) + v_{ave}\frac{W_{choke}^2}{C_{L1}^2} \tag{8.16}$$

so that the choked flow, W_{choke}, emerges as:

$$W_{choke} = C_{L1}C_v\sqrt{\frac{K_m}{C_{L1}^2 + K_m C_v^2}} \times \sqrt{\frac{p_1 - r_{vap}\,p_{vap}(T_1) - \frac{g}{v_{ave}}(z_{v1} - z_1)}{v_{ave}}} \tag{8.17}$$

As a result, if there is a possibility of the valve throat pressure falling to or near the vapour pressure of the liquid, equation (8.4) calculating mass flow should be modified to the more general form

$$W = \min \left[W_{choke}, C_T \sqrt{\frac{p_1 - p_4 + \dfrac{g}{v_{ave}}(z_1 - z_4)}{v_{ave}}} \right] \tag{8.18}$$

where W_{choke} is the choked flow for the temperature and pressure conditions, given by equation (8.17).

8.4 Cavitation during liquid flow

The onset of cavitation for a globe valve occurs with the onset of choking, as was noted in Section 7.5.2. Hence it will be predicted when $W = W_{choke}$.

Cavitation in a rotary valve occurs before choking, when the throat pressure equals the vapour pressure at valve inlet temperature, and equation (7.11) applies. This is of precisely the same form as equation (7.19), but with K_m replaced by K_c and r_{vap} set to unity. This similarity of form enables us to repeat the development set out in Section 8.3 to determine the flow at choking. Thus we reach the following equation for cavitating flow, W_{cav}:

$$W_{cav} = C_{L1} C_v \sqrt{\frac{K_c}{C_{L1}^2 + K_c C_v^2}} \times \sqrt{\frac{p_1 - p_{vap}(T_1) - \dfrac{g}{v_{ave}}(z_{v1} - z_1)}{v_{ave}}} \tag{8.19}$$

A flow greater than W_{cav} will imply that cavitation is occurring. But note that W_{cav} will depend on valve opening as a result of its dependence on C_v and K_c, so it will be necessary to calculate a value of W_{cav} for each valve opening.

8.5 Example: calculation of liquid flow

Consider the mass flow through the pipework shown in Figure 8.2.

The pipe is standard commercial pipe of diameter of 250 mm and length 60 m. It has a smoothly tapering entrance, two right-angled bends as shown, and rises 16 m just before the valve. The discharge vessel is 10 m after the valve, and is positioned 4 m above it. The valve is a 6-inch conventional-disc butterfly, with full-open coefficients: $C_V^* = 1750$ US gall min^{-1} psi$^{-(1/2)}$, $K_M = 0.3$, $K_C = 0.25$. The supply vessel contains water at 10 bar and 120°C, while the discharge vessel has a pressure of 4 bar.

It is known that the design flow is 165 kg/s. However, in this transient, the valve is initially 27° open, but after 10 seconds the valve is ramped over a period of half a minute to 54° open. Calculate the flow through the valve as a function of time, and estimate whether it becomes choked and whether cavitation occurs.

Calculation

Pipe cross-sectional area,

$$A = \frac{\pi}{4}(0.25)^2 = 0.0491\ \text{m}^2$$

From steam tables, specific volume,

$$v = 0.00106\ \text{m}^3/\text{kg}.$$

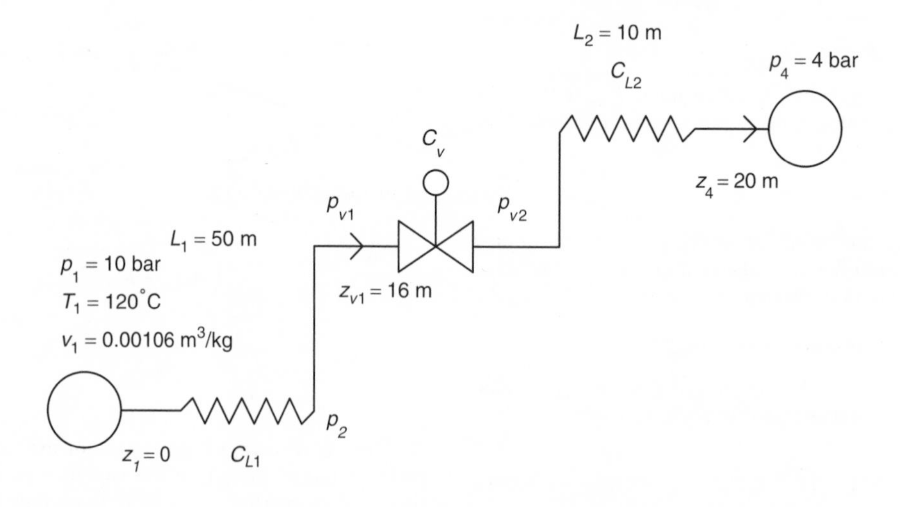

Figure 8.2

Pipe velocity at design flow,

$$c_D = \frac{vW_D}{A} = \frac{0.00106 \times 165}{0.0491} = 3.562 \text{ m/s}$$

Viscosity of water at 120°C, $\mu = 2.3 \times 10^{-4}$ Pa. s.

Reynolds number at design conditions,

$$(N_{RE})_D = \frac{c_D D}{v\mu} = \frac{3.562 \times 0.25}{1.06 \times 10^{-3} \times 2.3 \times 10^{-4}}$$
$$= 3.65 \times 10^6$$

Since the Reynolds number is greater than 10^5, the friction factor, f, will have reached a constant value (see Chapter 4, Section 4.4, and Figure 4.1). Using equation (4.29) for commercial pipe, the friction factor emerges as $f = 0.0045$.

Velocity head drop across the 50 m of pipe before the valve follows from equation (4.46) as

$$K_{L1} = 4f\frac{L_1}{D} = 4 \times 0.0045 \times \frac{50}{0.25} = 3.6$$

The fact that the entrance is smoothly tapered allows us to put $K_{con} \approx 0$ (see Chapter 4, Section 4.10), but we need to add the velocity-head drop from the two right-angle bends, $K_b = 0.75$ from Table 4.1, to produce the total velocity-head drop, K_{L1T}, between the upstream vessel and the valve: $K_{L1T} = K_{L1} + 2 \times 0.75 = 5.1$. Calculate line conductance upstream of valve using the velocity-head to conductance equation (4.52):

$$C_{L1} = A\sqrt{\frac{2}{K_{L1T}}} = 0.03074 \text{ m}^2$$

Velocity head drop across the 10 m of pipe downstream of valve:

$$K_{L2} = 4f\frac{L_2}{D} = 4 \times 0.0045 \times \frac{10}{0.25} = 0.72$$

Calculate line conductance downstream of valve,

$$C_{L2} = A\sqrt{\frac{2}{K_{L2}}} = 0.08181 \text{ m}^2$$

Valve conductance at fully open in SI units (see Equation (7.4)):

$$C_V = 2.3837 \times 10^{-5} C_V^* = 2.3837 \times 10^{-5} \times 1750$$
$$= 0.04171 \text{ m}^2$$

Initial valve travel, $x, = 27°/90° = 0.3$, final valve travel $= 54°/90° = 0.6$.

Calculate valve opening for general butterfly valve from $y = 1 - \cos((\pi/2)x)$ (equation (7.32), and valve conductance from $C_v = yC_V(\text{m}^2)$.

Total conductance is given by:

$$\frac{1}{C_T^2} = \frac{1}{C_{L1}^2} + \frac{1}{C_v^2} + \frac{1}{C_{L2}^2} \text{ (cf. equation (4.82))}$$

Given the valve travel, x, use equation (7.20) to find K_m/K_M, and equation (7.14) to find K_c/K_C. Hence given the values of K_M, K_C (0.3 and 0.25, respectively), calculate K_m and K_c at each value of x.

Figure 8.3 shows the calculated values of x, y, K_m and K_c against time.

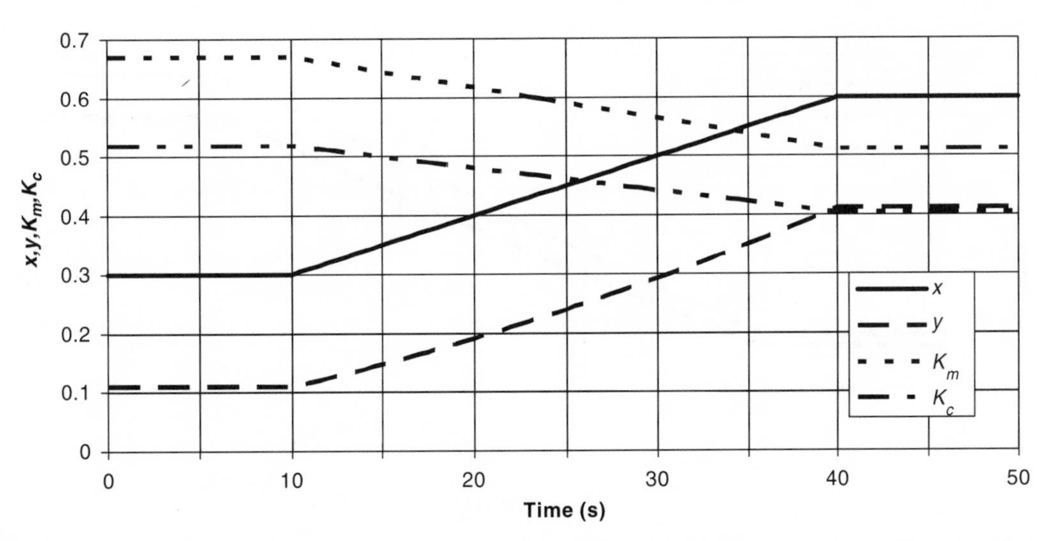

Figure 8.3 Showing transients for valve travel, x, valve opening, y, and choking and cavitation coefficients K_m and K_c.

Figure 8.4 shows the calculated values of C_v and C_T over the transient.

It may be observed from Figure 8.4 that $C_T \approx C_v$ at low values of valve opening, when almost all the pressure drop is across the valve. However, as the flow increases, the proportion of pressure drop across the line increases, and so $C_T < C_v$.

From steam tables, the vapour pressure of water at 120°C, $p_{vap} = 198\,530$ Pa, while the pressure at the critical point for water is 22.12 MPa. Hence the ratio of vapour pressure to critical pressure, $r_{pvc} = 0.1985\text{ MPa}/22.12\text{ MPa} = 0.00898$.

Vapour pressure fraction, r_{vap}, from correlation equation (7.17) emerges as $r_{vap} = 0.9324$ so $r_{vap}\, p_{vap}(T_1) = 185109$ Pa.

Supply vessel pressure $p_1 = 1\,000\,000$ Pa, discharge vessel pressure, $p_2 = 400\,000$ Pa.

With $v = v_{ave}$, use equation (8.17) to calculate the choking flow, (8.19) to calculate the cavitating flow and (8.18) to calculate the actual flow through line and valve. Figure 8.5 shows the result.

It is clear from Figure 8.5 that the actual flow remains less than the choked flow, $W < W_{choke}$, until roughly halfway through the transient, but after this time the actual flow is limited to the choked value, $W = W_{choke}$. Cavitation is indicated throughout the transient by the fact that $W > W_{cav}$ at all times.

Examining the first part of the transient in more detail, we see from Figure 8.6 that the flow gradually approaches the choking limit as the valve is opened,

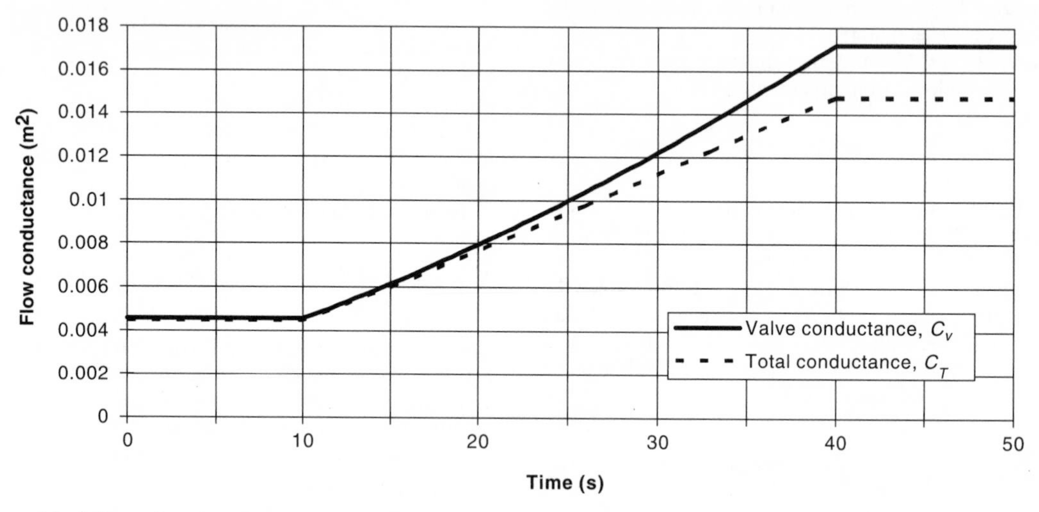

Figure 8.4 Valve and total conductances versus time.

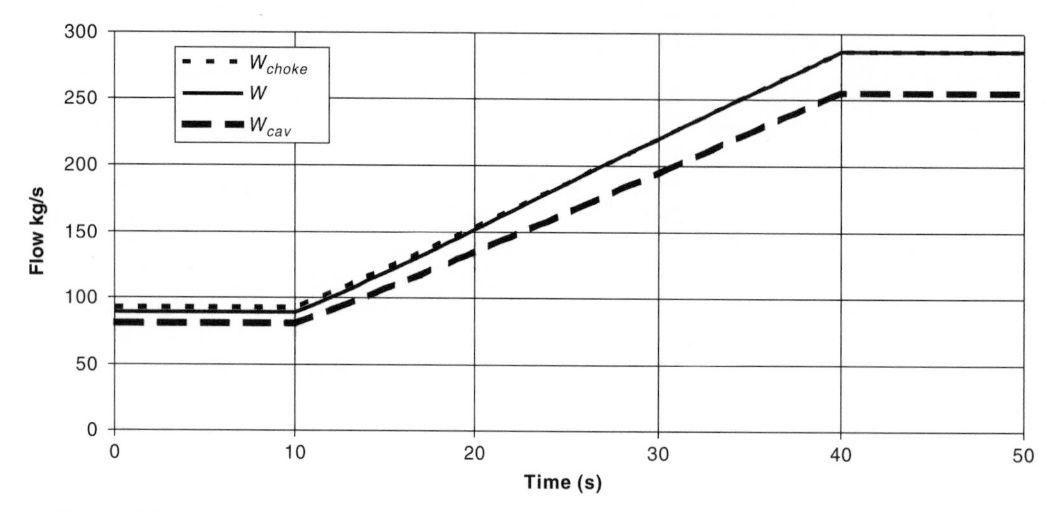

Figure 8.5 Actual flow, compared with choking flow and cavitating flow.

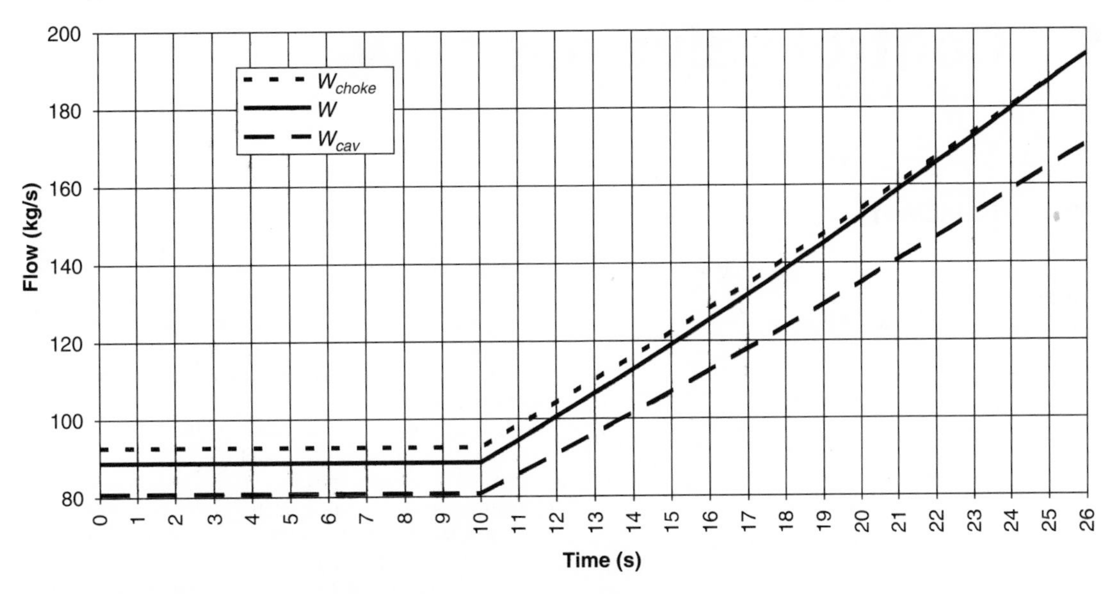

Figure 8.6 Pressures before and after the valve and in the valve throat.

with choking occurring after 26 seconds, at which time the valve travel is 0.46 and the valve opening is 0.25.

Finally, a calculation of the Reynolds number at the off-design flows shows it to be in excess of 1.96×10^6 ($\gg 10^5$) at all times, thus confirming that a constant value of friction factor, f, was in order.

The fact that the valve is cavitating at all times indicates a fault with the installation, since the valve is likely to undergo considerable wear. This would be avoided if the valve were repositioned at a lower level, 13 m or less above the supply vessel, rather than 16 m above it.

Obviously the example was designed to display the range of calculations needed with a control valve carrying a liquid, and is to that extent artificial. But it should not be thought that similar conditions are never found in practice: it is quite possible for a large process plant to contain such an instance of poor installation, particularly if it has undergone a number of modifications over the years, at the hands of different teams with different objectives.

9 Control valve gas flow

9.1 Introduction

Manufacturers have found that gas flow at high valve pressure ratios, p_2/p_1, may be calculated using the same equation as for liquid flow:

$$W = C_v\sqrt{\frac{\Delta P}{v_1}} = C_v\sqrt{\frac{p_1}{v_1}\left(1 - \frac{p_2}{p_1}\right)} \quad (9.1)$$

where the valve conductance, $C_v(\mathrm{m}^2)$, is the same for liquids and gases. (An explanation indicating why this equation may be used for gas flow at high-pressure ratios is given at the end of Section 9.2.) At very low pressure ratios, the flow of gas becomes choked and independent of downstream pressure. Here flow may be characterized by the equation:

$$W = C_g\sqrt{\frac{p_1}{v_1}} \quad (9.2)$$

where the limiting gas conductance, $C_g(\mathrm{m}^2)$, must be measured for each valve and depends on the properties of the gas being passed. Thus we have in equations (9.1) and (9.2) simple algorithms for calculating the gas flow through the control valve at either end of the range of possible pressure ratios. Unfortunately the equations are insufficient to define the point when choked flow begins and they do not indicate how flow should be calculated at intermediate pressure ratios.

Manufacturers often use a modified form of equation (9.1) as the basis for calculating gas flow throughout the pressure range up to choking. However, it is possible, as an alternative, to exploit the fact that the control valve passing gas has strong similarities to the gas nozzle. A comprehensive set of equations is available to describe the flow of gas through an ideal, frictionless nozzle. These equations cover the flow at intermediate pressure ratios as well as at each end of the range, thus providing automatically a smooth linkage between the two extreme modes of flow. However, the ideal-nozzle equations will be valid only over the section of the valve where the flow is accelerating and hence fluid frictional effects are low. In practice this means as far as the point of maximum constriction, which, following nozzle practice, we have called the throat of the valve. The increased friction associated with the diffusing flow downstream of the throat renders ideal nozzle theory a poor approximation from that point on. Fortunately it is possible to formulate a relationship based on the measured conductances, C_v and C_g, to link the throat pressure, p_t, to the valve exit pressure, p_2.

It is important for its practical use that any model of gas flow through the control valve shall require no more data than are customarily provided by valve manufacturers. Given the valve inlet and outlet pressures, the inlet specific volume, the valve opening and the type of gas being passed, the method presented in the main part of this chapter will require only the two conductances, C_v and C_g, both of which are supplied as a matter of course by manufacturers of control valves or else may be derived from the corresponding pair of values provided after units conversion.

9.2 Representing the first section of the control valve as a nozzle

The control valve shares the following features with the nozzle: a convergent inlet section, a divergent outlet section and a central region, the throat, where the flow area is restricted. Since the process of converting pressure head into velocity head in a converging flow is inherently very efficient, the frictional losses may be expected to be small as far as the throat in both the nozzle and the control valve. However, there is a tendency for fluid to break away from the walls of any diverging passage and flow backwards in the direction of the falling pressure gradient. Thus it is in the divergent section that most of the frictional loss will occur in a nozzle. The more complex geometry downstream of the valve's throat means that losses in a control valve will be even more significant. The space around the expanding stream will experience a pronounced eddying motion, the energy for which must be supplied from the main body of flow, which implies a large frictional loss. It follows that nozzle concepts may be applied directly only as far as the throat of the control valve.

Let us reconsider Figure 7.3, which represents the valve as a nozzle with a convergent and a divergent section. Conditions at the entrance to the nozzle are

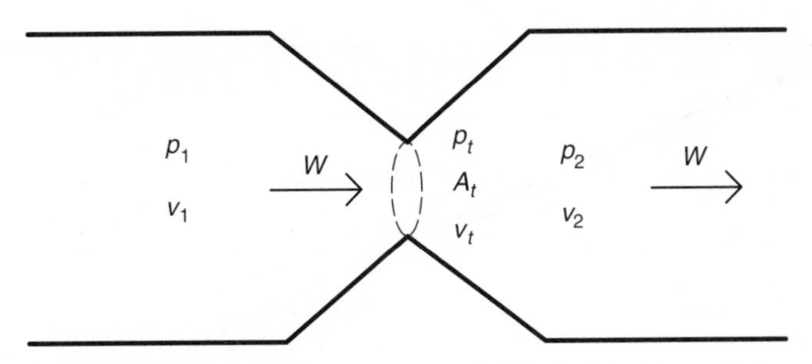

Figure 7.3 Flow constriction, representative of nozzle and of valve.

given the subscript '1', those at the selected downstream station are given the subscript '2', while those at the throat are given the subscript 't'.

The gas flow through the control valve as a whole will be very close to adiabatic, and, as just noted, the expansion as far as the valve throat will incur only a small frictional loss, implying a process that is approximately isentropic. Accordingly we may substitute the ratio of the gas's specific heats, $\gamma = c_p/c_v$ for the polytropic index, n, in the flow equations (5.25) and (5.26) derived for an ideal nozzle in Chapter 5. Combining those two equations gives the mass flow, W, as:

$$W = A_t\sqrt{2\frac{\gamma}{\gamma-1}\frac{p_1}{v_1}\left(\left(\frac{p_t}{p_1}\right)^{2/\gamma} - \left(\frac{p_t}{p_1}\right)^{(\gamma+1)/\gamma}\right)} \tag{9.3}$$

Following the procedure of Section 5.3, we may differentiate equation (9.3) with respect to throat pressure ratio and set the result to zero to show that the maximum flow occurs when the throat pressure ratio has decreased to a critical value given by:

$$\frac{p_{tc}}{p_1} = \left(\frac{2}{\gamma+1}\right)^{\gamma/(\gamma-1)} \tag{9.4}$$

at which point the maximum or choked flow, W_c, is given by:

$$W_c = A_t\sqrt{\frac{p_1}{v_1}\gamma\left(\frac{2}{\gamma+1}\right)^{(\gamma+1)/(\gamma-1)}} \tag{9.5}$$

No further reduction in pressure downstream of the throat will cause any further increase in speed in the throat, nor in the mass flow. The maximum mass flow will coincide with sonic flow in the throat.

It is possible to force the incompressible flow equation to match compressible flow equation (9.3) by introducing an expansion factor, $Y = Y(\gamma, p_t/p_1)$, into equation (9.1), modified to replace the downstream pressure, p_2, with the valve throat pressure, p_t:

$$W = YA_t\sqrt{2\frac{p_1}{v_1}\left(1-\frac{p_t}{p_1}\right)} \tag{9.6}$$

where division of equation (9.6) by equation (9.3) gives the expansion factor as:

$$Y = \sqrt{\frac{\gamma}{\gamma-1}\left(\frac{\left(\frac{p_t}{p_1}\right)^{2/\gamma} - \left(\frac{p_t}{p_1}\right)^{(\gamma-1)/\gamma}}{1-\frac{p_t}{p_1}}\right)} \tag{9.7}$$

This function is plotted against the pressure ratio, p_t/p_1, for a variety of specific-heat ratios, γ, in Figure 9.1 from which it may be seen that $Y \to 1$ for high values of the valve throat to inlet pressure ratio, p_t/p_1.

Hence at high-pressure ratios, $Y \approx 1$, so that equation (9.6) may be rewritten

$$W = A_t\sqrt{2}\sqrt{\frac{p_1}{v_1}\left(1-\frac{p_t}{p_1}\right)} \tag{9.8}$$

Comparing equation (9.8) with equation (9.1) applied as far as the throat, it is clear that they will take the same form if we define a conductance from the valve inlet to the valve throat, C_{vt} (m^2), that takes the value:

$$C_{vt} = A_t\sqrt{2} \tag{9.9}$$

While nozzle concepts have been applied above to the valve as far as the throat, we need to know the throat pressure ratio before they will prove their use in calculating the gas flow through the valve. The next section will show how the throat pressure ratio, p_t/p_1, may be related to the valve pressure ratio, p_2/p_1.

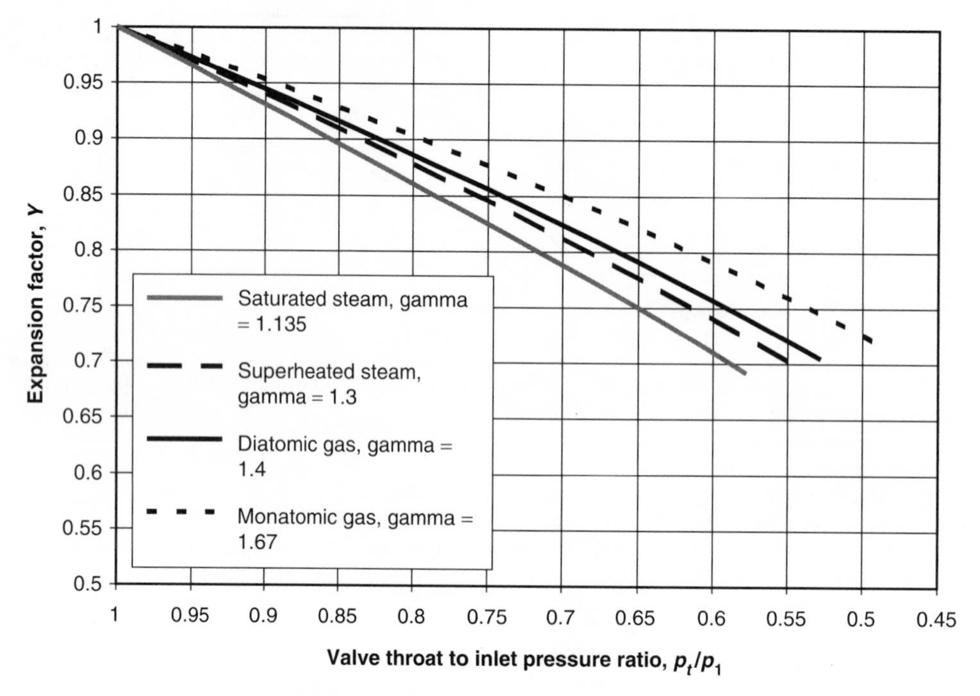

Figure 9.1 Expansion factor, Y, versus valve throat to inlet pressure ratio.

9.3 The relationship between throat ratio and the valve pressure ratio at high valve pressure ratios, p_2/p_1

The development of this section assumes that the gas flow may be modelled by incompressible-flow equations at high-pressure ratios, and is accordingly very similar to the development of Chapter 7, Section 7.3. If the valve pressure ratio, p_2/p_1, is high, we should be able to use either equation (9.1) or equation (9.8) to calculate the gas flow. Hence the following equality should hold:

$$C_{vt}\sqrt{\frac{p_1}{v_1}\left(1-\frac{p_t}{p_1}\right)} = C_v\sqrt{\frac{p_1}{v_1}\left(1-\frac{p_2}{p_1}\right)} \tag{9.10}$$

We may rearrange equation (9.10) to give the throat pressure ratio as a function of the valve pressure ratio:

$$\frac{p_t}{p_1} = \frac{1}{\left(\frac{C_{vt}}{C_v}\right)^2}\left(\frac{p_2}{p_1}-1\right)+1 \tag{9.11}$$

Let us define a friction coefficient for gas at high-pressure ratios, C_{fgh}, by

$$C_{fgh} = \frac{C_{vt}}{C_v} \tag{9.12}$$

so that

$$\frac{p_t}{p_1} = \frac{1}{C_{fgh}^2}\left(\frac{p_2}{p_1}-1\right)+1 \tag{9.13}$$

The friction coefficient determines whether or not pressure will rise downstream of the throat, and by how much. Pressure recovery will occur if $C_{fgh} < 1$, but if $C_{fgh} > 1$, the pressure will fall throughout the valve. The throat pressure and the valve outlet pressure will be equal when the friction coefficient takes its isobaric value, $C_{fgh} = 1$, at which condition $p_t = p_2$.

The valve conductance, C_v (m^2), will be available from manufacturer's data (it may be necessary to use the US to SI units-conversion factors given in Appendix 3). However, before we can calculate the friction coefficient, C_{fgh}, at high-pressure ratios, we need to know the valve conductance to the throat, C_{vt}, which, as we have seen from equation (9.9), depends on the throat area, A_t. This last parameter may be inferred from the limiting gas conductance, C_g.

9.4 Deriving a value for throat area, A_t, from the limiting gas conductance, C_g

Equation (9.2) holds at low valve pressure ratios, when flow is choked and where the value of C_g comes from manufacturer's experimental data. But from our model of the valve as an ideal nozzle undergoing isentropic, choked flow, equation (9.5) should also apply. Since the two equations should give the same flow, it follows that the SI limiting gas conductance C_g must be given by

$$C_g = A_t\sqrt{\gamma\left(\frac{2}{\gamma+1}\right)^{(\gamma+1)/(\gamma-1)}} \tag{9.14}$$

indicating (i) that C_g is independent of conditions and geometry downstream of the throat, as the physics of sonic flow would suggest, and (ii) that $C_g = C_g(\gamma)$ is dependent on specific-heat ratio, γ, and hence on the type of gas being passed. In most cases the manufacturer will base his quoted C_g-value on tests using air, a mixture predominantly of diatomic gases for which $\gamma = 1.4$, implying that the quoted C_g-value will be C_g (1.4). Hence we may calculate the effective throat area by inserting $\gamma = 1.4$ into equation (9.14):

$$A_t = 1.4604C_g(1.4) \tag{9.15}$$

Given the valve conductance, C_v, it is now possible to calculate the friction coefficient at high-pressure ratios, C_{fgh} from equations (9.9), (9.12) and (9.15).

9.5 Correlation of the friction coefficient at high-pressure ratios with the cavitation coefficient

The condition for cavitation in liquid flow, equations (7.11), may be re-expressed in the form:

$$\frac{p_{tcav}}{p_1} = \frac{1}{K_c}\left(\frac{p_2}{p_1} - 1\right) + 1 \tag{9.16}$$

which bears a strong resemblance to equation (9.13), suggesting that the friction coefficient, C_{fgh}, should be strongly correlated with the cavitation coefficient, K_c. Assuming a relationship of the form

$$(C_{fgh})^m \propto K_c \tag{9.17}$$

a least-squares straight line was fitted to the graph of log C_{fgh} versus log K_c for 70 valves varying in size from 0.5 inches to 72 inches (data taken from the *Fisher Control Valve Handbook*). This exercise yielded an optimal value of $m = 1.81$, which is very close to the value of $m = 2.0$ suggested by equation (9.13). In fact the correlation coefficients, r, found by fitting least-squares straight lines to C^m_{fgh} versus K_c first for $m = 1.81$ and then for $m = 2.0$ were almost identical at the high value, $r = 0.97$. Figure 9.2 shows the squared correlation:

$$C^2_{fgh} = 1.8817K_c \tag{9.18}$$

plotted against the valve data.

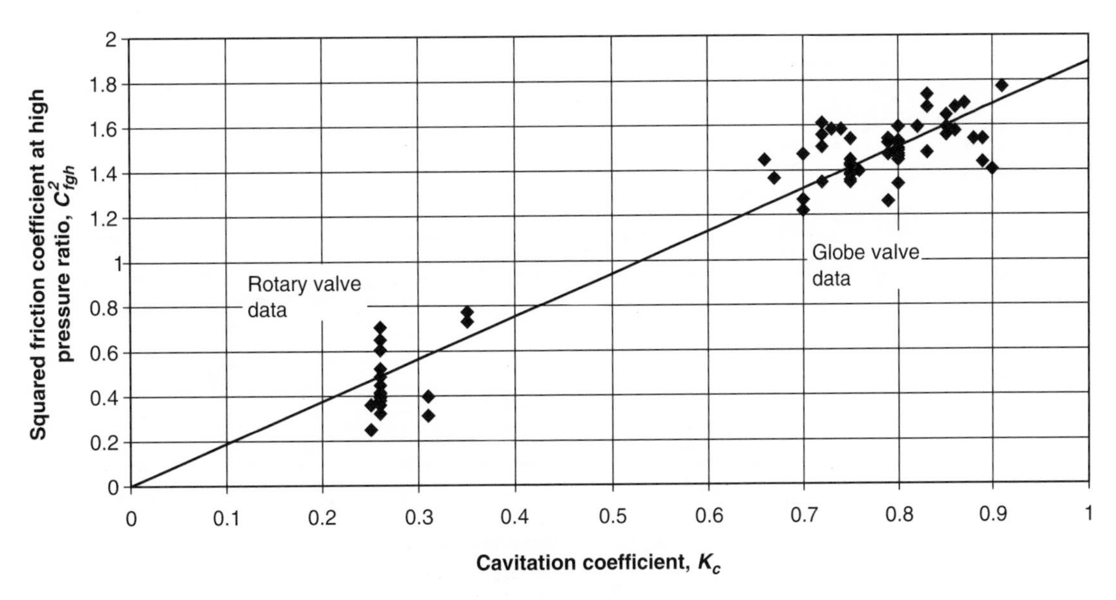

Figure 9.2 Correlation between the squared friction coefficient and the cavitation coefficient.

The linear relationship between the squared friction coefficient and the cavitation coefficient implies that

$$\frac{C_{fgh}^2}{C_{FGH}^2} = \frac{K_c}{K_C} \tag{9.19}$$

where the capital subscripts imply the value of the parameter when the valve is fully open.

The value of K_c has been found to independent of valve travel for a globe valve, reflecting the fact that the geometry downstream of the valve throat is unaffected by valve travel. Hence $K_c/K_C = 1$ and hence we may expect $C_{fgh}^2(x) = C_{FGH}^2$ for all valve travels, x. For rotary valves, however, the geometry downstream of the effective valve throat changes with valve travel, causing the cavitation constant, K_c, to change also. K_c may be related to valve travel, x, by piecewise linear equations (equation (7.14)). It follows from equation (9.19) that the same relationship should apply to the ratio of squared friction coefficients at high-pressure ratios:

$$\begin{aligned} \frac{C_{fgh}^2(x)}{C_{FGH}^2} &= 2.308 && \text{for } 0 \le x \le 0.15 \\ &= 2.5385 - 1.5385x && \text{for } 0.15 < x \le 1.0 \end{aligned} \tag{9.20}$$

9.6 The relationship between throat and valve pressure ratios when the valve pressure ratio is low

Equation (9.13) may be rearranged to give the valve pressure ratio as a function of the throat pressure ratio:

$$\frac{p_2}{p_1} = C_{fgh}^2\left(\frac{p_t}{p_1} - 1\right) + 1 \tag{9.21}$$

The derivation of this equation in Section 9.3 implies that it will be valid when the valve pressure ratio is high. The validity of equation (9.21) at low pressure ratios may be tested against direct data presented in the *Fisher Control Valve Handbook* for the low valve pressure ratios associated with the onset of choking. The Fisher data are in the form of a graph of $\Delta p/p_1(= 1 - p_2/p_1)$ versus C_1^* $(= C_g^*/C_v^*$, see Appendix 3) at the onset of choking, which may be converted into a relationship between $(p_2/p_1|_{choke})$ versus C_{fgh}^2 under the assumption that the data refer to the passage of air, which is a diatomic gas, with $\gamma = 1.4$. The critical throat pressure ratio, p_{tc}/p_1, for a diatomic gas is 0.5283 (equation (9.4)), which condition may be inserted into equation (9.21) to allow us to plot the variation of $(p_2/p_1|_{choke})$ with C_{fgh}^2 and compare the result with the direct data. See Figure 9.3.

We see that equation (9.21) is approximately valid up to the isobaric value of the friction coefficient, $C_{fgh} = 1$, where the prediction that $(p_2/p_1|_{choke}) = p_{tc}/p_1 = 0.5283$ is confirmed well by the direct data. But there is a growing divergence for values of C_{fgh} significantly greater than 1 and at the associated low valve pressure ratios. Here it should be noted that all globe valves operate with $C_{fgh} > 1$ (see Figure 9.2). To improve the match, we note first that the isobaric data point, $(p_2/p_1|_{choke}) = p_{tc}/p_1$ at $C_{fgh} = 1$, will be reproduced for any value of q by an equation of

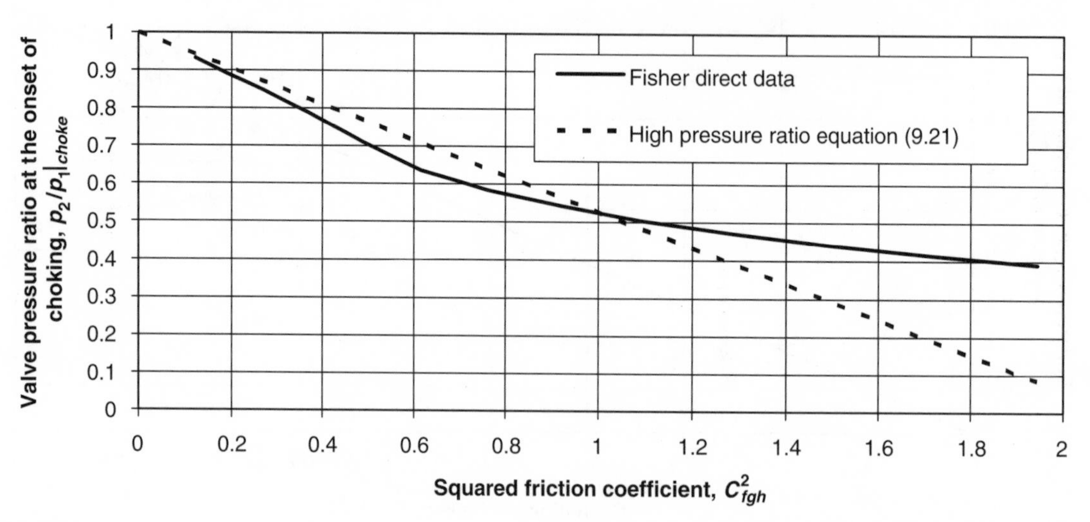

Figure 9.3 Valve pressure ratio at the onset of choking versus squared friction coefficient: comparison of equation (9.21) with Fisher direct data.

the form

$$\frac{p_2}{p_1} = C_{fgh}^{q}\left(\frac{p_t}{p_1} - 1\right) + 1 \qquad (9.22)$$

It is found, further, that the data at $C_{fgh} > 1$ are reproduced very well with $q = 0.75$, while those at very low C_{fgh} values are matched well with $q = 2$, the index previously shown to be valid at high valve pressure ratios. Introducing a smoothing function to link the two regions, the modified equations for very low valve pressure ratios become

$$\begin{aligned}\frac{p_2}{p_1} &= C_{fgh}^{2}\left(\frac{p_t}{p_1} - 1\right) + 1 \qquad \text{for } C_{fgh}^2 \leq 0.1\\ &= EC_{fgh}^{0.75}\left(\frac{p_t}{p_1} - 1\right) + 1 \quad \text{for } 0.1 < C_{fgh}^2 < 0.9\\ &= C_{fgh}^{0.75}\left(\frac{p_t}{p_1} - 1\right) + 1 \qquad \text{for } C_{fgh}^2 > 0.9\end{aligned} \qquad (9.23)$$

where the expression, E, is given by:

$$E = -1.25(C_{fgh}^{3.25} - C_{fgh}^{2} - 0.9C_{fgh}^{1.25} + 0.1) \qquad (9.24)$$

The improved match to the direct data is shown in Figure 9.4.

On inverting equation set (9.23) and introducing the critical pressure ratio limit in the valve throat, $p_{tc}(\gamma)/p_1$, the corresponding equations for the throat pressure ratio in terms of the valve pressure ratio when the latter is low become:

$$\frac{p_t}{p_1} = f_{lpr}\left(\frac{p_2}{p_1}, C_{fgh}\right) \qquad (9.25)$$

where f_{lpr} is the low-pressure-ratio function:

$$\begin{aligned}f_{lpr}\left(\frac{p_2}{p_1}, C_{fgh}\right)\\ &= \max\left(\frac{p_{tc}}{p_1}(\gamma), \frac{1}{C_{fgh}^2}\left(\frac{p_2}{p_1} - 1\right) + 1\right)\\ &\qquad \text{for } C_{fgh}^2 \leq 0.1\\ &= \max\left(\frac{p_{tc}}{p_1}(\gamma), \frac{1}{EC_{fgh}^{0.75}}\left(\frac{p_2}{p_1} - 1\right) + 1\right)\\ &\qquad \text{for } 0.1 < C_{fgh}^2 < 0.9\\ &= \max\left(\frac{p_{tc}}{p_1}(\gamma), \frac{1}{C_{fgh}^{0.75}}\left(\frac{p_2}{p_1} - 1\right) + 1\right)\\ &\qquad \text{for } C_{fgh}^2 \geq 0.9\end{aligned} \qquad (9.26)$$

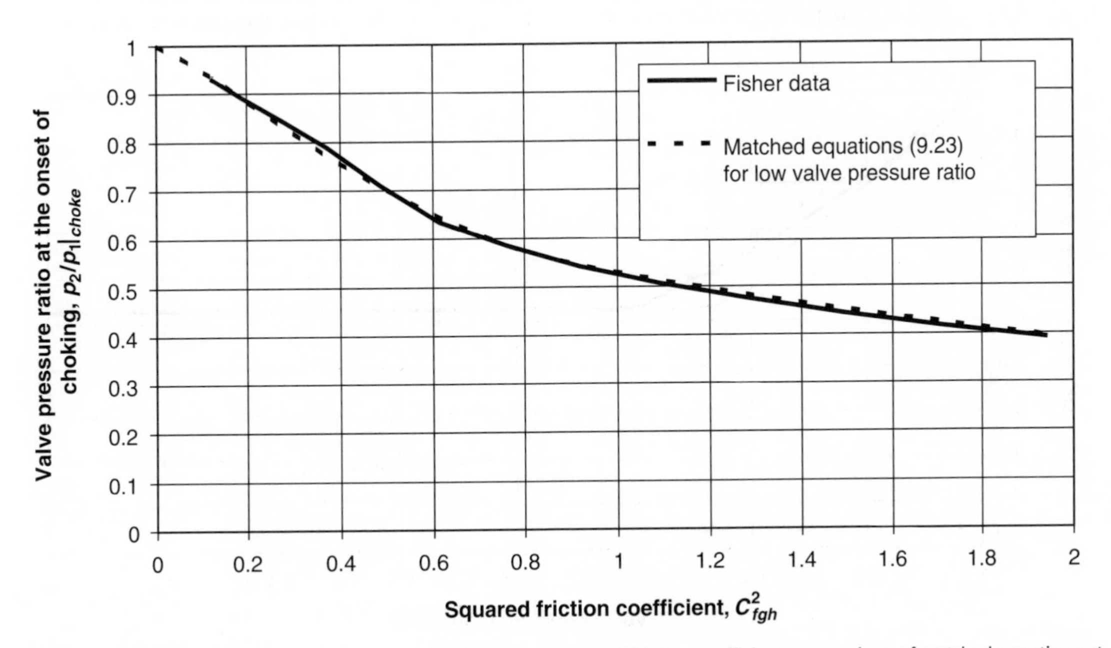

Figure 9.4 Valve pressure ratio at the onset of choking versus squared friction coefficient: comparison of matched equation set (9.23) with Fisher direct data.

9.7 Relating throat and exit pressure ratios throughout the pressure ratio range

Using equation (9.13) and adding the limiting throat pressure ratio at sonic flow, we may define the high-pressure ratio function $f_{hpr}(p_2/p_1, C_{fgh})$ by:

$$f_{hpr}\left(\frac{p_2}{p_1}, C_{fgh}\right) = \max\left(\frac{p_{tc}}{p_1}(\gamma), \frac{1}{C_{fgh}^2}\left(\frac{p_2}{p_1} - 1\right) + 1\right) \quad (9.27)$$

Using equations (9.25) and (9.27), the throat pressure ratio will be given at the extremities of the range of valve pressure ratios by:

$$\frac{p_t}{p_1} = f_{hpr}\left(\frac{p_2}{p_1}, C_{fgh}\right) \text{ when } \frac{p_2}{p_1} \to 1.0$$
$$= f_{lpr}\left(\frac{p_2}{p_1}, C_{fgh}\right) \text{ when } \frac{p_2}{p_1} \to \left.\frac{p_2}{p_1}\right|_{choke} \quad (9.28)$$

We may expect the behaviour at intermediate values of valve pressure ratio to be approximated reasonably well by a linear combination of the two functions, with the weighting of each dependent on the distance of the valve pressure ratio from the ends of the range. Hence we postulate that the throat pressure ratio may be found from

$$\frac{p_t}{p_1} = f_{comb}\left(\frac{p_2}{p_1}, C_{fgh}\right) \quad (9.29)$$

where the combination function, f_{comb}, is given by:

$$f_{comb} = \frac{\dfrac{p_2}{p_1} - \left.\dfrac{p_2}{p_1}\right|_{choke}}{1 - \left.\dfrac{p_2}{p_1}\right|_{choke}} f_{hpr} + \frac{1 - \dfrac{p_2}{p_1}}{1 - \left.\dfrac{p_2}{p_1}\right|_{choke}} f_{lpr} \quad \text{for } \frac{p_2}{p_1} > \left.\frac{p_2}{p_1}\right|_{choke} \quad (9.30)$$
$$= \frac{p_{tc}}{p_1}(\gamma) \quad \text{for } \frac{p_2}{p_1} \le \left.\frac{p_2}{p_1}\right|_{choke}$$

and where $(p_2/p_1|_{choke})$ is found by substituting $(p_t/p_1 = p_{tc}/p_1(\gamma))$ into equation set (9.23).

Figures 9.5 and 9.6 compare the expressions f_{lpr}, f_{hpr} and f_{comb} plotted against valve pressure ratio, p_2/p_1, for the cases when $C_{fgh} = 0.7$, typical of a fully open rotary valve and $C_{fgh} = 1.25$, typical of a globe valve. A diatomic gas is assumed in both cases, so that $\gamma = 1.4$.

It is clear that the functions, f_{lpr} and f_{comb}, produce similar values throughout the range of valve pressure ratios. Inspection of equation set (9.26) shows that f_{lpr} and f_{comb} must be identical for $C_{fgh}^2 < 0.1$. At higher values of the friction coefficient, $f_{lpr} \to 1.0$ as $p_2/p_1 \to 1.0$, so that the function, f_{lpr}, will be

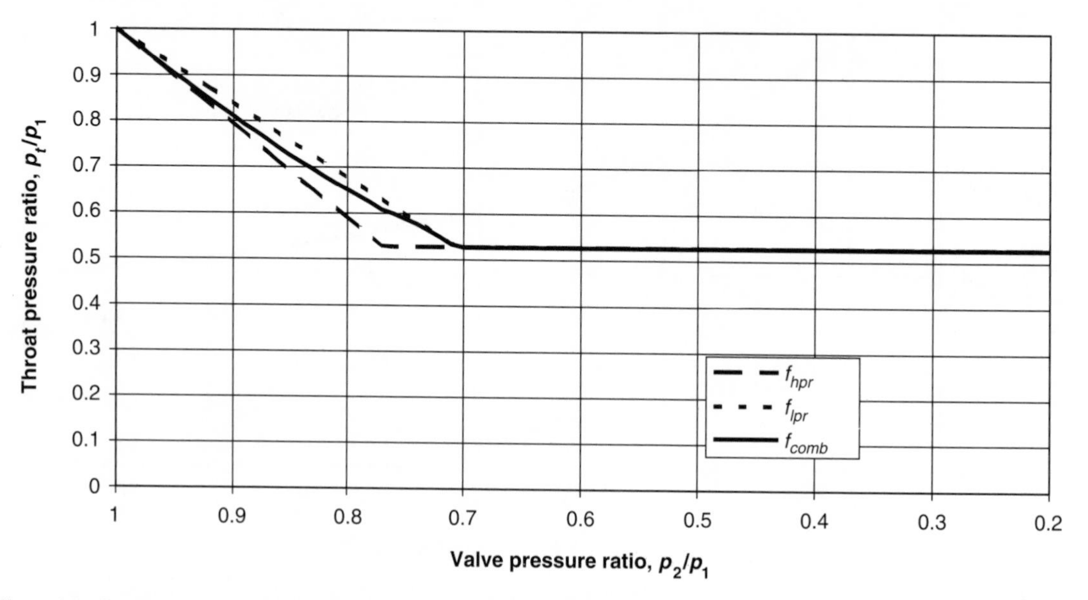

Figure 9.5 Pressure ratio relationships, f_{lpr}, f_{hpr} and f_{comb}, for a friction coefficient, $C_{fgh} = 0.7$.

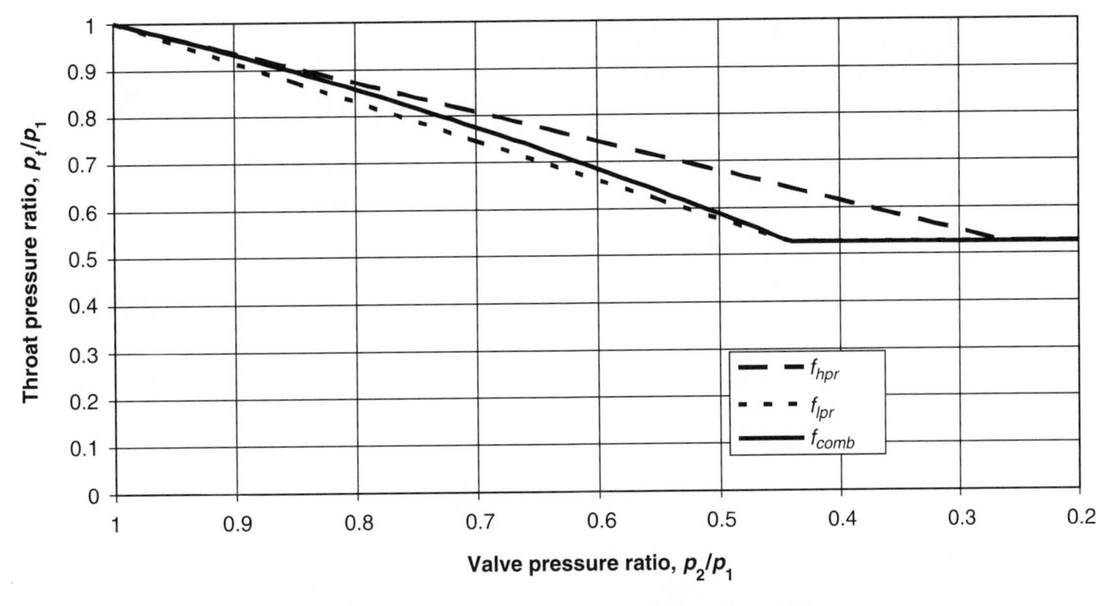

Figure 9.6 Pressure ratio relationships, f_{lpr}, f_{hpr} and f_{comb}, for a friction coefficient, $C_{fgh} = 1.25$.

valid at very high valve pressure ratios as well as low. We may find the maximum disparity between the two functions by forming the difference function, $f_{comb} - f_{lpr}$, differentiating this with respect to valve pressure ratio, p_2/p_1, and setting the result to zero. This procedure reveals that the maximum difference occurs when the valve pressure ratio takes the intermediate value:

$$\frac{p_2}{p_1} = \frac{1}{2}\left(1 + \left.\frac{p_2}{p_1}\right|_{choke}\right) \tag{9.31}$$

at which point the difference between f_{lpr} and f_{comb} is given by:

$$(f_{comb} - f_{lpr})|_{\max} = \frac{1}{4}\frac{p_{tc}}{p_1}\left(\frac{E_1}{C_{fgh}^{1.25}} - 1\right) \tag{9.32}$$

where E_1 is identical to E (see equation (9.24)) for $0.1 < C_{fgh}^2 < 0.9$, but takes the value of unity for higher friction coefficients. Equation (9.32) is plotted in Figure 9.7, which indicates that the maximum difference in predicted throat pressure ratio over the likely range of control valve C_{fgh}-values is ~0.035. This relatively small difference implies that there will be many cases where the function f_{lpr} may be used as an adequate approximation to f_{comb} throughout the range of valve pressure ratios.

9.8 Flow at partial valve openings

We may define the (fractional) valve opening, y, as the ratio of the current flow, W, to the flow, $W|_{\max}$, that would pass at fully open, given the same upstream conditions and throat pressure:

$$y = \frac{W}{W|_{\max}} \tag{9.33}$$

Combining equation (9.3) with equation (9.15), the gas flow through the control valve with limiting gas conductance, C_g (1.4), at the current valve opening is given by:

$$W = 2.065 C_g(1.4) \times \sqrt{\frac{\gamma}{\gamma-1}\frac{p_1}{v_1}\left(\left(\frac{p_t}{p_1}\right)^{2/\gamma} - \left(\frac{p_t}{p_1}\right)^{(\gamma+1)/\gamma}\right)} \tag{9.34}$$

which will reach a maximum value when the throat pressure ratio falls to its minimum physically realizable value, $p_t/p_1 = (p_{tc}/p_1(\gamma))$. For the same upstream conditions and throat pressure, the flow will clearly be greatest when $C_g(1.4)$ is greatest, and this will occur when the valve is fully open, i.e.:

$$C_g(1.4) = C_g(1.4)|_{y=1.0} = C_G(1.4) \tag{9.35}$$

where the capital subscript has been used to indicate the parameter value when the valve is fully open. It

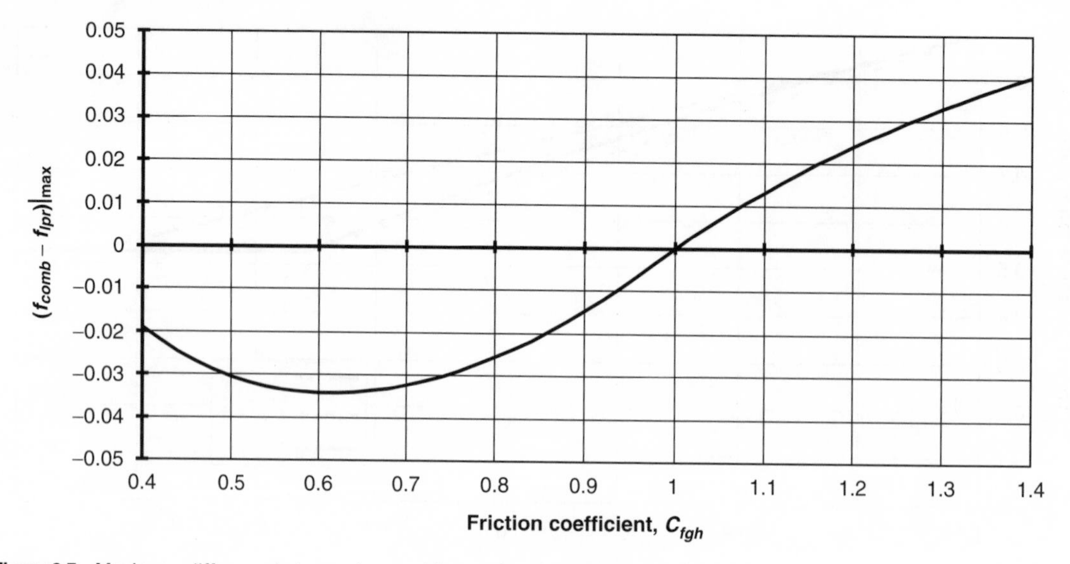

Figure 9.7 Maximum difference between f_{comb} and f_{ipr} as a function of friction coefficient, C_{fgh}.

follows that:

$$W|_{\max} = 2.065 C_G(1.4) \times \sqrt{\frac{\gamma}{\gamma-1}\frac{p_1}{v_1}\left(\left(\frac{p_t}{p_1}\right)^{2/\gamma} - \left(\frac{p_t}{p_1}\right)^{(\gamma+1)/\gamma}\right)} \tag{9.36}$$

Substituting from equations (9.34) and (9.36) into equation (9.33) gives the valve opening as the ratio of the current C_g-value to its value at fully open:

$$y = \frac{C_g(1.4)}{C_G(1.4)} \tag{9.37}$$

Substituting from equation (9.37) back into equation (9.34) gives the flow at valve opening, y, as:

$$W = 2.065 y C_G(1.4) \times \sqrt{\frac{\gamma}{\gamma-1}\frac{p_1}{v_1}\left(\left(\frac{p_t}{p_1}\right)^{2/\gamma} - \left(\frac{p_t}{p_1}\right)^{(\gamma+1)/\gamma}\right)} \tag{9.38}$$

9.9 Summary of the nozzle-based model for gas flow through the control valve

We assume that the following are given: the valve inlet pressure, p_1, the outlet pressure, p_2, the inlet specific volume, v_1, the valve travel, x, the fully open valve coefficient, C_V, the fully open, limiting gas conductance for air, $C_G(1.4)$.

The valve conductance to the throat at fully open may be evaluated using equations (9.9) and (9.15) as:

$$\begin{aligned} C_{VT} &= 1.4604\sqrt{2}C_G(1.4) \\ &= 2.065 C_G(1.4) \end{aligned} \tag{9.39}$$

Hence the friction coefficient at high-pressure ratio takes the fully open value, C_{FGH}:

$$C_{FGH} = \frac{C_{VT}}{C_V} \tag{9.40}$$

The friction coefficient at high-pressure ratio is the same at all valve openings for a globe valve, so that in such a case $C_{fgh}(x) = C_{FGH}$ for all x.

For a rotary valve, equation (9.20) is used to find the friction coefficient, $C_{fgh}(x)$, at a valve travel, x,:

$$\begin{aligned} \frac{C_{fgh}^2(x)}{C_{FGH}^2} &= 2.308 && \text{for } 0 \le x \le 0.15 \\ &= 2.5385 - 1.5385x && \text{for } 0.15 < x \le 1.0 \end{aligned} \tag{9.20}$$

Once the friction coefficient, C_{fgh}, has been determined, the valve pressure ratio at choking, $(p_2/p_1|_{choke})$, may be calculated by substituting the critical throat pressure ratio, $p_t/p_1 = (p_{tc}/p_1(\gamma))$, into equation set (9.23):

$$\left.\frac{p_2}{p_1}\right|_{choke} = C_{fgh}^2\left(\frac{p_{tc}}{p_1} - 1\right) + 1 \quad \text{for } C_{fgh}^2 \le 0.1$$
$$= EC_{fgh}^{0.75}\left(\frac{p_{tc}}{p_1} - 1\right) + 1 \quad \text{for } 0.1 < C_{fgh}^2 < 0.9 \qquad (9.41)$$
$$= C_{fgh}^{0.75}\left(\frac{p_{tc}}{p_1} - 1\right) + 1 \quad \text{for } C_{fgh}^2 > 0.9$$

where E is given by:

$$E = -1.25(C_{fgh}^{3.25} - C_{fgh}^2 - 0.9C_{fgh}^{1.25} + 0.1) \qquad (9.24)$$

The functions $f_{lpr}(p_2/p_1, C_f)$ and $f_{hpr}(p_2/p_1, C_f)$ may also be evaluated from equation sets (9.26) and (9.27), respectively.

$$f_{lpr}\left(\frac{p_2}{p_1}, C_f\right) = \max\left(\frac{p_{tc}}{p_1}(\gamma), \frac{1}{C_{fgh}^2}\left(\frac{p_2}{p_1} - 1\right) + 1\right) \quad \text{for } C_{fgh}^2 \le 0.1$$
$$= \max\left(\frac{p_{tc}}{p_1}(\gamma), \frac{1}{EC_{fgh}^{0.75}}\left(\frac{p_2}{p_1} - 1\right) + 1\right) \quad \text{for } 0.1 < C_{fgh}^2 < 0.9 \qquad (9.26)$$
$$= \max\left(\frac{p_{tc}}{p_1}(\gamma), \frac{1}{C_{fgh}^{0.75}}\left(\frac{p_2}{p_1} - 1\right) + 1\right) \quad \text{for } C_{fgh}^2 \ge 0.9$$

$$f_{hpr}\left(\frac{p_2}{p_1}, C_{fgh}\right) = \max\left(\frac{p_{tc}}{p_1}(\gamma), \frac{1}{C_{fgh}^2}\left(\frac{p_2}{p_1} - 1\right) + 1\right) \qquad (9.27)$$

Then the throat pressure ratio, p_t/p_1, may be found from equations (9.29) and (9.30).

$$\frac{p_t}{p_1} = f_{comb}\left(\frac{p_2}{p_1}, C_{fgh}\right) \qquad (9.29)$$

where

$$f_{comb} = \frac{\dfrac{p_2}{p_1} - \left.\dfrac{p_2}{p_1}\right|_{choke}}{1 - \left.\dfrac{p_2}{p_1}\right|_{choke}} f_{hpr} + \frac{1 - \dfrac{p_2}{p_1}}{1 - \left.\dfrac{p_2}{p_1}\right|_{choke}} f_{lpr} \quad \text{for } \frac{p_2}{p_1} > \left.\frac{p_2}{p_1}\right|_{choke} \qquad (9.30)$$
$$= \frac{p_{tc}}{p_1}(\gamma) \quad \text{for } \frac{p_2}{p_1} \le \left.\frac{p_2}{p_1}\right|_{choke}$$

Finally the flow through the valve is found from equation (9.38):

$$W = 2.065 y C_G(1.4) \times \sqrt{\frac{\gamma}{\gamma - 1}\frac{p_1}{v_1}\left(\left(\frac{p_t}{p_1}\right)^{2/\gamma} - \left(\frac{p_t}{p_1}\right)^{(\gamma+1)/\gamma}\right)} \qquad (9.38)$$

9.10 Worked example using the nozzle-based calculational model

Argon is being vented from a lagged upstream vessel initially at 0.251 MPa, 120°C to a storage vessel held at atmospheric pressure, 0.101 MPa. The short connecting pipe contains a 12-inch butterfly valve, for which the valve flow coefficient at full-open is given in US units as 6900 US gallons of water at 60°F per minute per psi$^{1/2}$, while the gas sizing coefficient at full-open is given in US units as 110 500 scf/h/psia. From an initial valve travel of 30%, the valve is ramped to 75% over 30 seconds. The valve travel remains unchanged for the next 40 seconds, but then the valve is ramped shut over the final 30 seconds of the transient. The upstream pressure, p_1, is assumed to be falling isentropically according to the following equation:

$$p_1 = 0.101 + 0.15\mathrm{e}^{-(t/60)} \text{ (MPa)}$$

where t is the time from the beginning of the transient in seconds. Neglecting the frictional losses in the pipe, calculate the initial mass flow through the valve, determine the maximum flow rate, and find the time at which the flow through the valve ceases to be choked.

Solution

Assuming that the C_g-value given refers to a diatomic gas, we may convert the valve coefficient and the limiting gas coefficient at fully open into the valve conductance and limiting gas conductance using the conversion factors given at the end of Appendix 3:

$$C_V = 2.3837 \times 10^{-5} C_V^* = 2.3837 \times 10^{-5} \times 6900 = 0.1645\,\mathrm{m}^2$$

$$C_G(1.4) = 4.02195 \times 10^{-7} C_G^* = 4.02195 \times 10^{-7} \times 110500 = 0.0444\,\text{m}^2$$

The conductance to the throat at fully open is then

$$C_{VT} = 2.065 \times C_G(1.4) = 2.065 \times 0.0444 = 0.09177\,\text{m}^2$$

The friction coefficient at high-pressure ratio at fully open is thus

$$C_{FGH} = C_{VT}/C_V = 0.09177 \div 0.1645 = 0.5580, \quad \text{so that } C_{FGH}^2 = 0.3113$$

Argon is a monatomic gas, with a specific-heat ratio, $\gamma = 1.67$. The throat pressure ratio at choking is given by equation (9.4) as:

$$p_{tc}/p_1 = (2/(1.67+1))^{(1.67)/(1.67-1)} = 0.4867$$

Argon's molecular weight is $w = 40$, and the specific volume follows from the equation of state: $v_1 = R_w T_1/p_1$ where $R_w = R/w$. Hence the specific volume at time zero is $v_1(0) = (8314/40) \times 393 \div 251\,000 = 0.3254\,\text{m}^3/\text{kg}$. Since the Argon in the upstream vessel is assumed to be expanding isentropically, the specific volume at later times will follow the equation:

$$v_1(t) = v_1(0)\left(\frac{p_1(0)}{p_1(t)}\right)^{1/\gamma}$$

Neglecting the thickness of the butterfly disc, the valve opening, y, for a butterfly valve may related to the valve travel, x, by the equation:

$$y = 1 - \cos\left(\frac{\pi}{2}x\right)$$

Hence when $x = 0.3$, $y = 0.109$.

At the initial valve travel of $x = 0.3$, the squared friction coefficient will be $C_{fgh}^2(x = 0.3) = (2.5385 - 1.5385 \times 0.3) \times C_{FGH}^2 = 2.077 \times 0.3113 = 0.6466$ (equation (9.20)). Thus $C_{fgh} = 0.8041$ at time zero. Once the friction coefficient, C_{fgh}, has been calculated, the parameter E of equation (9.24) may be evaluated:

$$E = -1.25 \times (0.8041^{3.25} - 0.8041^2 - 0.9 \times 0.8041^{1.25} + 0.1) = 0.9245$$

The valve pressure ratio at choking for conditions at time zero may be calculated from equation (9.23) with its throat pressure ratio at critical: $p_t/p_1 = p_{tc}/p_1 = 0.4867$. Thus $(p_2/p_1|_{choke}) = 0.9245 \times 0.8041^{0.75} \times (0.4867 - 1) + 1 = 0.597$. The actual valve pressure ratio at time zero is $p_2/p_1 = 101\,000 \div 251\,000 = 0.4024$, which is less than the choked value, so that the flow through the valve must be choked initially. As a result, both f_{lpr} and f_{hpr} (equations (9.26) and (9.27)) will come out at the critical value, 0.4867, as will f_{comb} (equation (9.30)).

The flow at time zero is therefore found by putting $p_t/p_1 = p_{tc}/p_1 = 0.4867$ into equation (9.38):

$$W = 2.065 \times 0.109 \times 0.0444 \times \sqrt{\frac{1.67}{1.67-1}\,\frac{251\,000}{0.3254}\left(\begin{array}{c}0.4867^{2/(1.67)} \\ -0.4867^{(1.67+1)/(1.67)}\end{array}\right)} = 4.514\,\text{kg/s}$$

The flow rate at later times may be calculated using a procedure similar to that just outlined, making due allowance for the falling upstream pressure and varying valve opening. The calculated flow rate over the transient is shown in Figure 9.8, from which it can be

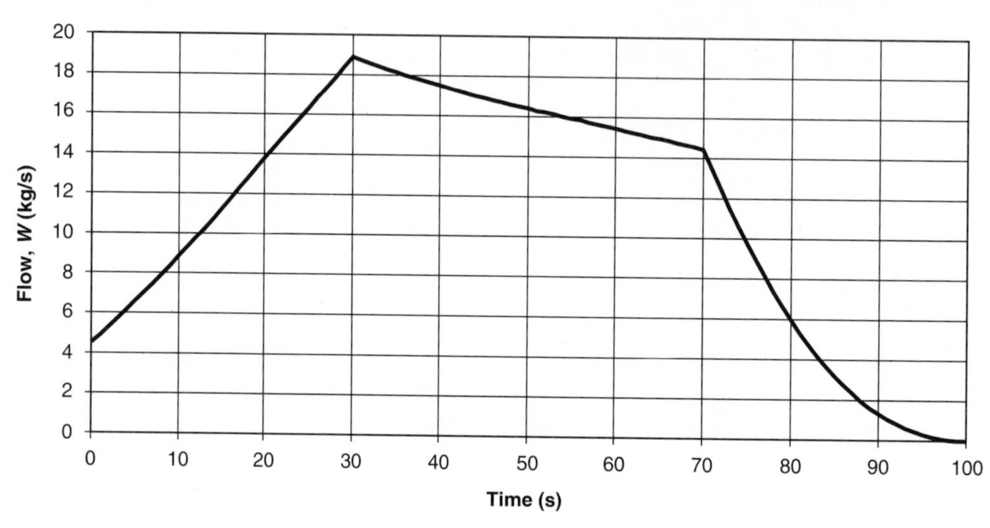

Figure 9.8 Worked example: mass flow rate versus time.

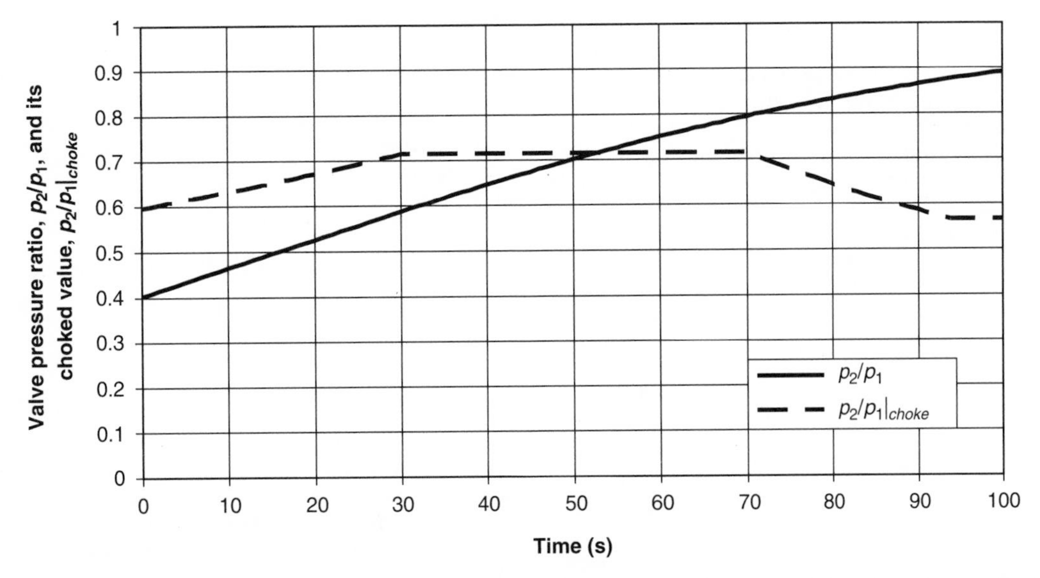

Figure 9.9 Worked example: transient behaviour of the calculated valve pressure ratio, p_2/p_1, and the valve pressure ratio at the onset of choking, $(p_2/p_1|_{choke})$.

seen that the maximum flow rate occurs at 30 seconds, when the flow is 18.89 kg/s.

We may deduce the time at which flow ceases to be choked by plotting both the actual valve pressure ratio, p_2/p_1, and its value at choking, $(p_2/p_1|_{choke})$, against time, and noting the moment that the two curves cross. Figure 9.9 shows the result, from which we may see that the flow is initially choked, but ceases to be so after 53 seconds.

This example was calculated under the assumption that the friction in the pipe was negligible, and that all the frictional pressure drop occurred across the valve. In the normal plant situation, however, significant friction losses are incurred as the gas flows through the pipework. This complication to the process of finding the flow is dealt with by a variety of methods in Chapter 10.

9.11 Other models for gas flow

9.11.1 The Fisher Universal Gas Sizing Equation (FUGSE)

The Fisher Universal Gas Sizing equation (FUGSE), given in the *Fisher Control Valve Handbook*, is a long-established method for finding gas flow through the valve, and is still widely used. The FUGSE maintains the liquid-flow algorithm for high-pressure ratios, but takes into account the choking at sonic flow induced by very low valve pressure ratios. The two distinct modes of behaviour are 'blended' using a sine function, the argument of which is chosen to be proportional to the square root of the fractional pressure drop, $\sqrt{\Delta p/p_1} = \sqrt{1 - p_2/p_1}$. The resulting equations for flow in the US units of standard cubic feet per hour are:

$$Q_{scfh} = \sqrt{\frac{520}{G_a T_1^*}} C_g^* p_1^* \sin\theta \tag{9.42}$$

$$\theta = \min\left(\frac{\pi}{2}, \frac{59.64}{C_1^*}\sqrt{\frac{\Delta p^*}{p_1^*}}\right) \tag{9.43}$$

where the variables marked with asterisks are in US units. Converted to SI units (see Appendix 3), the FUGSE gives the mass flow, W, in kg/s as:

$$W = C_g \sqrt{\frac{p_1}{v_1}} \sin\theta \tag{9.44}$$

$$\theta = \min\left(\frac{\pi}{2}, \frac{1.0063}{C_1}\sqrt{1 - \frac{p_2}{p_1}}\right) \tag{9.45}$$

where C_1 is the dimensionless ratio of the limiting gas conductance to the valve conductance:

$$C_1 = \frac{C_g}{C_v} \tag{9.46}$$

The angle, θ, will be small when the valve pressure ratio, p_2/p_1, is high, allowing application of the approximation $\sin\theta \approx \theta$, which transforms equations (9.44) and (9.45) essentially into equation (9.1). At very high-pressure drops, on the other hand, the limit of equation (9.45) is invoked, so that $\theta = \pi/2$ and $\sin\theta = 1$, which results in equation (9.44) reducing to equation (9.2). But while the FUGSE will be valid at either end of the pressure drop range, the fact that the sine function joins the extremes in a smooth manner does not guarantee that the formula will be accurate at intermediate values. In fact there appears to be no basis for choosing the sine function apart from mathematical convenience. As a result, the FUGSE turns out to be relatively poor at predicting the onset of choked flow in the valve, particularly at the higher values of the valve conductance ratio, C_1, at which many control valves operate.

The valve conductance ratio, C_1, is related to the friction coefficient at high valve pressure ratios, C_{fgh}, as follows:

$$C_{fgh} = \frac{C_{vt}}{C_v} = \frac{C_{vt}}{C_g}\frac{C_g}{C_v} = \frac{C_1}{C_{1t}} \tag{9.47}$$

where we have introduced the new coefficient, C_{1t}, which may be regarded as the conductance ratio for the valve as far as the throat:

$$C_{1t} = \frac{C_g}{C_{vt}} \tag{9.48}$$

Substituting for the limiting gas conductance, C_g, from equation (9.14) and for the conductance as far as the throat, C_{vt}, from equation (9.9) transforms equation (9.48) into

$$C_{1t} = \frac{C_g}{C_{vt}} = \frac{A_t\sqrt{\gamma\left(\frac{2}{\gamma+1}\right)^{(\gamma+1)/(\gamma-1)}}}{A_t\sqrt{2}} = \sqrt{\frac{\gamma}{2}\left(\frac{2}{\gamma+1}\right)^{(\gamma+1)/(\gamma-1)}} \tag{9.49}$$

which embodies the interesting result that the conductance ratio as far as the throat, C_{1t}, depends solely on the specific-heat ratio, γ, and not at all on the valve type, its geometry or the valve travel. Physically, these last three have been bundled up in the effective throat area, A_t, and have been cancelled out. Values of C_{1t} for commonly encountered values of γ are given in Table 9.1, which also lists for easy reference the critical throat pressure ratios calculated from equation (9.4).

From equation (9.47), the valve conductance ratio, C_1, may be written as:

$$C_1 = C_{1t}C_{fgh} \tag{9.50}$$

where, from Table 9.1, C_{1t} will be a constant for any given gas being passed. Hence

$$\frac{C_1}{C_I} = \frac{C_{fgh}}{C_{FGH}} \tag{9.51}$$

Table 9.1 p_{tc}/p_1 and C_{1t} values for different types of gas

Type of gas	γ	$\frac{P_{tc}}{P_1}$	C_{1t}
Dry saturated steam	1.135	0.5774	0.4494
Superheated steam	1.3	0.5457	0.4718
Polyatomic gas	1.333	0.5398	0.4760
Diatomic gas	1.4	0.5283	0.4842
Monatomic gas	1.67	0.4867	0.5138

where

$$C_I = \frac{C_G}{C_V} \tag{9.52}$$

is the conductance ratio at fully open, and so the valve conductance ratio will take on the relative characteristics of the friction coefficient, C_{fgh}. In particular, C_1 will be invariant with valve travel for globe valves, but will decrease with valve travel, x, for rotary valves according to the analogue of equation (9.20):

$$\begin{aligned}\frac{C_1^2}{C_I^2} &= 2.308 && \text{for } 0.0 \le x \le 0.15\\ &= 2.5385 - 1.5385x && \text{for } 0.15 < x \le 1.0\end{aligned} \tag{9.53}$$

The onset of choking is predicted from equation (9.45) when

$$\frac{1.0063}{C_1}\sqrt{1 - \frac{p_2}{p_1}} = \frac{\pi}{2} \tag{9.54}$$

or

$$\begin{aligned}\left.\frac{p_2}{p_1}\right|_{choke} &= 1 - \left(\frac{1}{2} \times \frac{\pi}{1.0063}\right)^2 C_1^2\\ &= 1 - 2.4366C_1^2\end{aligned} \tag{9.55}$$

If the FUGSE is used to model flow of a non-diatomic gas through the valve, it should be borne in mind that the value of C_g is dependent on the specific-heat ratio, γ, according to equation (9.14), and that the value of C_g given will be $C_g(1.4)$. Hence, substituting from equation (9.15) into equation (9.14), the C_g at any other specific-heat ratio, γ, will be given by:

$$C_g(\gamma) = 1.4604\sqrt{\gamma\left(\frac{2}{\gamma+1}\right)^{(\gamma+1)/(\gamma-1)}}\, C_g(1.4) \tag{9.56}$$

Appendix 4 gives a detailed comparison of the FUGSE with the nozzle-based model of gas flow through the valve, where it is shown that the FUGSE can be expected to produce reasonably accurate predictions of flow for rotary valves, especially at high valve openings. For example, using the FUGSE to compute the example given in Section 9.10 produces an almost identical mass-flow transient. It is likely to perform somewhat less well in situations of higher downstream friction ($C_{fgh} > 1.0$) associated with globe valves and with rotary valves at low travels, especially as the valve pressure ratio approaches its choked value. Nevertheless, the FUGSE has the undoubted merit of relative simplicity.

9.11.2 Approximate calculation of valve gas flow through modifying the liquid-flow equation

The nozzle-based model summarized in Section 9.9 gives an explicit description of gas flow through the valve up to and including the sonic condition. However, it will be found (Chapter 10) that an iterative procedure is necessary to solve the flow equations when the valve is inserted in a line of significant flow resistance. On the other hand, while the liquid-flow equation (7.3) is simple, it cannot be applied to model gas flow unless the pressure ratio across the valve is within a few percentage points of unity. But it turns out to be possible to adjust the liquid-flow equation to extend its range of use for lower pressure ratios through modifying the specific volume term and adopting a maximum flow based on the sonic limit.

Since the flow through the valve is governed by the area of the throat, it is reasonable to surmise that equation (7.3) will produce its best results when the specific volume term refers to the throat. The incompressible nature of liquids means that the inlet specific volume and the throat specific volume will be identical, so that the surmise is answered by equation (7.3) in its present form for liquids. However, we must make a change to cater for gases. Assuming a perfectly adiabatic expansion through the valve as far as the throat, the specific volume of the gas at the throat, v_t, will be given by

$$v_t = v_1 \left(\frac{p_1}{p_t}\right)^{1/\gamma} \tag{9.57}$$

We may borrow from the analysis of the early part of this chapter to find the ratio p_t/p_1 in the way summarized in Section 9.9: all the steps of that section are carried out except for the final calculation of flow from equation (9.38), which is replaced by:

$$W = C_v \sqrt{\frac{p_1 - p_2}{v_t}} \tag{9.58}$$

Such a formulation makes no allowance for critical flow, and so we need to impose the limit on mass flow implied by equation (9.2). Hence the modified liquid-flow model gives mass flow as:

$$W = \min\left[C_g\sqrt{\frac{p_1}{v_1}},\, C_v\sqrt{\frac{p_1 - p_2}{v_t}}\right] \tag{9.59}$$

Note that the value of C_g must be appropriate to the type of gas being passed, and should be calculated from equation (9.56).

We may generate a comparison between the modified liquid-flow model and the nozzle-based model by taking out the factor, $C_g\sqrt{p_1/v_1}$ from both equation (9.59) and equation (9.34). The modified liquid-flow function, f_{LG}, emerges as:

$$f_{LG} = \frac{W}{C_g\sqrt{\dfrac{p_1}{v_1}}} = \min\left[1.0,\, \frac{1}{C_1}\sqrt{\left(\frac{p_t}{p_1}\right)^{1/\gamma}\left(1 - \frac{p_2}{p_1}\right)}\right] \tag{9.60}$$

where equation (9.46) has been used to combine C_v and C_g and equation (9.57) has been used to replace the specific volume ratio with an equivalent pressure ratio. p_t/p_1 and p_2/p_1 may be related in the way explained in Section 9.9.

The equivalent nozzle function, f_{NV}, emerges from equation (9.34) as:

$$f_{NV} = 2.065\sqrt{\frac{\gamma}{\gamma - 1}\left(\left(\frac{p_t}{p_1}\right)^{2/\gamma} - \left(\frac{p_t}{p_1}\right)^{(\gamma+1)/\gamma}\right)} \quad \text{for } \frac{p_t}{p_1} > \frac{p_{tc}}{p_1}$$
$$= 1.0 \quad \text{for } \frac{p_t}{p_1} \leq \frac{p_{tc}}{p_1} \tag{9.61}$$

where the choked flow limit has been imposed.

The two functions, f_{LG} and f_{NV} may be compared by plotting them against valve pressure ratio with friction coefficient, C_{fgh}, as parameter. Figure 9.10 shows the plots for C_{fgh}-values of 0.7, 1.0 and 1.25. $C_{fgh} = 0.7$ is typical of a rotary valve near fully open while $C_{fgh} = 1.0$ is the isobaric value, typical of a rotary valve near fully closed; $C_{fgh} = 1.25$ is representative of a globe valve.

The graph shows that the modified liquid-flow model can give results that are reasonably close to those of the nozzle-based model, although the approach to choked flow is predicted to happen at higher valve pressure

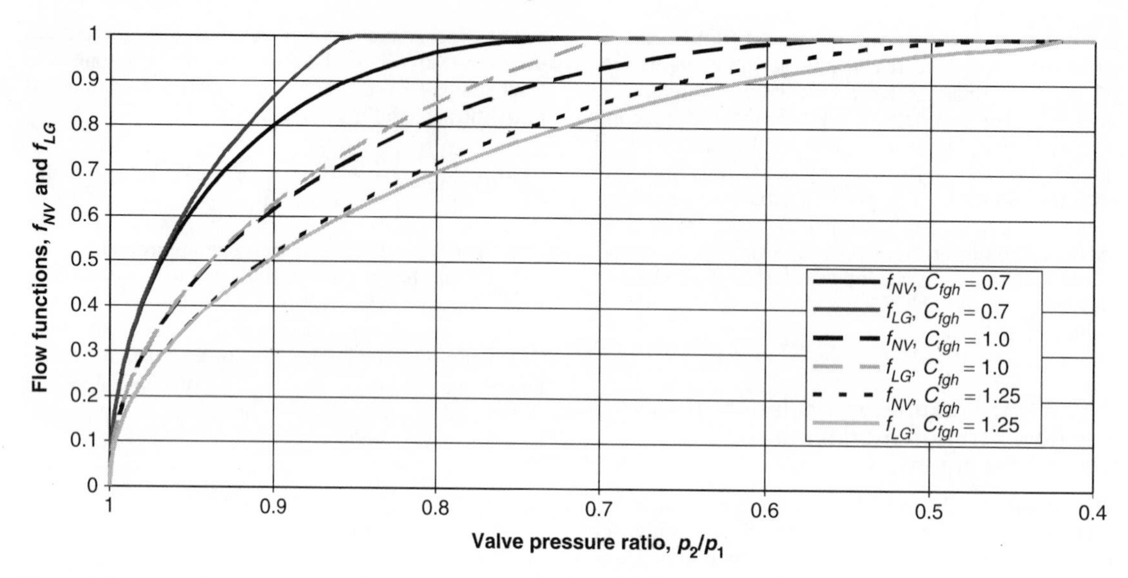

Figure 9.10 Comparison between nozzle-based function, f_{NV}, and modified liquid-flow function, f_{LG}.

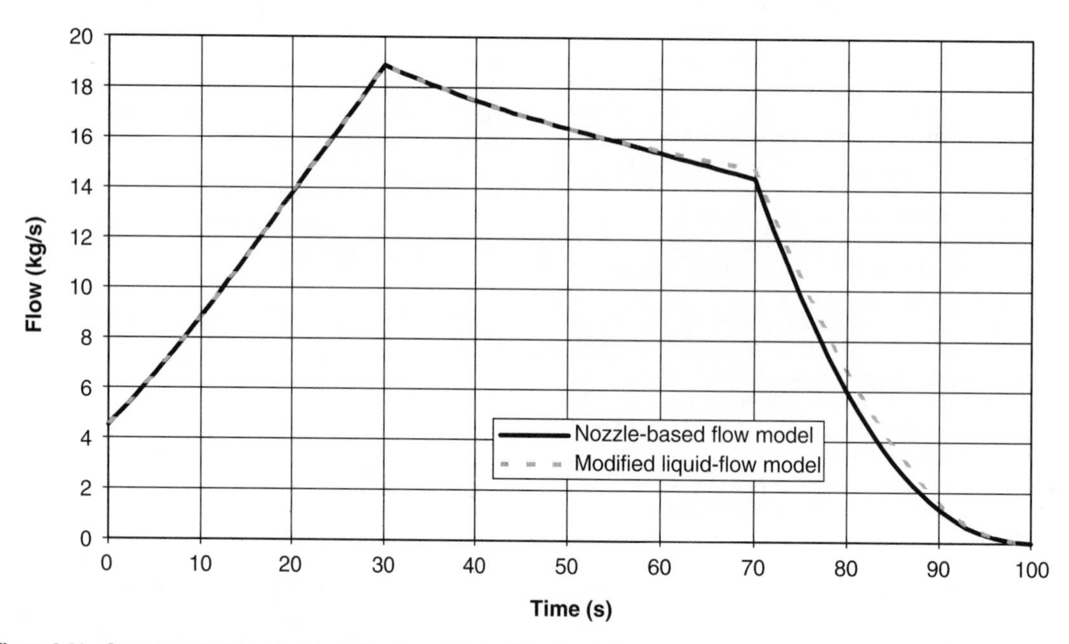

Figure 9.11 Comparison of flow transients for the example of Section 9.10.

ratios with the modified liquid-flow model when the friction coefficient, C_{fgh}, is low. Nevertheless, using the modified liquid-flow model to compute the example of Section 9.10 produces a flow transient that is very similar to that predicted by the nozzle-based model, as is shown in Figure 9.11.

9.12 Bibliography

Driskell, L.R. (1970). Sizing valves for gas flow, *ISA Transactions*, **9**, 325–331.

Faires, V.M. (1957). *Thermodynamics*, Macmillan, New York, pp. 375–404.

Fisher Controls (1977). *Control Valve Handbook*, 2nd edition. Fisher-Rosemount Ltd., Knight Road, Strood, Rochester, Kent, England, ME2 2EZ, pp. 60–78.

Lockey, J. (1966). *The Thermodynamics of Fluids*, Heinemann, London, pp. 153–180.

Rogers G. and Mayhew, Y. (1992). *Engineering Thermodynamics, Work and Heat Transfer*, 4th edition, Longman Scientific and Technical, Harlow, John Wiley & Son, New York, pp. 418–433, pp. 478–479.

Singh, M.G. Elloy, J-P. Mezencev, R. and Munro, N. (1980). *Applied Industrial Control, an Introduction*, Pergamon Press, Oxford, pp. 375–411.

Stephens, A.D. (1975). Stability and optimization of a methanol converter, *Chemical Engineering Science*, **30**, 11–19.

Streeter, V.L. and Wylie, E.B. (1983). *Fluid Mechanics*, First SI metric edition., McGraw-Hill, Singapore, pp. 82–111, 243–248, 262–276.

Webber, N.B. (1965). *Fluid Mechanics for Civil Engineers*, Spon's Civil Engineering Series, London, pp. 104–109.

Wherry, T.C. Peebles, J.R. McNeese, P.M. Teter, P.O. Worsham, B.S. and Young, R.M. (1983). Process control, in *Perry's Chemical Engineers' Handbook*, 6th edition, eds Perry, R.H., Green, D.W and Maloney, J.O., McGraw-Hill, New York, pp. 22.79–22.89.

10 Gas flow through the installed control valve

10.1 Introduction

The remarks made in Section 8.1 translate across to the gas flow cases more or less word for word, except that the methods of Chapter 9 must now be allied to those of Chapter 6 in order to calculate gas flow through line and valve. But the more complicated equations for both line flow and valve flow render explicit solutions to the full set of equations impossible. Two implicit methods, the Velocity-Head Implicit Method (VHIM) and the Smoothed Velocity-Head Implicit Method (SVHIM), will be presented, where the solution process has been reduced to iteration on a single variable. The SVHIM is judged to be more accurate because it deals with the compressible-flow valve equations at all times.

An approximate method, the Average Specific Volume Approximation Method (ASVAM), is also presented, based on the Long-Pipe Approximation described in Chapter 6. This approximate method retains much of the accuracy of the SVHIM, but has the advantage of yielding a direct estimate of flow.

Examples will be given where the various methods are applied and results compared.

10.2 Gas flow through an installed valve – Velocity-Head Implicit Method (VHIM)

10.2.1 VHIM without choking

Consider Figure 10.1, which shows the case of a control valve installed in a gas pipeline, with significant effective lengths of piping upstream and downstream of the valve. A valve of limiting gas conductance C_g (m^2) and liquid conductance C_v (m^2) is installed in a pipeline leading from a supply vessel at pressure p_1 (Pa) and specific volume v_1 (m^3/kg) to a receiving vessel at pressure p_4 (Pa). The effective length upstream of the valve is L_1 (m) and the effective length downstream of the valve is L_2 (m).

The methodology of Chapter 6 is applied in a fairly straightforward way in VHIM to estimate the flow as follows. First estimate the number of velocity heads, K_{T1}, dropped over the upstream section of pipe using equation (6.58) from Section 6.6, repeated below:

$$K_T = 4f\frac{L}{D} + K_{con} + \sum_i K_{bi} \tag{6.58}$$

and then repeat the exercise to find the number of velocity heads, K_{T2}, dropped over the section of pipe downstream of the valve. Evaluate the number of velocity heads, K_v, lost across the valve from equation (7.36), repeated below:

$$K_v = \frac{\pi^2 D^4}{8C_v^2} \tag{7.36}$$

where

$$C_v = yC_V \tag{7.25}$$

Then find the total number of velocity heads, K_T, dropped across pipe and valve from

$$K_T = K_{T1} + K_v + K_{T2} \tag{10.1}$$

Assigning the number (6.43a) to equation (6.43), modified by setting $K_T = 4fL/D$, find the critical ratio of specific volumes, $(v_3/v_2|_{crit})$, by solving the implicit equation:

$$\frac{\gamma+1}{2\gamma}\left(1-\left(\frac{v_3}{v_2}\bigg|_{crit}\right)^2\right) + \frac{\gamma+1}{\gamma}\ln\left(\frac{v_3}{v_2}\bigg|_{crit}\right) + K_T = 0 \tag{6.43a}$$

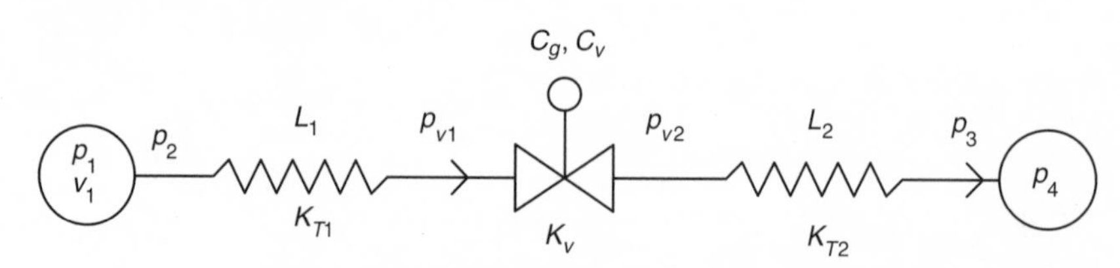

Figure 10.1 Control valve installed in a gas pipeline.

Use the result to determine the critical pressure ratio, p_{3c}/p_2:

$$\frac{p_{3c}}{p_2} = \frac{\left(\left.\frac{v_2}{v_3}\right|_{crit}\right)}{1 + \frac{\gamma - 1}{2}\left(1 - \left(\left.\frac{v_2}{v_3}\right|_{crit}\right)^2\right)} \tag{6.46}$$

Now solve the equations listed in Section 6.4, except that the term $4fL/D$ in equation (6.27) is replaced once more by K_T, the equivalent form for velocity heads dropped:

$$\mathbf{M}_2 = \sqrt{\frac{2}{\gamma - 1}\left(\left(\frac{p_1}{p_2}\right)^{(\gamma-1)/\gamma} - 1\right)} \tag{6.6}$$

$$p_3 = \max\left(p_4, p_2\frac{p_{3c}}{p_2}\right) \tag{6.51}$$

$$\frac{v_3}{v_2} = \frac{1}{(\gamma - 1)\mathbf{M}_2^2} \times \left[\sqrt{\left(\frac{p_3}{p_2}\right)^2 + 2(\gamma - 1)\mathbf{M}_2^2 + (\gamma - 1)^2\mathbf{M}_2^4} - \left(\frac{p_3}{p_2}\right)\right] \tag{6.32}$$

$$\left(\frac{2 + (\gamma - 1)\mathbf{M}_2^2}{2\gamma\mathbf{M}_2^2}\right)\left(1 - \frac{v_2^2}{v_3^2}\right) - \frac{\gamma + 1}{\gamma}\ln\left(\frac{v_3}{v_2}\right) - K_T = 0 \tag{6.27a}$$

where equation (6.27a) is equation (6.27) with the substitution $K_T = 4fL/D$. The solution sequence is as laid out in Section 6.4, whereby an initial guess of p_2 is made, and then refined estimates are generated until equation (6.27a) is satisfied.

10.2.2 Detecting the onset of sonic flow in the valve using VHIM

No allowance has been made above for sonic flow being reached in the valve, and the consequent choking that will occur. Whether sonic flow occurs will depend on the values of the pressures, p_{v1}, p_{v2}, immediately upstream and downstream of the valve. Once these are known, the ratio of the valve throat pressure to valve inlet pressure may be found using the method summarized in Section 9.9 of Chapter 9. Assuming for the moment that the flow in the valve is not choked, values for p_{v1} and p_{v2} may be found as follows.

Equations (6.32) and (6.27a) apply to a general downstream point, and hence may be modified to refer first to the point in the line immediately before the valve, where we use the subscript '$v1$', and then to the point immediately following the valve, given the subscript '$v2$'.

For the upstream point, equation (6.27a) becomes:

$$\left(\frac{2 + (\gamma - 1)\mathbf{M}_2^2}{2\gamma\mathbf{M}_2^2}\right)\left(1 - \frac{v_2^2}{v_{v1}^2}\right) - \frac{\gamma + 1}{\gamma}\ln\left(\frac{v_{v1}}{v_2}\right) - K_{T1} = 0 \tag{10.2}$$

The variables $\mathbf{M}_2$ (and also v_2) will be known from the solution of the overall flow equations, and hence equation (10.2) may be solved iteratively for the ratio, v_{v1}/v_2. This specific volume ratio may then be used in equation (10.3) below, derived from equation (6.30) to solve for valve upstream pressure, p_{v1}:

$$\frac{p_{v1}}{p_2} = \frac{1 + \frac{\gamma - 1}{2}\mathbf{M}_2^2\left(1 - \left(\frac{v_{v1}}{v_2}\right)^2\right)}{\frac{v_{v1}}{v_2}} \tag{10.3}$$

For the downstream point, the equations to be solved in a similar manner for v_{v2}/v_2 and then p_{v2} are (10.4) and (10.5) below:

$$\left(\frac{2 + (\gamma - 1)\mathbf{M}_2^2}{2\gamma\mathbf{M}_2^2}\right)\left(1 - \frac{v_2^2}{v_{v2}^2}\right) - \frac{\gamma + 1}{\gamma}\ln\left(\frac{v_{v2}}{v_2}\right) - K_{T1v} = 0 \tag{10.4}$$

$$\frac{p_{v2}}{p_2} = \frac{1 + \frac{\gamma - 1}{2}\mathbf{M}_2^2\left(1 - \left(\frac{v_{v2}}{v_2}\right)^2\right)}{\frac{v_{v2}}{v_2}} \tag{10.5}$$

where K_{T1v} is the velocity-head drop across upstream pipe and valve:

$$K_{T1v} = K_{T1} + K_v \tag{10.6}$$

As was the case with equation (10.2), the highly nonlinear form of equation (10.4) requires that the solution for the ratio v_{v2}/v_2 be iterative.

The ratio, p_t/p_{v1}, of valve throat pressure to valve inlet pressure is then found using the method set out in Chapter 9, Section 9.9. This is summarized in the equations below, modified where appropriate to conform to the valve labels given in Figure 10.1.

Derived coefficients for the fully open valve are:

$$C_{VT} = 2.065C_G(1.4) \tag{9.39}$$

$$C_{FGH} = \frac{C_{VT}}{C_V} \tag{9.40}$$

The friction coefficient at high-pressure ratio is the same at all valve openings for a globe valve, so that in such a case $C_{fgh}(x) = C_{FGH}$ for all x. For a rotary valve, on the other hand, the friction coefficient, $C_{fgh}(x)$, at a valve travel, x, is found from:

$$\frac{C_{fgh}^2(x)}{C_{FGH}^2} = 2.308 \qquad \text{for } 0 \le x \le 0.15$$
$$= 2.5385 - 1.5385x \qquad \text{for } 0.15 < x \le 1.0 \tag{9.20}$$

The valve pressure ratio at choking, $p_{v2}/p_{v1}|_{choke}$, may be calculated from:

$$\left.\frac{p_{v2}}{p_{v1}}\right|_{choke} = C_{fgh}^2\left(\frac{p_{tc}}{p_{v1}} - 1\right) + 1 \quad \text{for } C_{fgh}^2 \le 0.1$$
$$= EC_{fgh}^{0.75}\left(\frac{p_{tc}}{p_{v1}} - 1\right) + 1 \quad \text{for } 0.1 < C_{fgh}^2 < 0.9$$
$$= C_{fgh}^{0.75}\left(\frac{p_{tc}}{p_{v1}} - 1\right) + 1 \quad \text{for } C_{fgh}^2 > 0.9 \tag{10.7}$$

where E is given by:

$$E = -1.25(C_{fgh}^{3.25} - C_{fgh}^2 - 0.9C_{fgh}^{1.25} + 0.1) \tag{9.24}$$

The functions $f_{lpr}(p_{v2}/p_{v1}, C_{fgh})$ and $f_{hpr}(p_{v2}/p_{v1}, C_{fgh})$ are evaluated from

$$f_{lpr}\left(\frac{p_{v2}}{p_{v1}}, C_{fgh}\right) = \max\left(\frac{p_{tc}}{p_{v1}}(\gamma), \frac{1}{C_{fgh}^2}\left(\frac{p_{v2}}{p_{v1}} - 1\right) + 1\right) \quad \text{for } C_{fgh}^2 \le 0.1$$
$$= \max\left(\frac{p_{tc}}{p_{v1}}(\gamma), \frac{1}{EC_{fgh}^{0.75}}\left(\frac{p_{v2}}{p_{v1}} - 1\right) + 1\right) \quad \text{for } 0.1 < C_{fgh}^2 < 0.9 \tag{10.8}$$
$$= \max\left(\frac{p_{tc}}{p_{v1}}(\gamma), \frac{1}{C_{fgh}^{0.75}}\left(\frac{p_{v2}}{p_{v1}} - 1\right) + 1\right) \quad \text{for } C_{fgh}^2 \ge 0.9$$

$$f_{hpr}\left(\frac{p_{v2}}{p_{v1}}, C_{fgh}\right) = \max\left(\frac{p_{tc}}{p_{v1}}(\gamma), \frac{1}{C_{fgh}^2}\left(\frac{p_{v2}}{p_{v1}} - 1\right) + 1\right) \tag{10.9}$$

Finally the throat pressure ratio, p_t/p_{v1}, is found from

$$\frac{p_t}{p_{v1}} = f_{comb}\left(\frac{p_{v2}}{p_{v1}}, C_{fgh}\right) \tag{10.10}$$

where

$$f_{comb} = \frac{\dfrac{p_{v2}}{p_{v1}} - \left.\dfrac{p_{v2}}{p_{v2}}\right|_{choke}}{1 - \left.\dfrac{p_{v2}}{p_{v1}}\right|_{choke}} f_{hpr} + \frac{1 - \dfrac{p_{v2}}{p_{v1}}}{1 - \left.\dfrac{p_{v2}}{p_{v1}}\right|_{choke}} f_{lpr} \quad \text{for } \frac{p_{v2}}{p_{v1}} > \left.\frac{p_{v2}}{p_{v1}}\right|_{choke}$$
$$= \frac{p_{tc}}{p_{v1}}(\gamma) \quad \text{for } \frac{p_{v2}}{p_{v1}} \le \left.\frac{p_{v2}}{p_{v1}}\right|_{choke} \tag{10.11}$$

Sonic flow and choking will be indicated if

$$\frac{p_t}{p_{v1}} \le \frac{p_{tc}}{p_{v1}} = \left(\frac{2}{\gamma + 1}\right)^{\gamma/(\gamma-1)} \tag{10.12}$$

If inequality (10.12) is not satisfied, then choking is not occurring, and the values derived from the analysis of Section 10.2.1 will hold, and no more work needs to be done. If, however, inequality (10.12) is satisfied, then further calculations need to be carried out, as detailed in Section 10.2.3 below.

10.2.3 Calculating the flow and pipe conditions when valve flow is sonic in the VHIM

A new method of solution is required when the valve becomes choked. If the valve exit pressure has fallen below the value needed to produce choking, any variation below that pressure will have no effect on the flow, nor on upstream conditions. A considerable measure of decoupling occurs between upstream parameters and downstream parameters, although the two sections of pipe will carry the same flow, of course. In the calculation, the upstream conditions are determined first, and then the downstream conditions, subject to the constraint that the two pipe sections are linked by a common flow and a common stagnation enthalpy.

When the valve is choked, the flow through the valve will obey equation (9.2), which, in the notation of Figure 10.1, becomes simply for the sonic case:

$$W = C_g\sqrt{\frac{p_{v1}}{v_{v1}}} \tag{10.13}$$

But the flow through the valve is the same as the flow past station '2', given by combining equations (6.52) and (6.53) as

$$W = \sqrt{\gamma}A\mathbf{M}_2\sqrt{\frac{p_2}{v_2}} \tag{10.14}$$

so we may eliminate W from equations (10.13) and (10.14) to give the ratio of pressures at station '2' and valve inlet in terms of the ratio of specific volumes at the corresponding points:

$$\frac{p_{v1}}{p_2} = \gamma \frac{A^2}{C_g^2} \mathbf{M}_2^2 \frac{v_{v1}}{v_2} \tag{10.15}$$

The ratio of pressures may be eliminated by combining equation (10.15) with equation (10.3) to give the following relationship between Mach number at station '2' and the ratio of specific volumes at valve inlet and station '2':

$$\mathbf{M}_2^2 = \frac{1}{\left(\frac{v_{v1}}{v_2}\right)^2 \left(\gamma \frac{A^2}{C_g^2} + \frac{\gamma - 1}{2}\right) - \frac{\gamma - 1}{2}} \tag{10.16}$$

We may rewrite the first of the expressions used in the velocity-head equation (10.2) as:

$$\frac{2 + (\gamma - 1)\mathbf{M}_2^2}{2\gamma \mathbf{M}_2^2} = \frac{1}{\gamma}\left(\frac{1}{\mathbf{M}_2^2} + \frac{\gamma - 1}{2}\right) \tag{10.17}$$

Equation (10.16) is valid for the sonic case, and so may be substituted into equation (10.17) to produce:

$$\frac{2 + (\gamma - 1)\mathbf{M}_2^2}{2\gamma \mathbf{M}_2^2}$$

$$= \frac{1}{\gamma}\left(\left(\frac{v_{v1}}{v_2}\right)^2 \left(\gamma \frac{A^2}{C_g^2} + \frac{\gamma - 1}{2}\right) - \frac{\gamma - 1}{2} + \frac{\gamma - 1}{2}\right)$$

$$= \left(\frac{v_{v1}}{v_2}\right)^2 \left(\frac{A^2}{C_g^2} + \frac{\gamma - 1}{2\gamma}\right) \tag{10.18}$$

Hence the velocity-head equation (10.2) when the flow through the valve is sonic is:

$$\left(\frac{A^2}{C_g^2} + \frac{\gamma - 1}{2\gamma}\right)\left(\left(\frac{v_{v1}}{v_2}\right)^2 - 1\right) - \frac{\gamma + 1}{\gamma} \ln\left(\frac{v_{v1}}{v_2}\right) - K_{T1} = 0 \tag{10.19}$$

This equation may now be solved for the single unknown (v_{v1}/v_2). That done, the following steps may be followed to find the flow and the upstream thermodynamic variables.

(1) Find $\mathbf{M}_2$ by back-substitution of (v_{v1}/v_2) into equation (10.16).
(2) Given $\mathbf{M}_2$ find pressure, p_2, from a rearrangement of equation (6.6):

$$p_2 = p_1 \left(\frac{2}{2 + (\gamma - 1)\mathbf{M}_2^2}\right)^{\gamma/(\gamma-1)} \tag{10.20}$$

(3) Determine v_2 from equation (6.7), repeated below:

$$v_2 = v_1 \left(\frac{p_1}{p_2}\right)^{1/\gamma} \tag{6.7}$$

(4) Determine v_{v1} from a knowledge of (v_{v1}/v_2) and v_2.
(5) Determine the valve inlet pressure, p_{v1}, using equation (10.15).
(6) Find the mass flow, W, from equation (10.13) or (10.14).

Downstream section of pipe

Once choking has occurred, the pressures downstream of the throat become decoupled from the upstream pressures. The upstream and downstream sections of the valve and piping system are now linked by their common mass flow and their common stagnation enthalpy. It will be shown how the downstream thermodynamic variables may be found once the flow and the thermodynamic variables upstream of the valve have been calculated. To do so, we will need to solve iteratively a set of equations dependent on the pressure immediately downstream of the valve, p_{v2}.

We may calculate at the outset the choked pressure ratio, p_{3c}/p_{v2}, from equations analogous to (6.43a) and (6.46), namely:

$$\frac{\gamma + 1}{2\gamma}\left(1 - \left(\left.\frac{v_3}{v_{v2}}\right|_{crit}\right)^2\right) + \frac{\gamma + 1}{\gamma} \ln\left(\left.\frac{v_3}{v_{v2}}\right|_{crit}\right) + K_{T2} = 0 \tag{10.21}$$

$$\frac{p_{3c}}{p_{v2}} = \frac{\left(\left.\frac{v_{v2}}{v_3}\right|_{crit}\right)}{1 + \frac{\gamma - 1}{2}\left(1 - \left(\left.\frac{v_{v2}}{v_3}\right|_{crit}\right)^2\right)} \tag{10.22}$$

Whereas equations (6.43a) and (6.46) apply to the complete pipe and valve system, equations (10.21) and (10.22) apply to the pipework downstream of the valve (hence the use of K_{T2} to characterize the frictional loss, rather than K_T).

We may calculate the valve outlet temperature, T_{v2}, using the fact that the stagnation enthalpy will remain constant between the supply vessel and the valve outlet because the expansion is adiabatic (even though not reversible, and hence not isentropic):

$$h_{v2} + \tfrac{1}{2}c_{v2}^2 = h_1 \tag{10.23}$$

Now the valve outlet velocity, c_{v2}, may be derived by continuity from the mass flow through the valve, already calculated:

$$c_{v2} = \frac{v_{v2}W}{A} \tag{10.24}$$

where A is the area of the pipe immediately downstream of the valve, and the specific volume, v_{v2}, follows from the characteristic gas equation as:

$$v_{v2} = \frac{ZR_w T_{v2}}{p_{v2}} \tag{10.25}$$

Assuming also that enthalpy is equal to specific heat multiplied by the temperature, $h = c_p T$, allows us to rewrite equation (10.23) as:

$$\frac{1}{2}Z^2R_w^2\left(\frac{W}{Ap_{v2}}\right)^2 T_{v2}^2 + c_p T_{v2} - c_p T_1 = 0 \tag{10.26}$$

Only the positive root of this quadratic is physically meaningful:

$$T_{v2} = \frac{-c_p + \sqrt{c_p^2 + 2c_p Z^2 R_w^2 \left(\frac{W}{Ap_{v2}}\right)^2 T_1}}{Z^2R_w^2\left(\frac{W}{Ap_{v2}}\right)^2} = \frac{c_p}{ZR_w}\,\frac{\left[\sqrt{1 + 2\frac{ZR_w}{c_p}ZR_w\left(\frac{W}{Ap_{v2}}\right)^2 T_1} - 1\right]}{ZR_w\left(\frac{W}{Ap_{v2}}\right)^2} \tag{10.27}$$

The specific heat may be eliminated by using the fact that

$$\frac{c_p}{ZR_w} = \frac{\gamma}{\gamma - 1} \tag{10.28}$$

so that the valve outlet temperature emerges as

$$T_{v2} = \frac{\gamma}{\gamma - 1}\frac{A^2p_{v2}^2}{ZR_wW^2} \times \left[\sqrt{1 + 2\frac{\gamma-1}{\gamma}\frac{ZR_wW^2}{A^2p_{v2}^2}T_1} - 1\right] \tag{10.29}$$

Note that the solution to this equation requires us to guess the pressure immediately downstream of the valve, p_{v2}. However, it is possible to iterate on this variable according to the scheme presented next.

Once the temperature is calculated from equation (10.29), the specific volume at the valve outlet, v_{v2}, may then be calculated using equation (10.25), and the valve outlet velocity, c_{v2}, from equation (10.24). Next we may calculate the Mach number at valve outlet, $\mathbf{M}_{v2}$, from

$$\mathbf{M}_{v2} = \frac{c_{v2}}{\sqrt{\gamma p_{v2} v_{v2}}} \tag{10.30}$$

Then we apply the equations of Section 6.4 to the downstream pipe section:

$$p_3 = \max\left(p_4, p_{v2}\frac{p_{3c}}{p_{v2}}\right) \tag{10.31}$$

$$\frac{v_3}{v_{v2}} = \frac{1}{(\gamma-1)\mathbf{M}_{v2}^2} \times \left[\sqrt{\left(\frac{p_3}{p_{v2}}\right)^2 + 2(\gamma-1)\mathbf{M}_{v2}^2 + (\gamma-1)^2\mathbf{M}_{v2}^4} - \left(\frac{p_3}{p_{v2}}\right)\right] \tag{10.32}$$

$$\left(\frac{2+(\gamma-1)\mathbf{M}_{v2}^2}{2\gamma\mathbf{M}_{v2}^2}\right)\left(1 - \frac{v_{v2}^2}{v_3^2}\right) - \frac{\gamma+1}{\gamma}\ln\left(\frac{v_3}{v_{v2}}\right) - K_{T2} = 0 \tag{10.33}$$

The overall solution method is now very similar to that presented in Section 6.4. First guess p_{v2}, then calculate in turn T_{v2}, v_{v2}, c_{v2}, $\mathbf{M}_{v2}$, p_3, and finally v_3/v_{v2}, and iterate on p_{v2} until equation (10.33) is satisfied. At this point the flow and all the important thermodynamic variables upstream and downstream of the valve will have been calculated.

It should be noted, however, that the calculation scheme has required up to six single-variable iterative loops to solve:

- equation (10.21) for $(v_3/v_{v2}|_{crit})$,
- equation set (6.6), (6.51), (6.32) and (6.27a) for the pressure, p_2, just inside the pipe entrance,
- equation (10.2) for the specific volume ratio, v_{v1}/v_2,
- equation (10.4) for the specific volume ratio, v_{v2}/v_2,
- equation (10.19) for the specific volume ratio, v_{v1}/v_2 when the valve is choked,
- equation set (10.29), (10.25), (10.30), (10.31), (10.32) and (10.33) to find p_{v2} when the valve is choked.

Although single-variable iterative loops should not cause a large problem for computer simulation, the overall solution scheme is relatively complex.

10.3 Gas flow through an installed valve – Smoothed Velocity-Head Implicit Method (SVHIM)

10.3.1 SVHIM without choking

The VHIM is able to deal with both choking at the end of the pipe and choking of the valve. Nevertheless, it possesses some drawbacks. First, the characterization of valve flow in the subsonic regime using a velocity-head equivalent is based on incompressible flow rather than the compressible flow that actually occurs with a gas: the C_g value is not used at all in the subsonic region, and reliance on the C_v value will inevitably introduce a degree of error. Secondly, there is a need to perform a check for sonic flow through the valve, and the fact that two nonlinear, implicit equations need to be solved introduces complication and potentially extra problems of convergence. Finally, the transition from subsonic to sonic flow in the valve causes a potential discontinuity in the calculation as the incompressible formulation, based on a C_v value, is changed to a compressible formulation based on a C_g value.

SVHIM seeks to circumvent these problems by attacking the full set of line and valve equations at the outset. It is again necessary to solve the equations for subsonic and sonic flow in the valve separately, but the transition should now be smooth. The necessary equations for the subsonic case will now be listed in a logically consistent solution sequence. The basis of the method is first to guess and then to iterate on the value of the ratio of specific volumes v_{v1}/v_2.

Upstream section of pipe

Given a guessed value of the ratio of specific volumes v_{v1}/v_2, we may calculate the Mach number at station '2' from a rearrangement of equation (10.2):

$$\mathbf{M}_2 = \sqrt{\frac{1-\left(\dfrac{v_2}{v_{v1}}\right)^2}{\gamma K_{T1} + (\gamma+1)\ln\left(\dfrac{v_{v1}}{v_2}\right) - \dfrac{1}{2}(\gamma-1)\left(1-\left(\dfrac{v_2}{v_{v1}}\right)^2\right)}} \tag{10.34}$$

We may then calculate the pressure at station '2' from equation (10.20), repeated below:

$$p_2 = p_1\left(\frac{2}{2+(\gamma-1)\mathbf{M}_2^2}\right)^{\gamma/(\gamma-1)} \tag{10.20}$$

A knowledge of the pressure at station '2' enables us to calculate the specific volume from equation (6.7), repeated below:

$$v_2 = v_1\left(\frac{p_1}{p_2}\right)^{1/\gamma} \tag{6.7}$$

It also allows us to calculate the pressure at valve inlet using equation (10.3), repeated below:

$$\frac{p_{v1}}{p_2} = \frac{1+\dfrac{\gamma-1}{2}\mathbf{M}_2^2\left(1-\left(\dfrac{v_{v1}}{v_2}\right)^2\right)}{\dfrac{v_{v1}}{v_2}} \tag{10.3}$$

Combining the calculated value of v_2 with the guessed value of v_{v1}/v_2 gives the value of the specific volume at valve inlet, v_{v1}.

We may now use equation (10.14) to calculate the mass flow at station '2', W, which is the mass flow at all points throughout the pipe and valve:

$$W = \sqrt{\gamma}A\mathbf{M}_2\sqrt{\frac{p_2}{v_2}} \tag{10.14}$$

Valve

The mass flow through the valve is given by

$$W = C_g\sqrt{\frac{p_{v1}}{v_{v1}}}f_{NV}\left(\frac{p_t}{p_{v1}}\right) \tag{10.35}$$

where the nozzle-valve function f_{NV} comes from equation (9.61):

$$f_{NV}\left(\frac{p_t}{p_{v1}}\right) = 2.065\sqrt{\frac{\gamma}{\gamma-1}\left(\left(\frac{p_t}{p_{v1}}\right)^{2/\gamma} - \left(\frac{p_t}{p_{v1}}\right)^{(\gamma+1)/\gamma}\right)} \quad \text{for } \frac{p_t}{p_{v1}} > r_c$$

$$= 1 \quad \text{for } \frac{p_t}{p_{v1}} \le r_c \tag{10.36}$$

where r_c is the critical pressure ratio, given by

$$r_c = \frac{p_{tc}}{p_{v1}} = \left(\frac{2}{\gamma+1}\right)^{\gamma/(\gamma-1)} \tag{10.37}$$

Given the flow through valve and pipe from equation (10.34), we will wish to find the throat pressure, p_t. But the non-integer exponentiation in the nozzle-valve function, f_{NV}, means that this will require the iterative solution of a nonlinear, implicit equation. Fortunately, however, f_{NV} may be approximated to excellent accuracy by the function, f_{NVA}, where:

$$f_{NVA} = \sqrt{1-\left(\frac{\dfrac{p_t}{p_{v1}}-r_c}{1-r_c}\right)^2} \quad \text{for } \frac{p_t}{p_{v1}} > r_c$$

$$= 1 \quad \text{for } \frac{p_t}{p_{v1}} \le r_c \tag{10.38}$$

Using this function instead of f_{NV} in equation (10.35) and equating the valve flow to the flow at station '2' allows us to solve for the valve throat to valve inlet pressure ratio:

$$\frac{p_t}{p_{v1}} = r_c + (1 - r_c)\sqrt{\max\left(0, 1 - \frac{W^2}{C_g^2}\frac{v_{v1}}{p_{v1}}\right)} \tag{10.39}$$

The maximum function has been included in equation (10.39) to ensure that no attempt is made to take the square root of a negative number, while also accounting for the physical limitation that the throat pressure to inlet pressure ratio cannot drop below the critical value. We may now use a rearrangement of equation (9.23) together with equation (9.24) to find the valve outlet pressure, p_{v2}:

$$\frac{p_{v2}}{p_{v1}} = B\frac{p_t}{p_{v1}} + 1 - B \tag{10.40}$$

where:

$$\begin{aligned} B &= C_{fgh}^2 && \text{for } C_{fgh}^2 < 0.1 \\ &= \frac{1}{E}C_{fgh}^{0.75} && \text{for } 0.1 \le C_{fgh}^2 \le 0.9 \\ &= C_{fgh}^{0.75} && \text{for } C_{fgh}^2 > 0.9 \end{aligned} \tag{10.41}$$

and

$$E = -1.25(C_{fgh}^{3.5} - C_{fgh}^2 - 0.9C_{fgh}^{1.25} + 0.1) \tag{9.24}$$

It is possible to use this formulation, based on the use of the low-pressure-ratio function, f_{lpr}, for simplicity because f_{lpr} is a good approximation to the combination function, f_{comb}, as demonstrated in Chapter 9, Section 9.7.

Downstream section of pipe

Once the pressure immediately downstream of the valve has been calculated, the corresponding temperature may be calculated also, using equation (10.29), which applies generally to adiabatic flow, irrespective of whether the valve flow is sonic or subsonic:

$$T_{v2} = \frac{\gamma}{\gamma - 1}\frac{A^2 p_{v2}^2}{ZR_w W^2} \times \left[\sqrt{1 + 2\frac{\gamma - 1}{\gamma}\frac{ZR_w W^2}{A^2 p_{v2}^2}T_1} - 1\right] \tag{10.29}$$

The specific volume, v_{v2}, then follows from the characteristic gas equation:

$$v_{v2} = \frac{ZR_w T_{v2}}{P_{v2}} \tag{10.42}$$

The velocity at valve outlet may be found from

$$c_{v2} = \frac{v_{v2}W}{A} \tag{10.43}$$

Thus we may calculate the Mach number at valve outlet, $\mathbf{M}_{v2}$, from

$$\mathbf{M}_{v2} = \frac{c_{v2}}{\sqrt{\gamma p_{v2} v_{v2}}} \tag{10.44}$$

We may now apply the equations of Section 6.4 to the downstream pipe section, starting with the pressure in the pipe outlet, P_3:

$$p_3 = \max\left(p_4, p_{v2}\frac{p_{3c}}{p_{v2}}\right) \tag{10.31}$$

$$\frac{v_3}{v_{v2}} = \frac{1}{(\gamma - 1)\mathbf{M}_{v2}^2} \times \left[\sqrt{\left(\frac{p_3}{p_{v2}}\right)^2 + 2(\gamma - 1)\mathbf{M}_{v2}^2 + (\gamma - 1)^2\mathbf{M}_{v2}^4} - \left(\frac{p_3}{p_{v2}}\right)\right] \tag{10.32}$$

Constraint equation

All the above calculations are subject to the downstream velocity head constraint:

$$\left(\frac{2 + (\gamma - 1)\mathbf{M}_{v2}^2}{2\gamma\mathbf{M}_{v2}^2}\right)\left(1 - \frac{v_{v2}^2}{v_3^2}\right) - \frac{\gamma + 1}{\gamma}\ln\left(\frac{v_3}{v_{v2}}\right) - K_{T2} = 0 \tag{10.33}$$

The procedure requires iteration on v_{v1}/v_2 until equation (10.33) is satisfied.

10.3.2 Allowing for sonic flow in the valve using SVHIM

The full set of the above equations ceases to be valid when the flow in the valve goes sonic, since at this point there is a physical decoupling of the conditions upstream and downstream of the valve. In particular, the relationship between the ratios of throat and valve outlet pressures to valve inlet pressure given in equation (10.40) will hold no longer. Hence it will be necessary to detect the onset of sonic flow, which will occur when p_t/p_{v1}, as calculated from equation (10.39), is equal to the critical pressure ratio, $r_c = (2/(\gamma + 1))^{\gamma/(\gamma - 1)}$. We may then use the procedure outlined in Section 10.2.3. to calculate flow and conditions when the flow in the valve has become sonic. This is exactly the same procedure as used in VHIM, but now there will be no discontinuity to mark the transition from subsonic to sonic flow in the valve since equation (10.13) is the limiting case of equation (10.35).

10.4 Gas flow through an installed valve – Average Specific Volume Approximation Method (ASVAM)

This final method is based on the approximate methods for calculating compressible flow in a pipe, as described in Sections 6.8, 6.9 and Appendix 2, and in a valve, as described in Section 9.11.2. The major approximation made is that the specific volume of the gas in large stretches of pipework can be represented adequately by a notional average specific volume. The benefit of ASVAM is that it can be programmed to avoid implicit loops, and hence the associated problems of convergence in the main equations, while maintaining a very reasonable accuracy.

The equations underlying ASVAM will now be set down for the plant arrangement of Figure 10.1. First we calculate the frictional loss in velocity heads in the same way as laid down for VHIM in Section 10.2 for the upstream section of pipe, the downstream section and the valve (equations (6.58), (7.36) and (7.25)), and then find the total frictional head loss, K_T, from equation (10.1).

Having found K_T, we may estimate the critical pressure ratio, P_{3c}/P_1, that will lead to sonic flow at the exit from the downstream section of pipe from polynomial equation (6.66):

$$\frac{p_{3c}}{p_1} = \pi_0 + \pi_1(\ln K_T) + \pi_2(\ln K_T)^2 + \pi_3(\ln K_T)^3 + \pi_4(\ln K_T)^4 \qquad (6.66)$$

where the coefficients, $\pi_i = \pi_i(\gamma)$, are given for commonly encountered gases in Appendix 2. We may then determine the ratio of the downstream pipe exit pressure to the supply vessel pressure from equation (6.63), repeated below:

$$\frac{p_3}{p_1} = \max\left[\frac{p_4}{p_1}, \frac{p_{3c}}{p_1}\right] \qquad (6.63)$$

and hence also the actual pipe outlet pressure, p_3.

Having derived the ratio of supply vessel to pipe outlet pressures, p_3/p_1, we may estimate the notional average specific volume throughout pipe and valve from equation (6.62), namely:

$$v_{ave} = v_1 \frac{\gamma+1}{\gamma} \frac{\left(1 - \dfrac{p_3}{p_1}\right)}{\left(1 - \left(\dfrac{p_3}{p_1}\right)^{(\gamma+1)/\gamma}\right)} \qquad (6.62)$$

We shall use the long-pipe approximation to the compressible flow equation (cf. equation (6.65)) in order to calculate the flow before sonic conditions are reached in the valve:

$$W = b_0 C_T \sqrt{\frac{p_1 - p_3}{v_{ave}}} \qquad (10.45)$$

Here C_T is the total conductance, given by the conductance equation given in Chapter 8:

$$\frac{1}{C_T^2} = \frac{1}{C_{L1}^2} + \frac{1}{C_v^2} + \frac{1}{C_{L2}^2} \qquad (8.5)$$

in which the upstream and downstream conductances are given in terms of their respective velocity head drops by

$$C_{Li} = A\sqrt{\frac{2}{4f\dfrac{L_{eff\,i}}{D}}} = A\sqrt{\frac{2}{K_{Ti}}} \quad \text{for } i = 1, 2 \qquad (10.46)$$

An equivalent conversion between total velocity heads, K_T, and total conductance, C_T, is simply:

$$C_T = \sqrt{\frac{2}{4f\dfrac{L_{eff}}{D}}} = A\sqrt{\frac{2}{K_T}} \qquad (10.47)$$

The coefficient, b_0, is given as a function of the frictional drop, K_T, and the pressure ratio, p_4/p_1, by:

$$b_0\left(K_T, \frac{p_4}{p_1}\right) = b_0(K_T, 0.95) + m_b(K_T)\left(\frac{p_3}{p_1} - 0.95\right) \qquad (6.69)$$

where

$$b_0(K_T, 0.95) = \beta_0 + \beta_1 \ln K_T + \beta_2(\ln K_T)^2 + \beta_3(\ln K_T)^3 + \beta_4(\ln K_T)^4 \qquad (6.67)$$

$$m_b(K_T) = \mu_0 + \mu_1(\ln K_T) + \mu_2(\ln K_T)^2 + \mu_3(\ln K_T)^3 + \mu_4(\ln K_T)^4 + \mu_5(\ln K_T)^5 \qquad (6.68)$$

where the coefficients, $\beta_i = \beta_i(\gamma)$, $\mu_i = \mu_i(\gamma)$, are given for commonly encountered gases in Appendix 2. But before calculating the flow through valve and pipe we must allow for the possibility that the flow through the valve becomes sonic. Sonic flow represents the maximum mass flow that the valve will pass for any given conditions of valve inlet pressure, specific volume and valve opening:

$$W_{v\max} = C_g\sqrt{\frac{p_{v1}}{v_{v1}}} \qquad (10.48)$$

Hence there is need to find the pressure, p_{v1}, and specific volume, v_{v1}.

We calculate the valve inlet pressure, p_{v1}, by assuming that the general relationship of flow to pressure drop of equation (10.45) applies both to the whole pipeline and to the upstream pipeline. Rearranging (10.45) for the complete pipe gives:

$$W = b_0 C_T \sqrt{p_1} \sqrt{\frac{1 - \dfrac{p_3}{p_1}}{v_{ave}}} \tag{10.49}$$

while the equivalent formulation for upstream section of pipe gives:

$$W = b_{up\,0} C_{L1} \sqrt{p_1} \sqrt{\frac{1 - \dfrac{p_3}{p_1}}{v_{upave}}} \tag{10.50}$$

Here $b_{up\,0}$ is the value of the flow coefficient applicable over the upstream pipe section, and v_{upave} is the corresponding average specific volume. Solving equations (10.49) and (10.50) gives:

$$\frac{p_{v1}}{p_1} = 1 - \frac{v_{upave}}{v_{ave}} \left(\frac{b_0}{b_{up\,0}}\right)^2 \left(\frac{C_T}{C_{L1}}\right)^2 \left(1 - \frac{p_3}{p_1}\right) \tag{10.51}$$

A problem with this equation is that it is implicit, since both v_{upave} and $b_{up\,0}$ depend on p_{v1}/p_1. v_{upave} may be calculated by applying equation (6.62) to the upstream section of pipe:

$$v_{upave} = v_1 \frac{\gamma + 1}{\gamma} \frac{\left(1 - \dfrac{p_{v1}}{p_1}\right)}{\left(1 - \left(\dfrac{p_{v1}}{p_1}\right)^{(\gamma+1)/\gamma}\right)} \tag{10.52}$$

while $b_{up\,0}$ may be found by applying equations (6.67) to (6.69):

$$b_{up\,0} = b_0\left(K_{T1}, \frac{p_{v1}}{p_1}\right) = b_0(K_{T1}, 0.95) + m_b(K_{T1})\left(\frac{p_{v1}}{p_1} - 0.95\right) \tag{10.53}$$

where

$$b_0(K_{T1}, 0.95) = \beta_0 + \beta_1 \ln K_{T1} + \beta_2(\ln K_{T1})^2 + \beta_3(\ln K_{T1})^3 + \beta_4(\ln K_{T1})^4 \tag{10.54}$$

$$m_b(K_{T1}) = \mu_0 + \mu_1 \ln K_{T1} + \mu_2(\ln K_{T1})^2 + \mu_3(\ln K_{T1})^3 + \mu_4(\ln K_{T1})^4 + \mu_5(\ln K_{T1})^5 \tag{10.55}$$

However, we may make initial estimates of v_{upave} and $b_{up\,0}$ by assuming they take the values obtaining over the whole pipe and valve, i.e.:

$$v_{upave}|_1 = v_{ave} \tag{10.56}$$

$$b_{up\,0}|_1 = b_0 \tag{10.57}$$

We may then use these to derive an initial estimate, $(p_{v1}/p_1)|_1$, of p_{v1}/p_1, from equation (10.50). Then successively better estimates may be made of v_{upave} and $b_{up\,0}$ in order to improve the estimate of p_{v1}/p_1 according to the progression:

$$v_{upave}|_n = v_1 \frac{\gamma + 1}{\gamma} \frac{\left(1 - \left.\dfrac{p_{v1}}{p_1}\right|_{n-1}\right)}{\left(1 - \left(\left.\dfrac{p_{v1}}{p_1}\right|_{n-1}\right)^{(\gamma+1)/\gamma}\right)} \tag{10.58}$$

$$b_{up\,0}|_n = b_0(K_{T1}, 0.95) + m_b(K_{T1})\left(\left.\frac{p_{v1}}{p_1}\right|_{n-1} - 0.95\right) \tag{10.59}$$

$$\left.\frac{p_{v1}}{p_1}\right|_n = 1 - \frac{v_{upave}|_n}{v_{ave}} \left(\frac{b_0}{b_{up\,0}|_n}\right)^2 \left(\frac{C_T}{C_{L1}}\right)^2 \left(1 - \frac{p_3}{p_1}\right) \tag{10.60}$$

Clearly this procedure could continue with successively better estimates of the ratio p_{v1}/p_1, the flow coefficient, $b_{up\,0}$, and the upstream specific volume, v_{upave}, being generated with each iteration. The possibility of an unbounded number of iterations may be avoided if the number of iterations is fixed in advance, and usually a good degree of accuracy will be achieved by limiting n to, say, 4. At this point the specific volume at valve entry may be estimated by assuming an isentropic expansion over the upstream pipe:

$$v_{v1} = v_1 \left(\frac{p_1}{p_{v1}}\right)^{1/\gamma} \tag{10.61}$$

Having now found p_{v1} and v_{v1}, we may write down the final expression for the flow by making the calculation of equation (10.45) subject to a maximum flow defined by equation (10.48):

$$W = \max\left[C_g \sqrt{\frac{p_{v1}}{v_{v1}}}, b_0 C_T \sqrt{\frac{p_1 - p_3}{v_{ave}}}\right] \tag{10.62}$$

10.5 Example: calculation of gas flow

The example is based on that covered in Section 6.7, but with a valve inserted in the line. It will be seen that it provides a demanding test of the calculation methods, with the flow reaching sonic conditions at times, both in the valve and at the pipe outlet. The calculations will be performed first using SVHIM, since this takes most complete account of the pipe and valve characteristics and therefore provides the standard. Then VHIM and ASVAM will be applied, and their calculations compared with SVHIM.

Consider the flow of air from a large reservoir vessel at 1.101 MPa and 20°C through 10 m of horizontal Schedule 40 steel pipe (inside diameter = 52.5 mm), three standard elbows and a control valve to the atmosphere, assumed at 0.101 MPa. The length of pipe upstream of the valve is 4 m and contains one standard elbow. The pipe inlet is abrupt.

The control valve is a 2-inch globe valve, with a linear characteristic, a full-travel liquid sizing coefficient $C_V^* = 65.3$ US gall/min/psi$^{1/2}$ and a full-travel gas sizing coefficient $C_G^* = 2280$ scf/h/psia. The control valve travel is initially 100%, but this is decreased by a 5% per second ramp starting at time = 5 seconds to 5% open at time = 24 seconds. The control valve is then maintained in this position to the end of the transient at time = 30 seconds.

Calculate the mass flow and the thermodynamic conditions at various points in the pipe as functions of time.

10.5.1 SVHIM

SVHIM is considered to be the most accurate of the methods presented, and has been selected on this basis for the first analysis of the problem so as to bring out the most important features.

The pipe in the problem has the general configuration shown in Figure 10.1. We may set down the following data.

$p_1 = 1\,101\,000$ Pa, $p_4 = 101\,000$ Pa, $T_1 = 293$ K

$\gamma = 1.4$. Molecular weight = 29, hence $R_w = 8314/29 = 286.69$ J/(kg K).

Reservoir vessel specific volume, $v_1 = R_w T_1 / p_1 = 0.076294\,\text{m}^3/\text{kg}$.

Pipe cross-sectional area, $A = \pi \times (0.0525)^2/4 = 0.002165\,\text{m}^2$.

Assume that the Reynolds number is greater than 100 000 at all times, and so the Fanning friction factor may be taken as $f = 0.0045$ from equations (4.28) and (4.29). Upstream pipe section frictional loss in velocity heads, K_{T1}, is given by equation (6.58): $K_{T1} = 4 \times 0.0045 \times (4/0.0525) + 0.5 + 0.75 = 2.6214$ (dimensionless).

Downstream pipe section frictional loss:

$K_{T2} = 4 \times 0.0045 \times (6/0.0525) + 2 \times 0.75 = 3.5571$ (dimensionless).

Valve limiting gas conductance at full valve travel:

$C_G = 4.02195 \times 10^{-7} C_G^* = 0.000917\,\text{m}^2$.

Valve liquid conductance at full valve travel: $C_V = 2.3837 \times 10^{-5} C_V^* = 0.0015566\,\text{m}^2$.

Valve friction coefficient, C_{fgh}, (independent of travel for a globe valve):

$C_{FGH} = (2.065 C_G / C_V) = 1.2162$ (dimensionless).

Since $C_{FGH}^2 > 0.9$, $B = C_{FGH}^{0.75} = 1.158$ in equation (10.41).

Valve critical pressure ratio, $r_c = (2/(\gamma+1))^{\gamma/(\gamma-1)} = 0.5283$

Linear valve, so valve opening, $y = x$, valve travel.

Valve limiting gas conductance at valve travel, x, $C_g = y C_G$.

Valve effective throat area (needed for calculation of velocity in valve, given the flow rate and specific volume) at fully open follows from equation (9.15) as:

$A_T = 1.4604 C_G (1.4) = 0.001339\,\text{m}^2$.

The calculations laid out in Section 10.3 were carried out based on the data above, and the results are summarized in the four graphs, Figures 10.2 to 10.5.

The introduction of the valve into the pipe described in Section 6.7 has decreased the flow rate, even at fully open, from the value 2.626 kg/s calculated in Section 6.7 to 2.28 kg/s now. However, the flow at the outlet from the pipe is still sonic, and hence the pressure just inside the pipe outlet, p_3, is greater than the atmospheric pressure that exists just outside the pipe, p_4. As the valve is closed, however, the pressure drop across it increases, until, 17 seconds into the transient, the throat to inlet pressure ratio falls to the critical value needed for sonic flow in the valve. At this point we have the interesting phenomenon that the flow is sonic in the valve throat, then reduces to subsonic at the valve outlet, only to accelerate to sonic velocity at the pipe outlet. This is shown most clearly in Figure 10.3, which plots the Mach numbers at various points in the pipe. For about 3 seconds in the middle of the transient, the Mach number at the throat of the valve is equal to unity, as is the Mach number at the pipe outlet:

$\mathbf{M}_t = \mathbf{M}_3 = 1$.

Then as the flow is further constricted, the pressure at the pipe outlet, p_3, falls below the critical pressure and subsonic flow becomes established at that point.

It is shown in Figure 10.4 that the valve inlet pressure, p_{v1}, rises, as the flow falls, towards the pipe inlet pressure, p_2, which, in its turn, approaches the pressure in the upstream vessel, p_1. The valve outlet pressure, p_{v2}, falls towards the pipe outlet pressure, p_3, which becomes equal to the atmospheric discharge

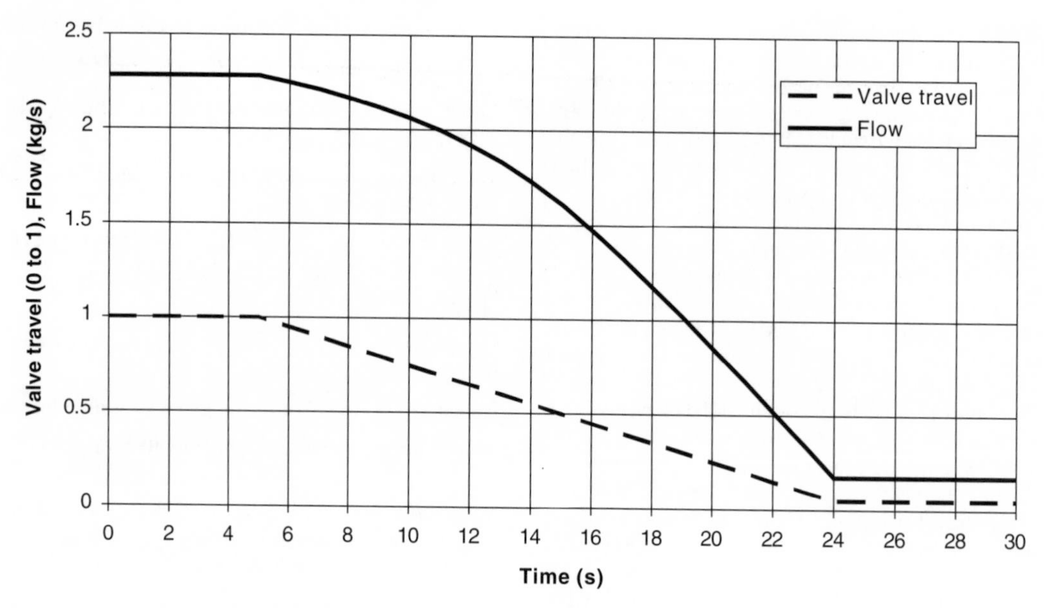

Figure 10.2 Mass flow transient (SVHIM).

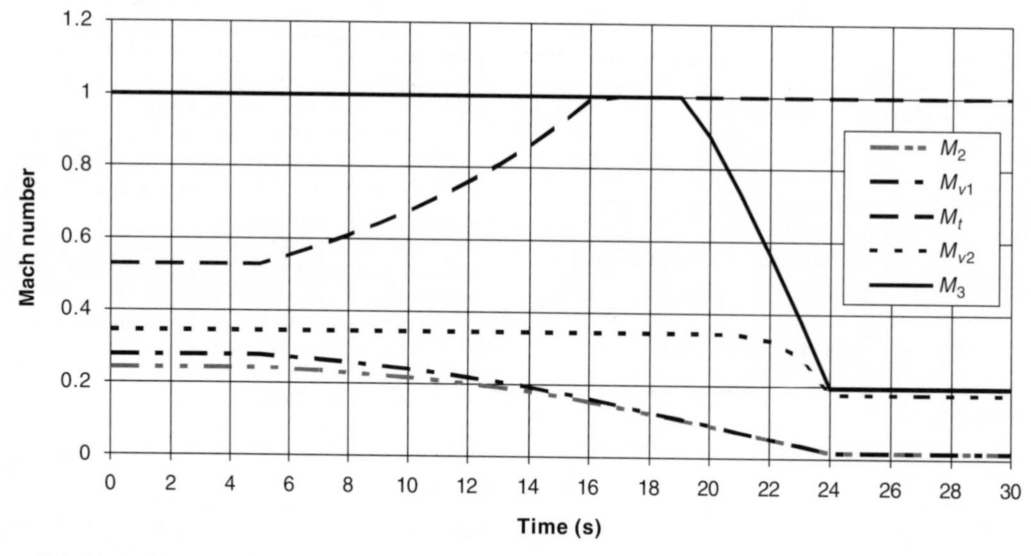

Figure 10.3 Mach numbers at various locations in the pipe (SVHIM).

pressure, p_4. Hence almost the whole pressure drop occurs across the valve at low valve openings.

Once the flow in the valve has gone sonic, the flow is given by (see equation (10.10)):

$$W = yC_G\sqrt{\frac{p_{v1}}{v_{v1}}}$$

The variations in p_{v1} and v_{v1} are small, so this is almost linear in valve opening, y. Therefore only a small degree of curvature in the flow transient may be observed in Figure 10.2 after sonic flow has begun.

The temperature transients are plotted in Figure 10.5, with both the pipe outlet temperature and the valve throat temperature undergoing large temperature swings. In each case the lowest temperature (about −30°C) is associated with sonic air flow.

The average temperature through pipe and valve at the end of the transient is approximately the same as

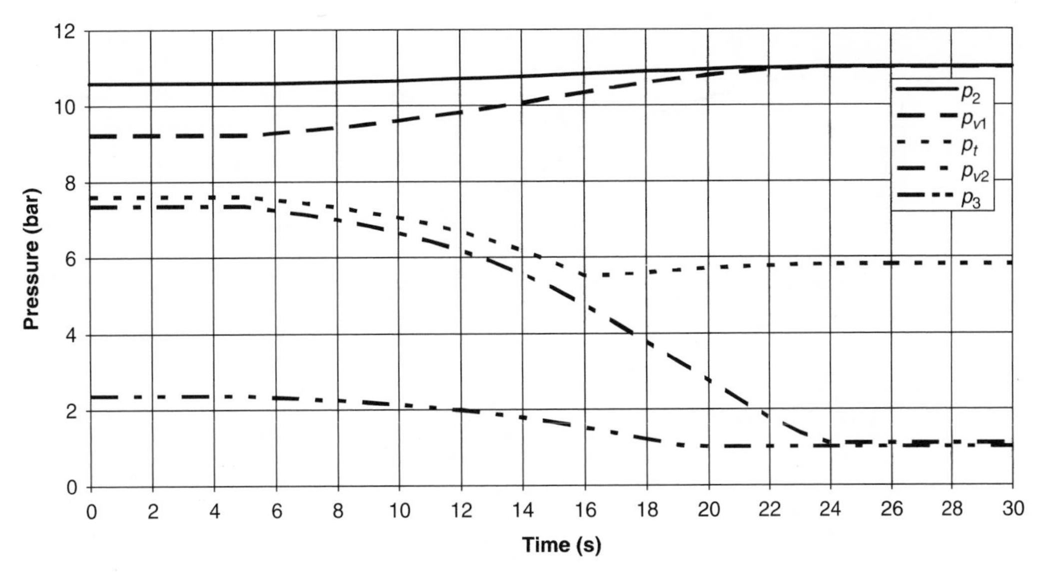

Figure 10.4 Pressures at various locations in the pipe (SVHIM).

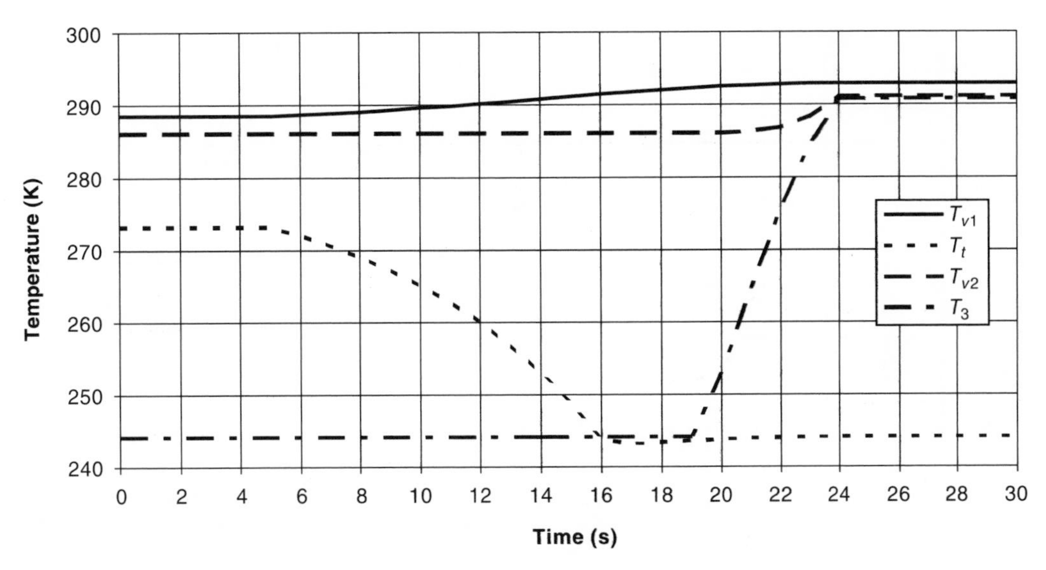

Figure 10.5 Temperatures at various locations in the pipe (SVHIM).

it was at the beginning: ~270 K. Hence the viscosity is $\mu \approx 1.7 \times 10^{-5}$ Pa s. The Reynolds number is given by equation (4.26) as

$$N_{RE} = \frac{WD}{A\mu} = \frac{0.1740 \times 0.0525}{0.002165 \times 1.7 \times 10^{-5}}$$

$$= 2.48 \times 10^5$$

This is greater than 100 000, and thus confirms that the Fanning friction factor, f, may be regarded as constant and equal to 0.0045 throughout the transient.

10.5.2 VHIM

The method relies on characterizing the pipe and the subsonic valve by frictional losses in velocity heads. The valve frictional loss is given from equations (7.25) and (7.36) as

$$K_v = \frac{1}{y^2}\frac{\pi^2 D^4}{8C_V^2} = \frac{1}{y^2}\frac{\pi^2(0.0525)^4}{8(0.0015566)^2} = \frac{3.8683}{y^2}$$

and the inverse dependence on the square of the valve opening means that it increases by more than two

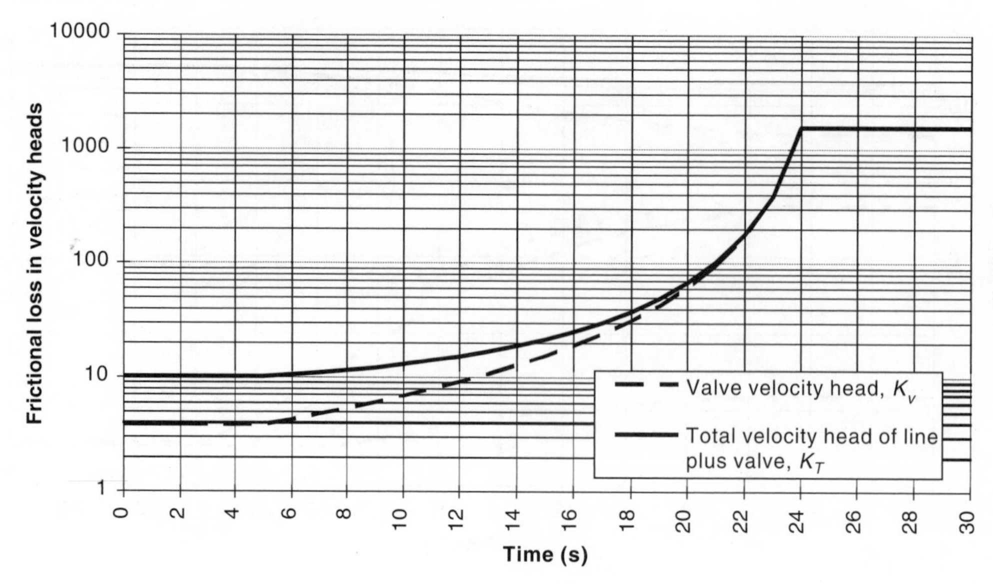

Figure 10.6 Frictional loss in velocity heads (VHIM).

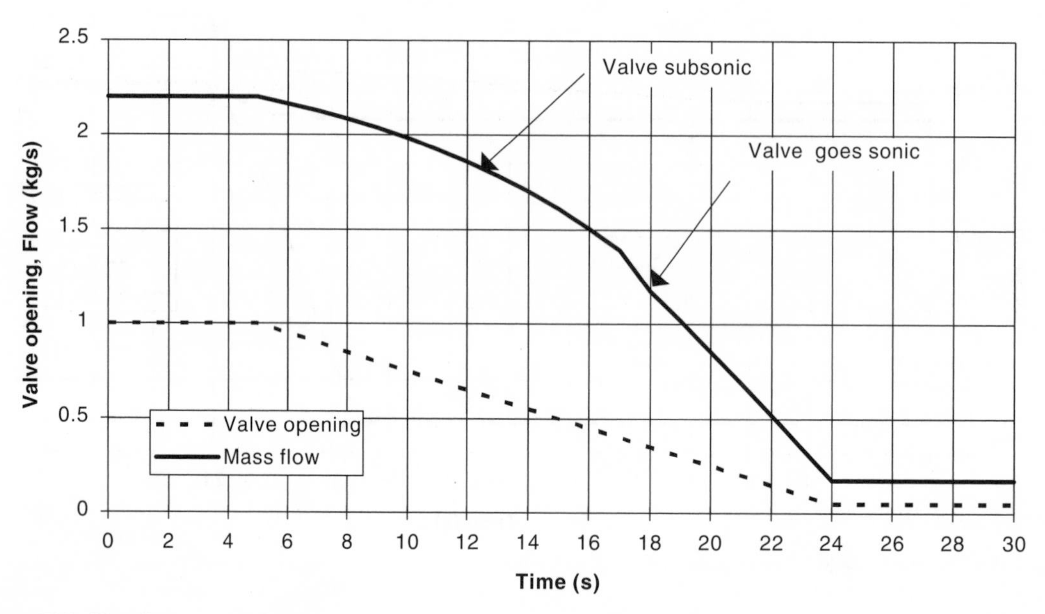

Figure 10.7 Mass flow transient (VHIM).

orders of magnitude over the transient, as shown in Figure 10.6. The figure also shows how the frictional loss through the valve begins at less than 40% of the total loss when the valve is fully open, but has reached nearly 100% of the total loss when the valve nears closure.

VHIM calculates the valve throat to valve inlet pressure ratio to be greater than the critical value until 17 seconds into the transient, indicating that the flow through the valve is subsonic to this time. But this pressure ratio then falls below critical, indicating choked flow in the valve, and necessitating the change in the calculational method described in Section 10.2.3 for sonic flow under VHIM. Graph 10.7 shows the mass flow transient.

It is noticeable that a small discontinuity is introduced between 17 and 18 seconds, caused by the

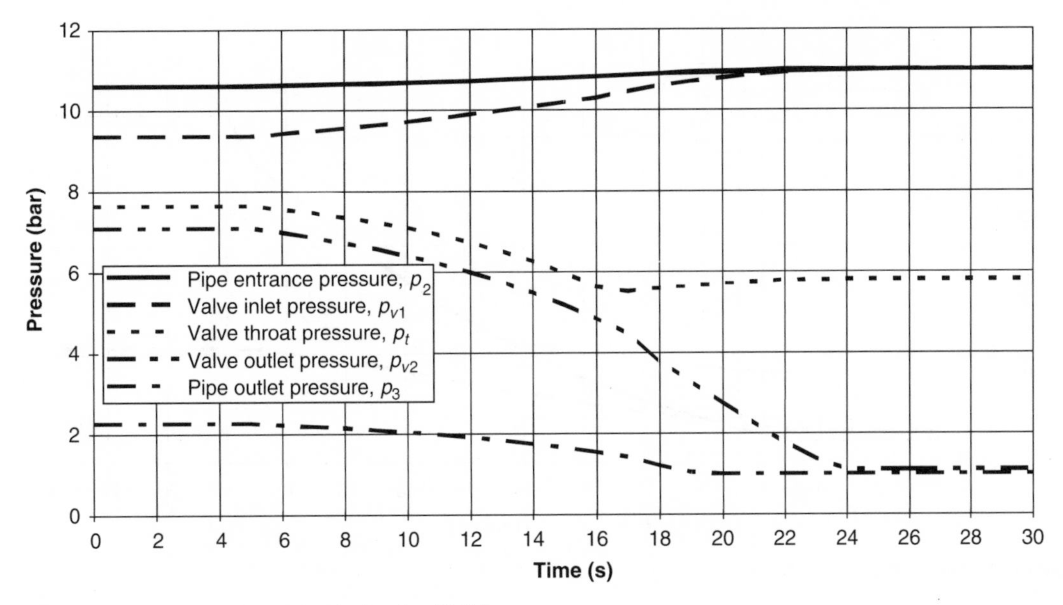

Figure 10.8 Pressures at various points in the pipe (VHIM).

change in the characterization of the valve capacity from the liquid coefficient, C_v, to the gas coefficient, C_g. However, the absolute value of the mass flow stays very close to that calculated by SVHIM throughout the transient. The largest difference occurs when the valve is fully open, when VHIM calculates 2.2 kg/s as opposed to the 2.28 kg/s calculated by SVHIM.

The pressure transient calculated by VHIM is shown in Figure 10.8, which matches closely that plotted in Figure 10.4 for SVHIM. In particular, it will be seen that the pipe outlet pressure, p_3, has reduced to atmospheric, p_4, by 20 seconds, in agreement with SVHIM, indicating that the outlet flow is subsonic at this time.

10.5.3 ASVAM

Applying this approximate method required the estimation of suitable values for the flow coefficient, b_0, which is a function of the total frictional loss in velocity heads, K_T, and the pressure ratio, p_4/p_1. The polynomial approximations of equations (6.66) to (6.69), used with the appropriate polynomial constants from Appendix 2, gave the values of b_0 shown in Figure 10.9. As can be seen, the variation over the transient is not great, even though K_T varies by a factor of over 150 through the transient, as has been noted already. The flow coefficient, $b_{up\,0}$, applicable to the upstream section of the pipe is also shown in Figure 10.9. $b_{up\,0}$ depends on the frictional loss, K_{T1}, and the pressure ratio, p_{v1}/p_1 over the upstream section. At the beginning of the transient, $b_{up\,0} \approx b_0$ although the two diverge as K_T increases substantially with time.

The resulting flow transient calculated by ASVAM is shown in Figure 10.10 compared with that of VHIM. The flows are almost identical over the subsonic region up to time = 16 seconds, and come back together again after 18 seconds. It is noticeable that the discontinuity between subsonic valve flow and sonic valve flow that characterizes VHIM disappears under ASVAM. This is because of the transition in ASVAM takes a very simple form, namely the maximum selection of equation (10.62).

Figure 10.11 compares the flow transient calculated by ASVAM with the standard transient calculated by SVHIM. ASVAM, like VHIM, relies implicitly on the C_v value to characterize valve flow in the subsonic region via the calculation of K_v and then K_T. As a result, ASVAM underestimates the flow by about 3% at the beginning of the transient in the same way as VHIM. But ASVAM produces essentially the same value as SVHIM for flow by time = 15 seconds. In fact, ASVAM predicts sonic flow in the valve at time = 16 seconds, a second in advance of SVHIM, but the difference in flow is very small.

The valve inlet pressure, p_{v1}, has been calculated on the basis of the fourth estimate of p_{v1}/p_1, $n = 4$ in equation (10.60). The valve inlet pressure, p_{v1}, and the valve outlet pressure, p_3, calculated under ASVAM trace a very similar path to that shown

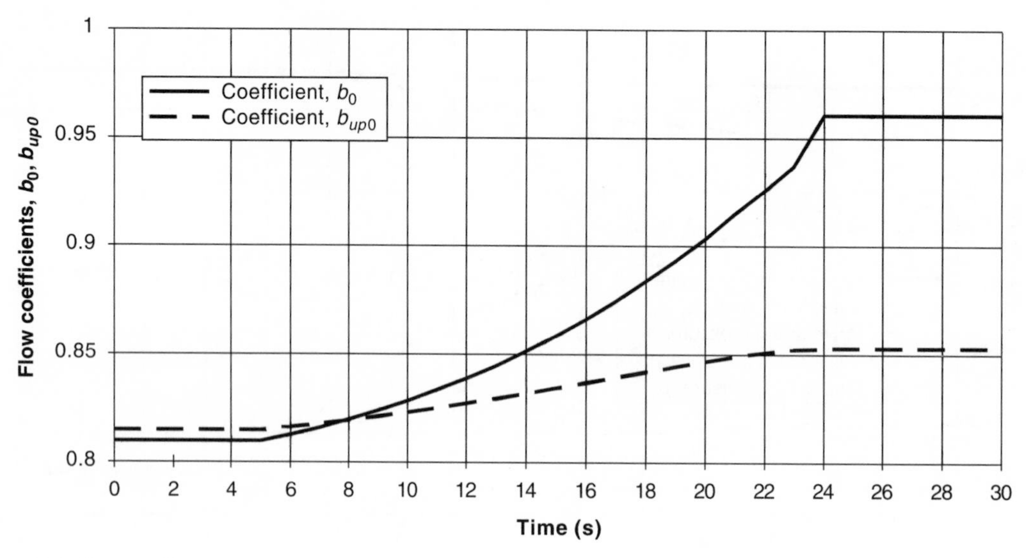

Figure 10.9 Flow coefficient, b_0 (ASVAM).

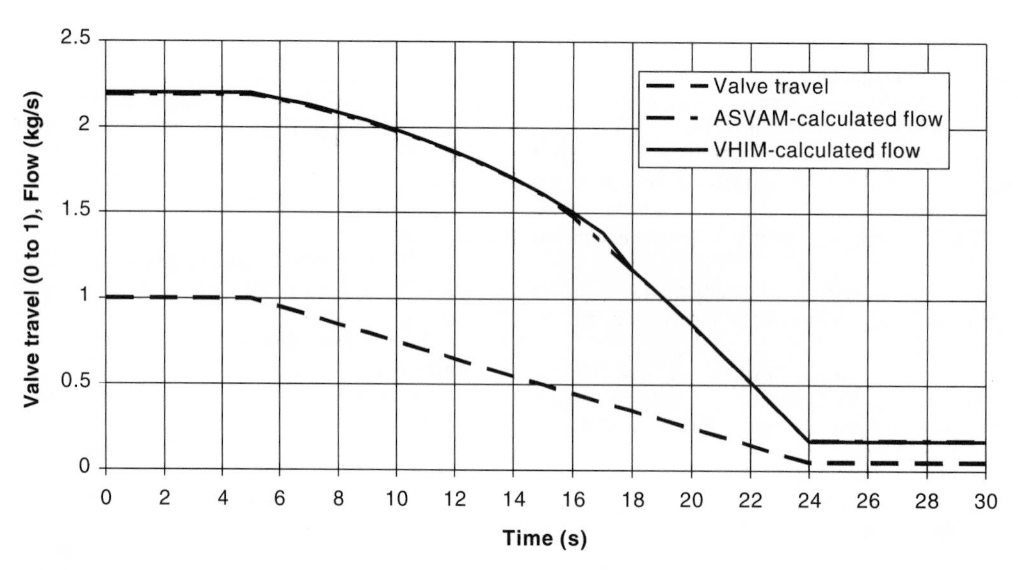

Figure 10.10 Mass flow transient, ASVAM compared with VHIM.

in Figure 10.4 for the more accurate SVHIM (see Figure 10.12). The transient for pipe outlet pressure, p_3, is correctly predicted in general terms through using the polynomial correlation of critical pressure ratio p_{3c}/p_1 with total frictional loss in velocity heads, K_T, in equation (6.6). In particular, the pipe outlet pressure, p_3, is calculated to have reduced to atmospheric, p_4, by 20 seconds, which indicates that the outlet flow is subsonic at this time, in agreement with SVHIM.

10.5.4 Simplified Average Specific Volume Method with constant b_0: SASVAM

The relatively small variation in the flow coefficient, b_0, over the transient, from 0.81 to 0.96 (see Figure 10.9) prompts the question whether it would be possible to assume a constant, mid-range value and maintain reasonable accuracy. The answer is yes, and such a value may be calculated either as the average value of b_0 in the transient, found from Figure 10.9,

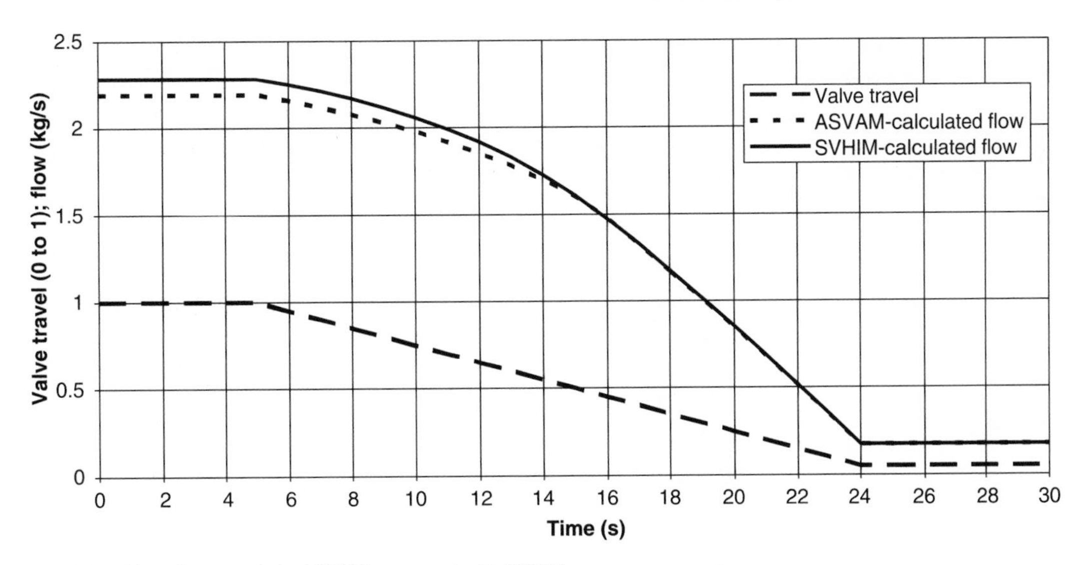

Figure 10.11 Mass flow transient, ASVAM compared with SVHIM.

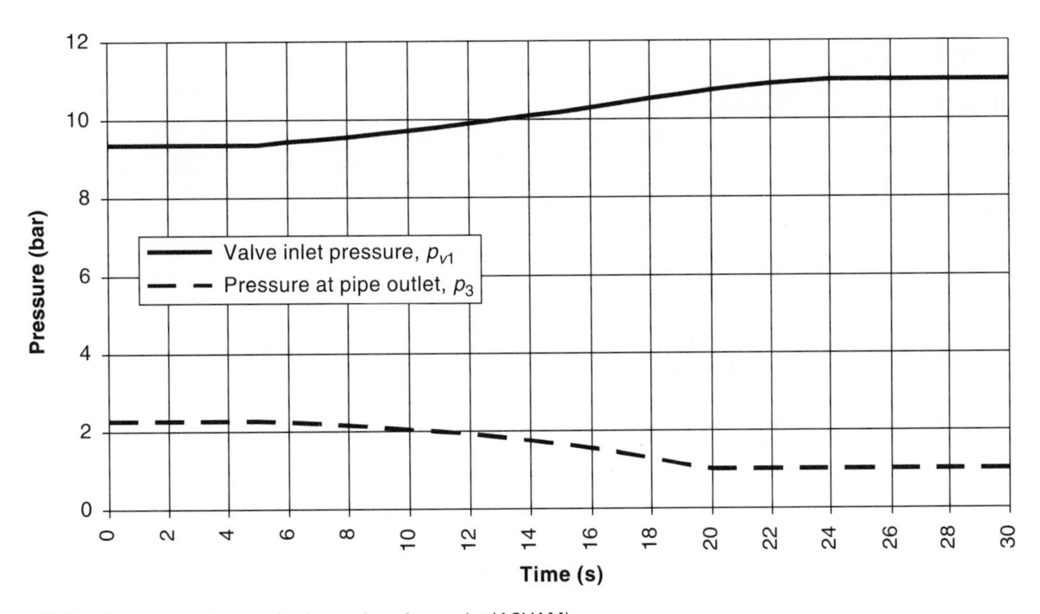

Figure 10.12 Pressure upstream of valve and at pipe outlet (ASVAM).

or from a reference or 'flowsheet' set of accurately calculated values.

The latter course has been adopted in this example. The following equations may be used to calculate a fixed value of b_0 (see Section 6.8):

$$\bar{b}_0 = \frac{\overline{W}}{\overline{C}_T}\sqrt{\frac{\overline{v}_{ave}}{\overline{p}_1 - \overline{p}_3 - \dfrac{g(\overline{z}_3 - \overline{z}_1)}{\overline{v}_{ave}}}} \tag{10.63}$$

(cf. equation (6.60)), where

$$\overline{v}_{ave} = \overline{v}_1 \frac{\gamma + 1}{\gamma} \frac{\left(1 - \dfrac{\overline{p}_3}{\overline{p}_1}\right)}{\left(1 - \left(\dfrac{\overline{p}_3}{\overline{p}_1}\right)^{(\gamma+1)/\gamma}\right)} \tag{10.64}$$

(cf. equation (6.62)).

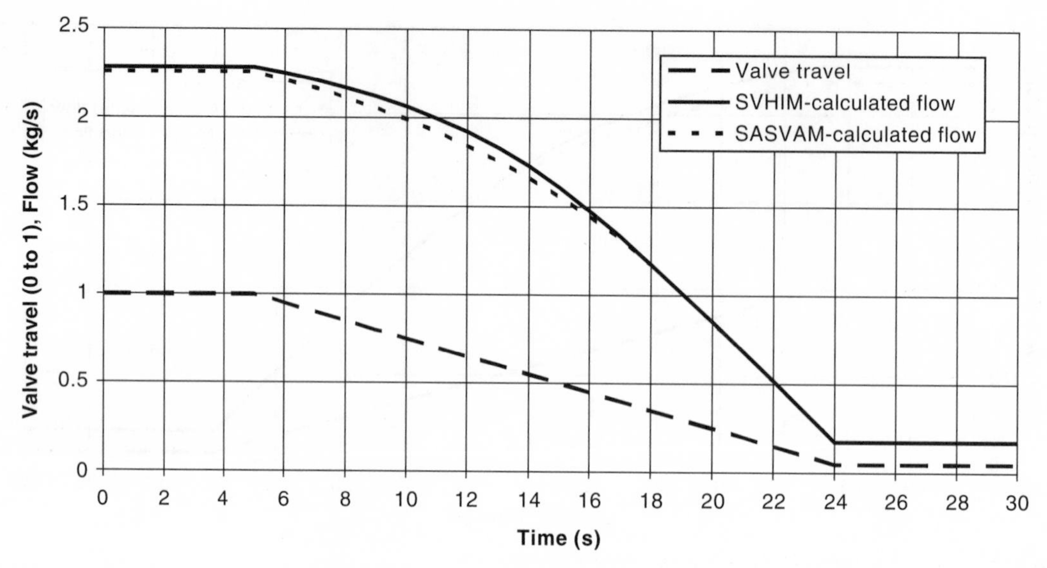

Figure 10.13 Mass flow transient, SASVAM compared with SVHIM.

Note that the calculation of the valve inlet pressure, p_{v1}, needed for determining sonic flow, is made on the basis of the first estimate of the ratio p_{v1}/p_1.

We shall assume that the values calculated by SVHIM for a 70% valve travel are available, namely:

$$\overline{p}_3 = 200124\,\text{Pa}$$

$$\overline{W} = 1.91971\,\text{kg/s}$$

$$\overline{C}_T = 0.000816\,\text{m}^2$$

The average specific volume at this condition is found to be $\overline{v}_{ave} = 0.1131\,\text{m}^3/\text{kg}$, and the value of b_0 is $\overline{b}_0 = 0.8333$. Replacing the variable value of b_0 with this constant value in ASVAM, but making no other changes, produces the mass flow transient shown in Figure 10.13.

The agreement with the standard case of SVHIM is generally very good. The flow when the valve is fully open is 2.26 kg/s, within 1% of the SVHIM value. SASVAM predicts the onset of sonic flow in the valve at 16 seconds, which is close to the SVHIM value of just after 17 seconds.

10.6 Discussion

The example of Section 10.5 presented a severe test for the methods of calculating gas flow through pipe and valve, since the overall pressure ratio was greater than 10:1, causing choking to occur first in the downstream pipe, then in both valve and downstream pipe, and finally in the valve alone, as the flow was reduced by a factor of more than 13. All the methods catered effectively for sonic flow occurring in the downstream pipe and in the valve.

SVHIM provides the most complete treatment of both valve and pipe, and is considered to have given the most accurate calculation of flow and thermodynamic conditions. However, this method requires an iterative solution, which is a disadvantage in a dynamic simulation, where the flow is likely to need calculating many hundreds or thousands of times, possibly over a wide range of conditions. Non-convergence is an ever-present danger for which the modeller must always be on his guard when the solution is iterative.

VHIM is based on a simple transfer to the pipe-plus-valve case of the method outlined in Section 6.4 for calculating flow in a pipe. It will be less accurate than SVHIM in the subsonic-valve region because only the liquid valve coefficient, C_v, is used in this flow regime, rather than the more representative gas coefficient, C_g. This causes a small discontinuity to occur when sonic flow conditions are met in the valve, and the C_v characterization is superseded by a characterization based on C_g. The loss in accuracy compared with SVHIM is 3% or less, but VHIM retains the disadvantage that it requires an iterative solution.

ASVAM is an essentially explicit approximation to VHIM. ASVAM benefits from the extensive off-line computations carried out to define both the critical pressure ratio, p_{3c}/p_1, as a function of the frictional loss in velocity heads, K_T, and also the shape of the b_0 versus K_T and p_4/p_1 surface. It is much quicker and less complicated than VHIM as a result, while remaining just about as accurate, always within about a percentage point or so of SVHIM. It suffers the same lack of accuracy as VHIM in the subsonic-valve region for identical reasons, but the transition to sonic flow in the valve is handled more smoothly because of the simple form of the transition equation.

SASVAM benefits from the previously generated function of p_{3c}/p_1 versus K_T, but uses a central, constant value of b_0. Obviously the accuracy of this method will depend on the extent of the variation in the 'true' value of b_0 over the transient. In general, one might expect SASVAM applied with a constant value, $\bar{b}_0$, that is well-selected, to give an accuracy of 5% or better, depending on the size and variation in both K_T and p_4/p_1. Large values of K_T and small values of p_4/p_1 favour higher accuracies.

All the accuracy figures above have been quoted against the standard of SVHIM, and these may be put in context by the consideration that no method can guarantee to calculate the flow on real plant to better than about 3%. Taking this into account, the example calculated would appear to suggest that there is not much to choose between VHIM, SVHIM or ASVAM on accuracy grounds. Nevertheless, of the iterative methods, we do expect SVHIM's use of the C_g value as well as the C_v value in the subsonic regime to make it somewhat more accurate than VHIM. Further, SVHIM introduces no discontinuity in the transition to sonic flow in the valve.

ASVAM has the distinct advantage that it avoids convergence problems, since its only iteration occurs in a minor loop, where the number of passes may be fixed in advance. ASVAM is therefore a very attractive option for the modeller.

SASVAM, where b_0 is given a constant value, may well carry only a small accuracy penalty if the ranges of K_T and p_4/p_1 over the transient are relatively restricted. It is the simplest of the four methods considered.

11 Accumulation of liquids and gases in process vessels

11.1 Introduction

In this chapter we will apply to typical process vessels the equations for the conservation of mass and of energy that were derived in Chapter 3, Sections 3.4, 3.5 and 3.6. We will begin with the simplest system, namely accumulation of liquid at a constant temperature, and build up to consider more complicated systems where both a liquid and a gas are present, and where the temperature of each phase varies. In each case a full set of equations will be developed, and the solution cycle will be outlined.

11.2 Accumulation of liquid in an open vessel at constant temperature

A version of such a system has already been covered in the example described in Chapter 2, Section 2.2. The accumulation of liquid in an open system at a constant temperature is about the simplest system one can expect to meet on a process plant, and Figure 11.1 sets out the basic features.

Figure 11.1 shows an open vessel being fed with W_1 kg/s of liquid at ambient temperature from an upstream source at pressure p_1 Pa and specific volume v_1 m^3/kg. The vessel is discharging W_2 kg/s to a downstream point at pressure p_2 Pa.

Atmospheric pressure is denoted p_{at} Pa. The vessel contains m kg of liquid, with a volume V m^3.

Since the whole system is at ambient temperature, the specific volume of the vessel contents will be the same as the specific volume of the upstream source: $v = v_1$. The volume of the vessel contents will be related to their mass by:

$$V = mv \tag{11.1}$$

The level, l, in the vessel may be found by dividing the volume by the base cross-sectional area, A, of the vessel:

$$l = \frac{V}{A} \tag{11.2}$$

The pressure at the inlet and outlet of the vessel is then

$$p = p_{at} + \frac{gl}{v} \tag{11.3}$$

The inflow will depend on the upstream pressure and specific volume, p_1 and v_1, and on the pressure at the entry to the vessel, p. For a horizontal pipe containing a valve, the flow will be given by

$$W_1 = C_{T1}\sqrt{\frac{p_1 - p}{v_1}} \tag{11.4}$$

where C_{T1} is the total conductance in m^2 of the upstream line and valve, assumed given. See Chapter 8, equations (8.4) and (8.5). (It has been assumed that the flow through the valve is not choked: if choking is a possibility, then equations (8.17) and (8.18) need to be used.)

In a similar way, the outflow will be given by:

$$W_2 = C_{T2}\sqrt{\frac{p - p_2}{v}} \tag{11.5}$$

where C_{T2} is the combined conductance (m^2) of the downstream line and valve, again assumed given.

The mass of liquid in the vessel will obey the mass conservation equation (3.28):

$$\frac{dm}{dt} = W_1 - W_2 \tag{11.6}$$

We now have in equations (11.1) to (11.6) a full set of equations describing this simple system. The equations may now be integrated numerically to produce values of $m(t)$ for all future times, t, according to the following cycle. From the initial mass, $m(0)$, the initial volume and level of the vessel contents may be found, allowing the pressure at the base of the vessel to be determined. Once this pressure has been found, the inflow and the outflow may be calculated, and equation (11.6) integrated one timestep to produce the next value of the mass, m. This allows a recalculation of the vessel variables at this later time instant, which enables the integration procedure to march forward.

11.3 Accumulation of gas in a vessel at constant temperature

This is the equivalent gas system to the liquid system analysed above. Figure 11.2 shows a well-insulated

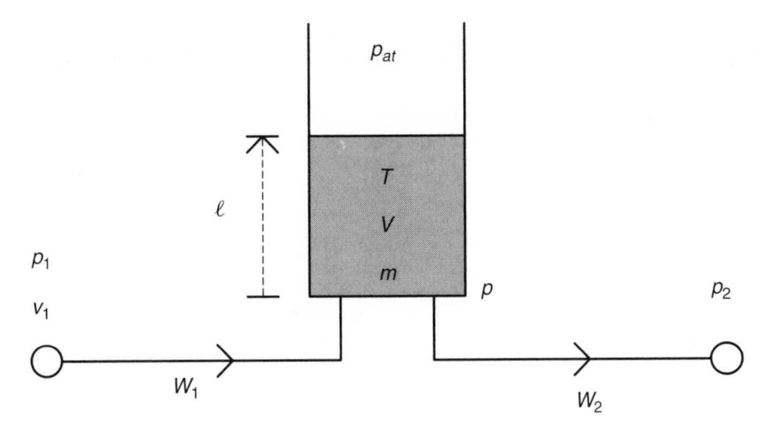

Figure 11.1 Liquid accumulation in an open vessel at constant temperature.

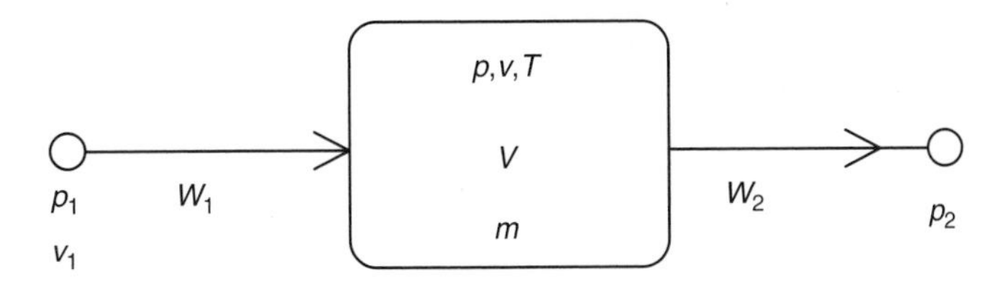

Figure 11.2 Gas vessel fed with inlet and outlet streams.

vessel of volume, V m^3, containing m kg of gas at temperature T. The vessel receives an inlet stream of W_1 kg/s of gas at the same temperature and discharges a flow of W_2 kg/s.

The specific volume may be calculated directly from

$$v = \frac{V}{m} \tag{11.7}$$

while the vessel pressure will obey the equation of state given in Chapter 3:

$$pv = \frac{ZRT}{w} \tag{3.2}$$

Hence the pressure is given by

$$p = \frac{mZRT}{wV} \tag{11.8}$$

Following the dimensional analysis of Appendix 2, Section A2.2, the mass flow, W_1, may be written generally as

$$W_1 = f_{flow}\left(p_1, v_1, \frac{p}{p_1}, A_1, K_{T1}, \gamma\right) \tag{11.9}$$

where

p_1 is the upstream pressure (Pa),
v_1 is the upstream specific volume (m^3/kg),
p is the pressure in the vessel (Pa),
A_1 is the cross-sectional area of the upstream pipe (m^2),
K_{T1} is the number of velocity heads characterizing the frictional drop over the upstream pipe (dimensionless); by equation (6.61), K_{T1} is related to the upstream conductance, C_{L1} (m^2), by $K_{T1} = 2A_1^2/C_{L1}^2$,
and γ is the specific-heat ratio of the gas (dimensionless).

Specific realizations of the compressible flow function, f_{flow}, are given in Chapters 6 and 10.

The vessel pressure constitutes the upstream pressure for the discharge flow, and so W_2 will have the form:

$$W_2 = f_{flow}\left(p, v, \frac{p_2}{p}, A_2, K_{T2}, \gamma\right) \tag{11.10}$$

The mass of gas in the vessel will obey the same differential equation as derived in the previous section, namely

$$\frac{dm}{dt} = W_1 - W_2 \tag{11.11}$$

The system may be regarded as 'solved' as long as the constants, T, γ, A_1 and A_2 are given and provided the following boundary conditions are also known at each timestep:

$$p_1, v_1, p_2, K_{T1}, K_{T2}$$

The initial value $m(0)$ allows the initial values $v(0)$ and $p(0)$ to be calculated, which in turn allow $W_1(0)$ and $W_2(0)$ to be calculated; knowledge of $W_1(0)$ and $W_2(0)$ allows numerical integration of equation (11.11), producing a later value of m, and thus allowing the cyclic process of solution to continue indefinitely.

11.4 Use of kilogram-moles in modelling the accumulation of a mixture of gases

The modelling of the gas accumulation process above has used kilograms as the unit. However, it is also possible to use the kilogram-mole (abbreviation: kmol) as the unit of accumulation, which is often more convenient when a mixture of gases is being considered.

The mass in kmol, M, is found by dividing the mass in kg, m, by the molecular weight, w:

$$M = \frac{m}{w} \tag{11.12}$$

Substituting this relationship into equation (11.8) gives

$$p = \frac{ZRTM}{V} \tag{11.13}$$

Equation (11.13) indicates that the pressure inside a vessel of given volume, V, and temperature, T, depends only on the mass in kilogram-moles, *irrespective of how many gases are present*, provided the compressibility factor, Z, is the same for all those gases. In fact, the compressibility factor will be close to unity for a large number of gases at moderately low pressures. If we assume that the gas contents of the vessel is made up from a number, n, of gases, each contributing a mass of M_i kmol, then the total mass in kmol is given by

$$M = M_1 + M_2 + \cdots + M_i + \cdots M_n = \sum_{i=1}^{n} M_i \tag{11.14}$$

where

$$M_i = \frac{m_i}{w_i} \tag{11.15}$$

Substituting for M from equation (11.14) into equation (11.13) gives:

$$p = \frac{ZRT}{V}M = \frac{ZRT}{V}M_1 + \frac{ZRT}{V}M_2 + \cdots + \frac{ZRT}{V}M_i + \cdots \frac{ZRT}{V}M_n \tag{11.16}$$

Thus the pressure may be considered to be made up of the sum of all the partial pressures:

$$p = p_1 + p_2 + \cdots + p_i + \cdots + p_n \tag{11.17}$$

where the partial pressure corresponding to the ith gas is defined as

$$p_i = \frac{ZRT}{V}M_i \tag{11.18}$$

Equation (11.17) is a statement of Dalton's law on partial pressures. A rearrangement of equation (11.16) gives the alternative formulation:

$$\begin{aligned} V &= \frac{ZRT}{p}M = \frac{ZRT}{p}M_1 + \frac{ZRT}{p}M_2 + \cdots \\ &\quad + \frac{ZRT}{p}M_i + \cdots \frac{ZRT}{p}M_n \\ &= V_1 + V_2 + \cdots + V_i + \cdots + V_n \end{aligned} \tag{11.19}$$

indicating that the volume may be considered to be the sum of the partial volumes, where the partial volume V_i is defined as

$$V_i = \frac{ZRT}{p}M_i \tag{11.20}$$

The partial volume represents the volume that M_i kmol of component i would occupy alone when kept at temperature T and pressure p. If we divide equation (11.20) by the total volume, $V = (ZRT/p)M$, we achieve an expression for the molar fraction, λ_i:

$$\lambda_i = \frac{M_i}{M} = \frac{V_i}{V} \tag{11.21}$$

It may be noted that the molar fraction is identical to the volume fraction for *a mixture of gases with the same compressibility factor*.

The mass fraction, μ_i, is the ratio of the mass of component i to the total mass of gas:

$$\mu_i = \frac{m_i}{m} \tag{11.22}$$

We may relate the mass fraction to the molar fraction using the following transformations based on equations (11.15) and (11.21):

$$\begin{aligned} \mu_i &= \frac{m_i}{m} = \frac{m_i}{\sum_{i=1}^{n} m_i} = \frac{M_i w_i}{\sum_{i=1}^{n} M_i w_i} \\ &= \frac{\frac{M_i}{M} w_i}{\sum_{i=1}^{n} \frac{M_i}{M} w_i} = \frac{\lambda_i w_i}{\sum_{i=1}^{n} \lambda_i w_i} \end{aligned} \tag{11.23}$$

Conversely, we may express the molar fraction in terms of the mass fraction using the transformations below based on equations (11.15) and (11.22):

$$\lambda_i = \frac{M_i}{M} = \frac{M_i}{\sum_{i=1}^{n} M_i} = \frac{\dfrac{m_i}{w_i}}{\sum_{i=1}^{n} \dfrac{m_i}{w_i}} = \frac{\dfrac{m_i}{m}\dfrac{1}{w_i}}{\sum_{i=1}^{n} \dfrac{m_i}{m}\dfrac{1}{w_i}} = \frac{\dfrac{\mu_i}{w_i}}{\sum_{i=1}^{n} \dfrac{\mu_i}{w_i}} \tag{11.24}$$

The effective molecular weight, w, of the mixture of gases in the vessel is found by combining equations (11.12), (11.14) and (11.15):

$$\frac{m}{w} = M = \sum_{i=1}^{n} M_i = \sum_{i=1}^{n} \frac{m_i}{w_i} \tag{11.25}$$

so that, dividing by the mass in kg, m, gives:

$$w = \frac{1}{\sum_{i=1}^{n} \dfrac{\mu_i}{w_i}} \tag{11.26}$$

An alternative but equivalent formulation in terms of the molar fraction, λ_i, may be derived from the same equations:

$$Mw = m = \sum_{i=1}^{n} m_i = \sum_{i=1}^{n} M_i w_i \tag{11.27}$$

Hence, dividing throughout by the mass in kmol, M, yields

$$w = \sum_{i=1}^{n} \lambda_i w_i \tag{11.28}$$

The effective molecular weight, w, represents the mass in kilograms of one kilogram-mole of the mixture of gases.

Since the kilogram-mole of a general substance has a mass of 1 kg multiplied by the molecular weight, it follows that any specific property of the substance expressed per kilogram will need to be multiplied by the molecular weight if it is to be re-expressed on a per kilogram-mole basis. Hence, for example, the enthalpy per kilogram-mole, $\hat{h}$, J/kmol, will be the specific enthalpy in J/kg, h, multiplied by the molecular weight:

$$\hat{h} = wh \tag{11.29}$$

Similarly for specific internal energy, specific entropy and specific volume:

$$\begin{aligned} \hat{u} &= wu \\ \hat{s} &= ws \\ \hat{v} &= wv \end{aligned} \tag{11.30}$$

where $\hat{u}$ is the specific internal energy in J/kmol, while u is in J/kg, $\hat{s}$ is the specific entropy in J/(kmol K), while s is in J/(kg K) and $\hat{v}$ is the specific volume in m^3/kmol, while v is in m^3/kg.

For a mixture of gases, each of which has the same compressibility factor, Z, the specific volume of the mixture may be found from equation (3.2), where the molecular weight, w, is taken as the effective molecular weight, as found from (11.26) or (11.28):

$$v = \frac{ZRT}{pw} \tag{11.31}$$

Given the mass fraction, μ_i, of a flow, we may calculate the ith component's mass flow in kg/s, W_i, from the total flow, W, as

$$W_i = \mu_i W \tag{11.32}$$

The flow in kmol/s of component i, F_i, is therefore

$$F_i = \frac{W_i}{w_i} = \frac{\mu_i}{w_i} W \tag{11.33}$$

The form of equation (11.9) for gas flow requires that the specific-heat ratio, γ, be available. It is possible to calculate an effective specific-heat ratio for a gas-mixture as follows. The specific heat at constant pressure for the mixture, c_p, will obey the equation:

$$mc_p = \sum_i m_i c_{pi} \tag{11.34}$$

so that

$$c_p = \sum_i \frac{m_i}{m} c_{pi} = \sum_i \mu_i c_{pi} \tag{11.35}$$

Similarly, the specific heat at constant volume for the mixture, c_v, will be

$$c_v = \sum_i \mu_i c_{vi} \tag{11.36}$$

Hence the specific-heat ratio for the mixture, γ, will be given by:

$$\gamma = \frac{c_p}{c_v} = \frac{\sum_i \mu_i c_{pi}}{\sum_i \mu_i c_{vi}} \tag{11.37}$$

The specific heats for near-ideal gases are given in terms of the number of degrees of freedom, molecular weights and compressibility factor by equations (3.21) and (3.23). Thus equation (11.37) becomes, for near-ideal gases,

$$\gamma = \frac{\sum_i \mu_i \left(1 + \frac{N_{Fi}}{2}\right) Z_i \frac{R}{w_i}}{\sum_i \mu_i \frac{N_{Fi}}{2} Z_i \frac{R}{w_i}} \tag{11.38}$$

in which the characteristics of each gas have been given the subscript, i, and the characteristic gas constant for each gas has been expanded according to $R_{wi} = R/w_i$. If the pressures are moderate, then $Z_i \approx Z \approx 1$, and may be cancelled along with the universal gas constant, R:

$$\gamma = \frac{\sum_i \frac{\mu_i}{w_i} \left(1 + \frac{N_{Fi}}{2}\right)}{\sum_i \frac{\mu_i}{w_i} \frac{N_{Fi}}{2}} \tag{11.39}$$

which may be further rearranged as follows:

$$\begin{aligned} \gamma &= \frac{\sum_i \frac{\mu_i}{w_i} \frac{N_{Fi}}{2}}{\sum_i \frac{\mu_i}{w_i} \frac{N_{Fi}}{2}} + \frac{\sum_i \frac{\mu_i}{w_i}}{\sum_i \frac{\mu_i}{w_i} \frac{N_{Fi}}{2}} \\ &= 1 + \frac{2}{\dfrac{\sum_i \frac{\mu_i}{w_i} N_{Fi}}{\sum_i \frac{\mu_i}{w_i}}} \\ &= 1 + \frac{2}{\sum_i \lambda_i N_{Fi}} \end{aligned} \tag{11.40}$$

where equation (11.24) was used to transform a mass fraction into a molar fraction. Equation (11.40) may be compared with equation (3.26), to which it degenerates in the case of a single gas constituent. For example, suppose that the mixture consisted of equal molar fractions of a monatomic gas ($N_{F1} = 3$, $\gamma_1 = 1.67$, see equations (3.22) and (3.26)) and a diatomic gas ($N_{F2} = 5$, $\gamma_1 = 1.4$); equation (11.40) predicts a combined specific-heat ratio of $\gamma = 1.5$.

11.4.1 Application to the accumulation of gas in a vessel of constant volume

Let us now consider the system of Figure 11.2. Given the initial masses $M_i(0)$ in kmol for the components $i = 1$ to n of the gas mixture in the vessel, the initial pressure in the vessel may be calculated from equations (11.14) and (11.13). A knowledge of the initial masses will also allow the effective specific-heat ratio, γ, to be calculated. Once γ has been determined, the outflow from the vessel, W_2, in kg/s may be calculated from equation (11.10). Equation (11.33) may then be applied to calculate the component gas outflows in kmol/s:

$$F_{2,i} = \frac{\mu_i}{w_i} W_2 \qquad \text{for } i = 1 \text{ to } n \tag{11.41}$$

In a similar way, if the pressure, p_1, temperature, T_1, and mass fraction, $\mu_{1,i}$ at the upstream point are given, then we may calculate the upstream specific volume from

$$v_1 = \frac{ZRT_1}{p_1 w_1} \tag{11.42}$$

where the effective molecular weight of the upstream mixture is given by:

$$w_1 = \frac{1}{\sum_{i=1}^{n} \frac{\mu_{1,i}}{w_i}} \tag{11.43}$$

Knowing the upstream specific volume, we may calculate the total inflow in kg/s from equation (11.9), and the inflow in kg/s of component i will be given by

$$W_{1,i} = \mu_{1,i} W_1 \tag{11.44}$$

The inflow in kmol/s is then given by

$$F_{1,i} = \frac{W_{1,i}}{w_i} = \frac{\mu_{1,i}}{w_i} W_1 \qquad \text{for } i = 1 \text{ to } n \tag{11.45}$$

When kmol units are used to characterize the accumulation of a mixture of gases, the mass balance of equation (11.11) is replaced by the molar balance for each component:

$$\frac{dM_i}{dt} = F_{1,i} - F_{2,i} \qquad \text{for } i = 1 \text{ to } n \tag{11.46}$$

where $F_{1,i}$ and $F_{2,i}$ are the inlet and outlet flows of component i in kmol/s. Since we have been given the initial values $M_i(0)$ for components $i = 1$ to n, equation (11.46) may be integrated numerically to produce the subsequent values of the masses in kmol at all later times.

11.5 Gas accumulation with heat exchange

Now we add heat transfer to the system of Figure 11.2 and produce Figure 11.3.

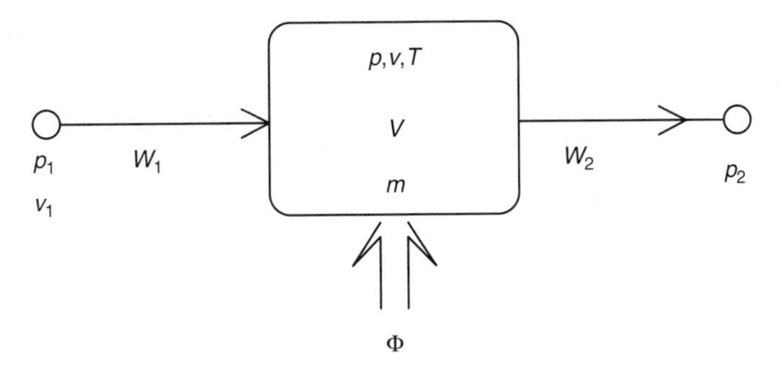

Figure 11.3 Gas accumulation with heat exchange.

We will assume that the inflow contains a mixture of gases, and so we will choose to work with kmol units.

Mass balance

The mass balance equations derived in Section 11.4 will apply as before, namely:

$$\frac{dM_i}{dt} = F_{1,i} - F_{2,i} \qquad \text{for } i = 1 \text{ to } n \qquad (11.46)$$

$$F_{1,i} = \frac{\mu_{1,i}}{w_i} W_1 \qquad \text{for } i = 1 \text{ to } n \qquad (11.45)$$

$$F_{2,i} = \frac{\mu_i}{w_i} W_2 \qquad \text{for } i = 1 \text{ to } n \qquad (11.41)$$

$$W_1 = f_{flow}\left(p_1, v_1, \frac{p}{p_1}, A_1, K_{T1}, \gamma_1\right) \qquad (11.47)$$

$$W_2 = f_{flow}\left(p, v, \frac{p_2}{p}, A_2, K_{T2}, \gamma\right) \qquad (11.10)$$

(The subscript '1' has been given to the average upstream specific-heat ratio in equation (11.47), which may be different in transient conditions from the average specific-heat ratio of the mixture in the vessel. Equation (11.47) is otherwise identical to equation (11.9).)

Pressure

The pressure, needed in equations (11.47) and (11.10) will obey equation (11.13) as before

$$p = \frac{ZRTM}{V} \qquad (11.13)$$

but now it will be necessary to calculate the temperature, T, at each time instant, as well as the mass in kmol, M. We will need to carry out an energy balance to calculate the instantaneous temperature value.

Energy balance

It will almost always be possible on a process plant to neglect kinetic and potential energies in the energy balance, so that we may apply equation (3.43) from Chapter 3, Section 3.5:

$$m\frac{du}{dt} = \Phi - P + W_1(h_1 - u) - W_2 pv \qquad (3.43)$$

The mass in this equation is expressed in kg, the flows in kg/s and the specific internal energy, the specific enthalpy in J/(kg K) and the specific volume in m^3/kg. We may convert to a kmol basis by dividing mass and mass flows by the average molecular weight, w, and by multiplying by w the specific internal energy, specific enthalpy and specific volume. Note that the energy input and power output (J) are unaffected:

$$M\frac{d\hat{u}}{dt} = \Phi - P + F_1(\hat{h}_1 - \hat{u}) - F_2 p\hat{v} \qquad (11.48)$$

In the case we are considering, there is no mechanical power extracted, and so we may set $P = 0$. In line with our choice of kmol units for the gas mixture, we will make the assumption that it acts as a gas with $Z = 1$ or a near-ideal gas (Z = constant). We may then substitute

$$\frac{d\hat{u}}{dt} = \frac{d\hat{u}}{dT}\frac{dT}{dt} \qquad (11.49)$$

into equation (11.48), and so obtain the necessary differential equation for temperature:

$$\frac{dT}{dt} = \frac{\Phi + F_1(\hat{h}_1 - \hat{u}) - F_2 p\hat{v}}{M\dfrac{d\hat{u}}{dT}} \qquad (11.50)$$

We may re-express the term $p\hat{v}$ by using the equation of state for the gas mixture, equation (3.2):

$$p\hat{v} = p(wv) = wpv = ZRT \qquad (11.51)$$

Hence, equation (11.50) may be re-expressed as:

$$\frac{dT}{dt} = \frac{\Phi + F_1(\hat{h}_1 - \hat{u}) - F_2 ZRT}{M\frac{d\hat{u}}{dT}} \tag{11.52}$$

Physically, the heat input will be dependent on the difference between the temperature of gas in the vessel and the temperature of the heating medium:

$$\Phi = UA_H(T_H - T) \tag{11.53}$$

where U is the overall heat transfer coefficient, W/(m^{2}K), A_H is the heat transfer area, m^2, and T_H (K) is the temperature of the heating medium. Thus it will be necessary to know the value of the gas temperature in order to calculate how much heat is being supplied. (There will normally be a feedback effect on the temperature of the heating medium also, since the heat supplied to the gas is the same as the heat taken from the heating medium. However, we will assume for simplicity that adequate compensatory action has been taken, and so there is no need for us to model the heating subsystem in this explanatory case.)

Sufficient equations have now been provided for the time-marching process of numerical integration to proceed, given the initial conditions of the masses of the gas components, $M_i(0)$, and the initial temperature, $T(0)$.

11.6 Liquid and gas accumulation with heat exchange

Figure 11.4 shows a vessel that is receiving and discharging both liquid and gas flows, each of which consists of a single component. The liquid is assumed to lie under an inert blanket of gas, such as nitrogen. The gas will not dissolve or condense in the liquid and we will assume that the liquid does not evaporate into the gas. The liquid in the vessel is being heated, and there will be a degree of heat transfer across the liquid–gas interface to the gas.

We will need to carry out mass and energy balances on both the liquid and the gas. We will choose to use kg units of mass for the liquid and kmol units for the gas.

Mass balance for the liquid

The equations follow those developed in Section 11.2, and may be written down using the notation of Figure 11.4 as:

$$\frac{dm_L}{dt} = W_{L1} - W_{L2} \tag{11.54}$$

$$W_{L1} = C_{T1}\sqrt{\frac{p_{L1} - p_L}{v_{L1}}} \tag{11.55}$$

$$W_{L2} = C_{T2}\sqrt{\frac{p_L - p_{L2}}{v_L}} \tag{11.56}$$

$$p_L = p_G + \frac{gl}{v_L} \tag{11.57}$$

$$l = \frac{V_L}{A} \tag{11.58}$$

$$V_L = m_L v_L \tag{11.59}$$

The liquid specific volume will have a known dependency on the temperature:

$$v_L = v_L(T) \tag{11.60}$$

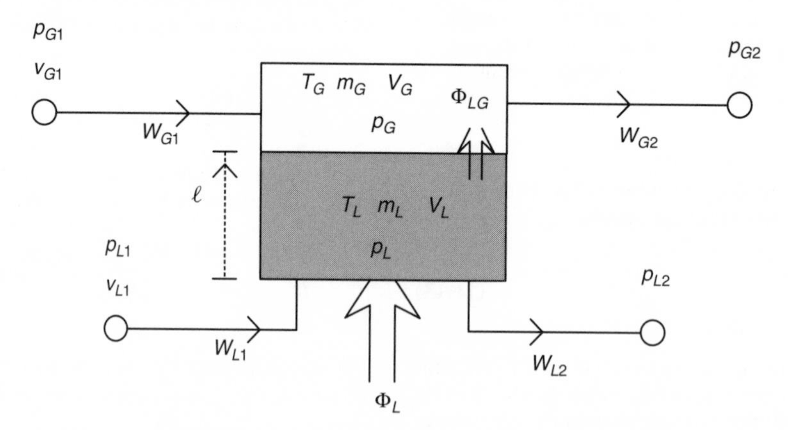

Figure 11.4 Liquid and gas accumulation with heat exchange.

Energy balance for the liquid

Applying equation (3.43) from Chapter 3, Section 3.5, we have for the liquid:

$$m_L \frac{du_L}{dt} = \Phi - P + W_{L1}(h_{L1} - u_L) - W_{L2} p_L v_L \tag{11.61}$$

where p_L is the pressure at the base of the vessel. Now for a liquid, the specific internal energy is almost independent of pressure, so that

$$\frac{du_L}{dt} \approx \frac{du_L}{dT_L}\frac{dT_L}{dt} \tag{11.62}$$

The heat input to the liquid will obey an equation similar to (11.53):

$$\Phi_L = U_{HL} A_H (T_H - T_L) \tag{11.63}$$

where U_{HL} is the heat transfer coefficient between the heating medium and the liquid. Similarly, the heat passed from the liquid to the gas will be proportional to the temperature difference between them:

$$\Phi_{LG} = U_{LG} A (T_L - T_G) \tag{11.64}$$

where U_{LG} is the heat transfer coefficient for heat transfer from the liquid to the gas, and A is the heat transfer area, equal to the cross-sectional area of the vessel. The net heat input to be used in equation (11.61) will be

$$\Phi = \Phi_L - \Phi_{LG} \tag{11.65}$$

When the level of the liquid rises, the liquid will do mechanical work against the gas pressure, as explained in Chapter 3, Section 3.6. From equation (3.47), the rate of performing work will be:

$$P = p_G \frac{dV_L}{dt} \tag{11.66}$$

Using (11.59) and (11.60), this may be re-expressed as:

$$P = p_G v_L \frac{dm_L}{dt} + p_G m_L \frac{dv_L}{dT_L}\frac{dT_L}{dt} \tag{11.67}$$

Using the expression for dm_L/dt given by equation (11.54) allows equation (11.67) to be written:

$$P = p_G v_L (W_{L1} - W_{L2}) + p_G m_L \frac{dv_L}{dT_L}\frac{dT_L}{dt} \tag{11.68}$$

Substituting back into the energy balance equation (11.61) gives the following equation in the temperature differential, dT_L/dt:

$$\begin{aligned} m_L \frac{du_L}{dT}\frac{dT_L}{dt} &= \Phi_L - \Phi_{LG} - p_G v_L (W_{L1} - W_{L2}) \\ &\quad - p_G m_L \frac{dv_L}{dT}\frac{dT_L}{dt} + W_{L1}(h_{L1} - u_L) \\ &\quad - W_{L2} p_L v_L \end{aligned} \tag{11.69}$$

Rearranging yields

$$\begin{aligned} &\left(\frac{du_L}{dT} + p_G \frac{dv_L}{dT}\right) m_L \frac{dT_L}{dt} \\ &\quad = \Phi_L - \Phi_{LG} + W_{L1}(h_{L1} - (u_L + p_G v_L)) \\ &\quad\quad - W_{L2} v_L (p_L - p_G) \end{aligned} \tag{11.70}$$

Two points may now be made:

(i) the gas pressure and the liquid pressure will differ only by the hydrostatic head, and thus in many cases $p_L - p_G \approx 0$;

(ii) the pv term in the enthalpy definition is usually very much smaller than the specific internal energy term, u, for a liquid, so that little error will be introduced by assuming

$$h_L \approx u_L + p_G v_L \tag{11.71}$$

The specific heat of the liquid, c_{pL}, at constant pressure is given by:

$$\begin{aligned} c_{pL} &= \left(\frac{dh_L}{dT}\right)_p = \frac{du_L}{dT_L} + p_L \frac{dv_L}{dT_L} \\ &\approx \frac{du_L}{dT_L} + p_G \frac{dv_L}{dT_L} \end{aligned} \tag{11.72}$$

Hence we may rearrange equation (11.70) to give the differential of liquid temperature as:

$$\frac{dT_L}{dt} = \frac{\Phi_L - \Phi_{LG} + W_{L1}(h_{L1} - h_L)}{m_L c_{pL}} \tag{11.73}$$

Mass balance for the gas

The equations for mass balance follow those of Section 11.5:

$$\frac{dM_G}{dt} = F_{G1} - F_{G2} \tag{11.74}$$

$$F_{G1} = \frac{W_{G1}}{w} \tag{11.75}$$

$$F_{G2} = \frac{W_{G2}}{w} \tag{11.76}$$

$$W_{G1} = f_{flow}\left(p_{G1}, v_{G1}, \frac{p_G}{p_{G1}}, A_{G1}, K_{TG1}, \gamma_{G1}\right) \tag{11.77}$$

$$W_{G2} = f_{flow}\left(p_G, v_G, \frac{p_{G2}}{p_G}, A_{G2}, K_{TG2}, \gamma\right) \tag{11.78}$$

Gas pressure

$$p_G = \frac{ZRT_G M_G}{V_G} \tag{11.79}$$

Volume and specific volume

$$V_G = V - V_L \tag{11.80}$$

$$\hat{v}_G = \frac{V_G}{M_G} \tag{11.81}$$

Energy balance for the gas

The energy balance follow the form of equation (11.48), which gives for the notation of Figure 11.4:

$$M_G \frac{d\hat{u}_G}{dt} = \Phi_{LG} - P_G + F_{G1}(\hat{h}_{G1} - \hat{u}_G) - F_{G2} p_G \hat{v}_G \tag{11.82}$$

The only mechanical power exchange is between the liquid and the gas, and the mechanical power taken from the liquid, P, is added to the gas. Hence

$$P_G = -P \tag{11.83}$$

where P is given by equation (11.67), in which dm_L/dt may be found from equation (11.54) and dT_L/dt from equation (11.73).

Provided that the gas is ideal, near-ideal or a real gas in most process conditions of pressure and temperature, $\hat{u}_G = \hat{u}_G(T_G)$, and we have

$$\hat{c}_v = \frac{d\hat{u}_G}{dT_G} \tag{11.84}$$

where $\hat{c}_v$ is the specific heat at constant volume of the gas, expressed in J/(kmol K). As a result, equation (11.82) may be rearranged to give the differential of the gas temperature:

$$\frac{dT_G}{dt} = \frac{\Phi_{LG} + P + F_{G1}(\hat{h}_{G1} - \hat{u}_G) - F_{G2} p_G v_G}{M_G \hat{c}_v} \tag{11.85}$$

We now have enough equations to 'solve' the system: given initial values of the state variables, liquid mass, liquid temperature, gas mass and gas temperature, we may integrate the equations in a time-marching way to give values of the state variables at all future times. The solution sequence is as follows.

1. Calculate the liquid specific volume from the known liquid temperature, then find the liquid volume, given the liquid mass, and then the gas volume by subtraction from the vessel volume.
2. Given the gas mass and temperature, use the gas volume to determine the gas pressure. Once the gas pressure has been established, find the liquid pressure at the liquid inlet and outlet, then the liquid flows, and finally the liquid mass differential.
3. Determine the gas specific volume from the gas volume and gas mass. Determine the gas flows and hence the gas mass differential.
4. Find the heat fluxes given the liquid and gas temperatures, and then find the liquid temperature differential.
5. Given the liquid mass and liquid temperature differentials, find the mechanical power absorbed from the liquid by the gas. Then calculate the gas temperature differential.
6. Take the four differentials for liquid mass, liquid temperature, gas mass and gas temperature and integrate forward one timestep.
7. Go back to step 1 and repeat the cycle.

12 Two-phase systems: boiling, condensing and distillation

12.1 Introduction

The last example of Chapter 11 dealt with the case where a liquid and a gas were present together, but the gas was inert with respect to the liquid. However, there are a large number of systems on a process plant where a liquid and its own vapour are present together: evaporators, condensers, steam drums, deaerators, refrigeration systems, stills and distillation columns. These systems exist in a state of vapour–liquid equilibrium, and their behaviour is significantly different from the gas–liquid system dealt with in Chapter 11, Section 11.6. The liquid and its vapour will have the same temperature, and it will not be possible to decouple the mass and energy balance equations for the liquid from those of the vapour. The way to obtain the necessary time differentials explicitly is to use the Method of Referred Derivatives.

12.2 Description of single component boiling/condensing: boiling model

Figure 12.1 is a schematic representation of a boiling/condensing vessel subject to a heat input (which will be negative in the case of condensing). Liquid and vapour flow into the vessel, and a liquid flow and a vapour flow are discharged by the vessel.

It will be assumed that the contents of each phase are well mixed and that the vapour and the liquid are in vapour–liquid equilibrium, so that they share a common temperature.

Our description of the boiling vessel must be sufficient to determine the mass of liquid, the temperature and pressure in the vessel, together with the specific volume of the vapour and the liquid. As well as these, we would like to know the mass of the vapour. Perhaps surprisingly, all these variables may be calculated from a knowledge of just two variables: the mass of liquid and the temperature. Accordingly we shall choose these two as our state variables.

Liquid mass balance

Following the approach detailed in Chapter 3, Section 3.4, we begin by carrying out a mass balance on the liquid space:

$$\frac{dm_L}{dt} = W_{L1} - W_{L2} - W_B \tag{12.1}$$

where
m_L is the mass of liquid (kg),
W_{L1} is the flow of liquid into the vessel (kg/s),
W_{L2} is the flow of liquid leaving the vessel (kg/s),
W_B is the rate of boiloff (condensation when negative) kg/s.

Vapour mass balance

Similarly, we may carry out a mass balance on the vapour space:

$$\frac{dm_V}{dt} = W_{V1} - W_{V2} + W_B \tag{12.2}$$

where
m_V is the mass of vapour (kg),
W_{V1} is the flow of vapour into the vessel (kg/s),
W_{V2} is the flow of vapour leaving the vessel (kg/s).

We may note immediately that although determining the vessel inlet and outlet flows in equations (12.2) should be a straightforward matter, calculating the boiloff rate is not so easy.

Energy balance on the vessel contents

We now follow the method set down in Section 3.5 to carry out an energy balance on the contents of the vessel. Using equation (3.34) as the basis, and ignoring kinetic and potential energy terms as negligible, we may write

$$\frac{dE}{dt} = \Phi + W_{L1}h_{L1} + W_{V1}h_{V1} - W_{L2}h_L - W_{V2}h_V \tag{12.3}$$

where
Φ is the heat input (W),
h_{L1} and h_{V1} are the enthalpies of the liquid and vapour flows entering the vessel (J/kg),
h_L and h_V are the enthalpies of the liquid and the vapour inside the vessel (J/kg),
and the energy of the contents of the vessel, E(J), is given by

$$E = m_L u_L + m_V u_V \tag{12.4}$$

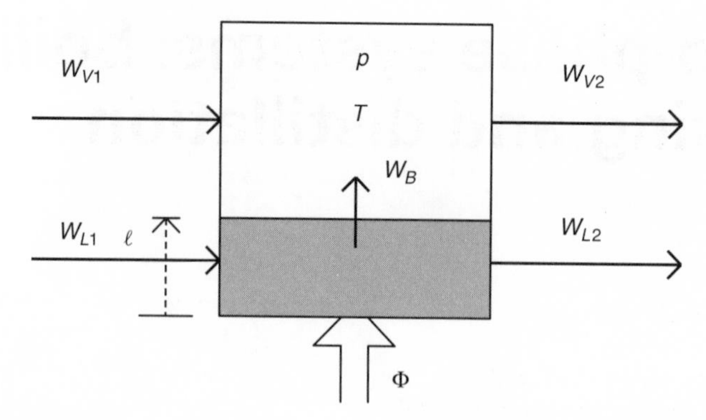

Figure 12.1 Boiling/condensing vessel.

Here u_L and u_V are the specific internal energies of the liquid and the vapour (J/kg), each of which is a function of temperature only in the state of vapour–liquid equilibrium. Hence when we differentiate E in equation (12.3) we produce an equation in the three state variable derivatives, dm_L/dt, dm_V/dt and dT/dt:

$$u_L \frac{dm_L}{dt} + u_V \frac{dm_V}{dt} + \left(m_L \frac{du_L}{dT} + m_V \frac{du_V}{dT} \right) \frac{dT}{dt} = \Phi + W_{L1} h_{L1} + W_{V1} h_{V1} - W_{L2} h_L - W_{V2} h_V \quad (12.5)$$

We may eliminate the mass derivatives by using equations (12.1) and (12.2), so that the temperature derivative emerges as:

$$\frac{dT}{dt} = [\Phi + W_{L1}(h_{L1} - u_L) + W_{V1}(h_{V1} - u_V) - W_{L2}(h_L - u_L) - W_{V2}(h_V - u_V) - W_B(u_V - u_L)] \div \left[m_L \frac{du_L}{dT} + m_V \frac{du_V}{dT} \right] \quad (12.6)$$

Volumes of liquid and vapour

The volumes of the liquid and the vapour together must always equal the total volume of the vessel, i.e.

$$V_L + V_V - V_T = 0 \quad (12.7)$$

where V_L is the volume of the liquid in the vessel (m^3) and V_T is the total volume of the vessel (m^3). Since

$$V_L = v_L m_L \quad (12.8)$$

where $v_L = v_L(T)$ is the specific volume of the liquid, we may re-express (12.7) as

$$V_V + v_L m_L - V_T = 0 \quad (12.9)$$

Pressure in the gas space

The pressure must obey both its characteristic equation and the condition of vapour–liquid equilibrium. Using equation (3.1) from Chapter 3, the first may be expressed in general form

$$p = f_1(v_V, T) \quad (12.10)$$

where v_V is the specific volume of the vapour. Since $v_V = V_V/m_V$, where V_V is the volume of the vapour space (m^3), equation (12.10) may be rewritten as:

$$p = f(m_V, V_V, T) \quad (12.11)$$

The condition of vapour–liquid equilibrium means that pressure will be a defined function of temperature:

$$p = K(T) \quad (12.12)$$

Combining equations (12.11) and (12.12) gives

$$K(T) - f(m_V, V_V, T) = 0 \quad (12.13)$$

Applying the Method of Referred Derivatives

We now have in (12.1) and (12.6) equations in our chosen state variables liquid mass and temperature. But the fact that we do not know the boiloff rate prevents us from solving these equations without the additional differential equation (12.2) for the mass of vapour and the algebraic equations for volume and pressure, (12.9) and (12.13). We may now use the Method of Referred Derivatives, outlined in Chapter 2, Section 2.11 (and in Chapter 18, Section 18.7 in further detail), to put these algebraic constraints into the appropriate form to be used in the differential equations defining the state variables.

Equations (12.9) and (12.13) represents a system of simultaneous, nonlinear equations of the form given by equation (2.93) in Section 2.10 of Chapter 2, namely

$$\mathbf{g}(\mathbf{z}(t), \mathbf{x}(t), \mathbf{u}(t)) = \mathbf{0} \quad (2.93)$$

where in this case the vector of unknown algebraic variables, $\mathbf{z}$, is:

$$\mathbf{z} = \begin{bmatrix} m_V \\ V_V \end{bmatrix} \tag{12.14}$$

the vector of states, $\mathbf{x}$, is given by

$$\mathbf{x} = \begin{bmatrix} m_L \\ T \end{bmatrix} \tag{12.15}$$

and there is no input vector, $\mathbf{u}$. Differentiating equation (2.93) with respect to time for the case when there is no input vector gives:

$$\frac{\partial \mathbf{g}}{\partial \mathbf{z}}\frac{d\mathbf{z}}{dt} + \frac{\partial \mathbf{g}}{\partial \mathbf{x}}\frac{d\mathbf{x}}{dt} = \mathbf{0} \tag{12.16}$$

(cf. equation (2.97) of Section 2.11), which allows the derivative of the algebraic unknowns to be referred to the derivatives of the state variables. We carry out this procedure on equations (12.9) and (12.13) to give:

$$\frac{dV_V}{dt} + v_L\frac{dm_L}{dt} + m_L\frac{dv_L}{dT}\frac{dT}{dt} = 0 \tag{12.17}$$

and

$$\left(\frac{dK}{dT} - \frac{\partial f}{\partial T}\right)\frac{dT}{dt} - \frac{\partial f}{\partial m_V}\frac{dm_V}{dt} - \frac{\partial f}{\partial V_V}\frac{dV_V}{dt} = 0 \tag{12.18}$$

Thus we have now referred the time derivatives dm_V/dt and dV_V/dt to the time derivatives of the state variables, dm_L/dt and dT/dt. We now have in equations (12.1), (12.2), (12.6), (12.17) and (12.18) a set of five linear, simultaneous equations in the five unknowns:

$$\frac{dT}{dt}, \frac{dm_L}{dt}, \frac{dm_V}{dt}, \frac{dV_V}{dt} \text{ and } W_B$$

so the solution is now guaranteed. Furthermore, we have already separated out expressions for dm_L/dt, dm_V/dt, dT/dt, and dV_V/dt in equations (12.1), (12.2), (12.6), and (12.17) respectively. Accordingly, we may substitute for these quantities into equation (12.18) to give an equation in a single unknown, namely the boiloff rate, W_B.

First let us use equation (12.17) to eliminate dV_V/dt from equation (12.18):

$$\left(\frac{dK}{dT} + \frac{\partial f}{\partial V_V}m_L\frac{dv_L}{dT} - \frac{\partial f}{\partial T}\right)\frac{dT}{dt} - \frac{\partial f}{\partial m_V}\frac{dm_V}{dt} + \frac{\partial f}{\partial V_V}v_L\frac{dm_L}{dt} = 0 \tag{12.19}$$

To simplify the algebra, we shall rewrite equation (12.6) in dT/dt as:

$$\frac{dT}{dt} = \frac{\alpha_1}{\alpha_2} - \frac{u_V - u_L}{\alpha_2}W_B \tag{12.20}$$

where

$$\alpha_1 = \Phi + W_{L1}(h_{L1} - u_L) + W_{V1}(h_{V1} - u_V) - W_{L2}(h_L - u_L) - W_{V2}(h_V - u_V) \tag{12.21}$$

and

$$\alpha_2 = m_L\frac{du_L}{dT} + m_V\frac{du_V}{dT} \tag{12.22}$$

Substituting for dm_L/dt from (12.1), for dm_V/dt from (12.2) and for dT/dt from equation (12.20) into equation (12.19) gives the solution for the boiloff rate, W_B, as:

$$W_B = \frac{\dfrac{\alpha_1}{\alpha_2}\left(\dfrac{dK}{dT} + \dfrac{\partial f}{\partial V_V}m_L\dfrac{dv_L}{dT} - \dfrac{\partial f}{\partial T}\right) - \dfrac{\partial f}{\partial m_V} \times (W_{V1} - W_{V2}) + \dfrac{\partial f}{\partial V_V}v_L(W_{L1} - W_{L2})}{\dfrac{1}{\alpha_2}\left(\dfrac{dK}{dT} + \dfrac{\partial f}{\partial V_V}m_L\dfrac{dv_L}{dT} - \dfrac{\partial f}{\partial T}\right)(u_V - u_L) + \dfrac{\partial f}{\partial m_V} + \dfrac{\partial f}{\partial V_V}v_L} \tag{12.23}$$

Once W_B is established, the derivatives of the two state variables may then be found:

$$\frac{dm_L}{dt} = W_{L1} - W_{L2} - W_B \tag{12.1}$$

$$\frac{dT}{dt} = \frac{\alpha_1}{\alpha_2} - \frac{u_V - u_L}{\alpha_2}W_B \tag{12.20}$$

Given a starting condition, $[m_L(0), T(0)]$, it is now possible to integrate equations (12.1) and (12.20) for all future times. The calculational route is to integrate to evaluate temperature, T, and liquid mass, m_L:

$$T(t) = T(t_0) + \int_{t_0}^{t}\frac{dT}{dt}\,dt \tag{12.24}$$

$$m_L(t) = m_L(t_0) + \int_{t_0}^{t}\frac{dm_L}{dt}\,dt \tag{12.25}$$

Once the temperature is known, the following thermodynamic variables may be calculated algebraically:

$$\begin{aligned} u_L &= u_L(T) \\ u_V &= u_V(T) \\ v_L &= v_L(T) \\ p &= K(T) \\ v_V &= f_v(p, T) \end{aligned} \tag{12.26}$$

where the function $f_v(p, T)$ is derivable from equation (12.10).

The volume of the liquid mass is given by

$$V_L = v_L m_L \tag{12.27}$$

and the volume of the vapour mass is consequently

$$V_V = V_T - v_L m_L \tag{12.28}$$

The mass of the vapour is then related algebraically to the mass of liquid:

$$m_V = \frac{V_V}{v_V} = \frac{V_T}{v_V} - \frac{v_L}{v_V} m_L \tag{12.29}$$

[It would be possible, as an alternative, to integrate for the mass of vapour based on equation (12.2). However, it is generally a good thing to reduce the number of integrations whenever possible, especially since in this case the time constant associated with the vapour space may well be short, adding unnecessary stiffness to the simulation.]

12.3 Functions used in the modelling of vapour–liquid equilibrium

We may normally use the modified ideal gas law of equation (3.2) as the function f_1:

$$p = f_1(v_V, T) = \frac{ZR_w T}{v_V} \tag{12.30}$$

so that the companion function f_v is simply:

$$v_V = f_v(p, T) = \frac{ZR_w T}{p} \tag{12.30a}$$

Multiplying both sides of equation (12.30a) by the vapour mass, m_V, and noting that $V_V = v_V m_V$ gives the required function, f:

$$p = \frac{m_V Z R_w T}{V_V} = f(m_V, V_V, T) \tag{12.31}$$

Hence

$$\begin{aligned} \frac{\partial f}{\partial m_V} &= \frac{ZR_w T}{V_V} \\ \frac{\partial f}{\partial V_V} &= -\frac{m_V Z R_w T}{V_V^2} \\ \frac{\partial f}{\partial T} &= \frac{m_V R_w}{V_V}\left(Z + T\frac{dZ}{dT}\right) \end{aligned} \tag{12.32}$$

Here R is the universal gas constant (=8314 J/kgmol K), w is the molecular weight of the fluid, $R_w = R/w$ is the characteristic gas constant and Z is the compressibility factor, which will be a function of temperature in conditions of boiling or condensing.

A convenient and accurate representation of the vapour pressure curve is the Antoine equation:

$$p = K(T) = \exp\left(A + \frac{B}{C + T}\right) \tag{12.33}$$

Hence

$$\frac{dK}{dT} = -\frac{B}{(C+T)^2} K \tag{12.34}$$

The Riedel equation for vapour pressure gives greater accuracy by relying on four constants rather than three:

$$p = K(T) = \exp\left(A + \frac{B}{T} + C \ln T + DT^6\right) \tag{12.35}$$

in which case

$$\frac{dK}{dT} = \left(-\frac{B}{T^2} + \frac{C}{T} + 6DT^5\right) K \tag{12.36}$$

Specific internal energy and specific volume functions may be provided by look-up tables (e.g. from the highly accurate steam tables for boiling/condensing water) or by polynomial approximations fitted over the pressure/temperature range of interest. For control engineering purposes, even low-order polynomials will give ample accuracy: e.g. third-order polynomials will give better than 0.1% accuracy over the range 40 to 100 bar for the specific internal energies and specific volumes of water and steam. Low-order polynomial approximations have the advantages that (i) the programming is simpler, since there is no need to invoke a look-up subroutine, and (ii) it is an easy matter to derive analytic expressions for the derivatives with respect to temperature: du_L/dT, du_V/dT and dv_L/dT.

Because the boiling model described above is fully explicit, it is rapid and easy to apply. It has been used with success in modelling steam drums and boiler recirculation loops, direct-contact feedwater heaters, high-pressure feedwater heaters, deaerator towers and vessels, condensers and refrigeration loops. The application of the boiling model to a steam drum and recirculation loop will be discussed in the next section because of its importance for the modelling of power stations.

12.4 Application of the boiling model to a steam drum and recirculation loop

In a recirculation boiler system, see Figure 12.2, the feedwater is normally fed through sparge pipes into the steam drum, where the downward-pointing spargers encourage the formation of a subcooled region. A mixture of subcooled and saturated water is then drawn

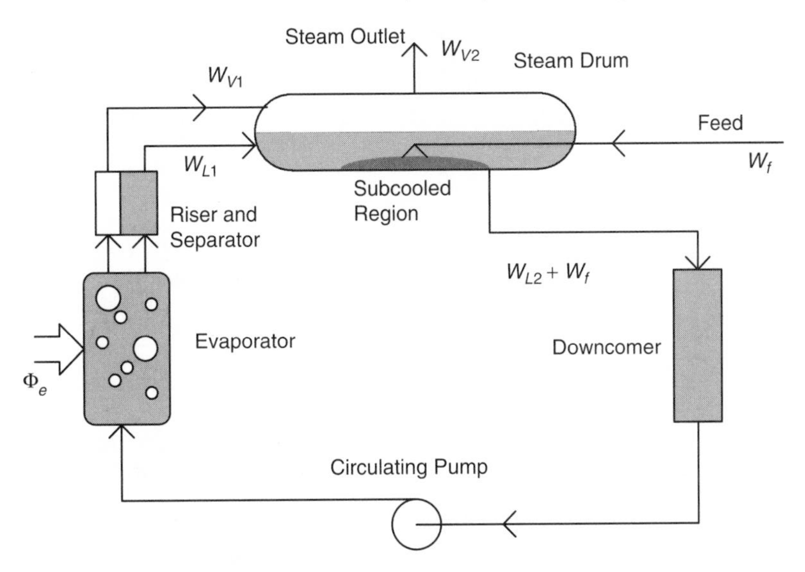

Figure 12.2 Schematic of boiler recirculation loop.

from the bottom of the drum into the downcomer pipework by the circulating pump, which supplies the evaporator. The heat is added to the evaporator causing steam to be formed. A two-phase flow emerges from the evaporator with a steam quality (ratio of steam mass flow to total mass flow) of about 10%. This two-phase flow passes up the riser pipework and into a set of cyclone separators, which disentrain the water so that it flows into the liquid space while allowing the steam to pass into the vapour space. Saturated steam is then further dried by a set of steam-scrubbers before being drawn off from the top of the steam drum.

It is possible to obtain a very good representation for control engineering purposes by decomposing the boiler recirculation loop, into four notional lumped sections:

- (i) the subcooled region and downcomer pipework,
- (ii) the evaporator,
- (iii) the riser and cyclone separators,
- (iv) the steam drum.

The subcooled region and downcomer pipework

To a good approximation, we may treat the subcooled region as having a constant volume of the order of 10% of the liquid volume in the drum. It may be treated as separate from the equilibrium thermodynamics of the steam drum. It may be regarded as an extension of the downcomer, with its volume added to that of the downcomer pipework in order to work out a downcomer time delay.

The downcomer flow, W_d, will be determined by the pump characteristic, as it pumps from the drum at pressure, p, to the evaporator at pressure, p_e. A steady-state flow balance at downcomer inlet gives the outlet water flow from the notional steam drum

$$W_{L2} = W_d - W_f \tag{12.37}$$

while a steady-state energy balance at the inlet to the downcomer gives the enthalpy, h_{di}, at the inlet to the downcomer as

$$h_{di} = \frac{W_f h_f + W_{L2} h_{L2}}{W_d} \tag{12.38}$$

The enthalpy, h_{do}, at the outlet of the downcomer will be delayed by a time delay, τ_d:

$$h_{do}(t) = h_{di}(t - \tau_d) \tag{12.39}$$

where τ_d is the time delay associated with the downcomer, dependent on downcomer pipework volume and downcomer flow.

The evaporator

While a rigorous treatment of the evaporator would make due allowance for its distributed nature, very useful and surprisingly accurate results are achieved by treating it as a lumped-parameter system, using the boiling model presented in Section 12.2. This may be used to determine the temperature and pressure inside the evaporator, and the split between water and steam.

The riser and cyclone separators

The mass flow between the evaporator and the steam drum may be calculated by the proportionality (cf. equation (4.81)):

$$W_r \propto \sqrt{p_e - p} \tag{12.40}$$

Despite the fact that this takes no account of variations in the specific volume of the riser flow, this simple model has proved adequate in modelling a number of commercial-size boiler systems.

The steam quality at the inlet to the riser is given by

$$q_{in} = \frac{m_{Ve}}{m_{Le} + m_{Ve}} \quad (12.41)$$

while the riser outlet quality is delayed by the riser time delay, τ_r:

$$q_{out}(t) = q_{in}(t - \tau_r) \quad (12.42)$$

The steam flow to the drum is then given by

$$W_{V1} = q_{out} W_r \quad (12.43)$$

and the liquid flow to the drum is

$$W_{L1} = (1 - q_{out})W_r \quad (12.44)$$

The cyclone separators are regarded as completely efficient in separating the steam and water flows.

The steam drum

The saturated contents of the steam drum may be modelled directly by the boiling model of Section 12.2.

The subcooled region is modelled separately along with the downcomer pipework. As far as the drum model is concerned, its sole contribution comes in determining liquid level, where it adds a constant term to the liquid volume. As mentioned previously, the subcooled region may make up typically about 10% or less of the water volume of the drum, and so its effect on steam drum dynamics is second-order only.

Of more concern is the effect of the storage of thermal energy in the drum metalwork and the transfer of heat between the drum metal and the drum contents. Using a lumped parameter model of the drum metal, an energy balance yields:

$$\frac{d}{dt}(m_m c_{pm} T_m) = -\Phi \quad (12.45)$$

or

$$\frac{dT_m}{dt} = -\frac{\Phi}{m_m c_{pm}} \quad (12.46)$$

where m_m is the metalwork mass (kg), c_{pm} is the specific heat of the metal (J/(kgK)) and the heat transferred, Φ, (W) is given by

$$\Phi = K_m A_m (T_m - T) \quad (12.47)$$

Here A_m is the metalwork area (m^2) and K_m, the heat transfer coefficient between the two-phase drum contents and the drum metal, has been calculated as about 10 000 W/(m^2K). The inclusion of the effect of the stored energy in the drum metalwork will have an appreciable damping effect on the temperature and pressure transients calculated.

The relatively simple, lumped-parameter system model described above has been tested against and used in earnest to analyse the behaviour of the boiler recirculation loops of a number of power stations. It has been found to give excellent quantitative predictions of all the variables whose trends are important for control engineering purposes, namely steam drum pressure and temperature, feedwater flow, steam production, downcomer flow and, very important, drum water level.

12.5 Continuous distillation in a distillation column

Figure 12.3 shows a distillation column used to separate the components of a feed stream on a continuous basis. The column contains many plates, each of which holds a liquid mixture in equilibrium with the vapour mixture above it. Some of the liquid emerging from the bottom of the column is heated in the reboiler, and the resulting vapour or 'boil-up' is re-introduced to the final plate, providing the source of heat for the column as a whole. Hot vapour rises from each plate into the next plate up, passing via bubble-caps or valve-caps into the liquid held on the tray above and thus transferring heat to that liquid and causing evaporation. Vapour is taken from the top of the column and passed through a condenser. Some of the resultant condensate is taken off as the top product, but some is recycled to the top plate as 'reflux'. The liquid reflux is fed onto the top tray, which then overflows via a downcomer onto the plate below it, initiating a continuous cascade of liquid from tray to tray down the column.

Because there is heating at the bottom of the column and cooling at the top, there will be a temperature gradient down the column, with lower temperatures occurring at the upper plates of the column. As a result, the 'lighter' components (i.e. those with a relatively low boiling point) will predominate in the upper part of the column, particularly in the vapour, while the 'heavier' components (those with a higher boiling point) will predominate in the lower part of the column, particularly in the liquid. Two concentration gradients will establish themselves in the column, one for the liquid phase and one for the vapour phase. In each case the top plate will possess the highest concentration of lighter components, while the bottom plate will contain the highest concentration of heavier components. The bottom product is taken from the liquid on the lowest plate.

The feed is introduced to a plate somewhere about the middle of the column, and the lighter components

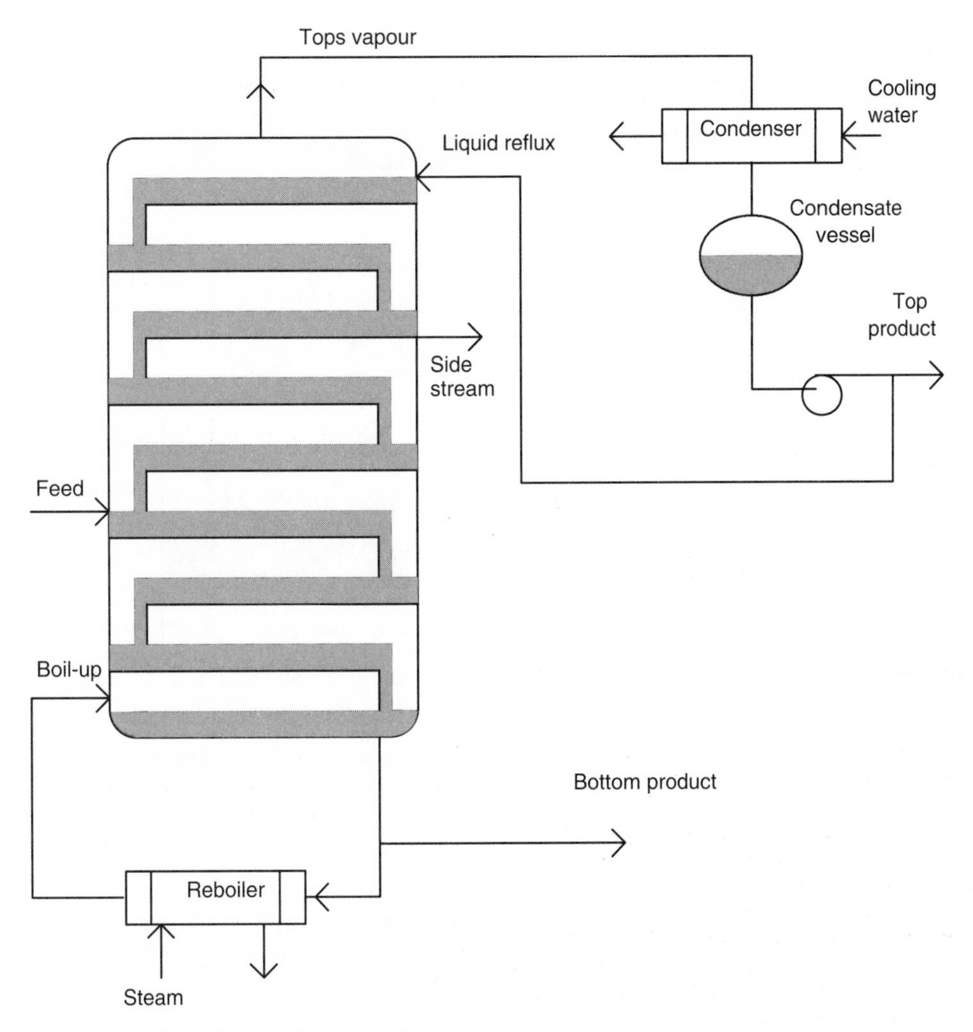

Figure 12.3 Schematic of a distillation column.

of the feed will tend to be taken up the column as vapour, while the heavier components will tend to move down the column as liquid. The plates above the feed are known as the 'rectifying' section of the column, while the plates below are known as the 'stripping' section. It is possible for the column to receive more than one feed stream, in which case the terms 'rectifying' and 'stripping' are less well-defined.

The distillation column produces two main product streams: the top product or 'tops', which is rich in the lighter components, and the bottom product or 'bottoms', which is rich in the heavier components. Sometimes one or more sidestreams are also taken at intermediate points on the column. The sum of the product flows will be equal to the sum of the feed flows in the steady state.

12.6 Mathematical model of the distillation plate

A schematic of the distillation plate is given in Figure 12.4. In many texts the terms 'plate' and 'tray' are used interchangeably, but we will use the convention that the 'plate' consists of a 'tray', which holds the liquid, and a 'vapour space', which holds the vapour. The ith plate receives a liquid flow L_{i+1} via the downcomer attached to the tray of the $(i + l)$th plate above. It also receives a vapour flow E_{i-1} from the $(i - 1)$th plate below it through bubble-caps or valve caps. (E is used to denote *evaporation* rate – the letter V for 'vapour' has already been claimed for volume.) The boiloff from the liquid on the ith tray into the plate's vapour space is denoted B_i, while the vapour flow from

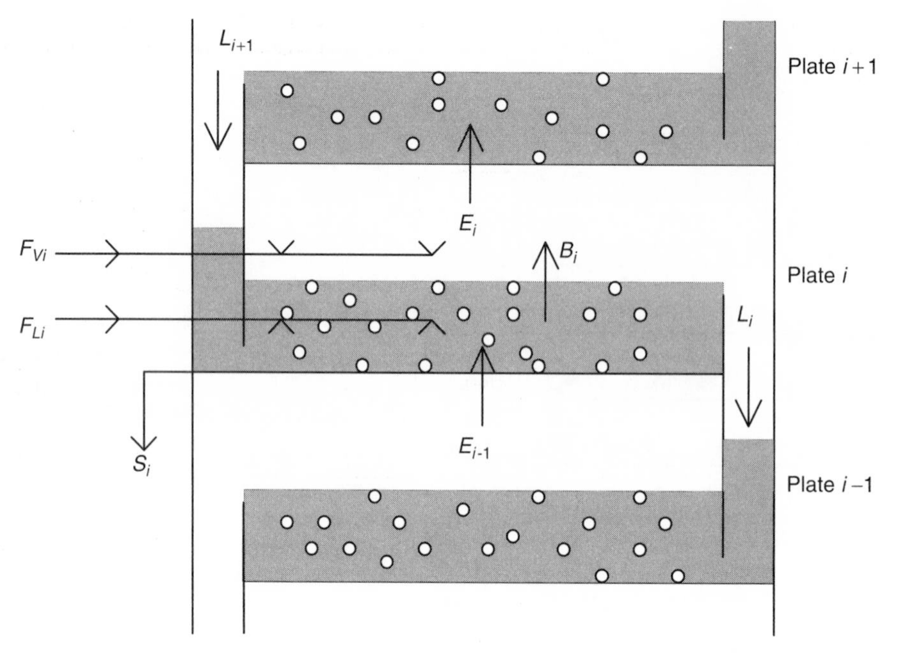

Figure 12.4 Distillation plate, showing liquid and vapour flows.

the ith plate into the $(i+1)$th plate above is labelled E_i. A liquid flow L_i will pass from the ith to the $(i-1)$th plate below.

It is possible for the ith plate to possess either a sidestream, S_i, or a feedstream, F_{Li} for liquid feed, F_{Vi} for vapour feed. Common sense dictates that sidestream and feedstream should not both be present on the same plate on a sensibly run distillation column, but both feed- and side-streams have been shown in Figure 12.4 for completeness, and the general plate analysed will be assumed to contain both.

We will now derive the equations necessary to describe the dynamic behaviour of a distillation plate. The modelling procedure for multi-component boiling has strong similarities to the situation of single component boiling condensing already described in Section 12.2. We shall wish to know the masses of all the components in the liquid phase and in the vapour phase, the temperature of the plate, the pressure in the plate, the specific volume of the liquid components and the specific volume of the vapour mixture as a whole. This information is sufficient to allow us to calculate the mole fractions in the vapour and the liquid and the inter-plate flows.

The choice of state variables follows the example of single component boiling. It is perhaps surprising that we can calculate all the variables listed above from a knowledge of only the component liquid masses in the plate and the plate temperature. These are thus the obvious state variables. It is customary and natural to use kmol units to analyse the distillation column because the essence of the process is that it deals with at least two distinct chemical components. Thus all masses will be in kmol and all flows in kmol/s.

Liquid mass balance for the ith *plate*

Assuming there are n components in the feed, then the mass balance over the liquid yields for the most general case:

$$\frac{dM_{Lij}}{dt} = x_{i+1,j}L_{i+1} + y_{i-1,j}E_{i-1} + z_{Lij}F_{Li} - B_{ij} - x_{ij}L_i - x_{ij}S_i \qquad \text{for } j = 1, \ldots n \tag{12.48}$$

where:

M_{Lij} is the mass of component j in the liquid phase in plate i (kmol),
L_i is the total liquid flow from plate i to plate $i-1$ (kmol/s),
E_i is the total vapour flow from plate i to plate $i+1$ (kmol/s),
F_{Li} is the liquid feed flow to plate i (kmol/s),
S_i is the total sidestream flow extracted from plate i (kmol/s),
B_{ij} is the boiloff rate of component j from the liquid in plate i (kmol/s),
x_{ij} is the mole fraction of component j in the liquid phase in plate i,

y_{ij} is the mole fraction of component j in the vapour phase in plate i,
z_{Lij} is the mole fraction of component j in the liquid feed to plate i.

The liquid mole fraction is given by

$$x_{ij} = \frac{M_{Lij}}{M_{Li}} \qquad \text{for } j = 1, \ldots, n \tag{12.49}$$

where the total liquid mass in the ith plate, M_{Li}, is given by:

$$M_{Li} = \sum_{j=1}^{n} M_{Lij} \tag{12.50}$$

We have by differentiation of (12.49):

$$\frac{dx_{ij}}{dt} = \frac{1}{M_{Li}}\left(\frac{dM_{Lij}}{dt} - x_{ij}\frac{dM_{Li}}{dt}\right) \qquad \text{for } j = 1, \ldots, n \tag{12.51}$$

We may carry out a total mass balance on the liquid in plate i to find dM_{Li}/dt:

$$\frac{dM_{Li}}{dt} = L_{i+1} + E_{i-1} + F_{Li} - B_i - L_i - S_i \tag{12.52}$$

where the total boiloff rate in plate i, B_i, is given by:

$$B_i = \sum_{j=1}^{n} B_{ij} \tag{12.53}$$

Vapour mass balance on plate i

A component mass balance on the vapour space yields

$$\frac{dM_{Vij}}{dt} = z_{Vij}F_{Vi} + B_{ij} - y_{ij}E_i \qquad \text{for } j = 1, \ldots, n \tag{12.54}$$

where
M_{Vij} is the mass of component j in the vapour of plate i (kmol),
F_{Vi} is the feed flow of vapour to plate i (kmol/s),
z_{Vij} is the mole fraction of component j in the vapour feed to plate i.

Energy balance for plate i

It is sensible to take an energy balance over the plate as a whole because the liquid and the vapour will share the same temperature, as a consequence of their equilibrium state. Following the methods of Chapter 3, Section 3.5, the total energy of the components on the plate is essentially the total internal energy, since the kinetic and potential energy may be ignored. The total internal energy of the liquid and vapour, U_i is given by

$$U_i = \sum_{j=1}^{n} (\hat{u}_{Lij}M_{Lij} + \hat{u}_{Vij}M_{Vij}) \tag{12.55}$$

where
$\hat{u}_{Lij}$ is the specific internal energy of component j in the liquid on plate i (J/kmol),
$\hat{u}_{Vij}$ is the specific internal energy of component j in the vapour on plate i (J/kmol).

Since both $\hat{u}_{Lij}$ and $\hat{u}_{Lij}$ will be functions of temperature alone, it is possible to differentiate equation (12.55) with respect to time to obtain:

$$\frac{dU_i}{dt} = \frac{dT_i}{dt}\sum_{j=1}^{n}\left(M_{Lij}\frac{d\hat{u}_{Lij}}{dT_i} + M_{Vij}\frac{d\hat{u}_{Vij}}{dT_i}\right) + \sum_{j=1}^{n}\left(\hat{u}_{Lij}\frac{dM_{Lij}}{dt} + \hat{u}_{Vij}\frac{dM_{Vij}}{dt}\right) \tag{12.56}$$

Carrying out the energy balance over the plate based on the energy flows into and out of the plate gives (cf. equation (3.38)):

$$\begin{aligned}\frac{dU_i}{dt} = {} & \Phi_i + L_{i+1}\sum_{j=1}^{n} x_{i+1,j}\hat{h}_{Li+1,j} \\ & + E_{i-1}\sum_{j=1}^{n} y_{i-1,j}\hat{h}_{Vi-1,j} + F_{Li}\sum_{j=1}^{n} z_{Lij}\hat{h}_{FLij} \\ & + F_{Vi}\sum_{j=1}^{n} z_{Vij}\hat{h}_{FVij} - L_i\sum_{j=1}^{n} x_{ij}\hat{h}_{Lij} \\ & - E_i\sum_{j=1}^{n} y_{ij}\hat{h}_{Vij} - S_i\sum_{j=1}^{n} x_{ij}\hat{h}_{Lij}\end{aligned} \tag{12.57}$$

where
$\hat{h}_{Lij}$ is the enthalpy of liquid component j on plate i (J/kmol),
$\hat{h}_{Vij}$ is the enthalpy of vapour component j on plate i (J/kmol),
$\hat{h}_{FLij}$ is the enthalpy of liquid component j in the feed to plate i (J/kmol),
$\hat{h}_{FVij}$ is the enthalpy of vapour component j in the feed to plate i (J/kmol),
Φ_i is the heat input to plate i, usually coming from or going to the metal of the plate and the surrounding shell (W).

Equating the right-hand sides of equations (12.56) and (12.57) and using the component mass balances of equations (12.48) and (12.54) produces the equation for the rate of change of plate temperature:

$$\frac{dT_i}{dt} = \begin{bmatrix} \Phi_i + L_{i+1}\sum_{j=1}^{n} x_{i+1,j}(\hat{h}_{Li+1,j} - \hat{u}_{Lij}) \\ + E_{i-1}\sum_{j=1}^{n} y_{i-1,j}(\hat{h}_{Vi-1,j} - \hat{u}_{Lij}) \\ + F_{Li}\sum_{j=1}^{n} z_{Lij}(\hat{h}_{FLij} - \hat{u}_{Lij}) \\ - L_i\sum_{j=1}^{n} x_{ij}(\hat{h}_{Lij} - \hat{u}_{Lij}) \\ - S_i\sum_{j=1}^{n} x_{ij}(\hat{h}_{Lij} - \hat{u}_{Lij}) \\ + F_{Vi}\sum_{j=1}^{n} z_{Vij}(\hat{h}_{FVij} - \hat{u}_{Vij}) \\ - E_i\sum_{j=1}^{n} y_{ij}(\hat{h}_{Vij} - \hat{u}_{Vij}) \\ - \sum_{j=1}^{n} B_{ij}(\hat{u}_{Vij} - \hat{u}_{Lij}) \end{bmatrix} \div \left[\sum_{j=1}^{n}\left(M_{Lij}\frac{d\hat{u}_{Lij}}{dT_i} + M_{Vij}\frac{d\hat{u}_{Vij}}{dT_i}\right)\right] \tag{12.58}$$

Plate liquid and vapour volumes

The volume of the liquid and the volume of the vapour will always equal the total volume of plate i:

$$V_{Li} + V_{Vi} - V_{Ti} = 0 \tag{12.59}$$

Using Amagat's law for liquid molar volumes at moderate pressures, we may write:

$$V_{Li} = \sum_{j=1}^{n} \hat{v}_{Lij} M_{Lij} \tag{12.60}$$

where $\hat{v}_{Lij} = \hat{v}_{Lij}(T_i)$ is the specific volume of liquid component j on plate i (m^3/kmol). Hence equation (12.59) becomes:

$$\sum_{j=1}^{n} \hat{v}_{Lij} M_{Lij} + V_{Vi} - V_{Ti} = 0 \tag{12.61}$$

Plate pressure

The partial pressure of each component in the vapour space of the plate will obey an equation of state of the general form of equation (12.11):

$$p_{ij} = f_j(M_{Vij}, T_i, V_{Vi}) \quad \text{for } j = 1, \ldots, n \tag{12.62}$$

where f_j is the state function describing the behaviour of component j, and V_{Vi} is the volume occupied by the vapour on plate i. But the partial pressure must also conform to the condition for vapour–liquid equilibrium, which relates it to the liquid fraction on each plate and the temperature:

$$p_{ij} = \gamma_{ij} K_j(T_i) x_{ij} \quad \text{for } j = 1, \ldots, n \tag{12.63}$$

where

$K_j(T)$ is the vapour pressure curve for component j, and

$\gamma_{ij} = \gamma_{ij}(x_{i1}, x_{i2}, \ldots, x_{in}, T_i)$ is the activity of component j on plate i, which will depend, in general, on the liquid fractions of all the components on the plate, and on the plate temperature. But note that the activity will be equal to unity for an ideal mixture.

By Dalton's law, the mole fraction of each component is given by:

$$y_{ij} = \frac{p_{ij}}{p_i} \quad \text{for } j = 1, \ldots, n \tag{12.64}$$

where the total plate pressure, p_i, is given by:

$$p_i = \sum_{j=1}^{n} p_{ij} \tag{12.65}$$

Combining equations (12.62) and (12.63) gives:

$$\gamma_{ij} K_j(T_i) x_{ij} - f_j(M_{Vij}, T_i, V_{Vi}) = 0 \quad \text{for } j = 1, \ldots, n \tag{12.66}$$

Applying the Method of Referred Derivatives

We have in equations (12.48) and (12.58) a set of equations in our chosen state variables, but the fact that we do not know the boiloff rate prevents us from solving these equations without the additional differential equation (12.54) and the algebraic equations (12.61) and (12.66). So now we use the Method of Referred Derivatives explained in Chapter 2, Section 2.11 to put the algebraic constraints into the appropriate differential form.

Equations (12.61) and (12.66) represents a system of simultaneous, nonlinear equations of the form given by equation (2.93) in Section 2.10 of Chapter 2, namely

$$\mathbf{g}(\mathbf{z}(t), \mathbf{x}(t), \mathbf{u}(t)) = \mathbf{0} \tag{2.93}$$

where in this case the vector of unknown algebraic variables, $\mathbf{z}$, contains the vapour component masses and the volume of the vapour space:

$$\mathbf{z} = \begin{bmatrix} M_{Vi1} \\ M_{Vi2} \\ \vdots \\ M_{Vin} \\ V_{Vi} \end{bmatrix} \tag{12.67}$$

and the vector of states, $\mathbf{x}$, for plate i is given by

$$\mathbf{x} = \begin{bmatrix} M_{Li1} \\ M_{Li2} \\ \vdots \\ M_{Lin} \\ T_i \end{bmatrix} \tag{12.68}$$

and there is no input vector, $\mathbf{u}$. Differentiating equation (2.93) with respect to time for the case when there is no input vector gives:

$$\frac{\partial \mathbf{g}}{\partial \mathbf{z}}\frac{d\mathbf{z}}{dt} + \frac{\partial \mathbf{g}}{\partial \mathbf{x}}\frac{d\mathbf{x}}{dt} = \mathbf{0} \tag{12.69}$$

To apply this procedure, we differentiate equations (12.61) and (12.66) with respect to time:

$$\frac{dV_{Vi}}{dt} + \frac{dT_i}{dt}\sum_{j=1}^{n} M_{Lij}\frac{d\hat{v}_{Lij}}{dT_i} + \sum_{j=1}^{n} \hat{v_{Lij}}\frac{dM_{Lij}}{dt} = 0 \tag{12.70}$$

and

$$\begin{aligned}
&\left(K_j x_{ij}\frac{d\gamma_{ij}}{dT_i} + \gamma_{ij}x_{ij}\frac{dK_j}{dT_i} - \frac{\partial f_j}{dT_i}\right)\frac{dT_i}{dt} + K_j\gamma_{ij}\frac{dx_{ij}}{dt} \\
&\quad + K_j x_{ij}\sum_{k=1}^{n}\frac{\partial \gamma_{ij}}{\partial x_{ik}}\frac{dx_{ik}}{dt} - \frac{\partial f_j}{\partial M_{Vij}}\frac{dM_{Vij}}{dt} \\
&\quad - \frac{\partial f_j}{\partial V_{Vi}}\frac{dV_{Vi}}{dt} = 0 \qquad \text{for } j = 1, \ldots, n
\end{aligned} \tag{12.71}$$

We may eliminate $\dfrac{dV_{Vi}}{dt}$ from equation (12.71) using equation (12.70):

$$\begin{aligned}
&\left(K_j x_{ij}\frac{d\gamma_{ij}}{dT_i} + \gamma_{ij}x_{ij}\frac{dK_j}{dT_i} + \frac{\partial f_j}{\partial V_{Vi}}\sum_{m=1}^{n} M_{Lim}\right. \\
&\quad \times \left.\frac{d\hat{v}_{Lim}}{dT_i} - \frac{\partial f_j}{dT_i}\right)\frac{dT_i}{dt} + K_j\gamma_{ij}\frac{dx_{ij}}{dt} \\
&\quad + K_j x_{ij}\sum_{k=1}^{n}\frac{\partial \gamma_{ij}}{\partial x_{ik}}\frac{dx_{ik}}{dt} - \frac{\partial f_j}{\partial M_{Vij}}\frac{dM_{Vij}}{dt} \\
&\quad + \frac{\partial f_j}{\partial V_{Vi}}\sum_{k=1}^{n}\hat{v}_{Lik}\frac{dM_{Lik}}{dt} = 0 \quad \text{for } j = 1, \ldots, n
\end{aligned} \tag{12.72}$$

We may eliminate $\dfrac{dx_{ij}}{dt}$ using equation (12.51):

$$\begin{aligned}
&\left(K_j x_{ij}\frac{d\gamma_{ij}}{dT_i} + \gamma_{ij}x_{ij}\frac{dK_j}{dT}\right. \\
&\quad \left.+ \frac{\partial f_j}{\partial V_{vi}}\sum_{m=1}^{n} M_{Lim}\frac{d\hat{v}_{Lim}}{dT} - \frac{\partial f_j}{dT_i}\right)\frac{dT_i}{dt} \\
&\quad + \frac{K_j x_{ij}}{M_{Li}}\sum_{k=1}^{n}\frac{\partial \gamma_{ij}}{\partial x_{ik}}\frac{dM_{Lik}}{dt} \\
&\quad - \frac{K_j x_{ij}}{M_{Li}}\left(\gamma_{ij} + \sum_{m=1}^{n}\frac{\partial \gamma_{ij}}{\partial x_{im}}x_{im}\right)\frac{dM_{Li}}{dt} \\
&\quad + \frac{\partial f_j}{\partial V_{Vi}}\sum_{k=1}^{n}\hat{v}_{Lik}\frac{dM_{Lik}}{dt} + \frac{K_j\gamma_{ij}}{M_{Li}}\frac{dM_{Lij}}{dt} \\
&\quad - \frac{\partial f_j}{\partial M_{Vij}}\frac{dM_{Vij}}{dt} = 0 \qquad \text{for } j = 1, \ldots n
\end{aligned} \tag{12.73}$$

Each of the derivatives

$$\frac{dT_i}{dt},\ \frac{dM_{Lij}}{dt},\ \frac{dM_{Li}}{dt}\left(= \sum_{j=1}^{n}\frac{dM_{Lij}}{dt}\right) \text{ and } \frac{dm_{Vij}}{dt}$$

has a dependence on the component boiloff rates, B_{ij}, $j = 1$ to n, and it is these boiloff rates that we wish to calculate by means of the n equations (12.73).

The several flows into and out of the distillation plate make the form of the differential equations rather 'busy', but it is possible to simplify the algebra without any loss of information by grouping together the expressions that are independent of the boiloff rates. Hence we may rewrite equation (12.58) as:

$$\frac{dT_i}{dt} = \frac{\alpha_1^{(i)}}{\alpha_2^{(i)}} - \frac{\sum_{j=1}^{n}(\hat{u}_{Vij} - \hat{u}_{Lij})B_{ij}}{\alpha_2^{(i)}} \tag{12.74}$$

where $\alpha_1^{(i)}$ is a grouped expression dependent on i only:

$$\begin{aligned}
\alpha_1^{(i)} &= \Phi_i + L_{i+1}\sum_{j=1}^{n} x_{i+1,j}(\hat{h}_{Li+1,j} - \hat{u}_{Lij}) \\
&\quad + E_{i-1}\sum_{j=1}^{n} y_{i-1,j}(\hat{h}_{Vi-1,j} - \hat{u}_{Lij}) \\
&\quad + F_{Li}\sum_{j=1}^{n} z_{Lij}(\hat{h}_{FLij} - \hat{u}_{Lij}) - L_i\sum_{j=1}^{n} x_{ij}(\hat{h}_{Lij} - \hat{u}_{Lij}) \\
&\quad - S_i\sum_{j=1}^{n} x_{ij}(\hat{h}_{Lij} - \hat{u}_{Lij}) + F_{Vi}\sum_{j=1}^{n} z_{Vij}(\hat{h}_{FVij} - \hat{u}_{Vij}) \\
&\quad - E_i\sum_{j=1}^{n} y_{ij}(\hat{h}_{Vij} - \hat{u}_{Vij})
\end{aligned} \tag{12.75}$$

$\alpha_1^{(i)}$ may be regarded as the net external heating to plate i.

The expression $\alpha_2^{(i)}$ is the total specific heat capacity at constant volume of the vapour and liquid on plate

i, and thus depends only on i:

$$\alpha_2^{(i)} = \sum_{m=1}^{n} \left(M_{Lim} \frac{d\hat{u}_{Lim}}{dT_i} + M_{Vim} \frac{d\hat{u}_{Vim}}{dT_i} \right) \quad (12.76)$$

The liquid component balance of equation (12.48) may be rewritten:

$$\frac{dM_{Lij}}{dt} = \alpha_3^{(ij)} - B_{ij} \qquad \text{for } j = 1, \ldots, n \quad (12.77)$$

where $\alpha_3^{(ij)}$ is an expression dependent on both i and j, representing the net *external* flow of component j into the liquid of the ith plate:

$$\alpha_3^{(ij)} = x_{i+1,j}L_{i+1} + y_{i-1,j}E_{i-1} + z_{Lij}F_{Li} - x_{ij}L_i - x_{ij}S_i \quad \text{for } j = 1, \ldots n \quad (12.78)$$

The total liquid balance of equation (12.52) may be re-expressed as:

$$\frac{dM_{Li}}{dt} = \alpha_4^{(i)} - \sum_{j=1}^{n} B_{ij} \quad (12.79)$$

where $\alpha_4^{(i)}$ depends only on i, and represents the net *external* flow into the liquid of the ith plate:

$$\alpha_4^{(i)} = L_{i+1} + E_{i-1} + F_{Li} - L_i - S_i \quad (12.80)$$

The component vapour balance of equation (12.54) may be rewritten:

$$\frac{dM_{Vij}}{dt} = \alpha_5^{(ij)} + B_{ij} \qquad \text{for } j = 1, \ldots, n \quad (12.81)$$

where $\alpha_5^{(ij)}$ is the net *external* flow of component j into the vapour of the ith plate, and is thus dependent on both i and j:

$$\alpha_5^{(ij)} = z_{Vij}F_{Vi} - y_{ij}E_i \qquad \text{for } j = 1, \ldots, n \quad (12.82)$$

It is also helpful to take groupings from equation (12.73):

$$\alpha_6^{(ij)} = K_j x_{ij} \frac{d\gamma_{ij}}{dT_i} + \gamma_{ij} x_{ij} \frac{dK_j}{dT_i} + \frac{\partial f_i}{\partial V_{Vi}} \sum_{m=1}^{n} M_{Lim} \frac{d\hat{v}_{Lim}}{dT_i} - \frac{\partial f_j}{dT_i} \qquad \text{for } j = 1, \ldots, n \quad (12.83)$$

$$\alpha_7^{(ij)} = \gamma_{ij} + \sum_{m=1}^{n} \frac{\partial \gamma_{ij}}{\partial x_{im}} x_{im} \qquad \text{for } j = 1, \ldots, n \quad (12.84)$$

Substituting into equation (12.73) and rearranging gives the required set of equations in the boiloff rates, B_{ij}:

$$\begin{aligned}
&\frac{\alpha_6^{(ij)}}{\alpha_2^{(i)}} \sum_{k=1}^{n} (\hat{u}_{Vik} - \hat{u}_{Lik}) B_{ik} + \frac{K_j x_{ij}}{M_{Li}} \sum_{k=1}^{n} \frac{\partial \gamma_{ij}}{\partial x_{ik}} B_{ik} \\
&\quad - \frac{K_j x_{ij}}{M_{Li}} \alpha_7^{(ij)} \sum_{k=1}^{n} B_{ik} + \frac{\partial f_j}{\partial V_{Vi}} \sum_{k=1}^{n} \hat{v}_{Lik} B_{ik} \\
&\quad + \left(\frac{K_j \gamma_{ij}}{M_{Li}} + \frac{\partial f_j}{\partial M_{Vij}} \right) B_{ij} \\
&= \frac{\alpha_6^{(ij)}}{\alpha_2^{(i)}} \alpha_1^{(i)} + \frac{K_j x_{ij}}{M_{Li}} \sum_{m=1}^{n} \frac{\partial \gamma_{ij}}{\partial x_{im}} \alpha_3^{(im)} - \frac{K_j x_{ij}}{M_{Li}} \alpha_7^{(ij)} \alpha_4^{(i)} \\
&\quad + \frac{\partial f_j}{\partial V_{Vi}} \sum_{m=1}^{n} \hat{v}_{Lim} \alpha_3^{(im)} + \frac{K_j \gamma_{ij}}{M_{Li}} \alpha_3^{(ij)} - \frac{\partial f_j}{\partial M_{Vij}} \alpha_5^{(ij)} \\
&\quad \text{for } j = 1, \ldots, n
\end{aligned} \quad (12.85)$$

Equation set (12.85) contains n linear, simultaneous equations in the n unknown boiloff rates, B_{ij}, for $j = 1, \ldots n$. It may be rewritten in the matrix form:

$$\mathbf{A}^{(i)} \mathbf{b}^{(i)} = \mathbf{c}^{(i)} \quad (12.86)$$

where $\mathbf{b}^{(i)}$ is the $n \times 1$ vector of component boiloff rates on plate i:

$$\mathbf{b}^{(i)} = \begin{bmatrix} B_{i1} \\ B_{i2} \\ \vdots \\ B_{in} \end{bmatrix} \quad (12.87)$$

$\mathbf{c}^{(i)}$ is the $n \times 1$ vector associated with plate i:

$$\mathbf{c}^{(i)} = \begin{bmatrix} c_1^{(i)} \\ c_2^{(i)} \\ \vdots \\ c_n^{(i)} \end{bmatrix} \quad (12.88)$$

where the element on the jth row of $\mathbf{c}^{(i)}$ is given by:

$$\begin{aligned}
c_j^{(i)} &= \frac{\alpha_6^{(ij)}}{\alpha_2^{(i)}} \alpha_1^{(i)} + \frac{K_j x_{ij}}{M_{Li}} \sum_{m=1}^{n} \frac{\partial \gamma_{ij}}{\partial x_{im}} \alpha_3^{(im)} \\
&\quad - \frac{K_j x_{ij}}{M_{Li}} \alpha_7^{(ij)} \alpha_4^{(i)} + \frac{\partial f_j}{\partial V_{Vi}} \sum_{m=1}^{n} \hat{v}_{Lim} \alpha_3^{(im)} \\
&\quad + \frac{K_j \gamma_{ij}}{M_{Li}} \alpha_3^{(ij)} - \frac{\partial f_j}{\partial M_{Vij}} \alpha_5^{(ij)}
\end{aligned} \quad (12.89)$$

The $n \times n$ matrix $\mathbf{A}^{(i)}$ associated with plate i is the sum of the two $n \times n$ matrices:

$$\mathbf{A}^{(i)} = \mathbf{D}^{(i)} + \Lambda^{(i)} \quad (12.90)$$

The element at the intersection of row j and column k of the matrix $\mathbf{D}^{(i)}$ is given by:

$$d_{jk}^{(i)} = \frac{\alpha_6^{(ij)}}{\alpha_2^{(i)}}(\hat{u}_{Vik} - \hat{u}_{Lik}) + \frac{K_j x_{ij}}{M_{Li}}\frac{\partial \gamma_{ij}}{\partial x_{ik}} - \frac{K_j x_{ij}}{M_{Li}}\alpha_7^{(ij)} + \frac{\partial f_j}{\partial V_{Vi}}\hat{v}_{Lik} \qquad (12.91)$$

while the matrix, $\Lambda^{(i)}$, is a diagonal matrix, with diagonal elements:

$$\lambda_{jj}^{(i)} = \frac{K_j \gamma_{ij}}{M_{Li}} + \frac{\partial f_j}{\partial M_{Vij}} \qquad (12.92)$$

Solution sequence

Matrix equation (12.86) is *linear*, and its solution is therefore guaranteed as

$$\mathbf{b}^{(i)} = (\mathbf{A}^{(i)})^{-1}\mathbf{c}^{(i)} \qquad (12.93)$$

Equation (12.93) is, in fact, a multicomponent generalization of the single component boiling/condensing equation (12.23).

For binary distillation, it is a simple matter to write an explicit solution for the boiloff rates for the two components on plate i:

$$B_{i1} = \frac{c_1^{(i)}a_{22}^{(i)} - c_2^{(i)}a_{12}^{(i)}}{a_{11}^{(i)}a_{22}^{(i)} - a_{12}^{(i)}a_{21}^{(i)}} \qquad (12.94)$$

$$B_{i2} = \frac{c_2^{(i)}a_{11}^{(i)} - c_1^{(i)}a_{21}^{(i)}}{a_{11}^{(i)}a_{22}^{(i)} - a_{12}^{(i)}a_{21}^{(i)}} \qquad (12.95)$$

where $a_{jk}^{(i)} = d_{jk}^{(i)} + \lambda_{jk}^{(i)}$ is the element at the intersection of row j and column k of the matrix $\mathbf{A}^{(i)}$. In a similar way, for ternary distillation, the boiloff rates for the three components on plate i may be found from Cramer's Rule as:

$$B_{i1} = \frac{c_1^{(i)}(a_{22}^{(i)}a_{33}^{(i)} - a_{23}^{(i)}a_{32}^{(i)}) - c_2^{(i)}(a_{12}^{(i)}a_{33}^{(i)} - a_{13}^{(i)}a_{32}^{(i)}) + c_3^{(i)}(a_{12}^{(i)}a_{23}^{(i)} - a_{13}^{(i)}a_{22}^{(i)})}{\det \mathbf{A}^{(i)}} \qquad (12.96)$$

$$B_{i2} = \frac{a_{11}^{(i)}(c_2^{(i)}a_{33}^{(i)} - c_3^{(i)}a_{23}^{(i)}) - a_{21}^{(i)}(c_1^{(i)}a_{33}^{(i)} - c_3^{(i)}a_{13}^{(i)}) + a_{31}^{(i)}(c_1^{(i)}a_{23}^{(i)} - c_2^{(i)}a_{13}^{(i)})}{\det \mathbf{A}^{(i)}} \qquad (12.97)$$

$$B_{i3} = \frac{a_{11}^{(i)}(c_3^{(i)}a_{22}^{(i)} - c_2^{(i)}a_{32}^{(i)}) - a_{21}^{(i)}(c_3^{(i)}a_{12}^{(i)} - c_1^{(i)}a_{32}^{(i)}) + a_{31}^{(i)}(c_2^{(i)}a_{12}^{(i)} - c_1^{(i)}a_{22}^{(i)})}{\det \mathbf{A}^{(i)}} \qquad (12.98)$$

$$\det \mathbf{A}^{(i)} = a_{11}^{(i)}(a_{22}^{(i)}a_{33}^{(i)} - a_{23}^{(i)}a_{32}^{(i)}) - a_{21}^{(i)}(a_{12}^{(i)}a_{33}^{(i)} - a_{13}^{(i)}a_{32}^{(i)}) + a_{31}^{(i)}(a_{12}^{(i)}a_{23}^{(i)} - a_{13}^{(i)}a_{22}^{(i)}) \qquad (12.99)$$

Explicit solutions cease to be computationally efficient as the number of components in the distillation process rises: in fact the situation is already marginal when $n = 3$. But it is possible to use a direct method such as Gauss elimination or an indirect, iterative method, each of which will generate a rapid solution. The fact that the equations are linear means that there will never be a problem with convergence, as is possible with nonlinear equations.

Once the component boiloff rates have been found, it becomes possible to calculate the derivatives of the liquid masses and temperature using equations (12.77) and (12.74):

$$\frac{dM_{Lij}}{dt} = \alpha_3^{(ij)} - B_{ij} \qquad \text{for } j = 1, \ldots, n \qquad (12.77)$$

$$\frac{dT_i}{dt} = \frac{\alpha_1^{(i)}}{\alpha_2^{(i)}} - \frac{\sum_{j=1}^{n}(\hat{u}_{Vij} - \hat{u}_{Lij})B_{ij}}{\alpha_2^{(i)}} \qquad (12.74)$$

Thus given a plate starting condition $[M_{Li1}(0), M_{Li2}(0), \ldots, M_{Lin}(0), T_i(0)]$, it is possible to integrate forward for all future times, as was the case for single component boiling/condensing.

Once the component masses are known, it is possible to calculate the mole fractions x_{ij} from equations (12.49) and (12.50). If the temperature is known also, the activity of each component, γ_{ij}, may be found from

$$\gamma_{ij} = \gamma_{ij}(x_{i1}, x_{i2}, \ldots, x_{in}, T_i) \qquad \text{for } j = 1, \ldots, n \qquad (12.100)$$

A knowledge of the plate temperature allows the following thermodynamic variables to be calculated:

$$\begin{aligned} \hat{u}_{Lij} &= \hat{u}_{Lij}(T_i) \\ \hat{u}_{Vij} &= \hat{u}_{Vij}(T_i) \qquad \text{for } j = 1, \ldots, n \\ \hat{v}_{Lij} &= \hat{v}_{Lij}(T_i) \end{aligned} \qquad (12.101)$$

The partial pressure of each component may now be found using equation (12.63). This allows the mole fraction of the components in the vapour space to be calculated using equations (12.64) and (12.65).

The volume of the liquid on the tray, V_{Li}, may be found from equation (12.60), and this allows the volume of the vapour, V_{Vi}, to be found from subtraction using equation (12.59). A knowledge of the volume of the vapour, the temperature and the partial pressure of each component allows the mass of each vapour component to be calculated using equation (12.62).

Knowledge of the total mass and total volume of the liquid on the tray allows the overall liquid specific volume in m³/kmol, $\hat{v}_{Li}$, to be calculated:

$$\hat{v}_{Li} = \frac{V_{Li}}{M_{Li}} \qquad (12.102)$$

and a similar calculation is possible for the overall vapour specific volume once the vapour volume and total vapour mass have been found:

$$\hat{v}_{Vi} = \frac{V_{Vi}}{M_{Vi}} \quad (12.103)$$

The effective molecular weight, w_{Li}, of the liquid on the ith tray follows from equation (11.28) of Chapter 11, Section 11.4 as

$$w_{Li} = \sum_{j=1}^{n} x_{ij} w_j \quad (12.104)$$

where w_j is the molecular weight of component j. Similarly, the effective molecular weight, w_{Vi}, of the vapour in the ith plate is given by:

$$w_{Vi} = \sum_{j=1}^{n} y_{ij} w_j \quad (12.105)$$

The effective molecular weights may be used to transform the specific properties from a per-kmol basis to a per-kg basis. For example, the specific volumes in m^3/kg are:

$$v_{Li} = \frac{\hat{v}_{Li}}{w_{Li}}, \qquad v_{Vi} = \frac{\hat{v}_{Vi}}{w_{Vi}} \quad (12.106)$$

from equation (11.30).

12.7 Functions used in the modelling of the distillation plate

It will normally be sufficient to use the ideal-gas relationship for the partial pressure function, f_j, for component j on plate i:

$$p_{ij} = f_j(M_{Vij}, T_i, V_{Vi}) = \frac{M_{Vij} R T_i}{V_{Vi}} \quad (12.107)$$

which yields the following partial derivatives:

$$\frac{\partial f_j}{\partial M_{Vij}} = \frac{R T_i}{V_{Vi}}, \qquad \frac{\partial f_j}{\partial V_{Vi}} = -\frac{M_{Vij} R T_i}{V_{Vi}^2}, \qquad \frac{\partial f_j}{\partial T_i} = \frac{R M_{Vij}}{V_{Vi}} \quad (12.108)$$

The vapour pressure curves, $K_j(T_i)$, may be represented by either Antoine or Riedel equations, as explained for the single-component case in Section 12.3. Similarly table lookups or low-order polynomial expressions may be used to describe the dependence on temperature (only) of the specific volume of each liquid component and the specific internal energy of each component in both the liquid and the vapour phase.

For an ideal solution, all the activity coefficients are equal to unity, and the partial derivatives are identically zero. Thus a useful simplification to the algebra occurs if it is possible to make this approximation in the actual case on plant. In general, however, the activity coefficients, γ_{ij}, are dependent on the liquid mole fractions and the plate temperature. They are calculated using correlations derived from experiments with binary mixtures. Margules, van Laar, Wilson and UNIQUAC correlations are in wide use. The Margules correlation is algebraically the simplest of these, and it offers an accuracy comparable to the van Laar correlation. Although there are circumstances where the additional complexity of the Wilson or UNIQUAC correlations are justified for design work, the Margules correlation is always likely to be sufficient for control engineering purposes.

The Margules correlation gives the two activities for a binary mixture (in plate i) about a reference temperature, T_0, as:

$$\begin{aligned} \ln \gamma_{i1}(T_0) &= x_{i2}^2 (A_{12} + 2x_{i1}(A_{21} - A_{12})) \\ \ln \gamma_{i2}(T_0) &= x_{i1}^2 (A_{21} + 2x_{i2}(A_{12} - A_{21})) \end{aligned} \quad (12.109)$$

where A_{12} and A_{21} are the binary interaction parameters for components 1 and 2. It may be seen by substituting $x_{i1} \to 1$ and $x_{i2} = 1 - x_{i1} \to 0$ into equation (12.109) that the activity of component 1, γ_{i1}, approaches unity when only an infinitesimal amount of component 2 is present in the liquid mixture. Conversely, substituting $x_{i1} \to 0$ and $x_{i2} = 1 - x_{i1} \to 1$ into equation (12.109) shows that the activity for component 1 is given by $\gamma_{i1} = e^{A_{12}}$ when component 1 is diluted to an infinite extent in component 2. Similarly, the activity for component 2, γ_{i2}, varies between unity and $e^{A_{21}}$ as the composition of the mixture changes from purely component 2 to purely component 1. Values of A_{12} and A_{21} for several chemical mixtures are found in Perry *et al.* (1984). Taking the case of the binary distillation of methanol (component 1) from water (component 2) at 1 atmosphere, $A_{12} = 0.7923$ and $A_{21} = 0.5434$. Hence γ_{i1} changes from 2.2085 to 1 and γ_{i2} from 1.0 to 1.7219 as the methanol content of the mixture changes from 0.0 to 1.0. These activity coefficients are seen to deviate significantly from unity at low concentrations of the individual components.

It is possible to extend the Margules correlation to multicomponent mixtures: although the expressions become more complicated, they depend still on the same, binary interaction parameters. See Perry (1963), Wohl (1946, 1953), Severns (1955), Chien and Null (1972). For ternary distillation, the activity coefficient at the reference temperature is given for component 1 by:

$$\ln \gamma_{i1}(T_0) = x_{i2}^2(A_{12} + 2x_{i1}(A_{21} - A_{12})) + x_{i3}^2(A_{13} + 2x_{i1}(A_{31} - A_{13})) + x_{i2}x_{i3} \times \begin{bmatrix} \frac{1}{2}(A_{21} + A_{12} + A_{31} + A_{13} - A_{23} - A_{32}) \\ + x_{i1}(A_{21} - A_{12} + A_{31} - A_{13}) \\ + (x_{i2} - x_{i3})(A_{23} - A_{32}) \end{bmatrix} \quad (12.110)$$

Similar equations for $\ln \gamma_{i2}$ and $\ln \gamma_{i3}$ may be obtained by changing subscripts from 1 to 2, from 2 to 3 and from 3 to 1; if the subscripts are changed once, the equation for $\ln \gamma_{i2}$ is obtained, and if they are changed twice, the equation for $\ln \gamma_{i3}$.

The Margules, binary-interaction parameters will normally be valid over a range of temperatures, corresponding usually to the terminal temperatures of a binary separation process operating at 1 atmosphere. If the distillation column is to be operated at a significantly different pressure, then allowance may need to be made for the resultant change in the temperature regime. The temperature dependency of the activity coefficient for component j may be represented by the form:

$$\ln \gamma_{ij}(T) = \ln \gamma_{ij}(T_0) + G_j\left(\frac{1}{T} - \frac{1}{T_0}\right) \quad (12.111)$$

where
T is the mid-point of the new temperature range (K),
T_0 is the mid-point of the reference temperature range (K).
and G_j is a constant for component j, with units of temperature, K.

G_j is given to reasonable accuracy by:

$$G_j = \frac{D_j}{R} \quad (12.112)$$

where D_j is the average partial heat of solution of component j (J/kmol) over the range T_0, to T, and R is the universal gas constant, $= 8314\,\text{J/(kmol K)}$. D_j is negative if heat is evolved when two liquids mix, and in this case the activity coefficient rises with temperature. However, many mixtures of organic liquids take up heat when mixing occurs, in which case D_j is positive, and the activity coefficient falls with temperature. Substituting from equation (12.109) into equation (12.111) gives the activity for component 1 at new temperature, T, as:

$$\ln \gamma_{i1}(T) = x_{i2}^2(A_{12} + 2x_{i1}(A_{21} - A_{12})) + G_1\left(\frac{1}{T} - \frac{1}{T_0}\right) \quad (12.113)$$

As the dilution of component 1 nears infinite, so $x_{i1} \to 0$, and we have

$$\ln \gamma_{i1}(T) \longrightarrow A_{12}(T) = A_{12}(T_0) + G_1\left(\frac{1}{T} - \frac{1}{T_0}\right) \quad (12.114)$$

Similarly, as the dilution of component 2 nears infinite, so $x_{i2} \to 0$, and we have

$$\ln \gamma_{i2}(T) \longrightarrow A_{21}(T) = A_{21}(T_0) + G_2\left(\frac{1}{T} - \frac{1}{T_0}\right) \quad (12.115)$$

Hence two new binary interaction coefficients have been derived for use with the new temperature range centred on T.

Although the development of the equations has allowed for the activity coefficients, γ_{ij} to be functions of temperature, we have seen that the Margules correlation is intended to be valid over a range of temperatures. Hence, if the Margules correlation is used, we may set the temperature derivatives of the activity coefficients to zero:

$$\frac{\partial \gamma_{ij}}{\partial T_i} = 0 \qquad \text{for all } i \text{ and } j \quad (12.116)$$

The partial derivatives with respect to liquid composition for a binary mixture are found by differentiating equation (12.109):

$$\begin{aligned} \frac{\partial \gamma_{il}}{\partial x_{i1}} &= 2x_{i2}^2(A_{21} - A_{12})\gamma_{i1} \\ \frac{\partial \gamma_{i1}}{\partial x_{i2}} &= 2x_{i2}(A_{12} + 2x_{i1}(A_{21} - A_{12}))\gamma_{i1} \\ \frac{\partial \gamma_{i2}}{\partial x_{i1}} &= 2x_{i1}(A_{21} + 2x_{i2}(A_{12} - A_{21}))\gamma_{i2} \\ \frac{\partial \gamma_{i2}}{\partial x_{i2}} &= 2x_{i1}^2(A_{12} - A_{21})\gamma_{i2} \end{aligned} \quad (12.117)$$

The partial derivatives with respect to liquid composition for a ternary mixture are found by differentiating equation (12.110):

$$\frac{\partial \gamma_{i1}}{\partial x_{i1}} = \gamma_{i1}[2x_{i2}^2(A_{21} - A_{12}) + 2x_{i3}^2(A_{31} - A_{13}) + x_{i2}x_{i3}(A_{21} - A_{12} + A_{31} - A_{13})]$$

$$\frac{\partial \gamma_{i1}}{\partial x_{i2}} = \gamma_{i1} \times \left\{ \begin{array}{l} 2x_{i2}(A_{12} + 2x_{i1}(A_{21} - A_{12})) + x_{i2}x_{i3}(A_{23} - A_{32}) \\ +x_{i3} \left[\begin{array}{l} \frac{1}{2}(A_{21} + A_{12} + A_{31} + A_{13} - A_{23} - A_{32}) \\ +x_{i1}(A_{21} - A_{12} + A_{31} - A_{13}) \\ +(x_{i2} - x_{i3})(A_{23} - A_{32}) \end{array} \right] \end{array} \right\}$$

$$\frac{\partial \gamma_{i1}}{\partial x_{i3}} = \gamma_{i1} \times \left\{ \begin{array}{l} 2x_{i3}(A_{13} + 2x_{i1}(A_{31} - A_{13})) + x_{i2}x_{i3}(A_{32} - A_{23}) \\ +x_{i2} \left[\begin{array}{l} \frac{1}{2}(A_{21} + A_{12} + A_{31} + A_{13} - A_{23} - A_{32}) \\ +x_{i1}(A_{21} - A_{12} + A_{31} - A_{13}) \\ +(x_{i2} - x_{i3})(A_{23} - A_{32}) \end{array} \right] \end{array} \right\} \quad (12.118)$$

Once again the derivatives $\partial \gamma_{i2}/\partial x_{ij}$, $\partial \gamma_{i3}/\partial x_{ij}$ for $j = 1, 2, 3$ may be found by rotating the subscripts in j.

12.8 Modelling the distillation column as a whole

Flow between plates

As shown in Figure 12.4, the liquid on tray i flows over a weir in order to reach tray (i-1), and we may use the Francis weir equation in order to relate the liquid flow rate in kmol/s to the height of the liquid on the tray:

$$L_i = 1.848 \frac{l_w(\theta_i - \theta_w)^{1.5}}{\hat{v}_{Li}} \quad (12.119)$$

where l_w is the length of the weir (m), θ_w is the height of the weir (m) and θ_i is the height of the liquid on the tray (m), given by:

$$\theta_i = \frac{V_{Li}}{A_{tray}} \quad (12.120)$$

where A_{tray} is the tray area in m^2.

The vapour flow will depend on both the difference in vapour-space pressure in adjacent plates and the head of liquid above the point of vapour entry into the plate above. For a sieve plate or valve-cap plate, the point of vapour entry is the bottom of the tray, but for a bubble-cap plate the point of entry is the lower edge of the bubble-cap. For a sieve or valve-cap plate, the vapour flow will have the form:

$$E_i = \frac{C_i}{w^{(i)}} \sqrt{\frac{p_{i+1} + g\dfrac{\theta_{i+1}}{v_{Li+1}} - p_i}{v_{Vi}}} \quad (12.121)$$

where C_i is the plate conductance between plate i and plate $i + 1$, with units of m^2, v_{Vi} is the overall specific volume of the vapours on plate i, while $w^{(i)}$ is the overall molecular weight of the vapours on plate i, given using equation (11.28) as

$$w^{(i)} = \sum_{j=1}^{n} \lambda_{ij} w_j \quad (12.122)$$

Here w_j is the molecular weight of component j, and λ_{ij} is the mole fraction of component j on plate i:

$$\lambda_{ij} = \frac{M_{Vij}}{\sum_{j=1}^{n} M_{Vij}} \quad (12.123)$$

Normally design calculations or plant data will be available from which C_i can be calculated.

Column efficiency

The theory developed in Section 12.6 assumed that the plate acts as a perfect equilibrium stage for separation. However, in practice it is found that this is only an approximation. While the temperature of the vapour leaving the plate is likely to be the same as that of the liquid leaving the plate, the difference in compositions of the vapour and the liquid is often not as great as implied by equation (12.63), and the plate has therefore not been such an efficient separation stage. The deviation from equilibrium is accounted for partly by gross physical phenomena such as carryover of liquid droplets in the vapour leaving the plate, but a more fundamental reason is the restricted rate of mass transfer between the vapour and liquid phases. A column efficiency, η_c, may be defined as the ratio of the number of plates needed in theory, N_T, to the number needed in practice, N_P, to achieve the desired separation:

$$\eta_c = \frac{N_T}{N_P} \quad (12.124)$$

A typical efficiency figure is 70%, although the efficiency can fall as low as 40%. The model may account for this less-than-perfect separation by using a reduced number, $\eta_c N_P$, of plates, but with the height of each plate and each weir upgraded by a factor of $1/\eta_c$ in order to maintain the overall volumes constant.

Metalwork thermal capacity

The thermal capacity of the metalwork is important for distillation dynamics in the same way as it was for steam drum dynamics. Allowance has been made for its inclusion by providing the energy balance equation for each plate with a heat flux variable, Φ_i. We may follow the procedure outlined in equations (12.45) to (12.47) to calculate the temperature of the metalwork associated with each plate and hence the heat flux,

Φ_i. Adapting equations (12.46) and (12.47) for the ith distillation plate yields:

$$\frac{dT_{mi}}{dt} = -\frac{\Phi_i}{m_{mi}c_{pmi}} \qquad (12.125)$$

$$\Phi_i = K_{mi}A_{mi}(T_{mi} - T_i) \qquad (12.126)$$

where

m_{mi} is the mass of the metalwork associated with the ith plate (kg),

c_{pmi} is the specific heat of the metal at plate i (J/(kgK)),

A_{mi} is the surface area of the metal exposed to plate i (m^2),

K_{mi} is the heat transfer coefficient between the metalwork and the two-phase plate contents ($W/(m^2K)$).

The analogy with the steam drum situation suggests that the value of K_{mi} will be very high – of the order of 10 000 $W/(m^2K)$.

Modelling the reboiler

The heating medium in a reboiler will often be condensing steam, in which case the reboiler may be viewed as a pair of two-phase systems: a single-component condenser exchanging heat with a multicomponent boiler. See Figure 12.5.

The steam condenser may be modelled by the equations developed in Section 12.2. Similarly, the multicomponent boiler will obey the equations developed in Section 12.6, after stipulating that there is only a liquid flow into the reboiler, which we may designate 'plate 0'.

The metalwork of the reboiler will obey the equation:

$$\frac{dT_{m0}}{dt} = -\frac{1}{m_{m0}c_{pm0}}(\Phi + \Phi_0) \qquad (12.127)$$

The heat passed from the reboiler metalwork to the steam/water at temperature T is given by:

$$\Phi = K_m A_m (T_{m0} - T) \qquad (12.128)$$

where the heat transfer coefficient, K_m, and the area, A_m, refer to the steam side of the reboiler (cf. equation (12.47)). The heat passed to the multicomponent mixture is given by:

$$\Phi_0 = K_{m0}A_{m0}(T_{m0} - T_0) \qquad (12.129)$$

where the nomenclature of equation (12.126) has been used, with $i = 0$.

Modelling the condenser and condensate vessel

The multicomponent mixture on the distillate side of the condenser and in the condensate vessel will obey the equations developed in Section 12.6. Thus the distillate side of the condenser and the condensate vessel together may be lumped together as 'plate $N + 1$', subject to an inlet flow of vapour only. However, assuming the cooling medium is water, the temperature of the water will vary along the condensing heat exchanger, as will the temperature of the metalwork. An accurate representation will require the water-side to be represented by a partial differential equation to account for this. (See Chapter 2, Section 2.6 for a discussion of the issue and Chapter 20, Section 20.2 for more detail.) The distributed aspects of the condenser and condensate vessel may therefore be represented as in Figure 12.6, where the temperature of

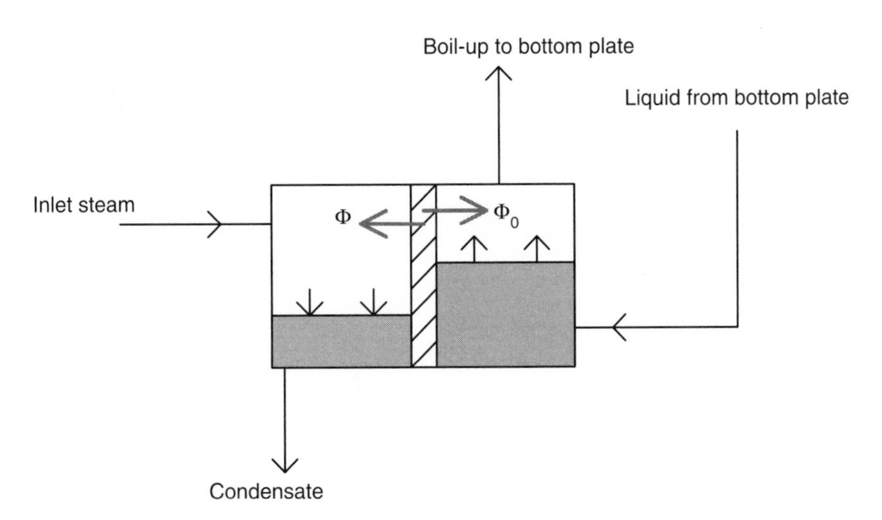

Figure 12.5 Schematic of steam reboiler.

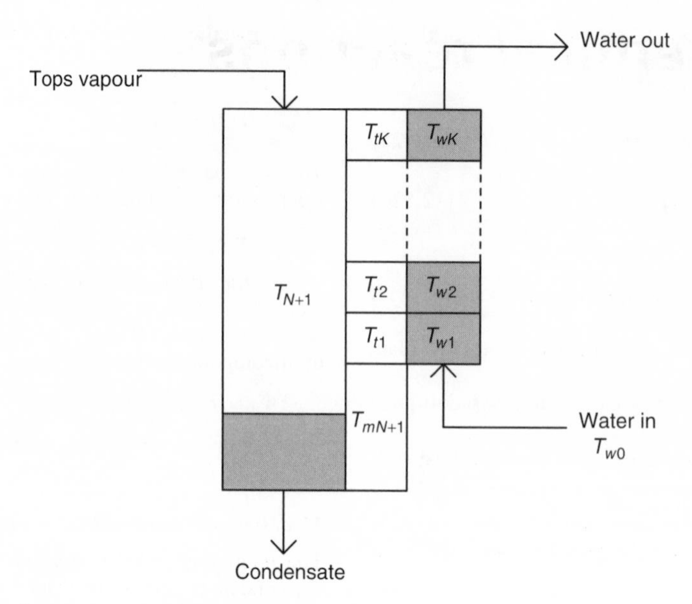

Figure 12.6 Schematic of condenser and condensate vessel.

the tube metal, T_{tk}, $k = 1, \ldots, K$, varies with distance along the condenser. A single temperature is sufficient for the metalwork in the condensing vessel.

The heat passed to the multicomponent mixture will be the sum of the contributions from the condenser tube and the rest of the metalwork:

$$\Phi_{N+1} = K_C \sum_{k=1}^{K} A_{tk}(T_{tk} - T_{N+1}) + K_{mN+1}A_{mN+1} \times (T_{mN+1} - T_{N+1}) \tag{12.130}$$

where

K_C is the heat transfer coefficient on the distillate side of the condenser (W/(m^{2}K)),

A_{tk} is the heat transfer area on the distillate side (m^2),

K_{mN+1} is the heat transfer coefficient for the condensate vessel (W/(m^{2}K)),

A_{mN+1} is the heat transfer area of the condensate vessel (m^2),

T_{N+1} is the temperature of the condensing vapour and condensed liquid (K),

T_{mN+1} is the temperature of the metalwork in the condensate vessel.

The term Φ_{N+1} will normally be negative since the design value should cause condensing.

12.9 Bibliography

Chien and Null, (1972) *American Institute of Chemical Engineers Journal*, **18**, 1177.

Knowles, J.B. (1990). *Simulation and Control in Electrical Power Stations*, Research Studies Press, Taunton, and John Wiley & Sons, New York.

Martin, E.N. (1970). The design and development of process control schemes using analogue and scale modelling techniques, *Chemical Age of India*, **21**, 184–194.

Perry, J.H. (1963). In *Chemical Engineers' Handbook*, 4th edition, edited by R.H. Perry, C.H. Chilton and S.D. Kirkpatrick, McGraw-Hill, New York. Chapter 13: Distillation.

Perry, R.H., Green, D.W. and Maloney, J.O. (eds) (1984). *Perry's Chemical Engineers' Handbook*, 6th edition, McGraw-Hill, New York. Chapter 13: Distillation; Chapter 18: Liquid-gas systems.

Severns, (1955). *American Institute of Chemical Engineers Journal*, **1**, 401.

Thomas, P.J. (1981). Dynamic simulation of multicomponent distillation processes, *Industrial Engineering Chemistry, Process Design and Development*, **20**, pp 166–168.

Thomas, P.J. (1997). Using the Method of Referred Derivatives to simulate the dynamics of systems in vapour-liquid equilibrium, *Trans. Inst. MC*, **19**, 23–37.

Wohl, (1953). *Chemical Engineering Progress*, **49**, 218.

Wohl, (1946). *Transactions of the American Institute of Chemical Engineers*, **42**, 215.

13 Chemical reactions

13.1 Introduction

Chemical reactions are at the heart of many process plants, where the changes devised to occur between molecules are scaled up to an industrial level. It is necessary to be able to predict how kilogram-mole quantities of products are formed, how fast those products are produced, and how fast the reagents are used up. Every chemical reaction is accompanied by an intake or release of heat energy, and the control engineer will need to be able to calculate the rate at which this occurs.

This chapter will explain the principles underlying chemical reactions, and it will go on to generalize these principles to the case of several concurrent reactions with large numbers of reagents and products. Then we shall extend to the case of chemical reaction the principles of mass balance and energy balance presented in Chapter 3. Finally we shall explain in detail how to simulate a gas reactor and a continuous stirred tank reactor (CSTR).

13.2 The reaction at the molecular and kilogram-mole levels

We will introduce the principles of chemical reaction with reference to the '*ABC*' reaction, where chemicals A and B react together to form chemical C. The reaction may be described by a 'chemical equation' of the form:

$$aA + bB = cC \tag{13.1}$$

where A represents one molecule of chemical A, B is one molecule of chemical B and C is one molecule of chemical C, while a, b and c are the stoichiometric coefficients. The meaning of chemical equation (13.1) is that a molecules of chemical A will combine with b molecules of chemical B to form c molecules of the chemical C.

A simple example involving just two molecules of reactants is the formation of phosgene from carbon monoxide and chlorine:

$$CO + Cl_2 = COCl_2 \tag{13.2}$$

In this case, $a = b = c = 1$. Another simple example is the reaction between carbon monoxide and oxygen, where two molecules of carbon monoxide combine with one molecule of oxygen to produce two molecules of carbon dioxide:

$$2CO + O_2 = 2CO_2 \tag{13.3}$$

Here $a = 2$, $b = 1$ and $c = 2$. Notice that in both these cases the number of molecules after the reaction is less than the number before the reaction, although the total mass will not have changed. In general, the total number of molecules in the system may be increased, decreased or left unchanged as a result of the reaction.

Chemical equation (13.1) will retain its validity if both sides are multiplied by any number we wish, since all that is then postulated is a number of similar, concurrent reactions. If we choose to multiply both sides of equation (13.1) by the number of molecules in a kilogram-mole, $N_{Ak} = 6.023 \times 10^{26}$ (Avogadro's number $\times$ 1000), then we have:

$$aN_{Ak}A + bN_{Ak}B = cN_{Ak}C \tag{13.4}$$

The quantity $N_{Ak}A$ in chemical equation (13.4) represents 6.023×10^{26} molecules of chemical A, i.e. a kilogram-mole's worth of chemical A. Similarly, $N_{Ak}B$ represents a kilogram-mole's worth of chemical B, and $N_{Ak}C$ represents a kilogram-mole's worth of chemical C. Let us use curly brackets, { }, to denote one kilogram-mole's worth of chemical, so that

$$\begin{aligned} \{A\} &= N_{Ak}A \\ \{B\} &= N_{Ak}B \\ \{C\} &= N_{Ak}C \end{aligned} \tag{13.5}$$

Chemical equation (13.4), may now be rewritten in terms of kilogram-mole quantities:

$$a\{A\} + b\{B\} = c\{C\} \tag{13.6}$$

Clearly equation (13.6) preserves the stoichiometric relationships of equation (13.1). Chemical equation (13.6) tells us that whenever c kmol of C are produced in the reaction defined by chemical equation (13.1), a kmol of chemical A and b kmol of chemical B are used up.

Chemical equation (13.6) may be multiplied by any number, n, in just the same way as equation (13.1):

$$an\{A\} + bn\{B\} = cn\{C\} \tag{13.7}$$

This implies that whenever cn kmol of C are produced in the reaction defined by chemical equation (13.1), an

kmol of chemical A and bn kmol of chemical B are used up.

13.3 Reaction rate relationship for the different chemical species in the reaction

Let us suppose that a system contains initial masses M_A kmol of chemical A, M_B kmol of chemical B and M_C kmol of chemical C. Let the reaction defined by equation (13.6) begin, so that after a time interval, δt, an additional δM_C kmol of chemical C have been formed. There is no loss of generality if we assume that δM_C is the mathematical product of the stoichiometric coefficient, c, and an increase in mass δM_R (kmol) defined by:

$$\delta M_C = c\,\delta M_R \tag{13.8}$$

We will now consider the significance of the term, δM_R. Recalling chemical equation (13.7), let us make the dimensionless multiplier, n, numerically equal to the magnitude of δM_R:

$$n = \delta M_R \tag{13.9}$$

so that equation (13.7) becomes:

$$a\,\delta M_R\{A\} + b\,\delta M_R\{B\} = c\,\delta M_R\{C\} \tag{13.10}$$

We may now apply to chemical equation (13.10) the interpretation we applied to chemical equation (13.7), namely that the production of $c\,\delta M_R$ kmol of chemical C has necessitated the using up of $a\,\delta M_R$ kmol of chemical A and $b\,\delta M_R$ kmol of chemical B, that is to say

$$\begin{aligned} \delta M_A &= -a\,\delta M_R \\ \delta M_B &= -b\,\delta M_R \end{aligned} \tag{13.11}$$

It will be seen by comparing equations (13.8) and (13.11) that

$$-\frac{1}{a}\,\delta M_A = -\frac{1}{b}\,\delta M_B = \frac{1}{c}\,\delta M_C = \delta M_R \tag{13.12}$$

Thus δM_R emerges from equation (13.12) as an important, normalized variable characterizing the progress of the reaction. It has units of kilogram-moles, but does not refer directly to any of the chemical species involved in the reaction, neither reactants nor products. Because of its importance, we choose to give M_R a special name, namely the 'kilogram-moles of reaction', with the units 'kmol rxn'. δM_R is the increase in the kilogram-moles of reaction, with the same units as M_R. It is clear from equation (13.12) that an increase of one kmol rxn implies that c kmol of chemical C will have been formed while a kmol of chemical A and b kmol of chemical B will have been used up.

If we divide (13.12) by the time interval, δt, and let $\delta t \to 0$, we come to the following relationship between the rates of changes of the various chemical kmol masses:

$$-\frac{1}{a}\frac{dM_A}{dt} = -\frac{1}{b}\frac{dM_B}{dt} = \frac{1}{c}\frac{dM_C}{dt} = \frac{dM_R}{dt} \tag{13.13}$$

We see that the term dM_R/dt is the normalized reaction rate (kmol rxn/s).

The relationship of equation (13.13) may be generalized to a chemical equation involving any number, N, of reactants and products. Let us extend the scope of chemical equation (13.6) by replacing the alphabetic symbols by alphanumerics:

$$\sum_{i=1}^{N} a_i A_i = 0 \tag{13.14}$$

Here N is the number of reagents and products involved in the reaction and a_i are a new set of stoichiometric coefficients, defined so that the product values are positive but the reagent values are negative. For the reaction defined by equation (13.6), $N = 3$ and the new stoichiometric coefficients are:

$$\begin{aligned} a_1 &= -a \\ a_2 &= -b \\ a_3 &= c \end{aligned} \tag{13.15}$$

We have chosen to define the chemicals of equation (13.14) directly in kmol quantities for convenience, so that for the reaction described by equation (13.6), we have:

$$\begin{aligned} A_1 &= \{A\} \\ A_2 &= \{B\} \\ A_3 &= \{C\} \end{aligned} \tag{13.16}$$

Using the generalized notation, we may state the normalized reaction rate for a reaction involving N reactants and products as

$$\frac{dM_R}{dt} = \frac{1}{a_i}\frac{dM_i}{dt}\bigg|_r \tag{13.17}$$

where we have introduced the notation $(dM_i/dt|_r)$ to signify the rate of change of the mass of chemical species i from the reaction alone, so as to exclude the effects on mass change of flows in and out of the system.

13.4 Reaction rates

Equation (13.17) gives us information on the relative sizes of the reaction rates, $(dM_i/dt|_r)$, for the chemical components involved in the reaction, but it gives no indication of how fast the reaction is proceeding in absolute terms. Information on the speed of the reaction must usually be found by chemical experiment, where measurements are made of the rate of disappearance of the reactants and the rate of creation of the product. These measurements are normally referred to unit volume in order to allow them to be scaled up to a different size of reaction vessel. The resultant figure is often called, loosely, the 'reaction rate', although what is actually quoted is a *reaction rate density,* the rate of reaction per unit volume. We will prefer to use the more precise term 'reaction rate density', r, and define it as the normalized rate of reaction per unit volume:

$$r = \frac{1}{a_i V} \frac{dM_i}{dt}\bigg|_r = \frac{1}{V}\frac{dM_R}{dt} \tag{13.18}$$

r will have units of kmol rxn/(m^{3}s). In the general case when chemical equation (13.1) describes a reversible reaction, there will be two competing processes taking place: the forward reaction, leading to an increase in product C, and a backward reaction, whereby product C is broken back down into the initial reagents A and B. Theory and experiment indicate that the reaction rate density for the reaction of chemical equation (13.1) will take the form:

$$r = k[A]^{\alpha}[B]^{\beta}[C]^{\gamma} - k'[A]^{\alpha'}[B]^{\beta'}[C]^{\gamma'} \tag{13.19}$$

where
- k is the forward 'velocity constant',
- k' is the backward 'velocity constant',
- and the square brackets denote the concentration of the bracketted chemical in kmol/m^3:

$$[A] = \frac{M_A}{V}$$
$$[B] = \frac{M_B}{V} \tag{13.20}$$
$$[C] = \frac{M_C}{V}$$

The velocity constants, k and k', are independent of concentrations for an ideal gas, but are found to depend to a degree on concentrations when the gas mixture is not ideal or when the reaction takes place in a liquid.

For many simple reactions, the powers associated with the forward and the backward reactions are the same as the stoichiometric coefficients of the reagents, i.e.

$$\alpha = a, \beta = b, \gamma = 0$$
$$\alpha' = 0, \beta' = 0, \gamma' = c \tag{13.21}$$

However, in the general, non-ideal case, the powers are related to the stoichiometric coefficients only by the condition:

$$\frac{\alpha - \alpha'}{a} = \frac{\beta - \beta'}{b} = \frac{\gamma' - \gamma}{c} > 0 \tag{13.22}$$

The velocity constants depend on temperature according to the Arrhenius equation:

$$k = k_{\infty} \exp\left(-\frac{E}{RT}\right) \tag{13.23}$$

where
- R is the universal gas constant, with a value of 8314 J/(kmol K),
- T is the absolute temperature (K),
- k_{∞} is a constant for the reaction, known as the 'frequency factor', and
- E is the activation energy, which is the amount of energy needed to start the reaction, even when the final energy release should favour the reaction. E is a constant for the reaction, and has values typically in the range $4 \times 10^7 < E < 2 \times 10^8$ (J/kmol).
- The units of k, k' and k_{∞} will be the same, but dependent on the powers in equation (13.19), $\alpha, \beta \ldots$ etc.

Equation (13.23) may be used to extrapolate from a known velocity constant at temperature T_1 (K) to a new velocity constant at temperature T_2 (K):

$$k(T_2) = k(T_1)\exp\left(\frac{E}{R}\left(\frac{1}{T_1} - \frac{1}{T_2}\right)\right) \tag{13.24}$$

13.5 Generalization for multiple reactions

When there is more than one reaction taking place at the same time, we need to introduce a further subscript to identify the reaction. For the case of M reactions involving N chemicals as reagents and products, there will be M chemical equations similar in form to chemical equation (13.14):

$$\begin{aligned} a_{11}A_1 + a_{21}A_2 + \cdots + a_{N1}A_N &= 0 \\ a_{12}A_1 + a_{22}A_2 + \cdots + a_{N2}A_N &= 0 \\ &\vdots \\ a_{1M}A_1 + a_{2M}A_2 + \cdots + a_{NM}A_N &= 0 \end{aligned} \tag{13.25}$$

or

$$\sum_{i=1}^{N} a_{ij}A_i = 0 \qquad \text{for } j = 1, \ldots, M \tag{13.26}$$

where a_{ij} is the stoichiometric coefficient of the ith component in the jth reaction.

The reaction rate relationship of equation (13.17) may be generalized to any number of reactants and products undergoing several reactions. Denoting by dM_{Rj}/dt the normalized reaction rate for reaction j, equation (13.17) becomes:

$$\frac{dM_{Rj}}{dt} = \frac{1}{a_{ij}} \frac{dM_i}{dt}\bigg|_j \quad \text{for } j = 1, \ldots, M \tag{13.27}$$

where $(dM_i/dt|_j)$ is the rate of change of mass of chemical i caused by reaction j.

The reaction rate density for reaction j in units of kmol rxn/(m^{3}s) is then:

$$r_j = \frac{1}{V} \frac{dM_{Rj}}{dt} \tag{13.28}$$

Equation (13.28) is the generalization to j reactions of equation (13.18). The equivalent generalization of equation (13.19) gives the reaction rate, r_j, in terms of the concentrations of all the chemical species present:

$$r_j = k_j \prod_{i=1}^{N} [A_i]^{\alpha_{ij}} - k'_j \prod_{i=1}^{N} [A_i]^{\alpha'_{ij}} \quad \text{for } j = 1 \text{ to } M \tag{13.29}$$

where $\alpha_{ij}, \alpha'_{ij}$ are the powers associated with the forward and backward reaction j, which will obey the generalized version of equation (13.22), namely:

$$\frac{\alpha'_{ij} - \alpha_{ij}}{a_{ij}} = \frac{\alpha'_{1j} - \alpha_{1j}}{a_{1j}} > 0 \quad \text{for } 1 \le i \le N, \quad \text{and for } j = 1, \ldots M \tag{13.30}$$

Meanwhile the forward and backward velocity constants will have the generalized form:

$$k_j = k_{\infty j} \exp\left(-\frac{E_j}{RT}\right)$$
$$k'_j = k'_{\infty j} \exp\left(-\frac{E'_j}{RT}\right) \quad \text{for } j = 1 \text{ to } M \tag{13.31}$$

13.6 Conservation of mass in a bounded volume

Consider Figure 13.1, which depicts a quite general bounded volume containing either a gas or a liquid mixture, with a single input and a single output stream.

Let us suppose that there are N chemical species in the volume undergoing M reactions. The presence of chemical reactions obliges us to extend the principle of mass conservation set out in Chapter 3, Section 3.4, which we may now set out as:

> the rate of change of mass of species i in kmol/s *equals* the inlet flow of species i in kmol/s *plus* the number of kmol of species i produced per second by reaction *minus* the number of kmol of species i eliminated per second by reaction *minus* the outlet flow of species i in kmol/s.

This may be expressed mathematically as:

$$\frac{dM_i}{dt} = F_{1,i} + V(a_{i1}r_1 + a_{i2}r_2 + \cdots a_{ij}r_j + \cdots a_{iM}r_M) - F_{2,i} \quad \text{for } i = 1 \text{ to } N \tag{13.32}$$

or

$$\frac{dM_i}{dt} = F_{1,i} + V\sum_{j=1}^{M} a_{ij}r_j - F_{2,i} \quad \text{for } i = 1 \text{ to } N \tag{13.33}$$

where

$F_{1,i}$ is the inflow of species i (kmol/s),
$F_{2,i}$ is the outflow of species i (kmol/s),
a_{ij} is the stoichiometric coefficient for species i in reaction j,

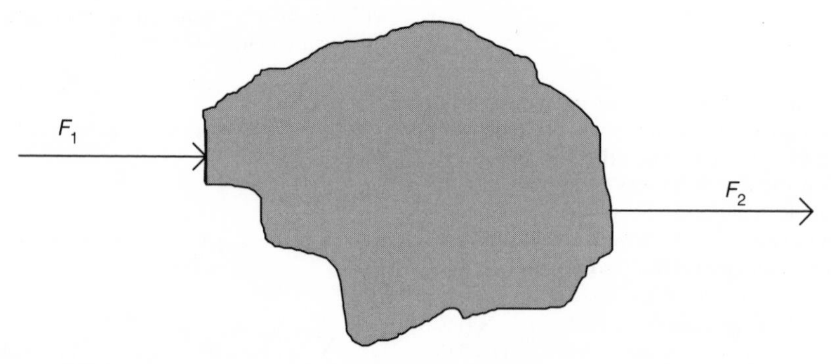

Figure 13.1 Bounded volume undergoing reactions.

r_j is the reaction rate density for reaction j (kmol rxn/(m^{3}s)), taking the form of equation (13.29), and

V is the volume of the fluid in the vessel (m^3).

13.7 Conservation of energy in a fixed volume

Here we extend the treatment of Chapter 3, Section 3.5 to cover the case of chemical reaction.

Considering Figure 13.2, conservation of energy requires that

> the change of energy in the fixed volume *equals* the external heat input *minus* the work output from the fixed volume *plus* the work done on the fixed volume by the incoming fluid *minus* the work done by the outgoing fluid *plus* the energy brought in by the incoming fluid *minus* the energy leaving with the outgoing fluid.

We may neglect the changes in kinetic energy and potential energy as explained in Section 3.5 and Appendix 1. In addition, the multiplicity of chemical species makes it sensible to work with kmol units. Accordingly, we may modify equation (3.29) for the case of N chemical species undergoing reaction to:

$$\delta E = \Phi\,\delta t - P\,\delta t + \sum_{i=1}^{N} F_{1,i}\,\delta t\,\hat{u}_{1,i} - \sum_{i=1}^{N} F_{2,i}\,\delta t\,\hat{u}_{2,i} + p_1 \sum_{i=1}^{N} F_{1,i}\,\delta t\,\hat{v}_{1,i} - p_2 \sum_{i=1}^{N} F_{2,i}\,\delta t\,\hat{v}_{2,i} \quad (13.34)$$

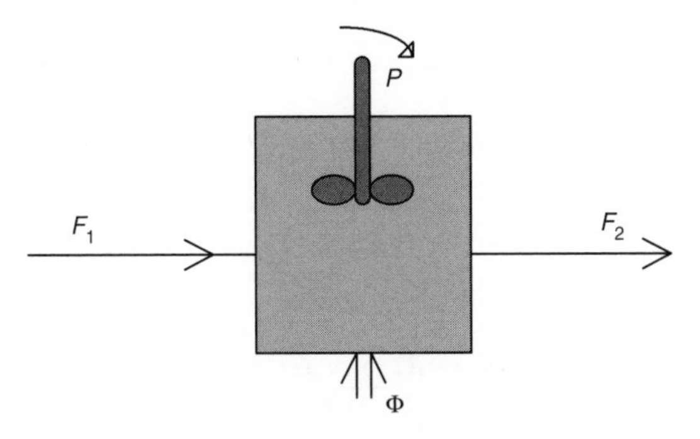

Figure 13.2 Fixed volume subject to a heat input and stirring.

where:

E is the energy contained in the fixed volume (J),

Φ is the heat flux into the fixed volume (W),

P is the mechanical power abstracted from the fixed volume (W); our sign convention means that P will be negative when mechanical power is added by a stirrer,

$F_{1,i}$, $F_{2,i}$ are the inlet and outlet mass flows of species i (kmol/s),

$\hat{u}_{1,i}$, $\hat{u}_{2,i}$ are the inlet and outlet specific internal energies for species i (J/kmol),

p_1, p_2 are the pressures of the inlet flow and the outlet flow (Pa).

Note that $\sum_{i=1}^{N} F_{1,i}\delta t\, v_{1,i} = \delta V_1$, the volume of fluid introduced in time δt, and similarly $\sum_{i=1}^{N} F_{2,i}\,\delta t\, v_{2,i} = \delta V_2$ the volume of fluid leaving in time δt.

Dividing equation (13.34) by the time interval, δt, and letting $\delta t \to 0$ allows us to write the energy equation as the differential equation:

$$\frac{dE}{dt} = \Phi - P + \sum_{i=1}^{N} F_{1,i}(\hat{u}_{1,i} + p_1\hat{v}_{1,i}) - \sum_{i=1}^{N} F_{2,i}(\hat{u}_{2,i} + p_2\hat{v}_{2,i}) \quad (13.35)$$

Using the definition of specific enthalpy, $\hat{h} = \hat{u} + p\hat{v}$, equation (13.35) may be simplified to:

$$\frac{dE}{dt} = \Phi - P + \sum_{i=1}^{N} F_{1,i}\hat{h}_{1,i} - \sum_{i=1}^{N} F_{2,i}\hat{h}_{2,i} \quad (13.36)$$

Ignoring the kinetic and potential energy terms as insignificant, the energy contained in the system is given by:

$$E = \sum_{i=1}^{N} M_i\hat{u}_i \quad (13.37)$$

Differentiating equation (13.37) with respect to time gives:

$$\frac{dE}{dt} = \sum_{i=1}^{N} \hat{u}_i\frac{dM_i}{dt} + \sum_{i=1}^{N} M_i\frac{d\hat{u}_i}{dt} \quad (13.38)$$

Equation (13.38) may be combined with the mass balance of equation (13.33) and then with equation (13.36) to give:

$$\sum_{i=1}^{N} \hat{u}_i\left(F_{1,i} + V\sum_{j=1}^{M} a_{ij}r_j - F_{2,i}\right) + \sum_{i=1}^{N} M_i\frac{d\hat{u}_i}{dt} = \Phi - P + \sum_{i=1}^{N} F_{1,i}\hat{h}_{1,i} - \sum_{i=1}^{N} F_{2,i}\hat{h}_{2,i} \quad (13.39)$$

so that

$$\sum_{i=1}^{N} M_i\frac{d\hat{u}_i}{dt} = \Phi - P + \sum_{i=1}^{N} F_{1,i}(\hat{h}_{1,i} - \hat{u}_i) - \sum_{i=1}^{N} F_{2,i}(\hat{h}_{2,i} - \hat{u}_i) - V\sum_{i=1}^{N}\sum_{j=1}^{M} \hat{u}_i a_{ij} r_j \quad (13.40)$$

Reversing the order of the summation of the last term on the right-hand side of equation (13.40) gives:

$$V\sum_{i=1}^{N}\sum_{j=1}^{M}\hat{u}_i a_{ij} r_j = V\sum_{j=1}^{M} r_j \sum_{i=1}^{N} a_{ij}\hat{u}_i \qquad (13.41)$$

Now $\sum_{i=1}^{N} a_{ij}\hat{u}_i$ is the change in specific internal energy brought about by reaction j, as it progresses by one kmol of reaction. It is convenient to define the term ΔU_j:

$$\Delta U_j = \sum_{i=1}^{N} a_{ij}\hat{u}_i \qquad (13.42)$$

where ΔU_j is thus the reaction-induced change in specific internal energy produced as reaction j advances by one kilogram-mole of reaction. We shall refer to it as the 'internal energy of reaction' for short, with units of J/kmol rxn; it is a parameter of the jth reaction only. ΔU_j will be positive for an endothermic reaction, where the specific internal energies of the products are greater than the specific internal energies of the reagents; conversely, ΔU_j will be negative for an exothermic reaction, where heat will be released because the specific internal energies of the products are less than the specific internal energies of the reactants. The value of ΔU_j may be found experimentally using a bomb calorimeter (see Appendix 5).

Equation (13.38) may be rewritten using equations (13.41) and (13.42) as

$$\sum_{i=1}^{N} M_i\frac{d\hat{u}_i}{dt} = \Phi - P + \sum_{i=1}^{N} F_{1,i}(\hat{h}_{1,i} - \hat{u}_i) - \sum_{i=1}^{N} F_{2,i}(\hat{h}_{2,i} - \hat{u}_i) - V\sum_{j=1}^{M} r_j\Delta U_j \qquad (13.43)$$

As explained in Section 3.5 of Chapter 3, specific internal energy may be regarded as a function of temperature alone for most process fluids, and so we may use the expansion:

$$\frac{d\hat{u}_i}{dt} = \frac{d\hat{u}_i}{dT}\frac{dT}{dt} \qquad (13.44)$$

Accordingly, the rate of change of temperature is given by:

$$\frac{dT}{dt} = \frac{\Phi - P + \sum_{i=1}^{N} F_{1,i}(\hat{h}_{1,i} - \hat{u}_i) - \sum_{i=1}^{N} F_{2,i}(\hat{h}_{2,i} - \hat{u}_i) - V\sum_{j=1}^{M} r_j\Delta U_j}{\sum_{i=1}^{N} M_i\frac{d\hat{u}_i}{dT}} \qquad (13.45)$$

The effect of movement of one or more boundaries of the volume under consideration will follow the treatment of Section 3.6 of Chapter 3. This has importance in the case of a liquid reaction in a tank supplied with an inert cover-gas, where the level of the liquid may change. The way in which this additional term is integrated into the analysis will be shown in Section 13.11.

13.8 The internal energy of reaction and the enthalpy of reaction

The chemical literature does not usually provide data on the internal energy of reaction, ΔU_j, preferring to list instead the reaction-induced change in specific enthalpy, ΔH_j, known as the 'enthalpy of reaction' for short. (ΔH_j is sometimes also called the 'heat of reaction', but this form of words will not be used in this book to avoid confusion with ΔU_j, which would seem to have an equal right to the term.) ΔH_j is defined for reaction j as:

$$\Delta H_j = \sum_{i=1}^{N} a_{ij}\hat{h}_i \qquad (13.46)$$

where $\hat{h}_i = \hat{u}_i + p\hat{v}_i$. ΔH_j is normally quoted at a standard temperature of 298.15 K.

The two terms, ΔU_j and ΔH_j, are different, although they are normally of similar magnitude. The difference between the two will be very small for liquids because of the low value of the pv term compared with the specific internal energy term. As for gases, equation (3.2) gives for ideal gases at the same temperature, T,

$$p\hat{v}_i = wpv_i = RT \qquad (13.47)$$

Thus we may substitute $\hat{h}_i = \hat{u}_i + RT$ into (13.46), so that:

$$\Delta H_j = \sum_{i=1}^{N} a_{ij}\hat{u}_i + RT\sum_{i=1}^{N} a_{ij} = \Delta U_j + RT\sum_{i=1}^{N} a_{ij} \qquad (13.48)$$

The term $\sum_{i=1}^{N} a_{ij}$ is the algebraic sum of the stoichiometric coefficients for reaction j, and since the stoichiometric coefficients for products are positive while those of the reactants are negative, there is a fair degree of cancelling inherent in $\sum_{i=1}^{N} a_{ij}$. However, $\Delta U_j = \Delta H_j$ only when the number of kilogram-moles of products formed by the reaction equals the number of kilogram-moles of reactants used up in doing so, i.e. when the reaction exhibits neither a molecular contraction nor a molecular expansion. Normally it will be necessary to find ΔU_j from the quoted value of ΔH_j using equation (13.48).

A note of caution is in order when looking up the enthalpy of reaction in chemical engineering textbooks. In setting down the chemical equation for a reaction, it is the ratios of the stoichiometric coefficients that must be constant, rather than their absolute values. Hence the same reaction may be described equally validly by chemical equations that are multiples of each other. However, the enthalpy of reaction, ΔH_j depends explicitly on the stoichiometric coefficients, $a_{ij}, i = 1, \ldots, N$, chosen to define the jth chemical reaction, and is thus tied to the absolute values of these coefficients. So while a reaction may be described validly by a second chemical equation that is any factor times the first (valid) chemical equation, it is important to realize that the enthalpy of reaction associated with the first equation needs to be multiplied by the same factor to find the enthalpy of reaction to be used alongside the second chemical equation. For example, suppose that reaction j is described by equation (13.26), and that the corresponding enthalpy of reaction is ΔH_j. It would be quite possible and legitimate to choose an alternative representation for reaction j, with a new set of stoichiometric coefficients equal to, say, half the old coefficients: $a'_{ij} = \frac{1}{2}a_{ij}$:

$$\sum_{i=1}^{N} a'_{ij} A_i = 0 \tag{13.49}$$

In such a case, the new enthalpy of reaction, $\Delta H'_j$, would be given by:

$$\begin{aligned} \Delta H'_j &= \sum_{i=1}^{N} a'_{ij} \hat{h}_i \\ &= \frac{1}{2} \sum_{i=1}^{N} a_{ij} \hat{h}_i \\ &= \frac{1}{2} \Delta H_j \end{aligned} \tag{13.50}$$

Thus the enthalpy of reaction has been halved because the stoichiometric coefficients have all been halved. The same cautionary remarks apply equally to the internal energy of reaction, ΔU_j, which is also tied to the absolute values of the stoichiometric coefficients used in the chemical equation describing the reaction.

Sometimes the enthalpy of reaction is quoted per kilogram-mole of one of the reactants, particularly if that reactant is a fuel. Let us suppose that we are dealing with the '*ABC*' reaction discussed in Sections 13.2 and 13.3, and that the enthalpy of reaction is given as ΔH_A in J/kmol of A. This implies that ΔH_A joules are released by the reaction when M_A decreases by one kilogram-mole, i.e.

$$\delta M_A = -1 \tag{13.51}$$

But by equation (13.12) the increase in the kilogram-moles of reaction is given by:

$$\delta M_R = -\frac{1}{a} \delta M_A \tag{13.52}$$

so that, substituting from equation (13.51),

$$\delta M_R = \frac{1}{a} \tag{13.53}$$

Hence ΔH_A joules will be released by the reaction when M_R increases by $1/a$ kmol rxn, and it follows that $a\,\Delta H_A$ joules will be released when M_R increases by 1 kmol rxn. That is to say:

$$\Delta H = a\,\Delta H_A \text{ J/kmol rxn} \tag{13.54}$$

In general terms, similar reasoning leads to the conclusion that

$$\Delta H_j = -a_{ij} \Delta H_{ij} \tag{13.55}$$

where ΔH_{ij} is the enthalpy of reaction for reaction j, when referred to chemical i and in units of J/kmol of chemical i, and a_{ij} is the stoichiometric coefficient for chemical i, positive for products and negative for reactants.

For example, the burning of carbon monoxide is governed by the chemical equation:

$$2\{CO\} + \{O_2\} = 2\{CO_2\} \tag{13.56}$$

where the curly brackets have been retained to emphasize that we are dealing with kmol quantities. Summing the stoichiometric coefficients gives:

$$\sum_{i=1}^{N} a_{ij} = -2 - 1 + 2 = -1 \tag{13.57}$$

implying a molecular contraction. The enthalpy of reaction at the standard temperature of 25°C/298.15 K is listed in Perry *et al.* (1984) as -282.989×10^6 J/(kmol CO), but the stoichiometric coefficient for CO in equation (13.56) is -2. Hence applying equation (13.55) gives:

$$\begin{aligned} \Delta H_j &= -(-2) \times -282.989 \times 10^6 \\ &= -565.979 \times 10^6 \text{ J/kmol rxn} \end{aligned} \tag{13.58}$$

Having found ΔH_j, we may compute ΔU_j at 298.15 K using equation (13.48) as:

$$\Delta U_j = \Delta H_j - RT \sum_{i=1}^{N} a_{ij}$$

$$= -565.979 \times 10^6 - 8314 \times 298.15 \times -1 \quad (13.59)$$
$$= -565.979 \times 10^6 + 2.479 \times 10^6$$
$$= -563.500 \times 10^6 \text{ J/(kmol rxn)}$$

An example where $\Delta U_j = \Delta H_j$ is the burning of methane, governed by the chemical equation:

$$\{CH_4\} + 2\{O_2\} = \{CO_2\} + 2\{H_2O\} \quad (13.60)$$

The burning will take place at a high temperature, and so that the physical state of the water produced will be steam, so that both reagents and products will be gaseous. Summing the stoichiometric coefficients algebraically gives:

$$\sum_{i=1}^{N} a_{ij} = -1 - 2 + 1 + 2 = 0 \quad (13.61)$$

Thus there will be neither a molecular expansion nor a molecular contraction. Accordingly, $\Delta U_j = \Delta H_j$.

The enthalpy of reaction, ΔH_j, may be found experimentally using an open system calorimeter. The basis of this procedure is described in Appendix 5, where it is contrasted with the method of measuring ΔU_j.

13.9 The effect of temperature on ΔU and ΔH

It is customary for chemists to quote either ΔH or ΔU at a standard temperature, namely $T_0 = 298.15$ K, but since temperature affects the specific internal energy of any given chemical species, we need to allow for the effect of temperature on both ΔH and ΔU.

Following equation (13.42), ΔU is given for a single reaction at temperature T by:

$$\Delta U(T) = \sum_{i=1}^{N} a_i \hat{u}_i(T) \quad (13.62)$$

But the specific internal energy at temperature T is related to the specific internal energy at temperature T_0 by:

$$\hat{u}_i(T) = \hat{u}_i(T_0) + \hat{c}_{vi}(T - T_0) \quad (13.63)$$

where $\hat{c}_{vi}$ is the average specific heat at constant volume of chemical i over the temperature range T_0 to T, in J/(kmol K). Substituting from equation (13.63) into equation (13.62) gives:

$$\Delta U(T) = \sum_{i=1}^{N} a_i \hat{u}_i(T_0) + (T - T_0) \sum_{i=1}^{N} a_i \hat{c}_{vi}$$
$$= \Delta U(T_0) + (T - T_0) \sum_{i=1}^{N} a_i \hat{c}_{vi} \quad (13.64)$$

A parallel development may be followed for the enthalpy of reaction:

$$\Delta H(T) = \sum_{i=1}^{N} a_i \hat{h}_i(T) \quad (13.65)$$

But the specific enthalpy at temperature T is related to the specific enthalpy at temperature T_0 by:

$$\hat{h}_i(T) = \hat{h}_i(T_0) + \hat{c}_{pi}(T - T_0) \quad (13.66)$$

where $\hat{c}_{pi}$ is the average specific heat at constant pressure of chemical i over the temperature range T_0 to T, in J/(kmol K). Substituting from equation (13.66) into equation (13.62) gives:

$$\Delta H(T) = \sum_{i=1}^{N} a_i \hat{h}_i(T_0) + (T - T_0) \sum_{i=1}^{N} a_i \hat{c}_{pi}$$
$$= \Delta H(T_0) + (T - T_0) \sum_{i=1}^{N} a_i \hat{c}_{pi} \quad (13.67)$$

We may take equation (13.64) further if we are dealing with gases that are ideal or near-ideal, in which case equation (3.21) relates the specific heat to the number of degrees of freedom (see Chapter 3, Section 3.3.1):

$$c_{vi} = \frac{N_{Fi}}{2} Z R_w = \frac{N_{Fi}}{2} \frac{ZR}{w_i} \quad (13.68)$$

where

c_{vi} is the specific heat of chemical i in J/(kg K),

N_{Fi} is the number of degrees of freedom for chemical i, = 3 for a monatomic gas, = 5 for a diatomic gas, = 6 for a polyatomic gas,

w_i is the molecular weight of chemical i,

R is the universal gas constant = 8314 J/(kmol K), and

Z is the compressibility factor, assumed the same for all gases (dimensionless).

Now the specific heat on a kmol basis is given by:

$$\hat{c}_{vi} = w_i c_{vi} = \frac{N_{Fi}}{2} ZR \quad (13.69)$$

using the result of equation (13.68). Hence we may substitute into equation (13.64) to give the internal energy of reaction at temperature T:

$$\Delta U(T) = \Delta U(T_0) + \frac{ZR(T - T_0)}{2} \sum_{i=1}^{N} a_i N_{Fi} \quad (13.70)$$

The enthalpy of reaction for a gas reaction may now be found using equation (13.48), where we have omitted the subscript j and added the temperature identifier for clarity:

$$\Delta H(T) = \Delta U(T) + ZRT \sum_{i=1}^{N} a_i \tag{13.71}$$

Let us now take some examples, in all of which we shall assume a compressibility factor of unity: $Z = 1$. First, consider the oxidation of carbon monoxide to carbon dioxide described by chemical equation (13.56), and assume that it takes place at 1250 K. Carbon monoxide is a diatomic gas, as is oxygen, while carbon dioxide is polyatomic. Hence, using the stoichiometric coefficients of equation (13.56), we have:

$$\sum_{i=1}^{N} a_i N_{Fi} = -2 \times 5 - 1 \times 5 + 2 \times 6 = -3 \tag{13.72}$$

Hence, using the value of the internal energy of reaction at the standard temperature of 298.15 K calculated in the previous section, we have:

$$\begin{aligned} \Delta U(1250) &= -563.501 \times 10^6 + \tfrac{1}{2} \times 8314 \\ &\quad \times (1250 - 298.15) \times -3 \qquad (13.73) \\ &= -563.501 \times 10^6 - 11.871 \times 10^6 \\ &= -575.372 \times 10^6 \text{ J/kmol rxn} \end{aligned}$$

The algebraic sum of the stoichiometric coefficients used in chemical equation (13.56) is -1 (see equation (13.57)), so that using equation (13.71), we have:

$$\begin{aligned} \Delta H(1250) &= -575.372 \times 10^6 + 8314 \\ &\quad \times 1250 \times -1 \qquad (13.74) \\ &= -575.372 \times 10^6 - 10.393 \times 10^6 \\ &= -585.765 \times 10^6 \text{ J/kmol rxn} \end{aligned}$$

As a further example, suppose that the methane combustion process described by chemical equation (13.60) takes place at 1500 K. Methane, carbon dioxide and steam are polyatomic, while oxygen is diatomic. Hence, using the stoichiometric coefficients of equation (13.60), we have:

$$\sum_{i=1}^{N} a_i N_{Fi} = -1 \times 6 - 2 \times 5 + 1 \times 6 + 2 \times 6 = -2 \tag{13.75}$$

Because the number of gas molecules is the same after as before the reaction, the internal energy of reaction is the same as the enthalpy of reaction, given in Perry *et al.* (1984) at the standard temperature of 298.15 K as -802.320×10^6 J/(kmol rxn). Hence:

$$\begin{aligned} \Delta U(1500) &= -802.320 \times 10^6 + \tfrac{1}{2} \times 8314 \\ &\quad \times (1500 - 298.15) \times -2 \\ &= -802.320 \times 10^6 - 9.992 \times 10^6 \qquad (13.76) \\ &= -812.312 \times 10^6 \text{ J/kmol rxn} \\ &= \Delta H(1500) \end{aligned}$$

It may be seen that the variations with temperature of both the internal energy of reaction and the enthalpy of reaction are often rather small. For many modelling applications a constant value will give sufficient accuracy.

As noted above, enthalpies of reaction are quoted at a standard temperature of 298.15 K. A realistic, physical interpretation of this standard implies that a fuel would be burnt at a high temperature and the products of combustion would then be cooled to the standard temperature. But the problem is that any water vapour present would condense at 298.15 K for any reasonable combustion pressure. This would present us with a difficulty in extrapolating the enthalpy of reaction at this standard temperature to the combustion temperatures met in the real world, since we would need to extrapolate across the condensing boundary, and would need to take the latent heat of steam into account. To counter this, chemical engineering texts quote two enthalpies of reaction for combustion reactions where water is produced as a by-product. The higher enthalpy of reaction applies to the situation where the steam condenses, giving up further heat to the surroundings, while the lower enthalpy of reaction corresponds to the artificial situation where the water produced is assumed to be kept in its vapour phase all the way down to 25°C. Since the lower enthalpy of reaction discounts the loss of the latent heat of steam on condensing, we may extrapolate from the lower enthalpy directly to the higher temperatures that will actually exist in a realistic combustion, where the water will certainly exist in its vapour phase.

13.10 Continuous reaction in a gas reactor

Let us consider the case of several gas reactions taking place inside a reaction vessel, which will have a fixed volume, V. In general, there may be N chemicals in the reactor, undergoing M reactions. It is assumed that there is sufficient turbulence and diffusion within the gas vessel for perfect mixing to occur. Heat may be given to or taken from the reactor.

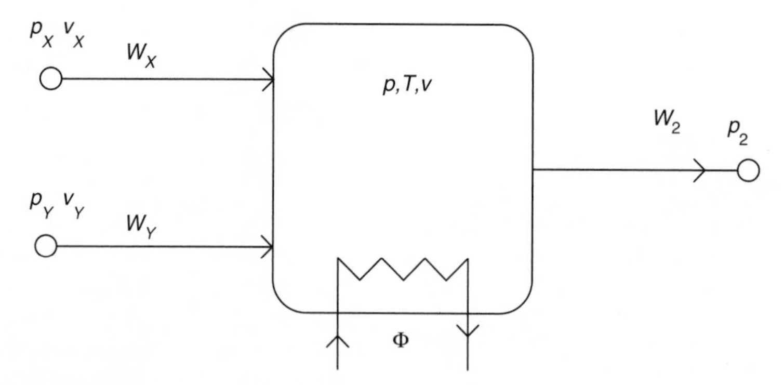

Figure 13.3 Gas reaction vessel.

Figure 13.3 shows two separate inlet feeds, X and Y, and one outlet stream. We will allow for the general case where each inlet stream may contain all N chemical species, reagents and products. Of course, each stream will contain normally a distinct subset only of the reactants.

Inlet flows

We shall assume that the mole fractions of the inlet streams are known, λ_{Xi} and λ_{Yi}, so that we may use equation (11.28) from Chapter 11, Section 11.4, to give the average molecular weights, w_X and w_Y, of the inlet streams:

$$w_X = \sum_{i=1}^{N} \lambda_{Xi} w_i \qquad w_Y = \sum_{i=1}^{N} \lambda_{Yi} w_i \tag{13.77}$$

where w_i is the molecular weight of chemical species i. Given the pressures and temperatures of these inlet streams, and assuming the gas mixtures are ideal, the specific volumes in m^3/kg may be evaluated using (11.31):

$$v_X = \frac{ZRT_X}{p_X w_X} \qquad v_Y = \frac{ZRT_Y}{p_Y w_Y} \tag{13.78}$$

The inflows of gas in kg/s will depend on the upstream pressures and specific volumes and on the pressure in the reaction vessel, in the way described in Chapter 11, Section 11.3, equation (11.9):

$$W_X = f_{flow}\left(p_X, v_X, \frac{p}{p_X}, A_X, K_{TX}, \gamma_X\right) \qquad W_Y = f_{flow}\left(p_Y, v_Y, \frac{p}{p_Y}, A_Y, K_{TY}, \gamma_Y\right) \tag{13.79}$$

where

A_X and A_Y are the cross-sectional areas (m^2) of the pipes bringing in feed-streams X and Y,

K_{TX} and K_{TY} are the velocity-head drops associated with the frictional resistance in the pipes bringing in feed-streams X and Y,

γ_X and γ_Y are the specific-heat ratios of gases X and Y, and

f_{flow} is the generalized gas flow function; specific realizations are given in Chapters 6 and 10.

The corresponding flows in kmol/s for each chemical species, i are found using equations (11.45) and (11.23):

$$F_{Xi} = \frac{\lambda_{Xi}}{\sum_{i=1}^{N} \lambda_{Xi} w_i} W_X \qquad F_{Yi} = \frac{\lambda_{Yi}}{\sum_{i=1}^{N} \lambda_{Yi} w_i} W_Y \qquad \text{for } i = 1 \text{ to } N \tag{13.80}$$

Chemical concentrations in the reaction vessel

The concentrations are given in kmol/m^3 by dividing the masses in kmol by the volume of the vessel:

$$[A_i] = \frac{M_i}{V} \qquad \text{for } i = 1 \text{ to } N \tag{13.81}$$

The mole fractions inside the reaction vessel are given by:

$$\lambda_i = \frac{M_i}{M} \tag{13.82}$$

where M in equation (13.82) is the total mass in kmol:

$$M = \sum_{i=1}^{N} M_i \tag{13.83}$$

Specific volume

The total mass in kg is given by:

$$m = \sum_{i=1}^{N} w_i M_i \tag{13.84}$$

Since the reactor volume is fixed, we may find the specific volume in m³/kg as

$$v = \frac{V}{m} \tag{13.85}$$

Pressure

Pressure is given by the modified gas law of equation (11.13):

$$p = \frac{ZRTM}{V} \tag{11.13}$$

where it is assumed that the value of Z applies to all gases in the reaction vessel. Normally we assume that the gases are ideal, and take $Z = 1$.

Outlet flow

The total outlet flow in kg/s is given by:

$$W_2 = f_{flow}\left(p, v, \frac{p_2}{p}, A_2, K_{T2}, \gamma\right) \tag{13.86}$$

where

A_2 is the cross-sectional area of the outlet pipe (m²),
K_{T2} is the number of velocity heads characterizing the frictional loss over the outlet pipe (dimensionless), and
γ is the specific-heat ratio of the gas in the reaction vessel, which may be evaluated using equation (11.40) from Section 11.4:

$$\gamma = 1 + \frac{2}{\sum_i \lambda_i N_{Fi}} \tag{11.40}$$

in which N_{Fi} is the number of degrees of freedom associated with the molecular structure of each of the gas components.

We may then derive the outflow in kmol/s of each chemical species using equations (11.41) and equation (11.23):

$$F_{2i} = \frac{\lambda_i}{\sum_{i=1}^{N} \lambda_i w_i} W_2 \qquad i = 1 \text{ to } N \tag{13.87}$$

Reactions and reaction rate densities

The reaction rate densities will be given by equation (13.29), repeated below:

$$r_j = k_j \prod_{i=1}^{N} [A_i]^{\alpha_{ij}} - k'_j \prod_{i=1}^{N} [A_i]^{\alpha'_{ij}} \qquad \text{for } j = 1 \text{ to } M \tag{13.29}$$

where it is assumed that the powers associated with the forward and backward reaction $j, \alpha_{ij}, \alpha'_{ij}$, are known, while the velocity constants for each forward and backward reaction are dependent on temperature, following equation (13.31):

$$k_j = k_{\infty j} \exp\left(-\frac{E_j}{RT}\right)$$

$$k'_j = k'_{\infty j} \exp\left(-\frac{E'_j}{RT}\right) \qquad \text{for } j = 1 \text{ to } M \tag{13.31}$$

with frequency factors, $k_{\infty j}$ and $k'_{\infty j}$, and activation energies, E_j and E'_j, all known.

Mass balance

We may now make use of equation (13.33), to find the mass balances for the chemical components in the reactor:

$$\frac{dM_i}{dt} = F_{Xi} + F_{Yi} + V \sum_{j=1}^{M} a_{ij} r_j - F_{2i} \qquad \text{for } i = 1 \text{ to } N \tag{13.88}$$

Energy balance

The energy balance follows the procedure laid down in Section 13.7, and we make use of equation (13.45), which gives the rate of change of temperature. In the absence of mechanical work by a stirrer, equation (13.45) becomes:

$$\frac{dT}{dt} = \frac{\Phi + \sum_{i=1}^{N} F_{Xi}(\hat{h}_{Xi} - \hat{u}_i) + \sum_{i=1}^{N} F_{Yi}(\hat{h}_{Yi} - \hat{u}_i) - \sum_{i=1}^{N} F_{2i}(\hat{h}_{2i} - \hat{u}_i) - V \sum_{j=1}^{M} r_j \Delta U_j}{\sum_{i=1}^{N} M_i \frac{d\hat{u}_i}{dT}} \tag{13.89}$$

where the terms $\hat{h}_{Xi}, \hat{h}_{Yi}$ are the specific enthalpies of the chemical species in the two inlet streams in J/kmol.

We will assume that the enthalpies of reaction ΔH_j for $j = 1, \ldots, M$ are available, at the standard temperature $T_0 = 298.15$ K. We may evaluate the internal energies of reaction ΔU_j for $j = 1, \ldots, M$ using equation (13.48), repeated below:

$$\Delta H_j = \Delta U_j + RT\sum_{i=1}^{N} a_{ij} \qquad (13.48)$$

Then the energies of reaction at temperature T may be related to their values at temperature $T_0 = 298.15\,\text{K}$ through equation (13.68):

$$\Delta U_j(T) = \Delta U_j(T_0) + \frac{R(T-T_0)}{2}\sum_{i=1}^{N} a_{ij}N_{Fi}$$

$$\text{for } j = 1 \text{ to } M \qquad (13.90)$$

There is often a significant thermal inertia contained in the walls of a reaction vessel. Accordingly, the heat input, Φ, will comprise in the general case the heat transferred from the wall of the vessel as well as the heat coming from a heating medium such as process steam. The heat supplied will depend on the temperature difference between the vessel wall and the gas and the temperature difference between the heating medium and the gas:

$$\Phi = K_H A_H (T_H - T) + K_m A_m (T_m - T) \qquad (13.91)$$

where K_H is the overall heat transfer coefficient, W/(m^{2}K), A_H is the heat transfer area, m^2, and T_H is the temperature of the heating medium, while K_m, A_m and T_m are the heat transfer coefficient, surface area and temperature of the metalwork, respectively. For simplicity we have treated the heater as a lumped parameter system, with the heating medium characterized by a single temperature, T_H.

Reaction vessel metalwork temperature

Assuming that the vessel is properly lagged, negligible heat will pass to the external environment, so an energy balance on the vessel metalwork gives:

$$\frac{d}{dt}(m_m c_{pm} T_m) = K_m A_m (T_m - T) \qquad (13.92)$$

Accordingly the rate of change of metalwork temperature is:

$$\frac{dT_m}{dt} = \frac{K_m A_m (T_m - T)}{m_m c_{pm}} \qquad (13.93)$$

Equation (13.93) describes the feedback effect of reaction temperature onto metalwork temperature. Equally, there will be a feedback effect onto the temperature of the heating medium, and that may require to be modelled. However, we shall assume for simplicity here that adequate compensatory action has been taken, so that there is no need to go on to model the heating subsystem.

Solution sequence

Given initial values of the masses of chemical species, M_i, $i = 1$ to N, and the temperature, T, it is possible to work out the starting pressure in the reaction vessel. This enables the total inflows to be calculated, and the component inflows may be calculated from our knowledge of the inlet compositions. Knowledge of the masses of chemical species, M_i, $i = 1$ to N, also allows a calculation of the overall specific volume of the reactor contents to be made, which, together with the pressure, allows the total outlet flow to be calculated. Further, kmol fractions, λ_i, may be found from M_i, and these enable the individual kmol flows to be found.

The concentrations in kmol/m^3 may be found also as soon as we know the masses, M_i. The concentrations may now be used along with the temperature to calculate the reaction rate densities, r_j, $j = 1$ to M. Now we know the inflow and outflow of each component, and the rate at which it is being produced or used up by the chemical reactions. Hence we may integrate the N mass-balance differential equations forward by one timestep.

A knowledge of the temperature inside the reactor allows the energies of reaction to be evaluated. Given also the vessel metalwork temperature and the temperature of the heating medium, the heat input may be found. The reactor temperature and the metalwork temperature differential equations may now be integrated forward by one timestep.

Having integrated the mass balance and the energy balance equations forward to give new values of the masses of N chemical species, M_i, the reactor temperature, T, and the metalwork temperature, T_m, we may repeat indefinitely the procedure outlined above, and the simulation problem may be regarded as solved.

13.11 Modelling a Continuous Stirred Tank Reactor (CSTR)

Figure 13.4 shows a liquid reaction vessel, subject to two feed streams and one outlet stream. It is the liquid analogue of the gas reactor described in Section 13.10. We shall again assume that there are M reactions involving N chemical species. The reactions occur continuously, and we shall assume that the stirrer effects perfect mixing. The gas above the liquid in the CSTR may consist of an inert gas blanket, perhaps under some form of pressure control, or it may be atmospheric air. The gas pressure may experience some small variability around the chosen pressure setpoint in the former case, while in the second case the pressure will undergo a slow, diurnal change of a few per cent at most.

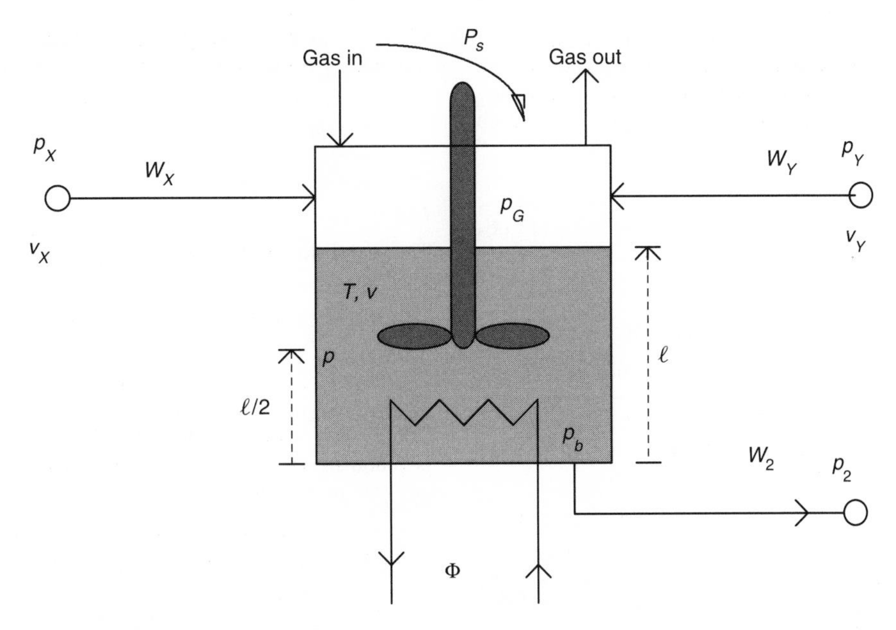

Figure 13.4 Continuous stirred tank reactor.

The analysis of the CSTR parallels closely that of the gas reactor, although differences arise from the fact that the reaction fluid is now a liquid. We shall need to allow for changes in the volume of the reaction fluid, and this will have effects on both the work done by the bulk volume and on the hydrostatic pressure.

Inlet flows

We shall assume that we have been given the temperatures of the inlet streams, T_X and T_Y and the mass fractions, μ_{Xi} and μ_{Yi}. The specific volume of each inlet stream will need to be calculated from the mass fractions. The simplest calculation follows Amagat's Law, which states that any liquid volume, V, is equal to the sum of the kmol masses, M_i, multiplied by the specific volumes, $\hat{v}_i$, in m^3/kmol:

$$V = \sum_{i=1}^{N} \hat{v}_i M_i \tag{13.94}$$

The overall specific volume is found by dividing by the mass in kg, m:

$$v = \frac{V}{m} = \frac{\sum_{i=1}^{N} \hat{v}_i M_i}{\sum_{i=1}^{N} w_i M_i} = \frac{\sum_{i=1}^{N} \hat{v}_i \frac{M_i}{M}}{\sum_{i=1}^{N} w_i \frac{M_i}{M}} = \frac{\sum_{i=1}^{N} \lambda_i \hat{v}_i}{\sum_{i=1}^{N} \lambda_i w_i} \tag{13.95}$$

where M is the total mass of all the components in kmol. Using the transformation between mass fraction and mole fraction of equation (11.24), the specific volume in m^3/kg emerges as simply the mass-fraction weighted average:

$$v = \sum_{i=1}^{N} \frac{\mu_i}{w_i} \hat{v}_i = \sum_{i=1}^{N} \mu_i v_i \tag{13.96}$$

Hence, allowing for the change in each component specific volume with temperature, we have:

$$\begin{aligned} v_X &= \sum_{i=1}^{N} \mu_{Xi} v_i(T_X) \\ v_Y &= \sum_{i=1}^{N} \mu_{Yi} v_i(T_Y) \end{aligned} \tag{13.97}$$

The inflows of liquid in kg/s will depend on the upstream pressures and specific volumes and on the pressure in the reaction vessel, in the way described in Chapter 11, Section 11.2. Hence, using equation (11.4), we have:

$$\begin{aligned} W_X &= C_{TX} \sqrt{\frac{p_X - p_G}{v_X}} \\ W_Y &= C_{TY} \sqrt{\frac{p_Y - p_G}{v_Y}} \end{aligned} \tag{13.98}$$

where C_{TX}, C_{TY} are the total conductances in m^2 of the upstream lines and valves, assumed known. The corresponding flows in kmol/s are found by dividing by the molecular weight, w_i, appropriate for each chemical species, i:

$$F_{Xi} = \frac{\mu_{Xi}}{w_i} W_X$$
$$F_{Yi} = \frac{\mu_{Yi}}{w_i} W_Y \tag{13.99}$$

Chemical concentrations in the reaction vessel

The equations are the same as for the gas reactor. Hence we have:

$$[A_i] = \frac{M_i}{V} \quad \text{for } i = 1 \text{ to } N \tag{13.81}$$

$$\lambda_i = \frac{M_i}{M} \tag{13.82}$$

$$M = \sum_{i=1}^{N} M_i \tag{13.83}$$

Liquid volume

Accurate prediction of the volume of a mixture of liquids will usually require experimental data to relate the masses in kmol to the volume in m^3. For example, Perry *et al.* (1984) give tabulations for several liquid mixtures found in process plants. In the absence of such data, the simplest relationship is the linear relationship of Amagat's law, already quoted as equation (13.94). The specific volume of liquid component i, $\hat{v}_i$ (m^3/kmol) will depend on the temperature, T, of the liquid in the CSTR, so that the liquid volume is given by.

$$V = \sum_{i=1}^{N} \hat{v}_i(T) M_i \tag{13.100}$$

Liquid specific volume

The same equations apply as for the gas reactor, namely:

$$m = \sum_{i=1}^{N} w_i M_i \tag{13.84}$$

$$v = \frac{V}{m} \tag{13.85}$$

Pressure

The pressure in Pa at the bottom of the CSTR is given by p_b:

$$p_b = p_G + p_l \tag{13.101}$$

where p_G is the pressure of the gas above the liquid (Pa) and p_l is the hydrostatic pressure due to the liquid level:

$$p_l = \frac{gl}{v} \tag{13.102}$$

Outler flow

The outlet flow in kg/s is given by:

$$W_2 = C_{T2}\sqrt{\frac{p_b - p_2}{v}} \tag{13.103}$$

where C_{T2} is the total conductance in m^2 of the downstream line and valve.

The outflow of each chemical component in kmol/s is found using the exactly same equations as for the gas reactor, namely:

$$F_{2i} = \frac{\lambda_i}{\sum_{i=1}^{N} \lambda_i w_i} W_2 \quad i = 1 \text{ to } N \tag{13.87}$$

Reaction rate densities

The equations for the reaction rate densities will be exactly the same as for the gas reactor, namely:

$$r_j = k_j \prod_{i=1}^{N} [A_i]^{\alpha_{ij}} - k'_j \prod_{i=1}^{N} [A_i]^{\alpha'_{ij}} \quad \text{for } j = 1 \text{ to } M \tag{13.29}$$

$$k_j = k_{\infty j} \exp\left(-\frac{E_j}{RT}\right)$$
$$k'_j = k'_{\infty j} \exp\left(-\frac{E'_j}{RT}\right) \quad \text{for } j = 1 \text{ to } M \tag{13.31}$$

Mass balance

We make use, once more, of equation (13.88), to find the mass balances for the chemical components in the reactor:

$$\frac{dM_i}{dt} = F_{Xi} + F_{Yi} + V\sum_{j=1}^{M} a_{ij} r_j - F_{2i} \quad \text{for } i = 1 \text{ to } N \tag{13.88}$$

Energy balance

The energy balance follows the procedure laid down in Section 13.7, and we make use of equation (13.43), which gives the rate of change of temperature. There are two differences from the gas-reactor case: (i) there is a stirrer present and (ii) the liquid volume can change. Hence we need to retain the power output term, P, so that equation (13.43) becomes for the case of two input streams, X and Y:

$$\sum_{i=1}^{N} M_i \frac{d\hat{u}_i}{dT}\frac{dT}{dt} = \Phi - P + \sum_{i=1}^{N} F_{Xi}(\hat{h}_{Xi} - \hat{u}_i) + \sum_{i=1}^{N} F_{Yi}(\hat{h}_{Yi} - \hat{u}_i) - \sum_{i=1}^{N} F_{2i}(\hat{h}_{2i} - \hat{u}_i) - V\sum_{j=1}^{M} r_j \Delta U_j \tag{13.104}$$

where the terms $\hat{h}_{Xi}$, $\hat{h}_{Yi}$ are the specific enthalpies of the chemical species in the two inlet streams in J/kmol.

The power abstracted from the liquid in the CSTR consists of two items: the rate, P_G, at which work is done against the cover-gas pressure and the net stirring power, P_s, experienced by the liquid in the reactor. The power expended in increasing the liquid volume against the cover-gas pressure, P_G, is found using equation (3.47) of Section 3.6:

$$P_G = p_G \frac{dV}{dt} \tag{13.105}$$

while the net power abstracted is found by subtracting the net stirring power, P_s:

$$P = p_G \frac{dV}{dt} - P_s \tag{13.106}$$

The rate of increase of liquid volume, dV/dt, is found by differentiating equation (13.100) with respect to time:

$$\frac{dV}{dt} = \sum_{i=1}^{N} \hat{v}_i \frac{dM_i}{dt} + \sum_{i=1}^{N} M_i \frac{d\hat{v}_i}{dT}\frac{dT}{dt} \tag{13.107}$$

Substituting from (13.106) and (13.107) into equation (13.104) gives

$$\frac{dT}{dt}\sum_{i=1}^{N} M_i \left(\frac{d\hat{u}_i}{dT} + p_G\frac{d\hat{v}_i}{dT}\right) = \Phi + P_s - \sum_{i=1}^{N} p_G\hat{v}_i \frac{dM_i}{dt} + \sum_{i=1}^{N} F_{Xi}(\hat{h}_{Xi} - \hat{u}_i) + \sum_{i=1}^{N} F_{Yi}(\hat{h}_{Yi} - \hat{u}_i) - \sum_{i=1}^{N} F_{2i}(\hat{h}_{2i} - \hat{u}_i) - V\sum_{j=1}^{M} r_j \Delta U_j \tag{13.108}$$

Now the rate of change of mass i in kmol is given by the mass balance equation (13.88). Accordingly, we may write

$$\sum_{i=1}^{N} p_G\hat{v}_i \frac{dM_i}{dt} = \sum_{i=1}^{N}\left(F_{Xi}\,p_G\hat{v}_i + F_{Yi}\,p_G\hat{v}_i + V p_G\hat{v}_i \sum_{j=1}^{M} a_{ij} r_j - F_{2i}\,p_G\hat{v}_i\right) = \sum_{i=1}^{N} F_{Xi}\,p_G\hat{v}_i + \sum_{i=1}^{N} F_{Yi}\,p_G\hat{v}_i - \sum_{i=1}^{N} F_{2i}\,p_G\hat{v}_i + V\sum_{j=1}^{M} r_j \sum_{i=1}^{N} a_{ij}\,p_G\hat{v}_i \tag{13.109}$$

We note also the definition of the internal energy of reaction given in equation (13.42), repeated below:

$$\Delta U_j = \sum_{i=1}^{N} a_{ij}\hat{u}_i \tag{13.42}$$

We may substitute from equations (13.109) and (13.42) into equation (13.108) to give:

$$\frac{dT}{dt}\sum_{i=1}^{N} M_i \left(\frac{d\hat{u}_i}{dT} + p_G\frac{d\hat{v}_i}{dT}\right) = \Phi + P_s + \sum_{i=1}^{N} F_{Xi}(\hat{h}_{Xi} - (\hat{u}_i + p_G\hat{v}_i)) + \sum_{i=1}^{N} F_{Yi}(\hat{h}_{Yi} - (\hat{u}_i + p_G\hat{v}_i)) - \sum_{i=1}^{N} F_{2i}(\hat{h}_{2i} - (\hat{u}_i + p_G\hat{v}_i)) - V\sum_{j=1}^{M} r_j \sum_{i=1}^{N} a_{ij}(\hat{u}_i + p_G\hat{v}_i) \tag{13.110}$$

Although the cover gas pressure, p_G, differs from the average pressure, p, and the base pressure, p_b, of the liquid in the CSTR only by the hydrostatic heads, the pressures may be significantly different in a large tank. For instance, if the cover gas is at atmospheric pressure, while the liquid is aqueous and has a level of 10 m, we have $p_G = \sim 1$ bar, while the pressure at the base, $p_b, = \sim 2$ bar, twice the gas pressure. Nevertheless, because we are dealing with liquids, the effect on specific enthalpy will normally be small because of the domination of the term for specific internal energy. Hence we will normally be justified in writing the specific enthalpy of liquid component i at the mid-point of the liquid as:

$$\hat{h}_i = \hat{u}_i + p\hat{v}_i \approx \hat{u}_i + p_G\hat{v}_i \quad (13.111)$$

while, similarly, the specific enthalpy at the base of the liquid may be written as:

$$\hat{h}_{2i} = \hat{u}_i + p_b\hat{v}_i \approx \hat{u}_i + p_G\hat{v}_i \approx \hat{h}_i \quad (13.112)$$

The specific heat at constant pressure for liquid component i is given by:

$$\hat{c}_{pi} = \left(\frac{d\hat{h}_i}{dT}\right)_p = \frac{d\hat{u}_i}{dT} + p\frac{d\hat{v}_i}{dT} \quad (13.113)$$

Hence we may re-express equation (13.110) as:

$$\frac{dT}{dt}\sum_{i=1}^{N} M_i\hat{c}_{pi} = \Phi + P_s + \sum_{i=1}^{N} F_{Xi}(\hat{h}_{Xi} - \hat{h}_i) + \sum_{i=1}^{N} F_{Yi}(\hat{h}_{Yi} - \hat{h}_i) - V\sum_{j=1}^{M} r_j \sum_{i=1}^{N} a_{ij}\hat{h}_i \quad (13.114)$$

We may use the definition of enthalpy of reaction given in equation (13.46) repeated below:

$$\Delta H_j = \sum_{i=1}^{N} a_{ij}\hat{h}_i \quad (13.46)$$

and then rearrange to give the time differential for temperature as:

$$\frac{dT}{dt} = \frac{\Phi + P_s + \sum_{i=1}^{N} F_{Xi}(\hat{h}_{Xi} - \hat{h}_i) + \sum_{i=1}^{N} F_{Yi}(\hat{h}_{Yi} - \hat{h}_i) - V\sum_{j=1}^{M} r_j \Delta H_j}{\sum_{i=1}^{N} M_i c_{pi}} \quad (13.115)$$

The fact that the volume of the liquid is able to change has led to a temperature differential framed exclusively in terms of enthalpies. This is in contrast to the situation with the gas reactor, where the reaction volume was fixed, and we dealt with both enthalpies and internal energies.

The enthalpies of reaction at temperature, T, are related to their values at the standard temperature, $T_0 = 298.15$ K using equation (13.70) as the basis:

$$\Delta H_j(T) = \Delta H_j(T_0) + (T - T_0)\sum_{i=1}^{N} a_i\hat{c}_{pi} \quad (13.116)$$

where $\hat{c}_{pi}$ is the specific heat at constant pressure.

Once again, the heat input, Φ, will comprise the heat transferred from the wall of the vessel as well as the heat exchanged with a heating medium such as process steam:

$$\Phi = K_H A_H (T_H - T) + K_m A_m (T_m - T) \quad (13.91)$$

Reaction vessel metalwork temperature

Again assuming that the vessel is properly lagged, the same equations apply to the CSTR as to the gas reactor:

$$\frac{d}{dt}(m_m c_{pm} T_m) = K_m A_m (T_m - T) \quad (13.92)$$

$$\frac{dT_m}{dt} = \frac{K_m A_m (T_m - T)}{m_m c_{pm}} \quad (13.93)$$

but note that here the area, A_m, is the area of the vessel wall in contact with the liquid.

Solution sequence

The solution sequence is very similar to that for the gas reactor. We shall assume that we know the initial conditions of the state variables, which are the masses in kmol of all the reagents and products, M_i, $i = 1$ to N, the temperature of the reaction liquid, T, and the temperature of the CSTR metalwork, T_m. We shall assume that the cover-gas pressure, p_G, and the temperature of the heating medium, T_H, are given as boundary conditions.

A knowledge of the cover-gas pressure and the inlet conditions allows the inlet flows to be calculated. The compositions of the inlet flows, assumed given, then allow us to work out the inlet flows of individual chemical species in kmol/s. Knowing the masses of the CSTR's chemical species and its liquid temperature, we may calculate the liquid volume and the overall specific volume. The height of the liquid may be deduced from the volume and then the hydrostatic pressure at the base of the CSTR. Having calculated the pressure and the specific volume, we may then calculate the total outlet flow. Further, kmol fractions, λ_i, may be found immediately from the kmol masses, M_i, and these enable the individual kmol flows to be found.

The concentrations in kmol/m^3 may be found also as soon as we know the kmol masses, M_i. The concentrations may now be used along with the temperature to calculate the reaction rate densities, r_j, $j = 1$ to M.

At this stage we know the inflow and outflow of each component, and the rate at which it is being produced or used up by the chemical reactions. Hence there is sufficient information to allow the N mass balance differential equations to be integrated forward by one timestep.

A knowledge of the temperature inside the reactor allows the enthalpies of reaction to be evaluated. Given also the vessel metalwork temperature and the temperature of the heating medium, the heat input may be found. Thus the reactor temperature and the metalwork temperature differential equations may now be integrated forward by one timestep.

We have now shown how a knowledge of the initial values of the state variables, M_i, $i = 1$ to N, T and T_m allows us to integrate the mass balance and the energy balance equations forward to give new values of the state variables at the next timestep. We may repeat indefinitely the procedure outlined above, and so derive values of the state variables at all future times.

13.12 Bibliography

Denbigh, K.G and Tumer, J.R. (1984). *Chemical Reactor Theory, an Introduction*, 3rd edition, Cambridge University Press, Cambridge.

Froment, G.F. and Bischoff, K.B. (1990). *Chemical Reactor Analysis and Design*, 2nd edition, John Wiley & Sons, New York.

Perry, R.H, Green, D.W. and Maloney, J.O. (eds) (1984). *Perry's Chemical Engineers' Handbook*, 6th edition, McGraw-Hill, New York. Chapter 3: Physical and Chemical Data.

Whitten, K.W, Gailey, K.D and Davis, R.E. (1992). *General Chemistry with Qualitative Analysis*, 4th edition, Saunders College Publishing, Harcourt Brace Jovanovich College Publishers, Fort Worth.

14 Turbine nozzles

14.1 Introduction

Chapter 5 gave a general introduction to nozzles, which were considered to be frictionless and to be supplied with gas at negligible velocity. However, an accurate model of a turbine requires us to consider the inefficiency introduced by frictional effects in nozzles. In addition, the turbine nozzle will be supplied with gas from the previous stage that may have an appreciable velocity, so that we need to develop a method of dealing with non-negligible inlet velocities.

The customary numbering convention for a turbine stage is to give the inlet station the number '0', the mid-stage station the number '1' and the stage outlet station the number '2'. This implies a turbine nozzle will have an inlet station of '0' and a discharge station of '1' immediately downstream of the nozzle. In some cases of sonic and supersonic nozzle flow, the conditions just inside the outlet of the nozzle will be different from those immediately downstream of the outlet. Hence we will refer to the station just inside the outlet of the nozzle as station 'N1'. These station identifiers will be used as subscripts to label conditions such as pressure and temperature. The general station downstream of the inlet will carry no subscript. Figure 14.1 illustrates a convergent-only nozzle with a parallel outlet section (the presence of which tends to promote a better-formed jet).

14.2 Velocity and enthalpy relationships in a turbine nozzle: nozzle efficiency

One form of the energy equation for general, steady-state flow is given by equation (4.7), derived in Section 4.2:

$$dq - dw - dh - c\,dc - g\,dz = 0 \qquad (4.7)$$

This equation applies to both incompressible and compressible flow, and no condition of reversibility is implied. Flow through a turbine nozzle is adiabatic, so that $dq = 0$; no work is done (until it reaches the turbine blade downstream of the nozzle), so that $dw = 0$; and the height difference is zero, so that $dz = 0$. Accordingly equation (4.7) simplifies to

$$dh + c\,dc = 0 \qquad (14.1)$$

or

$$d(h + \tfrac{1}{2}c^2) = 0 \qquad (14.2)$$

Thus the term $h + \frac{1}{2}c^2$ remains constant throughout the nozzle. This term is known as the 'stagnation enthalpy', since if the velocity were at any stage brought to zero without work being done, all its kinetic energy would be converted into enthalpy. This concept will be discussed further in Section 14.5. It follows from equation (14.2) that the increase in kinetic energy

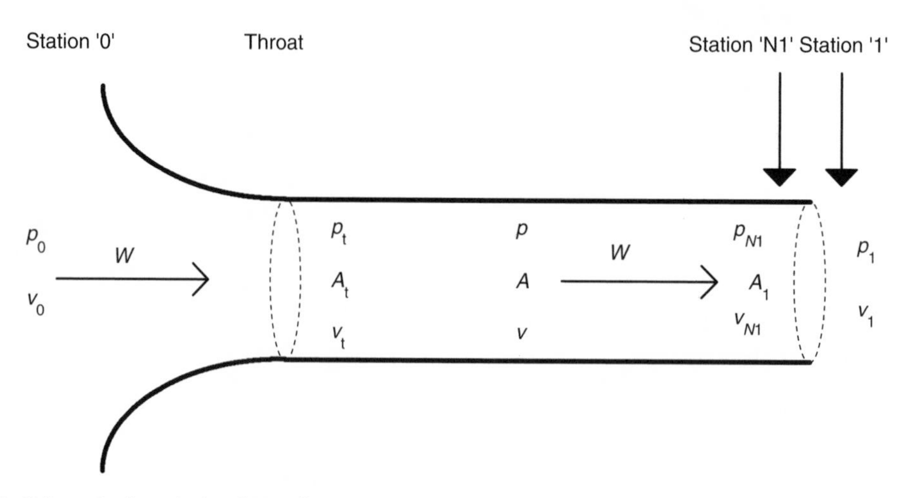

Figure 14.1 Schematic of nozzle showing stations.

at any downstream station comes at the expense of a decrease in enthalpy:

$$\tfrac{1}{2}c^2 - \tfrac{1}{2}c_0^2 = h_0 - h \tag{14.3}$$

The object of a turbine nozzle is to produce as much kinetic energy at the outlet, 'station 1', as possible from given inlet conditions and pressure drop. Although every effort will be made to reduce friction in the nozzle of a turbine by matching the nozzle inlet angle to the expected direction of incoming flow, by guiding the flow in the nozzle using gentle curves and by applying a gentle gradient to the diverging section (if present), small frictional losses will inevitably remain. The maximum kinetic energy achievable comes when there is no friction and the process is perfectly isentropic. We shall use the subscript '1' to denote the outlet velocity and enthalpy at station '1', and the subscript '1s' to denote the same after an isentropic expansion. Thus:

$$\tfrac{1}{2}c_{1s}^2 - \tfrac{1}{2}c_0^2 = h_0 - h_{1s} \tag{14.4}$$

The efficiency of the nozzle, η_N, is the ratio of the actual kinetic energy produced to this theoretical maximum:

$$\eta_N = \frac{\tfrac{1}{2}c_1^2 - \tfrac{1}{2}c_0^2}{\tfrac{1}{2}c_{1s}^2 - \tfrac{1}{2}c_0^2} = \frac{h_0 - h_1}{h_0 - h_{1s}} \tag{14.5}$$

When the gas has a specific heat at constant pressure that is constant over the temperature range of interest, T_0 to T_{1s}, implying that the gas acts as an ideal or near-ideal gas over this temperature range, then the enthalpy change is related to the temperature change by

$$\begin{aligned} h_1 - h_0 &= \int_{T_0}^{T_1} \frac{dh}{dT}\,dT = \int_{T_0}^{T_1} c_p\,dT \\ &= c_p \int_{T_0}^{T_1} dT = c_p(T_1 - T_0) \end{aligned} \tag{14.6}$$

Hence the nozzle efficiency equation may be re-expressed in terms of outlet temperatures:

$$\eta_N = \frac{T_0 - T_1}{T_0 - T_{1s}} = \frac{1 - \dfrac{T_1}{T_0}}{1 - \dfrac{T_{1s}}{T_0}} \tag{14.7}$$

14.3 Dependence of the polytropic exponent on nozzle efficiency

We may re-express equation (14.5) in the difference form

$$\eta_N = \frac{\Delta h_N}{\Delta h_{Ns}} \tag{14.8}$$

where Δh_{Ns} is the change in enthalpy over the nozzle that would occur for perfectly isentropic conditions, while Δh_N is the smaller change in enthalpy that actually will have occurred because of the friction causing additional heating of the fluid. If we make the assumption that the nozzle efficiency is constant over the whole length of the nozzle, then we may rewrite (14.8) in differential form:

$$\eta_N = \frac{dh_N}{dh_{Ns}} \tag{14.9}$$

We may express the enthalpy differential for an isentropic expansion from pressure p to pressure $p + dp$ as:

$$dh_{Ns} = c_p\,dT_s \tag{14.10}$$

where $c_p = dh/dT$ is the specific heat at constant pressure and dT_s is the change in temperature over the differential unit of nozzle for the ideal case where there is no friction present. Similarly, we may express the enthalpy differential for a frictionally resisted expansion between the same pressures as

$$dh_N = c_p\,dT \tag{14.11}$$

where dT is the change in temperature over the differential unit of nozzle for a frictionally resisted expansion. We assume that the same value of specific heat, c_p, may be applied in equations (14.10) and (14.11). Hence, using the definition of nozzle efficiency, equation (14.9), together with equations (14.10) and (14.11), we may relate the two temperature differentials:

$$\frac{dT}{dT_s} = \eta_N \tag{14.12}$$

Now a gas undergoing an isentropic expansion will obey the relationship

$$pv^\gamma = \text{constant} \tag{14.13}$$

as well as conforming to the characteristic equation

$$pv = ZR_wT \tag{3.2}$$

where R_w is the characteristic gas constant and Z is the compressibility factor, here assumed constant. Combining (14.13) with (3.2) gives the temperature in terms of the pressure and the isentropic index, γ:

$$T = bp^{(\gamma-1)/\gamma} \tag{14.14}$$

where b is a constant. Differentiating (14.14) gives

$$\frac{dT_s}{T} = \frac{\gamma - 1}{\gamma}\frac{dp}{p} \tag{14.15}$$

where the subscript 's' indicates that this temperature differential corresponds to an isentropic expansion.

We may now use equation (14.12) along with equation (14.15) to yield the corresponding temperature differential for a frictionally resisted expansion from the same initial conditions (p, T):

$$\frac{dT}{T} = \eta_N \frac{\gamma - 1}{\gamma} \frac{dp}{p} \tag{14.16}$$

Integrating equation (14.16) between an upstream station '0' and any downstream station, with corresponding pressures p_0 and p and temperatures T_0 and T, gives:

$$\ln \frac{T_0}{T} = \eta_N \frac{\gamma - 1}{\gamma} \ln \frac{p_0}{p} \tag{14.17}$$

From the gas characteristic equation (3.2)

$$\frac{T_0}{T} = \frac{p_0 v_0}{p v} \tag{14.18}$$

so it follows that

$$\ln \left[\frac{p_0}{p} \left(\frac{v_0}{v} \right)^{\gamma/(\gamma - \eta_N(\gamma - 1))} \right] = 0 \tag{14.19}$$

and hence that

$$p_0 v_0^m - p v^m = 0 \tag{14.20}$$

where

$$m = \frac{\gamma}{\gamma - \eta_N(\gamma - 1)} \tag{14.21}$$

Since the downstream station was chosen arbitrarily, it follows that the required relationship between specific volume for a frictionally resisted, adiabatic expansion is simply:

$$p v^m = \text{constant} \tag{14.22}$$

where the polytropic exponent, m, is given by equation (14.21) as a function of the isentropic index, γ, and the efficiency, η_N.

The significance of the change in polytropic exponent with efficiency is brought out more clearly if we take the mth root of both sides of equation (14.22) to restate the equation defining the frictionally resisted, adiabatic expansion in the alternative form:

$$v p^{1/m} = \text{constant} \tag{14.23}$$

where the constant of equation (14.23) is the mth root of the constant used in (14.22). Equation (14.23) shows the importance of the inverse of m, which from equation (14.21) is given as:

$$m^{-1} = \eta_N \gamma^{-1} + (1 - \eta_N)(1)^{-1} \tag{14.24}$$

Equation (14.24) indicates that the inverse of m is a weighted average of the inverse of the isentropic exponent, γ, and the inverse of the isothermal exponent, namely unity, with the weighting of the inverses dependent linearly on the nozzle efficiency. Thus when the nozzle is perfectly efficient, the flow through the nozzle will be isentropic, but as the nozzle efficiency falls below unity, the expansion will be a mixture of isentropic and isothermal. Physically, the turbulence in the flow will increase and cause the temperature at the outlet to rise above its isentropic value. In the limit, as the nozzle efficiency approaches zero, the expansion will be entirely isothermal.

Equation (14.22) may be used along with the gas equation (3.2) to relate the temperature and specific volume ratios in an inefficient expansion to the pressure ratio:

$$\begin{aligned} \frac{T}{T_0} &= \left(\frac{p}{p_0} \right)^{(m-1)/m} \\ \frac{v}{v_0} &= \left(\frac{p}{p_0} \right)^{-1/m} \end{aligned} \tag{14.25}$$

Equation (14.22) will prove very useful in characterizing flow through a nozzle in non-ideal conditions, although we need on occasion to remember that it derives from the assumption that equations (14.8) and (14.9) are equivalent. Strictly speaking, they are not, because of the effect of the 'reheat factor'. The loss of kinetic energy over each differential element governed by equation (14.9) will reappear as an increase in enthalpy entering the next element, and this will tend to increase the speed of the gas over that part of the nozzle. We may quantify this effect by recalculating the efficiency over the nozzle using equation (14.7), utilizing the relationship between temperature ratio and pressure ratio given by equation (14.25):

$$(\eta_N)_{recalc} = \frac{1 - \dfrac{T_1}{T_0}}{1 - \dfrac{T_{1s}}{T_0}} = \frac{1 - \left(\dfrac{p_1}{p_0} \right)^{(m-1)/m}}{1 - \left(\dfrac{p_1}{p_0} \right)^{(\gamma-1)/\gamma}} \tag{14.26}$$

Ideally $(\eta_N)_{recalc}$ should be the same as η_N, but this turns out to be exactly true only at vanishingly small pressure drops. The calculation of equation (14.26) has been performed for $\gamma = 1.4$ over a wide range of pressure ratios for two nozzles with differential efficiencies of 90 and 95% (Figure 14.2).

It will be seen that the recalculated value of nozzle efficiency deviates increasingly from the true value at the lower pressure ratios, with the deviation being greater when the original efficiency value is lower. However, the errors introduced will normally be small even in off-design conditions because of the combinations of pressure ratio and nozzle efficiency that are likely to be encountered in practice.

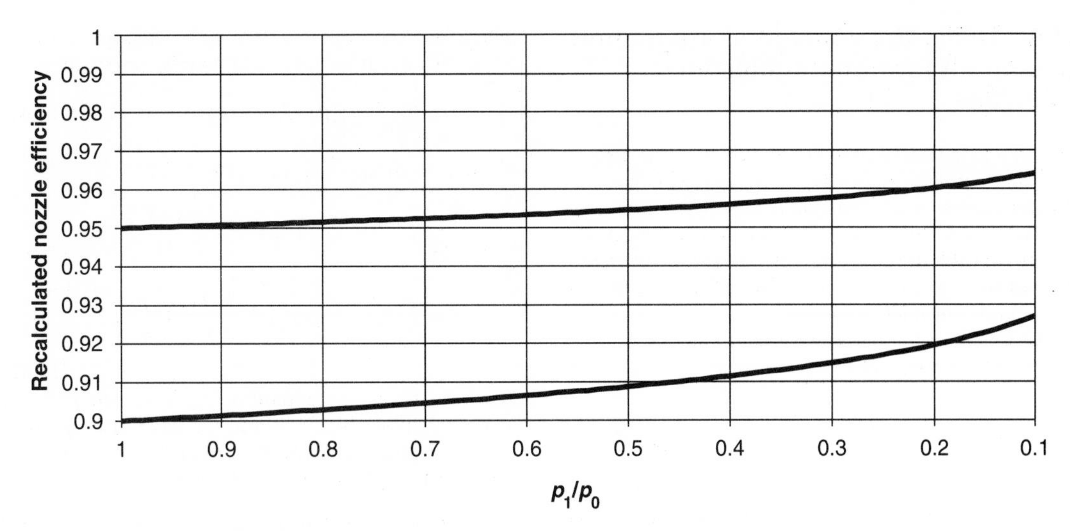

Figure 14.2 Recalculated nozzle efficiencies, compared with base values of 0.9 and 0.95.

14.4 Effect of nozzle efficiency on nozzle velocity

To evaluate the effect of a less-than-ideal nozzle on velocity, we rewrite equation (14.1), as

$$c\,dc = -dh \tag{14.27}$$

indicating that the fluid's acceleration will come directly from the drop in its enthalpy. Now enthalpy was defined in Chapter 3 as

$$h = u + pv \tag{3.6}$$

leading to the differential

$$dh = du + p\,dv + v\,dp \tag{14.28}$$

But the differential of specific entropy obeys the thermodynamic relation

$$T\,ds = du + p\,dv \tag{14.29}$$

so that equation (14.28) may be restated as:

$$dh = T\,ds + v\,dp \tag{14.30}$$

When no friction is present and the expansion is reversible and adiabatic, entropy is preserved. For this isentropic expansion, we may set $ds = 0$ in (14.30) to give:

$$dh_{Ns} = v\,dp \tag{14.31}$$

Therefore, the enthalpy change for a frictionally resisted expansion may be found by substituting from equation (14.9) into equation (14.31):

$$dh_N = \eta_N h_{Ns} = \eta_N v\,dp \tag{14.32}$$

Accordingly, the velocity for a frictionally resisted expansion through the nozzle is given by substituting from (14.32) into (14.27):

$$c\,dc = -\eta_N v\,dp \tag{14.33}$$

Since efficiency is assumed constant over the length of the nozzle, we arrive at the integral for velocity:

$$c^2 - c_0^2 = -2\eta_N \int_{p_0}^{p} v\,dp \tag{14.34}$$

This equation is very similar in form to that of equation (5.3) in Chapter 5, Section 5.2. Once more, we need to specify how specific volume, v, varies along the nozzle with pressure. Given the relationship between specific volume and pressure specified by equation (14.22), we may integrate equation (14.34) using the procedure outlined in Section 5.2.2 to give the velocity at a general downstream station in the nozzle as:

$$\begin{aligned} c^2 - c_0^2 &= 2\eta_N \frac{m}{m-1} p_0 v_0 \left(1 - \left(\frac{p}{p_0}\right)^{(m-1)/m}\right) \\ &= 2\frac{\gamma}{\gamma - 1} p_0 v_0 \left(1 - \left(\frac{p}{p_0}\right)^{(m-1)/m}\right) \end{aligned} \tag{14.35}$$

Here we have used the expression for m contained in equation (14.21) to eliminate η_N.

14.5 Using the concept of stagnation to account for non-neglible inlet velocities

As noted in the introduction to this chapter, the nozzles in a real turbine are likely to be fed with gas that already possesses an appreciable velocity. Up to now, we have deduced several useful results under the assumption that the inlet velocity is zero, but clearly we will need to extend our analysis if we are to deal with significant inlet velocities.

The equation for the downstream velocity, c, may be written from equation (14.35) as:

$$c^2 = 2\frac{\gamma}{\gamma - 1} p_0 v_0 \left(1 - \left(\frac{p}{p_0}\right)^{(m-1)/m}\right) + c_0^2 \tag{14.36}$$

We may set $p_0 v_0 = ZR_w T_0$ and also replace the pressure ratio term by the temperature ratio using equation (14.25):

$$c^2 = 2\frac{\gamma}{\gamma - 1} ZR_w T_0 \left(1 - \frac{T}{T_0}\right) + c_0^2 \tag{14.37}$$

This may be rearranged into the form:

$$c^2 = 2\frac{\gamma}{\gamma - 1} ZR_w \left(T_0 + \frac{\frac{1}{2}c_0^2}{\frac{\gamma}{\gamma - 1} ZR_w} - T\right) \tag{14.38}$$

We may combine the definition of $\gamma = c_p/c_v$ with the thermodynamic relationship between the specific heats at constant pressure (c_p) and constant volume (c_v) given in Chapter 3, Section 3.3:

$$c_p = c_v + ZR_w \tag{3.20}$$

to deduce that the specific heat at constant pressure for a gas is

$$c_p = \frac{\gamma}{\gamma - 1} ZR_w \tag{14.39}$$

Hence the term

$$\frac{1}{2}c_0^2 \Big/ \left(\frac{\gamma}{\gamma - 1} ZR_w\right)$$

represents the temperature rise that would occur if the incoming fluid were brought to rest, or stagnation, and its kinetic energy converted into enthalpy. It is natural to name this notional state the 'stagnation state', with a 'stagnation temperature', T_{0T}, at the nozzle inlet, given by:

$$T_{0T} = T_0 + \frac{\frac{1}{2}c_0^2}{c_p} \tag{14.40}$$

where the subscript 'T' has been added to indicate 'total'. Thus equation (14.38) may be rewritten

$$c^2 = 2\frac{\gamma}{\gamma - 1} ZR_w (T_{0T} - T) \tag{14.41}$$

We may rearrange equation (14.41) to:

$$\begin{aligned} c^2 &= 2\frac{\gamma}{\gamma - 1} ZR_w T_{0T} \left(1 - \frac{T_0}{T_{0T}} \frac{T}{T_0}\right) \\ &= 2\frac{\gamma}{\gamma - 1} p_{0T} v_{0T} \left(1 - \frac{T_0}{T_{0T}} \left(\frac{p}{p_0}\right)^{(m-1)/m}\right) \end{aligned} \tag{14.42}$$

Here the temperature ratio T/T_0 has been replaced using equation (14.25), while we have also assumed that the pressure and specific volume in the stagnation state are related to temperature by:

$$p_{0T} v_{0T} = ZR_w T_{0T} \tag{14.43}$$

Starting from the stagnation state, let us imagine that the actual gas temperature, pressure and speed at the entrance to the nozzle are reached by the same inefficient process of expansion that governs the subsequent flow through the nozzle. In this case the same polytropic index, m, will apply, so that we may use equation (14.25) to give the ratio of stagnation temperature to actual temperature in terms of the corresponding pressure ratio:

$$\frac{T_{0T}}{T_0} = \left(\frac{p_{0T}}{p_0}\right)^{(m-1)/m} \tag{14.44}$$

Hence equation (14.42) becomes:

$$\begin{aligned} c^2 &= 2\frac{\gamma}{\gamma - 1} p_{0T} v_{0T} \left(1 - \left(\frac{p_0}{p_{0T}}\right)^{(m-1)/m} \left(\frac{p}{p_0}\right)^{(m-1)/m}\right) \\ &= 2\frac{\gamma}{\gamma - 1} p_{0T} v_{0T} \left(1 - \left(\frac{p}{p_{0T}}\right)^{(m-1)/m}\right) \end{aligned} \tag{14.45}$$

It is highly convenient that equation (14.45) is of the same form as equation (14.36), with c_0^2 eliminated, and the inlet pressure and specific volume, p_0, v_0, replaced by their stagnation values, p_{0T}, v_{0T}.

[Note: the definition of the stagnation state offered above differs slightly from the conventional definition, where the expansion from the stagnation state to nozzle entrance is assumed to be isentropic. The relationship between conventional stagnation temperature and pressure and actual temperature and pressure may be found by replacing the polytropic index, m, in equation (14.44) by the isentropic index, γ. But the difference between m and γ will be small for the high nozzle efficiencies expected. Hence, given the same stagnation temperature, there will be little difference between the stagnation pressure found from equation (14.44) and the conventional stagnation pressure.]

Once the two thermodynamic variables, pressure and temperature, have been defined, it is possible to evaluate all the other thermodynamic variables characterizing the stagnation state. For example, stagnation specific volume emerges from equations (14.43) and (14.44) as

$$\frac{v_{0T}}{v_0} = \left(\frac{p_{0T}}{p_0}\right)^{(-1)/m} \tag{14.46}$$

The stagnation enthalpy follows directly by multiplying equation (14.40) throughout by the specific heat:

$$h_{0T} = h_0 + \tfrac{1}{2}c_0^2 \tag{14.47}$$

The mass flow at a general downstream station is given by the continuity equation

$$W = A\frac{c}{v} \tag{14.48}$$

which, using equations (14.45) and (14.46) becomes:

$$W = A\sqrt{2\frac{\gamma}{\gamma-1}\frac{p_{0T}}{v_{0T}}\left(\left(\frac{p}{p_{0T}}\right)^{2/m} - \left(\frac{p}{p_{0T}}\right)^{(m+1)/m}\right)} \tag{14.49}$$

14.6 Sonic flow

The remarks on sonic flow made at the beginning of Section 5.4.1 apply also to real nozzles, where there is a degree of friction present. The nozzle throat will pass flow at speeds up to and including sonic, but cannot support supersonic flow. Sonic flow will be reached when the ratio of throat pressure to inlet stagnation pressure has reached a critical value.

Nozzle efficiency for a well-designed convergent nozzle matched to its pressure drop can be expected to be better than 95%. For a well-designed, convergent–divergent nozzle, well matched to its pressure ratio, overall nozzle efficiency is likely to be 90% or better. In this case, the losses occur predominantly in the divergent section, which is not only longer, but also tends to allow the fluid to break away from the boundary wall to form loss-generating eddies. The convergent section of a convergent–divergent nozzle will have an efficiency similar to that of a convergent-only nozzle, i.e. in the high 90's%. Because of the disparity in efficiency between the two sections of a convergent–divergent nozzle, we will define a separate efficiency, η_{N0}, to cover the convergent section, which will have associated with it a polytropic index, m_0:

$$m_0 = \frac{\gamma}{\gamma - \eta_{N0}(\gamma-1)} \tag{14.50}$$

Clearly m_0 and m are identical for a convergent-only nozzle.

The efficiency of the nozzle will, in general, vary with inlet and outlet conditions, and this will lead to a different value of each of the polytropic indices, m and m_0, at each different set of conditions. The variation in efficiency has been investigated experimentally for a wide range of isentropic throat/discharge velocities, but the variation has been found to be small in practice for a convergent-only nozzle up to and including sonic velocity. Hence we may also assume that the efficiency of the convergent part of a convergent–divergent nozzle will stay approximately constant also. This matter is discussed further in both Section 14.7.1 and Appendix 6.

We will now develop an expression for the sonic speed experienced in the throat/outlet of a convergent-only nozzle and at the throat of a convergent–divergent nozzle when the expansion is frictionally resisted. Sonic conditions will exist in the throat when the velocity calculated by applying equation (14.45) to the convergent section of the nozzle has reached the local speed of sound, i.e.:

$$c_c = \sqrt{2\frac{\gamma}{\gamma-1}p_{0T}v_{0T}\left(1-\left(\frac{p_{tc}}{p_{0T}}\right)^{(m_{0c}-1)/m_{0c}}\right)}$$
$$= \sqrt{\gamma p_{tc}v_{tc}} \tag{14.51}$$

where the subscript 'tc' denotes critical conditions in the throat, and m_{0c} is the value of the polytropic index at critical conditions, corresponding to a convergent nozzle efficiency of η_{N0c}. Rearranging equation (14.51) gives

$$\frac{p_{0T}v_{0T}}{p_{tc}v_{tc}}\left(1-\left(\frac{p_{tc}}{p_{0T}}\right)^{(m_{0c}-1)/m_{0c}}\right) = \frac{\gamma-1}{2} \tag{14.52}$$

Substituting for the ratio of specific volumes using equation (14.46) gives:

$$\left(\frac{p_{0T}}{p_{tc}}\right)^{(m_{0c}-1)/m_{0c}}\left(1-\left(\frac{p_{tc}}{p_{0T}}\right)^{(m_{0c}-1)/m_{0c}}\right) = \frac{\gamma-1}{2} \tag{14.53}$$

Hence the ratio of the throat pressure to the stagnation inlet pressure at critical conditions is:

$$\frac{p_{tc}}{p_{0T}} = \left(\frac{2}{\gamma+1}\right)^{m_{0c}/(m_{0c}-1)} \tag{14.54}$$

We may see directly the effect of the inefficiency on the expansion by substituting for m_{0c} using equation (14.50):

$$\frac{p_{tc}}{p_{0T}} = \left(\frac{2}{\gamma+1}\right)^{\gamma/(\eta_{N\,0c}(\gamma-1))} = \left[\left(\frac{2}{\gamma+1}\right)^{\gamma/(\gamma-1)}\right]^{1/(\eta_{N\,0c})} = \left[\frac{p_{tc}}{p_{0T}}\bigg|_s\right]^{1/(\eta_{N\,0c})} \qquad (14.55)$$

where $(p_{tc}/p_{0T}|_s)$ is the critical pressure ratio at the throat for an ideal, isentropic expansion. The fact that the efficiency is less than unity means that the actual throat pressure needs to fall further than the ideal critical value to overcome the friction present, as one might expect on intuitive grounds. The value of the local speed of sound may be found by substituting from equations (14.55) and (14.46) into equation (14.51):

$$c_{tc} = \sqrt{\gamma p_{tc} v_{tc}} = \sqrt{\gamma \frac{p_{tc}}{p_{0T}} p_{0T} \frac{v_{tc}}{v_{0T}} v_{0T}} = \sqrt{\gamma \left(\frac{p_{tc}}{p_{0T}}\right)^{(m_{0c}-1)/m_{0c}} p_{0T} v_{0T}} = \sqrt{\frac{2}{\gamma+1}\gamma p_{0T} v_{0T}} \qquad (14.56)$$

which is the same speed as would have been reached in an ideal, isentropic expansion (see equation (5.58)).

The mass flow is given by:

$$W_c = A_t \frac{c_{tc}}{v_{tc}} = A_t \frac{c_{tc}}{v_{0T}} \left(\frac{p_{tc}}{p_{0T}}\right)^{1/m_{0c}} = A_t \frac{1}{v_{0T}} \left(\frac{2}{\gamma+1}\right)^{1/(m_{0c}-1)} \sqrt{\frac{2}{\gamma+1}\gamma p_{0T} v_{0T}} = A_t \sqrt{\left(\frac{2}{\gamma+1}\right)^{(m_{0c}+1)/(m_{0c}-1)} \gamma \frac{p_{0T}}{v_{0T}}} \qquad (14.57)$$

where A_t is the throat area. Using equation (14.50), we find that:

$$\frac{m_{0c}+1}{m_{0c}-1} = \frac{\gamma+1}{\gamma-1} + 2\frac{\gamma}{\gamma-1}\left(\frac{1-\eta_{N\,0c}}{\eta_{N\,0c}}\right) \qquad (14.58)$$

Hence we may rewrite equation (14.57) as:

$$W_c = A_t \sqrt{\left(\frac{2}{\gamma+1}\right)^{(\gamma+1)/(\gamma-1)} \gamma \frac{p_{0T}}{v_{0T}}} \times \left(\frac{2}{\gamma+1}\right)^{\gamma/(\gamma-1)((1-\eta_{N\,0c})/\eta_{N\,0c})} = W_{cs}\left(\frac{2}{\gamma+1}\right)^{\gamma/(\gamma-1)((1-\eta_{N\,0c})/\eta_{N\,0c})} \qquad (14.59)$$

where W_{cs} is the critical mass flow that would occur for an ideal, isentropic expansion through the nozzle (see equation (5.59)). For example, taking the case of superheated steam with $\gamma = 1.3$ and a nozzle efficiency as far as the throat of 0.95, the actual critical mass flow is about 3% less than the ideal critical flow.

It has been shown above that sonic flow will occur when the throat pressure has reached the critical value given in equation (14.54), but this begs the question of how we determine the throat pressure given only the inlet and outlet pressures for the nozzle, which will be the usual case. The answer may be stated easily for the case of the convergent-only nozzle, where the nozzle throat and the nozzle outlet are adjacent. Here the throat pressure will be the same as the outlet pressure until the ratio of the outlet to inlet pressures falls below the critical value, and thereafter the throat pressure will remain at the critical value. The case for the convergent–divergent nozzle is more complicated, and requires the fuller treatment given later in this chapter.

We shall discuss in Section 14.7 the nozzle efficiency for the convergent-only nozzle and hence calculate the outlet velocity and mass flow. We shall then develop similar methods for the convergent–divergent nozzle in Section 14.8.

14.7 The convergent-only nozzle

14.7.1 Estimating nozzle efficiency for a convergent-only nozzle

Estimating the efficiency of a nozzle depends ultimately on the results of tests on similar systems, and the manufacturer will supply a figure at the design point, either explicitly or from the thermodynamic data that accompany the turbine flowsheet. However, the control engineer will need to consider off-design conditions, including particularly the sonic limitations that may occur if the pressure drop falls below a critical value.

The convergent-only nozzle will usually be designed to work in the unchoked region, where the flow is always less than sonic. Provided the flow is not choked, the deviation from unity efficiency will be caused by skin-friction only. A simple application of the Fanning formula in its differential form to the frictional loss of energy along the nozzle gives:

$$\frac{dF}{dx} = \frac{4f}{D}\frac{c^2}{2} \tag{4.20}$$

where both the nozzle (effective) diameter, D, and the velocity, c, will need to vary with distance along the nozzle, x. If the friction factor, f, were constant, greater discharge speeds would lead to increased frictional losses and so to lower nozzle efficiencies. But experimental studies have shown a variety of behaviours for the nozzle efficiency as the isentropic discharge velocity increases towards sonic speed: it may rise by a few percentage points, it may exhibit a fall of a few percentage points and then recover, or it may stay substantially constant. The local Reynolds number, $N_{RE} = cD/v\mu$, seems to be the governing parameter. An increased discharge velocity and hence increased local Reynolds number appear to reduce the nozzle friction factor sufficiently to cancel out, more or less, the increase in frictional loss that would occur otherwise. Given the lack of a clear picture for the variation in efficiency, and the fact that the variation is likely to be small, we will assume that the efficiency of a convergent nozzle remains constant at its design value at off-design conditions up to sonic. (The reader is referred to Appendix 6, which contains a more extensive discussion of experimental data on nozzle efficiencies.)

The velocity of the gas at the point of discharge from a convergent-only nozzle cannot exceed the local speed of sound, no matter how low the exit pressure. But it is possible for gas that has already reached sonic velocity at the exit of a convergent-only nozzle to accelerate to supersonic speeds in the discharge manifold before reaching the turbine blade, provided there is space for the further expansion. In practice, any additional expansion would lose much of its additional velocity to turbulence if the discharge pressure were much less than the critical throat pressure, given by equation (14.54). But to make allowance for such a situation, we choose to assume that an additional expansion can occur down to a discharge pressure ratio of $r_{\text{lim}}(p_{tc}/p_{0T})$, where $r_{\text{lim}} \leq 1.0$ is an empirical discharge coefficient, dependent on the geometry of the manifold. For example, we might allow for a small extra expansion by putting $r_{\text{lim}} = 0.9$. Once again, the experimental evidence suggests that the efficiency of such an additional expansion is approximately the same as the design efficiency.

In summary, it is sufficient for the dynamic modeller to assume that the efficiency of the convergent nozzle remains constant at its design value down to a pressure ratio of $r_{\text{lim}}(p_{tc}/p_{0T})$:

$$\eta_{N0} = \eta_{N0c} = \eta_{N0D} \quad \text{for } \frac{p_1}{p_{0T}} \geq r_{\text{lim}}\frac{p_{tc}}{p_{0T}} \tag{14.60}$$

and hence

$$m_0 = m_{0c} = m_{0D} \quad \text{for } \frac{p_1}{p_{0T}} \geq r_{\text{lim}}\frac{p_{tc}}{p_{0T}} \tag{14.61}$$

14.7.2 Outlet velocity and mass flow in a convergent-only nozzle

The velocity of the gas reaching the turbine blade will increase as the pressure ratio falls down as far as $r_{\text{lim}}(p_{tc}/p_{0T})$, but it will remain constant thereafter. Thus it may be calculated using equation (14.45) as:

$$c_1 = \sqrt{2\frac{\gamma}{\gamma-1}p_{0T}v_{0T}\left(1-\left(\frac{p_1}{p_{0T}}\right)^{(m_0-1)/m_0}\right)} \quad \text{for } \frac{p_1}{p_{0T}} \geq r_{\text{lim}}\frac{p_{tc}}{p_{0T}}$$

$$= \sqrt{2\frac{\gamma}{\gamma-1}p_{0T}v_{0T}\left(1-\left(r_{\text{lim}}\frac{p_{tc}}{p_{0T}}\right)^{(m_{0c}-1)/m_{0c}}\right)} \quad \text{for } \frac{p_1}{p_{0T}} < r_{\text{lim}}\frac{p_{tc}}{p_{0T}} \tag{14.62}$$

where the subscript '1' refers to the discharge conditions.

While we have made allowance for a possible increase in velocity downstream of the nozzle, we should remember that the mass flow is still subject to an absolute choking limit. The mass flow under all conditions is given by combining equations (14.49) and (14.57):

$$W = A_t\sqrt{2\frac{\gamma}{\gamma-1}\frac{p_{0T}}{v_{0T}}\left(\left(\frac{p_1}{p_{0T}}\right)^{2/m_0}-\left(\frac{p_1}{p_{0T}}\right)^{(m_0+1)/m_0}\right)} \quad \text{for } \frac{p_1}{p_{0T}} \geq \left(\frac{2}{\gamma+1}\right)^{m_{0c}/(m_{0c}-1)}$$

$$= A_t\sqrt{\left(\frac{2}{\gamma+1}\right)^{(m_{0c}+1)/(m_{0c}-1)}\gamma\frac{p_{0T}}{\mu_{0T}}} \quad \text{for } \frac{p_1}{p_{0T}} < \left(\frac{2}{\gamma+1}\right)^{m_{0c}/(m_{0c}-1)} \tag{14.63}$$

where the throat area, A_t, is the same as the area at station 'N1' for a convergent-only nozzle: $A_t = A_1$.

The differentiation between the general polytropic index, m_0, and the value of the polytropic index at critical conditions, m_{0c}, has been retained for the sake of formal correctness, even though we shall regard them as equal, as explained in Section 14.7.1.

14.7.3 Nozzle efficiency in choked conditions for a convergent-only nozzle

The methods explained above allow us to calculate the outlet velocity and the mass flow through a convergent-only nozzle at all pressure ratios without the need to develop an expression for the efficiency of the nozzle deep into the choked region. However there are circumstances when such an expression can be useful (see Chapter 16), and we shall now perform the necessary analysis for the region

$$\frac{p_1}{p_{0T}} < r_{\mathrm{lim}} \frac{p_{tc}}{p_{0T}}$$

From equation (14.62), the velocity and hence the useful specific kinetic energy stays constant for this pressure ratio and lower:

$$\begin{aligned}\frac{1}{2}c_1^2 &= \frac{\gamma}{\gamma-1} p_{0T} v_{0T} \left(1 - \left(r_{\mathrm{lim}} \frac{p_{tc}}{p_{0T}}\right)^{(m_{0D}-1)/m_{0D}}\right) \\ &= \frac{\gamma}{\gamma-1} Z R_W T_{0T} \\ &\quad \times \left(1 - r_{\mathrm{lim}}^{(m_{0D}-1)/m_{0D}} \left(\frac{p_{tc}}{p_{0T}}\right)^{(m_{0D}-1)/m_{0D}}\right) \\ &\quad \text{for } \frac{p_1}{p_{0T}} \le r_{\mathrm{lim}} \frac{p_{tc}}{p_{0T}}\end{aligned} \tag{14.64}$$

where we have chosen to use equation (14.61) to replace m_{0c} with m_{0D}. Simplifying using equation (14.39) for specific heat and equation (14.54) for the throat critical pressure ratio produces:

$$\frac{1}{2}c_1^2 = c_p T_{0T} \left(1 - r_{\mathrm{lim}}^{(m_{0D}-1)/m_{0D}} \left(\frac{2}{\gamma+1}\right)\right) \tag{14.65}$$

Now from equation (14.40), the kinetic energy at nozzle inlet is:

$$\tfrac{1}{2}c_0^2 = c_p (T_{0T} - T_0) \tag{14.66}$$

Using equation (14.66) and also the definition of m in terms of nozzle efficiency, equation (14.50), allows us to re-express equation (14.65) as

$$\frac{1}{2}c_1^2 - \frac{1}{2}c_0^2 = c_p T_0 \left(1 - \frac{T_{0T}}{T_0} \frac{2}{\gamma+1} r_{\mathrm{lim}}^{\eta_{N\,0D}((\gamma-1)/\gamma)}\right) \tag{14.67}$$

Extending equation (14.5) to cover the additional expansion assumed to take place immediately downstream of the nozzle outlet, the effective nozzle efficiency is given by:

$$\begin{aligned}\eta_N &= \frac{\frac{1}{2}c_1^2 - \frac{1}{2}c_0^2}{h_0 - h_{1s}} = \frac{\frac{1}{2}c_1^2 - \frac{1}{2}c_0^2}{c_p T_0 \left(1 - \dfrac{T_{1s}}{T_0}\right)} \\ &= \frac{\frac{1}{2}c_1^2 - \frac{1}{2}c_0^2}{c_p T_0 \left(1 - \left(\dfrac{p_1}{p_0}\right)^{(\gamma-1)/\gamma}\right)}\end{aligned} \tag{14.68}$$

Hence

$$\eta_{N\,0} = \frac{1 - \dfrac{T_{0T}}{T_0} \dfrac{2}{\gamma+1} r_{\mathrm{lim}}^{\eta_{N\,0D}((\gamma-1)/\gamma)}}{1 - \left(\dfrac{p_1}{p_0}\right)^{(\gamma-1)/\gamma}} \tag{14.69}$$

For the case where it is assumed that the discharge speed cannot exceed the speed of sound in the nozzle throat, $r_{\mathrm{lim}} = 1$. If, in addition, the inlet velocity is negligible, then $T_{0T} = T_0$, and equation (14.69) simplifies to

$$\eta_{N\,0} = \frac{\gamma-1}{(\gamma+1)} \frac{1}{\left(1 - \left(\dfrac{p_1}{p_0}\right)^{(\gamma-1)/\gamma}\right)} \tag{14.70}$$

14.7.4 Nozzle efficiency over the whole range of pressure ratios for a convergent-only nozzle

Bringing together equations (14.60) and (14.69) gives the efficiency as

$$\begin{aligned}\eta_{N\,0} &= \eta_{N\,0D} \\ &\quad \text{for } \frac{p_1}{p_{0T}} \ge r_{\mathrm{lim}} \left(\frac{2}{\gamma+1}\right)^{m_{0D}/(m_{0D}-1)} \\ &= \frac{1 - \dfrac{T_{0T}}{T_0} \dfrac{2}{\gamma+1} r_{\mathrm{lim}}^{\eta_{N\,0D}((\gamma-1)/\gamma)}}{1 - \left(\dfrac{p_1}{p_0}\right)^{(\gamma-1)/\gamma}} \\ &\quad \text{for } \frac{p_1}{p_{0T}} < r_{\mathrm{lim}} \left(\frac{2}{\gamma+1}\right)^{m_{0D}/(m_{0D}-1)}\end{aligned} \tag{14.71}$$

As an example, let us take the case of a convergent-only nozzle being fed with initially stationary air at 20 bar and 400°C. The design efficiency is 0.97. The geometry of the discharge manifold is such that we may assume that the discharge velocity will continue increasing into the supersonic range until the discharge pressure ratio falls below 90% of the critical throat pressure ratio.

Air is essentially a mixture of two diatomic gases, oxygen and nitrogen, and hence we may take $\gamma = 1.4$. Since there is no inlet velocity, the stagnation values of temperature, pressure etc. are unchanged from the basic values. Treating air as a perfect gas with a compressibility factor of unity, we may use the characteristic gas equation to calculate the specific volume at nozzle inlet as

$$v_0 = \frac{8314 \times 673}{29 \times 20 \times 10^5} = 0.09647\,\text{m}^3/\text{kg}$$

Using equation (14.50) the value of the polytropic exponent for the convergent nozzle at design conditions is $m_{0D} = 1.4 \div (1.4 - 0.97(1.4 - 1)) = 1.3834$. Hence the critical throat pressure ratio is (equation 14.54)

$$\frac{p_{tc}}{p_{0T}} = \left(\frac{2}{1.4+1}\right)^{1.3834/(1.3834-1)} = 0.5180,$$

which is slightly below the isentropic critical throat pressure ratio, given by

$$\left(\frac{2}{1.4+1}\right)^{1.4/(1.4-1)} = 0.5283$$

Meanwhile, we may take $r_{\text{lim}} = 0.9$, so that the limiting pressure ratio for supersonic flow is $0.9 \times 0.518 = 0.4662$.

We may now evaluate equation (14.71) for this nozzle over the complete range of pressure ratios. The results are shown graphically in Figure 14.3. It is clear that the efficiency of the nozzle stays high until the pressure ratio falls to about 50%, after which there is a steady decline.

The convergent-only nozzle cannot produce the supersonic velocity and associated very high level of kinetic energy that would be available from a well-matched convergent–divergent nozzle. On the other hand, however, the convergent-only nozzle can offer a more robust efficiency curve if the turbine is required to work over a wide range of pressure ratios above critical. Convergent-only nozzles are often employed in the Rateau turbine. Such a turbine may be employed to power the boiler feedpump on a power station, where its ability to cope with widely varying steam plant conditions can prove essential to cope with markedly off-design conditions, which may result from load scheduling or from part of the plant being out for maintenance.

14.8 The convergent–divergent nozzle

14.8.1 Relationship between the throat pressure and the discharge pressure for a convergent–divergent nozzle

Figure 14.4 shows a convergent–divergent nozzle, intended to produce a supersonic outlet velocity. The pressure in the discharge manifold is denoted p_1. In most instances, $p_{N1} = p_1$, but this relationship will not always hold in all cases of supersonic flow.

Imagine the situation where the manifold pressure, p_1, is gradually decreased below the inlet pressure, p_0, causing the mass flow, W, to increase. Flow will be subsonic initially at all parts of the nozzle, so that the pressure at the exit from the nozzle will be identical with the discharge pressure: $p_{N1} = p_1$. Let the nozzle efficiency to the nozzle throat be η_{N0} and the overall nozzle efficiency be η_N, with corresponding expansion

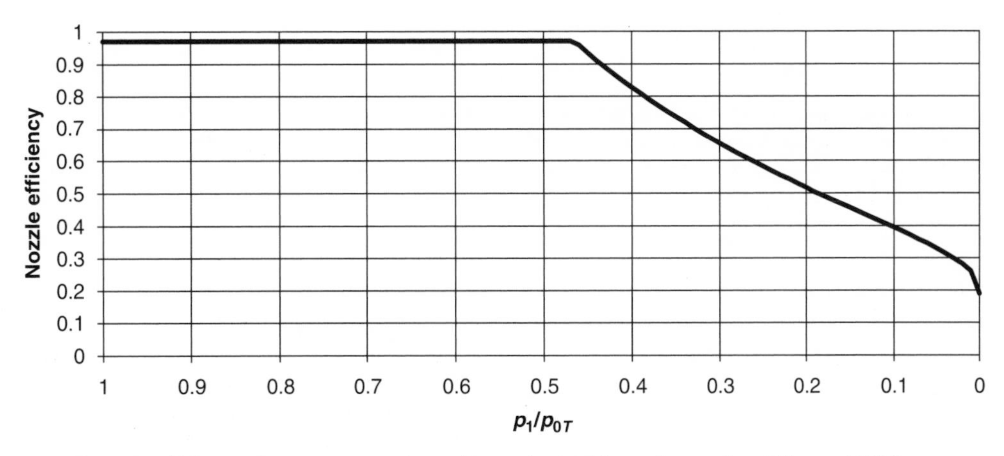

Figure 14.3 Example: efficiency of a convergent-only nozzle, passing initially stationary air at 20 bar and 400°C.

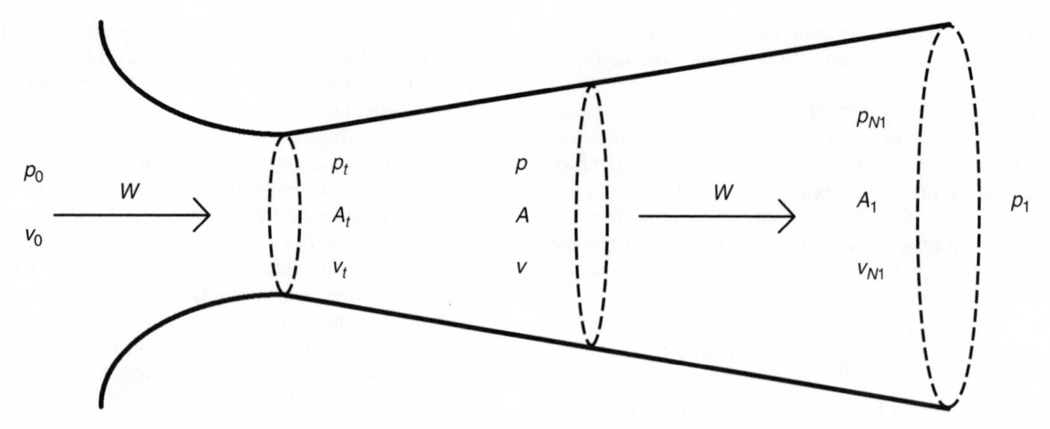

Figure 14.4 Convergent–divergent nozzle.

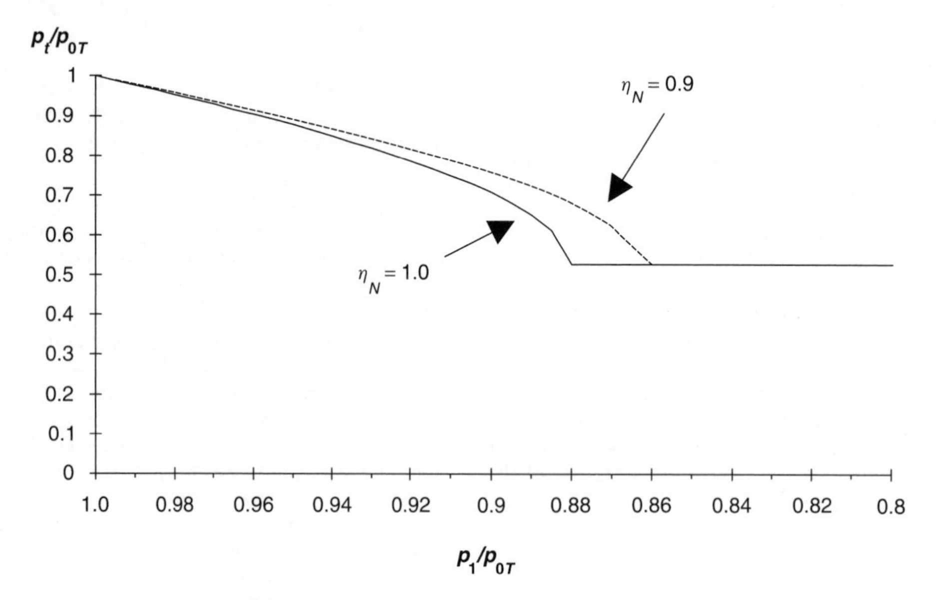

Figure 14.5 Variation in throat pressure ratio with exit pressure ratio.

indices, m_0 and m. In the steady state, the mass flow passing the throat will be the same as the mass flow at the exit from the nozzle. Equating these two flows using equation (14.49) gives

$$A_t\sqrt{2\frac{\gamma}{\gamma-1}\frac{p_{0T}}{v_{0T}}\left(\left(\frac{p_t}{p_{0T}}\right)^{2/m_0}-\left(\frac{p_t}{p_{0T}}\right)^{(m_0+1)/m_0}\right)}$$

$$=A_1\sqrt{2\frac{\gamma}{\gamma-1}\frac{p_{0T}}{v_{0T}}\left(\left(\frac{p_1}{p_{0T}}\right)^{2/m}-\left(\frac{p_1}{p_{0T}}\right)^{(m+1)/m}\right)} \tag{14.72}$$

Hence

$$\left(\frac{p_t}{p_{0T}}\right)^{2/m_0}-\left(\frac{p_t}{p_{0T}}\right)^{(m_0+1)/m_0}$$

$$=\left(\frac{A_1}{A_t}\right)^2\left(\left(\frac{p_1}{p_{0T}}\right)^{2/m}-\left(\frac{p_1}{p_{0T}}\right)^{(m+1)/m}\right) \tag{14.73}$$

Given the isentropic index, γ, the nozzle efficiency to the throat, η_{N0}, the overall nozzle efficiency, η_N, and the nozzle geometry to define A_1/A_t, it is possible to solve the implicit equation (14.73) for the throat pressure ratio, p_t/p_{0T}, in terms of the exit pressure ratio, p_1/p_{0T}. Figure 14.5 below shows the solution for a nozzle with an exit to throat area ratio of 1.5 for the two cases: (i) where the expansion is perfectly isentropic throughout the nozzle, with $\gamma = 1.4$ and (ii) where the expansion is isentropic as far as the throat, i.e. $\eta_{N0} = 1.0$, but then friction becomes significant in the divergent section, giving an overall nozzle efficiency, η_N, of 0.9.

The throat pressure ratio, p_t/p_{0T}, falls initially as the exit pressure ratio, p_1/p_{0T}, is reduced, but sonic conditions in the throat are soon reached for this nozzle: when the exit pressure ratio has come down to 0.88 for the perfectly isentropic case and 0.86 for the case where nozzle efficiency is 90%. After this point, the throat pressure stays constant no matter how much the exit pressure is reduced, and the mass flow rate through the nozzle stays constant.

Since the throat pressure becomes independent of the pressure in the discharge manifold once the former has reached a critical value, it is no longer sensible then to solve equation (14.73) for the throat pressure ratio for a given exit pressure ratio. Instead, we may use the nozzle mass flow equation (14.49) to examine the pressure at a general point downstream of the throat by setting the throat pressure ratio to its critical value. At this point, the mass flow will be at its critical value, W_c, given by equation (14.57). Equating this flow at the throat to the flow at a downstream point in the nozzle for an inefficient expansion using equation (14.49) gives:

$$A\sqrt{2\frac{\gamma}{\gamma-1}\frac{p_{0T}}{v_{0T}}\left(\left(\frac{p}{p_{0T}}\right)^{2/m}-\left(\frac{p}{p_{0T}}\right)^{(m+1)/m}\right)} = A_t\sqrt{\left(\frac{2}{\gamma+1}\right)^{(m_{0c}+1)/(m_{0c}-1)}\gamma\frac{p_{0T}}{v_{0T}}} \quad (14.74)$$

where m_{0c} is the value of the polytropic index for the convergent section of the nozzle at critical conditions. Hence the downstream pressure ratio, p/p_{0T}, is given in terms of the ratio of downstream area to throat area, A/A_t, by the implicit equation:

$$\left(\frac{p}{p_{0T}}\right)^{2/m}-\left(\frac{p}{p_{0T}}\right)^{(m+1)/m} = \left(\frac{A_t}{A}\right)^2\frac{\gamma-1}{2}\left(\frac{2}{\gamma+1}\right)^{(m_{0c}+1)/(m_{0c}-1)} \quad (14.75)$$

This equation may be used to generate the pressure profile downstream of the nozzle throat. It possesses two solutions at each downstream point in the nozzle, corresponding to the two cases:

(i) the discharge pressure ratio is equal to its upper critical value: $p_1/p_{0T} = (p_{1\,crit\,1})/p_{0T}$, so that the velocity in the throat just touches sonic, but the flow returns immediately without shock to subsonic downstream of the throat; and

(ii) the discharge pressure ratio is equal to its lower critical value: $p_1/p_{0T} = (p_{1\,crit\,2})/p_{0T}$. Here the velocity in the throat is sonic, and the velocity increases smoothly to supersonic downstream of the throat. The gas is accelerated in the supersonic state all the way to the end of the nozzle.

Both behaviours are displayed in Figure 14.6 for an ideal diatomic gas ($\gamma = 1.4$) and for a nozzle where the downstream radius increases linearly over the divergent section to give an exit-to-throat area ratio of 1.5. The figure shows both an ideal, isentropic case, where the overall nozzle efficiency is unity, and an inefficient case, where friction is negligible in the convergent section of the nozzle, but is significant in the divergent section, giving an overall nozzle efficiency of 90%.

Which behaviour occurs will depend on the pressure of the manifold into which the nozzle is discharging. If the manifold pressure is exactly the value of the either of the exit pressures shown in Figure 14.6, then the corresponding pressure profile will be followed. But the obvious question is: what happens when the pressure is between these pressures, or below the exit pressure corresponding to the supersonic profile? Theory and experiment have shown that the behaviour follows the general form of Figure 14.7.

Let us assume that the stagnation inlet pressure, p_{0T}, is kept constant. When the manifold pressure is reduced below p_{1a}, the value that just causes sonic flow in the throat, to a lower pressure p_{1b}, supersonic conditions are induced for some distance downstream of the throat. Then a stationary shock wave will occur, which causes the flow to change over a very short distance from supersonic to subsonic. As the pressure is further reduced to p_{1c}, the position of the shock wave migrates further downstream, but the behaviour is similar in kind, with a very rapid change from supersonic to subsonic speed. Once the manifold pressure is reduced below the throat critical pressure, p_{tc}, the shock wave is delayed until the exit of the nozzle and its character alters: the shock wave is no longer normal to the flow, but occurs obliquely, moving downstream at an angle from the inner periphery of the nozzle outlet. The velocity downstream of the nozzle will thus vary in the direction normal to the flow as well as in the direction of the flow, implying that flow becomes two-dimensional.

However, reducing the manifold pressure further to p_{1d} allows a smooth transition from the supersonic conditions in the nozzle to the manifold, and no shock occurs in the nozzle. This condition is the one aimed

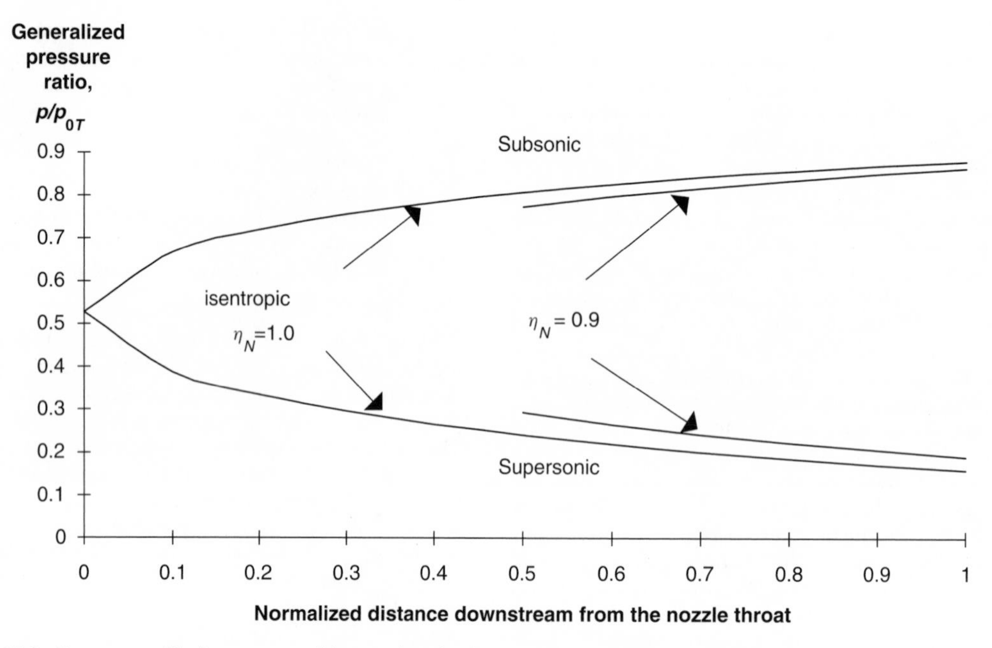

Figure 14.6 Pressure profile downstream of throat when the throat pressure is critical: subsonic and supersonic regimes (nozzle outlet corresponds to a normalized downstream distance of unity).

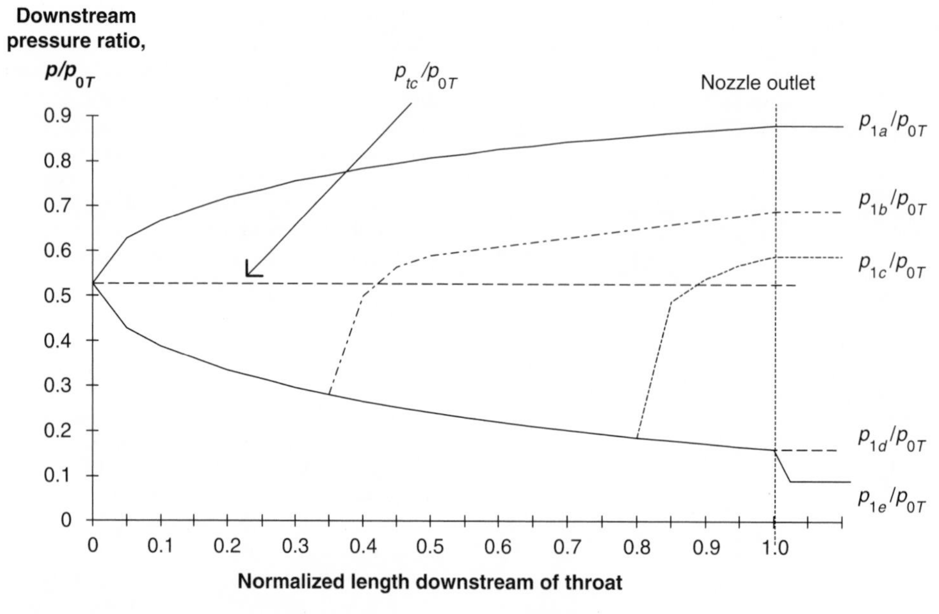

Figure 14.7 Pressure profiles downstream of throat when the throat pressure is critical: off-design cases.

at by turbine designers, and p_{1d} is the design manifold pressure. How well the supersonic speed is then utilized will depend on the design of the turbine stage.

When the manifold pressure is reduced further to p_{1e}, supersonic flow is again maintained in the nozzle, although no further increase in velocity occurs in the nozzle over and above that pertaining when the manifold pressure was p_{1d}. A manifold pressure of $p_{1e} < p_{1d}$ implies that the flow area of the manifold is greater than that of the nozzle outlet, and will permit expansion. Unless the area presented by the manifold to the flow is increased carefully in a taper (to become an extension of the divergent part of the nozzle, in effect), the resultant expansion will be uncontrolled and chaotic for any significant pressure drop below p_{1d}. In practical turbine applications, the flow leaving the nozzle will enter an associated blade, which will present essentially the same area as the nozzle outlet. An exception to this will be some cases of impulse stage (especially where the turbines consists of a single impulse stage), where one nozzle may supply several blades, which together will present a greater flow area. In such a case, a good deal of the kinetic energy resulting from the uncontrolled expansion may be used by the blades, provided the manifold pressure is not too far below the design pressure.

One important consequence of the discussion above is that sonic conditions will exist in the throat of the convergent–divergent nozzle as soon the discharge pressure ratio falls below the larger of the two roots of equation (14.75), the upper critical value, $p_{1\,crit\,1}/p_{0T}$. Although the velocity of the gas may increase downstream of the throat (given good pressure matching), the mass flow becomes choked at this point.

It may be observed that solving equation (14.75) for the upper critical value, $p_{1\,crit\,1}/p_{0T}$, has allowed us to transfer the choking criterion from the throat pressure ratio to the discharge pressure ratio.

14.8.2 Nozzle efficiencies for a convergent–divergent nozzle

By contrast with the convergent-only nozzle, we may expect the convergent–divergent nozzle to be designed to work in the region where flow is choked and the gas is accelerated to supersonic velocity in the divergent section – the presence of the divergent section would otherwise be disadvantageous, since it would act as a diffuser for subsonic flow, and reduce the kinetic energy of the outlet gas stream, the exact opposite of what is required. However, it is conceivable that the nozzle will be called upon occasionally to work at an off-design condition in the subsonic region.

We can expect the flowsheet to provide values of the pressure and temperature or enthalpy at the inlet and outlet of the nozzle at the design point. We may note also that the conditions just inside the nozzle outlet, station 'N1', will be the same as the conditions in the discharge manifold, station '1', at the design point. We may use these flowsheet values to deduce the overall nozzle efficiency at the design conditions using equation (14.5):

$$\eta_{ND} = \frac{h_{0D} - h_{1D}}{h_{0D} - h_{1sD}} = \frac{1 - \dfrac{T_{1D}}{T_{0D}}}{1 - \dfrac{T_{1sD}}{T_{0D}}} \qquad (14.76)$$

where the additional subscript 'D' indicates 'design'. (Note that the inlet quantities are the actual values, *not* the stagnation values). The downstream temperature for an isentropic expansion may be calculated from the expansion and gas equations:

$$\begin{aligned} p_0 v_0^\gamma &= p_1 v_{1s}^\gamma \\ \frac{p_0 v_0}{T_0} &= \frac{p_1 v_{1s}}{T_{1s}} \end{aligned} \qquad (14.77)$$

where the 's' subscript refers to isentropic conditions. We may eliminate specific volumes to give the isentropic temperature ratio in terms of the pressure ratio:

$$\frac{T_{1s}}{T_0} = \left(\frac{p_1}{p_0}\right)^{(\gamma-1)/\gamma} \qquad (14.78)$$

This expression may be used in equation (14.76) to find the overall nozzle efficiency at the design point.

We may assume that the nozzle will be designed to produce a supersonic discharge velocity, and that the design pressure ratio will coincide with the lower critical discharge pressure ratio for the design inlet conditions, i.e. $(p_1/p_{0T})|_D = (p_{1\,crit\,2}/p_{0T})|_D$. The nozzle will be choked at this point, and this implies that the pressure at the throat will be at its critical value. The associated throat critical pressure ratio is given by equation (14.55):

$$\left.\frac{p_{tc}}{p_{0T}}\right|_D = \left(\frac{2}{\gamma+1}\right)^{(1/\eta_{N\,0c})(\gamma/\gamma-1)} \qquad (14.79)$$

where $\eta_{N\,0c}$ is the efficiency of the convergent section of the nozzle at choking for the design inlet conditions. We may find $\eta_{N\,0c}$ as follows.

Since the design pressure ratio coincides with the lower critical discharge pressure ratio, it must satisfy equation (14.75), namely

$$\begin{aligned} &\left(\left.\frac{p_1}{p_{0T}}\right|_D\right)^{2/m_D} - \left(\left.\frac{p_1}{p_{0T}}\right|_D\right)^{(m_D+1)/m_D} \\ &= \left(\frac{A_t}{A_1}\right)^2 \frac{\gamma-1}{2}\left(\frac{2}{\gamma+1}\right)^{(m_{0c}+1)/(m_{0c}-1)} \end{aligned} \qquad (14.80)$$

where the indices, m_D and m_{0c}, are, respectively, the design value for the overall expansion and the choked value for the convergent section of the nozzle at the design inlet conditions:

$$m_D = \frac{\gamma}{\gamma - \eta_{ND}(1-\gamma)} \qquad (14.81)$$

and

$$m_{0c} = \frac{\gamma}{\gamma - \eta_{N\,0c}(1-\gamma)} \qquad (14.82)$$

Now the exponent $(m_{0c}+1)/(m_{0c}-1)$ may be expanded starting from equation (14.58):

$$\begin{aligned}\frac{m_{0c}+1}{m_{0c}-1} &= \frac{\gamma+1}{\gamma-1}+2\frac{\gamma}{\gamma-1}\left(\frac{1-\eta_{N\,0c}}{\eta_{N\,0c}}\right)\\ &= \frac{\gamma+1}{\gamma-1}-\frac{2\gamma}{\gamma-1}+\frac{1}{\eta_{N\,0c}}\frac{2\gamma}{\gamma-1}\\ &= \frac{1-\gamma}{\gamma-1}+\frac{1}{\eta_{N\,0c}}\frac{2\gamma}{\gamma-1}\\ &= -1+\frac{1}{\eta_{N\,0c}}\frac{2\gamma}{\gamma-1}\end{aligned} \tag{14.83}$$

Substituting back into equation (14.80) gives

$$\left(\left.\frac{p_1}{p_{0T}}\right|_D\right)^{2/m_D}-\left(\left.\frac{p_1}{p_{0T}}\right|_D\right)^{(m_D+1)/m_D} = \left(\frac{A_t}{A_1}\right)^2\frac{\gamma-1}{2}\frac{\gamma+1}{2}\left(\frac{2}{\gamma+1}\right)^{(1/\eta_{N\,0c})(2\gamma/(\gamma-1))} \tag{14.84}$$

Hence

$$\left(\frac{2}{\gamma+1}\right)^{(1/\eta_{N\,0c})(2\gamma/(\gamma-1))} = \frac{4}{\gamma^2-1}\left(\frac{A_1}{A_t}\right)^2\left(\left(\left.\frac{p_1}{p_{0T}}\right|_D\right)^{2/m_D}-\left(\left.\frac{p_1}{p_{0T}}\right|_D\right)^{(m_D+1)/m_D}\right) \tag{14.85}$$

Taking logs gives:

$$\frac{1}{\eta_{N\,0c}}\frac{2\gamma}{\gamma-1}\log\left(\frac{2}{\gamma+1}\right) = \log\left[\frac{4}{\gamma^2-1}\left(\frac{A_1}{A_t}\right)^2\left(\left(\left.\frac{p_1}{p_{0T}}\right|_D\right)^{2/m_D}-\left(\left.\frac{p_1}{p_{0T}}\right|_D\right)^{(m_D+1)/m_D}\right)\right] \tag{14.86}$$

so that the efficiency, $\eta_{N\,0c}$, of the convergent section of the nozzle at choking for the design inlet conditions emerges as:

$$\eta_{N\,0c} = \frac{2\gamma}{\gamma-1}\times\frac{\log\left[\dfrac{2}{\gamma+1}\right]}{\log\left[\dfrac{4}{\gamma^2-1}\left(\dfrac{A_1}{A_t}\right)^2\left(\left(\left.\dfrac{p_1}{p_{0T}}\right|_D\right)^{2/m_D}-\left(\left.\dfrac{p_1}{p_{0T}}\right|_D\right)^{(m_D+1)/m_D}\right)\right]} \tag{14.87}$$

At the upper critical discharge ratio, the divergent part of the nozzle will tend to reverse the action of the convergent part of the nozzle, and so it will see similar gas velocities, but in reverse order. However, as noted in Section 14.6, the divergent part of the nozzle will tend by its nature to lose more to friction than the convergent section. As a heuristic rule, we may assume that the fraction of energy lost to friction for the nozzle as a whole will be at least twice that lost in the convergent part of the nozzle at choking. Further, we will not expect the nozzle efficiency at the upper critical discharge ratio to better the efficiency at the lower critical discharge ratio. Pulling these two guidelines together, we may estimate the overall nozzle efficiency at the upper critical discharge ratio as

$$\eta_{N\,crit\,1} = \min(\eta_{N\,0c}^2, \eta_{ND}) \tag{14.88}$$

Given an estimate of overall nozzle efficiency, we may apply equation (14.75) once more and solve iteratively to find the upper critical pressure ratio for the design inlet conditions, $(p_{1\,crit\,1}/p_{0T}|_D)$:

$$\left(\left.\frac{p_{1\,crit\,1}}{p_{0T}}\right|_D\right)^{2/m_{crit\,1}}-\left(\left.\frac{p_{1\,crit\,1}}{p_{0T}}\right|_D\right)^{(m_{crit\,1}+1)/m_{crit\,1}} = \left(\frac{A_t}{A_1}\right)^2\frac{\gamma-1}{2}\left(\frac{2}{\gamma+1}\right)^{(m_{0c}+1)/(m_{0c}-1)} \tag{14.89}$$

where $m_{crit\,1}$ is the value of the polytropic index for the nozzle as a whole at the upper critical point.

Strictly speaking, the upper critical pressure ratio may differ somewhat from the design value if inlet conditions change. Off-design inlet conditions will induce a different speed of sound in the throat and hence a different efficiency for the convergent section of the nozzle. However, as discussed in Section 14.7.1, the change in efficiency is likely to be small, and the effect on the expansion exponent in critical conditions, m_{0c}, will be similarly small. The change in the upper critical pressure ratio derived from the solution of equation (14.89) can be expected thus to be second-order. A similar effect will be seen on the lower critical pressure ratio, and again we may neglect the change for the purposes of simulation. Hence we will use the approximations:

$$\begin{aligned}\frac{p_{1\,crit\,1}}{p_{0T}} &\approx \left.\frac{p_{1\,crit\,1}}{p_{0T}}\right|_D\\ \frac{p_{1\,crit\,2}}{p_{0T}} &\approx \left.\frac{p_{1\,crit\,2}}{p_{0T}}\right|_D\end{aligned} \tag{14.90}$$

for all anticipated inlet conditions.

Let us now turn our attention to the efficiency of an expansion taking place where the pressure ratio is less than the lower critical pressure. Such an additional expansion may occur if the pressure in the discharge manifold is lower than the pressure in the nozzle outlet (case *e* in Figure 14.7), but the gas will lose in turbulence most of the kinetic energy released if the discharge pressure is significantly below the nozzle outlet pressure. The situation is analogous to the case of supersonic flow downstream of a convergent-only nozzle discussed in the previous section. We may expect the design efficiency to hold down to a pressure ratio of $r_{\lim}(p_{1\,crit\,2}/p_{0T})$, where $r_{\lim}$ is an empirical discharge coefficient, $0 < r_{\lim} \leq 1$, and likely in practice to be close to unity.

We have now established methods of calculating overall nozzle efficiency above the upper critical pressure ratio and below the lower critical pressure ratio. However, we have not yet accounted for the very important situation where the pressure ratio falls between subsonic and the design value (pressure ratios between $p_{1\,crit\,1}/p_{0T}$ and $p_{1\,crit\,2}/p_{0T}$, corresponding to pressure profiles such as those terminating in p_{1b}/p_{0T} and p_{1c}/p_{0T} in Figure 14.7). In this case there will be a large shock loss that will predominate over the skin-friction loss. We will derive in the next section an expression for efficiency at the off-design points of this choked region.

14.8.3 Nozzle efficiency in off-design, choked conditions for a convergent–divergent nozzle

When the discharge pressure ratio falls between $p_{1\,crit\,1}/p_{0T}$, and $p_{1\,crit\,2}/p_{0T}$, shock causes an immediate reduction in velocity from supersonic to subsonic. We may calculate the discharge temperature in such a case by making use of the fact, noted in Section 14.2, that stagnation enthalpy stays constant throughout the nozzle, independent of whether the adiabatic expansion is reversible or irreversible (equation (14.2)). As a result, the stagnation enthalpy throughout the nozzle will remain equal to the stagnation enthalpy at the inlet, so that

$$h + \tfrac{1}{2}c^2 = h_{0T} \tag{14.91}$$

where h and c are the specific enthalpy and the fluid velocity at a general downstream station. The inlet stagnation enthalpy has a stagnation temperature associated with it given by:

$$T_{0T} = \frac{h_{0T}}{c_p} = \frac{h_0 + \frac{1}{2}c_0^2}{c_p} = T_0 + \frac{c_0^2}{2c_p} \tag{14.92}$$

Assuming that we may characterize the specific heat over the temperature range T_{0T} to T by a constant, average value, we may substitute into (14.91) to give

$$c^2 = 2c_p(T_{0T} - T) \tag{14.93}$$

From continuity

$$W = \frac{Ac}{v} \tag{14.94}$$

while the characteristic equation for a near-ideal gas (equation (3.2)) gives:

$$v = \frac{ZR_wT}{p} \tag{14.95}$$

where $R_w = R/w$ is the characteristic gas constant for the gas in question and Z is the compressibility factor, assumed constant over the region of interest.

Combining equations (14.93), (14.94) and (14.95) gives the following quadratic in the temperature at the general downstream station, T:

$$\frac{Z^2R_w^2}{p^2}\frac{W^2}{A^2}T^2 + 2c_pT - 2c_pT_{0T} = 0 \tag{14.96}$$

This has the following positive root:

$$T = \frac{c_p p^2}{Z^2R_w^2G^2}\left(\sqrt{1 + 2\frac{Z^2R_w^2G^2}{c_p p^2}T_{0T}} - 1\right) \tag{14.97}$$

where G is the mass velocity at a generalized downstream station, given by:

$$G = \frac{W}{A} \tag{14.98}$$

When sonic flow occurs in the throat as the result of an adiabatic expansion of a perfect gas, the mass velocity at the throat is given (from equation (14.57)) by:

$$\frac{W}{A_t} = \sqrt{\left(\frac{2}{\gamma+1}\right)^{(m_{0c}+1)/(m_{0c}-1)}\gamma\frac{p_{0T}}{v_{0T}}} \tag{14.99}$$

Since formally

$$\frac{W}{A} = \frac{W}{A_t}\frac{A_t}{A} \tag{14.100}$$

it follows that

$$G^2 = \gamma\left(\frac{2}{\gamma+1}\right)^{(m_{0c}+1)/(m_{0c}-1)}\frac{p_{0T}}{v_{0T}}\left(\frac{A_t}{A}\right)^2 \tag{14.101}$$

or, using $p_{0T}v_{0T} = ZR_wT_{0T}$,

$$G^2 = \gamma\left(\frac{2}{\gamma+1}\right)^{(m_{0c}+1)/(m_{0c}-1)}\frac{p_{0T}^2}{ZR_wT_{0T}}\left(\frac{A_t}{A}\right)^2 \tag{14.102}$$

Substituting back into equation (14.97) gives:

$$\frac{T}{T_{0T}} = \frac{c_p}{ZR_w}\frac{1}{\gamma}\left(\frac{2}{\gamma+1}\right)^{(m_{0c}+1)/(1-m_{0c})}\left(\frac{A}{A_t}\right)^2\left(\frac{p}{p_{0T}}\right)^2 \times\left[\sqrt{1+2\frac{ZR_w}{c_p}\gamma\left(\frac{2}{\gamma+1}\right)^{(m_{0c}+1)/(m_{0c}-1)}\left(\frac{A_t}{A}\right)^2\left(\frac{p_{0T}}{p}\right)^2}-1\right] \tag{14.103}$$

This equation can be simplified by using equation (14.39), repeated below:

$$c_p = \frac{\gamma}{\gamma-1}ZR_w \tag{14.39}$$

to eliminate the ratio c_p/R_w:

$$\frac{T}{T_{0T}} = \frac{1}{\gamma-1}\left(\frac{2}{\gamma+1}\right)^{(m_{0c}+1)/(1-m_{0c})}\left(\frac{A}{A_t}\right)^2\left(\frac{p}{p_{0T}}\right)^2 \times\left[\sqrt{1+2(\gamma-1)\left(\frac{2}{\gamma+1}\right)^{(m_{0c}+1)/(m_{0c}-1)}\left(\frac{A_t}{A}\right)^2\left(\frac{p_{0T}}{p}\right)^2}-1\right] \tag{14.104}$$

The actual, as opposed to stagnation, temperature ratio is given by:

$$\frac{T}{T_0} = \frac{T_{0T}}{T_0}\frac{T}{T_{0T}} \tag{14.105}$$

Equations (14.104) and (14.105) enable the temperature to be found at any point downstream of a choked throat in any nozzle of known geometry, once the pressure at that point has been specified. In practice, it is the nozzle discharge pressure that is most important. The conditions just inside the nozzle outlet are the same at those just downstream of the outlet down to and including a discharge pressure ratio of $p_{1\,crit\,2}/p_{0T}$ (see Figure 14.7), and so we may put $p = p_{N1} = p_1$, $T = T_{N1} = T_1$, as well as $A = A_1$ in equation (14.104). This will allow the discharge temperature to be found; a knowledge of the two thermodynamic variables pressure and temperature then allows any further thermodynamic variables to be determined at that point.

Having determined the temperature, nozzle efficiency may be determined from equation (14.7), which compares the actual outlet temperature ratio given by (14.104) and (14.105) with the isentropic outlet temperature ratio found from equation (14.78):

$$\eta_N = \frac{1-\dfrac{T_{0T}}{T_0}\dfrac{T_1}{T_{0T}}}{1-\dfrac{T_{1s}}{T_0}} = \frac{1-\dfrac{T_{0T}}{T_0}\dfrac{T_1}{T_{0T}}}{1-\left(\dfrac{p_1}{p_0}\right)^{(\gamma-1)/\gamma}} \tag{14.106}$$

14.8.4 The efficiency of a convergent–divergent nozzle down to just below the design pressure ratio

We may now combine the results of Sections 14.8.2 and 14.8.3 to give the efficiency of a convergent–divergent nozzle down to below design pressure ratio as:

$$\begin{aligned}\eta_N &= \min(\eta_{N\,0c}^2, \eta_{ND}) \quad \text{for } \frac{p_1}{p_{0T}} \geq \frac{p_{1\,crit\,1}}{p_{0T}}\\ &= \frac{1-\dfrac{T_{0T}}{T_0}\dfrac{T_1}{T_{0T}}}{1-\left(\dfrac{p_1}{p_0}\right)^{(\gamma-1)/\gamma}} \quad \text{for } \frac{p_{1\,crit\,2}}{p_{0T}} < \frac{p_1}{p_{0T}} < \frac{p_{1\,crit\,1}}{p_{0T}}\\ &= \eta_{ND}, \quad \text{for } r_{\lim}\frac{p_{1\,crit\,2}}{p_{0T}} \leq \frac{p_1}{p_{0T}} \leq \frac{p_{1\,crit\,2}}{p_{0T}}\end{aligned} \tag{14.107}$$

where

$$\eta_{N\,0c} = \frac{2\gamma}{\gamma-1} \times \frac{\log\left[\dfrac{2}{\gamma+1}\right]}{\log\left[\dfrac{4}{\gamma^2-1}\left(\dfrac{A_1}{A_t}\right)^2\left(\left(\dfrac{p_1}{p_{0T}}\bigg|_D\right)^{2/m_D}-\left(\dfrac{p_1}{p_{0T}}\bigg|_D\right)^{(m_D+1)/m_D}\right)\right]} \tag{14.87}$$

and

$$\frac{T_1}{T_{0T}} = \frac{1}{\gamma-1}\left(\frac{2}{\gamma+1}\right)^{(m_{0c}+1)/(1-m_{0c})}\left(\frac{A_1}{A_t}\right)^2\left(\frac{p_1}{p_{0T}}\right)^2 \times\left[\sqrt{1+2(\gamma-1)\left(\frac{2}{\gamma+1}\right)^{(m_{0c}+1)/(m_{0c}-1)}\left(\frac{A_t}{A_1}\right)^2\left(\frac{p_{0T}}{p_1}\right)^2}-1\right] \tag{14.108}$$

14.8.5 Discharge velocity and mass flow in a convergent–divergent nozzle

The discharge velocity is given from equation (14.45) as:

$$c_1 = \sqrt{2\frac{\gamma}{\gamma-1}p_{0T}v_{0T}\left(1-\left(\frac{p_1}{p_{0T}}\right)^{(m-1)/m}\right)} \quad \text{for } \frac{p_1}{p_{0T}} \geq r_{\lim}\frac{p_{1\,crit\,2}}{p_{0T}}$$

$$= \sqrt{2\frac{\gamma}{\gamma-1}p_{0T}v_{0T}\left(1-\left(r_{\lim}\frac{p_{1\,crit\,2}}{p_{0T}}\right)^{(m_D-1)/m_D}\right)}$$

$$\text{for } \frac{p_1}{p_{0T}} < r_{\lim}\frac{p_{1\,crit\,2}}{p_{0T}} \quad (14.109)$$

where p_1 is the pressure in the discharge manifold, and the index m is the function of overall nozzle efficiency, η_N, given by equation (14.21). The discharge velocity is limited through the use of the discharge coefficient, $r_{\lim}$, explained at the end of Section 14.8.2.

The mass flow is given by (14.49) and (14.57), and is subject to becoming choked as soon as the pressure ratio falls below the upper critical discharge ratio:

$$W = A_1\sqrt{2\frac{\gamma}{\gamma-1}\frac{p_{0T}}{v_{0T}}\left(\left(\frac{p_1}{p_{0T}}\right)^{2/m}-\left(\frac{p_1}{p_{0T}}\right)^{(m+1)/m}\right)}$$

$$\text{for } \frac{p_1}{p_{0T}} \geq \frac{p_{1\,crit\,1}}{p_{0T}} \quad (14.110)$$

$$= A_t\sqrt{\left(\frac{2}{\gamma+1}\right)^{(m_{0c}+1)/(m_{0c}-1)}\gamma\frac{p_{0T}}{v_{0T}}}$$

$$\text{for } \frac{p_1}{p_{0T}} < \frac{p_{1\,crit\,1}}{p_{0T}}$$

In equation (14.110), the overall nozzle index, m, takes the subcritical value:

$$m = \frac{\gamma}{\gamma - \min(\eta_{N\,0c}^2, \eta_{ND}) \times (\gamma-1)}$$

$$\text{for } \frac{p_1}{p_{0T}} \geq \frac{p_{1\,crit\,1}}{p_{0T}} \quad (14.111)$$

while the convergent section index, m_{0c}, is given by:

$$m_{0c} = \frac{\gamma}{\gamma - \eta_{N\,0c}(\gamma-1)} \quad (14.112)$$

14.8.6 Nozzle efficiency at discharge pressure ratios substantially below the lower critical ratio for a convergent–divergent nozzle

As was the case for the convergent-only nozzle, we have now devised methods for calculating the discharge velocity and mass flow without needing to evaluate the overall nozzle efficiency deep below the lower critical pressure ratio. However, a similar analysis to that contained in Section 14.7.3 may be applied to the convergent–divergent nozzle at discharge pressure ratios less than the lower critical pressure ratio. We may write the discharge velocity, c_1, at station '1', just downstream of the nozzle outlet as

$$\frac{1}{2}c_1^2 = c_pT_{0T}\left(1 - r_{\lim}^{(m_D-1)/m_D}\left(\frac{p_{1\,crit\,2}}{p_{0T}}\right)^{(m_D-1)/m_D}\right) \quad (14.113)$$

for $(p_1/p_{0T}) < r_{\lim}(p_{1\,crit\,2}/p_{0T})$ (cf. equation (14.65)). Hence the efficiency emerges from the same procedure as was used for the convergent-only nozzle as:

$$\eta_N = \frac{1 - \dfrac{T_{0T}}{T_0}r_{\lim}^{\eta_D((\gamma-1)/\gamma)}\left(\dfrac{p_{1\,crit\,2}}{p_{0T}}\right)^{\eta_D((\gamma-1)/\gamma)}}{1-\dfrac{T_{1s}}{T_0}} \quad (14.114)$$

or

$$\eta_N = \frac{1 - \dfrac{T_{0T}}{T_0}r_{\lim}^{\eta_D((\gamma-1)/\gamma)}\left(\dfrac{p_{1\,cirt\,2}}{p_{0T}}\right)^{\eta_D((\gamma-1)/\gamma)}}{1-\left(\dfrac{p_1}{p_0}\right)^{((\gamma-1)/\gamma)}}$$

$$\text{for } \frac{p_1}{p_{0T}} < r_{\lim}\frac{p_{1\,crit\,2}}{p_{0T}} \quad (14.115)$$

Adding this expression to the expressions summarized in Section 14.8.4 allows us to calculate the efficiency of a convergent–divergent nozzle over its full range.

14.8.7 Calculating the efficiency of a convergent–divergent nozzle over the full pressure range

To demonstrate the calculation of efficiency, let us take the case of a convergent–divergent nozzle being fed with initially stationary air at 20 bar and 400°C. The air is discharged at a temperature of 168°C into a manifold at a pressure of 3.69 bar. The ratio of outlet to throat area is 1.5. The geometry of the discharge manifold is such that the air coming out of the nozzle at the (supersonic) design speed will continue to be accelerated until the discharge pressure ratio reaches 90% of the design pressure ratio.

The first step is to calculate the efficiency at the design point. Since air is a mixture of essentially two diatomic gases, we may take $\gamma = 1.4$. The design pressure ratio is $3.69/20.0 = 0.1845$, and the overall nozzle efficiency follows from equations (14.76) and (14.78) as

$$\eta_{ND} = (1-(168+273)/(400+273)) \div (1-0.1845^{(1.4-1)/1.4}) = (1-0.655)\div(1-0.617) = 0.9$$

The value at the design point of the polytropic exponent m characterizing the overall expansion is therefore $m_D = 1.4 \div (1.4 - 0.9(1.4-1)) = 1.3462$.

We may now apply equation (14.87) to find the efficiency of the convergent section at choking for the design inlet conditions:

$$\frac{4}{\gamma^2-1}\left(\frac{A_1}{A_t}\right)^2\left(\left(\frac{p_1}{p_{0T}}\bigg|_D\right)^{2/m_D}-\left(\frac{p_1}{p_{0T}}\bigg|_D\right)^{(m_D+1)/m_D}\right)$$

$$=\frac{4}{1.4^2-1}(1.5)^2((0.1845)^{2/(1.3462)}$$

$$-(0.1845)^{(2.3462)/(1.3462)})=0.2683$$

Hence

$$\eta_{N\,0c}=\frac{2\times 1.4}{1.4-1.0}\frac{\log\left[\dfrac{2}{1.4+1}\right]}{\log[0.2683]}=0.97$$

Using equation (14.21), $m_{0c}=1.4\div(1.4-0.97(1.4-1))=1.3834$.

The subcritical overall nozzle efficiency is estimated by $\eta_N=\min(\eta_{N\,0c}^2,\eta_{ND})=\min(0.97^2,0.9)=0.9$. Hence the associated m-value is $m=m_D=1.3462$.

We now solve equation (14.75) for the nozzle discharge station '1' to find the upper value of the critical discharge pressure ratio, under the assumption that $m_{0c}=1.3834$ and $m=1.3462$. This gives the upper value of the critical discharge ratio as $p_{1\,crit\,1}/p_{0T}=0.87$. Above this value the nozzle efficiency is $\eta_N=0.9$.

We have already calculated the design pressure ratio, which can be assumed to be the same as lower critical discharge ratio: $(p_{1\,crit\,2})/p_{0T}=0.1845$.

The limiting value below which no further gas acceleration will take place is 90% of that pressure ratio

$$r_{\text{lim}}\frac{p_{1\,crit\,2}}{p_{0T}}=0.9\times 0.1845=0.16605$$

Between these values of pressure ratio, $\eta_N=\eta_D=0.9$.

We may apply equation (14.115) to calculate the efficiency at very low pressure ratios, i.e. less than

$$r_{\text{lim}}\frac{p_{1\,crit\,2}}{p_{0T}}=0.9\times 0.1845=0.16605$$

Turning to the choked region between $(p_{1\,crit\,1}/p_{0T})=0.87$ and $p_{1\,crit\,2}/p_{0T}=0.1845$, we may apply equations (14.104) and (14.105) to calculate the actual temperature ratio and equation (14.106) to calculate the overall nozzle efficiency.

The various expressions used to calculate efficiency over the different ranges of pressure ratio dovetail pretty well, but the transition is not perfect because of the implicit presence of the reheat factor discussed at the end of Section 14.3. Thus the choked calculation of efficiency (equations (14.104), (14.105) and (14.106) together) gives a value of $\eta_N=0.9020$ at $p_{1\,crit\,1}/p_{0T}=0.87$, as opposed to the estimate $\eta_N=\min(\eta_{N\,0c}^2,\eta_{ND})=\min(0.97^2,0.9)=0.9$; a slightly greater discrepancy occurs at $p_{1\,crit\,2}/p_{0T}=0.1845$, where the choked calculation gives $\eta_N=0.9203$ as opposed to the expected $\eta_N=\eta_D=0.9$. Similarly, the calculated value of efficiency at $r_{\text{lim}}(p_{1\,crit\,2}/p_{0T})=0.16605$ using (14.115) is $\eta_N=0.9213$, as opposed to $\eta_N=0.9$. Each of these discrepancies disappears

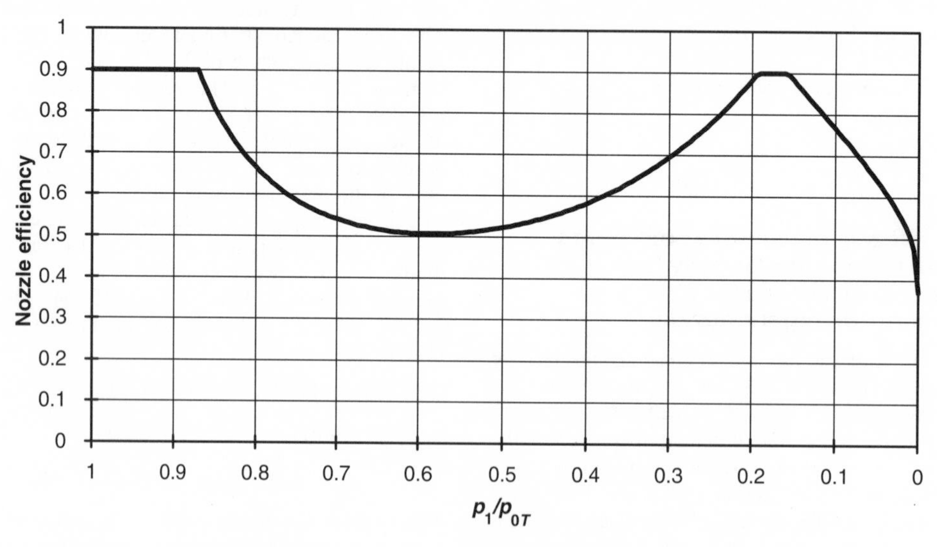

Figure 14.8 Nozzle efficiency for off-design conditions for a convergent–divergent nozzle. (outlet to throat area ratio = 1.5, diatomic gas).

within a change in pressure ratio of 0.01 or less. The effect has been smoothed out in Figure 14.8 by selecting the minimum of the two alternative efficiencies in the vicinity of each break-point in the pressure-ratio range. Figure 14.8 displays the nozzle efficiency calculated over the full pressure range.

The figure shows clearly the penalty for using a convergent–divergent nozzle away from its operating point: the efficiency of the nozzle can be halved or worse. Comparing Figure 14.8 with an equivalent graph for a convergent-only nozzle (Figure 14.3), it is clear that the pressure ratio for a convergent–divergent nozzle will need to be controlled much more tightly if good energy utilization is to be maintained. However, while a rise in the pressure ratio above the design point causes a sharp drop in efficiency, this is not tantamount to dropping over a cliff-edge; thus the situation is potentially controllable, as the control engineer would hope and expect.

The reader is referred also to Appendix 6, where these methods of calculating efficiency are compared with the experimental results for several convergent–divergent nozzles given in Keenan (1949). The analytical methods are shown to reproduce the main features of the variation in efficiency with pressure drop, particularly the pronounced dip between the upper and lower critical pressure ratios.

14.9 Bibliography

Faires, V.M. (1957). *Thermodynamics*, 3rd edition Macmillan, New York.

Giffen, E. and Crang, T.F. (1946). Steam flow in nozzles: velocity coefficient at low steam speeds, *Proceedings of the Institution of Mechanical Engineers*, **155**, 83–92.

Kearton, W.J. (1992, 1958). *Steam Turbine Theory and Practice*, Pitman and Sons, London.

Keenan, J.H. (1949). Reaction tests of turbine nozzles for supersonic velocities, *Transactions of the American Society of Mechanical Engineers*, **71**, 773–780.

Kraft, H. (1949). Reaction tests of turbine nozzles for subsonic velocities, *Transactions of the American Society of Mechanical Engineers*, **71**, 781–787.

Lockey, J. (1966). *The Thermodynamics of Fluids*, Heinemann Educational Books Ltd, London.

Rogers, G.F.C. and Mayhew, Y.R. (1992). *Engineering Thermodynamics: Work and Heat Transfer*, 3rd Edition, Longmans.

Streeter, V.L. and Wylie, E.B. (1983). *Fluid Mechanics, First SI Metric Edition*, McGraw-Hill, New York.

15 Steam and gas turbines

15.1 Introduction

Turbines are a common feature on process plant, where they provide economical drives for rotating machinery, particularly centrifugal and axial pumps and compressors, as well as electrical generators in power stations. They use the expansion of hot, high-pressure gas or steam to produce mechanical power.

A turbine will consist of one or more stages, each of which will contain a ring of nozzles followed by a ring of moving blades. The pressure drop across the nozzles causes a stream of gas to be directed at high velocity at the moving blades. The blades deflect the gas stream and in so doing experience a force which causes them to rotate. The turbine may be fed with high-pressure gas from more than one source, and it may discharge gas, at lower pressure and temperature, to a number of intermediate sinks – in theory there may be as many sinks as there are stages. The boundary conditions for simulation purposes are the gas conditions at the inlet to the turbine and the pressures in the discharge vessels. Any model of the turbine will then need to calculate the instantaneous power developed by the turbine and the flows and enthalpies of the discharge streams.

The gas volumes in the body of the turbine are very small, so that the establishment of flow in the turbine is very fast, with a response time measured in tenths of a second. Since these time constants are an order of magnitude less than the rotor time constant associated with turbine speed, it is permissible to treat the equations of fluid flow as algebraic, simultaneous equations. This simplification receives even greater justification when the model of the turbine and the machine it drives are interfaced with a model of the rest of the process plant, which will be invariably much slower still. It is only when the turbine is being considered in detail on its own, for example in designing a speed-control system, that some of the larger turbine pressure constants, such as those associated with the turbine inlet manifolds, will need to be considered.

15.2 The turbine stage

Figure 15.1 gives a plan view of a single nozzle–blade combination in a turbine stage, showing hot gas entering the stage at station '0' and being accelerated through a pair of fixed blades forming a nozzle to emerge at station '1' with increased velocity at an angle α_1 to the direction of wheel motion. The fast-moving gas then passes into the moving blade, which deflects the gas and experiences a force in so doing. The gas emerges from the stage at an angle α_2.

The boundary conditions we can expect when modelling a turbine stage will be

(i) the stage upstream pressure and another thermodynamic variable, such as specific enthalpy, from which pair any other thermodynamic variable, e.g. temperature and specific volume may be found,
(ii) the stage upstream inlet velocity, and
(iii) the stage downstream pressure.

We may apply the steady-state flow equation (4.7) to the complete turbine stage:

$$dq - dw - dh - c\,dc - g\,dz = 0 \qquad (4.7)$$

Dropping the terms dq and $g\,dz$ because the flow is adiabatic and experiences no or negligible height change, we may integrate between stage inlet (station '0') and outlet (station '2') to obtain:

$$h_0 + \tfrac{1}{2}c_0^2 - h_2 - \tfrac{1}{2}c_2^2 - w = 0 \qquad (15.1)$$

where w is the specific work (J/kg). The calculation plan for a multistage turbine is first to find the mid-stage velocity, c_1, and then deduce the specific work, w, and the outlet velocity, c_2. The method of doing this forms the main content of this chapter. Once this has been achieved, however, the outlet specific enthalpy, h_2, may be found from a simple rearrangement of equation (15.1):

$$h_2 = h_0 + \tfrac{1}{2}c_0^2 - \tfrac{1}{2}c_2^2 - w \qquad (15.2)$$

Introducing a double-subscript notation, with the second subscript identifying the stage, the inlet velocity to the first stage of a turbine will be essentially zero, i.e.

$$c_{0,1} = 0 \qquad (15.3)$$

For subsequent stages, the inlet velocity will be equal to the outlet velocity of the previous stage:

$$c_{0,i} = c_{2,i-1} \quad \text{for } i > 1 \qquad (15.4)$$

The specific enthalpy at the inlet to the first stage of the turbine is part of the boundary condition. For

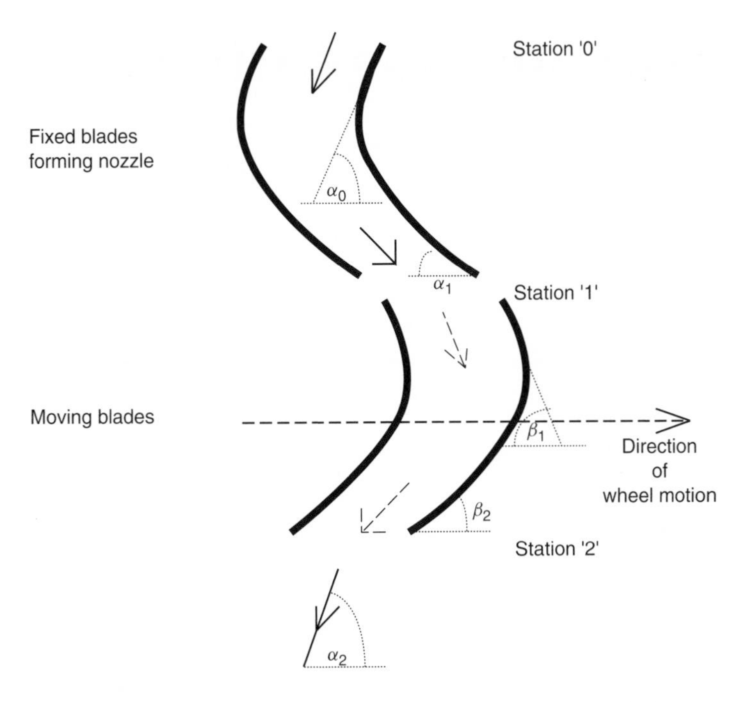

Figure 15.1 Plan view of turbine stage: single nozzle–blade combination.

subsequent stages

$$h_{0,i} = h_{2,i-1} \qquad \text{for } i > 1 \tag{15.5}$$

A knowledge of what type the stage is (impulse or reaction) allows us to estimate the mid-stage pressure, p_1. Knowing this pressure permits us to calculate both the mid-stage gas speed and the mass flow rate, W, through the nozzle and hence through the moving blade. The power produced by each stage is the product of the flow rate in kg/s multiplied by the specific work in J/kg:

$$P_i = W_i w_i \tag{15.6}$$

and the total power of the turbine is found by summing the powers of each of the stages:

$$P_T = \sum_i W_i w_i \tag{15.7}$$

The instantaneous speed of the turbine and its driven machine may now be found by putting $P_T = P_{in}$ into the differential equation for rotational speed, equation (3.50):

$$4\pi^2 JN\frac{dN}{dt} + 2\pi^2 N^2 \frac{dJ}{dt} = P_T - P_{out} - P_F \tag{15.8}$$

where J is the combined moment of inertia of turbine and driven machinery, while P_{out} and P_F are the power drawn from the driven machinery and the power lost to friction respectively. The moment of inertia will be constant in almost all cases, so that $dJ/dt = 0$ (exceptions will be fault conditions where part of the turbine shaft or its driven machine is lost during operation; also, for example, the case where the turbine is directly coupled to a solids–liquid separation centrifuge, and the centrifuge cake builds up slowly over time).

15.3 Stage efficiency and the stage polytropic exponent

We may define the stage efficiency, η_s, as the ratio of the actual specific work extracted from the stage, w (J/kg), to the maximum specific work that is possible to extract, w_{max} (J/kg):

$$\eta_s = \frac{w}{w_{max}} \tag{15.9}$$

Rearranging equation (15.1) gives the specific work performed in the stage, w, as:

$$w = h_0 - h_2 + \tfrac{1}{2}c_0^2 - \tfrac{1}{2}c_2^2 \tag{15.10}$$

In a multistage turbine the interstage velocities are equal, or nearly so, and so we may rewrite equation

(15.10) for all stages except the first as simply

$$w = h_0 - h_2 \tag{15.11}$$

The inlet velocity will be very small in the first stage, but the exit velocity may not. Nevertheless we may continue to use equation (15.11) as an approximation for the first stage because we can expect the term $c_2^2/2$ will be small compared with $h_0 - h_2$, although we should realize that this procedure of neglecting the unconverted kinetic energy will overestimate by a few per cent the specific work performed in the first stage.

The maximum work will be done when the outlet enthalpy is a minimum, and the minimum outlet enthalpy for an uncooled process will occur when the expansion is perfectly isentropic. Hence:

$$w_{\max} = h_0 - h_{2s} \tag{15.12}$$

where the subscript 's' implies an isentropic expansion: h_{2s} is the specific enthalpy that would be found at the outlet of the stage following an isentropic expansion from stage inlet to stage outlet. As a result

$$\eta_s = \frac{h_0 - h_2}{h_0 - h_{2s}} \tag{15.13}$$

This equation enables the stage efficiency to be calculated from the thermodynamic states at the inlet and outlet of the stage.

Equation (15.13) is of the same form as the equation derived in Section 14.2 for nozzle efficiency (equation (14.5)), and we may use a development similar to that given for the nozzle in Section 14.3 to relate the polytropic exponent for a stage-to-stage efficiency. While it is obviously true that the stage work is abstracted in the blade and not the nozzle, it is convenient to make the approximating assumption that the work is abstracted uniformly over the complete length of the turbine stage. This enables us to view the stage as a continuous sequence of differential elements, each of which obeys the energy equation (4.7), given above. Making the same assumptions as before, we may omit the terms $g\,dz$ and dq, and also $c\,dc$, under the assumption that the change in velocity over the complete stage is very small compared with the change in enthalpy. Then we may rearrange equation (4.7) into the form

$$dw = -dh \tag{15.14}$$

Equation (15.14) implies a stage where fluid enthalpy is converted continuously into mechanical work.

We now assume that the stage efficiency may be rewritten in the differential form

$$\eta_s = \frac{dw}{dw_s} \tag{15.15}$$

where dw_s is the work output from an ideal, isentropic differential unit. Using equation (15.14), this may be stated in terms of enthalpy differentials as

$$\eta_s = \frac{dh}{dh_s} \tag{15.16}$$

This is of exactly the same form as equation (14.9), although equation (15.16) refers to the complete stage rather than the nozzle. Precisely analogous arguments to those laid out in Section 14.3 may now be made in order to derive the relationship between specific volume and pressure over the complete stage. This relationship emerges as

$$pv^{m_s} = \text{constant} \tag{15.17}$$

where

$$m_s = \frac{\gamma}{\gamma - \eta_s(\gamma - 1)} \tag{15.18}$$

The caveat on reheat factor applied to equation (14.22) applies equally to equation (15.17), and the degree of approximation involved with equations (15.17) and (15.18) will increase as each of the stage pressure ratio and the stage efficiency falls. Nevertheless, this characterization has been found to be useful and accurate enough in practical simulation studies even when the turbine has been operating a long way from design conditions.

15.4 Reaction

Turbine stages may be classified either as 'impulse' or 'reaction'. An impulse stage is the simpler of the two conceptually, because all the expansion occurs in the nozzle, leaving the blade the single task of extracting the resultant kinetic energy. A turbine consisting of just one impulse stage is known as a de Laval turbine, named after the inventor of the first turbine of industrial significance, while a turbine made up of several impulse stages is known as a Rateau turbine. In a reaction stage, by contrast, a second expansion occurs over the blade, accompanied by a drop in enthalpy over the blade region. The most common design allows for 50% reaction, where the enthalpy drop over the blades is half the total enthalpy drop over the entire stage. A turbine consisting of several 50% reaction stages is known as a Parsons turbine. Now that the patents on the basic arrangements are long expired, it is common for manufacturers of power-station turbines to use combinations of both types of stage, with impulse stages at the high-pressure end, followed by 50% reaction stages to the exhaust.

The further expansion over the blade in the reaction stage causes the relative velocity of the gas leaving the blade to increase over its initial value at blade entry. This increase in kinetic energy occurs at the expense of enthalpy. As a result, we define the degree of reaction, ρ, as:

$$\% \text{ degree of reaction} = \frac{\text{Enthalpy drop in moving blades}}{\text{Enthalpy drop in the reaction stage}} \times 100$$

or

$$\rho = \frac{h_1 - h_2}{h_0 - h_2} \tag{15.19}$$

Note that the nozzle inlet enthalpy, h_0, is taken as the actual value, not the stagnation value in this definition. The expansion over the blades will, of course, be associated with a drop in pressure, and we may deduce its value by expressing the enthalpy drops in terms of temperature ratios and then pressure ratios as follows:

$$\rho = \frac{c_p(T_1 - T_2)}{c_p(T_0 - T_2)} = \frac{\dfrac{T_1}{T_0} - \dfrac{T_2}{T_0}}{1 - \dfrac{T_2}{T_0}} = \frac{\left(\dfrac{p_1}{p_0}\right)^{(m-1)/m} - \left(\dfrac{p_2}{p_0}\right)^{(m_s-1)/m_s}}{1 - \left(\dfrac{p_2}{p_0}\right)^{(m_s-1)/m_s}} \tag{15.20}$$

where c_p is the specific heat over the stage, m is the polytropic exponent governing the nozzle expansion, while m_s is the polytropic exponent governing the total stage expansion. Solving equation (15.20) for p_1/p_0 gives:

$$\frac{p_1}{p_0} = \left[\rho + (1 - \rho)\left(\frac{p_2}{p_0}\right)^{(m_s-1)/m_s}\right]^{m/(m-1)} \tag{15.21}$$

In fact, this pressure ratio is to a first approximation independent of the values of m and m_s when the stage pressure drop is only a small fraction of the stage inlet pressure, as may be seen by rearranging equation (15.21)

$$\rho = \frac{\left(\dfrac{p_1}{p_0}\right)^{(m-1)/m} - \left(\dfrac{p_2}{p_0}\right)^{(m_s-1)/m_s}}{1 - \left(\dfrac{p_2}{p_0}\right)^{(m_s-1)/m_s}} = \frac{\left(1 - \dfrac{(p_0 - p_1)}{p_0}\right)^{(m-1)/m} - \left(1 - \dfrac{(p_0 - p_2)}{p_0}\right)^{(m_s-1)/m_s}}{1 - \left(1 - \dfrac{(p_0 - p_2)}{p_0}\right)^{(m_s-1)/m_s}} \tag{15.22}$$

We may apply the binomial expansion and discard all but the first terms provided $(p_0 - p_2)/p_0 \ll 1$ (which of course implies $(p_0 - p_1)/(p_0 \ll 1)$:

$$\rho \approx \frac{\left(1 - \dfrac{m-1}{m}\dfrac{(p_0 - p_1)}{p_0}\right) - \left(1 - \dfrac{m_s - 1}{m_s}\dfrac{(p_0 - p_2)}{p_0}\right)}{1 - \left(1 - \dfrac{m_s - 1}{m_s}\dfrac{(p_0 - p_2)}{p_0}\right)} \tag{15.23}$$

and so

$$\rho \approx \frac{p_0 - p_2 - \dfrac{m_s}{m}\dfrac{m-1}{m_s - 1}(p_0 - p_1)}{p_0 - p_2} = \frac{p_0 - p_2 - \left(\dfrac{1 - \dfrac{1}{m}}{1 - \dfrac{1}{m_s}}\right)(p_0 - p_1)}{p_0 - p_2} \tag{15.24}$$

The nozzle exponent m is defined by equation (14.21), repeated below:

$$m = \frac{\gamma}{\gamma - \eta_N(\gamma - 1)} \tag{14.21}$$

Substituting into equation (15.24) using this and the analogous definition for the stage exponent, m, gives

$$\rho \approx \frac{p_0 - p_2 - \dfrac{\eta_N}{\eta_S}(p_0 - p_1)}{p_0 - p_2} = \frac{p_1\left(\dfrac{\eta_N}{\eta_S} + \left(1 - \dfrac{\eta_N}{\eta_S}\right)\dfrac{p_0}{p_1}\right) - p_2}{p_0 - p_2} \tag{15.25}$$

Provided p_1 and p_0 do not differ too greatly,

$$\frac{\eta_N}{\eta_S} + \left(1 - \frac{\eta_N}{\eta_S}\right)\frac{p_0}{p_1} \approx 1 \tag{15.26}$$

with the approximation gaining in validity when the stage efficiency and the nozzle efficiencies are reasonably close (in practice they are likely to be within roughly 15% of eachother.) Hence we may write:

$$\rho \approx \frac{p_1 - p_2}{p_0 - p_2} \tag{15.27}$$

As a rough rule of thumb, equation (15.27) is approximately valid for $(p_0 - p_2)/p_0 \leq \sim 0.3$ or $p_2/p_0 \geq \sim 0.7$. But the wider significance is that equation (15.27) is independent of the precise values of m and m_s, and hence we may calculate the mid-stage pressure ratio for off-design conditions using either the design values of m and m_s in equation (15.21), or indeed setting

$m \approx m_s \approx \gamma$. We are further helped by the fact that for the important case of a 50% reaction stage, the degree of reaction stays roughly constant even in the face of quite large excursions from the design conditions (see Appendix 7 for a justification of this statement).

15.5 Mid-stage pressure; nozzle discharge velocity; stage mass flow

The mid-stage pressure is needed if we are to calculate the nozzle discharge velocity and the nozzle flow. In fact, the mid-stage pressure is the same as the pressure at stage outlet for an impulse nozzle, while the mid-stage pressure for a reaction stage may be found from equation (15.21).

Once the mid-stage pressure is known, the nozzle discharge velocity, c_1, and mass flow, W, may be found using the methods of Chapter 14, see particularly Sections 14.7 for a convergent nozzle and Section 14.8 for a convergent–divergent nozzle. By continuity, the mass flow through the stage is the same as the mass flow through the nozzle.

15.6 Design conditions in an impulse stage

The force exerted on the blades in the direction of the turbine wheel motion is given by the momentum equation:

$$F = W\Delta c_w \tag{15.28}$$

where F is the force (N), W is the gas flow (kg/s) and Δc_w (m/s) is the change in gas velocity in the direction of the turbine wheel motion, sometimes known as the gas's 'whirl velocity'. The useful work done per second, or power, P, in watts, is found from multiplying the force by the distance moved by the blades in one second, i.e. the blade linear velocity, c_B (m/s):

$$P = Wc_B\Delta c_w \tag{15.29}$$

and the specific work, w (J/kg), is given by dividing by the mass flow rate:

$$w = \frac{P}{W} = c_B\Delta c_w \tag{15.30}$$

Note that the linear speed of the blades will increase steadily in the radial direction, from the root of the blade to its tip. Hence we have taken the blade velocity, c_B, as the average velocity, i.e. the velocity at the mid-length of the blade.

The only energy available to an impulse blade is the kinetic energy at the outlet from the inlet nozzle, and we may assume that a perfectly efficient impulse blade would convert all this energy into work, so that

$$w_{\max B} = \tfrac{1}{2}c_1^2 \tag{15.31}$$

Here $w_{\max B}$ is the maximum specific work that the blade can extract. Note that $w_{\max B}$ will be less than $w_{\max}$ as a result of the imperfect nature of the nozzle: a small fraction of the energy available to the stage will already have been lost as a result of nozzle friction. The efficiency of the blade, η_B, is defined as

$$\begin{aligned}\eta_B &= \frac{w}{w_{\max B}} \\ &= \frac{2c_B\Delta c_w}{c_1^2}\end{aligned} \tag{15.32}$$

To evaluate the term Δc_w, we need to examine the velocity diagram for the blade, as shown in Figure 15.2.

Figure 15.2 shows how the jets of hot gas or steam emerging from the nozzles are directed towards the turbine wheel at an angle α_1 to the direction of wheel

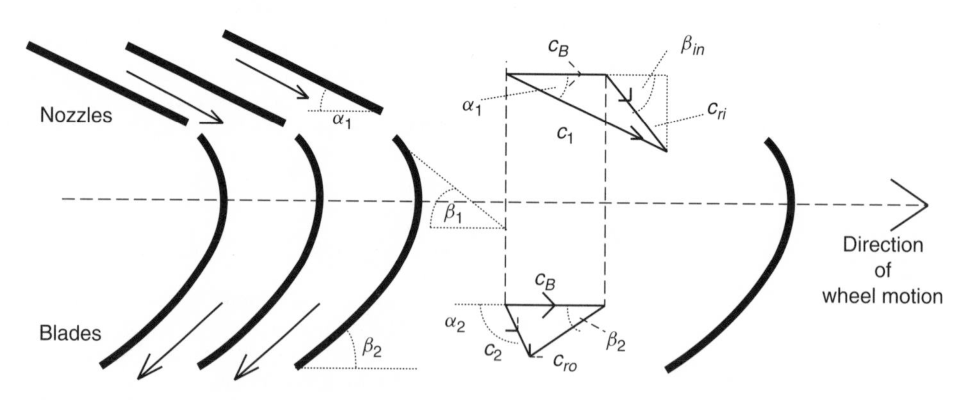

Figure 15.2 Diagram of impulse stage showing velocities.

motion. The value of α_1 is typically between 12° and 25°. Because the blade is moving at a velocity c_B to the right in the diagram, the gas jet approaches the blade at an angle β_{in} measured relative to the direction of wheel motion, where $\beta_{in} > \alpha_1$. In order to give entry without shock into the blade, the blade inlet angle, β_1, is made equal to the approach angle, β_{in}, at the design conditions. In an ideal impulse turbine with no frictional loss, the magnitude of the velocity relative to the blade is constant throughout the blade, and so $c_{ro} = c_{ri}$. It is normal in an impulse stage to maintain the blade height constant over the blade, so continuity of mass flow demands that the blade outlet angle, β_2, be made equal to the blade inlet angle, β_1. Figure 15.2 shows the velocity triangles at inlet and outlet from the blade. Because the blade speed is common to both triangles, it is convenient to use it as the base on which are superimposed both the inlet and the outline velocity triangles. Figure 15.3 shows such a superimposition.

From Figure 15.3, the change in velocity in the direction of wheel motion is given by

$$\Delta c_w = c_{ri} \cos \beta_{in} + c_{ro} \cos \beta_2 \tag{15.33}$$

Since for an impulse blade $c_{ro} = c_{ri}$, and the design conditions are

$$\beta_1 = \beta_{in}$$

and (15.34)

$$\beta_2 = \beta_1$$

it follows that at the design point

$$\Delta c_w = 2c_{ri} \cos \beta_{in} \tag{15.35}$$

From trigonometrical considerations in Figure 15.3,

$$c_{ri} \cos \beta_{in} = c_1 \cos \alpha_1 - c_B \tag{15.36}$$

so that

$$\Delta c_w = 2(c_1 \cos \alpha_1 - c_B) \tag{15.37}$$

Substituting into equation (15.32) gives the blade efficiency as

$$\eta_B = \frac{4c_B(c_1 \cos \alpha_1 - c_B)}{c_1^2} \tag{15.38}$$

Defining R_B as the ratio of blade average linear speed to mid-stage gas speed,

$$R_B = \frac{c_B}{c_1} \tag{15.39}$$

allows us to write

$$\eta_B = 4(R_B \cos \alpha_1 - R_B^2) \tag{15.40}$$

It is important to emphasize the limitations on the use of this equation. It is essentially a design equation and it represents the variation in blade efficiency with blade/gas speed ratio for a turbine impulse blade under the following conditions:

(i) the nozzle angle, α_1, is fixed;
(ii) the inlet angle, β_1, is varied continuously as the speed ratio changes so as to match the gas approach angle, β_{in};
(iii) the outlet angle is also varied continuously so as to keep permanent equality with the inlet angle: $\beta_2 = \beta_1 = \beta_{in}$.

Items (ii) and (iii) above are clearly only possible during the process of conceptual design.

Differentiating equation (15.40) with respect to speed ratio, R_B, and setting the result to zero gives the speed ratio for maximum efficiency as

$$R_B = \tfrac{1}{2} \cos \alpha_1 \tag{15.41}$$

At this optimum speed ratio, the efficiency is

$$\eta_B = 4\left(\frac{\cos^2 \alpha_1}{2} - \frac{\cos^2 \alpha_1}{4}\right) = \cos^2 \alpha_1 \tag{15.42}$$

This condition of maximum efficiency also fixes the values that should be designed for the blade inlet

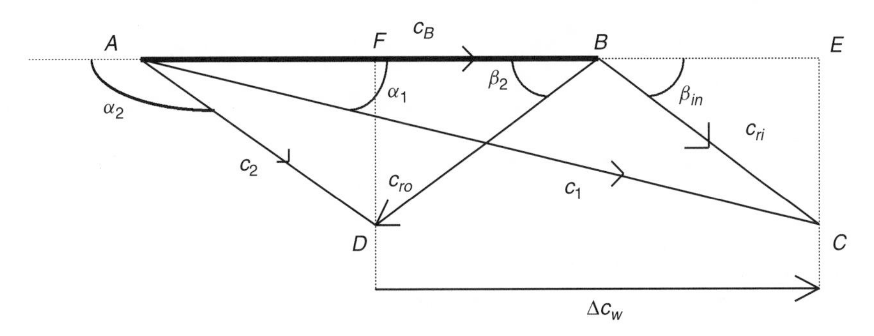

Figure 15.3 Inlet and outlet velocity triangles for an impulse blade.

and outlet angles, since from trigonometry applied to Figure 15.3,

$$\tan\beta_{in} = \frac{c_1 \sin\alpha_1}{c_1\cos\alpha_1 - c_B} = \frac{\sin\alpha_1}{\cos\alpha_1 - R_B} \qquad (15.43)$$

Substituting from (15.41) into (15.43) and adding the design conditions (15.34) gives

$$\beta_1 = \beta_2 = \beta_{in} = \tan^{-1}(2\tan\alpha_1) \qquad (15.44)$$

For example, if the nozzle angle is 20°, the inlet and outlet blade angles will each be 36°, and the optimum speed ratio will be 0.47, giving a blade efficiency of 0.883.

It is also evident that the absolute outlet velocity, c_2, emerges axially at maximum efficiency, i.e. $\alpha_2 = 90°$, since triangles ABD and EBC are congruent on the basis of two sides and included angle:

$$BE = c_1\cos\alpha_1 - c_B = c_1\cos\alpha_1 - \tfrac{1}{2}c_1\cos\alpha_1$$
$$= \tfrac{1}{2}c_1\cos\alpha_1 = AB$$
$$A\hat{B}D = \beta_2 = \beta_1 = \beta_{in} = E\hat{B}C$$
$$BD = c_{ro} = c_{ri} = BC \qquad (15.45)$$

The fact that the exit stream emerges axially demonstrates that all the kinetic energy associated with the direction of wheel motion has been abstracted from the gas stream, as one might expect at the point of greatest efficiency.

15.7 Off-design conditions in an impulse stage: blade efficiency and stage outlet velocity in the absence of blade and nozzle inlet loss

As indicated above, equation (15.40) can be used for design purposes only, since the blade inlet and outlet angles will be fixed once the turbine has been constructed. In particular, they will certainly not move in operation to adjust to the differing speed ratios that will occur in off-design conditions. To arrive at an expression for the blade efficiency of an operational impulse blade, we proceed in two steps. In this section we consider the geometry fixed at the optimum design conditions except for the blade inlet angle, which we will imagine we can vary with speed ratio so as always to give ideal, lossless gas entry. Then in the next section we will estimate the loss at blade entry due to a mismatched angle of approach and hence deduce a correction factor which to apply to the lossless calculation so as to arrive at the final blade efficiency for an operational blade. In a similar way, we shall calculate the stage outlet speed and direction, and then later add a loss-correction factor to find the inlet speed to the next stage.

Returning to Figure 15.3, the change in gas velocity in the direction of wheel motion is still given by equation (15.33), repeated below:

$$\Delta c_w = c_{ri}\cos\beta_{in} + c_{ro}\cos\beta_2 \qquad (15.33)$$

The conditions for the blade inlet and outlet angles are

$\beta_1 = \beta_{in}$, to give shockless entry as before, but now the blade exit geometry is fixed, so that

$$\beta_2 = \tan^{-1}(2\tan\alpha_1) = \text{constant} \qquad (15.46)$$

Equation (15.36) stills holds from trigonometry:

$$c_{ri}\cos\beta_{in} = c_1\cos\alpha_1 - c_B \qquad (15.36)$$

Further, applying Pythagoras's theorem to triangle BCE in Figure 15.3 produces an expression for the magnitude of the inlet velocity relative to the blades, c_{ri}.

$$c_{ri}^2 = c_1^2\sin^2\alpha_1 + (c_1\cos\alpha_1 - c_B)^2$$
$$= c_1^2\sin^2\alpha_1 + c_1^2\cos^2\alpha_1 - 2c_1c_B\cos\alpha_1 + c_B^2$$
$$= c_1^2 - 2c_1c_B\cos\alpha_1 + c_B^2 \qquad (15.47)$$

As before, we may take $c_{ro} = c_{ri}$ for an impulse blade, so that

$$c_{ro} = c_{ri} = c_1\sqrt{1 - 2R_B\cos\alpha_1 + R_B^2} \qquad (15.48)$$

It follows that we may re-express the change in velocity in the direction of wheel motion, using equations (15.33), (15.36) and (15.48), as:

$$\Delta c_w = c_1\left(\cos\alpha_1 - R_B + \cos\beta_2\sqrt{1 - 2R_B\cos\alpha_1 + R_B^2}\right) \qquad (15.49)$$

Blade efficiency is again the change in momentum in the direction of wheel motion divided by the available kinetic energy, as in equation (15.32). We shall call the value from this first step, where the gas is assumed to enter the blade with no loss, η_{Ba}. Substituting from (15.49) into (15.32) gives:

$$\eta_{Ba} = 2R_B\left(\cos\alpha_1 - R_B + \cos\beta_2\sqrt{1 - 2R_B\cos\alpha_1 + R_B^2}\right) \qquad (15.50)$$

Referring once more to Figure 15.3, trigonometry tells us that

$$FD = c_{ro}\sin\beta_2 \qquad (15.51)$$

and

$$AF = c_B - c_{ro}\cos\beta_2 \qquad (15.52)$$

Applying Pythagoras's theorem to the triangle AFD then gives the magnitude of the outlet velocity, c_2, as:

$$\begin{aligned} c_2^2 &= c_{ro}^2 \sin^2 \beta_2 + c_B^2 - 2c_B c_{ro} \cos \beta_2 + c_{ro}^2 \cos^2 \beta_2 \\ &= c_{ro}^2 - 2c_B c_{ro} \cos \beta_2 + c_B^2 \end{aligned} \tag{15.53}$$

Substituting from equation (15.48) gives:

$$\begin{aligned} c_2^2 &= c_1^2(1 - 2R_B \cos \alpha_1 + R_B^2) \\ &\quad - 2c_1 c_B \cos \beta_2 \sqrt{1 - 2R_B \cos \alpha_1 + R_B^2} + c_B^2 \\ &= c_1^2 \left(1 - 2R_B \left(\cos \alpha_1 + \cos \beta_2 \right.\right. \\ &\quad \left.\left. \times \sqrt{1 - 2R_B \cos \alpha_1 + R_B^2}\right) + 2R_B^2\right) \end{aligned} \tag{15.54}$$

Hence

$$c_2 = c_1 \sqrt{\begin{aligned} &1 - 2R_B \left(\cos \alpha_1 + \cos \beta_2\right. \\ &\left. \times \sqrt{1 - 2R_B \cos \alpha_1 + R_B^2}\right) + 2R_B^2 \end{aligned}} \tag{15.55}$$

From trigonometry, the sine of the outlet angle is given by:

$$\begin{aligned} \sin \alpha_2 &= \frac{FD}{AD} = \frac{c_{ro} \sin \beta_2}{c_2} \\ &= \sin \beta_2 \times \sqrt{\frac{1 - 2R_B \cos \alpha_1 + R_B^2}{\begin{aligned} &1 - 2R_B \left(\cos \alpha_1 + \cos \beta_2\right. \\ &\left. \times \sqrt{1 - 2R_B \cos \alpha_1 + R_B^2}\right) + 2R_B^2 \end{aligned}}} \end{aligned} \tag{15.56}$$

The outlet angle is then

$$\begin{aligned} \alpha_2 &= \sin^{-1}(\sin \alpha_2) &\quad \text{for } R_B \leq \frac{\cos \alpha_1}{2} \\ &= 180 - \sin^{-1}(\sin \alpha_2) &\quad \text{for } R_B > \frac{\cos \alpha_1}{2} \end{aligned} \tag{15.57}$$

where the inverse sine function is assumed to return the principal value of the angle.

15.8 Loss of kinetic energy caused by off-design angles of approach to moving and fixed blades

15.8.1 Loss of kinetic energy at the entry to a moving blade

Kinetic energy will be lost when the gas does not enter the moving blade smoothly, that is to say its angle of approach does not match the inlet angle of the blade. Figure 15.4 shows the very tip of the turbine blade receiving the inlet stream of gas for the two situations:

(i) the ratio of blade speed to gas speed is less than the optimum design ratio, so that $\beta_{in} < \beta_1$;
(ii) the ratio of blade speed to gas speed is greater than the optimum design ratio, so that $\beta_{in} > \beta_1$.

Consider first the situation in Figure 15.4(a), when the blade speed to gas speed ratio is lower than the design value. The incoming gas will strike the back face of the blade at the inlet. The main stream of gas will then flow immediately along the blade-tip face into the main

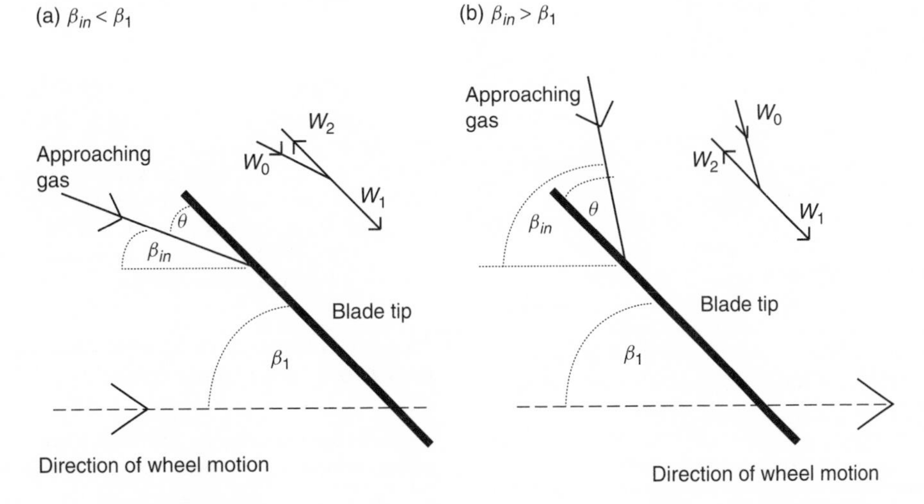

Figure 15.4 Showing the directions of approach to the tip of the moving blade for off-design speed ratios.

section of the blade, where the work will be extracted. However, a small fraction of the gas will be directed along the blade in the opposite direction. It will, of course, be constrained by the boundaries of the turbine, and continuity demands that it will eventually also flow towards the main section of the blade, but it will lose a good deal of its initial velocity. The diverted flow will cause turbulence to the incoming flow, converting kinetic energy into enthalpy as a result. We will use the fraction that would flow away in the unconstrained situation as our chosen measure of the turbulent loss of kinetic energy.

Assuming no energy loss due to the impact itself, the magnitude of the velocity will be unchanged before and after impact. Accordingly the velocity towards the main section of the blade and the velocity towards the edge of the blade will each be equal to the approach velocity. Let us apply the principle of the conservation of momentum along the blade tip. Since no force is applied in this direction, the momentum entering the control volume is equal to the momentum leaving, or in mathematical symbols:

$$\frac{c_{ri}A_0}{v}\cos\theta = \frac{c_{ri}A_1}{v} - \frac{c_{ri}A_2}{v} \tag{15.58}$$

where c_{ri} is the velocity of the incoming gas relative to the direction of wheel motion, θ is the angle of approach to the blade, v is the specific volume of the incoming gas, A_0 is the cross-sectional area of the incoming gas stream, A_1 is the cross-sectional area of the stream of gas passing along the blade towards the main blade section and A_2 is the cross-sectional area of the gas stream attempting to flow along the blade tip away from the main blade section.

Putting

$$\begin{aligned} W_0 &= \frac{c_{ri}A_0}{v} \\ W_1 &= \frac{c_{ri}A_1}{v} \\ W_2 &= \frac{c_{ri}A_2}{v} \end{aligned} \tag{15.59}$$

equation (15.58) becomes:

$$W_0\cos\theta = W_1 - W_2 \tag{15.60}$$

But continuity dictates that

$$W_0 = W_1 + W_2 \tag{15.61}$$

Accordingly, solving (15.60) and (15.61) gives the fraction of flow diverted to the outside of the blade as

$$\frac{W_2}{W_0} = \frac{1-\cos\theta}{2} \tag{15.62}$$

Considering now Figure 15.4(b), when the blade to gas speed ratio is higher than the design value, the angle of approach to the blade will be greater than the blade inlet value, so that the gas will no longer make its first incidence on the back of the first blade tip, but on the front of the next blade tip. Applying the conservation of momentum principle to the situation shown in Figure 15.4(b), this time along the back of the blade, produces the same equations (15.58) to (15.62).

From the geometry of Figure 15.4, it will be seen that the angle, θ, is given by

$$\theta = |\beta_1 - \beta_{in}| \tag{15.63}$$

where, from a consideration of the geometry of Figure 15.4

$$\begin{aligned} \beta_{in} &= \tan^{-1}\left(\frac{\sin\alpha_1}{\cos\alpha_1 - R_B}\right) && \text{for } R_B \le \cos\alpha_1 \\ &= 180 - \tan^{-1}\left(\frac{\sin\alpha_1}{R_B - \cos\alpha_1}\right) && \text{for } R_B > \cos\alpha_1 \end{aligned} \tag{15.64}$$

We will assume that some fraction e_1, $0 \le e_1 \le 1$ of the kinetic energy associated with the diverted flow will be lost, so that the fractional loss of kinetic energy will be:

$$\lambda_1 = e_1\frac{1-\cos\theta}{2} \tag{15.65}$$

This fraction of the kinetic energy possessed by the mid-stage gas stream will be converted into additional enthalpy. There is no mechanism for reconverting this enthalpy into kinetic energy in an impulse blade, so there will be no recovery of any energy lost at the blade entry in an impulse stage. A reasonable estimate may be $e_1 \approx 1$. The loss correction factor to be applied to the initial value of blade efficiency is simply $(1 - \lambda_1)$. The final calculation of blade efficiency for an operating impulse blade is thus

$$\eta_B = (1-\lambda_1)\eta_{Ba} \tag{15.66}$$

The situation for a reaction blade is different as a result of its nozzle action. The conversion of kinetic energy to enthalpy at the entrance to the blade will be followed by a balancing conversion back to kinetic energy by the reaction blade acting as a nozzle. This phenomenon is discussed in detail in Section 16.3 of the next chapter. Accordingly there is no need to consider inlet shock loss for a reaction blade.

15.8.2 Recovery of kinetic energy at the entry to a fixed blade (nozzle)

Figure 15.5 shows hot gas leaving the previous stage at an angle α_2 and entering the fixed blade forming the nozzle section of the next stage, which has a blade-inlet angle α_0. α_0 is normally 90° for both impulse and reaction fixed blades.

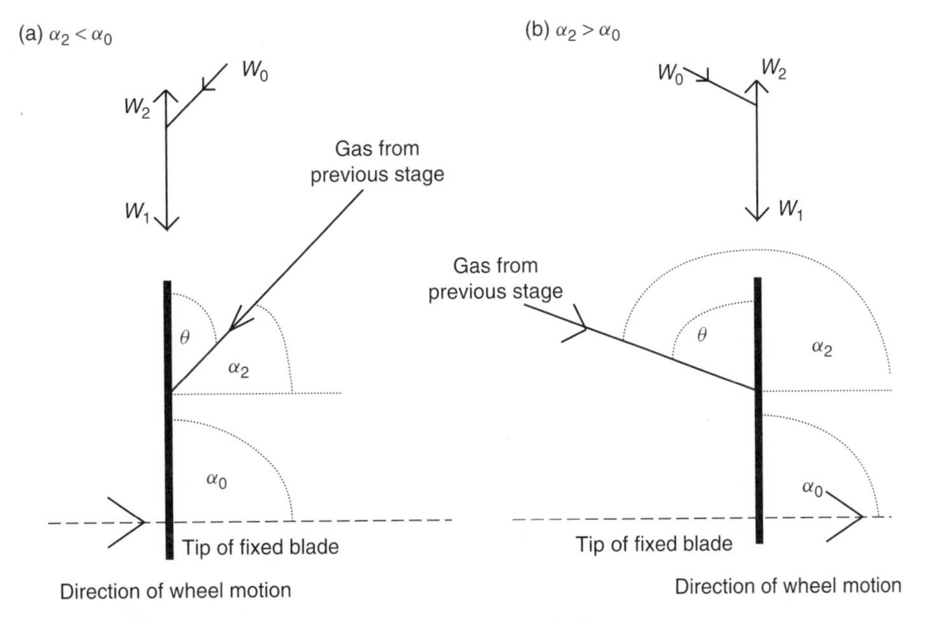

Figure 15.5 Entry into the fixed blade of the next stage at off-design speed ratios.

We may analyse the loss in kinetic energy in exactly the same way as was done in Section 15.8.1, and arrive at the conclusion that there will be a loss of kinetic energy at the entrance to the fixed blade/nozzle. However, the effect of the nozzle will be to reconvert essentially all the enthalpy gained back into kinetic energy. This phenomenon is discussed further in Section 16.2.

15.9 Off-design conditions in an impulse blade: typical corrections for kinetic energy losses

The uncorrected blade efficiency, η_{Ba}, is calculated from equation (15.50), and then the final efficiency figure is found from equation (15.66), thereby allowing for the kinetic energy loss at the entry to the moving blade. Let us consider the case of an impulse stage with a nozzle angle of 20°, with blade inlet and outlet angles optimized at 36°. Figure 15.6 shows a plot of η_{Ba} and η_B against R_B.

Considering the velocity diagram of Figure 15.3 once more, the limiting value of R_B at which no useful work will be done will occur when the angle of blade approach coincides with the outlet blade angle. There will then be no change in velocity over the blade, and the acceleration given to the turbine wheel by the gas in leaving will be balanced exactly by the retardation of the wheel at blade entry. The condition is

$$\beta_{in} = 180 - \beta_2 \tag{15.67}$$

Taking the tangent of both sides of equation (15.67) gives

$$\tan\beta_{in} = -\tan\beta_2 \tag{15.68}$$

Using equations (15.43) and (15.44), equation (15.68) becomes

$$\frac{\sin\alpha_1}{\cos\alpha_1 - R_B} = -2\tan\alpha_1 \tag{15.69}$$

Hence

$$\begin{aligned} R_B &= \cos\alpha_1 + \frac{\sin\alpha_1}{2\tan\alpha_1} \\ &= 1.5\cos\alpha_1 \end{aligned} \tag{15.70}$$

Thus the blade-to-gas speed ratio at which the blade will abstract no useful work is $R_B = 1.41$ when $\alpha_1 = 20°$. This is indicated in Figure 15.6 by the blade efficiency falling to zero at this point.

The shock correction factor $(1 - \lambda_1)$ remains close to unity at low speed ratios, but falls at high-speed ratios, where it causes a significant difference to arise between the uncorrected blade efficiency, η_{Ba}, and the blade efficiency corrected for inlet shock losses, η_B.

15.10 50% reaction stage: the design of the fixed blades (nozzles) and the moving blades

It is normal for a reaction stage to be designed to have to have an axial velocity, c_a, that is the same at the inlet

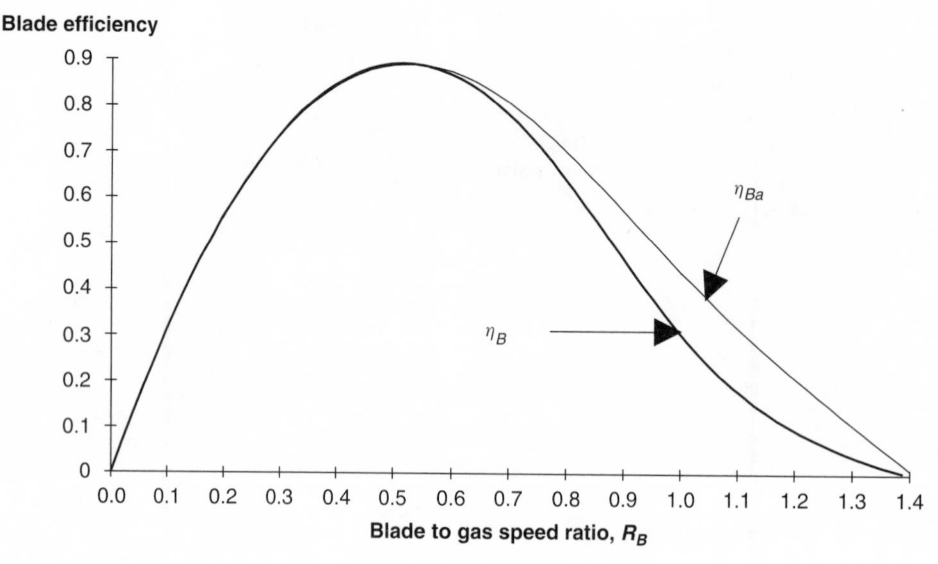

Figure 15.6 Blade efficiency for an operating impulse blade with nozzle angle = 20°.

to the moving blades as it is at the inlet to the fixed blades, which act purely as nozzles. Further, bearing in mind that the gas leaving one stage enters the next, the design usually provides for the velocity leaving the stage, c_2, to be the same as the velocity entering the (next) stage, c_0. Using subscript notation, $c_{0,i+1} = c_{2,i}$. Figure 15.7 below shows the inlet and outlet velocity triangles for the moving blade in a 50% reaction stage, with the blade speed, c_B, used as the common base.

Figure 15.7 differs from Figure 15.3 principally in the fact that the outlet velocity relative to the moving blade, c_{ro}, is greater than the inlet velocity relative to the moving blade, c_{ri}. The inlet velocity relative to the blade, c_{ri}, makes an angle β_{in} with the blade velocity, c_B, which will be the same as the moving blade inlet angle, β_1, at the design point, when shockless entry will occur, although this cannot be guaranteed away from the design conditions. Since the stage inlet and outlet velocities, c_0 and c_2, are equal, the power abstracted, P, is simply the difference in specific enthalpy at the stage inlet and outlet, multiplied by the mass flow:

$$P = W(h_0 - h_2) \tag{15.71}$$

But the power abstracted is also given from the velocity diagram as equation (15.29), repeated below:

$$P = Wc_B\Delta c_w \tag{15.29}$$

or

$$P = Wc_Bc_a(\cot\beta_{in} + \cot\beta_2) \tag{15.72}$$

Comparing equations (15.71) and (15.72), we see that

$$h_0 - h_2 = c_Bc_a(\cot\beta_{in} + \cot\beta_2) \tag{15.73}$$

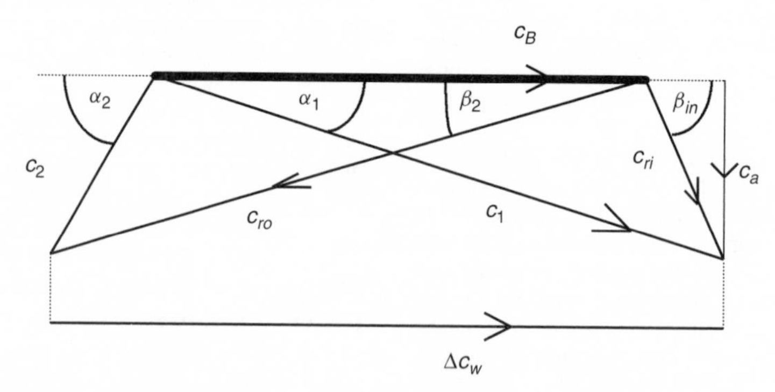

Figure 15.7 Inlet and outlet velocity triangles for the blade in a 50% reaction stage.

Now the increase in specific kinetic energy of the gas passing over the moving blades results from the decrease in specific enthalpy over the moving blades (see equation (14.3)). Hence

$$h_1 - h_2 = \tfrac{1}{2}c_{ro}^2 - \tfrac{1}{2}c_{ri}^2 \qquad (15.74)$$

But consideration of the geometry of Figure 15.7 shows that

$$\begin{aligned} c_{ri} &= c_a \operatorname{cosec} \beta_{in} \\ c_{ro} &= c_a \operatorname{cosec} \beta_2 \end{aligned} \qquad (15.75)$$

Substituting from (15.75) into (15.74) gives

$$\begin{aligned} h_1 - h_2 &= \tfrac{1}{2}c_a^2(\operatorname{cosec}^2 \beta_2 - \operatorname{cosec}^2 \beta_{in}) \\ &= \tfrac{1}{2}c_a^2(1 + \cot^2 \beta_2 - (1 + \cot^2 \beta_{in})) \\ &= \tfrac{1}{2}c_a^2(\cot \beta_2 - \cot \beta_{in})(\cot \beta_2 + \cot \beta_{in}) \end{aligned} \qquad (15.76)$$

At 50% reaction, we have from equation (15.19):

$$\frac{h_1 - h_2}{h_0 - h_2} = \frac{1}{2} \qquad (15.77)$$

so that, using (15.73) and (15.76), we have:

$$\frac{\tfrac{1}{2}c_a^2(\cot \beta_2 - \cot \beta_{in})(\cot \beta_2 + \cot \beta_{in})}{c_B c_a(\cot \beta_{in} + \cot \beta_2)} = \frac{1}{2} \qquad (15.78)$$

yielding the following equation for blade speed, c_B:

$$c_B = c_a(\cot \beta_2 - \cot \beta_{in}) \qquad (15.79)$$

But the geometry of Figure 15.7 shows that the blade velocity may also be expressed by the trigonometrical relations:

$$\begin{aligned} c_B &= c_a(\cot \alpha_1 - \cot \beta_{in}) \\ c_B &= c_a(\cot \beta_2 - \cot \alpha_2) \end{aligned} \qquad (15.80)$$

It follows from comparing equations (15.79) and (15.80) that for a 50% reaction stage with a constant axial velocity across the stage and equal stage inlet and outlet velocities:

$$\begin{aligned} \alpha_1 &= \beta_2 \\ \alpha_2 &= \beta_{in} \end{aligned} \qquad (15.81)$$

For shockless entry, the moving blade inlet angle needs to match the angle of gas entry:

$$\beta_1 = \beta_{in} \qquad (15.82)$$

In addition, the stipulation of equal absolute velocities at stage inlet and outlet implies that the angle of the absolute outlet velocity is equal to the inlet angle to the fixed blades of the next stage:

$$\alpha_0 = \alpha_2 \qquad (15.83)$$

An important consequence is that both the fixed and moving blades have the same cross-sectional shape (although the blade height will increase as the expansion proceeds along the turbine). Figure 15.8 shows the effect diagrammatically.

15.11 Blade efficiency at design conditions for a 50% reaction stage

Let the efficiency of the fixed blades, which act simply as nozzles, be η_N, and let the nozzle efficiency of the expansion that takes place in the moving blades be η_{BN}, defined by:

$$\eta_N = \frac{\tfrac{1}{2}c_1^2 - \tfrac{1}{2}c_0^2}{h_0 - h_{1s}} = \frac{h_0 - h_1}{h_0 - h_{1s}} \qquad (15.84)$$

$$\eta_{BN} = \frac{\tfrac{1}{2}c_{ro}^2 - \tfrac{1}{2}c_{ri}^2}{h_1 - h_{2sa}} = \frac{h_1 - h_2}{h_1 - h_{2sa}} \qquad (15.85)$$

where h_{1s} is the specific enthalpy that would result from an isentropic expansion over the fixed blade, while h_{2sa} is the specific enthalpy that would result from an isentropic expansion over the moving blade. The additional subscript 'a' in h_{2sa} connotes an isentropic expansion from mid-stage to stage outlet, as opposed to an expansion from stage inlet to stage outlet. At 50% reaction, equation (15.19) gives:

$$\rho = \frac{1}{2} = \frac{\eta_{BN}(h_1 - h_{2sa})}{\eta_N(h_0 - h_{1s}) + \eta_{BN}(h_1 - h_{2sa})} \qquad (15.86)$$

It follows that

$$\begin{aligned} \eta_N(h_0 - h_{1s}) &= \tfrac{1}{2}(c_1^2 - c_0^2) = \eta_{BN}(h_1 - h_{2sa}) \\ &= \tfrac{1}{2}(c_{ro}^2 - c_{ri}^2) \end{aligned} \qquad (15.87)$$

The maximum specific work, $w_{\max B}$, available for the moving blades to extract is the sum of the specific kinetic energy of the gas leaving the nozzle blades and the isentropic specific enthalpy drop across the moving blades:

$$w_{\max B} = \tfrac{1}{2}c_1^2 + h_1 - h_{2sa} \qquad (15.88)$$

From the symmetry of the velocity diagram, demonstrated in Section 15.10,

$$c_{ro} = c_1 \qquad (15.89)$$

while equation (15.47) still holds for the inlet velocity relative to the moving blades, c_{ri}:

$$c_{ri}^2 = c_1^2 - 2c_1 c_B \cos \alpha_1 + c_B^2 \qquad (15.47)$$

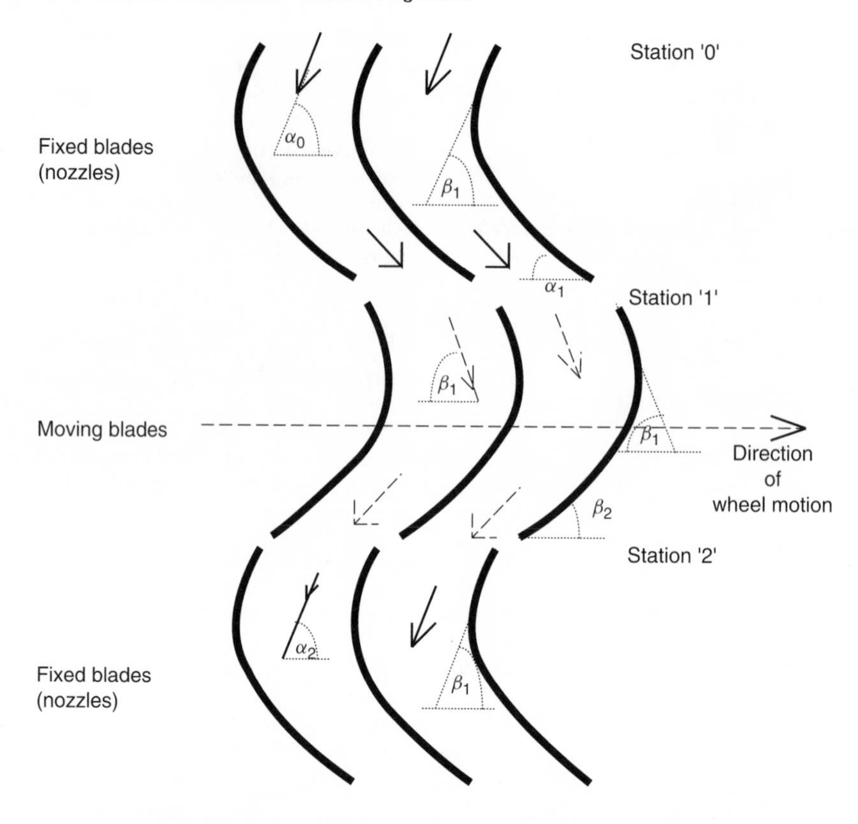

Figure 15.8 Schematic cross-section of 50% reaction stage.

Hence, combining equation (15.47) with equations (15.87) to (15.89) gives the maximum available specific work as

$$w_{\max B} = \frac{1}{2}\left(c_1^2 + \frac{1}{\eta_{BN}}(2c_1 c_B \cos\alpha_1 - c_B^2)\right)$$
$$= \frac{1}{2}c_1^2\left(1 + \frac{1}{\eta_{BN}}(2R_B \cos\alpha_1 - R_B^2)\right) \tag{15.90}$$

Meanwhile, the actual specific work extracted, w, is given by equation (15.31), repeated below:

$$w = c_B \Delta c_w \tag{15.31}$$

Now equation (15.33) still holds, as does equation (15.36), and so we may write the change in whirl velocity as:

$$\Delta c_w = c_1 \cos\alpha_1 + c_{r0}\cos\beta_2 - c_B \tag{15.91}$$

Further, by equations (15.81) and (15.89),

$$c_{ro}\cos\beta_2 = c_1 \cos\alpha_1 \tag{15.92}$$

and so the change in whirl velocity is:

$$\Delta c_w = 2c_1 \cos\alpha_1 - c_B \tag{15.93}$$

Hence the blade efficiency is given by:

$$\eta_B = \frac{w}{w_{\max B}} = \frac{c_B(2c_1\cos\alpha_1 - c_B)}{\frac{1}{2}c_1^2\left(1 + \frac{1}{\eta_{BN}}(2R_B\cos\alpha_1 - R_B^2)\right)}$$
$$= \frac{2(2R_B\cos\alpha_1 - R_B^2)}{1 + \frac{1}{\eta_{BN}}(2R_B\cos\alpha_1 - R_B^2)} \tag{15.94}$$

We may find the maximum blade efficiency for a given nozzle angle, α_1, by differentiating with respect to speed ratio, R_B, and setting the result to zero. (We will assume that the nozzle efficiency over the moving blades is unaffected by changes in speed ratio, R_B.) This yields the optimal speed ratio as

$$R_B = \cos\alpha_1 \tag{15.95}$$

with a corresponding maximum efficiency of

$$\eta_B = \frac{2\cos^2\alpha_1}{1 + \dfrac{\cos^2\alpha_1}{\eta_{BN}}} \tag{15.96}$$

The consequence of an optimal speed ratio on the angle of approach, β_{in}, is found by substituting from equation (15.95) into equation (15.43), repeated below:

$$\tan\beta_{in} = \frac{c_1 \sin\alpha_1}{c_1\cos\alpha_1 - c_B} = \frac{\sin\alpha_1}{\cos\alpha_1 - R_B} \tag{15.43}$$

so that

$$\tan\beta_{in} = \infty \tag{15.97}$$

implying

$$\beta_{in} = 90^\circ \tag{15.98}$$

It follows that the moving blade inlet angle, β_1, will be designed to be 90°. Further, by the design conditions (15.81), the stage outlet angle, α_2, will also be 90° at the optimal speed ratio. Thus the gas emerges axially at design conditions from the reaction stage in the same way as it did for an impulse stage.

15.12 Blade efficiency at off-design conditions for a 50% reaction stage

We need now to examine how the efficiency on an installed machine changes as the operating point moves away from the design point. We cannot use the design equation (15.94) for blade efficiency because it suffers from similar limitations to the design equation for the efficiency of an impulse blade. The equation contains two inherent assumptions that cannot be guaranteed away from the design point: first, that the moving blade angle, β_1, is adjusted continuously and kept equal at all times to the angle of gas approach, β_{in}, and second that the stage outlet angle, α_2, will be equal at all times to the angle of approach, β_{in}. It is technically more difficult to derive an expression for the off-design blade efficiency for a reaction stage than for an impulse stage because the fact that the reaction blade performs an expansion as well as a power extraction leads to a more complicated form for the maximum power, $w_{\max B}$, available for the moving blade to extract, namely

$$w_{\max B} = \tfrac{1}{2}c_1^2 + h_1 - h_{2sa} \tag{15.88}$$

Equation (15.88) contains a term not possessed by the equivalent equation (15.31) for the impulse blade, namely the enthalpy difference, $h_1 - h_{2sa}$, the magnitude of which depends on conditions upstream and downstream of the individual reaction stage. A complete analysis requires a consideration of all the reaction stages together, but it is nevertheless possible to derive an approximate expression for the blade efficiency in off-design conditions that depends only on the ratio of the blade speed to inlet gas speed. This may be done by making two assumptions for which justifications are presented in Appendix 7, namely that at off-design conditions

(i) the axial velocity at the outlet from the fixed blades remains equal to the axial velocity at the inlet to the fixed blades, and
(ii) the degree of reaction remains at 50%.

However the off-design point does not need to satisfy the design condition that the axial velocity of the gas emerging from the moving blade should be equal to the axial velocity at the outlet of the fixed blade.

As noted at the end of Section 15.8.1, there is no need to make an allowance for shock loss resulting from the angle of approach to the moving blade, β_{in}, being different from the design angle of approach, β_1, since the kinetic energy lost at this point is regained more or less in full as a result of the nozzle action of the reaction blade.

We may eliminate in equation (15.88) the isentropic drop in specific enthalpy over the moving blades, $h_1 - h_{2sa}$, by using equation (15.87):

$$w_{\max B} = \frac{1}{2}\left(c_1^2 + \frac{1}{\eta_{BN}}(c_1^2 - c_0^2)\right) \tag{15.99}$$

We may now use the axial velocity assumption listed above:

$$c_0 \sin\alpha_0 = c_1 \sin\alpha_1 \tag{15.100}$$

so that, noting also the design condition $\alpha_2 = \beta_1 = \alpha_0$ (equations (15.81), (15.82) and (15.83)), we may eliminate c_0 from equation (15.99)

$$w_{\max B} = \frac{1}{2}c_1^2\left(1 + \frac{1}{\eta_{BN}}\left(1 - \frac{\sin^2\alpha_1}{\sin^2\beta_1}\right)\right) \tag{15.101}$$

Meanwhile, the actual specific work abstracted, w, is given by the same equation as for the impulse blade, namely equation (15.31), repeated below:

$$w = c_B \Delta c_w \tag{15.31}$$

The change in whirl velocity is given by equation (15.33), as before:

$$\Delta c_w = c_{ri}\cos\beta_{in} + c_{ro}\cos\beta_2 \tag{15.33}$$

but since the design condition for a 50% reaction stage is $\alpha_1 = \beta_2$ (equation 15.80), Δc_w transforms to:

$$\Delta c_w = c_{ri}\cos\beta_{in} + c_{ro}\cos\alpha_1 \tag{15.102}$$

Equation (15.36) still holds:

$$c_{ri}\cos\beta_{in} = c_1\cos\alpha_1 - c_B \tag{15.36}$$

Meanwhile, the magnitude of the inlet velocity relative to the moving blades, c_{ri}, is given by equation (15.47) once more:

$$c_{ri}^2 = c_1^2 - 2c_1c_B\cos\alpha_1 + c_B^2 \tag{15.47}$$

The outlet velocity relative to the moving blades may be found from equation (15.87) as:

$$c_{ro}^2 = c_{ri}^2 + c_1^2 - c_0^2 \tag{15.103}$$

Substituting for c_{ri} from equation (15.47) and for c_0 from equation (15.100) with $\alpha_0 = \alpha_2 = \beta_1$ into equation (15.103) gives:

$$c_{ro}^2 = c_1^2\left(2 - \frac{\sin^2\alpha_1}{\sin^2\beta_1}\right) - 2c_1c_B\cos\alpha_1 + c_B^2 \tag{15.104}$$

Hence, making use of the moving blade to gas speed ratio, R_B, we may write:

$$c_{ro} = c_1\sqrt{2 - \frac{\sin^2\alpha_1}{\sin^2\beta_1} - 2R_B\cos\alpha_1 + R_B^2} \tag{15.105}$$

Collecting these expressions together, the actual specific work is given by:

$$w = c_B\left(c_1\cos\alpha_1 \times\left(1+\sqrt{2-\frac{\sin^2\alpha_1}{\sin^2\beta_1}-2R_B\cos\alpha_1+R_B^2}\right) - c_B\right) \tag{15.106}$$

Hence the efficiency of the 50% reaction blade is given by

$$\eta_B = \frac{w}{w_{\max B}} = \frac{c_B\left(c_1\cos\alpha_1\left(1+\sqrt{2-\frac{\sin^2\alpha_1}{\sin^2\beta_1}-2R_B\cos\alpha_1+R_B^2}\right)-c_B\right)}{\frac{1}{2}c_1^2\left(1+\frac{1}{\eta_{BN}}\left(1-\frac{\sin^2\alpha_1}{\sin^2\beta_1}\right)\right)} \tag{15.107}$$

or

$$\eta_B = \frac{2\left(R_B\cos\alpha_1\left(1+\sqrt{2-\frac{\sin^2\alpha_1}{\sin^2\beta_1}-2R_B\cos\alpha_1+R_B^2}\right)-R_B^2\right)}{1+\frac{1}{\eta_{BN}}\left(1-\frac{\sin^2\alpha_1}{\sin^2\beta_1}\right)} \tag{15.108}$$

Figure 15.9 shows blade efficiency at off-design conditions as a function of the ratio of blade speed to fixed blade gas speed for a 50% reaction stage with the following parameters: nozzle angle, $\alpha_1 = 20°$, blade inlet angle, $\beta_1 = 90°$ and blade nozzle efficiency, $\eta_{BN} = 1.0$. It is noticeable that the decline in efficiency with high blade to gas speed ratios is very much gentler than for an impulse blade.

The stage outlet velocity, c_2, is given by equation (15.53) once again:

$$c_2^2 = c_{ro}^2 - 2c_Bc_{ro}\cos\beta_2 + c_B^2 \tag{15.53}$$

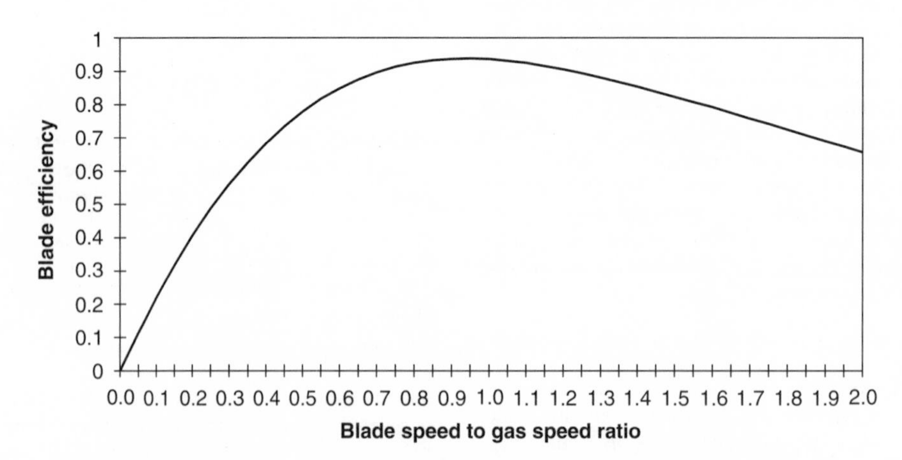

Figure 15.9 Blade efficiency for off-design conditions for a 50% reaction stage with a nozzle angle = 20°.

and substituting from equation (15.105) for c_{ro} gives:

$$c_2^2 = c_1^2\left(\left(2-\frac{\sin^2\alpha_1}{\sin^2\beta_1}\right)-2R_B\cos\alpha_1+R_B^2\right)-2c_1c_B\cos\beta_2$$
$$\times\sqrt{\left(2-\frac{\sin^2\alpha_1}{\sin^2\beta_1}\right)-2R_B\cos\alpha_1+R_B^2+c_B^2}$$
$$=c_1^2\left(\left(2-\frac{\sin^2\alpha_1}{\sin^2\beta_1}\right)-2R_B\left(\cos\alpha_1+\cos\beta_2\right.\right.$$
$$\left.\left.\times\sqrt{\left(2-\frac{\sin^2\alpha_1}{\sin^2\beta_1}\right)-2R_B\cos\alpha_1+R_B^2}\right)+2R_B^2\right) \quad (15.109)$$

Hence

$$c_2 = c_1 \times\sqrt{\left(2-\frac{\sin^2\alpha_1}{\sin^2\beta_1}\right)-2R_B\left(\cos\alpha_1+\cos\beta_2\times\sqrt{\left(2-\frac{\sin^2\alpha_1}{\sin^2\beta_1}\right)-2R_B\cos\alpha_1+R_B^2}\right)+2R_B^2} \quad (15.110)$$

15.13 The polytropic exponent for saturated steam

An isentropic expansion for superheated steam is well characterized by taking the exponent as $\gamma = 1.3$, but the exponent changes markedly when the steam becomes saturated. The value of γ appropriate in the saturated region will depend on the dryness fraction, x_0, at the start of the expansion, where x_0 will depend on the local thermodynamic properties at stage inlet, h_0, s_0, v_0, etc.:

$$x_0 = \frac{h_0 - h_f(p_0)}{h_g(p_0) - h_f(p_0)} = \frac{s_0 - s_f(p_0)}{s_g(p_0) - s_f(p_0)}$$
$$= \frac{v_0 - v_f(p_0)}{v_g(p_0) - v_f(p_0)}, \text{ etc.} \quad (15.111)$$

where the subscript f indicates saturated water and the subscript g indicates saturated steam. Since pressure is a unique function of temperature in the saturated region, each of the saturated variables: h_f, h_g, s_f, s_g, v_f, ... may be regarded as dependent on pressure only, as indicated.

Zeuner's equation for the saturated region gives the isentropic exponent as:

$$\gamma = 1.035 + 0.1x_0 \quad \text{for } 0.7 \le x_0 \le 1.0 \quad (15.112)$$

If the steam is superheated at the start of the expansion, then the value $\gamma = 1.3$ is used, the reason being that condensation will not occur immediately the saturation temperature is reached. Instead, the steam stays wholly in dry vapour form, when it is known as 'supersaturated'. This metastable state cannot be maintained indefinitely, however, and condensation will occur suddenly at some point after the steam has passed through the nozzle.

15.14 Calculation sequence for turbine simulation

The flows through a turbine with N stages is represented in Figure 15.10, which allows for the general case where hot gas or steam is taken off at the end of each stage, for example to provide heating to a boiler feedwater heater.

At any time instant, the boundary pressures, p_{up}, $p_{E,1}$, $p_{E,2}$, ..., $p_{E,i}$, ..., $p_{E,N}$, will be governed solely by the outputs of integrators and hence will be fixed. The same will apply to the upstream specific enthalpy, h_{up}, the upstream specific volume, v_{up}, and the turbine speed, N. As noted in the introduction to this chapter, the time constants associated with the establishment of fluid flow are small compared with the rotor time constant, and so we may assume the flow always to be in a steady state. The steady-state flow balance equations may be written from inspection of Figure 15.10 as:

$$W_{in} - W_1 = 0$$
$$W_i - W_{i+1} - W_{E,i} = 0 \quad \text{for } 1 \le i \le N-1$$
$$W_N - W_{E,N} = 0 \quad (15.113)$$

The inlet and outlet flows will face resistance from pipelines and valves and may be modelled using the methods of Chapters 6 and 10. The variables necessary to determine these flows are the upstream pressure, the upstream specific volume, the downstream pressure and the valve position. The internal mass flows, W_i, $i = 1, \ldots, N$, are calculated as the flows through each nozzle, and this procedure may be carried out iteratively at each timestep. Calculation of the overall performance of the turbine at each timestep follows the following sequence.

1. Provide a first estimate of the stage inlet pressures, $p_{0,i}$, $i = 1, \ldots, N$, and the last stage outlet pressure, $p_{2,N}$ (initially from best-estimates of the steady-state values, and then from the values at the last timestep).

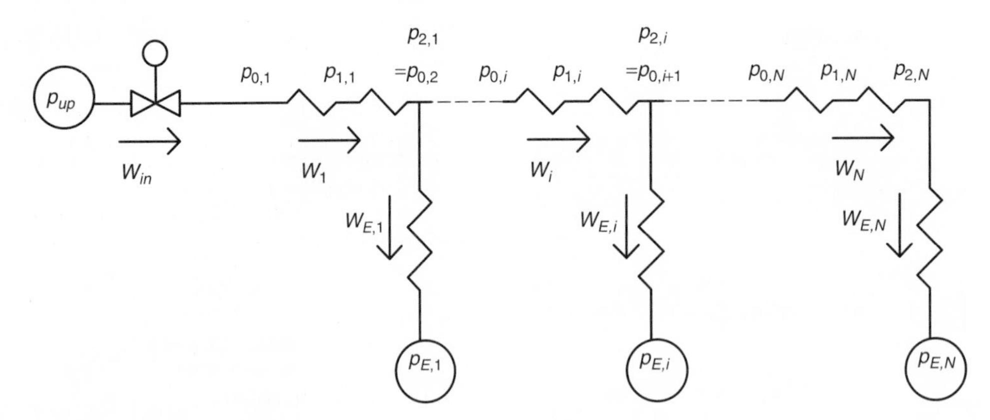

Figure 15.10 Flows through a turbine.

2. Calculate the mid-stage pressures, $p_{1,i}$. In the case of an impulse stage these will be the same as the inlet pressure to the next stage. The case of a reaction stage is more complex and is covered in Section 15.4 above.

3. Calculate the flow through the turbine control valve using the methods of Chapters 6 and 10. Enthalpy is unchanged by the throttling process, and so the specific enthalpy at the inlet to the first stage will be equal to the upstream specific enthalpy:

$$h_{0,1} = h_{up} \tag{15.114}$$

while the specific volume at the inlet to the first stage will reflect the fact that the temperature is kept constant:

$$v_{0,1} = v_{up}\frac{p_{up}}{p_{0,1}} \tag{15.115}$$

First stage

4. If the working fluid is steam, check whether it is superheated or saturated at the first stage by comparing the inlet specific enthalpy, $h_{0,1}$, with the specific enthalpy of saturated steam appropriate to the inlet pressure, $h_g(p_{0,1})$. If $h_{0,1} \leq h_g(p_{0,1})$, use equation (15.111) to estimate the initial dryness fraction and equation (15.112) to estimate the isentropic index, γ.

5. Since we now have available the upstream pressure and specific volume and the downstream pressure of the nozzle for the first stage, we may calculate its mass flow, W_1, and nozzle outlet velocity, c_1, using the methods of Chapter 14.

Steps 6 to 10 are common to the first and all subsequent stages

6. A knowledge of the instantaneous turbine speed, N, allows us to calculate the average blade speed for the stage from:

$$c_{B,i} = 2\pi N r_{B,i} \tag{15.116}$$

where $r_{B,i}$ (m) is the average blade radius for stage i.

7. Since we know the gas speed at the exit of the nozzle, we may calculate the ratio of gas speed to nozzle outlet speed, R_B. If the stage is an *impulse* stage we calculate the blade efficiency at off-design conditions using equations (15.50) and (15.66). The specific work of the stage will then be given by:

$$w = \tfrac{1}{2}c_1^2\eta_B \tag{15.117}$$

where the nozzle outlet velocity, c_1, has already been evaluated for the stage. If the stage is a *50% reaction* stage, then the specific work will be given by equation (15.106), repeated below:

$$w = c_B\left(c_1\cos\alpha_1 \times\left(1+\sqrt{2-\frac{\sin^2\alpha_1}{\sin^2\beta_1}-2R_B\cos\alpha_1+R_B^2}\right)-c_B\right) \tag{15.106}$$

8. We may at this point calculate the outlet velocity, c_2, from equation (15.55) for an impulse stage or from equation (15.110) for a 50% reaction stage.

9. Calculate the local specific enthalpy at the exit of the current stage/entry to the next stage using equation (15.2), repeated in subscripted form below:

$$h_{0,i+1} = h_{2,i} = h_{0,i} + \tfrac{1}{2}c_{0,i}^2 - \tfrac{1}{2}c_{2,i}^2 - w_i \tag{15.118}$$

10. After setting the inlet velocity of the next stage equal to the outlet velocity of the current stage:

$$c_{0,i+1} = c_{2,i} \tag{15.119}$$

we now have a full set of boundary conditions to calculate the performance of the next stage.

Next stage

11. If the working fluid is steam, check whether it is superheated or saturated at the entry to the stage by comparing the inlet specific enthalpy with the specific enthalpy of saturated steam appropriate to the inlet pressure, and if saturated use the methods outlined in step 4 to estimate the isentropic index, γ.

12. Estimate the stagnation values of pressure and specific volume at the inlet to the current stage using the methods outlined in Chapter 14, and hence determine the next stage mass flow and nozzle outlet velocity.

Second stage

13. Repeat steps 6 to 10 for the second stage.

All subsequent stages

14. Repeat steps 11, 12 and 6 to 10 for each subsequent stage in turn.

Turbine mass balance

15. Test whether the flow equations (15.113) are satisfied to the desired accuracy. If not, re-estimate the stage inlet pressures and the final stage outlet pressures and return to step 2.

Turbine power

16. When the flow equations are satisfied, calculate the power in each stage from equation (15.6) and the total power output of the turbine from equation (15.7):

$$P_i = W_i w_i \qquad (15.6)$$

$$P_T = \sum_i W_i w_i \qquad (15.7)$$

15.15 Bibliography

Faires, V.M. (1957). *Thermodynamics*, 3rd edition Macmillan, New York.

Kearton, W.J. (1922, 1958). *Steam Turbine Theory and Practice*, Pitman and Sons, London.

Lockey, J. (1966). *The Thermodynamics of Fluids*, Heinemann Educational Books Ltd, London.

Rogers, G.F.C. and Mayhew, Y.R. (1992). *Engineering Thermodynamics: Work and Heat Transfer*, 4th edition, Longmans.

16 Steam and gas turbines: simplified model

16.1 Introduction

The detailed approach used in the last chapter made proper allowance for the interstage velocities through the use of the notional stagnation state. But the size of the interstage kinetic energy term will usually be small relative to the specific enthalpy at the inlet to the next stage, which prompts the question: can we simplify the analysis and ignore interstage velocities? It is shown in Section 16.2 below that good accuracy may be achieved despite neglecting the interstage velocities in the calculation of stage-specific work. This opens up an approach to calculating turbine performance based on stage efficiencies, which may be calculated from the nozzle and blade efficiencies.

Having embarked on the quest of simplification, we show that the steam tables may be represented in the regions of interest to steam turbines by relatively simple approximating functions, examples of which are given. Further, it is shown that small changes in efficiency have little effect on the calculated mass flow, and it will often be sufficient for control engineering purposes to use the isentropic mass flow equation.

16.2 The effect of neglecting interstage velocities in modelling a real turbine stage: the approximate equivalence of kinetic energy and enthalpy at nozzle inlet

The sum of the specific enthalpy and specific kinetic energy at the stage outlet (station '2' of Figure 15.1) may be found by rearranging equation (15.2) to give:

$$h_2 + \tfrac{1}{2}c_2^2 = h_0 + \tfrac{1}{2}c_0^2 - w \tag{16.1}$$

Since the inlet to the present stage is the outlet from the previous stage, by introducing an additional subscript, i, to characterize the stage, we may re-express equation (16.1) as:

$$h_{0,i+1} + \tfrac{1}{2}c_{0,i+1}^2 = h_{0,i} + \tfrac{1}{2}c_{0,i}^2 - w_i \qquad \text{for } i \geq 2 \tag{16.2}$$

where $h_{0,i}$,$c_{0,i}$ are the specific enthalpy and velocity at the inlet to the current stage, the ith, and w_i is the specific work extracted from the current stage. The restriction on equation (16.2) is introduced for clarity, and follows from the fact that the inlet velocity to the first stage will be close to zero: $c_{0,1} \approx 0$, so that the equivalent equation for stage 1 is:

$$h_{0,2} + \tfrac{1}{2}c_{0,2}^2 = h_{0,1} - w_1 \tag{16.3}$$

Neglecting the interstage velocities implies putting $c_{0,i} = 0$ for all i, so that equations (16.2) and (16.3) would then be reduced to

$$h'_{0,i+1} = h'_{0,i} - w'_i \tag{16.4}$$

The primes have been added to the symbols for local specific enthalpy and stage specific work to indicate that we are dealing with approximate values. We will use the prime similarly to indicate the approximate values of the other local thermodynamic variables:

$$T'_{0,i},\ p'_{0,i},\ v'_{0,i}$$

In practice, the size of the kinetic energy terms neglected will be rather small compared with the enthalpy terms. For example, suppose the conditions at the second stage inlet of a steam turbine are 20 bar and 250°C, giving a local specific enthalpy from steam tables of 2904×10^3 J/kg. A typical value of inlet velocity to the second stage is 100 m/s, which implies a specific kinetic energy of 5×10^3 J/kg, only 0.17% of the specific enthalpy value. Similarly differences between stage inlet and outlet specific enthalpies are likely to dwarf the differences between the stage inlet and outlet velocities, which, with the exception of the first stage, are likely to be approximately equal.

We may gain a further insight into the implications of neglecting the interstage velocities by using the concept of stagnation specific enthalpy introduced in Section 14.5. Writing $h_{0T,i} = h_{0,i} + c_{0,i}^2/2$, equation (16.2) converts to:

$$h_{0T,i+1} = h_{0T,i} - w_i \tag{16.5}$$

Noting that the truly negligible inlet velocity to stage 1 implies

$$h_{0,1} = h'_{0,1} = h_{0T,1} \tag{16.6}$$

so that comparing equations (16.4) and (16.5) suggests that

$$h'_{0,i} \approx h_{0T,i} \qquad \text{for all } i \tag{16.7}$$

with the closeness of the approximation depending on how well w_i and w'_i match. Thus applying equation (16.4) to each stage in succession implies that we will overestimate the local inlet specific enthalpy to all stages after the first. Effectively, we will have raised the calculated local enthalpy at the stage inlet to its stagnation value. Similarly the neglect of interstage velocities will have raised the temperature estimated to its stagnation value: $T'_{0,i} \approx T_{0T,i}$.

It should be noted, however, that we will be unable to estimate the true inlet local temperature, $T_{0,i}$, from its stagnation equivalent without knowledge of the inlet velocity, which we have decided not to evaluate. The true value of inlet local temperature and the stagnation value are related by equation (14.40), repeated below with the addition of the stage subscript:

$$T_{0T,i} = T_{0,i} + \frac{\frac{1}{2}c_{0,i}^2}{c_p} \approx T'_{0,i} \tag{16.8}$$

and since we do not have any estimate for $c_{0,i}$ (apart from zero), we cannot disentangle the true local temperature, $T_{0,i}$. Moreover, we cannot, without a knowledge of $T_{0,i}$, perform the conversion between local pressure and stagnation pressure using equation (14.44), set down in subscript notation below:

$$\frac{p_{0T,i}}{p_{0,i}} = \left(\frac{T_{0T,i}}{T_{0,i}}\right)^{m/(m-1)} \tag{16.9}$$

But an accurate calculation of the performance of the nozzle requires a knowledge of both the stagnation and the local values of pressure. For example, the equation for mass flow for an impulse stage (e.g. equation (14.63)) will depend on the ratio $p_{1,i}/p_{0T,i}$, where the absence of a pressure drop over the blade implies $p_{1,i} = p_{2,i} = p_{0,i+1}$; unfortunately we will not have $p_{0,i+1}$ available, only the value $p'_{0,i+1}$, which will have to stand for both $p_{0,i+1}$ and $p_{0T,i+1}$. In fact, the assumption of zero interstage velocities means that we cannot distinguish between the stagnation and local values of any of the thermodynamic variables, and must be content with the approximations:

$$\begin{aligned} p_{0,i} &\approx p'_{0,i} \approx p_{0T,i} \\ T_{0,i} &\approx T'_{0,i} \approx T_{0T,i} \\ v_{0,i} &\approx v'_{0,i} \approx v_{0T,i} \end{aligned} \tag{16.10}$$

This will affect the calculations of both flow and specific work. Fortunately the effects are likely to be relatively small, however, because of the small size of the kinetic energy term. Further, there will be a degree of compensation in the case of errors in specific work: a specific work calculated too low in one stage will lead to a larger specific enthalpy at the entry to the next stage, tending to increase the specific work calculated for that stage; and *vice versa*.

16.3 Stage efficiency for an impulse stage

The concept of stage efficiency was introduced in Section 14.3, and combining equations (15.9) and (15.12) allows us to write:

$$\eta_s = \frac{w}{h_0 - h_{2s}} \tag{16.11}$$

In an impulse turbine there is no pressure drop over the blades and so the mid-stage and outlet specific enthalpies are the same. Thus for an isentropic expansion we have

$$h_{1s} = h_{2s} \tag{16.12}$$

Accordingly

$$\eta_s = \frac{w}{h_0 - h_{1s}} \tag{16.13}$$

The specific work, w, for an impulse stage may be written in terms of the blade efficiency, η_B, using equations (15.31) and (15.32):

$$w = \tfrac{1}{2}c_1^2\eta_B \tag{16.14}$$

Hence:

$$\eta_s = \frac{\frac{1}{2}c_1^2}{h_0 - h_{1s}}\eta_B \tag{16.15}$$

From equations (14.3), (14.4) and (14.5), the mid-stage velocity, c_1, is related to the nozzle efficiency by

$$\tfrac{1}{2}c_1^2 - \tfrac{1}{2}c_0^2 = \eta_N(h_0 - h_{1s}) \tag{16.16}$$

For the first stage, the inlet velocity, c_0, will be negligible, and so we may substitute back into equation (16.15) to give simply:

$$\eta_s = \eta_N\eta_B \tag{16.17}$$

For stages after the first, the stage inlet velocity will be approximately equal to the stage outlet velocity:

$$c_0 \approx c_2 \tag{16.18}$$

In addition, in an impulse stage $c_{ro} = c_{ri}$, so that from the velocity diagram, Figure 15.3, the stage outlet velocity, c_2, is related to the mid-stage velocity, c_1, by

$$c_2 \sin\alpha_2 = c_1 \sin\alpha_1 \tag{16.19}$$

and so

$$c_0 \approx c_1\frac{\sin\alpha_1}{\sin\alpha_2} \tag{16.20}$$

Accordingly equation (16.16) becomes

$$\frac{1}{2}c_1^2\left(1-\frac{\sin^2\alpha_1}{\sin^2\alpha_2}\right)\approx\eta_N(h_0-h_{1s}) \qquad (16.21)$$

or

$$\frac{1}{2}c_1^2\approx\frac{\sin^2\alpha_2}{\sin^2\alpha_2-\sin^2\alpha_1}\eta_N(h_0-h_{1s}) \qquad (16.22)$$

Substituting into equation (16.15) gives the stage efficiency for all impulse stages after the first as:

$$\eta_S=\frac{\sin^2\alpha_2}{\sin^2\alpha_2-\sin^2\alpha_1}\eta_N\eta_B \qquad (16.23)$$

Since we can expect $\alpha_2=90°$, this simplifies further to:

$$\eta_S=\frac{1}{\cos^2\alpha_1}\eta_N\eta_B \qquad (16.24)$$

We may evaluate the nozzle efficiency and the blade efficiency using the methods described in Chapters 14 and 15, and so it is possible to calculate the stage efficiency using equations (16.17) and (16.24).

Equation (15.42) gives the maximum blade efficiency for an impulse stage as $\cos^2\alpha_1$. Thus a second or later impulse stage operating at optimum blade conditions will have the same efficiency as its nozzle: $\eta_S=\eta_N$. This can lead to an apparently anomalous situation where the stage efficiency is greater than the blade efficiency. For example, let us take the case of an impulse stage where the nozzle outlet angle is $\alpha_1=20°$, so that the maximum blade efficiency is $\eta_B=0.883$. Suppose the nozzle efficiency is $\eta_N=0.95$, then the stage efficiency under optimum blade conditions ($R_B=\cos\alpha_1/2$) is $\eta_S=0.95$, so that $\eta_S>\eta_B$. The resolution of this apparent paradox lies in the less demanding definition of stage efficiency compared with blade efficiency. A finite outlet velocity is no bar to a stage efficiency of unity, since the definition of stage efficiency makes the assumption that the stage inlet and outlet velocities are equal. By contrast, a blade efficiency of unity requires all the kinetic energy to be used, and a finite outlet velocity will reduce the blade efficiency to a fractional value.

16.4 Stage efficiency for a reaction stage

The stage efficiency for a reaction stage is once again given by equation (16.11), but from equations (15.88) and (15.94) the specific work is now

$$w=\eta_B(\tfrac{1}{2}c_1^2+h_1-h_{2sa}) \qquad (16.25)$$

where h_1 is the mid-stage specific enthalpy (station '1' of Figure 15.8) and h_2a is the specific enthalpy following an isentropic expansion from station '1' to station '2'. Thus the stage efficiency is given by

$$\eta_S=\eta_B\frac{\frac{1}{2}c_1^2+h_1-h_{2sa}}{h_0-h_{2s}} \qquad (16.26)$$

Since the stagnation enthalpy stays constant in the nozzle (see Section 14.2), we may use equation (14.3) in equation (16.26) to produce the alternative statement:

$$\eta_S=\eta_B\frac{\frac{1}{2}c_0^2+h_0-h_{2sa}}{h_0-h_{2s}} \qquad (16.27)$$

If the reaction stage is the first stage in the turbine, then the stage inlet velocity will be negligible, in which case equation (16.27) becomes:

$$\eta_S=\eta_B\frac{h_0-h_{2sa}}{h_0-h_{2s}} \qquad (16.28)$$

For the stages following the first, the inlet velocity and the mid-stage velocity are related by equation (15.100), repeated below:

$$c_0\sin\alpha_0=c_1\sin\alpha_1 \qquad (15.100)$$

Equation (16.16) will still apply, linking mid-stage velocity to the frictionally resisted enthalpy drop over the nozzle:

$$\tfrac{1}{2}c_1^2-\tfrac{1}{2}c_0^2=\eta_N(h_0-h_{1s}) \qquad (16.16)$$

If we use equation (15.100) to eliminate the inlet velocity, c_0, from equation (16.16), the mid-stage velocity emerges as:

$$\frac{1}{2}c_1^2=\frac{\sin^2\alpha_0}{\sin^2\alpha_0-\sin^2\alpha_1}\eta_N(h_0-h_{1s}) \qquad (16.29)$$

Since we may expect $\alpha_0=\alpha_2$, equation (16.29) has exactly the same form as equation (16.22) developed in the previous section.

If, on the other hand, we eliminate the mid-stage velocity, c_1, instead, we achieve:

$$\frac{1}{2}c_0^2=\frac{\sin^2\alpha_1}{\sin^2\alpha_0-\sin^2\alpha_1}\eta_N(h_0-h_{1s}) \qquad (16.30)$$

Substituting from equation (16.30) into equation (16.27) gives the stage efficiency for a reaction stage after the first as:

$$\eta_S=\eta_B\frac{h_0-h_{2sa}}{h_0-h_{2s}}+\eta_N\eta_B\frac{\sin^2\alpha_1}{\sin^2\alpha_0-\sin^2\alpha_1}\frac{h_0-h_{1s}}{h_0-h_{2s}} \qquad (16.31)$$

In the normal case where $\alpha_0=90°$, the expression for stage efficiency simplifies to:

$$\eta_S=\eta_B\frac{h_0-h_{2sa}}{h_0-h_{2s}}+\eta_N\eta_B\tan^2\alpha_1\frac{h_0-h_{1s}}{h_0-h_{2s}} \qquad (16.32)$$

[It may be noted that equations (16.28) and (16.32), which cover cases with any degree of reaction, reduce to the impulse or zero-reaction equations when there is no expansion over the blade, so that $h_{2sa} = h_1, h_{2s} = h_{1s}$ and $\eta_N = (h_0 - h_1)/(h_0 - h_{1s})$. For example, equation (16.32) reduces to:

$$\eta_S = \eta_N\eta_B(1 + \tan^2\alpha_1) = \eta_N\eta_B\frac{1}{\cos^2\alpha_1} \qquad (16.33)]$$

Equations (16.28) and (16.32) require us to evaluate the enthalpy following a number of isentropic expansions: as the pressure drops from p_0 to p_1 over the nozzle, as the pressure drops from p_1 to p_2 over the blade, and as the pressure drops from p_0 directly to p_2 over the complete stage. But the isentropic expansion over the blade, with pressure dropping from p_1 to p_2, comes after a frictionally resisted expansion over the nozzle. Hence we need to develop a method of determining the enthalpy following both isentropic and frictionally resisted expansions. Methods of computing downstream specific enthalpy following both isentropic and frictionally resisted expansions will be discussed in the next section for both steam and gas.

Before we can make a start, we need the mid-stage pressure, and in this chapter's spirit of simplification, we may use the approximate equation (15.27), repeated below, to determine the mid-stage pressure from the inlet and outlet pressures:

$$\rho \approx \frac{p_1 - p_2}{p_0 - p_2} \qquad (15.27)$$

so that

$$p_1 = p_2 + \rho(p_0 - p_2) \qquad (16.34)$$

16.5 Evaluation of downstream enthalpies following isentropic and frictionally resisted expansions

We begin by developing a mathematical formula for the change in specific entropy over the expansion from general thermodynamic relationships. We will then consider how the resulting equation may be solved for the three cases of

(i) steam
(ii) a real gas
(iii) an ideal gas.

From equation (3.4) in Section 3.2, the change in specific entropy is given in terms of a heat increment, dq, added to a system at temperature, T:

$$T\,ds = dq \qquad (16.35)$$

But the first law of thermodynamics gives the term dq as:

$$dq = du + p\,dv \qquad (3.10)$$

Since specific enthalpy is given by equation (3.6) as $h = u + pv$, it follows that

$$dq = dh - v\,dp \qquad (16.36)$$

The specific volume, v, follows from equation (3.2) as:

$$v = \frac{ZR_wT}{p} \qquad (16.37)$$

Further, from the definition of specific heat at constant pressure (equation 3.18), we have:

$$dh = c_p\,dT \qquad (16.38)$$

Combining equations (16.35) to (16.38) allows us to write the differential of specific entropy as:

$$ds = \frac{c_p}{T}dT - \frac{ZR_w}{p}dp \qquad (16.39)$$

We will wish integrate this equation from general conditions (s_A, T_A, p_A) at the start of the expansion to general conditions (s_B, T_B, p_B) at the end:

$$\int_{s_A}^{s_B} ds = s_B - s_A = \int_{T_A}^{T_B}\frac{c_p}{T}dT - R_w\int_{p_A}^{p_B}\frac{Z}{p}dp \qquad (16.40)$$

where we have taken the term $R_w = R/w$ outside the integrand since it is independent of pressure: R is the universal gas constant and w is the molecular weight of the gas being expanded.

16.5.1 Evaluation of the entropy integral for steam

A problem arises when we attempt to integrate the right-hand side of equation (16.40) for steam, since we run into two difficulties immediately: the specific heat, c_p, and the compressibility factor, Z, are each fairly strong functions of both pressure and temperature. It turns out that an analytic integration of equation (16.40) cannot achieve accurate results over any extended region of the pressure–temperature plain. Instead, it is necessary to construct a two-dimensional table based on experiment to list specific volume, enthalpy and entropy against temperature over a range of pressures for superheated steam. A further table is needed to list the same thermodynamic properties against temperature, and, equivalently, pressure in the saturated region. These three tables together are

Table 16.1 Excerpt from a typical steam table for superheated steam

p (bar)		39			40	
T (°C)	*v* m³/kg	*h* kJ/kg	*s* kJ/kg K	*v* m³/kg	*h* kJ/kg	*s* kJ/kg K
400	0.0753	3215	6.782	0.0733	3214	6.769
410	0.0767	3239	6.817	0.0747	3237	6.803
420	0.0781	3262	6.850	0.0760	3260	6.837
430	0.0794	3285	6.884	0.0773	3283	6.870
440	0.0808	3308	6.916	0.0787	3307	6.903
450	0.0821	3331	6.948	0.0800	3330	6.935
460	0.0834	3354	6.980	0.0812	3353	6.967
470	0.0847	3377	7.011	0.0825	3376	6.998
480	0.0860	3400	7.042	0.0838	3399	7.029
490	0.0873	3423	7.072	0.0851	3422	7.059

known as the 'steam tables'. An excerpt from a typical steam table in the superheat region is listed in Table 16.1.

Specific enthalpy and specific entropy are measured relative to the properties of saturated water at its triple point: both specific enthalpy and specific entropy are taken by convention to be zero for saturated, liquid water at a temperature of 0.01°C.

Although the steam tables list steam properties against pressure and temperature, the fact that specific enthalpy and specific entropy are themselves thermodynamic variables means that it is possible to construe the steam tables as providing specific enthalpy as a tabular function of pressure and specific entropy. Alternatively, we may regard the tables as providing specific entropy as a tabular function of pressure and specific enthalpy. These two statements may be summarized mathematically by the two equations:

$$\begin{aligned} h &= h(p, s) \\ s &= s(p, h) \end{aligned} \tag{16.41}$$

We may apply this interpretation to find the specific enthalpy following an isentropic expansion, given the initial specific enthalpy, h_A, the initial pressure, p_A, and the final pressure, p_B. The specific entropy in state 'A' is found by first moving to the column where the pressure is p_A, finding the row at which the specific enthalpy is h_A, and then reading off the specific entropy, s_A, on the same row:

$$s_A = s(p_A, h_A) \tag{16.42}$$

To find the specific enthalpy in state 'B', we select the column for the final pressure, p_B, find the row where the specific entropy is the same as at the beginning of the expansion, s_A, and read off the specific enthalpy along this row. This will be the desired value following an isentropic expansion:

$$h_{Bs} = h(p_B, s_A) \tag{16.43}$$

where the additional subscript, s, signifies that the expansion was occurred isentropically.

We treat the real case where the expansion is frictionally resisted with efficiency, η, by first carrying out the tabular exercise for an isentropic expansion as above and then proceeding as follows. Since the real enthalpy drop will be the isentropic enthalpy drop multiplied by η, the actual specific enthalpy at the end of the expansion will be:

$$h_B = h_A - \eta(h_A - h_{Bs}) \tag{16.44}$$

The value of specific entropy at the end of the expansion, s_B, may be found by moving further down the p_B column until the row containing specific enthalpy, h_B, is found. The specific entropy, s_B, is then read off this row:

$$s_B = s(p_B, h_B) \tag{16.45}$$

Equation (16.45) is exactly analogous to equation (16.42).

In practice, the tabulation of steam properties in steam tables is not normally fine enough for us to be able to read off accurate values using the basic method of inspecting the existing columns and rows described above. Excellent accuracy may be achieved by using interpolation, however.

It is possible to computerize the steam tables and software packages have been written to do just this. Such packages should be rapid in execution, although the storage requirement will be substantial. One alternative is to store the data only for the regions of relevance to the turbine simulation. Another is to use some simple approximating functions, as explained in Section 16.6.

16.5.2 Evaluation of the entropy integral for a real gas

Gas turbines will pass a mixture of gases: normally dominated by air but also containing the products of hydrocarbon combustion. It is found that the specific heat is a function of temperature only, and the gas compressibility factor stays very close to unity, $Z \approx 1$, for all industrial applications. This latter fact allows the second term on the right-hand side of equation (16.40) to be integrated analytically:

$$\begin{aligned} s_B - s_A &= \int_{T_A}^{T_B} \frac{c_p}{T}\,dT - R_w \int_{p_A}^{p_B} \frac{dp}{p} \\ &= \int_{T_A}^{T_B} \frac{c_p}{T}\,dT - R_w \ln \frac{p_B}{p_A} \end{aligned} \tag{16.46}$$

We may express the remaining definite integral as the difference of integrals between 0 and T_B and 0 and T_A:

$$s_B - s_A = \int_0^{T_B} \frac{c_p}{T}\,dT - \int_0^{T_A} \frac{c_p}{T}\,dT - R_w \ln \frac{p_B}{p_A} \tag{16.47}$$

Let us now introduce a new thermodynamic variable, which is a function of temperature only:

$$\phi(T) = \int_0^{T} \frac{c_p}{T}\,dT \tag{16.48}$$

This variable is the temperature-dependent component of specific entropy, defined to be zero at a temperature of absolute zero. We will call it simply 'phi', J/(kgK). Substituting back into equation (16.47) gives:

$$s_B - s_A = \phi_B - \phi_A - R_w \ln \frac{p_B}{p_A} \tag{16.49}$$

It has been found more convenient to list phi rather than specific entropy in gas tables, which are thereby made independent of pressure.

The tables-based procedure for analysing a gas expansion is similar to that used for steam. Let us consider once more finding the specific enthalpy following an isentropic expansion, given the initial specific enthalpy, h_A, the initial pressure, p_A, and the final pressure, p_B.

Specific enthalpy is listed as a function of temperature, and we find the temperature row corresponding to the initial specific enthalpy, h_A. We now read off the value of phi, ϕ_A, in the same row. For an isentropic expansion, $s_B = s_A$, hence from equation (16.49), the value of phi at the end of the isentropic expansion is given by:

$$\phi_{Bs} = \phi_A + R_w \ln \frac{p_B}{p_A} \tag{16.50}$$

Thus ϕ_{Bs} will be less than ϕ_A for an expansion, and greater for a compression. We now find the temperature row containing the value calculated for ϕ_{Bs}, and take the value of specific enthalpy from the same row. This will be h_{Bs}.

In the real case where the expansion is frictionally resisted with efficiency, η, the actual specific enthalpy at the end of the expansion will be as given by equation (16.44):

$$h_B = h_A - \eta(h_A - h_{Bs}) \tag{16.44}$$

The value of phi after a frictionally resisted expansion, ϕ_B, will be found in the same temperature row as h_B.

Gas tables are generally more precise than steam tables because of the need to tabulate only against temperature, rather than temperature and pressure. However, interpolation may be needed to give the best possible accuracy.

The procedure above has been chosen because of its close relationship to the entropy relations used in steam tables. It may be noted in passing that an alternative procedure exists for applying the gas tables to expansion problems, based on the use of tabulated relative pressures. This procedure may be regarded as marginally more attractive because it replaces the calculation and subsequent addition of $R_w \ln(p_B/p_A)$ by a calculation of $p_r p_B/p_A$, where p_r is the tabulated relative pressure. An explanation of such a procedure will normally accompany the gas tables.

Finally, it should be noted that the modeller faces a very substantial task if he wishes to store significant portions of the gas tables for use in his computer simulation. Fortunately, the behaviour of the main gases employed in gas turbines is fairly close to ideal, and the control engineer may usually avoid the use of gas tables altogether and work with the very much simpler expressions arising from treating the turbine gas as a ideal gas. This option is explained in the next section.

16.5.3 Evaluation of the entropy integral for an ideal gas

Since both c_p and R_w are constant for an ideal gas and $Z = 1$, integrating equation (16.39) produces

$$\begin{aligned} s_B - s_A &= c_P \int_{T_A}^{T_B} \frac{dT}{T} - R_w \int_{P_A}^{P_B} \frac{dp}{p} \\ &= c_p \ln \frac{T_B}{T_A} - R_w \ln \frac{p_B}{p_A} \end{aligned} \tag{16.51}$$

Thus for an isentropic expansion, we have

$$c_p \ln\left(\frac{T_{Bs}}{T_A}\right) = R_w \ln\left(\frac{p_B}{p_A}\right) \tag{16.52}$$

so that

$$\frac{T_{Bs}}{T_A} = \left(\frac{p_B}{p_A}\right)^{(R_w/c_p)} \tag{16.53}$$

where T_{Bs} is the temperature following an isentropic expansion. Since, for an ideal gas (see equation (14.39)):

$$\frac{R_w}{c_p} = \frac{\gamma - 1}{\gamma} \tag{16.54}$$

equation (16.53) may be re-stated as:

$$\frac{T_{Bs}}{T_A} = \left(\frac{p_B}{p_A}\right)^{(\gamma-1)/\gamma} \tag{16.55}$$

This is the same result as would have been found by combining the equations $pv = R_w T$ and $pv^\gamma =$ constant, as we would expect.

Integrating the enthalpy differential between starting and end conditions gives:

$$h_{Bs} - h_A = c_p(T_{Bs} - T_A) = c_p T_A \left(\frac{T_{Bs}}{T_A} - 1\right) \tag{16.56}$$

so that the isentropic specific enthalpy is given by:

$$h_{Bs} = h_A - c_p T_A \left(1 - \left(\frac{p_B}{p_A}\right)^{(\gamma-1)/\gamma}\right) \tag{16.57}$$

The final enthalpy following a frictionally resisted expansion is given by:

$$\begin{aligned} h_B &= h_A - \eta(h_A - h_{Bs}) \\ &= h_A - \eta c_p T_A \left(1 - \left(\frac{p_B}{p_A}\right)^{(\gamma-1)/\gamma}\right) \end{aligned} \tag{16.58}$$

16.6 Analytic functions linking entropy and enthalpy for saturated and superheated steam

Although the functions $h(p, s)$ and $s(p, h)$ are tabulated in steam tables, unless a steam-tables software package is available, the modeller will face a very considerable workload if he attempts to store this information in a computer in the right form for retrieval for use in a simulation program. An alternative is to use analytic approximating functions.

Since they are thermodynamic variables, specific entropy and specific enthalpy may be regarded as functions of the two variables: temperature and pressure. However, pressure is a single-valued function of temperature along the saturated boundary, and so the specific entropy and enthalpy of saturated steam are themselves single-valued functions of temperature. Hence specific enthalpy may be plotted as a single-valued function of specific entropy in saturated conditions. This has been done in Figure 16.1.

The specific enthalpy of saturated water rises with specific entropy until it reaches a value of 2084 kJ/kg at a specific entropy of 4.43 kJ/(kgK) at the critical point, where the thermodynamic properties of saturated water and saturated steam are the same. The continuation of the graph constitutes the saturated steam line.

The equation set (16.41) requires the conversion from specific entropy at a given pressure to specific enthalpy at that pressure and vice versa. It turns out that we may derive a simple analytical expression for the rate of change of enthalpy with entropy. Then we may integrate from the base line of saturated steam conditions, along a line of constant pressure to find the appropriate value of specific enthalpy as a function of specific entropy. We are aided in these conversions by the fact that the enthalpy/entropy line for saturated steam is a fairly uncomplicated function, which may be approximated well by a low-order polynomial.

Returning to equation (16.40), it is clear that the second term on the right-hand side drops out when the integration proceeds along an isobar. Let us integrate at constant pressure, p, from the saturation line,

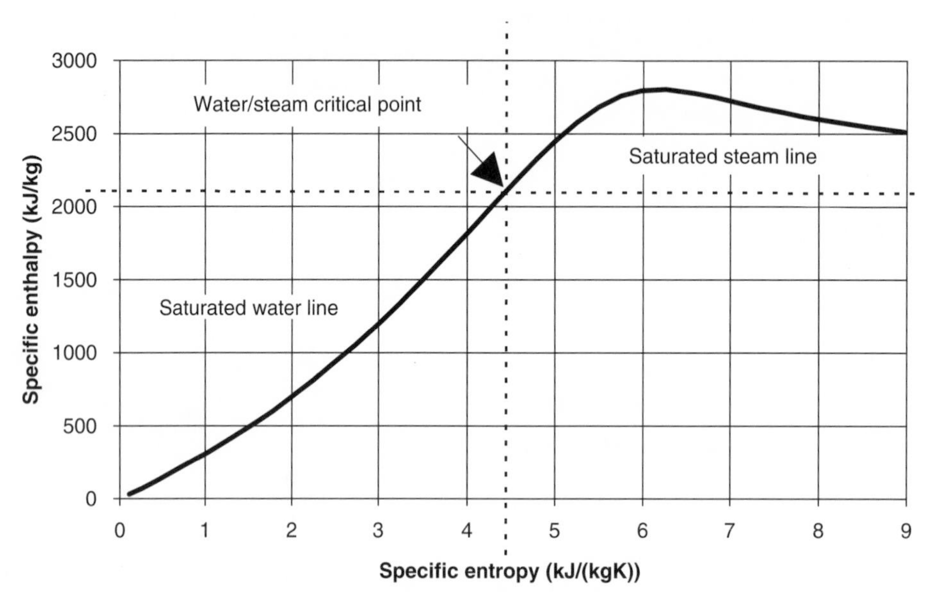

Figure 16.1 Specific enthalpy vs. specific entropy at saturation for water and steam.

where the temperature is $T_{sat} = T_{sat}(p)$ and the specific enthalpy is $s_g = s_g(T_{sat})$ to general end conditions, T, s. Equation (16.40) then becomes:

$$\int_{s_g}^{s} ds = \int_{T_{sat}}^{T} \frac{c_p}{T} dT \tag{16.59}$$

We shall now integrate this expression from the saturated h-versus-s line into the superheated region, and then consider separately the case of the wet-steam region.

16.6.1 Integrating into the superheated region

While for an ideal gas, c_p will have a constant value, c_p for superheated steam will depend on both temperature and pressure. We will assume that it is possible to fix an average value of specific heat for steam at each pressure that takes into account the sometimes substantial variations with temperature over the path of integration. Hence equation (16.59) becomes

$$\int_{s_g}^{s} ds = c_p \int_{T_{sat}}^{T} \frac{1}{T} dT \tag{16.60}$$

which produces the result:

$$s - s_g = c_p \ln\left(\frac{T}{T_{sat}}\right) \tag{16.61}$$

Meanwhile, we may also integrate equation (16.38) with respect to temperature from saturation conditions under the assumption that the specific heat is independent of temperature. This gives:

$$\begin{aligned} h - h_g &= c_p(T - T_{sat}) \\ &= c_p T_{sat}\left(\frac{T}{T_{sat}} - 1\right) \end{aligned} \tag{16.62}$$

where h_g is the specific enthalpy (J/kg) of saturated steam at the chosen pressure. We may now eliminate the temperature, T, from equations (16.61) and (16.62) to give the conversion equations:

$$h = h_g + c_p T_{sat}\left(\exp\left(\frac{s - s_g}{c_p}\right) - 1\right) \tag{16.63}$$

and

$$s = s_g + c_p \ln\left(1 + \frac{h - h_g}{c_p T_{sat}}\right) \tag{16.64}$$

It will be noted that these conversion equations depend on a temperature-invariant, average specific heat of steam at constant pressure, and such a figure will apply only at one temperature in the superheat region. Fortunately the specific heat terms in each equation undergo a significant degree of cancellation, with the result that the variations in specific heat due to temperature have a relatively small effect.

Let us assume that our estimate of average specific heat over the integration path contains an error, ε, so that $c_p \rightarrow c_p + \varepsilon$. Making use of the exponential expansion:

$$\exp(x) = 1 + x + \frac{x^2}{2!} + \frac{x^3}{3!} + \cdots \qquad \text{for } x^2 < \infty \tag{16.65}$$

in equation (16.63) gives:

$$\begin{aligned} h &= h_g + (c_p + \varepsilon)T_{sat} \\ &\quad \times \left(\frac{s - s_g}{c_p + \varepsilon} + \frac{(s - s_g)^2}{2(c_p + \varepsilon)^2} + \frac{(s - s_g)^3}{6(c_p + \varepsilon)^3} + \cdots\right) \\ &= h_g + T_{sat}(s - s_g) \\ &\quad \times \left(1 + \frac{1}{2}\frac{(s - s_g)}{(c_p + \varepsilon)} + \frac{1}{6}\frac{(s - s_g)^2}{(c_p + \varepsilon)^2} + \cdots\right) \end{aligned} \tag{16.66}$$

Since for steam turbine operation we may assume $c_p \geq 2000$ J/(kg K) and $s - s_g < 2000$ J/(kg K), squared and higher terms in (16.66) will contribute relatively little to the final figure, and may be neglected for the purpose of estimating the effect of the error in specific heat on the calculation of specific enthalpy. Assuming the error in average specific heat is fractional, we may expand using the binomial expansion:

$$\frac{1}{c_p + \varepsilon} = \frac{1}{c_p}\left(1 + \frac{\varepsilon}{c_p}\right)^{-1} \approx \frac{1}{c_p}\left(1 - \frac{\varepsilon}{c_p}\right) \tag{16.67}$$

Hence, substituting back into equation (16.66) gives:

$$h = h_g + T_{sat}(s - s_g) + T_{sat}\frac{(s - s_g)^2}{2c_p} - T_{sat}\frac{(s - s_g)^2}{2c_p}\frac{\varepsilon}{c_p} \tag{16.68}$$

Thus the error in specific heat causes an error of the order of

$$\frac{1}{2} T_{sat} \frac{(s - s_g)^2}{c_p^2} \varepsilon \tag{16.69}$$

in the calculation of specific enthalpy. Taking the case of superheated steam at 3 MPa and 550°C, the specific entropy is 7.374 kJ/(kg K) and the specific enthalpy is 3569 kJ/kg. The saturation temperature is $233.8 + 273.15 = 506.95$ K, the saturation specific entropy is 6.186 kJ/(kg K), and the saturation specific enthalpy is 2803 kJ/kg. Thus the specific heat averaged over the interval from the saturation line to the superheated conditions, 3 MPa, 550°C, is given by:

$$\begin{aligned} &(3569 \times 10^3 - 2803 \times 10^3)/(550 - 233.8) \\ &= 2422 \text{ J/(kg K)} = 2.422 \text{ kJ/(kg K)} \end{aligned}$$

Substituting these figures into the expression (16.69), we see that even an error as large as 25% in average specific heat will lead to an error in estimated specific enthalpy of about 37 kJ/kg, which represents an error in specific enthalpy of only about 1%.

A similar sensitivity analysis may be applied to equation (16.64) to show that quite sizeable errors in specific heat may be accommodated without causing large errors in calculated specific entropy.

16.6.2 Integrating into the wet-steam region

Here we begin not with equation (16.40), but with its precursor (16.39), repeated below:

$$ds = \frac{c_p}{T}\,dT - \frac{ZR_w}{P}\,dp \tag{16.39}$$

Substituting equation (16.38) into (16.39) and setting $dp = 0$ for variations along an isobar, we find the very simple relation:

$$dh = T\,ds \tag{16.70}$$

The integration of equation (16.70) along a line of constant pressure in the wet-steam region is extremely easy, because constant pressure implies constant temperature. Hence we have:

$$\int_{h_g}^{h} dh = T\int_{s_g}^{s} ds \tag{16.71}$$

which gives the simple conversion equations:

$$h = h_g + T(s - s_g) \tag{16.72}$$

and

$$s = s_g + \frac{h - h_g}{T} \tag{16.73}$$

The value of specific enthalpy in the wet region will be less than the specific enthalpy of saturated steam, being equal to the weighted average of the specific enthalpies of saturated steam and saturated water:

$$h = xh_g + (1 - x)h_f \tag{16.74}$$

Here x is the dryness fraction, which may be evaluated from equation (16.74) when the other quantities are available, or from the similar equation for specific entropy:

$$s = xs_g + (1 - x)s_f \tag{16.75}$$

16.6.3 Examples of analytic approximating functions

The following functions for the properties of saturated steam and water are approximately valid up to a pressure of 4 MPa/ temperature of 250°C, typical of the range needed to model the turbine driving a boiler feedpump in a power station. An Antoine formulation (see Section 12.3) gives:

$$T_c = \frac{-3858.3568}{\ln 10^{-5}p - 11.734649} - 229.21651 \tag{16.76}$$

where $T_c = T_{sat} - 273.15$ is the saturated temperature in degrees centrigrade and the pressure, p, is in pascals. Given the saturation temperature in °C, the following third-order polynomials give the specific entropy (J/(kgK)) of saturated steam, s_g, and saturated water, s_f:

$$s_g = 9157.7 - 24.9496T_c + 8.105556 \times 10^{-2}T_c^2 - 1.2192501 \times 10^{-4}T_c^3 \tag{16.77}$$

$$s_f = -0.2 + 15.122407T_c - 2.3580247 \times 10^{-2}T_c^2 + 3.1115684 \times 10^{-5}T_c^3 \tag{16.78}$$

The specific enthalpy of saturated steam (J/kg) is given over the same range of temperature and pressure by

$$h_g = 2.5016 \times 10^6 + 1.772778 \times 10^3 T_c + 1.11111111T_c^2 - 1.3786009 \times 10^{-2}T_c^3 \tag{16.79}$$

Average specific heat at constant pressure (J/(kgK)) may be approximated up to 4 MPa by the linear formulation:

$$c_p = 2000 + 1.5 \times 10^{-4}p \tag{16.80}$$

where the pressure, p, is in Pa.

The accuracy of these formulations may be judged by Tables 16.2 and 16.3.

Table 16.2 Saturated steam: approximating functions compared with steam tables

	Pressure (MPa)	*Saturated temperature (°C)*	*Saturated specific enthalpy (kJ/kg)*	*Saturated specific entropy (J/(kgK))*
Tables	0.5	151.8	2749	6822
Approximation		151.8	2748	6811
Tables	1.0	179.9	2778	6586
Approximation		179.9	2776	6583
Tables	2.0	212.4	2799	6340
Approximation		212.3	2796	6348
Tables	3.0	233.8	2803	6186
Approximation		233.8	2801	6197
Tables	4.0	250.3	2801	6070
Approximation		250.3	2799	6079

Table 16.3 Specific enthalpy and entropy for superheated steam: comparison of approximating function results with steam tables

	p *MPa*	*T* °*C*	*h* *kJ/kg*	*s* *J/kg K*	*T* °*C*	*h* *kJ/kg*	*s* *J/kg K*	*T* °*C*	*h* *kJ/kg*	*s* *J/kg K*
Tables	0.5	160	2768	6864	300	3065	7460	450	3377	7944
Approx.			2771	6857		3072	7448		3389	7928
Tables	1.0	200	2829	6695	350	3158	7301	500	3478	7761
Approx.			2828	6697		3162	7294		3487	7750
Tables	2.0	250	2904	6547	400	3248	7126	550	3578	7570
Approx.			2897	6560		3246	7129		3579	7568
Tables	3.0	250	2858	6289	400	3231	6921	550	3569	7374
Approx.			2848	6308		3228	6926		3567	7377
Tables	4.0	300	2963	6364	450	3330	6935	600	3674	7368
Approx.			2957	6375		3330	6936		3672	7370

Table 16.2 compares the results of equations (16.76), (16.77) and (16.79) with steam tables. It will be seen that there is a close correspondence. Table 16.3 uses the calculated results of Table 16.1 together with the enthalpy/entropy conversion equations (16.63) and (16.64). The value of specific heat needed for the approximations was found using equation (16.80).

Entropy/enthalpy and enthalpy/entropy conversions were carried out at several pressures in the range 0 to 4 MPa. First the value of specific entropy from the table was converted to give a calculated value of specific enthalpy at the given pressure, and then the value of specific enthalpy from the table was converted to give a calculated value of specific entropy. The calculations were performed at three points for each pressure: close to saturated conditions, then at a superheat of about 150°C and then at a superheat of about 300°C. It will be seen that the approximating functions produce values close to those given in the steam tables, despite the approximation inherent in assuming specific heat is independent of the degree of superheat.

The same form of equations may be used to model the steam conditions for the main turbine of a power station, where the pressure may be in the range 10 to 16 MPa. The greater spread of average specific heat at higher steam pressures may, however, require two or three values of specific heat to be used in equations (16.63) and (16.64) if the same level of accuracy is to be retained.

16.7 Specific volume at stage outlet

Once the overall stage efficiency has been calculated, it is possible to determine the specific volume at stage outlet from the inlet and outlet pressures and the specific volume at stage inlet. Here we use the equations developed in Section 15.3, namely:

$$pv^{m_s} = \text{constant} \tag{15.17}$$

where

$$m_s = \frac{\gamma}{\gamma - \eta_s(\gamma - 1)} \tag{15.18}$$

16.8 Simplifying the calculation of mass flow

The mass flow for a convergent-only nozzle is given by equation (14.63), repeated below:

$$W = A_t\sqrt{2\frac{\gamma}{\gamma-1}\frac{p_{0T}}{v_{0T}}\left(\left(\frac{p_1}{p_{0T}}\right)^{2/m_0} - \left(\frac{p_1}{p_{0T}}\right)^{(m_0+1)/m_0}\right)}$$

$$\text{for } \frac{p_1}{p_{0T}} \geq \left(\frac{2}{\gamma+1}\right)^{m_{0c}/(m_{0c}-1)}$$

$$= A_t\sqrt{\left(\frac{2}{\gamma+1}\right)^{(m_{0c}+1)/(m_{0c}-1)}\gamma\frac{p_{0T}}{v_{0T}}}$$

$$\text{for } \frac{p_1}{p_{0T}} < \left(\frac{2}{\gamma+1}\right)^{m_{0c}/(m_{0c}-1)} \tag{14.63}$$

where A_t is the throat (and also outlet) area, m_0 is the exponent of (frictionally resisted) expansion to the throat, while m_{0c} is the index of expansion at the point where the flow becomes choked, normally equal to m_0. Similarly, the mass flow through a convergent–divergent nozzle is given by equation (14.110), repeated below

$$W = A_1\sqrt{2\frac{\gamma}{\gamma-1}\frac{p_{0T}}{v_{0T}}\left(\left(\frac{p_1}{p_{0T}}\right)^{2/m} - \left(\frac{p_1}{p_{0T}}\right)^{(m+1)/m}\right)}$$

$$\text{for } \frac{p_1}{p_{0T}} \geq \frac{p_{1\,crit\,1}}{p_{0T}}$$

$$= A_t \sqrt{\left(\frac{2}{\gamma+1}\right)^{(m_{0c}+1)/(m_{0c}-1)} \gamma \frac{p_{0T}}{v_{0T}}}$$

$$\text{for } \frac{p_1}{p_{0T}} < \frac{p_{1\,crit\,1}}{p_{0T}} \quad (14.110)$$

where A_t is the area of the throat, at the junction of the convergent and divergent sections, A_1 is the area of the nozzle outlet and m is the index of expansion over the convergent–divergent nozzle as a whole. The upper critical discharge pressure ratio, $p_{1\,crit\,1}/p_{0T}$, is the larger solution of the two solutions, $p_{1\,crit\,1}/p_{0T}$ and $p_{1\,crit\,2}/p_{0T}$ of the implicit equation (see equation (14.75)):

$$\left(\frac{p_{1crit}}{p_{0T}}\right)^{2/m} - \left(\frac{p_{1crit}}{p_{0T}}\right)^{(m+1)/m} = \left(\frac{A_t}{A_1}\right)^2 \frac{\gamma-1}{2} \left(\frac{2}{\gamma+1}\right)^{(m_{0c}+1)/(m_{0c}-1)} \quad (16.81)$$

A simplification may be introduced as a result of the fact that all the nozzle efficiencies are likely to be near unity, at least near the design point. Hence we may use the approximation:

$$m \approx m_0 \approx m_{0c} \approx \gamma \quad (16.82)$$

Hence equation (14.63) for mass flow through a convergent-only nozzle becomes:

$$W = A_t \sqrt{2 \frac{\gamma}{\gamma-1} \frac{p_{0T}}{v_{0T}} \left(\left(\frac{p_1}{p_{0T}}\right)^{2/\gamma} - \left(\frac{p_1}{p_{0T}}\right)^{(\gamma+1)/\gamma}\right)}$$

$$\text{for } \frac{p_1}{p_{0T}} \geq \left(\frac{2}{\gamma+1}\right)^{\gamma/(\gamma-1)}$$

$$= A_t \sqrt{\left(\frac{2}{\gamma+1}\right)^{(\gamma+1)/(\gamma-1)} \gamma \frac{p_{0T}}{v_{0T}}}$$

$$\text{for } \frac{p_1}{p_{0T}} < \left(\frac{2}{\gamma+1}\right)^{\gamma/(\gamma-1)} \quad (16.83)$$

In a similar way, the simplified equations for mass flow through a convergent–divergent nozzle are:

$$W = A_1 \sqrt{2 \frac{\gamma}{\gamma-1} \frac{p_{0T}}{p_{0T}} \left(\left(\frac{p_1}{p_{0T}}\right)^{2/\gamma} - \left(\frac{p_1}{p_{0T}}\right)^{(\gamma-1)/\gamma}\right)}$$

$$\text{for } \frac{p_1}{p_{0T}} \geq \frac{p_{1\,crit\,1}}{p_{0T}}$$

$$= A_t \sqrt{\left(\frac{2}{\gamma+1}\right)^{(\gamma+1)/(\gamma-1)} \gamma \frac{p_{0T}}{v_{0T}}}$$

$$\text{for } \frac{p_1}{p_{0T}} < \frac{p_{1\,crit\,1}}{p_{0T}} \quad (16.84)$$

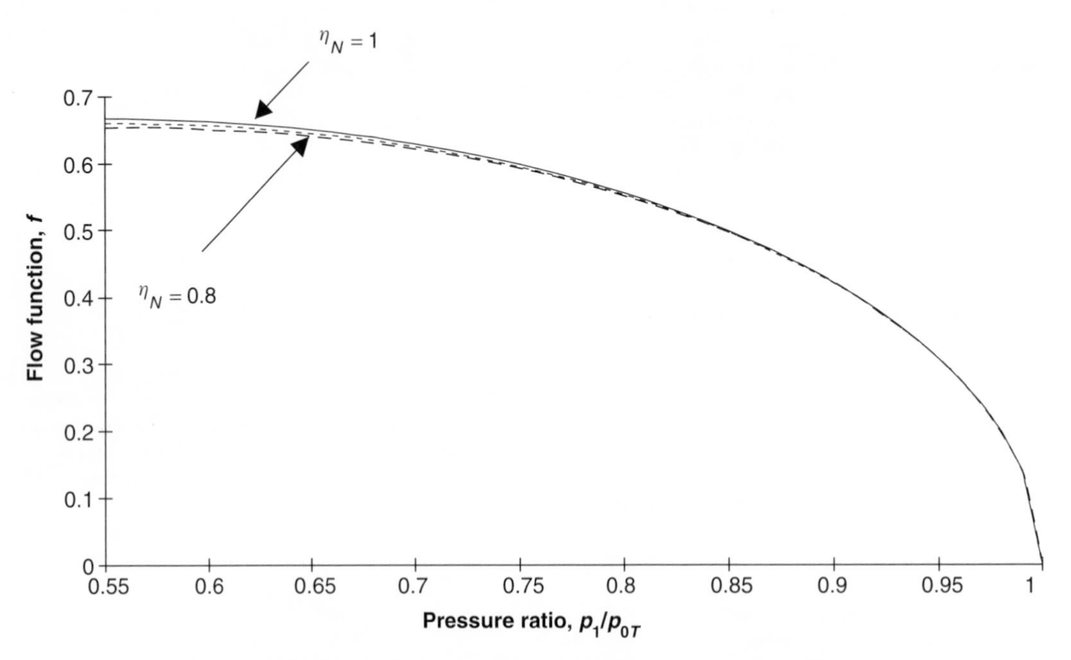

Figure 16.2 Behaviour of the mass flow function for a range of nozzle efficiencies: $\eta_N = 1.0$, 0.9 and 0.8.

where the upper critical discharge ratio, $p_{1\,cirt\,1}/p_{0T}$, is the larger solution of the equation:

$$\left(\frac{p_{1crit}}{p_{0T}}\right)^{2/\gamma} - \left(\frac{p_{1crit}}{p_{0T}}\right)^{(\gamma+1)/\gamma} = \left(\frac{A_t}{A_1}\right)^2 \frac{\gamma-1}{2}\left(\frac{2}{\gamma+1}\right)^{(\gamma+1)/(\gamma-1)} \quad (16.85)$$

Equation (16.85) possesses the advantage that it is independent of efficiency calculations. Hence $p_{1\,crit\,1}/p_{0T}$ may now be computed off-line from the main simulation in all cases where γ is constant. The only exception is the case of wet steam, where the effective value of γ depends on the dryness fraction, x.

The inaccuracy introduced into the flow calculation is likely to be small in practice. Let us consider the case of a convergent-only nozzle, and consider the effect of nozzle efficiencies of 90% and 80% – significantly less than we could reasonably expect in a well-designed turbine. Assuming the nozzle is passing superheated steam, we may take $\gamma = 1.3$. Computing the value of m from equation (14.21) we find that $m = 1.26$ for $\eta_N = 0.9$, while $m = 1.23$ for $\eta_N = 0.8$. We may now compare the behaviour of the mass flow function:

$$f\left(\frac{p_1}{p_{0T}}, \eta_N\right) = \sqrt{2\frac{\gamma}{\gamma-1}\left(\left(\frac{p_1}{p_{0T}}\right)^{2/m} - \left(\frac{p_1}{p_{0T}}\right)^{(m+1)/m}\right)} \quad (16.86)$$

at each efficiency with the base case where the expansion is isentropic ($\eta_N = 1$, $m = \gamma$). This is done in Figure 16.2, which shows what a small difference is made to the mass flow function even when the nozzle efficiency is dropped to the low value of 80%.

The maximum difference in the mass flow functions for the three cases shown in Figure 16.2 is about 2%.

16.9 Calculation sequence for the simplified turbine model

We shall now demonstrate how the simplified model may be applied to calculate the behaviour of the N stage turbine considered in Section 15.14. Figure 15.10, which depicts the turbine, is repeated below.

As before, we may regard the boundary pressures, $p_{up}, p_{E,1}, p_{E,2}, \ldots, p_{E,i}, \ldots, p_{E,N}$, as governed solely by the outputs of integrators (local mass and temperature), so that they will be fixed at any time instant. The same will apply to the upstream specific enthalpy, h_{up}, the upstream specific volume, v_{up}, and the turbine speed, N. The steady-state flow balance equations are:

$$\begin{aligned} W_{in} - W_1 &= 0 \\ W_i - W_{i+i} - W_{E,i} &= 0 \quad \text{for } 1 \le i \le N-1 \\ W_N - W_{E,N} &= 0 \end{aligned} \quad (15.113)$$

The variables necessary to determine the flows coming into the turbine and leaving the turbine are the upstream pressure, the upstream specific volume, the downstream pressure and the valve position. See Chapters 6 and 10 for details of calculational methods. The internal mass flows, W_i, $i = 1, \ldots, N$ are calculated as the flows through each nozzle, and this procedure will need to be carried out iteratively at each timestep. Calculation of the overall performance of the turbine at each timestep may follow the following sequence.

1. Provide a first estimate of the stage inlet pressures, $p_{0,i}$, $i = 1, \ldots, N$, and the last stage outlet pressure, $p_{2,N}$ (initially from a best estimate of the steady-state conditions, and then from the value at the last timestep).

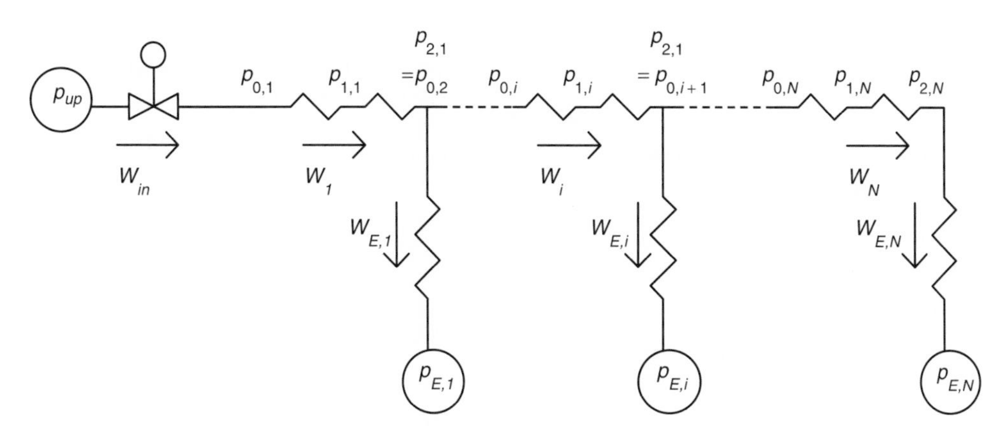

Figure 15.10 Flows through a turbine.

2. Calculate the mid-stage pressures, $p_{1,i}$. In the case of an impulse stage these will be the same as the inlet pressure to the next stage. Use equation (16.34) for a reaction stage.

3. Calculate the flow through the turbine control valve using the methods of Chapters 6 and 10. Enthalpy is unchanged by the throttling process, and so the specific enthalpy at the inlet to the first stage will be equal to the upstream specific enthalpy:

$$h_{0,1} = h_{up} \tag{16.87}$$

where the subscript 'up' has been used to denote the conditions upstream of the turbine control valve. Similarly the specific volume at the inlet to the first stage will reflect the fact that the temperature is kept constant:

$$v_{0,1} = v_{up}\frac{p_{up}}{p_{0,1}} \tag{16.88}$$

Stage calculations

4. If the working fluid is steam, calculate the specific entropy at stage inlet from the inlet specific enthalpy and the inlet pressure using either steam tables or equation (16.64). Check whether the steam is initially superheated or saturated by comparing the inlet specific entropy, s_0, with the saturated specific entropy at inlet pressure, $s_g(p_0)$. If $s_{0,1} \leq s_g(p_0)$, use equation (16.75) to estimate the initial dryness fraction and Zeuner's equation (15.112) to estimate the isentropic index, γ, for the first stage.

5. Calculate the mass flow through the first stage, which is equal to the mass flow through the first-stage nozzle, using equation (16.83) if the nozzle is convergent-only, or equation (16.84) if the nozzle is convergent–divergent.

6. Calculate the nozzle efficiency using the methods of Chapter 14.

7. If the first stage is being considered, determine the nozzle outlet velocity from equation (16.16) with $c_0 = 0$:

$$\tfrac{1}{2}c_1^2 = \eta_N(h_0 - h_{1s}) \tag{16.89}$$

If the stage under consideration is a second or subsequent stage, make approximate allowance for a non-zero inlet velocity by using equation (16.22), repeated below:

$$\frac{1}{2}c_1^2 \approx \frac{\sin^2\alpha_2}{\sin^2\alpha_2 - \sin^2\alpha_1}\eta_N(h_0 - h_{1s}) \tag{16.22}$$

8. A knowledge of the instantaneous turbine speed, N, allows us to calculate the average blade speed for the stage from:

$$c_B = 2\pi N r_B \tag{16.90}$$

where r_B (m) is the average blade radius for the stage. Hence we may calculate the blade speed to gas speed ratio at blade inlet, $R_B = c_B/c_1$, and use this figure to calculate nozzle efficiency. Equations (15.50) and (15.66) apply to an impulse stage, while equation (15.108) applies to a 50% reaction blade. Set $\eta_{BN} = \eta_N$ in equation (15.108).

9. If the stage is the first stage and is an impulse stage, calculate overall stage efficiency from equation (16.17):

$$\eta_s = \eta_N\eta_B \tag{16.17}$$

If the first stage is a reaction stage, use equation (16.28):

$$\eta_s = \eta_B\frac{h_0 - h_{2sa}}{h_0 - h_{2s}} \tag{16.28}$$

in which h_{2s} and h_{2sa} may be derived as discussed in Section 16.5.

10. If the stage is a later stage and is an impulse stage, calculate overall stage efficiency from equation (16.23):

$$\eta_s = \frac{\sin^2\alpha_2}{\sin^2\alpha_2 - \sin^2\alpha_1}\eta_N\eta_B \tag{16.23}$$

If the later stage is a reaction stage, calculate overall stage efficiency from equation (16.31):

$$\eta_s = \eta_B\frac{h_0 - h_{2sa}}{h_0 - h_{2s}} + \eta_N\eta_B\frac{\sin^2\alpha_1}{\sin^2\alpha_0 - \sin^2\alpha_1}\frac{h_0 - h_{1s}}{h_0 - h_{2s}} \tag{16.31}$$

11. Calculate the specific work using equation (16.11) rearranged:

$$w = \eta_s(h_0 - h_{2s}) \tag{16.91}$$

12. Calculate the stage outlet specific enthalpy as:

$$h_2 = h_0 - w \tag{16.92}$$

This will become the inlet specific enthalpy to the next stage.

13. Use equations (15.17) and (15.18) to derive the specific volume at stage outlet:

$$v_2 = v_0\left(\frac{p_0}{p_2}\right)^{1-\eta_s(\gamma-1)/\gamma} \tag{16.93}$$

This will become the inlet specific volume to the next stage.

14. Return to step 4 until the performance of all stages has been calculated for the current set of pressure estimates.

Flow iteration

15. Refine the pressure estimates iteratively until equation set (15.113) is satisfied to a good accuracy.

Turbine power

16. When the flow equations are satisfied, calculate the power in each stage from equation (15.6) and the total power output of the turbine from equation (15.7):

$$P_i = W_i w_i \tag{15.6}$$

$$P_T = \sum_i W_i w_i \tag{15.7}$$

At this stage a solution has been found for the current time instant.

16.10 Bibliography

Bain, R.W. (1964). *Steam Tables*, HMSO, Edinburgh.
Keenan, J.H., Chao, J. and Kaye, J. (1980). *Gas Tables, Second Edition (English Units)*, John Wiley and Sons.

17 Turbo pumps and compressors

17.1 Introduction

Machines for increasing the pressure of a flowing liquid are known as pumps, while those that increase the pressure of a flowing gas are known as compressors. Turbo pumps and turbo compressors use a rotating impeller to transfer kinetic energy to the fluid, and then use diffusion to convert most of this kinetic energy into potential or pressure energy. There are two major classes of machine for both liquids and gases: centrifugal and axial. The most common machine for pressurizing liquids is the centrifugal pump, but axial pumps have a place in applications where high-volume liquid flow is required but the pressure rise is small. Mixed-flow machines consitute a further class of liquid pump, and are a combination of centrifugal and axial flow pumps, with part of the liquid flow in the impeller being radial and part axial. The mixed-flow pump is used for high-volume flow at a greater head than a purely axial device could deliver. Compressors may be centrifugal or axial, with axial machines being used for larger volume flows (more than about three cubic metres per second actual volume flow at delivery).

The analyses of turbo pumps and compressors have much in common, although the compressible nature of the gas in a compressor makes it rather more difficult to model. One important feature is that the internal dynamics of both pumps and compressors will be very rapid in comparison with the process plant they are serving. Accordingly it is normally possible to use a steady-state model of each in the overall simulation.

17.2 Applying dimensional analysis to centrifugal and axial pumps

While not providing an explanation of the physical working of pumps, dimensional analysis provides a short-cut to understanding what are the key features affecting their behaviour. The first step is to select the variables that should be included in the analysis. Frictional losses will depend in part on the Reynolds number, which is itself a function of viscosity. However, the friction factor is almost independent of Reynolds number at the very high Reynolds numbers generally experienced by the highly turbulent flow through industrial turbo pumps, and hence we may neglect viscous effects with only a small loss of accuracy. The pressure rise across the pump is clearly important, but we shall prefer to use the composite variable $gH(= v\Delta p)$ in its stead. The chosen parameter set is therefore:

$$(gH, Q, N, P_D, D, v)$$

where

H is the head produced by the pump (m),
Q is the volume flow rate (m^3/s),
N is the rotational speed (rps),
P_D is the power demanded by the pump (W),
D is the impeller diameter (m),
v is the specific volume (m^3/kg)

We shall make the further assumption that gH and P_D are each dependent on (Q, N, D, v).

Pump head

Applying Rayleigh's method of dimensional analysis to the pump head, we write our test equation as:

$$gH = \text{const} \times Q^a N^b D^c v^d \tag{17.1}$$

Substituting dimensions yields:

$$[L^2T^{-2}] = [L^3T^{-1}]^a[T^{-1}]^b[L]^c[L^3M^{-1}]^d \tag{17.2}$$

Equating powers of M, L and T gives three equations in the four unknown indices, a, b, c and d:

$$\begin{aligned} 0 &= -d \\ 2 &= 3a + c + 3d \\ -2 &= -a - b \end{aligned} \tag{17.3}$$

The index, d, emerges unequivocally as zero, while the indices, b and c, are expressible in terms of a, which may take any value:

$$\begin{aligned} b &= 2 - a \\ c &= 2 - 3a \end{aligned} \tag{17.4}$$

Substituting back into equation (17.1) gives

$$\begin{aligned} gH &= \text{const} \times Q^a N^{2-a} D^{2-3a} \\ &= \text{const} \times N^2 D^2 \left(\frac{Q}{ND^3}\right)^a \end{aligned} \tag{17.5}$$

which implies that the following general relationship holds between head and flow :

$$\frac{gH}{N^2D^2} = \phi_1\left(\frac{Q}{ND^3}\right) \tag{17.6}$$

It should be noted that the specific volume does not enter into this equation, implying that the head produced for a given flow will be the same whatever the liquid being pumped–hence the preference of working in terms of head rather than pressure rise. It is possible to estimate the form of the function, ϕ_1, from a detailed analysis of the velocities of the impeller and liquid, making due allowance for losses, but an accurate determination requires experiment.

The function, ϕ_1, applies to pumps of different sizes, provided they are geometrically similar. Hence the form of the function could be determined from a model with impeller diameter, D_0, or one with diameter, D_1, or one with a diameter, D_2, etc. Equally, all the experimental points could be taken at a constant speed, N_0, or at a different constant speed, N_1, or at a third constant speed, N_2, etc. Theoretically, it should not matter what liquid is being pumped, although in reality the liquid should have a viscosity reasonably close to that of the ultimate process liquid.

Let us suppose we take the machine with impeller diameter, D_0, and run it at a fixed speed, N_0; by varying the downstream flow conductance, we may take a number of different readings of flow and corresponding head. These readings will be pairs of measurements of Q_0 and H_0, the flow and head at conditions (D_0, N_0). (It should be emphasized that Q_0 and H_0 are variables, not constants.) Now we move on to the next machine, with diameter, D_1. We run this at constant speed, N_1, and take a new set of readings, this time of Q_1 and H_1, where Q_1 and H_1 are again variables, not constant values. Moving to the third machine, we take measurements of the variables, Q_2 and H_2, at fixed conditions (D_2, N_2). We may, indeed, repeat the process with any available machine of similar geometry. We may then plot:

$$\frac{gH_0}{N_0^2 D_0^2} \text{ vs. } \frac{Q_0}{N_0 D_0^3} \quad \text{for } 0 \leq Q_0 \leq (Q_0)_{\max}$$

$$\frac{gH_1}{N_1^2 D_1^2} \text{ vs. } \frac{Q_1}{N_1 D_1^3} \quad \text{for } 0 \leq Q_1 \leq (Q_1)_{\max}$$

$$\frac{gH_2}{N_2^2 D_2^2} \text{ vs. } \frac{Q_2}{N_2 D_2^3} \quad \text{for } 0 \leq Q_2 \leq (Q_2)_{\max}$$

and equation (17.6) tells us that all the points should fall on the same curve. This theoretical result is found to be largely true in practice, although small discrepancies can occur as a result of the change in Reynolds number and hence friction when the scale-factor is large. The function, ϕ_1, for a centrifugal pump will have the general shape shown in Figure 17.1.

If we select a test point, $P = (x, y)$, on the curve of Figure 17.1, then, from the discussion above, the x-coordinate could be any of the following:

$$x = \frac{Q_0}{N_0 D_0^3}, \text{ or } x = \frac{Q_1}{N_1 D_1^3}, \text{ or } x = \frac{Q_2}{N_2 D_2^3}, \text{ etc.} \tag{17.7}$$

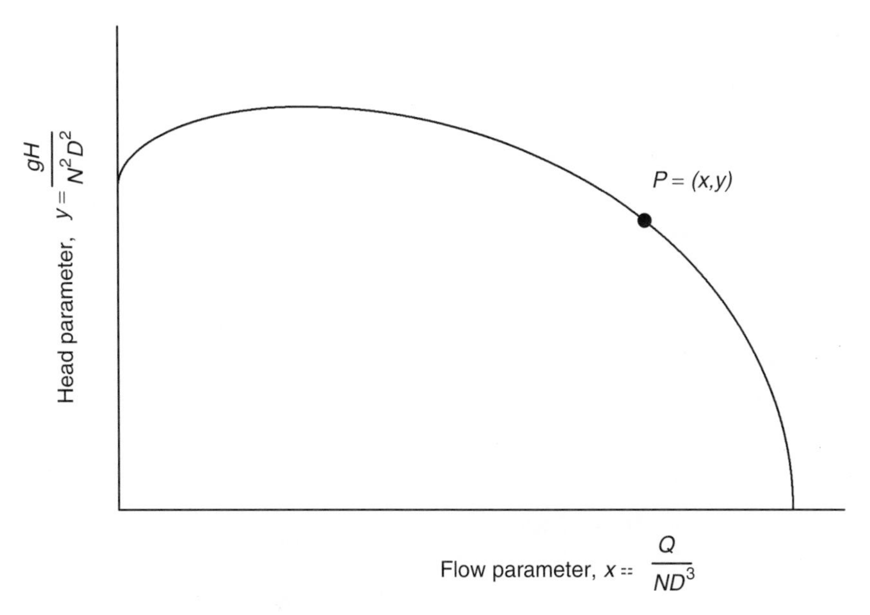

Figure 17.1 Generalized curve showing relationship between head and volume flow.

while the y-coordinate may be written as any of the following

$$y = \frac{gH_0}{N_0^2 D_0^2}, \text{ or } y = \frac{gH_1}{N_1^2 D_1^2}, \text{ or } y = \frac{gH_2}{N_2^2 D_2^2}, \text{ etc.} \tag{17.8}$$

Since the test point could be any point on the curve describing the behaviour of similar pumps, we may conclude that the following relations hold across all pumps in the set, at whatever speeds, flows and heads they are operating:

$$\frac{Q_0}{N_0 D_0^3} = \frac{Q_1}{N_1 D_1^3} = \frac{Q_2}{N_2 D_2^3} \tag{17.9}$$

and

$$\frac{H_0}{N_0^2 D_0^2} = \frac{H_1}{N_1^2 D_1^2} = \frac{H_2}{N_2^2 D_2^2} \tag{17.10}$$

Pump power

The test equation for the power demanded by the pump is:

$$P_D = \text{const} \times Q^a N^b D^c v^d \tag{17.11}$$

Substituting dimensions yields:

$$[ML^2T^{-3}] = [L^3T^{-1}]^a[T^{-1}]^b[L]^c[L^3M^{-1}]^d \tag{17.12}$$

Equating powers of M, L and T gives three equations in the four unknown indices, a, b, c and d:

$$\begin{aligned} 1 &= -d \\ 2 &= 3a + c + 3d \\ -3 &= -a - b \end{aligned} \tag{17.13}$$

The index, d, emerges unequivocally as -1, while the indices, b and c, are expressible in terms of a, which may take any value:

$$\begin{aligned} d &= -1 \\ b &= 3 - a \\ c &= 5 - 3a \end{aligned} \tag{17.14}$$

Substituting back into equation (17.11) gives

$$\begin{aligned} P_D &= \text{const} \times Q^a N^{3-a} D^{5-3a} v^{-1} \\ &= \text{const} \times \frac{N^3 D^5}{v}\left(\frac{Q}{ND^3}\right)^a \end{aligned} \tag{17.15}$$

which implies that the following general relationship holds between demanded power and flow :

$$\frac{vP_D}{N^3 D^5} = \phi_2\left(\frac{Q}{ND^3}\right) \tag{17.16}$$

Note that in this case specific volume of the fluid has an effect: working at the same speed, a pump can raise a flow of mercury to the same head as the same volume flow of water, but it will consume more power in doing so. The form of the function, ϕ_2, is found from experiment, usually on one machine, but theoretically on several machines, as described above for the determination of the general function, ϕ_1. Since a single function characterizes all the pumps in the set, the arguments set down above for the relationship between pump head and flow apply equally to the relationship between demanded power and flow. Hence we may write the following equation for pumps of different size, running at different speeds and pumping liquids of different specific volumes:

$$\frac{v_0 P_{D0}}{N_0^3 D_0^5} = \frac{v_1 P_{D1}}{N_1^3 D_1^5} = \frac{v_2 P_{D2}}{N_2^3 D_2^5} \tag{17.17}$$

Equations (17.9), (17.10) and (17.17) are known as the 'affinity laws'.

The useful pumping power, P_P, will be less than the power demanded, P_D, because of frictional losses. The pumping power is the power expended in lifting the mass flow, W (kg/s), to a height H (m):

$$P_P = WgH = \frac{QgH}{v} \tag{17.18}$$

Using equation (17.6), we may rewrite equation (17.18) as:

$$P_P = \frac{QN^2D^2}{v}\phi_1\left(\frac{Q}{ND^3}\right) \tag{17.19}$$

The pump efficiency, η_P, may be found by dividing the pumping power from equation (17.19) by the power demanded, given by equation (17.16):

$$\begin{aligned} \eta_P = \frac{P_P}{P_D} &= \frac{\frac{QN^2D^2}{v}\phi_1\left(\frac{Q}{ND^3}\right)}{\frac{N^3D^5}{v}\phi_2\left(\frac{Q}{ND^3}\right)} \\ &= \frac{Q}{ND^3}\frac{\phi_1\left(\frac{Q}{ND^3}\right)}{\phi_2\left(\frac{Q}{ND^3}\right)} \end{aligned} \tag{17.20}$$

which may be re-expressed as simply

$$\eta_P = \phi_3\left(\frac{Q}{ND^3}\right) \tag{17.21}$$

From equation (17.21), efficiency is simply a function of the dimensionless variable, $Q/(ND^3)$.

When simulating a particular pump on a process plant, the diameter of the pump impeller will, of course, be fixed, and so we may simplify the affinity laws to:

$$\frac{Q}{N} = \frac{Q_0}{N_0} \tag{17.22}$$

$$\frac{H}{N^2} = \frac{H_0}{N_0^2} \tag{17.23}$$

$$v\frac{P_D}{N^3} = v_0\frac{P_{D0}}{N_0^3} \tag{17.24}$$

where Q, H, N, P_D and v are general values and Q_0, H_0, N_0, P_{D0} and v_0 are reference or design values.

17.3 Pump characteristic curves

A set of empirical functions is needed to characterize any given pump. Rather than the generalized functions, ϕ, however, it is customary for manufacturers to provide instead (i) the characteristic of head versus volume flow at the design speed, together with either (ii) a curve showing power demand versus volume flow at the design speed and specific volume, or else (iii) a curve specifying pump efficiency versus volume flow at the design speed.

A set of curves typical of a large centrifugal pump are shown in Figure 17.2. Note that the second graph includes the pumping power curve for interest's sake, derived by application of equation (17.18). These manufacturers' graphs will have been derived from tests either on the pump itself, or else from tests on a pump of similar geometry.

Using the subscript '0' to denote conditions at the design speed and specific volume, we may regard the curve of head versus discharge rate as a graphical interpretation of the function:

$$H_0 = f_{P1}(Q_0) \tag{17.25}$$

and the power demand curve as a graphical interpretation of the function:

$$P_{D0} = f_{P2}(Q_0) \tag{17.26}$$

The efficiency curve will be a graphical interpretation of the function:

$$\eta_P = f_{P3}(Q_0) \tag{17.27}$$

To understand the effect of a change from the design speed, N_0, and design specific volume, v_0, we need to make use of the affinity equations. Substituting for Q_0 from equation (17.22) and for H_0 from equation (17.23) into equation (17.25) gives:

$$H\frac{N_0^2}{N^2} = f_{P1}\left(Q\frac{N_0}{N}\right) \tag{17.28}$$

Rearranging, the head vs. discharge characteristic at a speed different from the design speed is

$$H = \frac{N^2}{N_0^2} f_{P1}\left(Q\frac{N_0}{N}\right) \tag{17.29}$$

Similarly, substituting for Q_0 from (17.22) and for P_{D0} from equation (17.24) into equation (17.26) gives:

$$\frac{v}{v_0} P_D \frac{N_0^3}{N^3} = f_{P2}\left(Q\frac{N_0}{N}\right) \tag{17.30}$$

Accordingly the power demand at a speed different from the design speed is given by:

$$P_D = \frac{v_0}{v}\frac{N^3}{N_0^3} f_{P2}\left(Q\frac{N_0}{N}\right) \tag{17.31}$$

The pumping power at a speed different from the design speed is found by substituting for head, H, from (17.28) into equation (17.18):

$$P_P = \frac{g}{v} Q \frac{N^2}{N_0^2} f_{P1}\left(Q\frac{N_0}{N}\right) \tag{17.32}$$

To find the efficiency at any speed, N, we subsitute for Q_0 from (17.22) into equation (17.27):

$$\eta_P = f_{P3}\left(Q\frac{N_0}{N}\right) \tag{17.33}$$

It is clear from the above equations that the ratio of flow to normalized speed,

$$Q\frac{N_0}{N} = \frac{Q}{N/N_0}$$

is of fundamental importance in calculating the behaviour of the pump at any speed different from the design speed. It is further clear from equation (17.21) that the functions f_{P1}, f_{P2} and f_{P3} are interrelated, and that if two are specified, the third can be calculated. Concerning the choice of functions for modelling the behaviour of the pump, the head/discharge curve should always be the starting point. The efficiency curve is often used in calculating pump performance in preference to the demanded power curve, but there are grounds for sometimes choosing the latter instead, namely:

(i) the demanded power comes directly from test data;
(ii) the demanded power is usually a less complicated curve than the efficiency curve, and so may be modelled more accurately for the same order of polynomial fit or the same number of points in a look-up table;
(iii) if it is necessary to derive the input power at zero flow, e.g. for starting purposes, then a

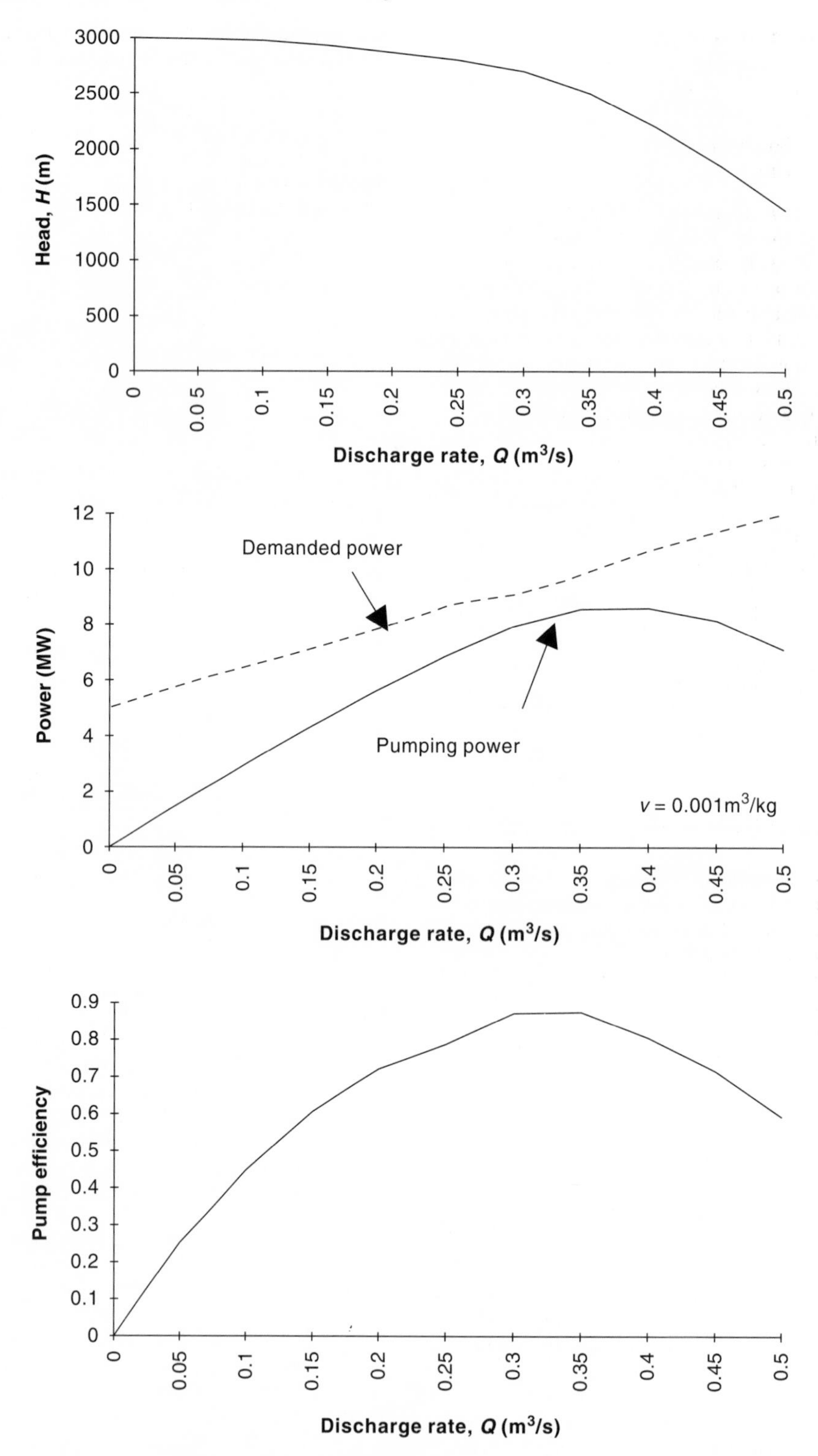

Figure 17.2 Typical pump characteristic curves at the design speed.

problem arises in using the expression:

$$P_D = \frac{P_P}{\eta_P} \qquad (17.34)$$

since both will be zero, and the answer indeterminate.

Each of the functions f_{P1}, f_{P2} and f_{P3} may be approximated for the purposes of dynamic simulation by a low-order polynomial. However, care is needed when fitting a polynomial to the head/discharge characteristic using a simple least-squares procedure, since it is important not to introduce a spurious reduction in head at low flows, which can lead to numerical problems caused by flow being a multivalued function of head. The numerical problem has an analogue in reality, since such a characteristic on a real process pump can induce an unstable mode of operation. In the past, centrifugal pumps displaying a postive slope dH/dQ at low flows were sometimes made and were prone to unstable, surging behaviour if the flow dropped to low values: a drop in flow would lead to a drop in head, which would lead to a further drop in flow, etc. It should be noted that suitable mechanical design of the pump, namely ensuring that the guide vanes are angled sufficiently backwards at the impeller exit, can eliminate this unstable mode of behaviour, and modern centrifugal pumps do not usually suffer from this problem.

17.4 Pump dynamics

The pump flow will establish itself extremely rapidly. The response of pump speed to changes in power input and output will also be rapid, of the order of a few seconds at most, but it may be essential to model this response, for instance if the model is to be used to design a pump speed-control system. Further, it may be desirable to use the numerical decoupling that including the additional state, namely pump speed, allows. This assumes particular importance if both the power supplied and the power demand are calculated from sets of nonlinear simultaneous equations, as will be the case for a pump powered by a turbine and supplying a complicated liquid-flow network.

The dynamics of the pump are derived from the principle of the conservation of energy applied to the system with boundaries drawn around the pump as shown in Figure 17.3.

An energy balance over a general bounded volume is given by equation (3.34), repeated below:

$$\frac{dE}{dt} = \Phi - P + W_1\left(h_1 + \frac{1}{2}c_1^2 + gz_1\right) - W_2\left(h_2 + \frac{1}{2}c_2^2 + gz_2\right) \qquad (3.34)$$

In this case the heat input, Φ, is zero and there is no mechanical power output, but a mechanical power supplied, P_S. Hence $P = -P_S$. There will be no significant difference in height over the pump, so $z_1 = z_2$. In addition, the almost incompressible nature of a liquid means that the specific volume will be essentially constant, allowing us to write:

$$v_1 = v_2$$

$$W_1 = \frac{Q}{v_1} = \frac{Q}{v_2} = W_2 = W \qquad (17.35)$$

$$c_1 = c_2$$

The energy of the pump system enclosed within the boundaries will be a combination of mechanical and thermal energy:

$$E = \tfrac{1}{2}J\omega^2 + mu = 2\pi^2 JN^2 + mu \qquad (17.36)$$

Hence, by differentiating E and using the conditions of equation (17.35) in equation (3.34), we may write:

$$4\pi^2 JN\frac{dN}{dt} + \frac{d(mu)}{dt} = P_s + W(h_1 - h_2) \qquad (17.37)$$

To understand the response of pump speed to changes in power input and in flow conditions, we substitute

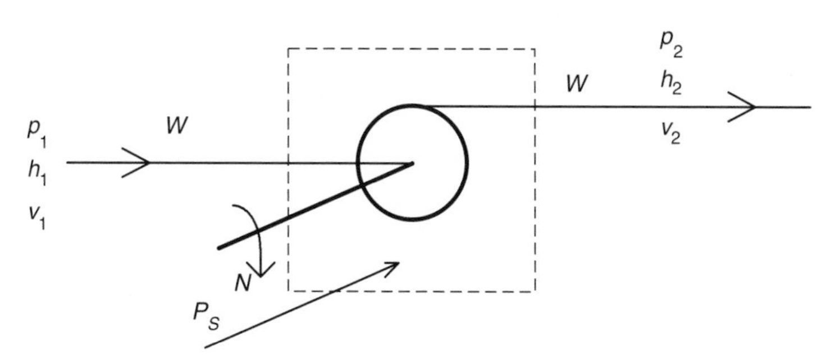

Figure 17.3 Showing energy boundaries around the pump.

$h = u + pv$ and $W = Q/v$ into equation (17.37):

$$4\pi^2 JN\frac{dN}{dt} + \frac{d(mu)}{dt} = P_S - Q(p_2 - p_1) - W(u_2 - u_1) \quad (17.38)$$

The left-hand side of equation (17.38) will be zero in the steady state, and the power supplied to the pump will be equal to the power demanded

$$P_S = P_D \quad (17.39)$$

so that at the steady-state condition equation (17.38) may be written :

$$0 = P_D - Q(p_2 - p_1) - W(u_2 - u_1) \quad (17.40)$$

Now

$$Q(p_2 - p_1) = vW\left(\frac{gH}{v}\right) = WgH = P_P \quad (17.41)$$

from equation (17.18), while

$$P_P = \eta_P P_D = \eta_P P_S \quad (17.42)$$

by the definition of pump efficiency (equation (17.21)), added to the steady-state condition (17.39). It follows that the expression

$$W(u_2 - u_1) = (1 - \eta_P)P_D = (1 - \eta_P)P_S \quad (17.43)$$

represents the frictional losses that cause an increase in the internal energy of the liquid being pumped. We may interpret equations (17.42) and (17.43) as indicating that in the steady state a fraction, η_P, of the power supplied goes into pumping power, while the remainder is lost to frictional heating. Assuming that this same process occurs in the dynamic state also allows us to decompose equation (17.38) into the mechanical and thermal energy equations:

$$4\pi^2 JN\frac{dN}{dt} = \eta_P P_S - Q(p_2 - p_1) \quad (17.44)$$

and

$$\frac{d(mu)}{dt} = (1 - \eta_P)P_S - W(u_2 - u_1) \quad (17.45)$$

The last equation could be used to find the outlet specific energy by putting

$$\frac{d(mu)}{dt} = m\frac{du_2}{dt}$$

and integrating; here m may be considered constant because of incompressibility and the specific internal energy may be considered that of the outlet. Outlet enthalpy may then be found from $h_2 = u_2 + p_2 v_2$. However, since process thermal time constants downstream of the pump will almost invariably be significantly longer than the time constants of the pump, it is simpler and usually sufficient to use the steady-state solution of equation (17.37) to give the pump outlet enthalpy, which is then:

$$h_2 = h_1 + \frac{P_S}{W} \quad (17.46)$$

Meanwhile equation (17.44) is in a form where it may be integrated from initial conditions of speed and pump flow. The efficiency, η_P, used above may be found either from equation (17.33), if the function, f_{P3}, is supplied, or else from the ratio of pumping power to demanded power (equations (17.31) and (17.32)) if the function, f_{P2}, is supplied.

17.5 Calculating the flow pumped through a pipe

Consider Figure 17.4, which shows a liquid of specific volume v_1 (m^3/kg) being pumped from a vessel at pressure p_1 (Pa) and height z_1 (m) up to a vessel at

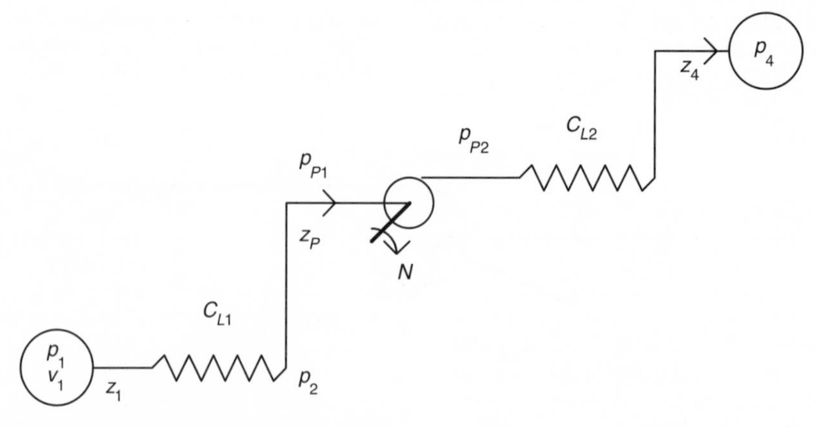

Figure 17.4 Liquid being pumped from one vessel to another.

pressure p_4 and height z_4. The conductance between the suction vessel and the pump (including pipe entrance effects) is C_{L1} (m^2), while the conductance between pump and discharge vessel is C_{L2} (m^2). The pump is situated at a height z_P (m), and is rotating at a speed N (rps). The pressure at pump inlet is p_{P1} (Pa), and is p_{P2} (Pa) at pump outlet.

Applying equation (4.81) to the pipe connecting the suction vessel to the pump, the mass flow is given by:

$$W = C_{L1}\sqrt{\frac{p_1 - p_{P1} - \dfrac{g(z_P - z_1)}{v_1}}{v_1}} \qquad (17.47)$$

Similarly, the flow from the pump exit to the discharge vessel is given by:

$$W = C_{L2}\sqrt{\frac{p_{P2} - p_4 - \dfrac{g(z_4 - z_P)}{v_1}}{v_1}} \qquad (17.48)$$

Here the specific volume, v_1, has been retained because the temperature rise across the pump is usually small. Squaring and adding the last two equations produces:

$$\left(\frac{1}{C_{L1}^2} + \frac{1}{C_{L2}^2}\right) v_1 W^2 = p_1 - p_4 - \frac{g(z_4 - z_1)}{v_1} + p_{P2} - p_{P1} \qquad (17.49)$$

or

$$\frac{v_1 W^2}{C_L^2} = p_1 - p_4 - \frac{g(z_4 - z_1)}{v_1} + \Delta p_P \qquad (17.50)$$

where C_L is the conductance characterizing the complete length of pipework between the suction and discharge vessel, given by

$$\frac{1}{C_L^2} = \frac{1}{C_{L1}^2} + \frac{1}{C_{L2}^2} \qquad (17.51)$$

(cf. equation (4.82)), and Δp_P is the pressure rise across the pump

$$\Delta p_P = p_{P2} - p_{P1} \qquad (17.52)$$

Equation (17.50) implies that, as far as the calculation of flow is concerned, it does not matter where the pump is situated in the line between the suction and discharge vessels. [Note that it does mattter in practice, and the pump will normally be situated as near the suction vessel as possible, so as to avoid the pump's inlet pressure falling below the vapour pressure of the liquid, when cavitation will reduce pumping efficiency and cause damage to the pump.]

The head developed across the pump at speed, N, is given by equation (17.28), and the corresponding pressure difference is:

$$\Delta p_P = \frac{g}{v_1}\frac{N^2}{N_0^2} f_{P1}\left(Q\frac{N_0}{N}\right) \qquad (17.53)$$

where N_0 is the design speed (rps) and the volume flow, Q is related to the mass flow, W, by:

$$Q = v_1 W \qquad (17.54)$$

Combining equations (17.50), (17.53) and (17.54) gives the following equation in volume flow rate, Q:

$$\frac{Q^2}{v_1 C_L^2} = p_1 - p_4 - \frac{g(z_4 - z_1)}{v_1} + \frac{gN^2}{v_1 N_0^2} f_{P1}\left(Q\frac{N_0}{N}\right) \qquad (17.55)$$

Solving equation (17.55) for a general, nonlinear function, f_{P1}, of head versus flow will require iteration. However, the function may be represented by a low-order polynomial

$$f_{P1}(Q_0) = a_0 + a_1 Q_0 + a_2 Q_0^2 + a_3 Q_0^3 + \cdots \qquad (17.56)$$

where a_i are constant coefficients. Sufficient accuracy is often obtained using a either a second-order or a third-order representation, allowing equation (17.55) to be re-expressed as:

$$\left[a_3\frac{g}{v_1}\frac{N_0}{N}\right] Q^3 + \left[a_2\frac{g}{v_1} - \frac{1}{v_1 C_L^2}\right] Q^2 + \left[a_1\frac{g}{v_1}\frac{N}{N_0}\right] Q + \left[a_0\frac{g}{v_1}\frac{N^2}{N_0^2} + p_1 - p_4 - \frac{g(z_4 - z_1)}{v_1}\right] = 0 \qquad (17.57)$$

Equation (17.57) may be solved either iteratively or, because it is a cubic, analytically. If a second-order expression for the pump characteristic, f_{P1}, is sufficient then $a_3 = 0$, and equation (17.57) becomes a quadratic, the solution of which is particularly easy.

17.6 Rotary compressors

The compressor performs for gases the function carried out by a pump for liquids: it adds energy to the gas to cause it to flow from one unit operation to the next. The pressure of the gas is raised in the process, so that the gas is 'compressed'. The compressor consists of one or more sections, with each section containing one or more stages, which are sometimes referred to as 'wheels'. Gas leaves the machine at the end

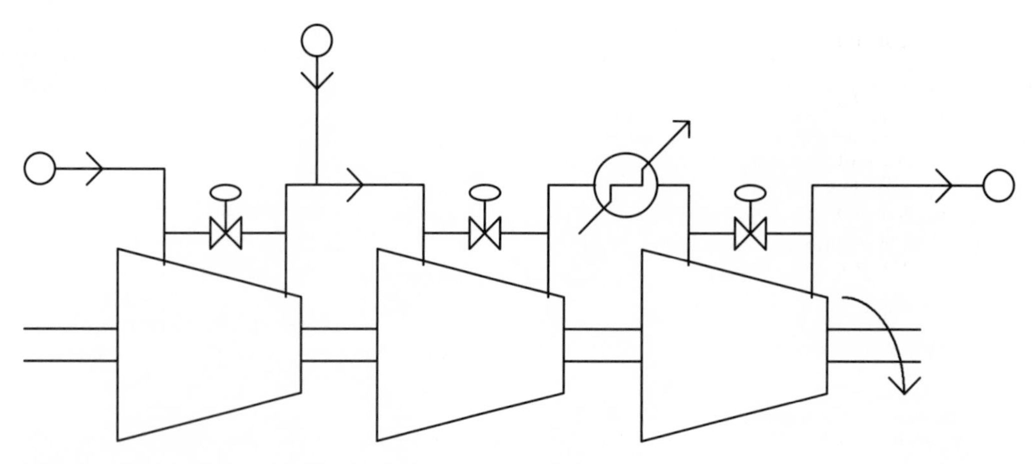

Figure 17.5 Schematic of a three-section compressor.

of each section and is either passed to another part of the process plant or else is fed to an intercooler prior to being returned to the machine for further compression. Alternatively, the end of one section and the beginning of the next may be defined by the addition of more gas from an external source. The general layout of a compressor containing three sections is shown in Figure 17.5. The figure illustrates also the 'kickback' flows from the section outlet to the section inlet that are a normal provision of an anti-surge control system.

The compressor acts like a turbine in reverse, compressing rather than expanding the gas that passes through it. An axial compressor, in particular, has a geometry similar to that of the axial-flow turbines described in Chapters 15 and 16. But a fundamental difference to the ease of analysis is that the flow in a compressor is diffusing, rather than accelerating. We can no longer rely on the blades to direct the flow as precisely as in a turbine because there is a natural tendency for diffusing gas to break away from the walls of the diverging, compressor passage and form eddies which flow back against the main stream. The result of this is that neither an axial nor a centrifugal compressor is amenable to the sort of geometrical treatment outlined in Chapter 15. This renders the task facing the designer more difficult, but, paradoxically, makes life easier for the control engineer on the process plant, since the greater difficulty in analysis forces the manufacturer to provide information on his compressor in the form of characteristic curves, usually based on test data. Using these curves to determine the flow through the compressor is usually a relatively straightforward matter.

Two separate sets of curves are possible, both arising from dimensional analysis. The basis for each will now be described.

17.7 Compressor characteristics based on polytropic head

17.7.1 Isentropic efficiency and isentropic head

The energy equation for general, steady-state flow was derived in Section 4.2 as:

$$dq - dw - dh - c\,dc - g\,dz = 0 \tag{4.7}$$

Flow through the compressor section will be adiabatic, so that $dq = 0$, and the height difference will be negligible, so that $dz = 0$. Further, the difference between the inlet and outlet velocities is small for both axial and centrifugal compressors so that $c\,dc \approx 0$. Accordingly the differential of specific work is given by:

$$-dw = dh \tag{17.58}$$

Integrating over the compressor section gives the specific work done on the gas, $-w$, as

$$-w = h_2 - h_0 \tag{17.59}$$

where the subscript '0' is taken to refer to the section inlet, while '2' refers to the section outlet. Further, the gases compressed may be assumed to have a constant specific heat over the temperature range of interest (this applies approximately even to refrigeration systems, where the vapour being compressed begins at saturation temperature). Thus equation (17.59) may be rewritten

$$-w = c_p(T_2 - T_0) = c_p T_0 \left(\frac{T_2}{T_0} - 1\right) \tag{17.60}$$

For most gases, the specific heat is given to good accuracy by equation (14.39) repeated below:

$$c_p = \frac{\gamma}{\gamma - 1} Z R_w \tag{14.39}$$

in which case equation (17.60) becomes

$$-w = \frac{\gamma}{\gamma - 1} Z R_w T_0 \left(\frac{T_2}{T_0} - 1 \right) \quad (17.61)$$

The specific work required will be a minimum when the temperature, T_2, is a minimum, and this will occur when the compression process is not only adiabatic but also reversible, that is isentropic. An isentropic compression is governed by the equation:

$$pv^{\gamma} = \text{constant} \quad (3.24)$$

Combining this with the characteristic gas equation

$$pv = Z R_w T \quad (3.2)$$

allows us to deduce the following relationship between temperature and pressure ratios for an isentropic compression:

$$\frac{T_{2s}}{T_0} = \left(\frac{p_2}{p_0} \right)^{(\gamma-1)/\gamma} \quad (17.62)$$

Substituting from equation (17.62) into equation (17.61) gives the isentropic specific work, $-w_s$, associated with compressing the gas from inlet pressure, p_0, to section outlet pressure, p_2, as:

$$-w_s = \frac{\gamma}{\gamma - 1} Z R_w T_0 \left(\left(\frac{p_2}{p_0} \right)^{(\gamma-1)/\gamma} - 1 \right) \quad (17.63)$$

The isentropic efficiency, η_s, is defined as the ratio of the isentropic specific work, $-w_s$, to the actual specific work, $-w$, associated with compressing the gas from inlet pressure, p_0, to section outlet pressure, p_2:

$$\eta_s = \frac{w_s}{w} \quad (17.64)$$

Compressor section efficiencies at the design point are in the mid-80s per cent for an axial compressor and in the high 70s per cent for a centrifugal compressor. Using equation (17.64) in conjunction with equation (17.59), the isentropic efficiency of the section may be expressed in terms of enthalpy differences as

$$\eta_s = \frac{h_{2s} - h_0}{h_2 - h_0} \quad (17.65)$$

(This may be contrasted with the inverse formulation used for turbine stage efficiency, equation (15.13).) Assuming the gas has a constant specific heat, the isentropic efficiency, η_s, takes the form

$$\eta_s = \frac{c_p(T_{2s} - T_0)}{c_p(T_2 - T_0)} = \frac{T_{2s} - T_0}{T_2 - T_0} \quad (17.66)$$

Thus an experimental determination of the isentropic efficiency, η_s, may be made by measuring the pressures and temperatures at the section inlet and section outlet, using equation (17.62) to determine the additional temperature, T_{2s}, and then applying equation (17.66).

Combining equations (17.63) and (17.64), the actual specific work of compression is given by:

$$-w = \frac{-w_s}{\eta_s} = \frac{1}{\eta_s} \frac{\gamma}{\gamma - 1} Z R_w T_0 \left(\left(\frac{p_2}{p_0} \right)^{(\gamma-1)/\gamma} - 1 \right) \quad (17.67)$$

The actual power needed for the compression is the product of the actual specific work and the flow rate of gas:

$$P = -Ww \quad (17.68)$$

We may use equation (17.67) to re-express the actual compression power in terms of isentropic specific work and isentropic efficiency:

$$P = -\frac{W w_s}{\eta_s} \quad (17.69)$$

We may compare this expression with the expression for the power demanded by the pump, P_D, found by combining equations (17.18) and (17.21):

$$P_D = \frac{W g H}{\eta_P} \quad (17.70)$$

where η_P is the pump efficiency. It is clear from inspection that the term $-w_s$ used in the compression power equation is analogous to the term gH, the pump head multiplied by the gravitational acceleration, used in the pumping power equation, and has the same dimensions. Hence it is common practice to refer to the isentropic specific work as the 'isentropic head', which is given the alternate symbol H_s:

$$H_s = -w_s \quad (17.71)$$

Note, however, that the compressor's isentropic head, H_s, has units of J/kg or m^2/s^2, in contrast to the metres that characterize the pump head, H.

While it is possible to use isentropic head as a characterizing variable for a compressor, most manufacturers who appeal to the concept of 'head' prefer to use a numerically similar parameter, the 'polytropic head'. The polytropic efficiency and polytropic head will be discussed in the next section.

17.7.2 Polytropic efficiency and polytropic head

The definition of isentropic efficiency, equation (17.64), is based on a ratio of isentropic specific work to actual specific work, across a complete section of the compressor. However, it is also possible to define a differential efficiency, assumed constant over the section, known as the polytropic efficiency, η_p, given by

$$\eta_p = \frac{dw_s}{dw} \tag{17.72}$$

We may use equation (17.58) to transform equation (17.72) into:

$$\eta_p = \frac{dh_s}{dh} = \frac{c_p\,dT_s}{c_p\,dT} = \frac{dT_s}{dT} \tag{17.73}$$

where we have also assumed that the specific heat is constant. The form of equation (17.73) enables us to use the mathematical treatment outlined in Section 14.3 for nozzle efficiency, except that the nozzle efficiency, η_N, used in that section will be replaced by the inverse of the polytropic efficiency, $1/\eta_p$. Thus the actual compression may be characterized by

$$pv^m = \text{constant} \tag{17.74}$$

where the exponent, m, is given by:

$$m = \frac{\eta_p \gamma}{1 - \gamma(1 - \eta_p)} \tag{17.75}$$

As a consequence of equation (17.74) and the characteristic gas equation (3.2), the actual temperature ratio across a compressor section operating at a polytropic efficiency, η_p, will be given by

$$\frac{T_2}{T_0} = \left(\frac{p_2}{p_0}\right)^{(m-1)/m} \tag{17.76}$$

where, from the definition of the exponent, m, in equation (17.75)

$$\frac{m-1}{m} = \frac{1}{\eta_p}\frac{\gamma - 1}{\gamma} \tag{17.77}$$

The expression for the temperature ratio in an actual compression may be found by substituting into equation (17.61) to give the actual specific work:

$$-w = \frac{\gamma}{\gamma - 1} Z R_w T_0 \left(\left(\frac{p_2}{p_0}\right)^{(m-1)/m} - 1\right) \tag{17.78}$$

We may use this new expression for the actual specific work together with equation (17.63) that gives the isentropic specific work in order to evaluate the isentropic efficiency:

$$\eta_s = \frac{\dfrac{\gamma}{\gamma - 1} Z R_w T_0 \left(\left(\dfrac{p_2}{p_0}\right)^{(\gamma-1)/\gamma} - 1\right)}{\dfrac{\gamma}{\gamma - 1} Z R_w T_0 \left(\left(\dfrac{p_2}{p_0}\right)^{(m-1)/m} - 1\right)} = \frac{\left(\left(\dfrac{p_2}{p_0}\right)^{(\gamma-1)/\gamma} - 1\right)}{\left(\left(\dfrac{p_2}{p_0}\right)^{1/\eta_p((\gamma-1)/\gamma)} - 1\right)} \tag{17.79}$$

From their definitions, we would expect the isentropic efficiency and the polytropic efficiency to be similar in value, and this is indeed the case. Nevertheless, it is possible to evaluate equation (17.79) over a range of pressure ratios for fixed values of polytropic efficiency, and thus highlight divergencies. Figure 17.6 shows the isentropic efficiency calculated for three typical values of polytropic efficiency over a range of pressure ratios.

It will be seen that the isentropic efficiency of the section is the same as the polytropic efficiency at unity pressure ratio, but falls away as the pressure

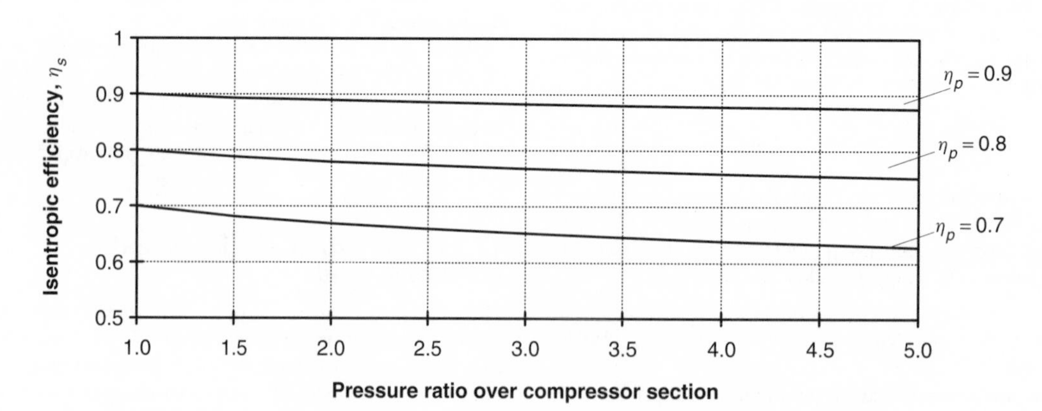

Figure 17.6 Isentropic efficiency against pressure ratio for polytropic efficiencies of 0.7, 0.8 and 0.9, with $\gamma = 1.4$.

ratio rises, the fall being rather more marked at lower polytropic efficiencies. The isentropic effficiency is between about 2% and 7% less than the polytropic efficiency, depending on the latter's value, for the normal range of pressure ratios found on industrial plant, namely 2.5 to 4.5.

The fact that the isentropic efficiency varies with pressure ratio for a constant value of polytropic efficiency has led some to regard the polytropic efficiency as a preferable foundation on which to base their analysis of the compressor section. Instead of taking the isentropic specific work as our ideal against which to measure the actual specific work, it is possible to devise a new ideal measure, the polytropic specific work, $-w_p$, defined so that its ratio to the actual specific work is the polytropic efficiency, η_p:

$$\eta_p = \frac{w_p}{w} \tag{17.80}$$

Combining equations (17.78) and (17.80), the polytropic specific work emerges as:

$$-w_p = \eta_p \frac{\gamma}{\gamma - 1} Z R_w T_0 \left(\left(\frac{p_2}{p_0} \right)^{(m-1)/m} - 1 \right) \tag{17.81}$$

or, using equation (17.77)

$$-w_p = \frac{m}{m - 1} Z R_W T_0 \left(\left(\frac{p_2}{p_0} \right)^{(m-1)/m} - 1 \right) \tag{17.82}$$

An experimental determination of m may be made after measuring the pressure and temperature at the section inlet, p_0, T_0 and outlet, p_2, T_2 and applying equation (17.76), which may be solved to give the formula:

$$m = \frac{\ln \dfrac{p_2}{p_0}}{\ln \dfrac{p_2}{p_0} \dfrac{T_0}{T_2}} \tag{17.83}$$

The polytropic efficiency may be found similarly by solving equations (17.76) and (17.77):

$$\eta_p = \frac{\gamma - 1}{\gamma} \frac{\ln \dfrac{p_2}{p_0}}{\ln \dfrac{T_2}{T_0}} \tag{17.84}$$

In fact, an experimental determination of the exponent, m, making use of equation (17.83) (or an equivalent) will introduce a small error due to the implicit inclusion of an imperfect model of compressibility effects. To compensate for this, manufacturers sometimes introduce an additional factor, f, into the equation for polytropic specific work, although in most cases f is so close to unity as to make negligible difference. Polytropic specific work is normally given the name 'polytropic head', H_p, based on the same reasoning used for isentropic head, and so we have the final form:

$$H_p = f \frac{m}{m - 1} Z R_w T_0 \left(\left(\frac{p_2}{p_0} \right)^{(m-1)/m} - 1 \right) \tag{17.85}$$

The power absorbed by the compressor section is given by equation (17.68). We may use equation (17.80) to express the power in terms of polytropic specific work/polytropic head and polytropic efficiency:

$$\begin{aligned} P &= -\frac{W w_p}{\eta_p} \\ &= \frac{W H_p}{\eta_p} \end{aligned} \tag{17.86}$$

where the final form uses the fact that $H_p = -w_p$.

It should be emphasized that the polytropic head is an idealization in the same way that the isentropic head was an idealization. Taken together with the polytropic efficiency, however, it provides a way of analysing compressor performance, as will be shown in the next section.

17.7.3 Dimensional analysis applied to the polytropic head: the use of a characteristic curve and the affinity laws

Since the polytropic head has the same units as the pump head times the acceleration due to gravity, it follows that we may apply the dimensional analysis of Section 17.2 to the compressor as well as the pump, with H_p replacing gH. This reveals the same independence of the specific volume of the fluid being passed, and the same dependence on volume flow rate at suction, Q, and speed, N.

The manufacturer will normally supply a set of curves for each compressor section (Figure 17.7), at the design speed with further curves at $+/-10\%$ or $+/-20\%$ speed. The surge line will be marked, indicating the region of unstable operation that must be avoided. The manufacturer should also specify the polytropic efficiency.

Because of the affinity laws, however, the modeller need work with only the characteristic giving polytropic head vs. volume flow at the design point, which may be stored in the form of a curve of Q_0 vs. H_{p0}. Then for any off-design point, he may calculate the current polytropic head, H_p, from the ratio of the inlet and outlet pressures and the inlet temperature using equation (17.85). Given the current operating speed, N, he may calculate the polytropic head at the design

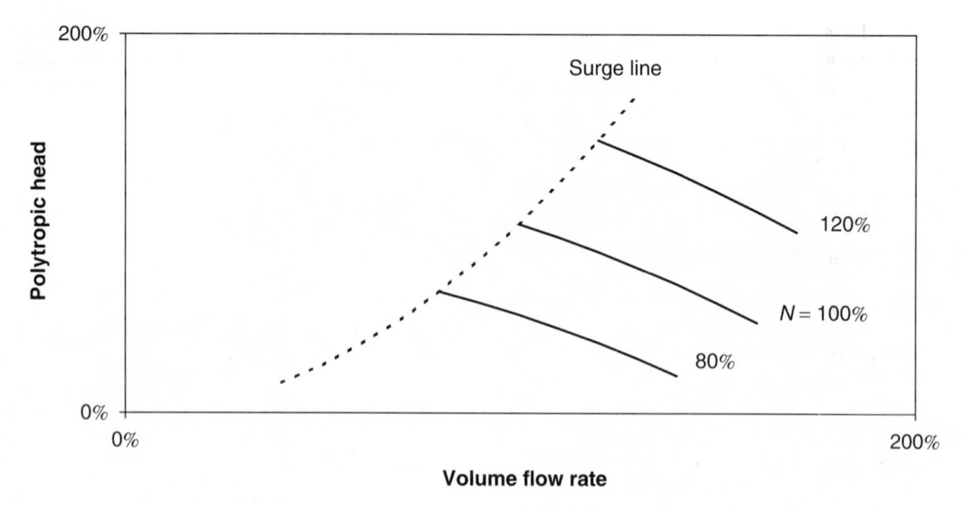

Figure 17.7 Typical manufacturer's curves of polytropic head vs. volume flow rate with speed as parameter.

speed from the affinity law (17.23):

$$H_{p0} = \frac{N_0^2}{N^2} H_p \tag{17.87}$$

Reference to the stored characteristic curve then gives the volumetric flow rate, Q_0, at the design speed and this allows the current suction volume flow to be calculated from:

$$Q = \frac{N}{N_0} Q_0 \tag{17.88}$$

The affinity laws are valid only for compressors where the Mach number (defined in terms of the exit velocity from the stage) is less than 0.8, or, under vacuum conditions, the mean free path does not approach the impeller dimensions. Outside these limits, no simple correlations with speed exist, and the full characteristics must be stored rather than a single curve for each section. Fortunately, most industrial compressors will stay within these limits.

17.8 Compressor characteristics based on pressure ratio

17.8.1 Dimensional analysis applied to compressor pressure ratio

Instead of considering polytropic head, it is possible to carry out an alternative dimensional analysis based on inlet and outlet pressures of a compressor section. We will begin by assuming that the compression process may be represented using the parameter set:

$$(p_{0T}, p_{2T}, T_{0T}, T_{2T}, v_{0T}, v_{2T}, W, N, D, \mu)$$

where the second subscript 'T' denotes the stagnation state, thus accounting for inlet and outlet velocities. Now the specific volumes follows from the characteristic gas equation (3.2) as:

$$v = \frac{ZR_w T}{p} \tag{17.89}$$

For compression gases, we may normally take the dimensionless compressibility factor, Z as unity or at least constant in the pressure range we are studying, and consider it no further. The dependence of specific volume on pressure and temperature means that we may eliminate this variable from our chosen parameter set. Further, it has been found convenient to use the grouping $R_w T$ instead of temperature alone. Thus the parameter set becomes:

$$(p_{0T}, p_{2T}, R_w T_{0T}, R_w T_{2T}, W, N, D, \mu)$$

Following the procedure outlined in Section 17.2, we write our test equation as:

$$p_{2T} = \text{const} \times p_{0T}^a (R_w T_{0T})^b (R_w T_{2T})^c W^d N^e D^f \mu^g \tag{17.90}$$

Substituting dimensions yields:

$$[ML^{-1}T^{-2}] = [ML^{-1}T^{-2}]^a [L^2T^{-2}]^b [L^2T^{-2}]^c \times [MT^{-1}]^d [T^{-1}]^e [L]^f [ML^{-1}T^{-1}]^g \tag{17.91}$$

Equating powers of M, L and T gives three equations in the seven unknowns, a, b, c, d, e, f and g:

$$\begin{aligned} 1 &= a + d + g \\ -1 &= -a + 2b + 2c + f - g \\ -2 &= -2a - 2b - 2c - d - e - g \end{aligned} \tag{17.92}$$

It has been found convenient to solve for a, b and f in terms of c, d, e and g:

$$a = 1 - d - g$$
$$b = \frac{d}{2} - c - \frac{e}{2} + \frac{g}{2} \quad (17.93)$$
$$f = e - 2d - g$$

Substituting back into equation (17.90) gives

$$p_{2T} = \text{const} \times p_{0T}^{1-d-g}(R_w T_{0T})^{(d/2-c-e/2+g/2)} \times (R_w T_{2T})^c W^d N^e D^{e-2d-g} \mu^g \quad (17.94)$$

or

$$p_{2T} = \text{const} \times p_{0T} \left(\frac{R_w T_{2T}}{R_w T_{0T}}\right)^c \left(\frac{W\sqrt{R_w T_{0T}}}{p_{0T} D^2}\right)^d \times \left(\frac{ND}{\sqrt{R_w T_{0T}}}\right)^e \left(\frac{\mu\sqrt{R_w T_{0T}}}{p_{0T} D}\right)^g \quad (17.95)$$

Equation (17.95) implies that the pressure ratio is some function, ϕ, of four dimensionless numbers:

$$\frac{p_{2T}}{p_{0T}} = \phi\left\{\left(\frac{R_w T_{2T}}{R_w T_{0T}}\right), \left(\frac{W\sqrt{R_w T_{0T}}}{p_{0T} D^2}\right), \times \left(\frac{ND}{\sqrt{R_w T_{0T}}}\right), \left(\frac{\mu\sqrt{R_w T_{0T}}}{p_{0T} D}\right)\right\} \quad (17.96)$$

The first dimensionless number obviously may be simplified by cancellation of R_w. The fourth dimensionless group may be simplified by substituting from the characteristic equation for an ideal gas: $p_{0T} = (R_w T_{0T}/v_{0T})$:

$$\frac{\mu\sqrt{R_w T_{0T}}}{p_{0T} D} = \frac{v_{0T}\mu\sqrt{R_w T_{0T}}}{R_w T_{0T} D} = \frac{v_{0T}\mu}{\sqrt{R_w T_{0T}} D} \quad (17.97)$$

The term $\sqrt{R_w T_{0T}}$ is proportional to the speed of sound at the inlet stagnation conditions, and is a velocity term. Hence, bearing in mind the definition of the Reynolds number:

$$N_{RE} = \frac{cD}{v\mu} \quad (4.33)$$

we may replace the fourth term by the Reynolds number. As a result, the pressure ratio is given by:

$$\frac{p_{2T}}{p_{0T}} = \phi\left\{\frac{T_{2T}}{T_{0T}}, \left(\frac{W\sqrt{R_w T_{0T}}}{p_{0T} D^2}\right), \left(\frac{ND}{\sqrt{R_w T_{0T}}}\right), N_{RE}\right\} \quad (17.98)$$

But the Reynolds number in most industrial cases will be so high and the flow so turbulent that changes in this parameter may normally be neglected. Further, the temperature ratio may be related to the pressure ratio and the polytropic efficiency by applying to the stagnation states equations (17.76) and (17.77) developed in the previous section:

$$\frac{T_{2T}}{T_{0T}} = \left(\frac{p_{2T}}{p_{0T}}\right)^{(m-1)/m} = \left(\frac{p_{2T}}{p_{0T}}\right)^{1/\eta_p((\gamma-1)/\gamma)} \quad (17.99)$$

In addition, the control engineer will normally be dealing with a machine that has already been designed and has a fixed value of diameter, D. Accounting for all these effects together, we may re-express the pressure ratio as:

$$\frac{p_{2T}}{p_{0T}} = \phi\left\{\eta_p, \left(\frac{W\sqrt{R_w T_{0T}}}{p_{0T}}\right), \left(\frac{N}{\sqrt{R_w T_{0T}}}\right)\right\} \quad (17.100)$$

The dependence on polytropic efficiency may be dispensed with if polytropic efficiency may be regarded as function of the second and third dimensionless groups:

$$\eta_p = f_{C2}\left\{\left(\frac{W\sqrt{R_w T_{0T}}}{p_{0T}}\right), \left(\frac{N}{\sqrt{R_w T_{0T}}}\right)\right\} \quad (17.101)$$

so that the pressure ratio emerges as:

$$\frac{p_{2T}}{p_{0T}} = f_{C1}\left\{\left(\frac{W\sqrt{R_w T_{0T}}}{p_{0T}}\right), \left(\frac{N}{\sqrt{R_w T_{0T}}}\right)\right\} \quad (17.102)$$

Figure 17.8 shows the form of a manufacturer's curve expressed in terms of pressure ratio. It is customary to plot the pressure ratio and the polytropic efficiency against the mass flow parameter, $(W\sqrt{R_w T_{0T}})/p_{0T}$, for a range of speed parameters, $N/\sqrt{R_w T_{0T}}$. Manufacturers often omit the value of R_w, so that the groups are no longer dimensionless. An alternative formulation is to use the volume flow at compressor inlet:

$$Q_{0T} = v_{0T} W = \frac{R_w T_{0T}}{p_{0T}} \quad (17.103)$$

Hence:

$$\frac{p_{2T}}{p_{0T}} = f_{C1}\left\{\left(\frac{Q_{0T}}{\sqrt{R_w T_{0T}}}\right), \left(\frac{N}{\sqrt{R_w T_{0T}}}\right)\right\} \quad (17.104)$$

$$\eta_p = f_{C2}\left\{\left(\frac{Q_{0T}}{\sqrt{R_w T_{0T}}}\right), \left(\frac{N}{\sqrt{R_w T_{0T}}}\right)\right\} \quad (17.105)$$

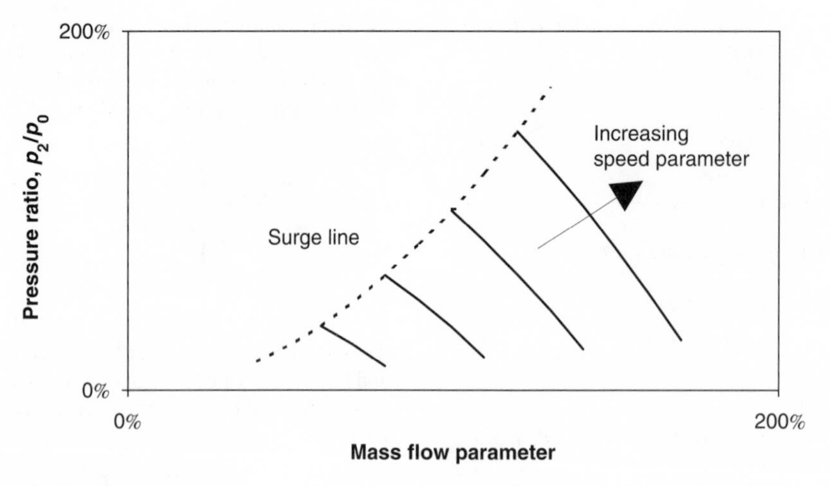

Figure 17.8 Manufacturer's curve expressed in terms of pressure ratio vs. mass flow parameter, $(W\sqrt{R_w T_{0T}})/(p_{0T})$, with speed parameter, $N/(\sqrt{R_w T_{0T}})$ as added variable.

17.8.2 Applying the pressure ratio and efficiency characteristics to estimating flow and section power

The formulation above used stagnation values of inlet and outlet pressure and temperature, but since the section inlet and outlet velocities will be small, little will be lost by using the actual values: $p_0 \approx p_{0T}$, $p_2 \approx p_{2T}$, $T_0 \approx T_{0T}$, $T_2 \approx T_{2T}$. Given the inlet temperature and the speed, the speed parameter, $N/\sqrt{R_w T_{0T}}$, may be calculated. Add to this a knowledge of the section's inlet and outlet pressures, from which to calculate p_2/p_0, and the flow parameter, $(W\sqrt{R_w T_{0T}})/p_{0T}$, may be read off the projection of the characteristic curve onto the horizontal axis, usually after interpolation for the speed parameter. Mass flow, W, may then be disentangled from the flow parameter. A knowledge of both the speed and the flow parameters allows the polytropic efficiency to be found from a map (assumed provided) of equation (17.101).

The power consumed in the section is the flow rate multiplied by the specific work. Using equations (17.68) and (17.78), we have

$$P = Ww = W\frac{\gamma}{\gamma-1}ZR_wT_0\left(\left(\frac{p_2}{p_0}\right)^{1/\eta_p((\gamma-1)/\gamma)} - 1\right) \tag{17.106}$$

17.9 Computing the performance of the complete compressor

We may represent the flows through the N stages of a general compressor by Figure 17.9. For each section of the compressor, there will be an external flow, $W_{E,i}$, a kickback flow, $W_{K,i}$, and a compressor section flow, $W_{C,i}$,. The last external flow, $W_{E,N+1}$, will be from the compressor to the discharge vessel, and some of the intermediate external flows may also be from the compressor to another part of the process.

The pressures, $p_{E,1}, p_{E,2}, \ldots, p_{E,N+1}$ in the vessels to which the compressor is connected will be determined in the simulation by the outputs of integrators (local mass and temperature) and so will be fixed at each timestep. The same applies to the specific volumes of the gas in these vessels. In addition, the speed of the compressor, N, will be the output of an integrator, and may be regarded as fixed at the current time.

Figure 17.9 represents a complex flow network, which may be solved iteratively at each timestep by guessing and then successively refining the estimates of the pressures and temperatures at the compressor nodes, $p_{0,i}, T_{0,i}, i = 1, 2, \ldots, N+1$. The correct values are those that bring the steady-state mass flows and energy flows into balance at each node.

Flow through the ith compressor stage

If the pressure-ratio method is to be used, then a knowledge of the compressor speed, N, the estimated inlet temperature, $T_{0,i}$, inlet pressure, $p_{0,i}$, and outlet pressure, $p_{2,i} = p_{0,i}$, make it possible to calculate first the speed parameter and then the mass flow, $W_{C,i}$, using the pressure ratio equation (17.102). The polytropic efficiency, $\eta_{p,i}$, may then be calculated from equation (17.101).

Alternatively, the same set of variables may be used to calculate the polytropic head at the current

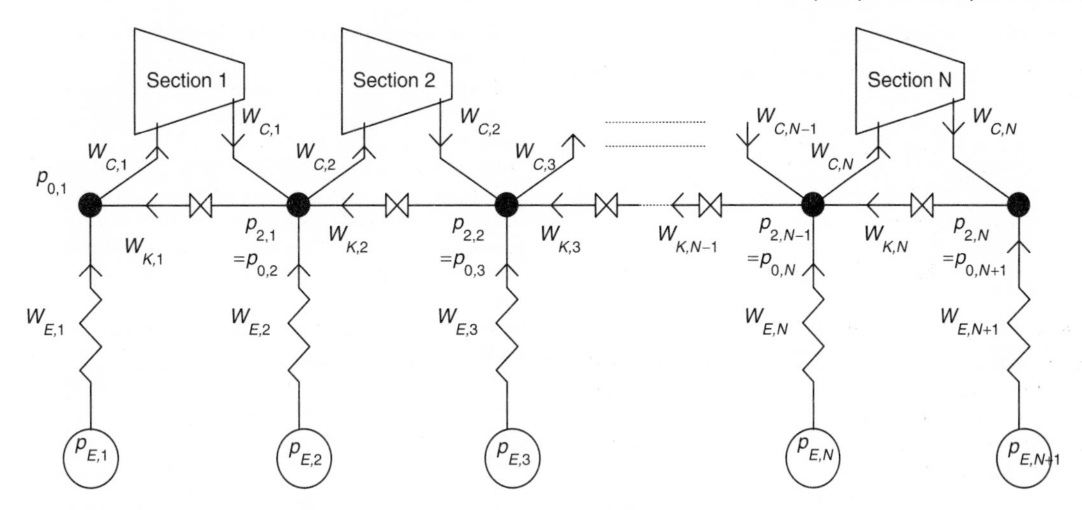

Figure 17.9 Schematic of flows through the compressor.

speed, H_p, from equation (17.85). The polytropic head at the design speed, H_{p0}, may then be calculated from equation (17.87), and then the suction volume flow at the design speed, Q_0, may be found from the H_{p0} vs. Q_0 characteristic. Then the suction volume flow at the current speed, Q, may be found from equation (17.88). Using the subscripts appropriate to Figure 17.9, the specific volume at suction may be calculated from

$$v_{0,i} = \frac{ZR_w T_{0,i}}{p_{0,i}} \tag{17.107}$$

and the mass flow is then $W_{C,i} = Q_{C,i}/v_{0,i}$.

The temperature at compressor section delivery is found from equation (17.76):

$$T_{2,i} = T_{0,i}\left(\frac{P_{2,i}}{T_{0,i}}\right)^{1/\eta_{pi}((\gamma-1)/\gamma)} \tag{17.108}$$

A constant value of polytropic efficiency is usually assumed at each stage if the polytropic head method is used. The error introduced to delivery temperature by using a constant value of efficiency in equation (17.108) is likely to be small.

Kickback flows

The specific volume at each node will have been found from equation (17.107). Hence the kickback flow may be determined using the flow functions developed in Chapters 6 and 10:

$$W_{k,i} = f_{flow}\left(p_{0,i+1}, v_{0,i+1}, \frac{P_{0,i}}{p_{0,i+1}}, A_{k,i}, K_{k,i}, \gamma\right) \tag{17.109}$$

where

$A_{K,i}$ is the cross-sectional area of the kickback pipe returning gas to the inlet of section i, and

$K_{K,i}$ is the number of velocity heads characterizing the frictional drop along the kickback pipe returning gas to the inlet of section i.

External flows into the compressor

The external flow from the suction vessel to the inlet node of the compressor may also be determined using the flow functions developed in Chapters 6 and 10:

$$W_{E,i} = f_{flow}\left(p_{E,i}, v_{E,i}, \frac{P_{0,i}}{P_{E,i}}, A_{E,i}, K_{E,i}, \gamma\right) \tag{17.110}$$

where

$A_{E,i}$ is the cross-sectional area of the external pipe carrying gas to the inlet of section i, and

$K_{E,i}$ is the number of velocity heads characterizing the frictional drop along the external pipe carrying gas to the inlet of section i.

External flows from the compressor

An external flow away from the compressor may be calculated as:

$$W_{E,i} = -W_{D,i}$$

$$W_{D,i} = f_{flow}\left(p_{0,i}, v_{0,i}, \frac{p_{E,i}}{p_{0,i}}, A_{E,i}, K_{E,i}, \gamma\right) \tag{17.111}$$

where $W_{D,i}$ is the absolute value of the discharge flow from node i.

Flow balance at each node

Performing an algebraic sum at each of the nodes gives:

$$W_{E,1} + W_{K,1} - W_{C,1} = 0$$

$$W_{E,i} + W_{K,i} + W_{C,i-1} - W_{K,i-1} - W_{C,i} = 0$$

$$\text{for } i = 2, 3, \ldots, N$$

$$W_{E,N-1} + W_{C,N} - W_{K,N} = 0 \quad (17.112)$$

Energy balance at each node

At each node, the estimated specific enthalpy at the inlet to the section is referred to the specific enthalpy of the first suction vessel, $h_{E,1}$, and may be calculated from the (successively refined) guess for the temperature at section inlet, $T_{0,i}$:

$$h_{0,i} = h_{E,1} + c_p(T_{0,i} - T_{E,1}) \quad (17.113)$$

Similarly, the specific enthalpy at the outlet of each compressor section is given by:

$$h_{2,i} = h_{E,1} + c_p(T_{2,i} - T_{E,1}) \quad (17.114)$$

where the outlet temperature, $T_{2,i}$, has been found using equation (17.108). The specific enthalpy of each of the suction vessels may be similarly referred to that of the first suction vessel:

$$h_{E,i} = h_{E,1} + c_p(T_{E,i} - T_{E,1}) \qquad \text{for } i \geq 2 \quad (17.115)$$

A steady-state energy balance at each of the nodes gives:

$$W_{E,1}h_{E,1} + W_{K,1}h_{0,2} - W_{C,1}h_{0,1} = 0$$

$$\max(0, W_{E,i})h_{E,i} + W_{K,i+1}h_{0,i+1} + W_{C,i-1}h_{2,i-1}$$
$$- (W_{K,i} + W_{C,i} + \max(0, -W_{E,i}))h_{0,i} = 0$$

$$\text{for } i = 2 \text{ to } N$$

$$W_{C,N}h_{2,N} - (W_{K,N} - W_{E,N+1})h_{0,N+1} = 0 \quad (17.116)$$

The maximum-seeking function has been introduced to allow for negative intermediate external flows, i.e. those leaving the compressor. In such case they will not bring enthalpy into the node, but will take it away.

Equation sets (17.112) and (17.116) represent $2N + 2$ nonlinear, simultaneous equations in the $2(N + 1)$ unknowns, $p_{0,i}, T_{0,i}, i = 1, 2, \ldots, N + 1$. An iterative solution may be made at each timestep. The calculation will begin with an initial set of guesses (perhaps the design values), and proceed to a solution at time 0. The resulting estimates of node pressures and temperatures may then be used as the starting values for the next timestep.

The power consumed in each section may be calculated either by equation (17.106) if the pressure-ratio method is being followed or by equation (17.86) if the polytropic head method is being used. The total power consumption is then the sum of the power consumptions of all the individual sections.

17.10 Bibliography

Bacon, D.H and Stephens, R.C. (1990). *Mechanical Technology*, 2nd edition, Heinemann Newnes, London.

Martin, E.N. (1979). 'The modelling of injectors, pumps and compressors in process simulation', *Transactions of the Institute of Measurement and Control*, **1**, No 2, 67–73.

Moore, R.L. (1989). *Control of Centrifugal Compressors*, Instrument Society of America.

Perry, R.H., Green, D.W. and Maloney, J.O. (eds) (1984). *Perry's Chemical Engineers' Handbook*, 6th edition, McGraw-Hill, New York. Chapter 2: Mathematics.

Rogers, G.F.C. and Mayhew, Y.R. (1992). *Engineering Thermodynamics, Work and Heat Transfer*, 4th edition, Longman Scientific and Technical.

Sayers, A.T. (1990). *Hydraulic and Compressible Flow Turbomachines*, McGraw-Hill, London.

Schultz, J.M. (1962). 'The polytropic analysis of centrifugal compressors', *Trans. ASME, J. Eng. for Power*, **84**, 69–82.

Stepanoff, A.J. (1955). *Turboblowers. Theory, Design and Application of Centrifugal and Axial Flow Compressors and Fans*, John Wiley, New York and Chapman & Hall, London.

Stepanoff, A.J. (1957). *Centrifugal and Axial Flow Pumps*, 2nd edition, Krieger Publishing Company, Krieger Drive, Malabar, Florida.

Streeter, V.L. and Wylie, E.B. (1983). *Fluid Mechanics, First SI Metric Edition*, McGraw-Hill, New York.

18 Flow networks

18.1 Introduction

The time constants associated with the settling of flow in a network are often very short compared with the other time constants of the plant being modelled, and in such cases it is justifiable to assume that the flows adjust instantaneously to the slower changes in the conditions (pressure, specific volume) occurring at the boundaries of the network. Further discussion of and justification for the assumption of instantaneous adjustment is given in Section 18.10 of this chapter.

Examples where this approach is valid are:

- where there are a number of conductances in series in a pipe transporting either a liquid or a gas, for example valves and bends;
- a system of pipes supplied with liquid by a pump;
- a network of pipes connected to a compressor;
- the serially connected nozzles and blades making up a steam turbine.

The assumption that the flows in the network are in a continuously evolving steady state brings the benefit that the model is rendered much less stiff thereby. Furthermore, the resulting simultaneous equations may be solved explicity in simple flow networks. In more complex networks, however, the resulting set of nonlinear, simultaneous equations requires a more sophisticated approach in order to bring about the desired savings in model execution time. This chapter considers both simple and complex networks and offers methods for solving the resulting equations efficiently.

18.2 Simple parallel networks

18.2.1 Liquids

Consider the simple parallel network shown in Figure 18.1, where there are N lines connecting a supply point at pressure p_1 (Pa) and height z_1 (m) with a downstream vessel at pressure p_2 and height z_2. If liquid is being carried, the flow in each line will conform to equation (4.81) given below with notation adapted to that of Figure 18.1:

$$W_i = C_i \sqrt{\frac{p_1 - p_2 - \dfrac{g(z_2 - z_1)}{v_1}}{v_1}} \qquad i = 1 \text{ to } N \tag{18.1}$$

where the C_i is the conductance (m^2) of each line, and it has been assumed that the specific volume stays constant at its inlet value, v_1, along each line.

The total flow is found by summation of both sides of equation (18.1):

$$W = \sum_{i=1}^{N} W_i = \left(\sum_{i=1}^{N} C_i\right) \sqrt{\frac{p_1 - p_2 - \dfrac{g(z_2 - z_1)}{v_1}}{v_1}} \tag{18.2}$$

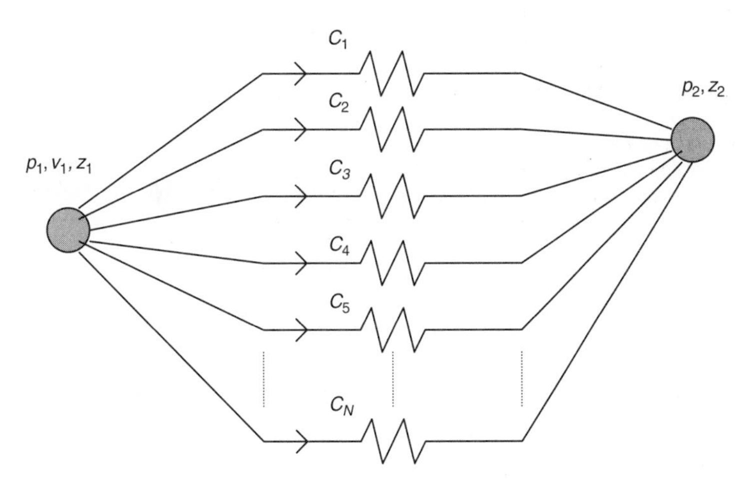

Figure 18.1 Simple parallel network.

We may compare this with a description of flow in terms of an overall conductance, C_T (m^2):

$$W = C_T \sqrt{\frac{p_1 - p_2 - \frac{g(z_2 - z_1)}{v_1}}{v_1}} \tag{18.3}$$

which demonstrates that for a simple parallel network the overall conductance is simply the sum of the individual line conductances:

$$C_T = \sum_{i=1}^{N} C_i \tag{18.4}$$

18.2.2 Gases

The long-pipe approximation for gas flow is given by equation (6.60), which may be tailored to the notation of Figure 18.1 as:

$$W_i = b_{0,i} C_i \sqrt{\frac{p_1 - p_2 - \frac{g(z_2 - z_1)}{v_{ave}}}{v_{ave}}} \quad i = 1 \text{ to } N \tag{18.5}$$

where v_{ave} is the isentropic average specific volume in all the lines, given by:

$$v_{ave} = v_1 \frac{\gamma + 1}{\gamma} \frac{\left(1 - \frac{p_2}{p_1}\right)}{\left(1 - \left(\frac{p_2}{p_1}\right)^{(\gamma+1)/\gamma}\right)} \tag{18.6}$$

(cf. equation (6.62)). Meanwhile $b_{0,i}$ is the adjustment parameter needed in this formulation to account for the compressibility of the gas in each line. An accurate characterization requires $b_{0,i}$ to be a function of the line conductance and the pressure ratio, but it is often possible, if the transient deviations from flowsheet values are not expected to be very great, to use a constant value of $b_{0,i}$ derived by applying equations (18.5) and (18.6) to the flowsheet conditions.

Summing equation (18.5) over all N shows that the conductances may be combined in essentially the same way as for the liquid case. The overall flow is given by:

$$W = b_{0,T} C_T \sqrt{\frac{p_1 - p_2 - \frac{g(z_2 - z_1)}{v_{ave}}}{v_{ave}}} \tag{18.7}$$

where the effective conductance is given by:

$$b_{0,T} C_T = \sum_{i=1}^{N} b_{0,i} C_i \tag{18.8}$$

18.3 Simple series network

18.3.1 Liquids

Consider the simple series network of Figure 18.2, containing N conductances, C_i, $i = 1, \ldots, N$ in series.

We will assume that the network is carrying liquid at constant temperature, so that the specific volume is the same all along the pipe. To retain p_1, z_1, v_1 and p_2, z_2 as the boundary conditions, the pressures and heights downstream of each conductance have been labelled $p_{m,i}, z_{m,i}$ ('m' for midway). We may expand the pressure and height differences between pipe inlet and outlet as follows:

$$\begin{aligned} p_1 - p_2 - \frac{g(z_2 - z_1)}{v_1} &= (p_1 - p_{m1}) - \frac{g(z_{m1} - z_1)}{v_1} \\ &+ (p_{m1} - p_{m2}) - \frac{g(z_{m2} - z_{m1})}{v_1} + \cdots \\ &+ (p_{m,i} - p_{m,i}) - \frac{g(z_{m,i} - z_{m,i})}{v_1} + \cdots \\ &+ (p_{m,N-2} - p_{m,N-1}) - \frac{g(z_{m,N-2} - z_{m,N-1})}{v_1} \\ &+ (p_{m,N-1} - p_2) - \frac{g(z_{m,N-1} - z_2)}{v_1} \end{aligned} \tag{18.9}$$

where we have used the fact that the specific volume throughout will be equal to the inlet specific volume, v_1.

The steady-state flow through each conductance will be the same, and given by an equation of the form of equation (18.1):

$$W = C_i \sqrt{\frac{p_{m,i} - p_{m,i+1} - \frac{g(z_{m,i+1} - z_{m,i})}{v_1}}{v_1}} \quad \text{for } i = 0 \text{ to } N \tag{18.10}$$

Substituting back into equation (18.9), noting that $z_{m,0} = z_1$ and $p_{m,0} = p_1$ while $z_{m,N} = z_2$ and $p_{m,N} = p_2$ gives:

$$\begin{aligned} p_1 - p_2 - \frac{g(z_2 - z_1)}{v_1} &= v_1 \frac{W^2}{C_1^2} + v_1 \frac{W^2}{C_2^2} + \cdots + v_1 \frac{W^2}{C_N^2} \\ &= v_1 W^2 \left(\sum_{i=1}^{N} \frac{1}{C_i^2} \right) \end{aligned} \tag{18.11}$$

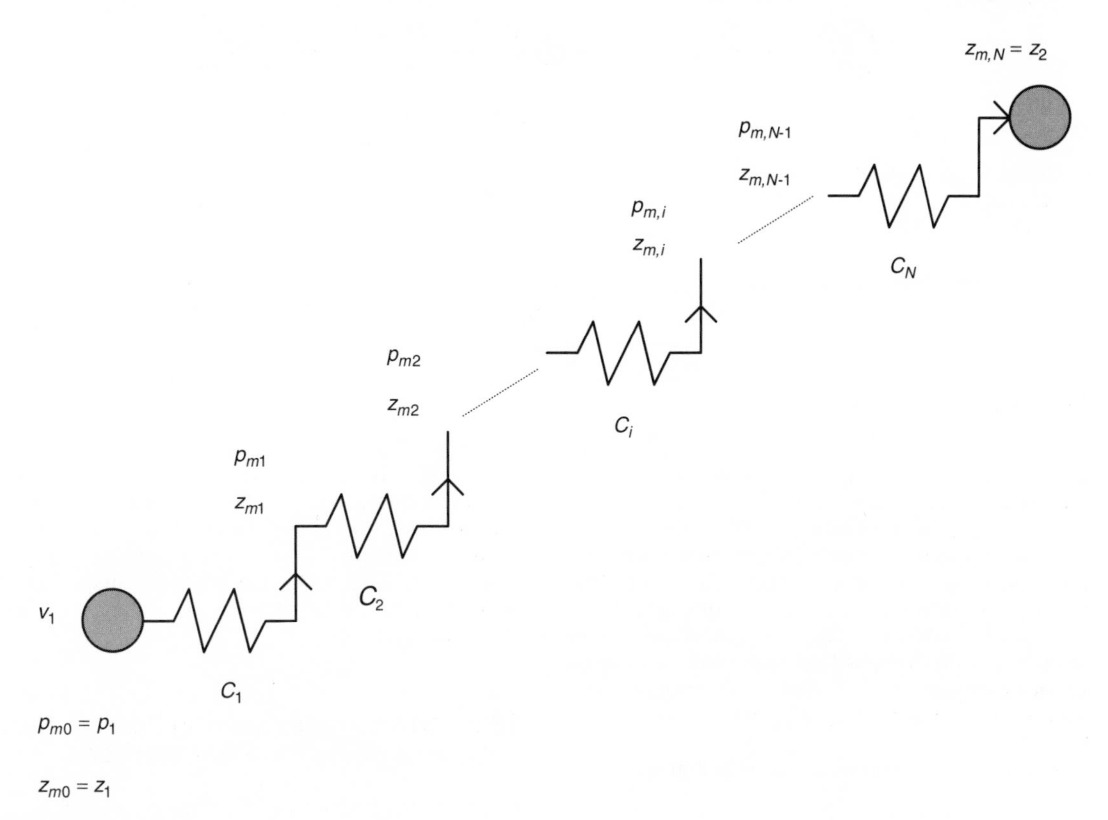

Figure 18.2 Simple series network.

Thus the flow may be written

$$W = C_T \sqrt{\frac{p_1 - p_2 - \dfrac{g(z_2 - z_1)}{v_1}}{v_1}} \tag{18.3}$$

where the overall conductance, C_T, for the simple series network is given by:

$$\frac{1}{C_T^2} = \sum_{i=1}^{N} \frac{1}{C_i^2} \tag{18.12}$$

18.3.2 Gases

Provided the overall pressure drop is not too great, the treatment given above for liquids may be transferred to gases, with the inlet specific volume replaced by the isentropic average specific volume over the line as a whole, v_{ave}, as given by equation (18.6), and the line conductances, C_i, replaced by the effective conductances, $b_{0,i}C_i$. As a result the overall flow equation is given by equation (18.7)

$$W = b_{0,T} C_T \sqrt{\frac{p_1 - p_2 - \dfrac{g(z_2 - z_1)}{v_{ave}}}{v_{ave}}} \tag{18.7}$$

where the overall effective conductance, $b_{0,T}C_T$, is given by:

$$\frac{1}{(b_{0,T}C_T)^2} = \sum_{i=1}^{N} \frac{1}{(b_{0,i}C_i)^2} \tag{18.13}$$

18.4 Complex networks

It will always be worth attempting to reduce the complexity of any extensive flow network by applying the rules for parallel and series networks as outlined in the previous two sections. However, there are many plant arrangements where this approach can only proceed

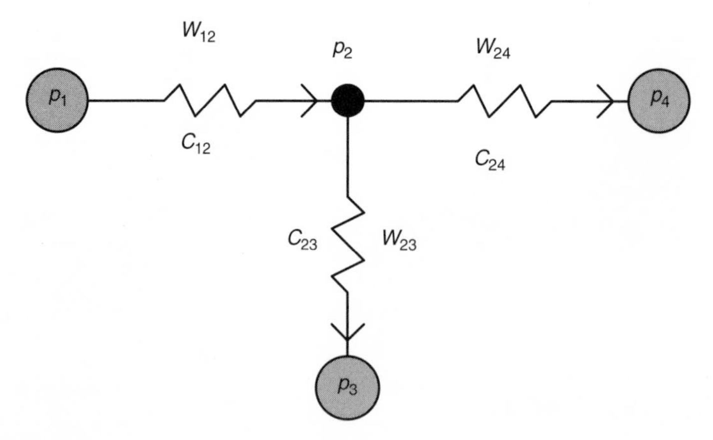

Figure 18.3 Liquid flow network.

a certain distance, leaving a residual set of connected flows that can be reduced no further.

An example of such an irreducible set is given in the flow network of Figure 18.3, assumed for simplicity to carry liquid at a constant temperature. This network can hardly be described as complicated, but it cannot be reduced further using either parallel or serial transformations, and we shall need to solve an implicit equation in order to find the intermediate pressure, p_2, and the flows, W_{12}, W_{23} and W_{24}, as will now be demonstrated.

A steady-state mass balance at node 2 gives

$$W_{12} - W_{23} - W_{24} = 0 \tag{18.14}$$

Since the pipes are carrying a liquid, flow equation (4.81) applies, so that the flow, W_{jk}, from node j to node k may be written as:

$$W_{jk} = C_{jk}\sqrt{\frac{p_j - p_k}{v_j}} \tag{18.15}$$

Substituting back into equation (18.14) under the assumption that the temperature and hence specific volume is invariant over the network gives

$$C_{12}\sqrt{p_1 - p_2} - C_{23}\sqrt{p_2 - p_3} - C_{24}\sqrt{p_2 - p_4} = 0 \tag{18.16}$$

Equation (18.16) gives the intermediate pressure, p_2, in terms of the boundary pressures, p_1, p_3 and p_4. However, it is impossible to write down an explicit algebraic expression for p_2, which implies that equation (18.16) is an implicit equation.

The problems encountered on the simple liquid system described above are met once again in accentuated form for example in the modelling of a steam turbine with several stages, as described in Chapters 15 and 16. Here it will be necessary to solve for a number of thermodynamic variables, including efficiencies and specific work, as well as for flows. Nevertheless the basic method of solving such a complex flow network is the same as for the relatively simple liquid network discussed in Section 2.10 of Chapter 2 and shown in Figure 2.3.

18.5 Strategy for solving flow networks using iterative methods

Invariably the flow network will form part of a larger model, which will define the boundary pressures. These will remain fixed during the process of iteration. The basic strategy is to use initial guesses of the pressures at the intermediate nodes and, in some cases, of the nodal temperatures or specific volumes, in order to calculate the flows and then to see if they balance. The guesses will then be refined successively using an algorithm such as Newton–Raphson until the flows balance to a predetermined tolerance.

Using the notation of Section 2.10, the implicit flow equations to be solved at each timestep will take the form:

$$\mathbf{g}(\mathbf{z}(t), \mathbf{x}(t), \mathbf{u}(t)) = \mathbf{0} \tag{2.93}$$

where $\mathbf{g}$ is the set of nonlinear, simultaneous equations representing the flow balance at each intermediate node, $\mathbf{z}$ is the vector of intermediate pressures and specific volumes (implicit variables), $\mathbf{x}$ is the vector of boundary pressures (state variables), and $\mathbf{u}$ is the vector of inputs. $\mathbf{x}$ and $\mathbf{u}$ will be defined at time t by the larger model and will remain unaltered during each iteration sequence.

Successively better estimates for $\mathbf{z}$ at each time, t, may be calculated using the Newton–Raphson algorithm:

$$\mathbf{z}^{(j+1)} = \mathbf{z}^{(j)} - \left[\frac{\partial \mathbf{g}(\mathbf{z}^{(j)}, \mathbf{x}(t), \mathbf{u}(t))}{\partial \mathbf{z}}\right]^{-1} \times \mathbf{g}(\mathbf{z}^{(j)}, \mathbf{x}(t), \mathbf{u}(t)) \tag{2.95}$$

where j is the iteration index. The integration algorithm for the model as a whole will not be permitted to march on until the iteration is complete at each timestep.

18.6 Modifying the flow equations to speed up the Newton–Raphson method

Applying the Newton–Raphson method to a flow network is not without its problems. These may be understood by exploring the method's application to a single implicit equation, where the algorithm will produce successive approximations to the root of the equation $g(z) = 0$ from the scalar version of equation (2.95), namely:

$$z^{(j+1)} = z^{(j)} + \frac{g(z^{(j)})}{\dfrac{dg(z^{(j)})}{dz}} \approx z^{(j)} + \frac{g(z^{(j)})}{\dfrac{\Delta g(z^{(j)})}{\Delta z}} \tag{18.17}$$

Inspection of equation (18.17) shows that difficulties will occur when the function $g(z)$ contains a discontinuity, which will lead to an inconsistent value of the numerical derivative. Just such a discontinuity may be introduced into a flow network by the presence of a non-return valve. Accordingly it will be preferable to ignore the action of the non-return valve if it is known that the flow will always turn out to be in the forward direction in the transient under consideration. Allowing for a small amount of artificial leakage is a way around problems of convergence if the pressure drop is expected to experience reversal at some point in the transient.

Problems of convergence can also arise when the function, while continuous, contains a region in which the derivative becomes either zero or infinite. It is clear from equation (18.17) that a zero derivative will lead to a next value of z that is unbounded. On the other hand, suppose that at a particular point generated by the algorithm, say the Jth, the derivative $dg(z^{(J)})/dz$ becomes infinite. The corresponding value of the numerical derivative, $\Delta g(z^{(J)})/\Delta z$, will be very large, with the result that, using equation (18.17), the value of the next estimate, $z^{(J+1)}$, will be little different from the current estimate, $z^{(J)}$. It may be seen that the region on and around $z^{(J)}$ becomes a potential trap for the method: if it 'lands' in such a region, the algorithm will proceed only very slowly away from it, and convergence will be extremely slow.

A problem for the modeller is that all the normal flow equations, liquid and gas, contain a point of inflexion at flow reversal, when the derivative takes an infinite value. Thus all the normal flow equations possess an 'inflexion region' from which, once there, the algorithm will emerge only slowly.

To illustrate the situation further, let us consider again the flow of liquid from node j to node k in a pipework system. Developing equation (18.15) gives:

$$W_{jk} = C_{jk}\sqrt{\frac{p_j}{v_j}}\sqrt{1 - \frac{p_k}{p_j}} \tag{18.18}$$

if the pressure at node j is greater than that at node k, while if the pressure gradient is in the opposite direction the flow from node j to node k will be given by:

$$\begin{aligned} W_{jk} &= -C_{jk}\sqrt{\frac{p_k - p_j}{v_k}} \\ &= -C_{jk}\sqrt{\frac{p_j}{v_j}}\sqrt{\frac{v_j}{v_k}}\sqrt{\frac{p_k}{p_j} - 1} \end{aligned} \tag{18.19}$$

Hence we may write:

$$W_{jk} = C_{jk}\sqrt{\frac{p_j}{v_j}}\, f_{liq,jk}\left(\frac{v_j}{v_k}, \frac{p_k}{p_j}\right) \tag{18.20}$$

where

$$\begin{aligned} f_{liq,jk}\left(\frac{v_j}{v_k}, \frac{p_k}{p_j}\right) &= \sqrt{1 - \frac{p_k}{p_j}} && \text{for } \frac{p_k}{p_j} \le 1 \\ &= -\sqrt{\frac{v_j}{v_k}}\sqrt{\frac{p_k}{p_j} - 1} && \text{for } \frac{p_k}{p_j} > 1 \end{aligned} \tag{18.21}$$

The function, $f_{liq,jk}$, is plotted in Figure 18.4 against p_k/p_j under the condition that the two nodal specific volumes are equal.

Figure 18.4 demonstrates that the curve contains an inflexion region centered on $p_k/p_j = 1$, the point of flow reversal.

To understand the effect of possible flow reversals occurring within a network of interconnected flows, let us return to the very simple liquid-flow network of Figure 18.3. As already noted, although the network is simple in construction, it cannot be reduced further

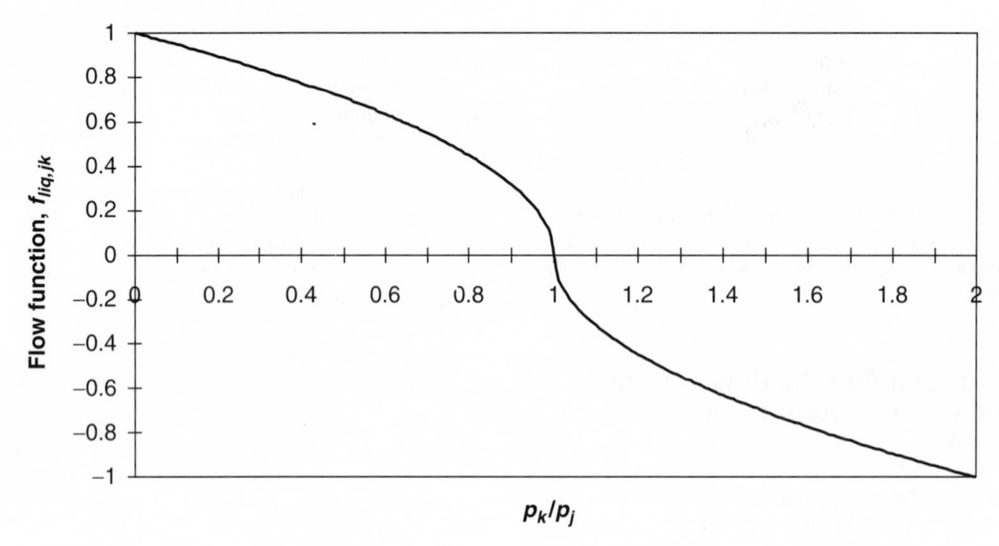

Figure 18.4 Liquid flow function versus pressure ratio.

using either parallel or serial transformations, and we shall need to solve an implicit equation in order to find the intermediate pressure, p_2, and the flows, W_{12}, W_{23} and W_{24}. From equation (18.20), the mass flow from node 1 to node 2 is

$$W_{12} = C_{12}\sqrt{\frac{p_1}{v_1}} f_{liq\,12}\left(\frac{v_1}{v_2}, \frac{p_2}{p_1}\right) \tag{18.22}$$

The mass flow from node 2 to node 3 may be written using the notation developed above as:

$$\begin{aligned} W_{23} &= C_{23}\sqrt{\frac{p_2}{v_2}} f_{liq\,23}\left(\frac{v_2}{v_3}, \frac{p_3}{p_2}\right) \\ &= C_{12}\sqrt{\frac{p_1}{v_1}} \times \frac{C_{23}}{C_{12}}\sqrt{\frac{v_1}{v_2}}\sqrt{\frac{p_2}{p_1}} f_{liq\,23}\left(\frac{v_2}{v_3}, \frac{p_3}{p_2}\right) \\ &= C_{12}\sqrt{\frac{p_1}{v_1}} \times \frac{C_{23}}{C_{12}} F_{liq\,23}\left(\frac{v_1}{v_2}, \frac{v_1}{v_3}, \frac{p_2}{p_1}, \frac{p_3}{p_1}\right) \end{aligned} \tag{18.23}$$

where:

$$\begin{aligned} F_{liq\,23} &= \sqrt{\frac{v_1}{v_2}}\sqrt{\frac{p_2}{p_1}} \times f_{liq\,23} \\ &= \sqrt{\frac{v_1}{v_2}}\sqrt{\frac{p_2}{p_1}} \times \sqrt{1 - \frac{p_3}{p_2}} \\ &= \sqrt{\frac{v_1}{v_2}}\sqrt{\frac{p_2}{p_1} - \frac{p_3}{p_1}} \qquad \text{for } \frac{p_3}{p_1} \le \frac{p_2}{p_1} \\ &= \sqrt{\frac{v_1}{v_2}}\sqrt{\frac{p_2}{p_1}} \times -\sqrt{\frac{v_2}{v_3}}\sqrt{\frac{p_3}{p_2} - 1} \\ &= -\sqrt{\frac{v_1}{v_3}}\sqrt{\frac{p_3}{p_1} - \frac{p_2}{p_1}} \qquad \text{for } \frac{p_3}{p_1} > \frac{p_2}{p_1} \end{aligned} \tag{18.24}$$

In a similar way, the flow from node 2 to node 4 is given by:

$$W_{24} = C_{12}\sqrt{\frac{p_1}{v_1}} \times \frac{C_{24}}{C_{12}} F_{liq\,24}\left(\frac{v_1}{v_2}, \frac{v_1}{v_4}, \frac{p_2}{p_1}, \frac{p_4}{p_1}\right) \tag{18.25}$$

where:

$$\begin{aligned} F_{liq\,24} &= \sqrt{\frac{v_1}{v_2}}\sqrt{\frac{p_2}{p_1} - \frac{p_4}{p_1}} \qquad \text{for } \frac{p_4}{p_1} \le \frac{p_2}{p_1} \\ &= -\sqrt{\frac{v_1}{v_4}}\sqrt{\frac{p_4}{p_1} - \frac{p_2}{p_1}} \qquad \text{for } \frac{p_4}{p_1} > \frac{p_2}{p_1} \end{aligned} \tag{18.26}$$

A steady-state mass balance at node 2 implies that the algebraic sum of the flows must be zero:

$$W_{12} - W_{23} - W_{24} = 0 \tag{18.27}$$

Substituting from equations (18.22), (18.23) and (18.25) and scaling by dividing throughout by the common term $C_{12}\sqrt{p_1/v_1}$ reduces equation (18.27) to the form:

$$f_{liq\,12} - \frac{C_{23}}{C_{12}} F_{liq\,23} - \frac{C_{24}}{C_{12}} F_{liq\,24} = 0 \tag{18.28}$$

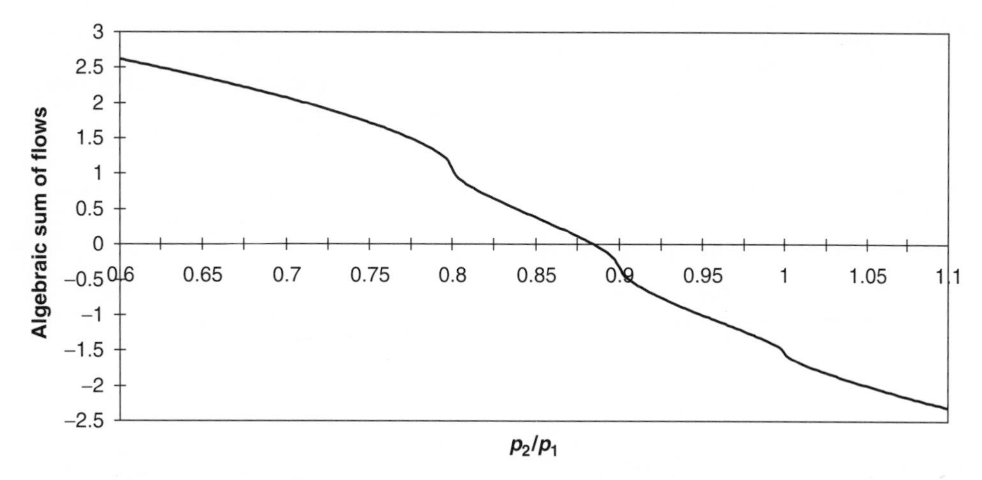

Figure 18.5 Scaled algebraic sum of the flows to node 2.

The problem caused by inflexion regions may be understood by evaluating equation (18.28) for a sample set of numerical values. The situation is simplified by taking the common case where the liquid temperature and hence specific volume are the same throughout the network:

$$v_1 = v_2 = v_3 = v_4 \tag{18.29}$$

Let us suppose that, in our illustrative example, the conductance in each downstream leg is double that in the upstream leg:

$$\begin{aligned} \frac{C_{23}}{C_{12}} &= 2 \\ \frac{C_{24}}{C_{12}} &= 2 \end{aligned} \tag{18.30}$$

The pressures, p_1, p_3 and p_4, are boundary conditions that can be assumed to be either the outputs of integrators or algebraic combinations of the outputs of integrators. They will therefore be known at any given time instant and will remain unchanged during the iteration cycle needed to solve the flow network equations. Let us suppose that at a particular time instant, the ratios of the downstream pressures to the upstream pressure have been found to be

$$\begin{aligned} \frac{p_3}{p_1} &= 0.8 \\ \frac{p_4}{p_1} &= 0.9 \end{aligned} \tag{18.31}$$

Equations (18.31), taken together with the assumption of constant specific volume (equation (18.29)), render equation (18.28) into the form

$$F\left(\frac{p_2}{p_1}\right) = 0 \tag{18.32}$$

where F is the scaled algebraic sum of flows. The flow function, F, is shown versus pressure ratio, p_2/p_1, in Figure 18.5. The solution to equation (18.32) is that value of pressure ratio, p_2/p_1, which causes the algebraic sum of the flows to be zero, and thus that which causes the curve shown in Figure 18.5 to cross the pressure-ratio axis. The Newton–Raphson iterative procedure begins with an initial guess of the solution value for p_2/p_1 and then makes successively better estimates.

It is clear from the figure that the inflexion regions contained in all of the individual flow functions have been carried across to the flow balance function more or less without modification. One inflexion region occurs for each flow in the flow balance equation, in this case at the pressure ratios:

$$\begin{aligned} p_2/p_1 &= 1.0 \quad (\text{i.e. } p_2 = p_1) \\ &= 0.9 \quad (\text{i.e. } p_2 = p_4) \\ &= 0.8 \quad (\text{i.e. } p_2 = p_3) \end{aligned} \tag{18.33}$$

These inflexion regions act rather like bunkers in golf, in which the unwary golfer can become trapped and may waste many shots attempting to get out. The inflexion 'bunkers' lie in wait for the successive estimates of pressure ratio produced by the algorithm. In the case under consideration, a well-chosen initial guess might mean that the traps at $p_2/p_1 = 0.8$ and 1.0 might be avoided altogether, but the trap at $p_2/p_1 =$ 0.9 lies very close to the solution and has a good chance of catching one of the estimates generated by the algorithm, leading to slow convergence.

A method of circumventing these problems is to modify the form of the flow function to remove its point of inflexion. For example, a linear approximation to the differential pressure function may be used in the neighbourhood of $p_k/p_j = 1$:

$$W_{jk} = C_{jk}\sqrt{\frac{p_j}{v_j}}\sqrt{1-\frac{p_k}{p_j}} \quad \text{for } \frac{p_k}{p_j} \le 0.98$$

$$= C_{jk}\sqrt{\frac{p_j}{v}} \times 7.07107\left(1-\frac{p_k}{p_j}\right)$$

$$\text{for } 0.98 \le \frac{p_k}{p_j} < 1.02 \qquad (18.34)$$

$$= C_{jk}\sqrt{\frac{p_j}{v_k}}\sqrt{\frac{p_k}{p_j}-1} \quad \text{for } \frac{p_k}{p_j} > 1.02$$

In addition the transition between upstream and downstream specific volumes may be smoothed, so that the transitional specific volume, v, used in equation (18.34) may be given by:

$$v = v_j \quad \text{for } \frac{p_k}{p_j} \le 1.0$$

$$= 50\left(v_j\left(1.02-\frac{p_k}{p_j}\right) + v_k\left(\frac{p_k}{p_j}-1\right)\right)$$

$$\text{for } 1.0 < \frac{p_k}{p_j} \le 1.02 \qquad (18.35)$$

$$= v_k \quad \text{for } \frac{p_k}{p_j} > 1.02$$

This formulation for transitional specific volume allows for a smooth transition at flow reversal, while giving a slightly greater weight to the accurate determination of forward flow.

The flow-balance function obtained by using the modified equations (18.34) and (18.35) is illustrated in Figure 18.6 for the example considered previously using the full set of equations. It is clear that the curve displayed is a good approximation to that of Figure 18.5, but has the important property that it avoids the inflexion traps.

The general problem of points of inflexion applies equally to gas flow networks. Take, for example, the steam flows in a steam turbine. Using the idealized, isentropic form given in Chapter 16, Section 16.8, the flow at pressure ratios in excess of the respective critical pressures for both a convergent-only and a convergent–divergent nozzle is given by (see equations (16.83) and (16.84)):

$$W = A_1\sqrt{2\frac{\gamma}{\gamma-1}\frac{p_0}{p_1}}\, f\left(\frac{p_1}{p_0}\right) \qquad (18.36)$$

where

$$f\left(\frac{p_1}{p_0}\right) = \sqrt{\left(\frac{p_1}{p_0}\right)^{2/\gamma} - \left(\frac{p_1}{p_0}\right)^{(\gamma+1)/\gamma}} \qquad (18.37)$$

and $A_1 = A_t$ for a convergent-only nozzle. The flow will exhibit a point of inflexion with infinite slope at a pressure ratio, $r = p_1/p_0 = 1$. This can be avoided by approximating the function of equation (18.37) by a linear approximation above a selected pressure ratio $r_s = (p_1/p_0|_s)$, such that the line passes through the point $(r_s, f(r_s))$ and through the point (1,0). This linear approximation will thus have the form:

$$f_{approx}(r) = \sqrt{r^{2/\gamma} - r^{(\gamma+1)/\gamma}} \qquad \text{for } r \le r_s$$

$$= \frac{r-1}{r_s-1}\sqrt{r_s^{2/\gamma} - r_s^{(\gamma+1)/\gamma}} \qquad \text{for } r > r_s \qquad (18.38)$$

A reasonable value for r_s may be 0.98, as was the case for the liquid flow approximation. Reverse flow in the

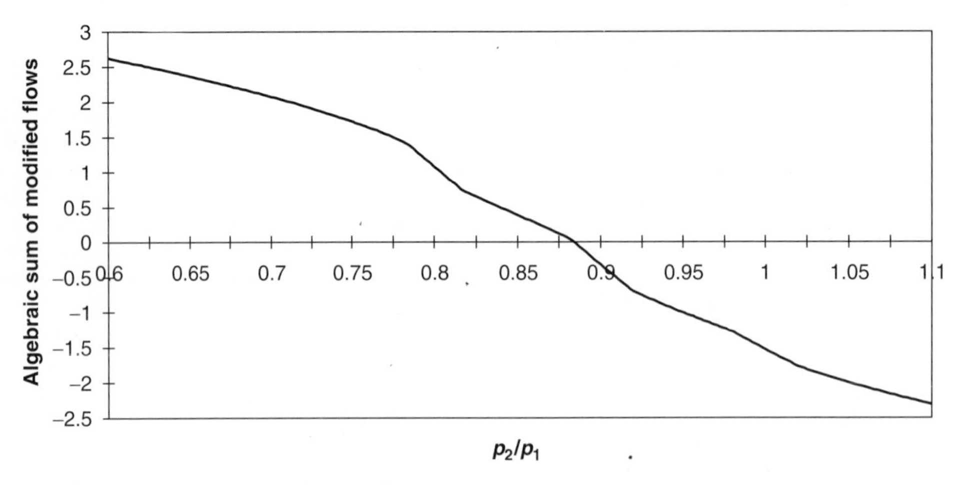

Figure 18.6 Scaled algebraic sum of the modified flows to node 2.

turbine will not be expected for many transients of interest, and in such cases the modeller will be content to use equation (18.37) with no upper limit on pressure ratio, $r = p_1/p_0$.

18.7 Solving the steady-state flow network using the Method of Referred Derivatives

18.7.1 Outline of the method

An alternative method of solving the implicit equations associated with complex flow networks is to use the Method of Referred Derivatives discussed in Section 2.11 of Chapter 2. The general set of nonlinear, simultaneous equations

$$\mathbf{g}(\mathbf{z}(t), \mathbf{x}(t), \mathbf{u}(t)) = \mathbf{0} \qquad (2.93)$$

may be differentiated with respect to time to give a new set of equations that are **linear** in the derivatives, $d\mathbf{z}(t)/dt$:

$$\frac{\partial \mathbf{g}}{\partial \mathbf{z}}\frac{d\mathbf{z}}{dt} = -\frac{\partial \mathbf{g}}{\partial \mathbf{x}}\frac{d\mathbf{x}}{dt} - \frac{\partial \mathbf{g}}{\partial \mathbf{u}}\frac{d\mathbf{u}}{dt} \qquad (2.98)$$

where the matrices,

$$\frac{\partial \mathbf{g}}{\partial \mathbf{z}}(t),\ \frac{\partial \mathbf{g}}{\partial \mathbf{x}}(t),\ \frac{\partial \mathbf{g}}{\partial \mathbf{u}}(t)$$

may be derived numerically or, in simpler cases, analytically. Because it is linear, equation (2.98) may be solved relatively easily for the derivatives, $d\mathbf{z}(t)/dt$, for example using Gauss elimination. The derivatives may then be integrated numerically (along with the state variables) from the initial conditions $\mathbf{z}(t_0)$ at time, t_0, to give the set of values, $\mathbf{z}(t)$, at any later time, t.

The method assumes the availability of a set of values, $\mathbf{z}(t_0)$, of the implicit variables (normally network nodal pressures and specific volumes) at the beginning of the transient, time t_0. Methods of finding $\mathbf{z}(t_0)$ are discusssed in the next section.

18.7.2 Determining the initial conditions for the implicit variables: Prior Transient Integration and Extended Prior Transient Integration

The initial conditions for the implicit variables, $\mathbf{z}(t_0)$, will be a solution of the equation

$$\mathbf{g}(\mathbf{z}(t_0), \mathbf{x}(t_0), \mathbf{u}(t_0)) = \mathbf{0} \qquad (18.39)$$

One way of finding the initial conditions, $\mathbf{z}(t_0)$, is to use an iterative method. Since this iteration would precede the transient to be considered, it would be off-line to the main simulation and thus avoid the drawback associated with the uncertain number of iterations needed to give convergence.

However, an option more in keeping with the Method of Referred Derivatives is that of 'Prior Transient Integration'. At an assumed time, t_{prior}, prior to the transient start time, we select an artificial set of states, $\mathbf{x}(t_{prior})$, and inputs, $\mathbf{u}(t_{prior})$, chosen solely on the basis that they enable a direct (i.e. non-iterative) solution for the variables $\mathbf{z}$ to be found at time t_{prior} from:

$$\mathbf{g}(\mathbf{z}(t_{prior}), \mathbf{x}(t_{prior}), \mathbf{u}(t_{prior})) = \mathbf{0} \qquad (18.40)$$

The states and inputs are then made to follow any convenient trajectory from the artificial conditions $\mathbf{x}(t_{prior})$, $\mathbf{u}(t_{prior})$ to their starting values, $\mathbf{x}(t_0)$, $\mathbf{u}(t_0)$, at the beginning of the main transient. It is simplest if the trajectory is chosen to be a ramp, so that the time differentials over the period t_{prior} to t_0 become:

$$\begin{aligned}\frac{d\mathbf{x}}{dt} &= \frac{1}{t_0 - t_{prior}}(\mathbf{x}(t_0) - \mathbf{x}(t_{prior}))\\ \frac{d\mathbf{u}}{dt} &= \frac{1}{t_0 - t_{prior}}(\mathbf{u}(t_0) - \mathbf{u}(t_{prior}))\end{aligned} \qquad (18.41)$$

Equation (2.98) may now be solved for $(d\mathbf{z}/dt)(t)$ and then the derivatives, $(d\mathbf{z}/dt)$, $(d\mathbf{x}/dt)$, $(d\mathbf{u}/dt)$, integrated from t_{prior} to t_0, at which time $\mathbf{x}(t_0)$, $\mathbf{u}(t_0)$, and, most importantly, $\mathbf{z}(t_0)$ will have been computed.

The above procedure has the advantage that it maintains the calculational structure of the main simulation programme. However, we may give ourselves rather more latitude in the process of finding the initial conditions, $\mathbf{z}(t_0)$, if we assume that some of the parameters in the constraint equation (2.93), hitherto regarded as constant, are now allowed to vary in the notional time interval, t_{prior} to t_0. Let the parameters now allowed to vary be denoted by the vector, $\mathbf{a}$, and let the constraint equation be modified accordingly to

$$\mathbf{h}(\mathbf{z}(t), \mathbf{x}(t), \mathbf{u}(t), \mathbf{a}(t)) = \mathbf{0} \qquad (18.42)$$

which on differentiating gives, once again, a system of linear equations in the unknown derivatives, $d\mathbf{z}/dt$:

$$\frac{\partial \mathbf{h}}{\partial \mathbf{z}}\frac{d\mathbf{z}}{dt} = -\frac{\partial \mathbf{h}}{\partial \mathbf{x}}\frac{d\mathbf{x}}{dt} - \frac{\partial \mathbf{h}}{\partial \mathbf{u}}\frac{d\mathbf{u}}{dt} - \frac{\partial \mathbf{h}}{\partial \mathbf{a}}\frac{d\mathbf{a}}{dt} \qquad (18.43)$$

(cf. equation (2.98)). Let us choose the values of $\mathbf{x}(t_{prior})$, $\mathbf{u}(t_{prior})$ and $\mathbf{a}(t_{prior})$ so that a direct solution, $\mathbf{z}(t_{prior})$, is possible for the equation:

$$\mathbf{h}(\mathbf{z}(t_{prior}), \mathbf{x}(t_{prior}), \mathbf{u}(t_{prior}), \mathbf{a}(t_{prior})) = \mathbf{0} \qquad (18.44)$$

Now assume that the selected constants follow a convenient trajectory such as a ramp:

$$\frac{d\mathbf{a}}{dt} = \frac{1}{t_0 - t_{prior}}(\mathbf{a}(t_0) - \mathbf{a}(t_{prior})) \qquad (18.45)$$

in the same way as the state and input variables, whose derivatives are given in equation (18.41). We may now solve equation (18.43) for $d\mathbf{z}/dt(t)$, and then integrate $(d\mathbf{z}/dt)$, $(d\mathbf{x}/dt)$, $(d\mathbf{u}/dt)$, $(d\mathbf{a}/dt)$ from t_{prior} to t_0 to find $\mathbf{z}(t_0)$. This method of finding the initial conditions brings in an extra set of integrations for the nominally constant parameters, and is therefore entitled 'Extended Prior Transient Integration'.

18.8 Worked example using the Method of Referred Derivatives: liquid flow network

Figure 2.3, repeated below, shows a network carrying water at a constant temperature, for which the specific volume may be taken as 0.001 m^3/kg. All the conductances are the same and equal to 2.45×10^{-4} m^2.

The network is assumed to form part of a larger model, to which it is interfaced through accumulators at nodes 3 and 5. The pressures, p_3, p_5, are assumed to depend on conditions both within the network and external to it, obeying the differential equations:

$$\begin{aligned} \frac{dp_3}{dt} &= k_3 W_{23} - c_3(p_3 - p_{3M}) \\ \frac{dp_5}{dt} &= k_5 W_{45} - c_5(p_5 - p_{5M}) \end{aligned} \qquad (18.46)$$

where p_{3M}, p_{5M} are pressures in the main part of the model. Pressures p_1, p_6 are imposed by external conditions, and are assumed to be under perfect pressure control, obeying the differential equations

$$\begin{aligned} \frac{dp_1}{dt} &= \frac{p_{1s} - p_1}{T_1} \\ \frac{dp_6}{dt} &= \frac{p_{6s} - p_6}{T_6} \end{aligned} \qquad (18.47)$$

where p_{1s}, p_{6s} are the pressure setpoints. The constant coefficients in equations (18.46) and (18.47) are given in Table 18.1.

Table 18.1 Constant coefficients used in worked example, equations (18.46) and (18.47)

T_1 (s)	T_6 (s)	c_3 (s^{-1})	c_5 (s^{-1})	k_3 (Pa kg^{-1}s)	k_5 (Pa kg^{-1}s)
50	50	0.05	0.0333	500	1000

The main model pressures, p_{3M}, p_{5M}, and the external pressure setpoints, p_{1s}, p_{6s}, undergo step changes at time 0, as summarized in Table 18.2.

Table 18.2 Step changes in main model pressures and external pressure setpoints

Time (s)	p_{1s} (kPa)	p_{3M} (kPa)	p_{5M} (kPa)	p_{6s} (kPa)
0	1000	900	800	600
0+	1050	850	820	680

Simulate the transient behaviour of the system for 100 seconds and determine the intermediate pressures, p_2, p_4, and the network flows, W_{12}, W_{23}, W_{24}, W_{45} and W_{46} at each time instant.

Solution

Pressures p_3, p_5 will be regarded as (slow) state variables, $\mathbf{x}$, and pressures p_1, p_6 will be regarded as input variables, $\mathbf{u}$. We may safely assume that the intermediate nodal pressures, p_2, p_4, will have fast dynamics

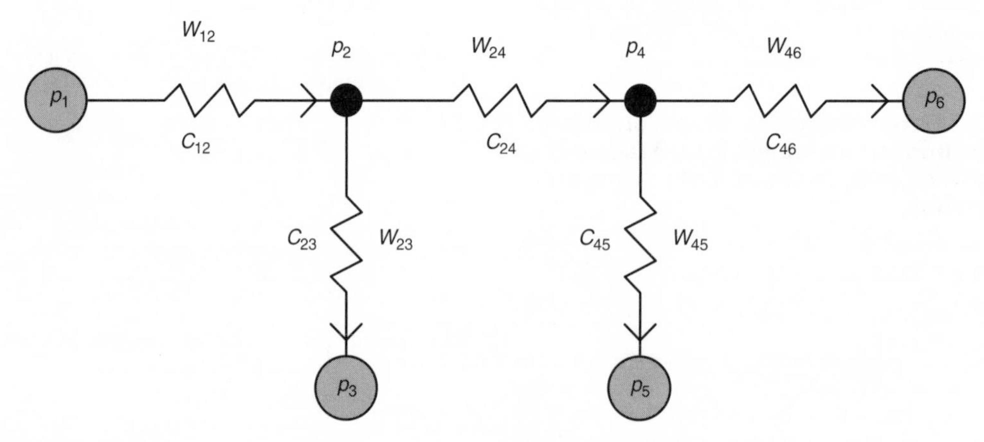

Figure 2.3 Liquid flow network.

compared with the other pressures and we will choose these to be the implicit variables, $\mathbf{z}$.

From inspection of Figure 2.3, the equations defining the steady-state flow balances are:

$$g_1(p_1, p_2, \ldots, p_6) = W_{12} - W_{23} - W_{24} = 0$$
$$g_2(p_1, p_2, \ldots, p_6) = W_{24} - W_{45} - W_{46} = 0 \quad (18.48)$$

Since the flow is liquid, the mass flow from node j to node k, W_{jk}, will obey:

$$W_{jk} = C_{jk}\sqrt{\frac{p_j}{v}}\sqrt{\left|1 - \frac{p_k}{p_j}\right|} \times \frac{1 - \dfrac{p_k}{p_j}}{\left|1 - \dfrac{p_k}{p_j}\right|} \quad (18.49)$$

where, in this network, each line conductance, C_{jk}, will be the same: $C_{jk} = C = 2.45 \times 10^{-4}\,\text{m}^2$.

Applying equation (2.98) to the implicit equation set (18.48), the derivatives of the intermediate pressures, $\mathbf{z} = (p_2, p_4)^T$, are given in terms of the derivatives of the interface pressures, $\mathbf{x} = (p_3, p_5)^T$, and the derivatives of the externally determined pressures, $\mathbf{u} = (p_1, p_6)^T$, by:

$$\begin{bmatrix} \dfrac{\partial W_{12}}{\partial p_2} - \dfrac{\partial W_{23}}{\partial p_2} - \dfrac{\partial W_{24}}{\partial p_2} & -\dfrac{\partial W_{24}}{\partial p_4} \\ \dfrac{\partial W_{24}}{\partial p_2} & \dfrac{\partial W_{24}}{\partial p_4} - \dfrac{\partial W_{45}}{\partial p_4} - \dfrac{\partial W_{46}}{\partial p_4} \end{bmatrix} \begin{bmatrix} \dfrac{dp_2}{dt} \\ \dfrac{dp_4}{dt} \end{bmatrix}$$

$$= \begin{bmatrix} \dfrac{\partial W_{23}}{\partial p_3} & 0 \\ 0 & \dfrac{\partial W_{45}}{\partial p_5} \end{bmatrix} \begin{bmatrix} \dfrac{dp_3}{dt} \\ \dfrac{dp_5}{dt} \end{bmatrix}$$

$$+ \begin{bmatrix} -\dfrac{\partial W_{12}}{\partial p_1} & 0 \\ 0 & \dfrac{\partial W_{46}}{\partial p_6} \end{bmatrix} \begin{bmatrix} \dfrac{dp_1}{dt} \\ \dfrac{dp_6}{dt} \end{bmatrix} \quad (18.50)$$

The partial differentials may be found analytically in this case. Differentiation of equation (18.49) gives:

$$\frac{\partial W_{jk}}{\partial p_j} = \frac{W_{jk}}{2(p_j - p_k)} \quad (18.51)$$

$$\frac{\partial W_{jk}}{\partial p_k} = \frac{-W_{jk}}{2(p_j - p_k)} \quad (18.52)$$

Generating the starting conditions using Prior Transient Integration

To produce a convenient set of conditions at time, t_{prior}, the start time for the prior transient, let us make the artificial assumptions:

(i) that the interface pressure, p_3, is equal to the starting value of the externally applied pressure, p_1, and
(ii) that the interface pressure, p_5, is equal to the starting value of the externally applied pressure, p_6.

This changes the effective layout of the network at time, t_{prior}, from that of Figure 2.3 to that of Figure 18.7.

The advantage of this simplified arrangement is that we may use parallel and serial transformations to generate the total conductance. Summing parallel conductances, the overall conductance between new combined node 1 and node 2 is simply $2C$, as is the overall conductance between node 4 and new combined node 6. The overall conductance, C_T, between new node 1 and new node 6 is found by the series transformation of summing the inverse squares (equation (18.4)). Hence:

$$\frac{1}{C_T^2} = \frac{1}{(2C)^2} + \frac{1}{C^2} + \frac{1}{(2C)^2} = \frac{6}{4C^2}$$

so that $C_T = \sqrt{\frac{2}{3}} \times C = 2.0004 \times 10^{-4}\,\text{m}^2$.

Applying equation (18.49) between new nodes 1 and 6 gives the total flow as

$$W = C_T\sqrt{\frac{p_1 - p_6}{v_1}} = 2.0004 \times 10^{-4}$$
$$\times \sqrt{\frac{10^6 - 0.6 \times 10^6}{0.001}} = 4.000833\,\text{kg/s}$$

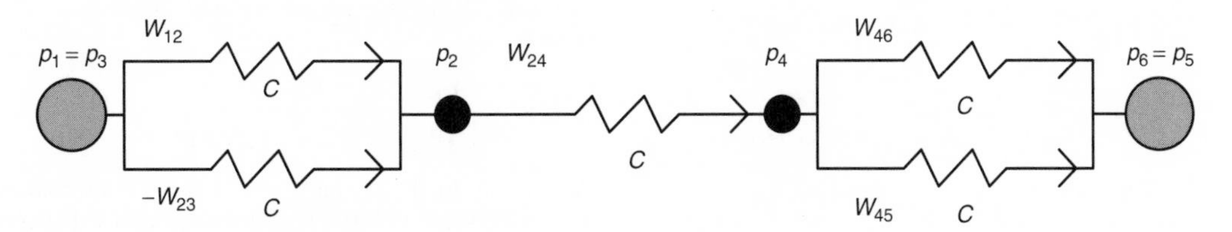

Figure 18.7 Layout of network at the beginning of the Prior Transient.

Knowing the flow, we may apply equation (18.49) to the section between new node 1 and node 2 to determine the intermediate pressure, p_2:

$$p_2 = p_1 - v\frac{W^2}{(2C)^2} = 1000 \times 10^3 - 0.001 \times \left(\frac{4.000833}{4.9 \times 10^{-4}}\right)^2 = 933\,333.3\,\text{Pa}$$

A similar procedure applied between node 4 and node 6 gives

$$p_4 = p_6 + v\frac{W^2}{(2C)^2} = 600 \times 10^3 + 0.001 \times \left(\frac{4.000833}{4.9 \times 10^{-4}}\right)^2 = 666\,666.7\,\text{Pa}$$

as the pressure at node 4 at time t_{prior}.

We now apply ramps to intermediate pressures p_3 and p_5 over an arbitrary time interval between t_{prior} and t_0, selected to be 100 seconds. Such a time interval implies that ramps of $dp_3/dt = -1000$ Pa/s and $dp_5/dt = 2000$ Pa/s are needed to bring p_3 from 1000 kPa to the starting value of 900 kPa and P_5 from 600 kPa to 800 kPa.

Generating the starting conditions using Extended Prior Transient Integration

As an alternative, let us allow ourselves the liberty of varying the constant conductances C_{23} and C_{45} over the period t_{prior} to t_0. The procedure has an obvious physical significance, in that the conductances are being treated as valves. Assuming $C_{23} = C_{45} = 0$ at $t = t_{prior}$ reduces the network to a single line of flow from node 1 to node 6, with an overall conductance found from summing the inverse squares of the individual conductances (equation (18.12)):

$$\frac{1}{C_T^2} = \frac{1}{C^2} + \frac{1}{C^2} + \frac{1}{C^2} = \frac{3}{C^2}$$

so that $C_T = \sqrt{\frac{1}{3}} \times C = 1.415 \times 10^{-4}\,\text{m}^2$.

Applying equation (18.49) between new nodes 1 and 6 gives the total flow as

$$W = C_T\sqrt{\frac{p_1 - p_6}{v_1}} = 1.415 \times 10^{-4} \times \sqrt{\frac{10^6 - 0.6 \times 10^6}{0.001}} = 2.829016\,\text{kg/s}$$

which may be used to determine the intermediate pressures, p_2 and p_4, at time t_{prior} in the way previously illustrated:

$$p_2 = p_1 - v\frac{W^2}{C^2} = 1000 \times 10^3 - 0.001 \times \left(\frac{2.829016}{2.45 \times 10^{-4}}\right)^2 = 866\,666.7\,\text{Pa}$$

$$p_4 = p_6 + v\frac{W^2}{C^2} = 600 \times 10^3 + 0.001 \times \left(\frac{2.829016}{2.45 \times 10^{-4}}\right)^2 = 733\,333.3\,\text{Pa}$$

Maintaining the boundary pressures at their t_0-values, we now apply ramps to C_{23} and C_{45} over the arbitrary period of 100 seconds assumed to elapse between t_{prior} and t_0 to take them both to their rightful value of $2.45 \times 10^{-4}\,\text{m}^2$ at t_0. Applying equation (18.43) to the constraint equations (18.48) gives

$$\begin{bmatrix} \frac{\partial W_{12}}{\partial p_2} - \frac{\partial W_{23}}{\partial p_2} - \frac{\partial W_{24}}{\partial p_2} & -\frac{\partial W_{24}}{\partial p_4} \\ \frac{\partial W_{24}}{\partial p_2} & \frac{\partial W_{24}}{\partial p_4} - \frac{\partial W_{45}}{\partial p_4} - \frac{\partial W_{46}}{\partial p_4} \end{bmatrix} \begin{bmatrix} \frac{dp_2}{dt} \\ \frac{dp_4}{dt} \end{bmatrix} = \begin{bmatrix} \frac{\partial W_{23}}{\partial C_{23}} & 0 \\ 0 & \frac{\partial W_{45}}{\partial C_{45}} \end{bmatrix} \begin{bmatrix} \frac{dC_{23}}{dt} \\ \frac{dC_{45}}{dt} \end{bmatrix} \quad (18.53)$$

where, from equation (18.49)

$$\frac{\partial W_{jk}}{\partial C_{jk}} = \sqrt{\frac{p_j}{v}}\sqrt{\left|1 - \frac{p_k}{p_j}\right|} \times \frac{1 - \frac{p_k}{p_j}}{\left|1 - \frac{p_k}{p_j}\right|} = \frac{W_{jk}}{C_{jk}} \quad \text{for } C_{jk} \neq 0 \quad (18.54)$$

Accuracy of the starting conditions found from Prior and Extended Prior Transient Integration

The pressure derivatives found from equation (18.50) for Prior Transient Integration and (18.54) for Extended Prior Transient Integration were integrated using the Euler method. In each case the error was measured in terms of the sum of the squared flow errors, E, at time t_0:

$$E = (W_{12} - W_{23} - W_{24})^2 + (W_{24} - W_{45} - W_{46})^2\,\text{kg}^2/\text{s}^2$$

A timestep of 0.5 seconds was used for both methods, with additional timesteps of 1 second and 0.1 seconds being used to test the sensitivity of the Prior Transient

Integration method. The pressures computed in this way were also supplied as starting guesses for an iterative method, which was used to refine the estimated starting conditions for pressures p_2 and p_4 at time t_0. Table 18.3 shows the comparisons.

The accuracy of both integration methods was very good, with pressure errors of the order of 0.01% or better for both integration methods for all the timesteps chosen. The accuracy of the the integration methods could be increased significantly by reducing the size of the timestep, with the sum of the squared flow errors being decreased by two orders of magnitude as the time interval was reduced from 1 second to 0.1 seconds in Prior Transient Integration.

Main transient

The initial conditions for the intermediate pressures found by the iterative method were used because of their availability and somewhat better accuracy. Euler integration with a timestep of 0.5 seconds was used to simulate 100 seconds of transient, and the resulting pressures and flows are shown graphically in Figures 18.8 and 18.9, respectively.

The first few seconds of the transient see a reversal of the pressure gradient between nodes 2 and 3, leading to a change in direction for flow W_{23}. This has an adverse effect on the sum of the squared flow errors (as is shown in Figure 18.10), which reaches a maximum of about $3 \times 10^{-3}\,\text{kg}^2/\text{s}^2$ at the point of flow reversal. However the error sum reduces subsequently to settle at a steady value of about $4.3 \times 10^{-4}\,\text{kg}^2/\text{s}^2$ for the rest of the transient.

The sum of the squared flow errors is a sensitive measure of error, especially in the vicinity of flow reversal. Table 18.4 compares the values of the intermediate pressures p_2 and p_4 found from integration

Table 18.3 Comparison of calculated starting pressures for transient

Method	*Euler timestep, (s)*	$p_2(t_0)$ *(kPa)*	$p_4(t_0)$ *(kPa)*	$E(t_0)$ *(kg/s)*^{2}
Prior Transient Integration	1.0	899.857	788.794	2.69×10^{-3}
	0.5	899.800	788.948	9.98×10^{-4}
	0.1	899.766	789.325	2.84×10^{-5}
Extended Prior Transient Integration	0.5	899.298	789.416	9.31×10^{-5}
Iterative method		899.757	789.394	1.05×10^{-11}

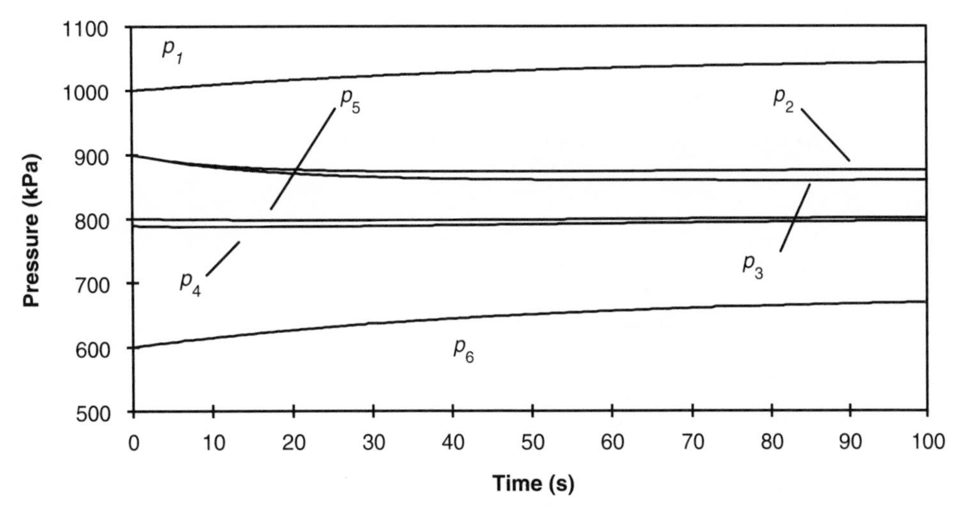

Figure 18.8 Pressures in worked example.

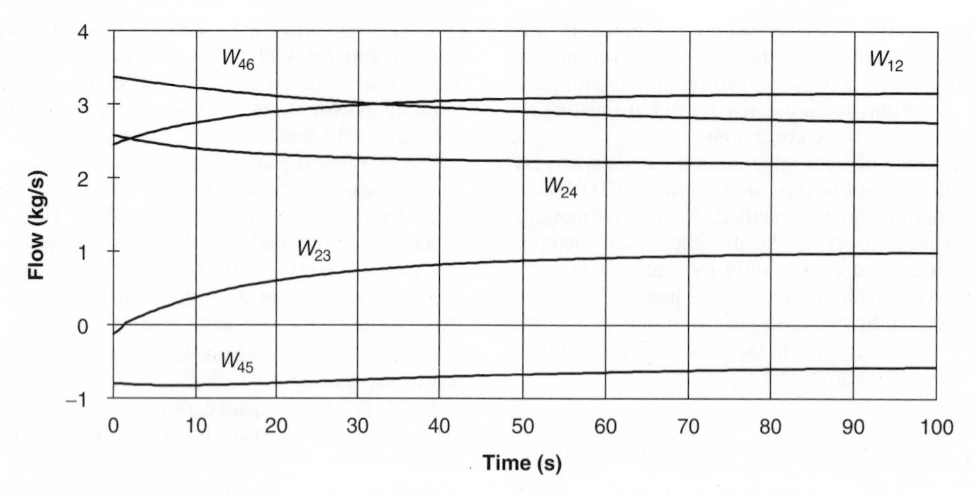

Figure 18.9 Flows in worked example.

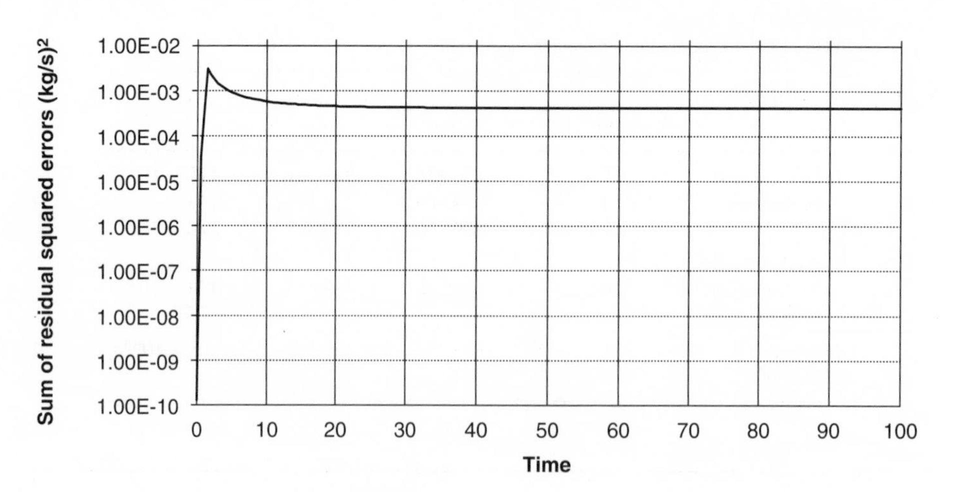

Figure 18.10 Sum of squared flow errors in worked example.

Table 18.4 Pressures computed by the Method of Referred Derivatives compared with the values found by iteration

Time into transient (seconds)	*Method*	p_2 *(kPa)*	p_4 *(kPa)*	*Flow error, E* (kg^2/s^2)
1.5	Referred Derivatives	896 299	788 934	3.15×10^{-3}
1.5	Iteration	896 273	788 927	1.11×10^{-14}
100	Referred Derivatives	876 227	796 084	4.25×10^{-4}
100	Iteration	875 822	796 009	5.04×10^{-26}

with values found using the time-consuming processes of iteration. The iterative solutions were based on the previously computed values for the boundary pressures, p_1, p_3, p_5, p_6, at 1.5 seconds (the time of maximum flow error) and 100 seconds. The close correspondence between the sets of values calculated for the intermediate pressures indicates that their transient behaviour will have been tracked closely by the Method of Referred Derivatives.

18.9 Avoiding problems at flow reversal with the Method of Referred Derivatives

It is interesting that the form of the flow equation (18.49) could cause problems at flow reversal for the Method of Referred Derivatives analogous to the problems discussed in Section 18.6 for the Newton–Raphson iterative method. Difficulties would arise if the upstream and downstream pressure became exactly equal at any time, as will now be shown. Combining equations (18.49) with (18.51) and (18.52) gives the partial differentials of flow with respect to upstream and downstream pressures as:

$$\frac{\partial W_{jk}}{\partial p_j} = \frac{1}{2} C_{jk} \sqrt{\frac{1}{v}} \frac{1}{\sqrt{p_j - p_k}} = -\frac{\partial W_{jk}}{\partial p_k} \qquad (18.55)$$

indicating infinite partial derivatives at the point of flow reversal, when $p_j = p_k$. Apart from the problem of the computer flagging an error for a division by zero, the effect of these very large, theoretically infinite, partial derivatives would be to force a spurious equality between (dp_j/dt) and (dp_k/dt) and so ensure $p_j = p_k$ at the next timestep, and, by extension, at all future times. That this is the case may be seen by expanding the top line of equation (18.50) and examining the situation when $j = 2$ and $k = 3$ and $p_2 = p_3$:

$$\frac{dp_2}{dt} = \frac{-\dfrac{\partial W_{23}}{\partial p_3}}{-\dfrac{\partial W_{12}}{\partial p_2} + \dfrac{\partial W_{23}}{\partial p_2} + \dfrac{\partial W_{24}}{\partial p_2}} \frac{dp_3}{dt} + \frac{\dfrac{\partial W_{12}}{\partial p_1}}{-\dfrac{\partial W_{12}}{\partial p_2} + \dfrac{\partial W_{23}}{\partial p_2} + \dfrac{\partial W_{24}}{\partial p_2}} \frac{dp_1}{dt} - \frac{\dfrac{\partial W_{24}}{\partial p_4}}{-\dfrac{\partial W_{12}}{\partial p_2} + \dfrac{\partial W_{23}}{\partial p_2} + \dfrac{\partial W_{24}}{\partial p_2}} \frac{dp_4}{dt} \qquad (18.56)$$

Letting $(\partial W_{23}/\partial p_2) = -(\partial W_{23}/\partial p_3) \rightarrow \infty$ in equation (18.56) causes $(dp_2/dt) \rightarrow (dp_3/dt)$. In practice, it is unlikely that the numerical procedure of integration will produce values of upstream and downstream pressure that are exactly the same. Nevertheless if the partial derivatives are evaluated from analytical expressions (18.51) and (18.52), it is probably wise to introduce a simple numerical trap to limit their size and forestall such an eventuality. A numerically estimated partial derivative will never equal infinity, of course.

A most complete method of avoiding infinite partial derivatives at flow reversal is to use a linear approximation in the vicinity of $p_j = p_k$, as suggested in Section 18.6, equation set (18.34). Assuming the upstream and downstream specific volumes are equal, differentiating the modified flow equations in the region $0.98 \leq p_k/p_j < 1.02$ gives:

$$\frac{\partial W_{jk}}{\partial p_j} = \frac{W_{jk}}{p_j - p_k} \frac{p_j + p_k}{2p_j} \qquad (18.57)$$

$$\frac{\partial W_{jk}}{\partial p_k} = \frac{-W_{jk}}{p_j - p_k} \qquad (18.58)$$

Since from equation set (18.34) $W_{jk} \propto p_j - p_k$ in the region either side of pressure equality, we see from equations (18.57) and (18.58) that the possibility of an infinite partial derivative has disappeared. Use of the linear formulation near flow reversal also decreases the sensitivity of flow to pressure errors. Figure 18.11 shows the result of replacing the flow function of equation (18.49) by that of equation set (18.34): the sum of the squared flow errors is maintained below 10^{-6} at all times, indicating that the use of the modified flow equations has led to greater internal consistency.

But it should be noted that this apparent advantage has been achieved as a result of solving a less demanding set of equations, which is likely to give a slightly poorer representation of the real world. Table 18.5 shows the results after 100 seconds found by using the Method of Referred Derivatives to solve the full flow equations compared with those achieved by using the same method to solve the modified flow equations. Both sets of results are also compared with the 'exact' solution. It will be seen that although the sum of the squared flow errors, E, is smaller when the modified flow equations are solved, the pressures computed by applying the Method of Referred Derivatives to the full flow equations are slightly closer overall to the 'exact' values. On the other hand, it was found that the pressures and flows calculated using the modified flow equations were almost identical to the transients depicted in Figures 18.8 and 18.9.

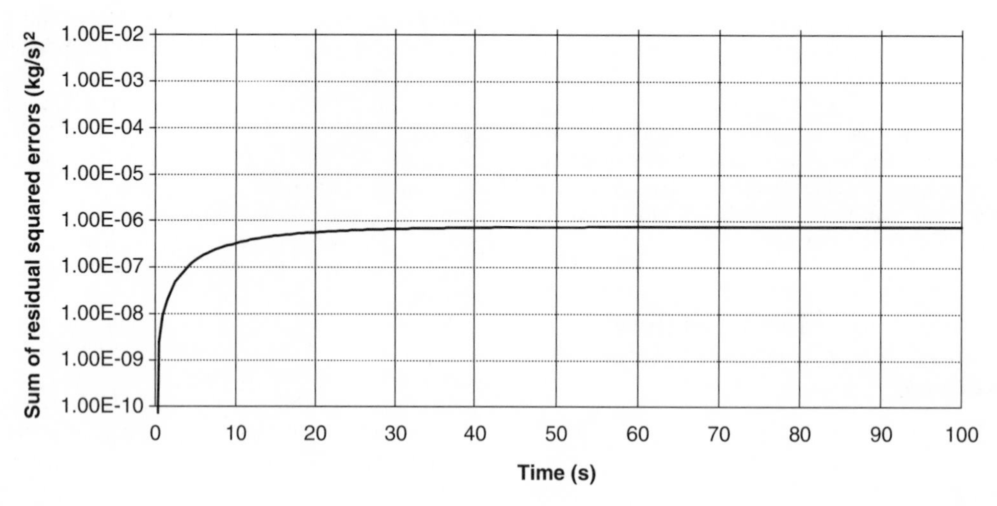

Figure 18.11 Sum of squared flow errors when a linear approximation is made to the flow function near flow reversal (equation 18.34).

Table 18.5 Intermediate pressures after 100 s found by applying the Method of Referred Derivatives to the full flow equations and to the modified flow equations. Comparison with 'exact' values

	$p_2(100)$ *(kPa)*	$p_4(100)$ *(kPa)*	*Flow error, E(100)* (kg^2/s^2)
Full flow equations. Method of Referred Derivatives ($\Delta t = 0.5$ s)	876 227	796 084	4.25×10^{-4}
'Exact' solution for computed boundary pressures at 100 s. Iteration using full flow equations	875 822	796 009	5.04×10^{-26}
Modified flow equations. Method of Referred Derivatives ($\Delta t = 0.5$ s)	875 730	794 118	7.66×10^{-7}

18.10 Liquid networks containing nodes with significant volume: allowing for temperature changes

In a network that is completely filled with either a liquid or a gas, ignoring any expansion or contraction in the metalwork, the volume, V, of fluid at each node will remain constant at all times. Writing

$$V = vm \tag{18.59}$$

constant volume implies that the derivative with respect to time is zero, so that:

$$\frac{dV}{dt} = v\frac{dm}{dt} + m\frac{\partial v}{\partial T}\frac{dT}{dt} + m\frac{\partial v}{\partial p}\frac{dp}{dt} = 0 \tag{18.60}$$

Here we have used the fact that specific volume, v, will be a function only of thermodynamic variables, temperature, T, and pressure, p. Equation (18.60) will be referred to as the 'Nodal Volume Conservation Equation'. Its implications are different for gas and liquid networks.

Applying the Nodal Volume Conservation Equation directly to a gas node would require a full solution for the pressure dynamics in the manner described in Chapter 11 for a process vessel of fixed volume. However, when the nodal volume is small, the gas pressure and temperature will reach equilibrium quickly, after which time $dT/dt = dp/dt = 0$. Hence, from equation (18.60), $dm/dt = 0$. While this last equation will be fully valid only after pressure and temperature have reached equilibrium, it will be acceptable as an approximate characterization of reality at all times provided the nodal volume is small enough to allow very rapid establishment of pressure and temperature equilibria. This is the basis for modelling gas flow in networks under the assumption that steady-state equations for mass balance are valid.

The situation for liquids is different as a result of the liquid specific volume being very nearly independent of pressure, i.e. $\partial v/\partial p \approx 0$. Substituting back into equation (18.60) gives the Nodal Volume Conservation Equation for Liquids as

$$v\frac{dm}{dt} + m\frac{\partial v}{\partial T}\frac{dT}{dt} = 0 \tag{18.61}$$

Note that this equation degenerates into the simple form $dm/dt = 0$ if the temperature stays constant, irrespective of the volume of the node. If, on the other hand, the nodal temperature is subject to change, the use of the steady-state mass balance will only be valid if the temperature dynamics are very fast. Temperature equilibrium will then be reached rapidly, after which time $dT/dt = 0$, and so, from (18.61), $dm/dt = 0$. But fast temperature dynamics require a small thermal inertia at the node, hence a small mass and hence a small volume. Accordingly the steady-state equations for mass balance will be valid for a liquid with a varying temperature only if all the network's nodes possess small volumes.

However, there may be occasions when the modeller will wish to deal with liquid nodes that not only experience significant changes in temperature but also have sizeable volumes. In such cases we can no longer assume simply that $dm/dt = 0$ but must use the Nodal Volume Conservation Equation for Liquids, equation (18.61) instead, which forces us to take account of the temperature dynamics at the node.

A degree of complication can arise in calculating the dynamic response of the nodal temperature when flows have the potential to reverse direction. To illustrate the point, let us consider the node shown in Figure 18.12, assumed to have a sizeable volume, e.g. as a result of lumping together the volume of several connecting pipes.

The node is normally supplied with inlet flows, $W_{up,m}$, from M nominally upstream nodes, and discharges outlet flows, $W_{d,n}$, to N nominally downstream nodes. $W_{up,m}$ is positive if the physical flow is entering the node and negative if it is leaving. Conversely $W_{d,n}$ is positive if the physical flow is leaving the node and negative when it enters. The pressures and enthalpies at the upstream nodes are denoted $P_{up,m}$, $h_{up,m}$, while the corresponding variables at the downstream nodes are denoted $p_{d,n}$, $h_{d,n}$. The node is assumed also to be heated with a heat flux, Φ, which may represent the exchange of heat between the liquid and the metalwork at the node.

The mass balance equation may be written simply:

$$\frac{dm}{dt} = \sum_{m=1}^{M} W_{up,m} - \sum_{n=1}^{N} W_{d,n} \tag{18.62}$$

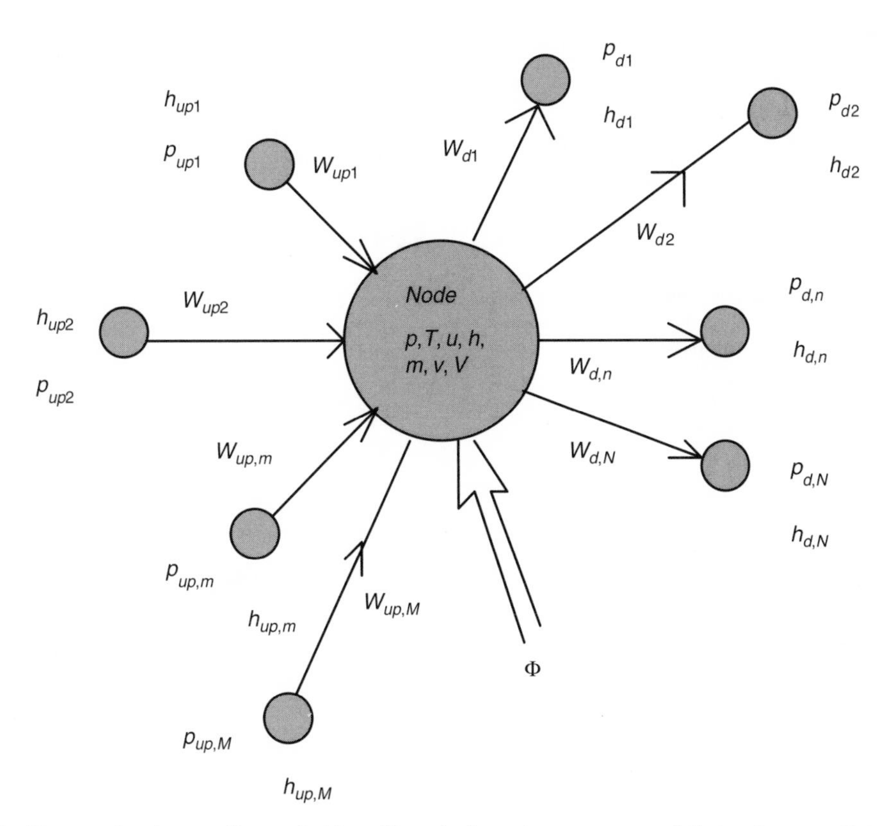

Figure 18.12 Diagram of node normally supplied from *M* nominally upstream sources and discharging normally to *N* nominally downstream sinks.

where the flows are functions, f_L, of the nodal pressures and specific volumes as given in equations (18.20) and (18.21) or in equations (18.34) and (18.35):

$$\begin{aligned} W_{up,m} &= f_L(p_{up,m}, v_{up,m}, p, v) \\ W_{d,n} &= f_L(p, v, p_{d,n}, v_{d,n}) \end{aligned} \tag{18.63}$$

An energy balance on the fixed volume of the node will produce an equation for the rate of change of temperature of the form of equation (3.44), repeated below:

$$\frac{dT}{dt} = \frac{\Phi - P + W_1(h_1 - u) - W_2 pv}{m\dfrac{du}{dT}} \tag{3.44}$$

where W_1 is the mass flow entering the fixed volume and W_2 is the mass flow leaving. However, applying (3.44) to the node of Figure 18.12 calls for special care to allow for possible flow reversals if W_1 or W_2 reverse and become negative. The appropriate equation is:

$$\frac{dT}{dt} = \left(\begin{array}{l} \Phi + \sum_{m=1}^{M} \max[W_{up,m}, 0](h_{up,m} - u) \\ + \sum_{n=1}^{N} \max[-W_{d,n}, 0](h_{d,n} - u) \\ -pv\left(\sum_{m=1}^{M} \max[-W_{up,m}, 0] \right. \\ \left. + \sum_{n=1}^{N} \max[W_{d,n}, 0] \right) \end{array} \right) \times \frac{1}{m\dfrac{du}{dT}} \tag{18.64}$$

The equation defining flows into and out of the node is found by substituting equations (18.62) and (18.64) into the Nodal Volume Conservation Equation for Liquids, equation (18.61), to give:

$$0 = \sum_{m=1}^{M} W_{up,m} - \sum_{n=1}^{N} W_{d,n} + \left(\begin{array}{l} \Phi + \sum_{m=1}^{M} \max[W_{up,m}, 0](h_{up,m} - u) \\ + \sum_{n=1}^{N} \max[-W_{d,n}, 0](h_{d,n} - u) \\ -pv\left(\sum_{m=1}^{M} \max[-W_{up,m}, 0] \right. \\ \left. + \sum_{n=1}^{N} \max[W_{d,n}, 0] \right) \end{array} \right) \times \frac{\dfrac{\partial v}{\partial T}}{v\dfrac{du}{dT}} \tag{18.65}$$

Given an initial temperature for the node, T, it is possible to find the specific internal energy, $u = u(T)$, and the specific volume, $v = v(T)$, and hence the mass: $m = V/v$. Equation (18.65), taken in conjunction with auxiliary equations (18.63), represents an implicit equation in the nodal pressure, p, which may be solved using the methods already outlined, either iteration or the Method of Referred Derivatives. The upstream and downstream flows, $W_{up,m}$ and $W_{d,n}$, may then be found, so that it becomes possible to calculate the right-hand side of the temperature differential equation (18.64). Equation (18.64) may then be integrated to find the temperature of the node at the next timestep. The process may then be repeated for the duration of the transient under consideration.

18.11 Bibliography

Thomas, P.J. (1997). The Method of Referred Derivatives: a new technique for solving implicit equations in dynamic simulation, *Trans. Inst.M.C.*, **19**, 13–21.

19 Pipeline dynamics

19.1 Introduction

The lengths of pipe connecting plant components are usually relatively short, so that the time constants associated with flow settling are short, allowing us to use steady-state flow equations within the dynamic simulation (as was our practice in Chapters 4, 6, 8, 10 and 18). But pipes connecting distant plants may be kilometres long, and in such cases we may no longer assert that at any given time the flow is the same at all points along the pipeline. In such a case the flow going into the line may be markedly different from the flow coming out at the other, leading to variable storage within the line. Such storage is known in the gas industry as 'linepack', and it may be exploited to help balance supply and demand. Other less desirable effects may occur: the rapid opening or closing of a valve may cause rapid oscillations in pressure and flow at various points along the pipeline, known as waterhammer when the fluid is a liquid, but also possible in gas systems.

A long pipeline is clearly a distributed system, and requires the use of partial differential equations in time and space to describe the dynamic behaviour of pressure and flow along its length. The basic equations derived in this chapter and their methods of solution apply equally to liquids and to gases.

19.2 Dynamic equations for a pipeline: the full equations

The general equations for flow in a pipe were derived in Chapter 3, Sections 3.8, 3.9 and 3.10. For a pipe of uniform cross-sectional area, the equations are:

Conservation of mass

$$\frac{\partial v}{\partial t} = v^2 \frac{\partial}{\partial x}\left(\frac{c}{v}\right) \tag{3.58}$$

Conservation of energy

$$\left(u + \frac{1}{2}c^2 + gx\sin\theta\right)\frac{\partial}{\partial t}\left(\frac{1}{v}\right) + \frac{1}{v}\frac{\partial}{\partial t}\left(u + \frac{1}{2}c^2\right)$$
$$= \frac{\phi}{A} - \frac{c}{v}\frac{\partial}{\partial x}\left(h + \frac{1}{2}c^2\right) - \left(h + \frac{1}{2}c^2 + gx\sin\theta\right)$$
$$\times \frac{\partial}{\partial x}\left(\frac{c}{v}\right) - \frac{c}{v}g\sin\theta \tag{3.69}$$

Conservation of momentum

$$\frac{\partial}{\partial t}\left(\frac{c}{v}\right) = -\frac{\partial}{\partial x}\left(p + \frac{c^2}{v}\right) - \frac{2c\,|c|}{vD}f - \frac{g}{v}\sin\theta \tag{3.73}$$

A conceptual scheme for solving these three equations was given at the end of Section 3.10. However, in practice it is usual to simplify the problem by ignoring changes in energy, and hence fluid temperature, and solving only the conservation equations for mass and momentum. This is a good approximation for liquids in pipelines. It is less good for gases, but nevertheless widely used (Fincham and Goldwater, 1979). While it will not capture fully the dynamic effects of a sudden pipe-break, for instance, it has been found to give reasonable results for slow transients (hours long) on buried natural gas pipelines, where the temperature changes within the gas are sufficient slow to be cancelled out by heat conduction between the pipe and the surrounding soil.

19.3 Development of the equation for conservation of mass

As noted in Chapter 3, specific volume depends in general on both temperature and pressure for gases and for liquids, although the pressure dependence is very much smaller in the latter case. But if we assume that the temperature is fixed, then specific volume will depend on pressure only:

$$v = v(p) \tag{19.1}$$

The pressure will vary along the length of the pipe and its value at any point will vary with time. This situation may be described algebraically by

$$p = p(x, t) \tag{19.2}$$

We may use the two relations above to find the partial differential of specific volume with respect to time, $\partial v/\partial t$, in the form:

$$\frac{\partial v}{\partial t} = \frac{dv}{dp}\frac{\partial p}{\partial t} \tag{19.3}$$

Now the bulk modulus of elasticity of the fluid, κ (Pa), is given by the ratio of the increase in its pressure to the fractional decrease in its specific volume:

$$\kappa = -v\frac{dp}{dv} \tag{19.4}$$

Further, physics tells us that the speed of sound in the fluid, c_{son}, may be expressed as

$$c_{son} = \sqrt{v\kappa} \tag{19.5}$$

Combining equations (19.4) and (19.5) allows us to relate the change in specific volume with respect to pressure to the speed of sound:

$$\frac{dv}{dp} = \left(\frac{dp}{dv}\right)^{-1} = -\frac{v}{\kappa} = -\frac{v^2}{v\kappa} = -\frac{v^2}{c_{son}^2} \tag{19.6}$$

Substituting from equations (19.3) and (19.6) into equation (3.58) relates the rate of change of pressure with time to the rate of change with distance of the composite quantity, c/v

$$\frac{\partial p}{\partial t} + c_{son}^2\frac{\partial}{\partial x}\left(\frac{c}{v}\right) = 0 \tag{19.7}$$

An alternative form of equation (19.7) is found by noting that the mass flow in kg/s is given by

$$W = \frac{Ac}{v} \tag{19.8}$$

so that for a pipe of uniform cross-section,

$$\frac{\partial W}{\partial x} = A\frac{\partial}{\partial x}\left(\frac{c}{v}\right) \tag{19.9}$$

Substituting into equation (19.7) gives

$$\frac{\partial p}{\partial t} + \frac{c_{son}^2}{A}\frac{\partial W}{\partial x} = 0 \tag{19.10}$$

19.4 Development of the equation for conservation of momentum

Using equation (19.8), equation (3.73) may be re-written as:

$$\frac{\partial}{\partial t}\left(\frac{W}{A}\right) + \frac{\partial p}{\partial x} + \frac{\partial}{\partial x}\left(v\frac{W^2}{A^2}\right) + \frac{2vW\,|W|}{A^2D}f + \frac{g}{v}\sin\theta = 0 \tag{19.11}$$

Now the term $(\partial/\partial x)(v(W^2/A^2))$ may be expanded as:

$$\frac{\partial}{\partial x}\left(v\frac{W^2}{A^2}\right) = 2v\frac{W}{A^2}\frac{\partial W}{\partial x} + \frac{W^2}{A^2}\frac{\partial v}{\partial x} \tag{19.12}$$

Further, under the same assumption of isothermal conditions used in the previous section, we may transform the rate of change of specific volume with distance into a rate of change of pressure:

$$\frac{\partial v}{\partial x} = \frac{dv}{dp}\frac{\partial p}{\partial x} = -\frac{v^2}{c_{son}^2}\frac{\partial p}{\partial x} \tag{19.13}$$

where equation (19.6) has been used in the final step. Hence equation (19.12) becomes

$$\frac{\partial}{\partial x}\left(v\frac{W^2}{A^2}\right) = 2v\frac{W}{A^2}\frac{\partial W}{\partial x} - \frac{v^2}{c_{son}^2}\frac{W^2}{A^2}\frac{\partial p}{\partial x} \tag{19.14}$$

Substituting from equation (19.14) into equation (19.11) gives the new form:

$$\frac{1}{A}\frac{\partial W}{\partial t} + \frac{\partial p}{\partial x}\left(1 - \frac{v^2}{c_{son}^2}\frac{W^2}{A^2}\right) + 2v\frac{W}{A^2}\frac{\partial W}{\partial x} + \frac{2vf}{A^2D}W\,|W| + \frac{g}{v}\sin\theta = 0 \tag{19.15}$$

19.5 Applying the Method of Characteristics to pipeline dynamics

We will preserve the physical information contained in equation (19.10), conservation of mass, and equation (19.15), conservation of momentum, if instead of treating the original equations, we deal with two distinct, linear combinations in their place. Let us multiply equation (19.10) by an arbitrary multiplier, λ, and add the result to equation (19.15):

$$\lambda\left(\frac{\partial p}{\partial t} + \frac{\partial p}{\partial x}\frac{1}{\lambda}\left(1 - \frac{v^2}{c_{son}^2}\frac{W^2}{A^2}\right)\right) + \frac{1}{A}\left(\frac{\partial W}{\partial t} + \frac{\partial W}{\partial x}\left(\lambda c_{son}^2 + 2v\frac{W}{A}\right)\right) + \frac{2vf}{A^2D}W\,|W| + \frac{g}{v}\sin\theta = 0 \tag{19.16}$$

Noting that the pressure and flow can be different at different locations and different time instants, i.e. $p = p(x, t)$ and $W = W(x, t)$, the total differentials with respect to time of pressure and flow are given by

$$\frac{dp}{dt} = \frac{\partial p}{\partial t} + \frac{\partial p}{\partial x}\frac{dx}{dt} \tag{19.17}$$

and

$$\frac{dW}{dt} = \frac{\partial W}{\partial t} + \frac{\partial W}{\partial x}\frac{dx}{dt} \tag{19.18}$$

Comparing equations (19.17) and (19.18) with the first two terms of equation (19.16), it is clear that equation (19.16) can be written in the much simpler form:

$$\lambda \frac{dp}{dt} + \frac{1}{A}\frac{dW}{dt} + \frac{2vf}{A^2 D} W|W| + \frac{g}{v}\sin\theta = 0 \tag{19.19}$$

provided the velocity, dx/dt, at which changes of pressure and mass flow propagate through the pipe obeys the relation

$$\frac{dx}{dt} = \frac{1}{\lambda}\left(1 - \frac{v^2}{c_{son}^2}\frac{W^2}{A^2}\right) = \left(\lambda c_{son}^2 + 2v\frac{W}{A}\right) \tag{19.20}$$

Equation (19.20) implies that λ must take the two values, λ_1, λ_2, that are the solutions of the quadratic

$$c_{son}^2 \lambda^2 + 2v\frac{W}{A}\lambda + \left(\frac{v^2}{c_{son}^2}\frac{W^2}{A^2} - 1\right) = 0 \tag{19.21}$$

which may be simplified by using equation (19.8) to convert the mass flow into velocity:

$$c_{son}^2 \lambda^2 + 2c\lambda + \left(\frac{c^2}{c_{son}^2} - 1\right) = 0 \tag{19.22}$$

The two solutions are:

$$\lambda_1 = \frac{-c + c_{son}}{c_{son}^2} \tag{19.23}$$

$$\lambda_2 = \frac{-c - c_{son}}{c_{son}^2} \tag{19.24}$$

Selecting the first root, equation (19.19) becomes:

$$\frac{-c + c_{son}}{c_{son}^2}\frac{dp}{dt} + \frac{1}{A}\frac{dW}{dt} + \frac{2vf}{A^2 D} W|W| + \frac{g}{v}\sin\theta = 0 \tag{19.25}$$

which is valid for the velocity found by substituting from equation (19.23) into equation (19.20):

$$\frac{dx}{dt} = c + c_{son} \tag{19.26}$$

A similar pair of equations may be found for the second root, namely

$$\frac{-c - c_{son}}{c_{son}^2}\frac{dp}{dt} + \frac{1}{A}\frac{dW}{dt} + \frac{2vf}{A^2 D} W|W| + \frac{g}{v}\sin\theta = 0 \tag{19.27}$$

Equation (19.27) is valid for the velocity:

$$\frac{dx}{dt} = c - c_{son} \tag{19.28}$$

The two equations, (19.25) and (19.27), taken together with their auxiliary equations (19.26) and (19.28), respectively, contain the same information about the pipeline as the original equations for conservation of mass and momentum, equations (19.10) and (19.15).

Starting at a set of locations along the pipeline, $x_i(t_0)$, $i = 1, 2, \ldots$ at time, t_0, equation (19.26) defines a set of forward trajectories or characteristic curves in distance and time, while equation (19.28) defines a set of backward characteristic curves starting at the same point:

$$\begin{aligned} x_i^+(t) &= x_i(t_0) + \int_{t_0}^{t} (c + c_{son})\, dt \\ &= x_i(t_0) + c_{son}(t - t_0) + \int_{t_0}^{t} c\, dt \\ x_i^-(t) &= x_i(t_0) + \int_{t_0}^{t} (c - c_{son})\, dt \\ &= x_i(t_0) - c_{son}(t + t_0) + \int_{t_0}^{t} c\, dt \end{aligned} \tag{19.29}$$

The speed of sound has been assumed constant over space and time because of the assumption of unvarying temperature, but the velocity of the fluid, c, will vary over space and time. Accordingly the characteristic curves defined by equations (19.29) will, in general, not be straight. However, the fluid velocity will normally be very much smaller than the sonic velocity. For example, the sonic velocity in natural gas pipelines is of the order of 400 m/s, while the speed of sound in liquid systems will be two or three times greater. By contrast the fluid velocity in natural gas pipelines and in oil pipelines is normally less than 10 m/s, so that $c/c_{son} < \sim 2.5\%$. As a result it is common practice on both liquid and gas pipelines to ignore the contribution of the fluid velocity and put $c = 0$ in equations (19.25) and (19.27) and (19.29). This has the beneficial effect of rendering the characteristics straight and invariant with time. This simplification is equivalent to ignoring the term $(\partial/\partial x)(v(W^2/A^2))$ in equation (19.11).

The fact that the characteristics are straight enables us to set up a rectangular grid of distance versus time as shown in Figure 19.1. Equation (19.25) will be valid along all the forward characteristics such as the lines marked '+' in Figure 19.1, while (19.27) will be valid along all the backward characteristics such as the lines marked '−' in the figure. Both equations will be valid at the intersection of a forward characteristic with a backward characteristic, and this enables us to find a time-marching solution as follows. Let us apply equations (19.25) and (19.27) with $c = 0$ to the marked characteristics and use a finite difference formulation to approximate the time derivatives. This

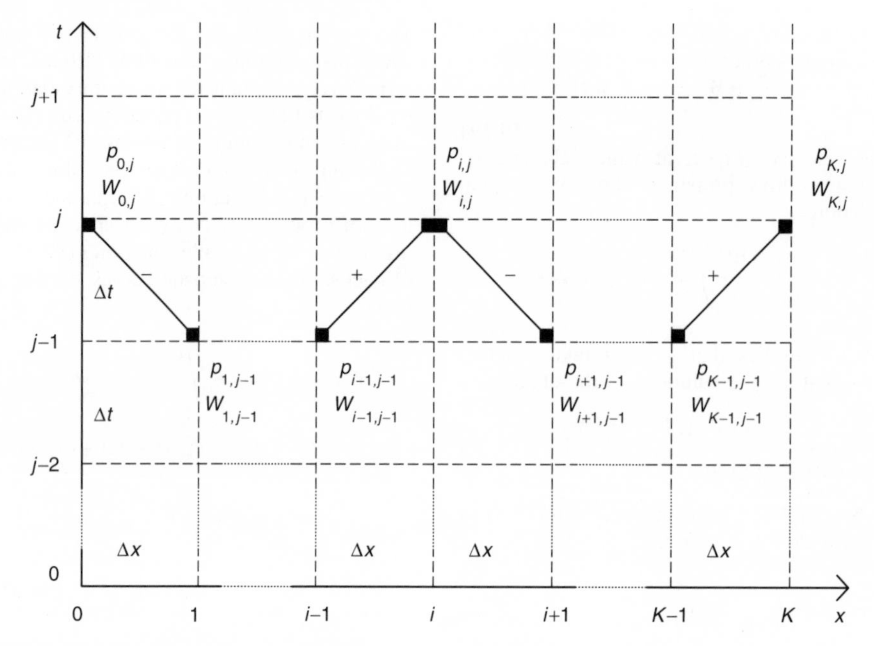

Figure 19.1 Rectangular grid for pipeline characteristics.

produces:

$$\frac{1}{c_{son}}\frac{p_{i,j}-p_{i-1,j-1}}{\Delta t}+\frac{1}{A}\frac{W_{i,j}-W_{i-1,j-1}}{\Delta t}$$
$$+\frac{2f}{A^2D}v_{i-1,j-1}W_{i-1,j-1}\left|W_{i-1,j-1}\right|$$
$$+\frac{g}{v_{i-1,j-1}}\sin\theta=0 \tag{19.30}$$

which is valid along the forward characteristic, and

$$-\frac{1}{c_{son}}\frac{p_{i,j}-p_{i+1,j-1}}{\Delta t}+\frac{1}{A}\frac{W_{i,j}-W_{i+1,j-1}}{\Delta t}$$
$$+\frac{2f}{A^2D}v_{i+1,j-1}W_{i+1,j-1}\left|W_{i+1,j-1}\right|$$
$$+\frac{g}{v_{i+1,j-1}}\sin\theta=0 \tag{19.31}$$

which is valid along the backward characteristic.

Let us assume that the values of the pressures and flows all along the pipe are known at the $(j-1)$th time instant, i.e. $p_{i,j-1}, i=0,1,\ldots,K$ and $W_{i,j-1}, i=0,1,\ldots,K$. If the fluid is a gas, we may evaluate the corresponding specific volumes, $v_{i,j-1}, i=0,1,\ldots,K$ from the characteristic equation (3.2), since we have assumed that the temperature along the line is approximately constant. The situation is simpler when the fluid is a liquid, since we may then assume a constant value of specific volume. Given these starting values, equations (19.30) and (19.31) represent a system of two linear, simultaneous equations in two unknowns, namely the pressure and flow at location i at the jth time instant, $p_{i,j}$ and $W_{i,j}$. An explicit solution is therefore possible, namely:

$$p_{i,j}=\frac{1}{2}(p_{i-1,j-1}+p_{i+1,j-1})+\frac{c_{son}}{2A}(W_{i-1,j-1}-W_{i+1,j-1})$$
$$+\Delta t\frac{c_{son}}{2}\left[\frac{2f}{A^2D}\left(v_{i+1,j-1}W_{i+1,j-1}\left|W_{i+1,j-1}\right|\right.\right.$$
$$\left.-v_{i-1,j-1}W_{i-1,j-1}\left|W_{i-1,j-1}\right|\right)$$
$$\left.+g\sin\theta\left(\frac{1}{v_{i+1,j-1}}-\frac{1}{v_{i-1,j-1}}\right)\right]$$
$$\text{for } j=1,2,\ldots,K-1 \tag{19.32}$$

$$W_{i,j}=\frac{1}{2}(W_{i-1,j-1}+W_{i+1,j-1})$$
$$+\frac{A}{2c_{son}}(p_{i-1,j-1}-p_{i+1,j-1})$$
$$-\Delta t\frac{A}{2}\left[\frac{2f}{A^2D}\left(v_{i-1,j-1}W_{i-1,j-1}\left|W_{i-1,j-1}\right|\right.\right.$$
$$\left.+v_{i+1,j-1}W_{i+1,j-1}\left|W_{1+1,j-1}\right|\right)$$
$$\left.+g\sin\theta\left(\frac{1}{v_{t-1,j-1}}+\frac{1}{v_{i+1,j-1}}\right)\right]$$
$$\text{for } j=1,2,\ldots,K-1 \tag{19.33}$$

Note that both the line pressures and, more unusually, the line flows are the outputs of numerical integrations, and may therefore be regarded as state variables. As with the other state variables in the simulation, calculation of their values at the current, jth instant presumes a knowledge of their values at the previous, $(j-1)$th instant.

Applying equations (19.32) and (19.33) systematically will allow the pressures and flows along the pipe to be determined for the jth time instant, with the exception of the pressures at flows at the boundaries, namely at $i = 0$ and $i = K$. The method of determining pressures and flows at these points will be discussed in the next section.

19.6 Interfacing the Method-of-Characteristics pipeline model to the rest of the process simulation: boundary conditions

Equations (19.32) and (19.33) apply to all interior points within the pipeline, so that the pressures and flows may be calculated for $i = 1, 2, 3, \ldots, K-1$. However, at the upstream boundary, when $i = 0$, no forward characteristic is present, as may be seen from inspecting Figure 19.1. At this location only the backward characteristic is available, so that only equation (19.31) is valid, which is a single equation containing the two unknowns of pressure and mass flow. A solution is possible, however, if the upstream conditions allow us to specify one of the following at $i = 0$:

(i) the pressure, independent of flow (for example if the pipe is connected directly to a tank of liquid or a gas pressure vessel);
(ii) the flow, independent of pressure (for example a reciprocating pump);
(iii) an additional relationship between pressure and flow (for example an upstream valve, or a pump feeding liquid forward from a liquid tank, or a compressor taking gas from an upstream pressure vessel).

In exactly the same way, at the end of the pipeline only the forward characteristic is available, so that only equation (19.30) is valid. Once again the Method of Characteristics produces a single equation in the two unknowns, pressure and mass flow. Once again an additional equation is needed to specify either the pressure, or the flow, or else a relationship between the two that is valid at $i = K$.

In general it may be necessary to solve two simultaneous equations in pressure and flow at each boundary. The same general principles apply irrespective of whether the fluid is a gas or a liquid, but the equations governing gas flow through line components (valves, compressors) are more complex than for the corresponding components in a liquid pipeline; there is also a need to allow for gas specific volume varying in both space and time. By contrast, the specific volume may be regarded as constant in calculating the transient behaviour of a liquid pipeline.

Regarding specific volume as constant in a liquid system may seem contradictory in view of the first equation given in this chapter, equation (3.58), where specific volume is assumed to vary in both time and space. The resolution to this apparent paradox is that the variation in specific volume for a liquid is sufficiently small for good results to be produced when a constant value is employed in the frictional and gravitational terms in equations (19.30) and (19.31). Numerical justification for this approach will be given in the example considered in Section 19.8.

The greater tractability of the equations when the fluid is a liquid means that it is often possible to produce explicit solutions to boundary problems. A number of boundary-condition examples will now be given where explicit solutions are possible. Most of these will involve liquid rather than gas flow on account of the simpler form of the valve equation and the fact that it is possible to regard the specific volume as approximately constant throughout the system. The simultaneous equations defining a gas-flow boundary are often highly nonlinear. But while direct, explicit solutions are not usually possible for gas pipeline boundaries (the example given in Section 19.6.6 is an exception), the principle of solving a pair of simultaneous equations in pressure and flow remains the same; it is just that now the solution must be iterative.

19.6.1 Constrained pressure at the inlet to a liquid pipeline

This is the situation when flow is being taken directly from a tank, as illustrated in Figure 19.2. The liquid in the tank will obey the mass-balance equation:

$$\frac{dm}{dt} = W_{IN} - W_0 \tag{19.34}$$

Assuming Euler integration, and using a subscript terminology, the mass in the tank at the jth time instant may be calculated from the mass and the flows at the $(j-1)$th time instant:

$$m_j = m_{j-1} + \Delta t(W_{IN,j-1} - W_{0,j-1}) \tag{19.35}$$

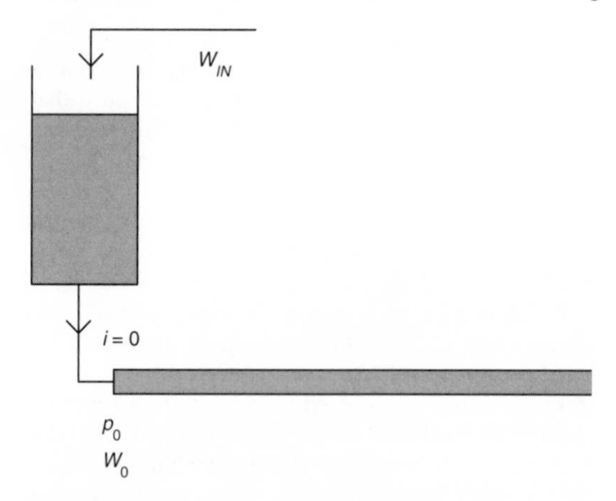

Figure 19.2 Liquid pipeline coupled directly to a feed tank.

The corresponding pressure at the outlet of the tank/inlet to the line will be calculated as:

$$p_{0,j} = \frac{g}{A_t} m_j = \frac{g}{A_t}(m_{j-1} + \Delta t(W_{IN,j-1} - W_{0,j-1})) \tag{19.36}$$

which illustrates that the pressure at the jth time instant may be calculated solely from values at the $(j-1)$th time instant.

To find the flow at the pipe inlet at time j, $W_{0,j}$, the backward characteristic, equation (19.31) may be applied with $i = 1$. Since the fluid is assumed to be a liquid at constant temperature, we will assign a constant value to specific volume, v. Thus $W_{0,j}$, is given in terms of the current pressure, $p_{0,j}$, and past flows by

$$W_{0,j} = W_{1,j-1} + \frac{A}{c_{son}}(p_{0,j} - p_{1,j-1}) - A\Delta t\left[\frac{2fv}{A^2 D} W_{1,j-1}\left|W_{1,j-1}\right| + \frac{g}{v}\sin\theta\right] \tag{19.37}$$

19.6.2 Constrained flow at the outlet of a liquid pipeline

This corresponds to the situation where a flow controller is used at the end of the line. The calculation may also be invoked to simulate approximately rapid control valve closure at the end of the line.

The procedure is to force the outlet flow to obey some externally determined function of time:

$$W_K = W_K(t) \tag{19.38}$$

and use the forward characteristic with i set to K. Once again we will use a single value of specific volume, v. Thus equation (19.30) gives:

$$p_{K,j} = p_{K-1,j-1} - \frac{c_{son}}{A}(W_{K,j} - W_{K-1,j-1}) - c_{son}\Delta t\left[\frac{2f}{A^2 D} v W_{K-1,j-1}\left|W_{K-1,j-1}\right| + \frac{g}{v}\sin\theta\right] \tag{19.39}$$

19.6.3 Valve at the inlet to a liquid pipeline

Consider Figure 19.3, where liquid from a tank is passing into a long pipeline. Assuming the valve is very close to the beginning of the pipeline, the flow

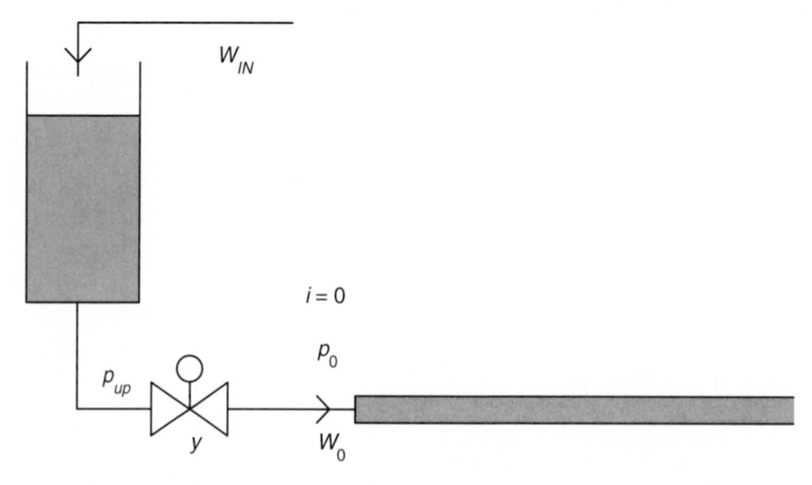

Figure 19.3 Interface of the pipeline with a terminal control valve.

through the valve will be the same as the flow at the start of the pipeline. The flow of liquid through the valve will obey equation (7.24) repeated below.

$$W = yC_V\sqrt{\frac{\Delta p}{v_1}} \tag{7.24}$$

where y is the fractional valve opening and C_V is the conductance (m^2) at fully open. At the jth time instant the flow through the valve will be

$$W_{0,j} = y_j C_V\sqrt{\frac{|p_{up,j} - p_{0,j}|}{v}} \tag{19.40}$$

in which the specific volume, v, is assumed the same throughout the system. The valve opening, y_j, and the upstream pressure, $p_{up,j}$, will be determined by the outputs of integrators in the main simulation, the latter in precisely the way described in Section 19.6.1. The valve relationship of equation (19.40) may be rewritten as:

$$p_{0,j} = p_{up,j} - \frac{v}{y_j^2 C_V^2} W_{0,j}^2 \quad \text{for } W_{0,j} \geq 0 \tag{19.41a}$$

$$p_{0,j} = p_{up,j} + \frac{v}{y_j^2 C_V^2} W_{0,j}^2 \quad \text{for } W_{0,j} < 0 \tag{19.41b}$$

Meanwhile along the backward characteristic, equation (19.31) will apply, which may be written for $i = 0$ and for constant specific volume in the form

$$-\frac{A}{c_{son}} p_{0,j} + W_{0,j} + \frac{A}{c_{son}} p_{1,j-1} - W_{1,j-1} + A\Delta t\left[\frac{2fv}{A^2 D} W_{1,j-1}\left|W_{1,j-1}\right| + \frac{g}{v}\sin\theta\right] = 0 \tag{19.42}$$

For positive flow, equations (19.41a) and (19.42) may be combined to give the quadratic in the pipe entry flow at the jth time instant, $W_{0,j}$:

$$\frac{A}{c_{son}} \frac{v}{y_j^2 C_V^2} W_{0,j}^2 + W_{0,j} - W_{1,j-1} + \frac{A}{c_{son}}(p_{1,j-1} - p_{up,j}) + A\Delta t \times \left[\frac{2fv}{A^2 D} W_{1,j-1}\left|W_{1,j-1}\right| + \frac{g}{v}\sin\theta\right] = 0 \tag{19.43}$$

Putting

$$X = \frac{A}{c_{son}} \frac{v}{y_j^2 C_V^2} \tag{19.44}$$

and

$$Y = -W_{1,j-1} + \frac{A}{c_{son}}(p_{1,j-1} - p_{up,j}) + A\Delta t\left[\frac{2f}{A^2 D} v W_{1,j-1}\left|W_{1,j-1}\right| + \frac{g}{v}\sin\theta\right] \tag{19.45}$$

equation (19.43) has the general solution

$$W_{0,j} = \frac{-1 \pm \sqrt{1 - 4XY}}{2X} \tag{19.46}$$

But for our stipulated case of $W_{0,j} \geq 0$, in view of the fact that X is always positive, it is clear first that the positive root must be chosen and secondly that

$$+\sqrt{1 - 4XY} \geq 1 \tag{19.47}$$

which implies that $Y \leq 0$.

For negative flow, equations (19.41b) and (19.42) may be combined to give the quadratic in the pipe entry flow at the jth time instant, $W_{0,j}$:

$$-\frac{A}{c_{son}} \frac{v}{y_j^2 C_V^2} W_{0,j}^2 + W_{0,j} - W_{1,j-1} + \frac{A}{c_{son}}(p_{1,j-1} - p_{up,j}) + A\Delta t\left[\frac{2fv}{A^2 D} W_{1,j-1}\left|W_{1,j-1}\right| + \frac{g}{v}\sin\theta\right] = 0 \tag{19.48}$$

which quadratic has the general solution, using the notation above:

$$W_{0,j} = \frac{1 \mp \sqrt{1 + 4XY}}{2X} \tag{19.49}$$

Since we have stipulated that the flow shall be negative, it is clear that the negative square root must be taken, and that

$$+\sqrt{1 + 4XY} \geq 1 \tag{19.50}$$

which implies that $Y > 0$. Since Y may be calculated from previously calculated values, it emerges as an important demarcation variable, with a negative value indicating positive flow and a positive value indicating negative flow. Thus we calculate Y and X first, which allow us to calculate $W_{0,j}$ from:

$$\begin{aligned} W_{0,j} &= \frac{-1 + \sqrt{1 - 4XY}}{2X} \quad \text{for } Y \leq 0 \\ &= \frac{1 - \sqrt{1 + 4XY}}{2X} \quad \text{for } Y > 0 \end{aligned} \tag{19.51}$$

Once $W_{0,j}$ has been found, $p_{0,j}$ may be found from equation (19.41a) if the flow is positive or (19.41b) if it is negative.

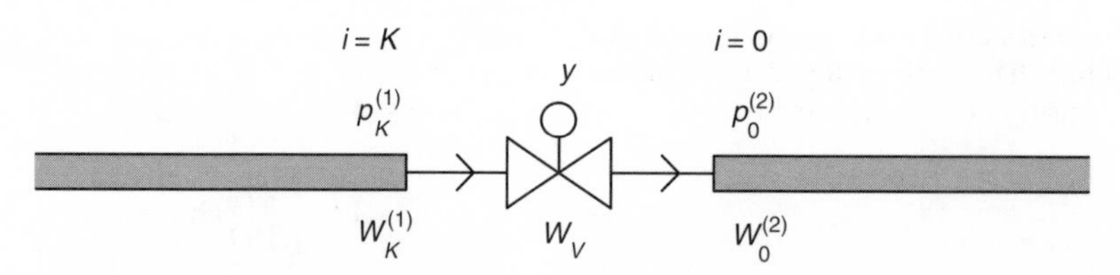

Figure 19.4 Valve located some distance along the pipeline.

19.6.4 In-line valve in a liquid pipeline

The situation is similar, but a little more complex when the valve occurs some distance down the line. Now we need to consider the upstream and downstream sections of pipeline separately. Let us give them the labels (1) and (2), respectively. Figure 19.4 illustrates the situation. A constant specific volume will be assumed once more.

The flow leaving the upstream section of the line will be the same as that passing through the valve, $W_{V,j}$, and the same as that entering the next section of the pipeline, i.e.

$$W_{V,j} = W_{K,j}^{(1)} = W_{0,j}^{(2)} \tag{19.52}$$

The valve equation gives:

$$W_{V,j} = y_j C_V \sqrt{\frac{\left|p_{K,j}^{(1)} - p_{0,j}^{(2)}\right|}{v}} \tag{19.53}$$

or

$$\frac{v}{y_j^2 C_V^2} W_{V,j}^2 + p_{0,j}^{(2)} - p_{K,j}^{(1)} = 0 \qquad \text{for } W_{V,j} \geq 0 \tag{19.54a}$$

$$-\frac{v}{y_j^2 C_V^2} W_{V,j}^2 + p_{0,j}^{(2)} - p_{K,j}^{(1)} = 0 \quad \text{for } W_{V,j} < 0 \tag{19.54b}$$

Again the fractional valve opening, y_j, will depend on the output of an integrator in the main simulation. The forward characteristic of pipe section (1) will apply, namely equation (19.30). Putting $i = K$ and putting $W_{V,j} = W_{K,j}^{(1)}$ gives

$$\frac{1}{c_{son}} \frac{p_{K,j}^{(1)} - p_{K-1,j-1}^{(1)}}{\Delta t} + \frac{1}{A} \frac{W_{V,j} - W_{K-1,j-1}^{(1)}}{\Delta t} + \frac{2fv}{A^2 D} W_{K-1,j-1}^{(1)} \left|W_{K-1,j-1}^{(1)}\right| + \frac{g}{v} \sin\theta = 0 \tag{19.55}$$

so that the pressure just upstream of the valve at time instant j is given by

$$p_{K,j}^{(1)} = p_{K-1,j-1}^{(1)} - \frac{c_{son}}{A} W_{V,j} + \frac{c_{son}}{A} W_{K-1,j-1}^{(1)} - c_{son} \Delta t \times \left[\frac{2fv}{A^2 D} W_{K-1,j-1}^{(1)} \left|W_{K-1,j-1}^{(1)}\right| + \frac{g}{v} \sin\theta\right] \tag{19.56}$$

Meanwhile, the backward characteristic of pipe section (2) will also apply, so that we may apply equation (19.31) with $i = 0$ and use equation (19.52) :

$$-\frac{1}{c_{son}} \frac{p_{0,j}^{(2)} - p_{1,j-1}^{(2)}}{\Delta t} + \frac{1}{A} \frac{W_{V,j} - W_{1,j-1}^{(2)}}{\Delta t} + \frac{2fv}{A^2 D} W_{1,j-1}^{(2)} \left|W_{1,j-1}^{(2)}\right| + \frac{g}{v} \sin\theta = 0 \tag{19.57}$$

so that the pressure just downstream of the valve at time instant j is given by

$$p_{0,j}^{(2)} = p_{1,j-1}^{(2)} + \frac{c_{son}}{A} W_{V,j} - \frac{c_{son}}{A} W_{1,j-1}^{(2)} + c_{son} \Delta t \times \left[\frac{2fv}{A^2 D} W_{1,j-1}^{(2)} \left|W_{1,j-1}^{(2)}\right| + \frac{g}{v} \sin\theta\right] \tag{19.58}$$

Substituting from equations (19.56) and (19.58) into equation (19.54a) gives a quadratic equation which may be solved for a positive flow through the valve, $W_{V,j}$:

$$\frac{A}{2c_{son}} \frac{v}{y_j^2 C_V^2} W_{V,j}^2 + W_{V,j} + \frac{A}{2c_{son}} \times \left(p_{1,j-1}^{(2)} - p_{K-1,j-1}^{(1)}\right) - \frac{1}{2}\left(W_{1,j-1}^{(2)} + W_{K-1,j-1}^{(1)}\right) + \Delta t \frac{A}{2} \left[\frac{2fv}{A^2 D} \left(W_{1,j-1}^{(2)} \left|W_{1,j-1}^{(2)}\right| + W_{K-1,j-1}^{(1)} \left|W_{K-1,j-1}^{(1)}\right|\right) + \frac{g}{v} \sin\theta\right] = 0 \tag{19.59}$$

When the flow through the valve is negative, we combine equations (19.54b), (19.56) and (19.58) to give:

$$-\frac{A}{2c_{son}}\frac{v}{y_j^2 C_V^2}W_{V,j}^2 + W_{V,j} + \frac{A}{2c_{son}} \times (p_{1,j-1}^{(2)} - p_{K-1,j-1}^{(1)}) - \frac{1}{2}(W_{1,j-1}^{(2)} + W_{K-1,j-1}^{(1)}) + \Delta t\frac{A}{2}\left[\frac{2fv}{A^2 D}\left(W_{1,j-1}^{(2)}\left|W_{1,j-1}^{(2)}\right| + W_{K-1,j-1}^{(1)}\left|W_{K-1,j-1}^{(1)}\right|\right) + \frac{g}{v}\sin\theta\right] = 0 \quad (19.60)$$

Equations (19.59) and (19.60) have the same general properties as equations (19.43) and (19.48), and the calculation of positive and negative flows follows the same procedure. Putting

$$X = \frac{A}{2c_{son}}\frac{v}{y_j^2 C_V^2} \quad (19.61)$$

and

$$Y = \frac{A}{2c_{son}}(p_{1,j-1}^{(2)} - p_{K-1,j-1}^{(1)}) - \frac{1}{2}(W_{1,j-1}^{(2)} + W_{K-1,j-1}^{(1)}) + \Delta t\frac{A}{2} \times \left[\frac{2fv}{A^2 D}\left(W_{1,j-1}^{(2)}\left|W_{1,j-1}^{(2)}\right| + W_{K-1,j-1}^{(1)}\left|W_{K-1,j-1}^{(1)}\right|\right) + \frac{g}{v}\sin\theta\right] \quad (19.62)$$

the flow through the valve may be calculated from

$$W_{V,j} = \frac{-1 + \sqrt{1 - 4XY}}{2X} \quad \text{for } Y \leq 0$$
$$= \frac{1 - \sqrt{1 + 4XY}}{2X} \quad \text{for } Y > 0 \quad (19.63)$$

The valve upstream and downstream pressures $p_{K,j}^{(1)}$ and $p_{0,j}^{(2)}$ may be found from equations (19.56) and (19.58) as soon as $W_{V,j} = W_{K,j}^{(1)} = W_{0,j}^{(2)}$ has been calculated.

19.6.5 Pump feeding the pipeline from an upstream tank

Conceptually this is very similar to the case of the upstream valve, in that the boundary condition at pipe entry relates flow to pressure difference. Once again we will assume that the specific volume, v, is the same throughout the system. Figure 19.5 illustrates such a pumped system.

The head, H (m), developed across the pump at speed N (rps), is related to the volume flow Q (m^3/s) by equation (17.28):

$$H\frac{N_0^2}{N^2} = f_{P1}\left(Q\frac{N_0}{N}\right) \quad (17.28)$$

where f_{P1} is the experimentally determined pump characteristic of head versus flow at the design speed,

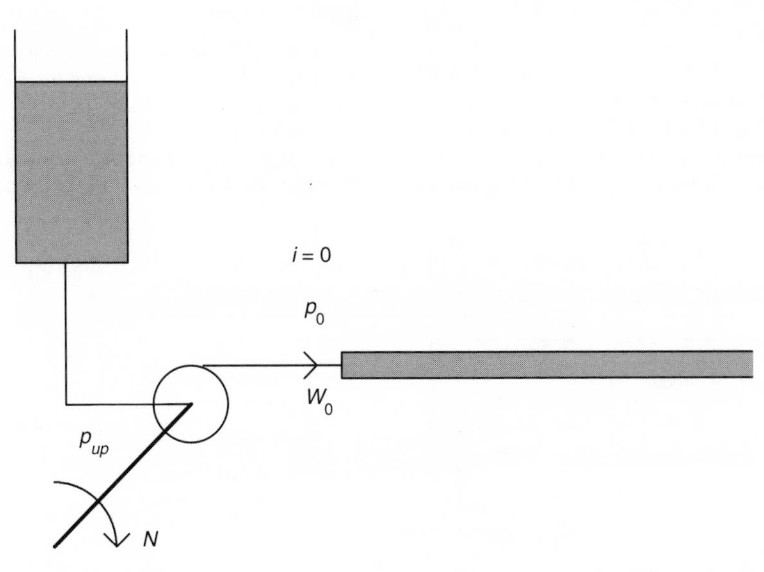

Figure 19.5 Pumping liquid into the line.

N_0, which may be represented by a polynomial, usually of low order:

$$f_{P1}(Q_0) = a_0 + a_1 Q_0 + a_2 Q_0^2 + a_3 Q_0^3 + \cdots \tag{17.56}$$

Equation (17.28) may be recast in terms of pressure rise across the pump, Δp_p, and mass flow, W, by using the relations $H = v\Delta p_P/g$ and $Q = vW$:

$$\Delta p_P = \frac{g}{v}\frac{N^2}{N_0^2} f_{P1}\left(vW\frac{N_0}{N}\right) \tag{19.64}$$

Applying equation (19.64) to the system depicted in Figure 19.5 for time instant j gives:

$$p_{0,j} = p_{up,j} + \frac{g}{v}\frac{N_j^2}{N_0^2} f_{P1}\left(vW_{0,j}\frac{N_0}{N_j}\right) \tag{19.65}$$

which assumes that the flow through the pump will be the same as the flow into the pipeline, $W_{0,j}$. The upstream pressure, $p_{up,j}$, and the pump speed, N_j, will be integrator outputs or dependent on integrator outputs, and may be assumed available in the way explained in Section 19.6.1.

Meanwhile the backward characteristic with $i = 0$ will apply, i.e. equation (19.42), repeated below:

$$-\frac{A}{c_{son}} p_{0,j} + W_{0,j} + \frac{A}{c_{son}} p_{1,j-1} - W_{1,j-1} + A\Delta t\left[\frac{2fv}{A^2 D} W_{1,j-1}\left|W_{1,j-1}\right| + \frac{g}{v}\sin\theta\right] = 0 \tag{19.42}$$

Equations (19.65) and (19.42) are two simultaneous equations which may be solved for the two unknowns, $W_{0,j}$ and $p_{0,j}$. This is a straightforward matter when the pump head/flow characteristic may be represented by a quadratic, in practice often a reasonable approximation. The delivery pressure is then given by combining equations (17.56) and (19.65):

$$p_{0,j} = p_{up,j} + \frac{g}{v}\frac{N_j^2}{N_0^2}\left(a_0 + a_1\left(vW_{0,j}\frac{N_0}{N_j}\right) + a_2\left(vW_{0,j}\frac{N_0}{N_j}\right)^2\right)$$
$$= p_{up,j} + a_0\frac{g}{v}\frac{N_j^2}{N_0^2} + a_1 g\frac{N_j}{N_0}W_{0,j} + a_2 g v W_{0,j}^2 \tag{19.66}$$

Substituting from equation (19.66) into equation (19.42) gives the following quadratic in mass flow into the pipeline, $W_{0,j}$:

$$-\frac{A}{c_{son}} a_2 g v W_{0,j}^2 + \left(1 - a_1 g\frac{A}{c_{son}}\frac{N_j}{N_0}\right)W_{0,j} + \frac{A}{c_{son}}(p_{1,j-1} - p_{up,j}) - W_{1,j-1} + A\,\Delta t \times\left[\frac{2fv}{A^2 D}W_{1,j-1}\left|W_{1,j-1}\right| + \frac{g}{v}\sin\theta\right] - \frac{A}{c_{son}} a_0 \frac{g}{v}\frac{N_j^2}{N_0^2} = 0 \tag{19.67}$$

The positive root should be selected. The pressure, $p_{0,j}$, then follows from equation (19.66).

19.6.6 Junction of two or more pipes: liquid or gas

Consider the junction of several pipes of different diameters, R of which normally carry fluid into the junction and S of which normally carry fluid away (Figure 19.6).

The forward characteristic will apply for the lines normally carrying fluid into the junction. Let us use

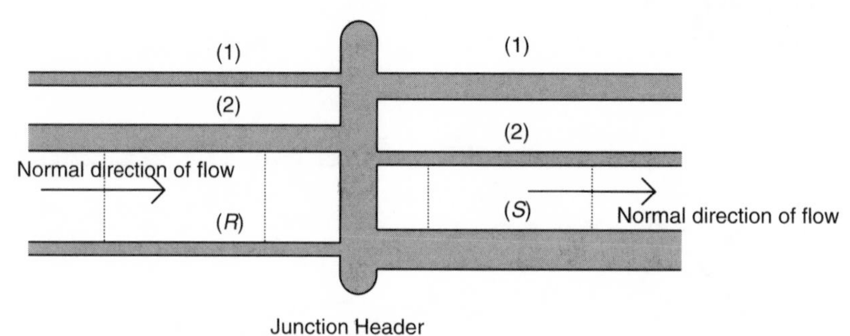

Figure 19.6 Schematic of a junction linking pipes of different diameters.

the superscript (A,r) to specify such a pipeline, where the letter A denotes that the pipeline is carrying fluid into the junction, while the letter r, which can take values $r = 1, 2, \ldots, R$, specifies which of the afferent pipelines is being referred to. We may use equation (19.30), setting the distance index $i = K$ to imply the end of each line:

$$\frac{1}{c_{son}} \frac{p_{K,j}^{(A,r)} - p_{K-1,j-1}^{(A,r)}}{\Delta t} + \frac{1}{A^{(A,r)}} \frac{W_{K,j}^{(A,r)} - W_{K-1,j-1}^{(A,r)}}{\Delta t} + \frac{2f^{(A,r)}}{(A^{(A,r)})^2 D^{(A,r)}} v_{K-1,j-1}^{(A,r)} W_{K-1,j-1}^{(A,r)} \left| W_{K-1,j-1}^{(A,r)} \right| + \frac{g}{v_{K-1,j-1}^{(A,r)}} \sin\theta^{(A,r)} = 0 \quad \text{for } r = 1, 2, \ldots, R \tag{19.68}$$

Conversely, the backward characteristic will apply for the lines that normally carry fluid away from the junction. Hence using the superscript (B, s), $s = 1, 2, \ldots, S$ to specify such an efferent pipeline, we may use equation (19.31), setting the distance index $i = 0$ to imply the beginning of each line:

$$-\frac{1}{c_{son}} \frac{p_{0,j}^{(B,s)} - p_{1,j-1}^{(B,s)}}{\Delta t} + \frac{1}{A^{(B,s)}} \frac{W_{0,j}^{(B,s)} - W_{1,j-1}^{(B,s)}}{\Delta t} + \frac{2f^{(B,s)}}{(A^{(B,s)})^2 D^{(B,s)}} v_{1,j-1}^{(B,s)} W_{1,j-1}^{(B,s)} \left| W_{1,j-1}^{(B,s)} \right| + \frac{g}{v_{1,j-1}^{(B,s)}} \sin\theta^{(B,s)} = 0 \quad \text{for } s = 1, 2, \ldots, S \tag{19.69}$$

We may solve equations (19.68) and (19.69) by noting

(i) the pressures at the exits of the lines bringing fluid to the junction will be the same as the pressures at the entry to the lines taking fluid away, i.e.

$$p_{K,j}^{(A,r)} = p_{0,j}^{(B,s)} = p_{JUNC,j} \quad \text{for } r = 1, 2, \ldots, R; s = 1, 2, \ldots, S \tag{19.70}$$

(ii) the algebraic sum of the flows at the junction is zero:

$$\sum_{r=1}^{R} W_{K,j}^{(A,r)} - \sum_{s=1}^{S} W_{0,j}^{(B,s)} = 0 \tag{19.71}$$

Carrying out a summation over all the incoming flows using equation (19.68) gives:

$$\sum_{r=1}^{R} W_{K,j}^{(A,r)} + p_{JUNC,j} \frac{1}{c_{son}} \sum_{r=1}^{R} A^{(A,r)} = \sum_{r=1}^{R} \left\{ W_{K-1,j-1}^{(A,r)} + \frac{A^{(A,r)}}{c_{son}} p_{K-1,j-1}^{(A,r)} - A^{(A,r)} \Delta t \times \left[\frac{2f^{(A,r)}}{(A^{(A,r)})^2 D^{(A,r)}} v_{K-1,j-1}^{(A,r)} W_{K-1,j-1}^{(A,r)} \left| W_{K-1,j-1}^{(A,r)} \right| + \frac{g}{v_{K-1,j-1}^{(A,r)}} \sin\theta^{(A,r)} \right] \right\} \tag{19.72}$$

Similarly, carrying out a summation of all the normally outgoing flows using equation (19.69) gives:

$$\sum_{s=1}^{S} W_{0,j}^{(B,s)} - p_{JUNC,j} \frac{1}{c_{son}} \sum_{s=1}^{S} A^{(B,s)} = \sum_{s=1}^{S} \left\{ W_{1,j-1}^{(B,s)} - \frac{A^{(B,s)}}{c_{son}} p_{1,j-1}^{(B,s)} - A^{(B,s)} \Delta t \times \left[\frac{2f^{(B,s)}}{(A^{(B,s)})^2 D^{(B,s)}} v_{1,j-1}^{(B,s)} W_{1,j-1}^{(B,s)} \left| W_{1,j-1}^{(B,s)} \right| + \frac{g}{v_{1,j-1}^{(B,s)}} \sin\theta^{(B,s)} \right] \right\} \tag{19.73}$$

Subtracting equation (19.73) from equation (19.72) and using equation (19.71) to eliminate the junction flows gives junction pressure as:

$$p_{JUNC,j} = c_{son} \left(\frac{1}{\sum_{r=1}^{R} A^{(A,r)} + \sum_{s=1}^{S} A^{(B,s)}} \right) \times \left[\sum_{r=1}^{R} \left\{ W_{K-1,j-1}^{(A,r)} + \frac{A^{(A,r)}}{c_{son}} p_{K-1,j-1}^{(A,r)} - A^{(A,r)} \Delta t \left[\frac{2f^{(A,r)}}{(A^{(A,r)})^2 D^{(A,r)}} v_{K-1,j-1}^{(A,r)} W_{K-1,j-1}^{(A,r)} \times \left| W_{K-1,j-1}^{(A,r)} \right| + \frac{g}{v_{K-1,j-1}^{(A,r)}} \sin\theta^{(A,r)} \right] \right\} - \sum_{s=1}^{S} \left\{ W_{1,j-1}^{(B,s)} - \frac{A^{(B,s)}}{c_{son}} p_{1,j-1}^{(B,s)} - A^{(B,s)} \Delta t \right.$$

$$\times \left[\frac{2f^{(B,s)}}{(A^{(B,s)})^2 D^{(B,s)}} v^{(B,s)}_{1,j-1} W^{(B,s)}_{1,j-1} \left| W^{(B,s)}_{1,j-1} \right| + \frac{g}{v^{(B,s)}_{1,j-1}} \sin\theta^{(B,s)} \right] \Bigg\} \Bigg] \quad (19.74)$$

Once the junction pressure at time instant j has been found, the various flows into and out of the junction may be found from equations (19.68) and (19.69).

While the need for generality has made equation (19.74) very 'busy' in terms of superscripts and subscripts, most junctions will connect just three or perhaps four pipes. Further, in the common case where the pipe diameters are all the same, the following simplifications apply:

$$D^{(A,r)} = D^{(B,s)}$$

$$A^{(A,r)} = A^{(B,s)} \qquad \text{for all } r \text{ and } s$$

$$\sum_{r=1}^{R} A^{(A,r)} + \sum_{s=1}^{S} A^{(B,s)} = (R+S)A \quad (19.75)$$

No allowance has been made above for additional, minor pipe inlet losses due to changes in pipe diameter. A simple way of including them is to use an effective pipe-length in place of the actual length, calculated from equation (6.59). (See Chapter 6, Section 6.6 and Chapter 4, Sections 4.9 and 4.10.) This will have the disadvantage of introducing a small error into the period calculated for any oscillations, although the effect will be very minor for long pipelines. An alternative method is to adjust the value of the friction factor in the adjoining pipe reaches.

19.7 Correcting the speed of sound for the elasticity of the pipe material

The effect of the elasticity of the wall is to cause a slowing of the sound waves, particularly when the fluid being carried is a liquid. Streeter and Wylie (1983) developed a correction formula that may be written:

$$c_{son} = \frac{\sqrt{v\kappa}}{\sqrt{1 + \beta \dfrac{\kappa}{E} \dfrac{2D}{D_0 - D}}} = \frac{c_{son\infty}}{\sqrt{1 + \beta \dfrac{\kappa}{E} \dfrac{2D}{D_0 - D}}} \quad (19.76)$$

where

$c_{son\infty}$ is the speed of sound in an infinite fluid medium (m/s),

κ is the bulk modulus of elasticity of the fluid (Pa),

E is the Young's modulus of elasticity of the wall material (Pa),

D is the internal diameter of the pipe (m),

D_0 is the external diameter of the pipe (m),

and the coefficient β lies in the range $0 \leq \beta \leq 1$. The coefficient β takes the value zero when the wall effects are negligible, but is unity when the pipeline contains expansion joints.

Take for example, a pipe with external diameter $D_0 = 0.762$ m, and internal diameter $D = 0.7366$ m. Let the pipe be steel so that the Young's modulus is $E = 2 \times 10^{11}$ Pa and let it incorporate expansion joints, so that $\beta = 1$. Suppose the flow being passed is water at 20°C, with $v = 0.001\ \text{m}^3/\text{s}$, $\kappa = 2.2 \times 10^9$ Pa.

$$c_{son\infty} = \sqrt{v\kappa} = \sqrt{10^{-3} \times 2.2 \times 10^9} = 1483\ \text{m/s}$$

$$c_{son} = \frac{c_{son\infty}}{\sqrt{1 + \beta \dfrac{\kappa}{E} \dfrac{2D}{D_0 - D}}}$$

$$= \frac{1483}{\sqrt{1 + 1 \times \dfrac{2.2 \times 10^9}{2 \times 10^{11}} \dfrac{2 \times 0.7366}{0.762 - 0.7366}}}$$

$$= \frac{1483}{\sqrt{1.638}} = 0.7813 \times 1483 = 1159\ \text{m/s}$$

In this case, the speed of sound in the pipeline, c_{son}, is a sizeable 22% down on the speed of sound in the infinite fluid medium, $c_{son\infty}$.

The difference between c_{son} and $c_{son\infty}$ is usually a lot less when the pipeline is carrying a gas. For example, suppose now that the same pipe is being used to pass natural gas, CH_4, at 5 MPa and 15°C/288K. Take $\gamma = 1.35$. The speed of sound in an infinite medium is $c_{son\infty} = \sqrt{\gamma R_w T} = \sqrt{1.35 \times (8314/16) \times 288} = 449.5$ m/s. The bulk modulus of elasticity for a gas is given by $\kappa = \gamma p = 1.35 \times 5 \times 10^6 = 6.75 \times 10^6$ Pa. Hence the speed of sound waves in the pipe is

$$c_{son} = \frac{c_{son\infty}}{\sqrt{1 + \beta \dfrac{\kappa}{E} \dfrac{2D}{D_0 - D}}}$$

$$= \frac{449.5}{\sqrt{1 + 1 \times \dfrac{6.75 \times 10^6}{2 \times 10^{11}} \dfrac{2 \times 0.7366}{0.762 - 0.7366}}}$$

$$= \frac{449.5}{\sqrt{1.00196}} = 0.999 \times 449.5 = 449.0\ \text{m/s}$$

In this case, c_{son} and $c_{son\infty}$ are almost the same.

19.8 Example of pipeline flow using the Method of Characteristics

The example is a variant of example 12.8, given in Section 12.8 of Streeter and Wylie (1983). Referring to Figure 19.7, water at 20°C flows from a reservoir with a height under perfect level-control at 100 m through a horizontal, 2 m-diameter pipeline that is 4800 m long. A valve with $C_V = 1.2229\,\text{m}^2$ is situated 1200 m down the line, and is left fully open. The line discharges to atmospheric pressure, with its exit flow under flow control, assumed to be perfect. The flow controller discharges a steady flow of 2626.6 kg/s for the first 10 seconds, then reduces the exit flow to half that amount linearly over the next 10 seconds. The friction factor is $f = 0.0055$, while the speed of sound in the pipe is 1200 m/s.

(i) Taking the mean value of the specific volume as $v = 0.0010013\,\text{m}^3/\text{s}$, calculate the flow through the in-line valve and the pressure transients just before the in-line and the final valve.
(ii) Provide justification for the use of a mean value of specific volume.
(iii) Assess the effect of linearly reducing the opening of the valve 1200 m down the line (the 'in-line valve') over 16 seconds from fully open to 20% open, starting 30 seconds from the beginning of the simulation.

Solution

The solution was programmed using a spread-sheet package.

(i) We begin by dividing the pipeline into eight reaches of $\Delta x = 600\,\text{m}$, two before the in-line valve and six after. We shall use the superscripts (1) and (2) to denote the sections upstream and downstream of the in-line valve, as shown in Figure 19.7.

Finding the initial steady state

Assuming an initial steady state, the flow will be the same in each section: $W_{i,0}^{(m)} = 2626.6\,\text{kg/s}$ for $m = 1, 2$ and all i.

The pressure drop across each reach will be the same and given by steady-state flow equation (4.40):

$$\Delta p_{i,0}^{(m)} = 2f\frac{\Delta x}{D}\frac{v}{A^2}(W_{i,0}^{(m)})^2$$

$$= 2 \times 0.0055 \times \frac{600}{2} \times \frac{0.0010013}{\left(\pi\frac{2^2}{4}\right)^2} \times 2626.6^2$$

$$= 2309.7\,\text{Pa}$$

The pressure drop across the in-line valve at fully open is given by

$$\Delta p_0 = v\frac{(W_{i,0}^{(m)})^2}{y^2 C_V^2}$$

$$= 0.0010013 \times \frac{2626.6^2}{1 \times 1.2229^2} = 4619.5\,\text{Pa}$$

The reservoir pressure is given for all times, j, including the initial time, by

$$p_{0,j}^{(1)} = \frac{gl}{v} = \frac{100 \times 9.8}{0.0010013} = 1078728\,\text{Pa}$$

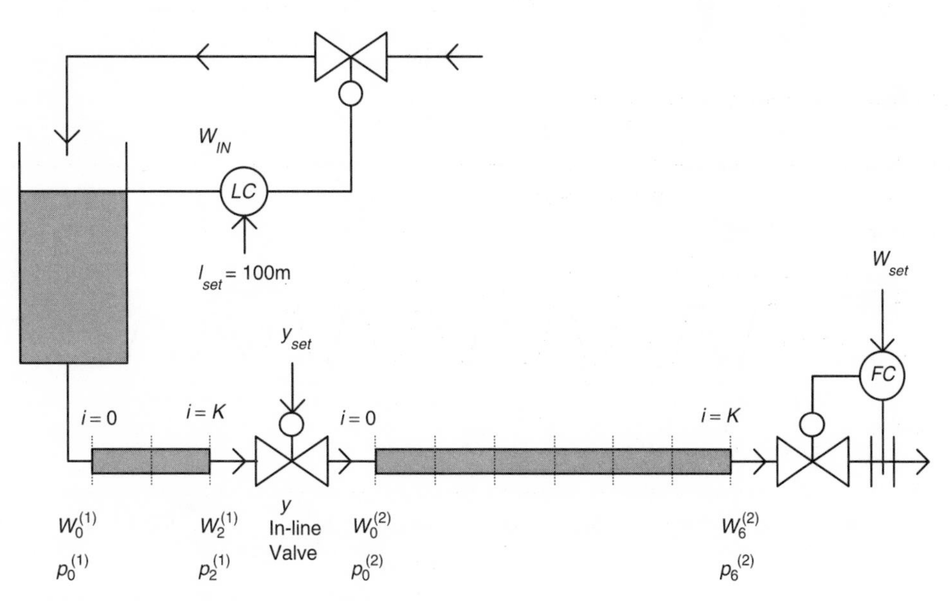

Figure 19.7 Schematic of liquid pipeline example.

Knowing the reservoir pressure and each of the pressure drops along the line allows us to calculate the pressure in each reach by subtraction.

Applying the Method of Characteristics

Having selected the length of each reach, Δx, the time-step is fixed and given by $\Delta t = \Delta x / c_{son} = 600/1200 = 0.5$ s.

The pressure at the inlet to pipe branch (1) is fixed, and so Section 19.6.1 applies. Equation (19.37) was programmed for flow 0 in branch (1), $W_0^{(1)}$. The end of branch (1) is joined to the beginning of branch (2) via an in-line valve and so the equations listed in Section 19.6.4 were programmed: equation (19.63) to give the valve flow, and hence also flow 2 in branch (1), $W_2^{(1)}$, and flow 0 in branch (2), $W_0^{(2)}$; equation (19.56) to give pressure 2 in branch (1), $p_2^{(1)}$ and (19.58) to give pressure 0 in branch (2), $p_0^{(2)}$, respectively. Finally, the flow out of branch (2), $W_6^{(2)}$, is fixed, and so Section 19.6.2 applies. Equation (19.39) was programmed for pressure 6 in branch (2), $p_6^{(2)}$.

The pressure and flow in each of the remaining, internal reaches were determined by programming the characteristic equations (19.32) and (19.33). These equations were applied to give pressure 1 and flow 1 in branch (1) and pressures and flows 1,2,3,4 and 5 in branch (2).

Simulation results

The simulation was run for 200 s (see Figures 19.8 and 19.9). Figure 19.8 shows the calculated flow through the in-line valve while Figure 19.9 shows the pressures just before the in-line valve and the final valve. Clearly the closure has caused a very oscillatory

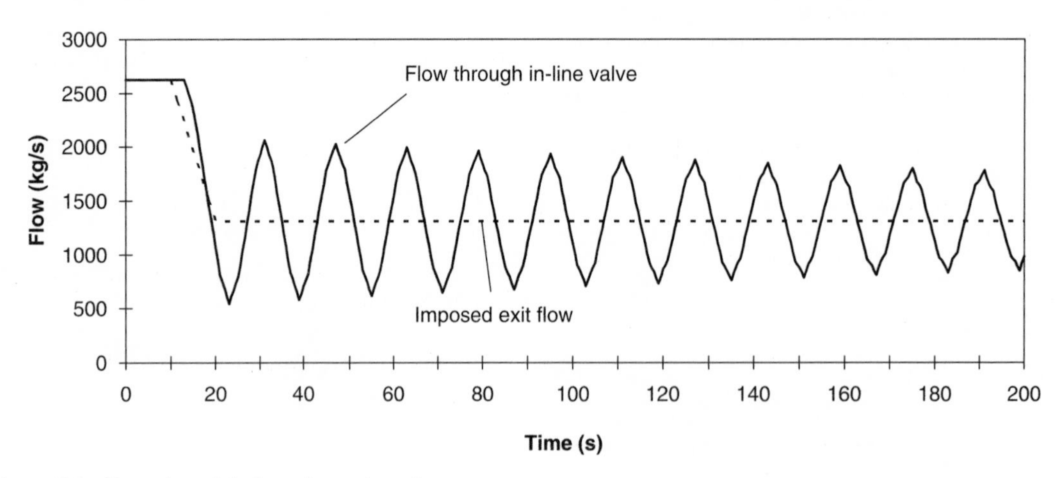

Figure 19.8 Flows through in-line valve and at exit.

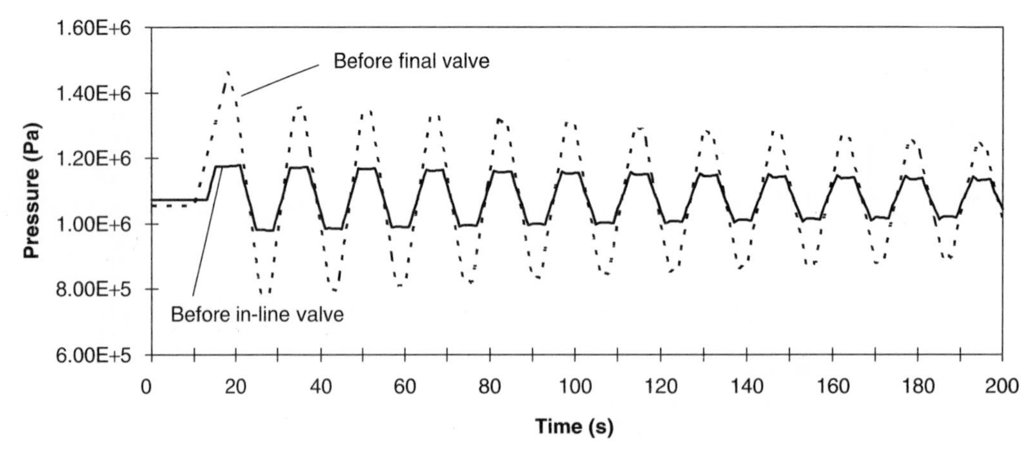

Figure 19.9 Pressure before in-line and final valves.

system response. The oscillations have a period of 16 s, which corresponds to the expected period of the system fundamental, namely $T = 4L/c_{son}$, and they decay only very slowly.

(ii) The mass in the ith reach will obey the equation

$$\frac{dm_i}{dt} = W_{i-1} - W_i \qquad (19.77)$$

which may be programmed using Euler integration as

$$m_{i,j} = m_{i,j-1} + \Delta t(W_{i-1,j-1} - W_{i,j-1}) \qquad (19.78)$$

The specific volume in the ith reach may then be found from

$$v_{i,j} = \frac{A\,\Delta x}{m_{i,j}} \qquad (19.79)$$

This was done for the reach undergoing the most major excursions in flow and pressure, namely the reach just before the final valve. Figure 19.10 shows the result.

Clearly the specific volume does change, but the maximum excursion is less than 0.0000003 m^3/kg, or less than 0.03%, confirming the validity of using a single, mean value of specific volume. (Similar values would be obtained by using steam tables to read off specific volume at 20°C as a function of pressure excursion.)

(iii) The in-line valve opening was programmed to reduce to 20%, and the simulation re-run. The resulting transient is shown in Figures 19.11 and 19.12. Clearly the valve has had a useful damping effect, restoring the pipeline to equilibrium at a new operating condition at the end of the period examined.

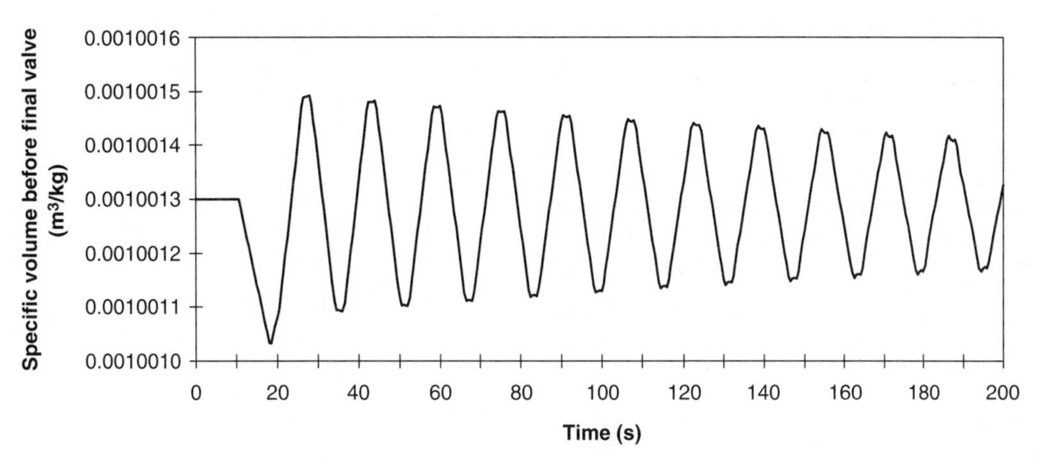

Figure 19.10 Specific volume before final valve.

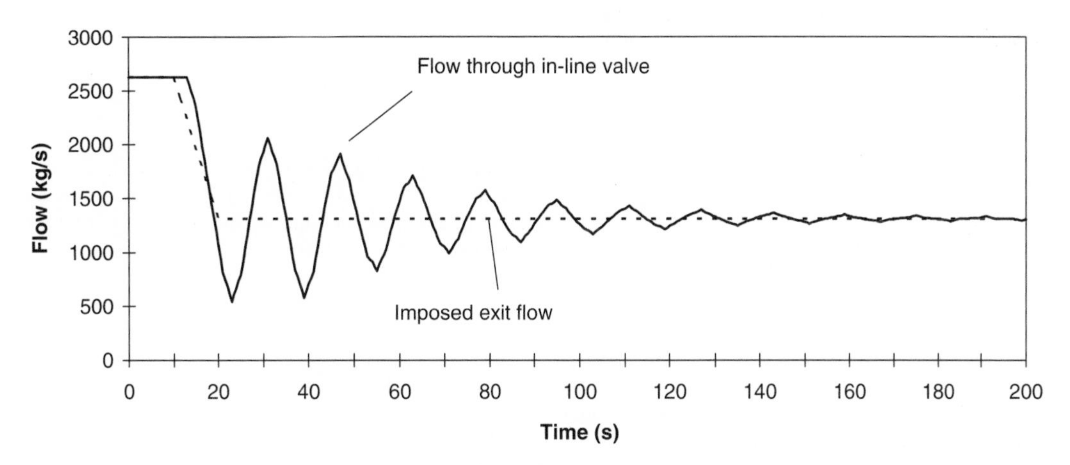

Figure 19.11 Flows when in-line valve is reduced to 20% opening.

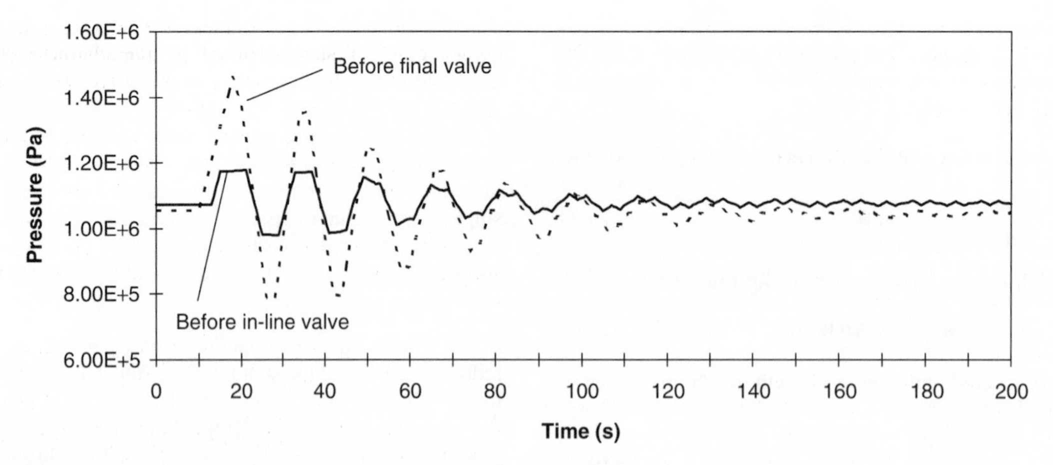

Figure 19.12 Pressures when in-line valve is reduced to 20% opening.

19.9 Finite differences

The method of characteristics is accepted as the most accurate way of solving hyperbolic partial differential equations such as those treated in this chapter. The method is particularly well suited to solving pipeline equations because neglecting the pipeline velocity by comparison with the speed of sound allows the use of a constant rectangular grid in x and t, which makes interfacing to the rest of the simulation program straightforward. Nevertheless mention should be made of a popular alternative method of solution, namely finite differencing of the space variable, and using the integration algorithm of the main program to integrate the time variables.

The two equations to be considered are the continuity equation (19.10) and the momentum equation (19.11). In line with the treatment using the method of characteristics, we will neglect the term $(\partial/\partial x)(v(W^2/A^2))$ in equation (19.11), using the justification contained in Section 19.5. Assuming a uniform pipe diameter, the two equations to be solved are thus:

$$\frac{\partial p}{\partial t} + \frac{c_{son}^2}{A}\frac{\partial W}{\partial x} = 0 \tag{19.10}$$

and

$$\frac{1}{A}\frac{\partial W}{\partial t} + \frac{\partial p}{\partial x} + \frac{2f}{A^2 D} vW\,|W| + \frac{g}{v}\sin\theta = 0 \tag{19.80}$$

Maudsley (1984) proposes central differences for the inner points, with forward and backward differences at the boundaries. The pipeline is split into K pipe reaches, each of which is assumed to have two state variables associated with it: a pressure and a flow, both of which will vary with time. Central differencing implies that the spatial derivatives will be approximated by subtracting the value of the state variable one reach upstream from the corresponding value one reach downstream and dividing the result by the distance between. Hence

$$\begin{aligned} \frac{\partial p_i}{\partial x} &\approx \frac{p_{i+1} - p_{i-1}}{2\Delta x} \\ \frac{\partial W_i}{\partial x} &\approx \frac{W_{i+1} - W_{i-1}}{2\Delta x} \end{aligned} \tag{19.81}$$

An advantage of this formulation is that both upstream and downstream disturbances may be transmitted in a natural way from pipe reach to pipe reach. Rearranging equations (19.10) and (19.80) and using central differencing for K pipe reaches gives:

$$\frac{dp_i}{dt} = \frac{c_{son}^2}{A}\frac{W_{i-1} - W_{i+1}}{2\Delta x} \tag{19.82}$$

$$\frac{dW_i}{dt} = A\frac{p_{i-1} - p_{i+1}}{2\Delta x} - A \times \left[\frac{2f}{A^2 D} v_i W_i\,|W_i| + \frac{g}{v_i}\sin\theta\right] \tag{19.83}$$

which equations are valid for $i = 1, 2, \ldots, K-1$. Central differencing is obviously not possible at either end of the pipeline, and so the boundary differentials with respect to time are calculated by:

$$\frac{dp_0}{dt} = \frac{c_{son}^2}{A}\frac{W_0 - W_1}{\Delta x} \tag{19.84}$$

$$\frac{dW_0}{dt} = A\frac{p_0 - p_1}{\Delta x} - A \times \left[\frac{2f}{A^2 D} v_0 W_0\,|W_0| + \frac{g}{v_0}\sin\theta\right] \tag{19.85}$$

$$\frac{dp_K}{dt} = \frac{c_{son}^2}{A}\frac{W_{K-1} - W_K}{\Delta x} \quad (19.86)$$

$$\frac{dW_K}{dt} = A\frac{p_{K-1} - p_K}{\Delta x} - A \times \left[\frac{2f}{A^2 D} v_K W_K |W_K| + \frac{g}{v_K}\sin\theta\right] \quad (19.87)$$

Only one of the two boundary pairs will be used at each boundary. If a pressure is the boundary condition, then only the flow differential will be used. Conversely, if a flow is the boundary condition, then only the pressure differential will be used.

Such a scheme has been found to attenuate higher frequency disturbances as they pass through the system, so that discontinuities tend to be smeared. Maudsley (1984) gives the number of finite difference segments, N_S, as a function of the highest frequency, f_R, that the engineer wishes the simulation to reproduce:

$$N_s \geq \pi \frac{f_R}{f_{fun}} \quad (19.88)$$

where $f_{fun} = c_{son}/4L$ is the fundamental frequency of the pipeline. It is usually sufficient if the simulation reproduces the third harmonic, so that $f_R/f_{fun} = 3$, which implies $N_S \geq 3\pi$, or, since N_S must be an integer:

$$N_s \geq 10 \quad (19.89)$$

a figure that is reassuringly close to the sort of value an experienced modeller would be likely to try based on past experience.

The selected integer value of N_S fixes the length of the pipe segment, $\Delta x = L/N_S$. Computed results should not be expected to hold outside the cone of influence in x,t space defined by the characteristic lines, as shown in the centre of Figure 19.1. Hence we must choose a timestep smaller than $\Delta x/c_{son}$:

$$\Delta t \leq \frac{\Delta x}{c_{son}} \quad (19.90)$$

The size of the timestep below that limit depends on the time integration algorithm. It is shown in Ames (1992) that stability problems arise from combining a forward-difference representation of the time differential with a central difference for the space differential. Accordingly the Euler integration method that was perfectly feasible with the method of characteristics is no longer a realistic choice. Maudsley (1984) reports good results when a fourth-order Runge–Kutta routine with variable timestep was used. The results were comparable with those produced by the method of characteristics, but interestingly the latter needed only half the number of pipeline reaches.

19.10 Bibliography

Ames, W.F. (1992). Numerical Methods for Partial Differential Equations, 3rd edition, Academic Press, Boston.

Fincham, A.E. and Goldwater, M.H. (1979). Simulation models for gas transmission networks, *Trans. Inst. M. C.*, **1**, 3–13.

Maudsley, D. (1984). Errors in the simulation of pressure transients in a hydraulic system, *Trans. Inst. M. C.*, **6**, 7–12.

Streeter, V.L. and Wylie, E.B. (1983). Fluid Mechanics, First SI Metric Edition, McGraw-Hill, New York.

20 Distributed components: heat exchangers and tubular reactors

20.1 Introduction

Until Chapter 19, we considered plant components where the important parameters could be defined at particular points in the process. For instance, our purposes were served adequately when the model calculated steam-drum temperature and pressure as 'point' parameters in Chapter 12 because these variables would represent conditions at all locations inside the steam drum. However, Chapter 19 showed how the flow in long pipelines could vary in space as well as time, making the pipeline a distributed system, the accurate analysis of which required the application of partial differential equations. In fact, systems do not have to be kilometres long to be classed as 'distributed'. For example, the temperature inside the tube of a heat exchanger a few metres long will vary continuously along its length, as will the temperature and the reaction rate inside a tubular reactor. Such plant components are by their nature distributed, and, as was the case for long pipelines, require the use of partial differential equations in time and space to describe their dynamic behaviour.

The fluids running through both components considered in this chapter will flow through a containment of uniform cross-section: the tubes and the shell in the heat exchanger, the catalyst beds in the tubular reactor. The general equations describing the dynamics of such fluid flows are derived in Chapter 3, Sections 3.8, 3.9 and 3.10. The general equations for chemical reaction dynamics are given in Chapter 13. This chapter will discuss how those equations may be applied, first to a typical heat exchanger and then to a tubular reactor.

20.2 General arrangement of a shell-and-tube heat exchanger

The heat exchanger is thermodynamically more efficient when its fluids run in counter-current mode, as illustrated in Figure 20.1, and accordingly this mode of

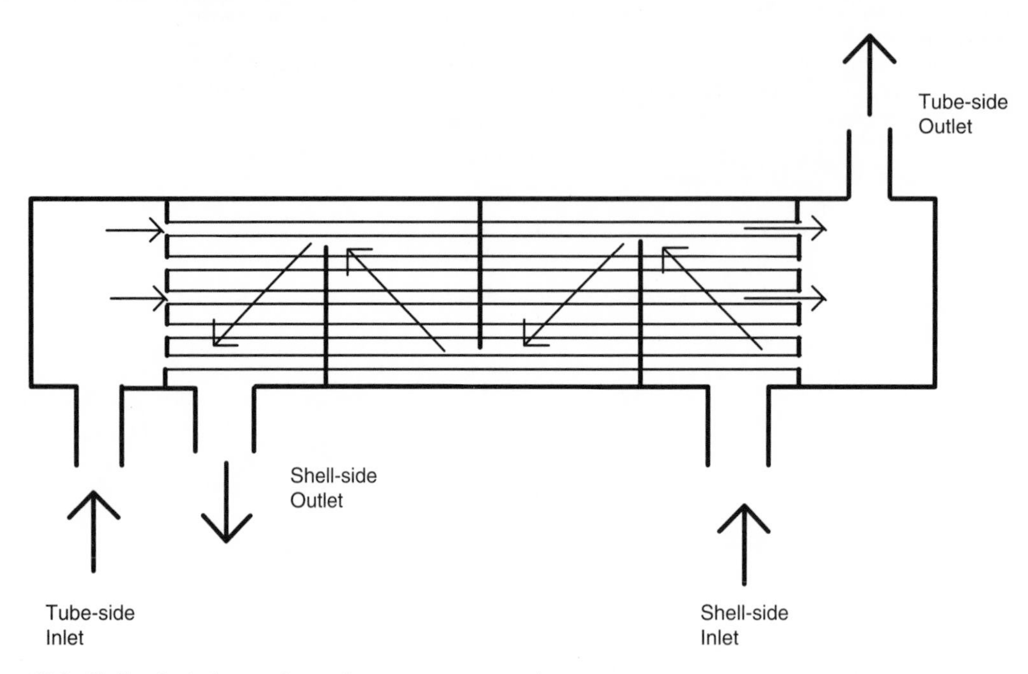

Figure 20.1 Shell and tube heat exchanger in counter-current mode.

operation is the one encountered most frequently. The tube fluid runs through a bank of tubes (in practice often numbered in hundreds) from left to right in the diagram, while the shell fluid runs in the general direction right to left. Baffles direct the flow of the shell fluid, so that its flow across the tubes is approximately cross-flow.

At each point in the heat exchanger, heat passes from the hotter fluid to the tube wall, and then from the tube wall to the cooler fluid. The tube wall will have its own dynamic response, and, just as it separates physically the two fluids, so it separates mathematically the calculations of the two fluid temperatures. Heat will also flow between the shell fluid and the shell wall, which will normally be heavily insulated to prevent heat flow to the environment. The shell wall will be relatively massive and cause a significant slowing of the response of the shell-side fluid outlet temperature.

20.3 Equations for flow in a duct subject to heat exchange

Mass balance

The dynamic mass balance for a duct of uniform cross-sectional area is given by equation (3.57), repeated below:

$$\frac{\partial}{\partial t}\left(\frac{1}{v}\right) = -\frac{\partial}{\partial x}\left(\frac{c}{v}\right) \tag{3.57}$$

The flow dynamics will be very much faster than the temperature dynamics that will be our primary concern. Hence we may assume that the flow is in an evolving steady state, which allows us to set the time differential to zero. But by equation (3.57), the space differential must also be zero:

$$0 = \frac{\partial}{\partial t}\left(\frac{1}{v}\right) = -\frac{\partial}{\partial x}\left(\frac{c}{v}\right) \tag{20.1}$$

Since the cross-sectional area of the heat exchanger duct, A, will be unchanged with distance, equation (20.1) implies that the mass flow will be the same at all points along the heat exchanger duct, as can be seen from:

$$0 = A\frac{\partial}{\partial x}\left(\frac{c}{v}\right) = \frac{\partial}{\partial x}\left(\frac{Ac}{v}\right) = \frac{\partial W}{\partial x} = 0 \tag{20.2}$$

It is normally reasonable for both liquids and gases to calculate the flow through each heat exchanger duct using a steady-state flow equation of the form:

$$W = C\sqrt{\frac{p_1 - p_2}{v}} \tag{20.3}$$

where C is the flow conductance in m^2 associated with the duct (shell or tube) of the heat exchanger being considered, while v is the mean specific volume (see Chapters 4 and 6). The value of C may be calculated from flowsheet conditions or from plant measurements. The specific volume may be updated as the simulation proceeds, but often a constant value will suffice.

Energy balance

Assuming the cross-sectional area of the duct stays constant with length, the appropriate form of the energy equation is given by equation (3.69), repeated below:

$$\begin{aligned}\left(u + \frac{1}{2}c^2 + gx\sin\theta\right)\frac{\partial}{\partial t}\left(\frac{1}{v}\right) + \frac{1}{v}\frac{\partial}{\partial t}\left(u + \frac{1}{2}c^2\right)\\ = \frac{\phi}{A} - \frac{c}{v}\frac{\partial}{\partial x}\left(h + \frac{1}{2}c^2\right)\\ - \left(h + \frac{1}{2}c^2 + gx\sin\theta\right)\frac{\partial}{\partial x}\left(\frac{c}{v}\right) - \frac{c}{v}g\sin\theta\end{aligned} \tag{3.69}$$

We may use the steady-state flow equation (20.1) to eliminate a number of terms and so give:

$$\frac{1}{v}\frac{\partial}{\partial t}\left(u + \frac{1}{2}c^2\right) = \frac{\phi}{A} - \frac{c}{v}\frac{\partial}{\partial x}\left(h + \frac{1}{2}c^2\right) - \frac{c}{v}g\sin\theta \tag{20.4}$$

Further, the assumption of an evolving steady state for flow implies that the flow becomes established instantaneously, so that at each instant we have

$$0 = \frac{\partial W}{\partial t} = A\frac{\partial}{\partial t}\left(\frac{c}{v}\right) = A\left(\frac{1}{v}\frac{\partial c}{\partial t} + c\frac{\partial}{\partial t}\left(\frac{1}{v}\right)\right) \tag{20.5}$$

Since $(\partial/\partial t)(1/v) = 0$ from equation (20.1), it follows that equation (20.5) implies $\partial c/\partial t = 0$ also. This result may be used to simplify equation (20.1) to

$$\frac{1}{v}\frac{\partial u}{\partial t} = \frac{\phi}{A} - \frac{c}{v}\frac{\partial}{\partial x}\left(h + \frac{1}{2}c^2\right) - \frac{c}{v}g\sin\theta \tag{20.6}$$

A further simplification results from the fact that the change in enthalpy with distance will be orders of magnitude higher than the change in kinetic energy. This statement will be valid for both liquids and gases. For example, consider the case of water at 20°C and 1 bar entering a heat exchanger tube with at a typical velocity of 1 m/s and emerging at 70°C. Neglecting the small pressure drop across the heat exchanger, we may construct Table 20.1, in which the outlet velocity is calculated from the inlet by the steady-state relationship: $c_{out} = (v_{out}/v_{in})c_{in}$.

Similarly, let us consider air at 1 bar flowing into a heat exchanger duct at 77°C/350 K and emerging

Table 20.1 Comparison of changes in enthalpy and kinetic energy: water case

	Inlet	*Outlet*	*Difference between inlet and outlet*
Temperature (°C)	20	70	50
Specific volume (m^3/kg)	0.0010017	0.0010227	2.1×10^{-5}
Velocity (m/s)	1.0	1.021	0.021
Specific kinetic energy (J/kg)	0.5	0.521	0.021
Specific enthalpy (J/kg)	84 000	293 000	209 000

Table 20.2 Comparison of changes in enthalpy and kinetic energy: air case

	Inlet	*Outlet*	*Difference between inlet and outlet*
Temperature (°C)	77	127	50
Specific volume (m^3/kg)	1.005	1.148	0.143
Velocity (m/s)	20	22.85	2.85
Specific kinetic energy (J/kg)	200	261	61
Specific enthalpy (J/kg)	350 700	401 200	50 500

at 127°C/400 K. The inlet velocity is assumed to be 20 m/s. Table 20.2 may be constructed. The change in enthalpy is massively larger than the change in kinetic energy in both instances: seven orders of magnitude for the water case, three orders for the air case considered. Thus the remaining kinetic energy term may be dropped, leaving (20.6) as:

$$\frac{\partial u}{\partial t} + c\frac{\partial h}{\partial x} = \frac{v}{A}\phi - cg\sin\theta \tag{20.7}$$

For a horizontal heat exchanger $0 = \theta = \sin\theta$, so that we arrive at the simpler form:

$$\frac{\partial u}{\partial t} + c\frac{\partial h}{\partial x} = \frac{v}{A}\phi \tag{20.8}$$

20.4 Equation for liquid flow in a duct subject to heat exchange

The specific internal energy and the specific enthalpy of liquids are strongly dependent on temperature and only very weakly on pressure, so that we may expand equation (20.8) to

$$\frac{du}{dT}\frac{\partial T}{\partial t} + c\frac{dh}{dT}\frac{\partial T}{\partial x} = \frac{v}{A}\phi \tag{20.9}$$

Since $(du/dT) \approx (dh/dt) = c_p$ for a liquid, we may re-express (20.9) as:

$$\frac{\partial T}{\partial t} + c\frac{\partial T}{\partial x} = \frac{v}{Ac_p}\phi \tag{20.10}$$

Since the temperature variable is a function of both distance and time, $T = T(x, t)$, a formal differention gives

$$\frac{dT}{dt} = \frac{\partial T}{\partial t} + \frac{\partial T}{\partial x}\frac{dx}{dt} \tag{20.11}$$

where dx/dt is the velocity with which temperature changes will propagate through the duct. A comparison of equations (20.10) and (20.11) indicates that temperature changes will propagate through the duct at the same velocity as the fluid, c.

20.5 Equation for gas flow in a duct subject to heat exchange

The specific internal energy and the specific enthalpy of an ideal gas are dependent on temperature alone, and so equation (20.9) is valid for a gas as well as a liquid. Further, we may note that the specific heats at constant volume and constant pressure for a gas are given by:

$$c_v = \frac{du}{dT} \tag{3.12}$$

$$c_p = \frac{dh}{dT} \tag{3.16}$$

Hence equation (20.9) may be reduced to:

$$\frac{\partial T}{\partial t} + \gamma c\frac{\partial T}{\partial x} = \frac{v}{Ac_v}\phi \tag{20.12}$$

where γ is the ratio of the specific heats:

$$\gamma = \frac{c_p}{c_v} \tag{3.25}$$

Comparing equation (20.12) with equation (20.11), we see that the temperature front in a gas duct subject to heat exchange will travel at a velocity of γc, somewhat faster than the speed of the flowing gas.

20.6 Application of the duct equations to the tube-side fluid

We will assume that all the tubes behave in a similar fashion, and so may lump them together for the purposes of calculation. Let us introduce the following subscripts: t to denote tube fluid, tw to denote tube wall and twt to denote a transfer from tube wall to tube fluid. The inside diameter of the tube will be written D_{ti} (the subscripts denoting 'tube' and 'inside'). Using this notation, the heat flux per unit length, ϕ_{twt}, from tube wall to tube fluid is the product of the heat transfer coefficient, K_{twt}, the temperature difference, $(T_{tw} - T_t)$ and the tube-wall surface area per unit length, $N_t \pi D_{ti}$, where N_t is the number of tubes:

$$\phi_{twt} = K_{twt} N_t \pi D_{ti} (T_{tw} - T_t) \tag{20.13}$$

The cross-sectional area of the inside of all the tube, $N_t A_t$, is given by $N_t A_t = N_t \pi D_{ti}^2/4$, allowing the energy equation (20.10) for the liquid to be written for the bank of tubes as

$$\frac{\partial T_t}{\partial t} + c_t \frac{\partial T_t}{\partial x} = 4 \frac{v_t}{D_{ti} c_{pt}} K_{twt} (T_{tw} - T_t) \tag{20.14}$$

where c_t is the velocity and c_{pt} is the specific heat of the liquid flowing through the tubes.

The corresponding equation for the gas is found by substituting into equation (20.12):

$$\frac{\partial T_t}{\partial t} + \gamma_t c_t \frac{\partial T_t}{\partial x} = 4 \frac{v_t}{D_{ti} c_{vt}} K_{twt} (T_{tw} - T_t) \tag{20.15}$$

where c_t is the velocity and c_{vt} is the specific heat at constant volume of the gas flowing through the tubes, while γ_t is its specific-heat ratio.

The heat transfer coefficient for tube flow will be proportional to the mass flow raised to the power 0.8. Hence if the flowsheet mass flow is W_{t0} and the flowsheet heat transfer coefficient is K_{twt0}, the heat transfer coefficient at any other tube-side flow, W_t, will be given by:

$$K_{twt} = K_{twt0} \left(\frac{W_t}{W_{t0}} \right)^{0.8} \tag{20.16}$$

20.7 Application of the duct equations to the shell-side fluid

The flow geometry of the shell side of the heat exchanger is complex and three-dimensional. Flow in each baffled section is bounded by the vertical baffles, by the cylindrical shell and by a very large number of cylindrical tubes. While the flow is in reality close to cross-flow, it is nevertheless possible to achieve good results for the purposes of control engineering studies by assuming the flow is parallel to the tube axis and applying equations (20.10) and (20.12).

Continuing with the sign convention that the heat flux is positive when heat flows into the fluid, the total heat flux per unit length for the shell-side fluid is now the sum of the heat flux from the tube wall to the shell-side fluid, ϕ_{tws}, and the heat flux from the shell wall to the shell-side fluid, ϕ_{sws}:

$$\phi_{tws} = K_{tws} N_t \pi D_{to} (T_{tw} - T_s) \tag{20.17}$$

$$\phi_{sws} = K_{sws} \pi D_{si} (T_{sw} - T_s) \tag{20.18}$$

where additional subscripts have been introduced: s to denote shell-side fluid, sw to denote shell wall, tws to denote a transfer from tube wall to shell-side fluid and sws to denote a transfer from the shell wall to the shell-side fluid. The outside diameter of a tube is written D_{t0}, while the inside diameter of the shell is written D_{si}.

Applying equation (20.10), the energy equation for the shell-side fluid when that fluid is a liquid becomes:

$$\frac{\partial T_s}{\partial t} + c_s \frac{\partial T_s}{\partial x} = \frac{v_s}{A_s c_{ps}} (K_{tws} N_t \pi D_{to} (T_{tw} - T_s) + K_{sws} \pi D_{si} (T_{sw} - T_s)) \tag{20.19}$$

where c_s is the velocity and c_{ps} is the specific heat of the shell-side liquid.

The corresponding equation when the shell side is carrying a gas is:

$$\frac{\partial T_s}{\partial t} + \gamma_s c_s \frac{\partial T_s}{\partial x} = \frac{v_s}{A_s c_{vs}} (K_{tws} N_t \pi D_{to} (T_{tw} - T_s) + K_{sws} \pi D_{si} (T_{sw} - T_s)) \tag{20.20}$$

where c_s is the velocity and c_{vs} is the specific heat at constant volume of the shell-side gas, while γ_s is its specific-heat ratio.

In both (20.19) and (20.20), A_s is the axial cross-sectional area of the shell-side that the shell-side fluid may pass through, given by:

$$A_s = \frac{\pi}{4} (D_{si}^2 - N_t D_{to}^2) \tag{20.21}$$

Note that if the heat exchanger is counter-current, the x-variable in the shell-side equations above is measured in the opposite sense from the x-variable in the tube-side equations.

Although the calculational method assumes flow parallel to the tubes and parallel to the shell, in reality the flow is close to cross-flow, when the heat transfer coefficient has a dependence on velocity raised to the power 0.6. Hence:

$$K_{tws} = K_{tws\,0}\left(\frac{W_s}{W_{s0}}\right)^{0.6} \tag{20.22}$$

$$K_{sws} = K_{sws\,0}\left(\frac{W_s}{W_{s0}}\right)^{0.6} \tag{20.23}$$

where $K_{tws\,0}$ is the heat transfer coefficient between tube wall and shell-side fluid at design conditions, when the shell-side mass flow is W_{s0}, while K_{tws} is the heat transfer coefficient between tube wall and shell-side fluid when the shell-side mass flow is at a different value, W_s. Similarly $K_{sws\,0}$ is the heat transfer coefficient between shell wall and shell-side fluid at design conditions, when the shell-side mass flow is W_{s0}, while K_{sws}, is the heat transfer coefficient between shell wall and shell-side fluid when the shell-side mass flow is at a different value, W_s.

20.8 Equations for the tube wall and the shell wall

Figure 20.2 is a schematic of a counter-current heat exchanger showing the heat flows. Neglecting heat conduction in a longitudinal direction, an energy balance on the tube wall element gives:

$$\frac{\partial}{\partial t}(m_{tw}\,\delta x\, c_{ptw} T_{tw}) = -\phi_{twt}\,\delta x - \phi_{tws}\,\delta x \tag{20.24}$$

where c_{ptw} is the specific heat of the tube wall material and m_{tw} is the mass of tube wall per unit length, given by

$$m_{tw} = \frac{N_t \pi}{4 v_{tw}}(D_{to}^2 - D_{ti}^2) \tag{20.25}$$

where v_{tw} is the specific volume of the tube wall. Hence the rate of change of tube wall temperature at each location is given by

$$\frac{\partial T_{tw}}{\partial t} = \frac{4 v_{tw}}{N_t \pi c_{ptw}(D_{to}^2 - D_{ti}^2)}(-\phi_{twt} - \phi_{tws}) \tag{20.26}$$

or

$$\frac{\partial T_{tw}}{\partial t} = \frac{4 v_{tw}}{c_{ptw}(D_{to}^2 - D_{ti}^2)}(K_{twt} D_{ti}(T_t - T_{tw}) + K_{tws} D_{to}(T_s - T_{tw})) \tag{20.27}$$

A similar procedure applied to the shell wall gives the equation

$$\frac{\partial T_{sw}}{\partial t} = \frac{4 v_{sw}}{f_b c_{psw}(D_{si}^2 - D_{so}^2)} K_{sws} D_{si}(T_s - T_{sw}) \tag{20.28}$$

where

v_{sw} is the specific volume of the shell wall,
c_{psw} is the specific heat of the shell wall, and
f_b is a multiplying factor to account for the additional metal contained in the baffles, assumed to be at the same temperature as the shell.

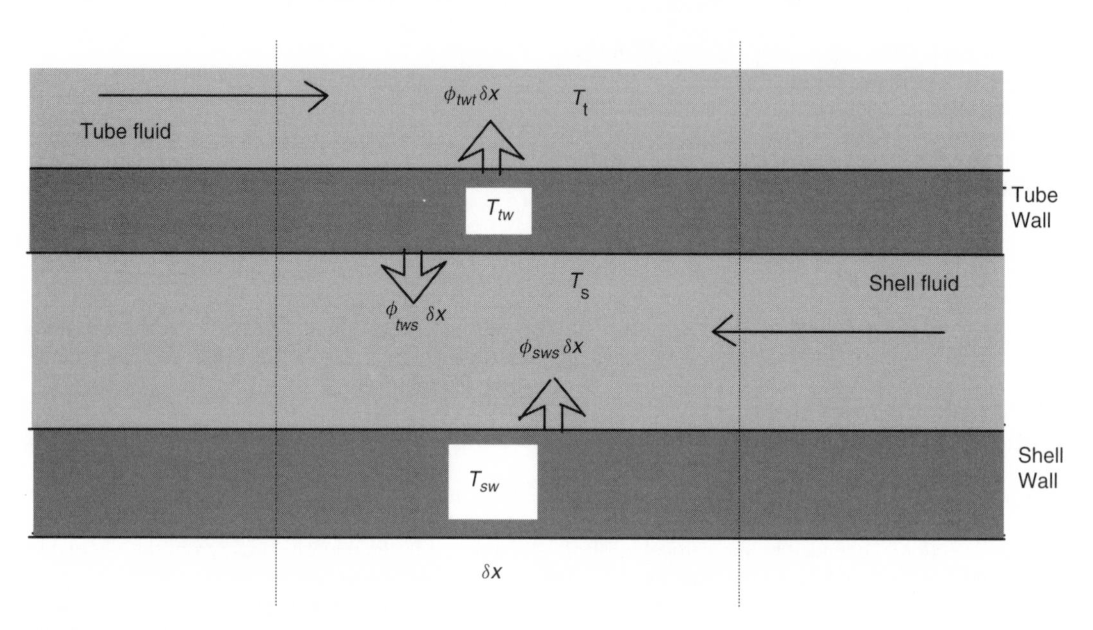

Figure 20.2 Schematic of heat exchanger element showing heat flows.

20.9 Solving the heat exchanger equations using spatial finite differences

As was noted in Chapter 2, Section 2.6, and Chapter 19, the most accurate method of solving a partial differential equation of the form (20.9) or (20.12) is the Method of Characteristics. However, when the heat exchanger is merely a part of a much larger simulation, it will often be easier to apply finite differences to the spatial variable. Consider the schematic of Figure 20.3, where the heat exchanger is divided into K sections.

Given that the modeller may wish to try a number of integration routines for the time differentials in the model as a whole, it is wise to choose one of the simplest finite difference representation of the partial differential, $\partial T/\partial x$, namely a backward difference measured relative to fluid flow. This will be stable when used in conjunction with the popular Euler integration of the time derivative, provided the timestep does not carry the calculation beyond the range of influence of the characteristics defined by the speed of propagation of temperature disturbances (see equation (20.35)). By contrast, using central differencing for the spatial variable will lead to numerical instability when used in conjunction with Euler integration of the time derivative, (Ames, 1992).

Assuming that the tube-side fluid is liquid, we may apply backward differencing for the x-variable to the liquid-flow equation (20.14) so that the ordinary time differential of temperature at the end of each section emerges as:

$$\frac{dT_{t,k}}{dt} = 4\frac{v_{t,k}}{D_{ti}c_{pt}}K_{twt}\left(T_{tw,k} - T_{t,k}\right) - c_{t,k}\left(\frac{T_{t,k} - T_{t,k-1}}{\Delta x}\right) \quad \text{for } k = 1, 2, \ldots K \tag{20.29}$$

Assuming, for example, that the shell-side fluid is a gas flowing in counter-current mode, the equation corresponding to (20.29) for the rate of change of temperature at the end of each section for the shell fluid is derived from equation (20.20):

$$\frac{dT_{s,k}}{dt} = \frac{v_{s,k}}{A_s c_{vs}}(K_{tws}N_t\pi D_{to}(T_{tw,k} - T_{s,k}) + K_{sws}\pi D_{si}(T_{sw,k} - T_{s,k})) - \gamma_s c_{s,k}\left(\frac{T_{s,k} - T_{s,k+1}}{\Delta x}\right) \quad \text{for } k = 0, 1, 2, \ldots, K-1 \tag{20.30}$$

where the velocity of the shell fluid is measured in the direction of shell flow.

We take the specific volume at each point as a function of temperature at that point and the prevailing pressure in the heat exchanger:

$$\begin{aligned} v_{t,k} &= v_{t,k}(T_{t,k}, p_t) \\ v_{s,k} &= v_{s,k}(T_{s,k}, p_s) \end{aligned} \tag{20.31}$$

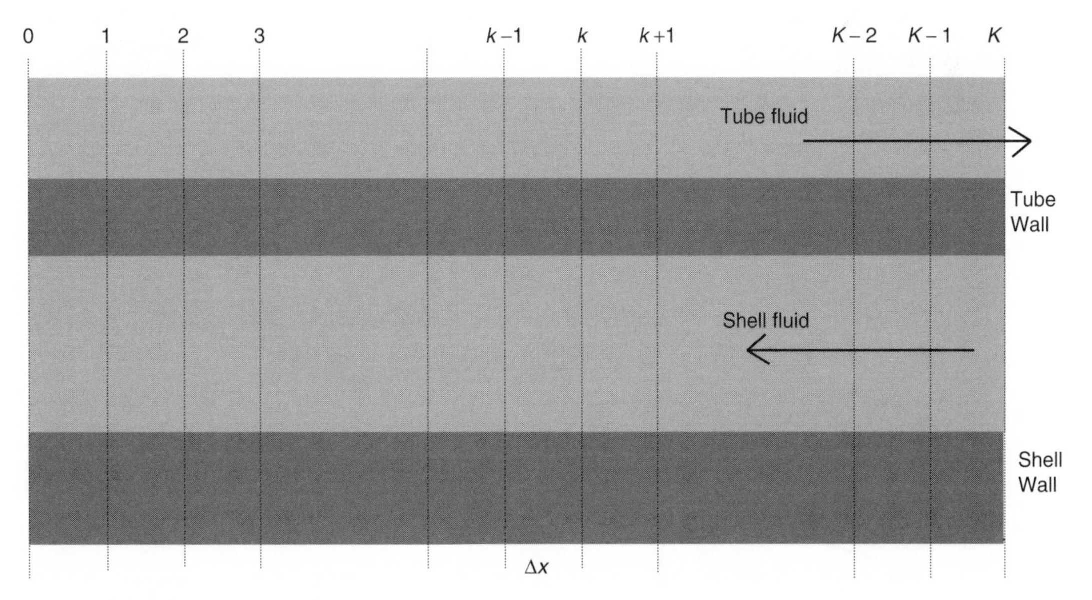

Figure 20.3 Sectionalized heat exchanger.

where $T_{t,k}$ and $T_{s,k}$ are the temperatures in segment k of the tube-side and shell-side fluids, and p_t and p_s are the average pressures on the tube and shell sides.

The velocity at each point then follows from continuity:

$$c_{t,k} = v_{t,k}\frac{W_t}{N_t A_t}$$
$$c_{s,k} = v_{s,k}\frac{W_s}{A_s} \tag{20.32}$$

where $v_{t,k}$ and $v_{s,k}$ are the specific volumes of the tube-side and shell-side fluids in the kth segment of the heat exchanger.

From equation (20.27), the rate of change of tube wall temperature at each point is:

$$\frac{dT_{tw,k}}{dt} = \frac{4v_{tw}}{c_{ptw}(D_{to}^2 - D_{ti}^2)}(K_{twt}D_{ti}(T_{t,k} - T_{tw,k}) + K_{tws}D_{to}(T_{s,k} - T_{tw,k})) \quad \text{for } k = 1, 2, \ldots, K \tag{20.33}$$

while the rate of change of shell wall temperature at each point follows from equation (20.28) as:

$$\frac{dT_{sw,k}}{dt} = \frac{4v_{sw}}{f_b c_{psw}(D_{si}^2 - D_{so}^2)}K_{sws}D_{si}(T_{s,k} - T_{sw,k}) \tag{20.34}$$

The numerical solution of the partial differential equations must not attempt to use a temperature at a distance beyond the range of influence determined by the characteristic velocity of temperature propagation. In this case, where the tubes are carrying a liquid and the shell a gas, the spatial step, Δx, and the timestep, Δt, must conform to a limit set by the highest velocity of temperature propagation:

$$\Delta t \leq \min_k \left[\frac{\Delta x}{c_{t,k}}, \frac{\Delta x}{\gamma c_{s,k}}\right] \tag{20.35}$$

Experience has shown that the maximum value of timestep consistent with accurate integration of the equations for the rest of the process plant is often about 1 second. For the water duct considered in Table 20.1, the velocity was 1 m/s, which implies that the spatial step could be about 1 m. Hence if the tubes are 10 m long, the number of sections, M, will be 10. Such a figure is typical for the discretization of distributed process components such as heat exchangers. It is gratifying that although the differential equations are based on the theoretical differential increment, δx, approaching zero, very sizeable spatial increments, Δx, are found to give accurate results in practice.

20.10 The tubular reactor

The tubular reactor is a relatively common component on chemical plants. The reactants enter at one end and the products leave from the other, with a continuous variation in the composition and the temperature of the reacting mixture in between. It is common for the feed to consist of a mixture of gases, as in the case of ammonia and methanol synthesis, where the feed gas passes through a densely packed catalyst bed which promotes a number of different reactions simultaneously.

In the following sections, we will analyse the dynamic behaviour of a gas-fed, adiabatic, catalytic-bed reactor, of the sort commonly used for methanol and ammonia production. Figure 20.4 gives a schematic representation of an element from such a

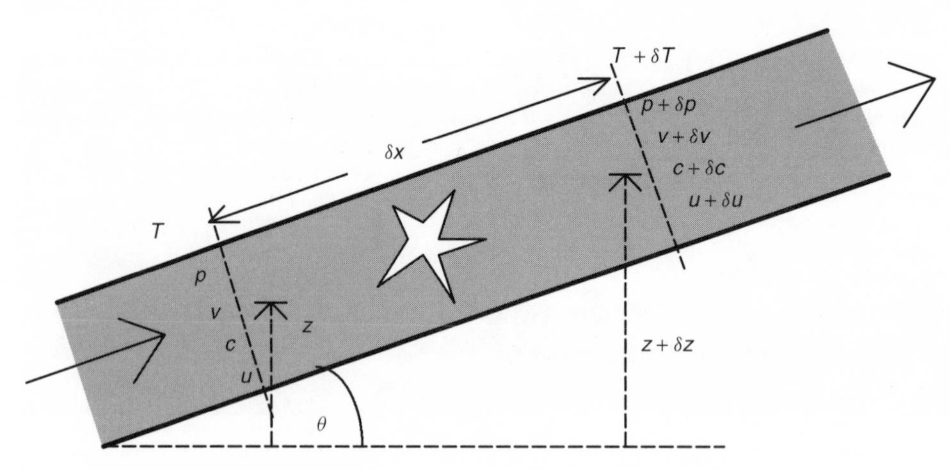

Figure 20.4 Element of catalyst bed inside a tubular reactor.

bed, which is assumed to promote M simultaneous reactions.

20.11 Mass balance for the gas flowing through the catalyst bed

The mass balance follows closely that given in Chapter 13, Section 13.6. The conservation of mass requires that

> the rate of change of mass of species i in kmol/s *equals* the inlet flow of species i in kmol/s *plus* the number of kmol of species i produced per second by all the reactions *minus* the number of kmol of species i eliminated per second by all the reactions *minus* the outlet flow of species i in kmol/s.

Expressed mathematically:

$$\frac{\partial}{\partial t}\left(\frac{\lambda_i A_g \delta_x}{\hat{v}_i}\right) = F_i + A\,\delta x \sum_{j=1}^{M} a_{ij} r_j - \left(F_i + \frac{\partial F_i}{\partial x}\,\delta x\right) \quad \text{for } i = 1, \ldots, n \tag{20.36}$$

or

$$\frac{\partial}{\partial t}\left(\frac{\lambda_i A_g}{\hat{v}_i}\right) = A \sum_{j=1}^{M} a_{ij} r_j - \frac{\partial F_i}{\partial x} \quad \text{for } i = 1, \ldots, n \tag{20.37}$$

where
- A_g is the cross-sectional area of the void at position x (m^2),
- A is the cross-sectional area of the catalyst bed at position x (m^2),
- $\hat{u}_i$ is the molar specific internal energy of component i (J/kmol),
- $r_j = r_j(T)$ is the reaction rate density (kmol rxn/(m^3s)), referred to the volume of the packed bed (the most likely reference volume for the experimental results on reaction rate),
- a_{ij} stoichiometric coefficient for species i undergoing reaction j (will be negative when species i is being eliminated by the reaction),
- n is the total number of chemical species in the feed and product gas streams,
- M is the number of separate reactions,
- λ_i is the mole fraction of chemical species i, given by $\lambda_i = F_i/F$,
- F_i is the flow of chemical component i (kmol/s),
- F is the total flow of all chemical components (kmol/s): $F = \sum_{i=l}^{n} F_i$.

The flow may be expected to settle to a new steady state significantly faster than the temperature of the bed, in a matter of seconds compared with minutes for the temperature of a typical catalyst bed. In this case case we may consider the mass throughout the bed to be in an evolving steady-state. The flow through the bed may be determined by an expression of the form of equation (20.3). Further, setting the partial differential with respect to time equal to zero in equation (20.37) gives the rate of change of flow for each chemical species as a function of distance along the bed only:

$$\frac{\partial F_i}{\partial x} = A \sum_{j=1}^{M} a_{ij} r_j \quad \text{for } i = 1, 2, \ldots, n \tag{20.38}$$

20.12 Energy balance for the gas flowing through the catalyst bed

The derivation of the energy equation for the catalyst bed follows a procedure very similar to that used in Section 3.9 to derive the energy equation for a pipe carrying a fluid. It is somewhat more complicated, however, because of the need to allow for both the presence of several chemical species in the inlet and outlet streams, and for heat conduction by the catalyst pellets. The intimate contact of gas with the catalyst pellets means that it is legitimate to assume that the gas and the catalyst share the same temperature at any given distance through the bed.

Heat conduction follows Fourier's law, namely that the heat passed along the bed in the x-direction, Φ_x (W), is proportional to the temperature gradient and to the cross-sectional area:

$$\Phi_x = -K_c A \frac{\partial T}{\partial x} \tag{20.39}$$

where K_c is the effective thermal conductivity of the catalyst bed (J/(ms)) and A is the cross-sectional area of the catalyst bed (m^2). The negative sign accounts for the fact that the heat travels from a higher to a lower temperature. The value of thermal conductivity of the catalyst will depend on the packing density, the gas being passed and the temperature. However, transient effects may be calculated adequately by regarding K_c as constant about any given operating point.

Conservation of energy requires that

> the change in energy in the bed element *equals* the energy brought into the bed element by the incoming fluid *minus* the energy leaving the bed element with the outgoing fluid *plus* the work done on the bed element by the incoming fluid *minus* the work done by the outgoing fluid *plus* the heat diffusing into the catalyst *minus* the heat diffusing out of the catalyst.

Applying this statement for a time interval δt yields the following equation:

$$\delta E = \sum_{i=1}^{n} F_i\,\delta t\left(\hat{u}_i + \frac{1}{2}w_i c^2 + w_i g z\right) - \left[\sum_{i=1}^{n} F_i\,\delta t\left(\hat{u}_i + \frac{1}{2}w_i c^2 + w_i g z\right) + \frac{\partial}{\partial x}\sum_{i=1}^{n} F_i\,\delta t\left(\hat{u}_i + \frac{1}{2}w_i c^2 + w_i g z\right)\delta x\right] + pA_g c\,\delta t - \left[pA_g c\,\delta t + \frac{\partial}{\partial x}(pA_g c\,\delta t)\,\delta x\right] - K_c A\frac{\partial T}{\partial x}\,\delta t - \left[-K_c A\frac{\partial T}{\partial x}\,\delta t + \frac{\partial}{\partial x} \times \left(-K_c A\frac{\partial T}{\partial x}\,\delta t\right)\delta x\right] \quad (20.40)$$

where

F_i is the flow of chemical species i (kmol/s),
$\hat{u}_i$ is the molar specific internal energy of species i (J/kmol),
w_i is the molecular weight of chemical species i,
A_g is the cross-sectional area of the void (m^2).

Cancelling terms in equation (20.40) gives:

$$\delta E = -\frac{\partial}{\partial x}\sum_{i=1}^{n} F_i\,\delta t\left(\hat{u}_i + \frac{1}{2}w_i c^2 + w_i g z\right)\delta x - \frac{\partial}{\partial x}(pA_g c\,\delta t)\,\delta x + \frac{\partial}{\partial x}\left(K_c A\frac{\partial T}{\partial x}\,\delta t\right)\delta x \quad (20.41)$$

We may now note that:

$$pA_g c = pv\frac{A_g c}{v} = pvW \quad (20.42)$$

where W is the flow of gas (kg/s). This equation may be developed further by multiplying top and bottom by the effective molecular weight, w, of the mixture of gases

$$pvW = pwv\frac{W}{w} = p\hat{v}F = p\hat{v}\sum_{i=1}^{n} F_i = \sum_{i=1}^{n}(F_i p\hat{v}) \quad (20.43)$$

where F is the total molar flow at location x. Equation (20.43) has made use of the effective molecular weight to convert mass flow to molar flow and specific quantities to specific molar quantities. (See Chapter 10, Section 10.4 for the basis of the conversions between kg and kmol units.) We may proceed further by noting that the molar specific volumes (m^3/kmol) of each of the constituent gases and of the mixture will be equal:

$$w_i v_i = \hat{v}_i = wv = \hat{v} \quad \text{for all } i \quad (20.44)$$

where v_i is the specific volume of gas species i in m^3/kg, while the circumflex indicates specific molar quantities. (The validity of equation (20.44) may be seen by examining the characteristic equations for both the gas mixture and for each of its gas species, which, using equation (3.2), may be written:

$$wv = \frac{ZR}{p}T \qquad w_i v_i = \frac{Z_i R}{p}T \quad (20.45)$$

Equation (20.44) follows provided the compressibility factors of constituents, Z_i, and mixture, Z, are the same.)

Hence

$$\sum_{i=1}^{n}(F_i p\hat{v}) = \sum_{i=1}^{n} F_i p\hat{v}_i \quad (20.46)$$

Combining equations (20.42), (20.43) and (20.46) gives the result:

$$pA_g c = \sum_{i=1}^{n} F_i p\hat{v}_i \quad (20.47)$$

Using the fact that $\hat{h}_i = \hat{u}_i + p\hat{v}_i$, we may substitute from equation (20.47) to simplify equation (20.41):

$$\delta E = -\frac{\partial}{\partial x}\sum_{i=1}^{n} F_i\,\delta t\left(\hat{h}_i + \frac{1}{2}w_i c^2 + w_i g z\right)\delta x + \frac{\partial}{\partial x}\left(K_c A\frac{\partial T}{\partial x}\,\delta t\right)\delta x \quad (20.48)$$

Dividing by δt and then letting $\delta t \to 0$ produces the partial differential equation:

$$\frac{\partial E}{\partial t} = -\frac{\partial}{\partial x}\sum_{i=1}^{n} F_i\left(\hat{h}_i + \frac{1}{2}w_i c^2 + w_i g z\right)\delta x + K_c A\frac{\partial^2 T}{\partial x^2}\,\delta x \quad (20.49)$$

The energy contained in the bed element, E, is the sum of the energy contained by the gas and that contained by the catalyst pellets. Allowing for internal energy, kinetic energy and potential energy for each of the gas species and for the catalyst pellets, we have:

$$E = \sum_{i=1}^{n}\frac{\lambda_i A_g\,\delta x}{\hat{v}_i}\left(\hat{u}_i + \frac{1}{2}wc^2 + wgz\right) + \frac{A\,\delta x}{v_c}(u_c + gz) \quad (20.50)$$

where u_c is the specific internal energy of the catalyst pellets in the bed (J/kg), v_c is the specific volume of the catalyst pellet packing (m^3/kg), while the effective molecular weight, w, given in terms of the molar fractions, λ_i, by equation (11.28), repeated below:

$$w = \sum_{i=1}^{n} \lambda_i w_i \tag{11.28}$$

and $\hat{u}$ (J/kmol) is the molar specific internal energy of the mixture, given by:

$$\hat{u} = \sum_{i=1}^{n} \lambda_i \hat{u}_i = \sum_{i=1}^{n} \lambda_i w_i u_i \tag{20.51}$$

For a catalyst bed fixed in space and with uniform cross-sectional areas, differentiation of equation (20.50) with respect to time gives:

$$\frac{\partial E}{\partial t} = \sum_{i=1}^{n} \frac{\lambda_i A_g \, \delta_x}{\hat{v}_i} \frac{\partial}{\partial t}\left(\hat{u}_i + \frac{1}{2}wc^2 + wgz\right) + \sum_{i=1}^{n}\left(\hat{u}_i + \frac{1}{2}wc^2 + wgz\right)\frac{\partial}{\partial t}\left(\frac{\lambda_i A_g \, \delta_x}{\hat{v}_i}\right) + \frac{A\,\delta x}{v_c}\frac{\partial u_c}{\partial t} \tag{20.52}$$

Combining equations (20.49) and (20.52) and cancelling the δx term gives the full energy equation as

$$\sum_{i=1}^{n} \frac{\lambda_i A_g}{\hat{v}_i} \frac{\partial}{\partial t}\left(\hat{u}_i + \frac{1}{2}wc^2 + wgz\right) + \sum_{i=1}^{n}\left(\hat{u}_i + \frac{1}{2}wc^2 + wgz\right)\frac{\partial}{\partial t}\left(\frac{\lambda_i A_g}{\hat{v}_i}\right) + \frac{A}{v_c}\frac{\partial t_c}{\partial t} = -\frac{\partial}{\partial x}\sum_{i=1}^{n} F_i\left(\hat{h}_i + \frac{1}{2}w_i c^2 + w_i gz\right) + K_c A\frac{\partial^2 T}{\partial x^2} \tag{20.53}$$

Assuming the mass at each point in the bed reaches a steady value immediately allows us to put $(\partial/\partial t)(\lambda_i A_g/\hat{v}_i) = 0$. Hence

$$\sum_{i=1}^{n} \frac{\lambda_i A_g}{\hat{v}_i} \frac{\partial}{\partial t}\left(\hat{u}_i + \frac{1}{2}wc^2 + wgz\right) + \frac{A}{v_c}\frac{\partial u_c}{\partial t} = -\sum_{i=1}^{n} F_i \frac{\partial}{\partial x}\left(\hat{h}_i + \frac{1}{2}w_i c^2 + w_i gz\right) - \sum_{i=1}^{n}\left(\hat{h}_i + \frac{1}{2}wc^2 + wgz\right)\frac{\partial F_i}{\partial x} + K_c A\frac{\partial^2 T}{\partial x^2} \tag{20.54}$$

Neglecting kinetic and potential energies for reasons similar to those given in the analysis of the heat exchanger duct (Section 20.3), and substituting for $\partial F_i/\partial x$ from equation (20.38), we may reduce equation (20.54) to:

$$\left(\sum_{i=1}^{n} \frac{\lambda_i A_g}{\hat{v}_i}\frac{d\hat{u}_i}{dT} + \frac{A}{v_c}\frac{du_c}{dT}\right)\frac{\partial T}{\partial t} = -\sum_{i=1}^{n} F_i \frac{d\hat{h}_i}{dT}\frac{\partial T}{\partial x} - A\sum_{i=1}^{n}\hat{h}_i \sum_{j=1}^{M} a_{ij} r_j + K_c A\frac{\partial^2 T}{\partial x^2} \tag{20.55}$$

Further simplification follows from the fact that any specific molar variable, let us say $\hat{\beta}$, is defined for the mixture of gases by

$$F\hat{\beta} = \sum_{i=1}^{n} F_i \hat{\beta}_i \tag{20.56}$$

so that

$$\hat{\beta} = \sum_{i=1}^{n} \frac{F_i}{F}\hat{\beta}_i = \sum_{i=1}^{n} \lambda_i \hat{\beta}_i \tag{20.57}$$

It follows that

$$\sum_{i=1}^{n} \lambda_i \frac{d\hat{u}_i}{dT} = \frac{d\hat{u}}{dT} = \hat{c}_v = wc_v \tag{20.58}$$

where $\hat{c}_v$ is the molar specific heat at constant volume of the gas mixture (J/(kmolK)), while c_v is the specific heat at constant volume of the gas mixture (J/(kgK)). It also follows that:

$$\sum_{i=1}^{n} F_i \frac{d\hat{h}_i}{dT} = F\frac{d\hat{h}}{dT} = F\hat{c}_p \tag{20.59}$$

Making use of the fact $\hat{v}_i = \hat{v} = wv$ and putting $c_{pc} = du_c/dt$ for the specific heat of the catalyst gives

$$\left(\frac{A_g}{v}c_v + \frac{A}{v_c}c_{pc}\right)\frac{\partial T}{\partial t} = -F\hat{c}_p\frac{\partial T}{\partial x} + K_c A\frac{\partial^2 T}{\partial x^2} - A\sum_{j=1}^{M} r_j \sum_{i=1}^{n}\hat{h}_i a_{ij} \tag{20.60}$$

The term $\sum_{i=l}^{n}\hat{h}_i a_{ij}$ is the enthalpy of reaction for reaction j:

$$\Delta H_j = \sum_{i=1}^{n}\hat{h}_i a_{ij} \tag{20.61}$$

which takes a positive value when the reaction is endothermic and a negative value when the reaction is exothermic (see Chapter 13, Section 13.8). Equation (20.60) may thus be written in the final form:

$$\frac{\partial T}{\partial t} = \frac{K_c A \dfrac{\partial^2 T}{\partial x^2} - F\hat{c}_p \dfrac{\partial T}{\partial x} - A \sum_{j=1}^{M} r_j \Delta H_j}{\dfrac{A_g}{v} c_v + \dfrac{A}{v_c} c_{pc}} \tag{20.62}$$

The partial derivative in temperature given by equation (20.62) may be compared with the derivative of the temperature in a reaction vessel of fixed volume, given by equation (13.45). The reader may notice that the partial derivative with respect to time of equation (20.62) depends on the enthalpy of each reaction, ΔH_j, whereas the total derivative of equation (13.45) depends on the internal energy of (each) reaction, ΔU_j, a different quantity, although numerically similar (see Chapter 13, Section 13.8). The reason for this apparent anomaly is that the mass flow has been considered to be in an evolving steady state in the catalyst bed reactor, whereas the analysis of the reaction vessel allows for unsteady mass flow.

20.13 Solving the temperature and conversion equations using finite differences

The catalyst bed is first sectionalized into cells in a similar way to the sectionalization of the heat exchanger (Figure 20.5).

The temperatures along the reactor, T_k, $k = 1, 2, \ldots, K$, are state variables, and the solution requires that we know their values at time t_0 – the same requirement as for all the other state variables in the simulation. The values, T_k, might be taken initially as the design values. Alternatively, approximate values could be chosen, and the model run to the steady state appropriate for the study under consideration.

We have assumed in Section 20.11 that the mass flows will reach an immediate equilibrium with the prevailing temperature of the catalyst bed. Hence the molar flows, F_i, $i = 1, \ldots, n$, of each chemical species at a distance x_k along the bed may be found at any time, t, by a spatial integration of the steady-state mass balance equation (20.38):

$$F_i(x_k) = F_i(0) + A \int_{x=0}^{x_k} \sum_{j=1}^{M} a_{ij} r_j(x)\, dx \qquad \text{for } i = 1, 2, \ldots, n \tag{20.63}$$

where $r_j(x)$ is the rate of reaction j at a distance, x, through the reactor. Equation (20.63) may be solved numerically using, for instance, the Euler equation:

$$F_i(k+1) = F_i(k) + \Delta x A \sum_{j=1}^{M} a_{ij} r_j(k) \qquad \text{for } k = 0, \ldots, K-1 \tag{20.64}$$

Here $r_j(k)$ denotes the rate of reaction j in cell k, which will depend on the cell's temperature, T_k,

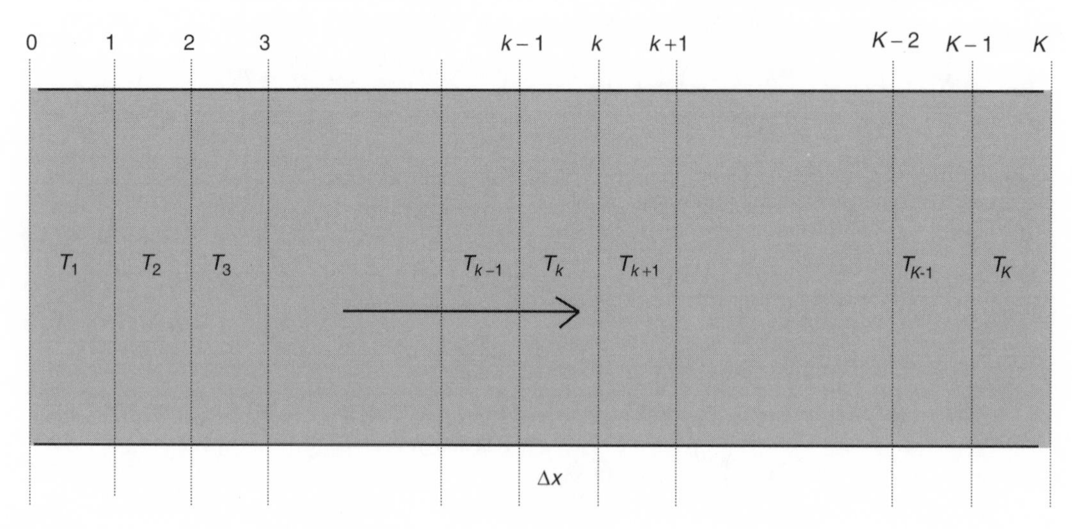

Figure 20.5 Flow through the catalyst bed.

and its chemical concentrations, $A_i(k)$. Hence $r_j(k) = r_j(T_k, A_i(k))$. (General expressions for the reaction rate are discussed in Chapter 13, Sections 13.4 and 13.5; the modeller will need to have access to design information on the reaction rate constants that characterize the particular process under consideration.) The chemical concentrations required to calculate the reaction rate in each cell may be found from the flows entering the cell:

$$A_i(k) = \frac{F_i(k)}{\sum_{i=1}^{n} F_i(k)} \qquad i = 1, \dots, n \qquad (20.65)$$

Equation (20.64), used together with the known cell temperatures and equation (20.65), enables the molar flows and the concentrations in each cell to be calculated. Putting $k = K$, in equation (20.65) allows $A_i(K)$ to be calculated, which represents the concentrations leaving the reactor.

It may be noted that the cell concentrations, $A_i(k)$, $k = 1, \dots, K$, are not state variables in themselves, but each is algebraically dependent on one or more cell temperatures, T_k, which are state variables. For example, $A_i(1)$ is dependent on T_1, $A_i(2)$ is dependent on T_1 and T_2, and so on, until $A_i(K)$ is dependent on all the cell temperatures, $T_1, T_2, \dots, T_k$.

The cell temperatures may be updated by solving equation (20.62) numerically. Replacing the space differentials by discrete approximations gives the time differential for cell temperature for every cell except the last as:

$$\frac{dT_k}{dt} = \frac{1}{\frac{A_g}{v}c_v + \frac{A}{v_c}c_{pc}} \left(K_c A \frac{1}{\Delta x} \left(\frac{T_{k+1} - T_k}{\Delta x} - \frac{T_k - T_{k-1}}{\Delta x} \right) - F\hat{c}_p \frac{T_k - T_{k-1}}{\Delta x} - A \sum_{j=1}^{M} r_j(k) \Delta H_j(T_k) \right)$$

$$\text{for } k = 1, 2, \dots, K-1 \qquad (20.66)$$

which may be simplified to:

$$\frac{dT_k}{dt} = \frac{1}{\frac{A_g}{v}c_v + \frac{A}{v_c}c_{pc}} \left(K_c A \frac{T_{k+1} - 2T_k + T_{k-1}}{\Delta x^2} - F\hat{c}_p \frac{T_k - T_{k-1}}{\Delta x} - A \sum_{j=1}^{M} r_j(k) \Delta H_j(T_k) \right)$$

$$\text{for } k = 1, 2, \dots, K-1 \qquad (20.67)$$

The temperature differential for the end cell may be written in terms of available cell temperatures as:

$$\frac{dT_k}{dt} = \frac{1}{\frac{A_g}{v}c_v + \frac{A}{v_c}c_{pc}} \left(K_c A \frac{T_k - 2T_{k-1} + T_{k-2}}{\Delta x^2} - F\hat{c}_p \frac{T_k - T_{k-1}}{\Delta x} - A \sum_{j=1}^{M} r_j(k) \Delta H_j(T_k) \right)$$

$$\text{for } k = K \qquad (20.68)$$

The reaction rates, $r_j(k) = r_j(T_k, A_i(k))$ in equations (20.67) and (20.68) may be computed from the cell's temperature, T_k, and its chemical concentrations, $A_i(k)$, with the latter having been computed in the course of the spatial integration. The enthalpy of reaction, ΔH_j, will have a weak dependence on cell temperature, T_k, although this may be neglected in many cases (see Section 13.9).

Equations (20.67) and (20.68) may now be integrated using the integration algorithm of the main simulation.

20.14 Bibliography

Ames, W. F. (1992). *Numerical Methods Partial Differential Equations, 3rd Edition*, Academic Press.

Smith, A. J. (1984). Solving distributed heat-exchanger models by the method of characteristics, *Trans. Inst. M. C.*, **6**, No. 2, 83–88.

Smith, G. D. (1965, 1974). *Numerical Solution of partial Differential Equations*, Oxford University Press.

Stephens, A. D. (1975). Stability and optimisation of a methanol converter, Chemical Engineering Science, **30**, 11–19.

21 Nuclear reactors

21.1 Introduction

The Calder Hall plant in the UK fed the first nuclear-generated power into a public electricity supply system in 1956, and nuclear energy has become a firmly established component of the world's electric power industry in the years since then. Worldwide, the fraction of electricity produced by nuclear power stations reached 17% in 1995. Within Europe, the UK generated 25% of its electricity from nuclear power stations, Germany generated 29%, Switzerland 40%, while France's nuclear contribution was 76%. Outside Europe, the USA generated 22% of its electricity from nuclear, while Japan generated 33% and Korea 36%. Thus the nuclear reactor must be regarded now as a common unit process in the power generation industry.

With nuclear fusion energy at the stage of research only, all the nuclear power referred to above was generated by fission reactors. Such reactors produce heat energy by causing neutrons to split both uranium and plutonium nuclei in a chain reaction that is sustained at a constant rate by the next-generation neutrons that are emitted at and soon after fission. The heat produced by the nuclear reactor is then used to raise steam in a conventional steam plant.

The fuel for commercial nuclear reactors is usually uranium, often enriched in the fissile uranium-235 isotope. A useful quantity of plutonium will also be formed during operation as a result of some of the neutrons being absorbed by the majority uranium-238 isotope. The subsequent fission of some of this plutonium *in situ* increases the energy yield of the original, uranium fuel. All truly commercial nuclear reactors to date have been 'thermal reactors', so-called because they use a moderator to slow the neutrons down to thermal energy levels so as to increase the probability of nuclear fission. The different types of thermal reactor result from different choices of moderator and coolant at the design stage. For example, the UK's Advanced Gas Reactor (AGR) uses carbon in the form of graphite as the moderator and carbon dioxide gas as the reactor coolant, while the Pressurized Water Reactor (PWR), designed in the USA but now in established use all over the world, uses hydrogen in the form of water, pressurized to remain in its liquid phase, as both moderator and coolant.

A more advanced design is the plutonium-fuelled 'fast reactor', where the higher neutron speed permitted by avoiding thermalization means that slightly more neutrons are produced at each fission event. These extra neutrons allow more of the non-fissile isotope, uranium-238, to be converted into plutonium. A carefully designed fast reactor may, during a reactor run, convert uranium-238 in the 'blanket' surrounding the core into slightly more plutonium than the reactor has used up in the fuel-core. Theoretically, this process allows access to all the energy inherent in natural uranium, instead of just the 1% or so available using thermal reactors. Inefficiencies in the fast reactor fuel cycle mean that not all the available energy is liberated in practice, but nevertheless uranium utilization can be increased very substantially: a factor of 50 to 60. Fast-reactor power stations of industrial scale have been built and run in the UK, France, Germany, Russia and Japan. So far, however, their large capital cost has meant that they have not been adopted commercially, despite their significant potential for additional energy supply in the long term.

This chapter will develop a model of the dynamics of a general nuclear fission reactor based on the assumption that the nuclear characteristics of the reactor can be represented by averages applicable over the whole of the reactor core. The resulting equations are often referred to as the 'point kinetics model'. With the correct choice of nuclear parameters, the model is applicable to both thermal and fast reactors. The model may be adapted to give an approximate representation of spatial variations in neutron density and hence power density. Finally, a method will be indicated for calculating both the fuel temperatures and the coolant temperatures at various heights in the reactor.

21.2 General description of a nuclear reactor

Figure 21.1 is a diagram of a typical nuclear reactor. The reactor fuel, normally uranium oxide in pellet form, is placed inside long, narrow sheaths to form fuel pins a metre or more in length. The cladding sheaths are made of either stainless steel or a zirconium alloy, typically 1 cm in diameter and half a millimetre thick. Fuel pins are gathered together and held rigidly inside a fuel assembly. The reactor core is then constructed using a large number of fuel assemblies, so that the total mass of uranium oxide is typically 100 te in a reactor of commercial size (taken to mean an electricity generation of roughly 1000 MWe, which at typical efficiencies implies a heat generation of roughly

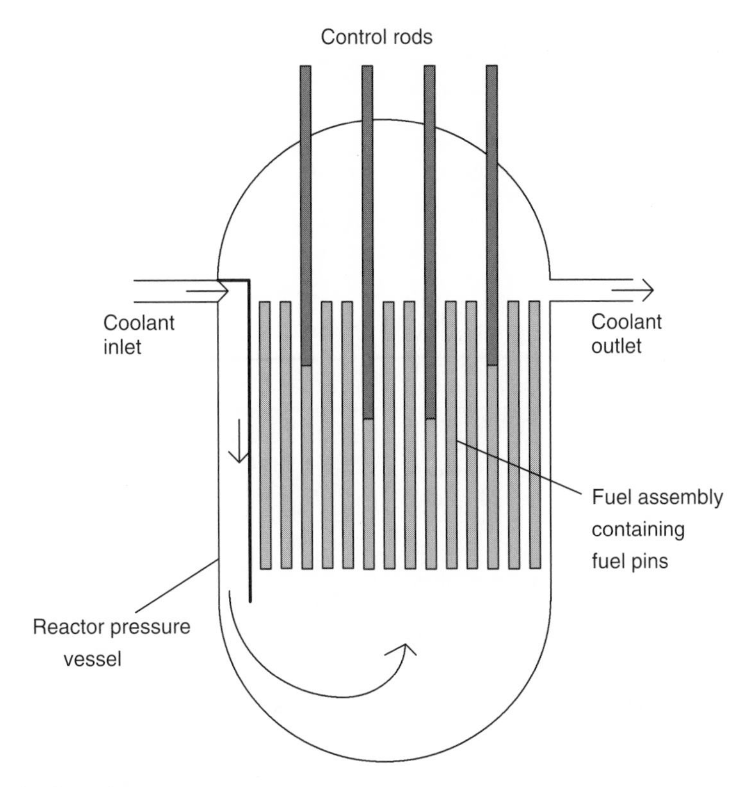

Figure 21.1 Schematic of a nuclear reactor.

3000 MWth). Reactor coolant, which may be a liquid, a gas or a vapour–liquid mixture, is pumped through the core. The coolant is constrained to run parallel to the fuel pins and to make intimate contact with their exterior cladding so that it may abstract efficiently the heat generated within the fuel. The neutron density and hence rate of heat generation are controlled by rods containing neutron-absorbing material, such as boron, silver, indium and cadmium. The control rods are raised or lowered as necessary in order to regulate the reactor's power output. The schematic of Figure 21.1 shows a configuration typical of a PWR, where clusters of control rods may move up and down within certain of the fuel assemblies.

21.3 The process of nuclear fission

Natural uranium is composed of two different isotopes: 99.3% is uranium-238, the nucleus of which contains 238 nucleons, comprising 92 protons and 146 neutrons, while the other 0.7% is made up of uranium-235, which contains 235 nucleons, comprising 92 protons and 143 neutrons. Being positively charged, all the protons in the nucleus exert a repulsive electrostatic force on one another. These forces are very strong and would drive the protons apart immediately if it were not for the even more powerful strong-nuclear force binding the nucleus together. This force has been likened to glue, since it is effective only over very small distances. In fact, both types of uranium nuclei, U-235 and U-238, are too large for the strong-nuclear force to hold them intact indefinitely, and a collection of uranium nuclei will decompose slowly into more stable isotopes by throwing off alpha particles (2 protons plus 2 neutrons) at sporadic intervals.

A more spectacular mode of decomposition occurs if a neutron, particularly a slow-moving neutron, happens to approach very close to a uranium-235 nucleus. The strong-nuclear force will pull the neutron into the nucleus at enormous speed, and its deceleration on arrival will send the resulting uranium-236 nucleus into spatial oscillation. The dominance of the strong-nuclear force will be upset, and the nucleus will usually split into two roughly equal parts as a result. Once these parts have become separated by more than about 10^{-15} m, the electrostatic repulsion will take over, causing the them to fly apart at very great velocities. This process is known as fission, and the energy of fission is contained mainly in the kinetic energy of the two fragments or fission products. In addition to the two large fission products, the fission will also release two or three neutrons. The process is illustrated diagrammatically in Figure 21.2.

The fissile uranium-235 nucleus may be fissioned by both slow and fast neutrons, whereas an uranium-238 nucleus can be fissioned only by a very fast neutron, corresponding to the extreme high end of the spectrum of neutron energy in a fast reactor. An uranium-238 nucleus may indeed capture a slow neutron coming close to it, but instead of fissioning, the resultant uranium-239 nucleus will undergo a series of radioactive decay processes over the next few days to become transformed into plutonium-239. The nucleus of plutonium-239 contains 94 protons and 145 neutrons, and its nuclear behaviour is very similar to that of uranium-235, in that it is fissile and it will release a comparable amount of energy together with 2 or 3 neutrons when fissioned.

About 2.4 neutrons on average are produced per fission inside a thermal nuclear reactor. Of these, about 1.2 will be lost by absorption into non-fissile nuclei present in the core, namely those of the majority uranium-238 isotope and those of the control rods and of structural materials. A further small number will be absorbed by uranium-235 nuclei without causing fission. The chain reaction will be maintained, however, provided one of the 2.4 neutrons produced goes on to split a uranium-235 nucleus.

The loss of neutrons to non-fissile absorption represents a significant problem for the reactor designer, particularly near the end of the reactor run, when the fuel is starting to become used up. While very careful attention to neutron economy may allow a reactor to be designed to run on natural uranium (e.g. the UK's Magnox and Canada's CANDU reactors), most commercial reactors use enriched uranium as the fuel, where the abundance of the fissile uranium-235 isotope has been increased from its natural percentage of 0.7% to about 3%.

21.4 Delayed neutrons

The fission products illustrated in Figure 21.2 will contain a mixture of neutrons and protons held together by the strong-nuclear force, and as such each fission product will constitute the nucleus of an element whose chemical properties will be determined by the number of protons it contains. Each fission-product nucleus will contain too many neutrons to be stable, however, and will undergo typically four stages of radioactive decay, with a succession of more stable nuclei being formed until full stability is reached. Usually the decay process will involve the emission of a negative beta particle or electron, accompanied by a gamma photon.

Six groups of fission products are particularly important to the control engineer. These decay to form daughter nuclei that are in a sufficiently excited state to throw off a neutron on formation. Such neutrons are called delayed neutrons, and the six groups of fission products are known as delayed neutron precursors. Delayed neutrons make up less than 1% of the total number of neutrons liberated by fission, but the fact that they are released on a much slower timescale than the 'prompt' neutrons liberated at the time of fission renders the control of nuclear reactors relatively easy.

We shall begin the process of constructing a mathematical model of the nuclear reactor by considering the behaviour of the precursor groups in more detail. Let C_i be the concentration of nuclei per m^3 of the ith precursor group at time, t. The rate of radioactive decay, R_i, will be proportional to the existing concentration,

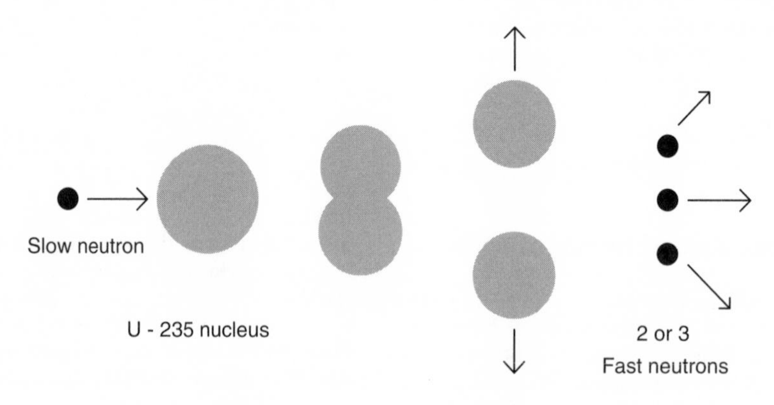

Figure 21.2 Diagram of the fission process.

thus obeying the equation:

$$R_i = \lambda_i C_i \qquad i = 1 \text{ to } 6 \qquad (21.1)$$

where λ_i is the constant of proportionality.

Every time a precursor nucleus decays, a daughter nucleus is formed, and a delayed neutron is ejected immediately from that daughter nucleus. It follows that the rate of production of delayed neutrons from that precursor group will be equal to the rate of decay of the precursor nuclei. Let dn_{di}/dt be the rate of production of delayed neutrons associated with precursor group i. This variable will obey the equation:

$$\frac{dn_{di}}{dt} = R_i = \lambda_i C_i \qquad i = 1 \text{ to } 6 \qquad (21.2)$$

The rate of production of all delayed neutrons is found by adding the contributions of all six precursor groups:

$$\frac{dn_d}{dt} = \sum_{i=1}^{6} \lambda_i C_i \qquad (21.3)$$

where n_d is the concentration of delayed neutrons per m^3.

21.5 Reactor multiplication factor, *k*

The multiplication factor for the reactor, k, is defined as:

$$k = \frac{\text{number of neutrons produced in the next generation}}{\text{number of neutrons produced in this generation}}$$

which may be expanded to

$$k = \frac{\text{no. of prompt neutrons + no. of delayed neutrons produced in the next generation}}{\text{number of neutrons produced in this generation}}$$

Hence we may regard the multiplication factor, k, as composed of a prompt part, k_p, and a delayed part, k_d:

$$k = k_p + k_d \qquad (21.4)$$

where

$$k_p = \frac{\text{number of prompt neutrons produced in the next generation}}{\text{number of neutrons produced in this generation}}$$

and

$$k_d = \sum_{i=1}^{6} k_{di} \qquad (21.5)$$

in which k_{di} is given by

$$k_{di} = \frac{\text{number of group } i \text{ delayed neutrons produced in the next generation}}{\text{number of neutrons produced in this generation}}$$

Since the total number of neutrons in the next generation will be proportional to k and the number of next-generation prompt neutrons will be proportional to k_p, it follows that the fraction of prompt neutrons in the next generation will be k_p/k. Similarly the fraction of next-generation delayed neutrons will be k_{di}/k, for $i = 1$ to 6. The delayed neutron fraction for group i is given the symbol β_i, so that

$$k_{di} = \beta_i k \qquad (21.6)$$

Substituting from equation (21.6) back into equation (21.5) gives

$$k_d = \sum_{i=1}^{6} \beta_i k = \beta k \qquad (21.7)$$

where β is the total fraction of delayed neutrons. Using equation (21.7) in equation (21.4) allows us to express k_p in terms of the total delayed-neutron fraction:

$$k_p = k(1 - \beta) \qquad (21.8)$$

We may also note that, since each delayed neutron is preceded by a precursor nucleus, it follows that

$$k_{di} = \frac{\text{number of group } i \text{ delayed neutrons precursor nuclei in the next generation}}{\text{number of neutrons produced in this generation}}$$

The reactor multiplication factor, k, depends on:

(i) fixed conditions, such as the reactor geometry, the form of the fuel and whether a moderator is present to thermalize the neutrons;
(ii) conditions that are varying slowly with time, like the degree of fuel burn-up; and
(iii) conditions that may vary rapidly, such as the position of the control rods and the temperature of the reactor.

The control engineer will be able to obtain data on the items above from the reactor designers in the first instance and then from reactor physicists employed to work on the reactor, who can be expected to refine the match to the operational reactor by collecting extensive operating data and performing experiments where necessary. For example, it is customary to perform experiments to calibrate the worth of each control rod. In this case the data is usually given in the form of a graph of control rod insertion distance versus reactivity, where reactivity, ρ, is the fractional deviation from unity of the multiplication factor, k:

$$\rho = \frac{k-1}{k} \tag{21.9}$$

Reactivity, ρ, is a dimensionless quantity, but sometimes it is assigned the dimensionless unit of 'Niles', where 1 Nile = 0.01. (The word 'Nile' originates from the terminology $\Delta k = k - 1$, and is a pun on the River Nile's well-known delta downstream of Cairo.) Alternatively, reactivity may also be referred to in dollars, the ratio of the reactivity to the delayed neutron fraction:

$$\text{Reactivity in dollars} = \frac{\rho}{\beta} \tag{21.10}$$

21.6 Absorption of neutrons and the production of prompt neutrons

The time between neutron generations is the time between a neutron being produced and the time it is absorbed, into either a fissile or non-fissile nucleus. In reality this time interval will vary between individual neutrons, but we will make the simplifying assumption that the lifetime of all neutrons may be characterized by the average neutron lifetime, l, which is typically 1 millisecond in a commercial thermal reactor. Neutrons are being 'born' continuously in a reactor, and we may assume that at any instant of time the neutrons have a uniform spread of all ages between 0 and l seconds. Let us divide the neutron lifetime, l, into a large number, M, of time intervals, δt, where

$$\delta t = \frac{l}{M} \tag{21.11}$$

Then, considering a typical region of the reactor core of volume $1\,\text{m}^3$, we divide the n neutrons (prompt and delayed) it contains at the start of the current generation into M groups, each of which contains n_G neutrons that are within δt of eachother in age, where

$$n_G = \frac{n}{M} \tag{21.12}$$

Making M very large implies $\delta t \to 0$ from equation (21.11), allowing us to consider that all the neutrons in each group have the same age. Since there will be about 10^{14} neutrons per m^3 in an operating reactor, there is no problem in making M very large in practice and still retaining enough neutrons in each group to ensure that the group's sample of neutrons retains the same statistical properties as the ensemble. In particular, the average lifetime of the neutrons in each group will be the same as the average lifetime of all the neutrons in the reactor.

Let us assign the name Group 1 to the group in the current generation that contains the neutrons that were born first. These are the oldest neutrons and will therefore 'die' first as a result of absorption. Group 2 neutrons were born second, make up the second oldest group, and so will die second. We may then continue this process of assignment until we encounter the youngest group of neutrons, which will be given the name Group M. Under our assumption that all neutrons live for the average neutron lifetime, all the neutrons in any given group will 'die' at the same time through absorption in either fuel or non-fuel material. This we call an 'absorption event'. Figure 21.3 shows how the M groups of the current generation come to the end of their lives in an absorption event and produce neutrons in the next generation. Most of the neutrons in any given group will be absorbed into nuclei that do not fission, but some will be absorbed by fissile nuclei (U-235 or Pu-239) and a fission will occur. The fissions will lead to the production of $k_p n_G$ prompt neutrons, as well as $k_d n_G$ delayed-neutron-precursor nuclei.

From Figure 21.3 we may see that an absorption event occurs every δt seconds, at which point n_G neutrons (delayed and prompt) are absorbed and $k_p n_G$ prompt neutrons are produced. Let the decrease in neutrons in an absorption event be called δn_a, where

$$\delta n_a = n_G \tag{21.13}$$

It follows from equations (21.11) and (21.12) that

$$n_G = \frac{n}{l}\delta t \tag{21.14}$$

Substituting this back into equation (21.13) gives:

$$\frac{\delta n_a}{\delta t} = \frac{n}{l} \tag{21.15}$$

Making M very large (which presents no problem, as has been noted) means that $\delta t \to 0$, allowing equation (21.15) to be replaced by the differential equation for the rate of loss of neutrons in the reactor by absorption:

$$\frac{dn_a}{dt} = \frac{n}{l} \tag{21.16}$$

Turning now to the prompt neutrons formed, let us name the increase in prompt neutrons immediately following an absorption event δn_p, so that we have

$$\delta n_p = k_p n_G \tag{21.17}$$

Substituting from equation (21.14) gives:

$$\frac{\delta n_p}{\delta t} = k_p \frac{n}{l} \tag{21.18}$$

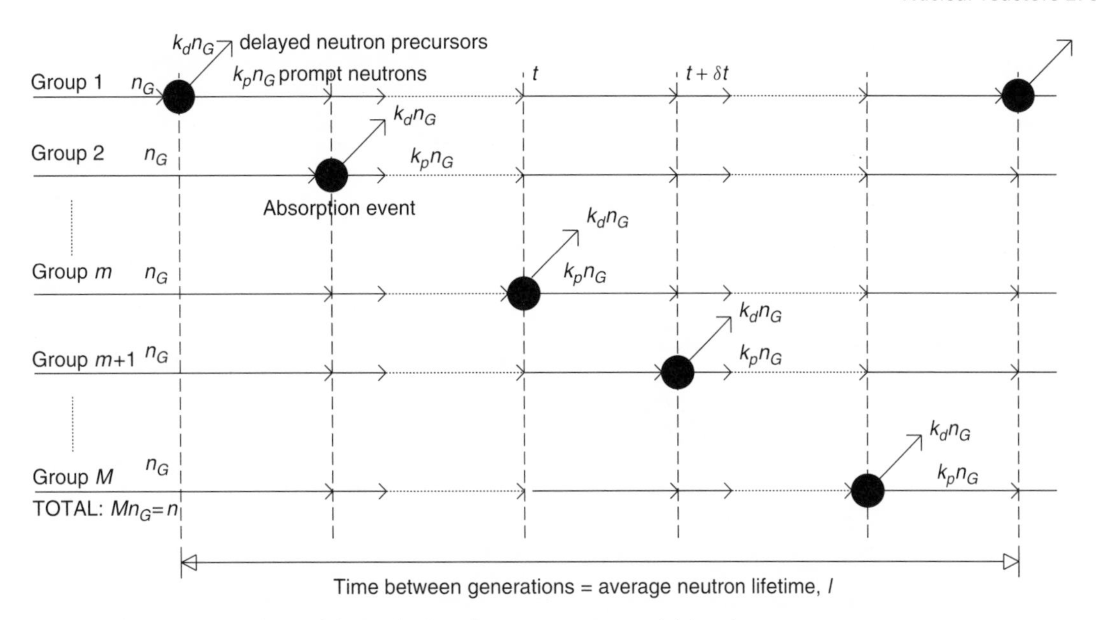

Figure 21.3 Neutron absorption and the production of prompt neutrons and delayed-neutron precursors.

As $\delta t \to 0$ as a result of making M very large produces the rate of production of prompt neutrons:

$$\frac{dn_p}{dt} = k_p \frac{n}{l} \tag{21.19}$$

We may substitute from equation (21.8) for k_p to give:

$$\frac{dn_p}{dt} = k(1-\beta)\frac{n}{l} \tag{21.20}$$

21.7 Overall neutron balance

The overall neutron balance may be expressed for a volume of 1 m^3 as

> the rate of change of the number of neutrons equals the rate of production of prompt neutrons *plus* the rate of production of delayed neutrons *plus* the rate of release of neutrons from the neutron source *minus* the rate of absorption of neutrons

The neutron source referred to above is a normal component of a nuclear reactor core, which produces a stream of neutrons, S, per second per m^3 of core, independent of the operating power of the reactor. Such a source is included both to initiate the chain reaction at first start-up and to ensure that the neutron flux detectors will always have a recognizably valid signal to detect even when the reactor is shut down.

The neutron balance may be expressed in symbols as

$$\frac{dn}{dt} = \frac{dn_p}{dt} + \frac{dn_d}{dt} + S - \frac{dn_a}{dt} \tag{21.21}$$

or, using, equations (21.3), (21.16) and (21.20):

$$\frac{dn}{dt} = k(1-\beta)\frac{n}{l} + \sum_{i=1}^{6}\lambda_i C_i + S - \frac{n}{l} \tag{21.22}$$

We may rearrange this as follows:

$$\begin{aligned}\frac{dn}{dt} &= (k-1-k\beta)\frac{n}{l} + \sum_{i=1}^{6}\lambda_i C_i + S \\ &= k\left(\frac{k-1}{k} - \beta\right)\frac{n}{l} + \sum_{i=1}^{6}\lambda_i C_i + S\end{aligned} \tag{21.23}$$

Hence, using the definition of reactivity given in equation (21.9), the rate of change of neutron density is:

$$\frac{dn}{dt} = k\frac{(\rho-\beta)}{l}n + \sum_{i=1}^{6}\lambda_i C_i + S \tag{21.24}$$

21.8 The balance for delayed neutron precursors

Let the production of the nuclei of delayed neutron precursor i due to absorption in a fission event be

δC_{fi}. We may see from Figure 21.3 that $k_d n_G$ delayed neutron precursor nuclei per m^3 are produced at every absorption event, which events occur every δt seconds. It follows that

$$\delta C_{fi} = k_{di} n_G \qquad i = 1 \text{ to } 6 \tag{21.25}$$

Following the procedure of Section 21.6, we may substitute $n_G = (n/l)\delta t$ and let $\delta t \to 0$ to give the rate of production of delayed neutron precursor nuclei per m^3 as

$$\frac{dC_{fi}}{dt} = k_{di}\frac{n}{l} = k\beta_i\frac{n}{l} \qquad i = 1 \text{ to } 6 \tag{21.26}$$

where equation (21.6) has been used in the last step.

The rate of increase of concentration of delayed neutron precursors is the rate of production minus the rate of decay:

$$\frac{dC_i}{dt} = \frac{dC_{fi}}{dt} - R_i \qquad i = 1 \text{ to } 6 \tag{21.27}$$

and using equations (21.1) and (21.26), we may re-express this as:

$$\frac{dC_i}{dt} = \frac{k\beta_i}{l}n - \lambda_i C_i \qquad i = 1 \text{ to } 6 \tag{21.28}$$

21.9 Summary of neutron kinetics equations; reactor power

The neutronic behaviour of the reactor is described by equations (21.24) and (21.28), repeated below.

$$\frac{dn}{dt} = k\frac{(\rho - \beta)}{l}n + \sum_{i=1}^{6}\lambda_i C_i + S \tag{21.24}$$

$$\frac{dC_i}{dt} = \frac{k\beta_i}{l}n - \lambda_i C_i \qquad i = 1 \text{ to } 6 \tag{21.28}$$

These equations may be integrated in a time-marching manner once starting values $n(t_0)$ and $C_i(t_0)$ are available.

[The derivation has been based on the variables, n, C_i and S, being understood as density terms, expressed per unit volume of the reactor core. However, it is possible to multiply both sides of each equation by the reactor core volume and so produce an equivalent set of equations in the total number of neutrons and delayed neutron precursors contained within the core.]

21.10 Values of delayed neutron parameters and the problem of stiffness

Table 21.1 gives values of delayed neutron parameters, β_i, λ_i, for thermal fission, while Table 21.2 gives the corresponding values for fast fission. Following nuclear physics convention, the tables also list the effective half-life of each precursor, based on the formula:

$$t_{1/2,i} = \frac{\ln 2}{\lambda_i} = \frac{0.6931}{\lambda_i} \tag{21.29}$$

Tables 21.1 and 21.2 are based on information supplied in Lewins, 1978 Chapter 2.

It may be seen from the tables that the spread of time constants defining the production of delayed neutrons, $\tau_i = 1/\lambda_i$, corresponds to a factor of over 200, and hence the neutronics equations are stiff according to the criteria of Chapter 2, Section 2.7. It is therefore common for control engineers interested in modelling the interaction of the reactor with the steam plant to

Table 21.1 Delayed neutron parameters for thermal fission

Isotope	*Group*	β_i	λ_i (s^{-1})	$t_{1/2,i}$ (s)
Uranium-235	1	0.000266	0.0127	54.58
	2	0.001492	0.0317	21.87
	3	0.001317	0.115	6.03
	4	0.002851	0.311	2.23
	5	0.000897	1.40	0.495
	6	0.000182	3.87	0.179
	Overall	0.00700	0.0784	8.84
Plutonium-239	1	0.000086	0.0129	53.73
	2	0.000637	0.0311	22.29
	3	0.000491	0.134	5.17
	4	0.000746	0.331	2.09
	5	0.000234	1.26	0.55
	6	0.000080	3.21	0.216
	Overall	0.00227	0.0683	10.14

Table 21.2 Delayed neutron parameters for fast fission

Isotope	*Group*	β_i	λ_i (s^{-1})	$t_{1/2,i}$ (s)
Uranium-235	1	0.000251	0.0127	54.58
	2	0.001406	0.0317	21.87
	3	0.001241	0.115	6.03
	4	0.002687	0.311	2.23
	5	0.000845	1.40	0.495
	6	0.000172	3.87	0.179
	Overall	0.00660	0.0784	8.84
Plutonium-239	1	0.000081	0.0129	53.73
	2	0.000594	0.0311	22.29
	3	0.000458	0.134	5.17
	4	0.000695	0.331	2.09
	5	0.000218	1.26	0.55
	6	0.000074	3.21	0.216
	Overall	0.00212	0.0682	10.17

approximate the behaviour of the six precursor groups by a smaller number of groups, usually between two and four effective groups. A new group with parameters, $\hat{\beta}_j$, $\hat{\lambda}_j$, may be formed from several precursor groups by a weighting approach applied to the time constants:

$$\hat{\beta}_j = \sum_{i=n_1}^{n_2} \beta_i \tag{21.30}$$

$$\hat{\tau}_j = \frac{1}{\hat{\beta}_j} \sum_{i=n_1}^{n_2} \beta_i \tau_i \tag{21.31}$$

or

$$\frac{1}{\hat{\lambda}_j} = \frac{1}{\hat{\beta}_j} \sum_{i=n_1}^{n_2} \frac{\beta_i}{\lambda_i} \tag{21.32}$$

This method of weighting has been used to combine all six groups into one in the 'Overall' rows in Tables 21.1 and 21.2.

To give another example, the number of delayed neutron groups for thermal fission of uranium-235 could be reduced from six to three by combining groups 3, 4, 5 and 6: $\hat{\beta}_3 = 0.005247$ and $\hat{\lambda}_3 = 0.2463$, implying a time constant $\hat{\tau}_3 = 4.06$ seconds, which will be comparable with other small time constants within a larger power plant simulation.

21.11 Relationship between neutron density, neutron flux and thermal power

The neutron flux is the product of the neutron density and the neutron speed. Assuming that all the neutrons are travelling at the same speed, equal to the average speed, c_n, the reactor flux averaged over the complete core, χ_{ave}, neutrons/m^2/s will be the product:

$$\chi_{ave} = nc_n \tag{21.33}$$

The assumption of a constant speed, c_n, implies that the reactor flux will be proportional to the neutron density, n. The fission rate of the reactor per m^3 of fuel, F, is given by the product of the neutron flux and the total cross-sectional area for fission:

$$F = \chi_{ave} N_f \sigma_f \tag{21.34}$$

where N_f is the number of fissile nuclei per m^3 of fuel in the core, while σ_f is the effective cross-sectional area for fission of each fissile nucleus (m^2). Approximately 32 pJ of heat energy will be liberated per fission, so that the power density, ξ (W/m^3), or instantaneous power released as heat per cubic metre of fuel, averaged over the complete core will be:

$$\begin{aligned}\xi &= 32 \times 10^{-12} F \\ &= 32 \times 10^{-12} \chi_{ave} N_f \sigma_f \\ &= 32 \times 10^{-12} N_f \sigma_f c_n n\end{aligned} \tag{21.35}$$

The effective cross-sectional area for fission of each fissile nucleus, σ_f, depends on the thermalization temperature, i.e. the temperature of the moderator, although small deviations from the setpoint temperature will have a negligible effect on σ_f. But the number of and composition of the fissile nuclei will change over the months of reactor operation, as U-235 is used up and Pu-239 is produced. Hence the quantity $N_f \sigma_f$ will change during the reactor run. For the rapid transients of interest to the control engineer, however, it will be sufficient to use a constant figure for $N_f \sigma_f$ appropriate to the current burn-up, so that

$$\xi = \alpha n \tag{21.36}$$

where $\alpha(= 32 \times 10^{-12} N_f \sigma_f c_n)$ is taken to be a constant, with units of watts.

Consider, for example, a PWR containing 80 te of uranium oxide enriched to 3%, with an average moderator temperature of 310°C. The effective specific volume of uranium oxide pellets will be 1.11×10^{-4} m^3/kg. The fission cross-section at a moderator temperature of 20°C/293 K, $\sigma_f(293)$, is 582×10^{-28} m^2 for U-235, while the corresponding figure for plutonium-239 is 743×10^{-28} m^2. Calculate $N_f \sigma_f$ at the start of life and at the end of life, and hence deduce the variation in the constant of proportionality, α.

At 3% enrichment, the mass of U-235-bearing UO_2 is $0.03 \times 80\,000 = 2400$ kg. The mass fraction of U-235 in each UO_2 molecule is $235/(235 + 2 \times 16) = 0.88$. Therefore the mass of U-235 in the core $= 0.88 \times 2400 = 2112$ kg. 1 kg-mol of U-235 has a mass of 235 kg and contains Avogadro's number of nuclei, namely 6.022×10^{26}. Therefore the number of U-235 nuclei in the whole core will be:

$$\frac{2112}{235} \times 6.022 \times 10^{26} = 5.412 \times 10^{27}$$

The volume of the fuel will be $1.11 \times 10^{-4} \times 80\,000 = 8.88$ m^2, so that $N_f = (5.412 \times 10^{27})/(8.88) = 6.09 \times 10^{26}$ nuclei/m^3 of fuel.

It is shown in Glasstone and Sesonske (1981) that, assuming complete thermalization, the fission cross-section at moderator temperature $T(K)$ is related to fission cross-section at T_0 by the equation:

$$\sigma_f(T) = g_{corr} \frac{1}{1.128} \left(\frac{T_0}{T}\right)^{1/2} \sigma_f(T_0) \tag{21.37}$$

where g_{corr} is a correction factor close to unity, dependent on both the fissile isotope and the operating temperature. For U-235 in a reactor with a moderator temperature of 310°C, $g_{corr} \approx 0.94$. Hence

$$\sigma_f(583) = 0.94 \times \frac{1}{1.128} \times \left(\frac{293}{583}\right)^{1/2} \times 582 \times 10^{-28} = 344 \times 10^{-28}\,\mathrm{m}^2$$

It follows that $N_f\sigma_f = 6.09 \times 10^{26} \times 344 \times 10^{-28} = 20.97\,\mathrm{m}^{-1}$.

Using the Maxwell–Boltzmann energy distribution (Glasstone and Sesonske, 1981), it may be shown that the average speed of perfectly thermalized neutrons is given by

$$c_n = 128\sqrt{T} \tag{21.38}$$

Hence at a moderator temperature of 310°C/583 K, $c_n = 128 \times \sqrt{583} = 3091$ m/s. Substituting into equation (21.35) gives the relationship between neutron density, n, and average power density, ξ, at the start of core life as

$$\begin{aligned} \xi &= 32 \times 10^{-12} \times 20.97 \times 3091n \\ &= 2.074 \times 10^{-6}n \end{aligned} \tag{21.39}$$

For a typical neutron density of $n = 1.6 \times 10^{14}$ neutrons/m^3, the average power density emerges as $\xi = 331.8 \times 10^6$ W/m^3. Since the core has a total volume of fuel of 8.88 m^3, the total power is then 2947 MWth.

At the end of the reactor run, the number of fissile nuclei, N_f, will be approximately halved, and only about 55% of the nuclei will be uranium-235. The rest of the fissile nuclei will be plutonium-239, for which σ_f (293) has been given above as 743×10^{-28} m^2. To apply equation (21.37) to plutonium-239 at a moderator temperature of 310°C/583 K, we need to set the correction factor, $g_{corr} \approx 1.45$. Thus

$$\sigma_f|_{Pu}(583) = 1.45 \times \frac{1}{1.128}\left(\frac{293}{583}\right)^{1/2} \times 743 \times 10^{-28} = 677 \times 10^{-28}\,\mathrm{m}^2$$

At the end of life, with both U-235 and Pu-239 present, the product $N_f\sigma_f$ becomes a composite variable, based on the characteristics of U-235 and Pu-239:

$$\begin{aligned} N_f\sigma_f &\approx \frac{6.09 \times 10^{26}}{2} \times 0.55 \times 344 \times 10^{-28} \\ &\quad + \frac{6.09 \times 10^{26}}{2} \times 0.45 \times 677 \times 10^{-28} \\ &= 15.04\,\mathrm{m}^{-1} \end{aligned}$$

Since the speed of the thermalized neutrons depends only on moderator temperature, the same average speed of 3091 m/s holds. Thus the relationship between neutron density and power density at the end of life is

$$\begin{aligned} \xi &= 32 \times 10^{-12} \times 15.04 \times 3091n \\ &= 1.488 \times 10^{-6}n \end{aligned} \tag{21.40}$$

Thus the constant of proportionality has decreased by about 30% over the life of the core. More accurate relationships between thermal power density and neutron density at different stages of the reactor run may be available from the design calculations or from plant-specific reactor physics data.

21.12 Spatial variations in neutron flux and power: centre-line and average reactor flux

There is a significant variation in neutron flux both radially and axially in a typical nuclear reactor of cylindrical shape. The radial variation will be of concern to those analysing the detailed behaviour of the reactor itself, particularly the phenomenon of slow, xenon-induced, spatial oscillations in power that may occur in large reactor cores such as are found in the Magnox and CANDU reactors. However, the control engineer concerned with analysing the behaviour of a larger power system of which the reactor is only one component will be concerned mainly with the input–output characteristics of the reactor. Assuming good mixing of the coolant at the outlet, it will then be sufficient to consider the axial variations in neutron flux only. This is the approach adopted here.

It is shown in Glasstone and Sesonske (1981) that in the steady state, the flux averaged over the core radius at height z, $\overline{\chi}(z)$, will obey the relationship.

$$\overline{\chi} = \overline{\chi}_0 \cos\frac{\pi z}{\varepsilon L} \tag{21.41}$$

where

- L is the height of the reactor (m),
- $\overline{\chi}_0$ is the radially averaged flux at the centre-line, $z = 0$, of the reactor (neutrons/m^2/s),
- ε is the elongation factor. Close to unity, ε makes allowance for the effective increase in reactor height produced by the neutron reflector. $\varepsilon = 1.0$ for a reactor with no reflector, while $\varepsilon \approx 1.2$ for most commercial reactors.

Since, from equation (21.33), the neutron flux is proportional to the neutron density, it follows that the radially averaged neutron density at any height, $\overline{n}_z(z)$, will bear the same relationship to the average neutron density at the centre line, $\overline{n}_0 = \overline{n}_z(0)$:

$$\overline{n}_z = \overline{n}_0 \cos \frac{\pi z}{\varepsilon L} \tag{21.42}$$

The variation in neutron density between the centre-line of a commercial reactor and the upper and lower extremities of the core is significant, as may be judged by substituting $z = 0.5L$ and $\varepsilon = 1.2$ into equation (21.42), from which it may be seen that $\overline{n}_z(0.5L)/\overline{n}_0 = 0.26$.

Since the neutronics equations (21.24) and (21.28) will produce a value for average neutron density over the core as a whole rather than the maximum, centre-line density, we need to convert equation (21.42) into an equivalent expression involving the average neutron density, n. This may be found by averaging $\overline{n}_z$ over the whole height of the reactor:

$$nL = \int_{z=-(L/2)}^{z=(L/2)} \overline{n}_z \, dz = \int_{z=-(L/2)}^{z=(L/2)} \overline{n}_0 \cos \frac{\pi z}{\varepsilon L} \, dz$$

$$= \frac{\varepsilon L}{\pi} \overline{n}_0 \left[\sin \frac{\pi z}{\varepsilon L} \right]_{(-L/2)}^{(L/2)} = \frac{2\varepsilon L}{\pi} \overline{n}_0 \sin \frac{\pi}{2\varepsilon} \tag{21.43}$$

where the last step has used the fact that $\sin(-x) = -\sin x$. Hence the radially averaged neutron density at the centre-line is given in terms of the overall-average neutron density by

$$\overline{n}_0 = \frac{\left(\frac{\pi}{2\varepsilon}\right)}{\sin \frac{\pi}{2\varepsilon}} n \tag{21.44}$$

Substituting into equation (21.42) gives the axial variation in neutron density as:

$$\overline{n}_z = n \frac{\left(\frac{\pi}{2\varepsilon}\right)}{\sin \frac{\pi}{2\varepsilon}} \cos \frac{\pi z}{\varepsilon L} \tag{21.45}$$

This distribution equation, which allows us to calculate the axial distribution of neutron density from the core-average neutron density, n, is assumed to hold transiently as well as in the steady state.

21.13 Flux and power in axial segments of the reactor core

Let us divide the reactor core into N equal axial segments, each of which will be $\Delta L = L/N$ long. See Figure 21.4.

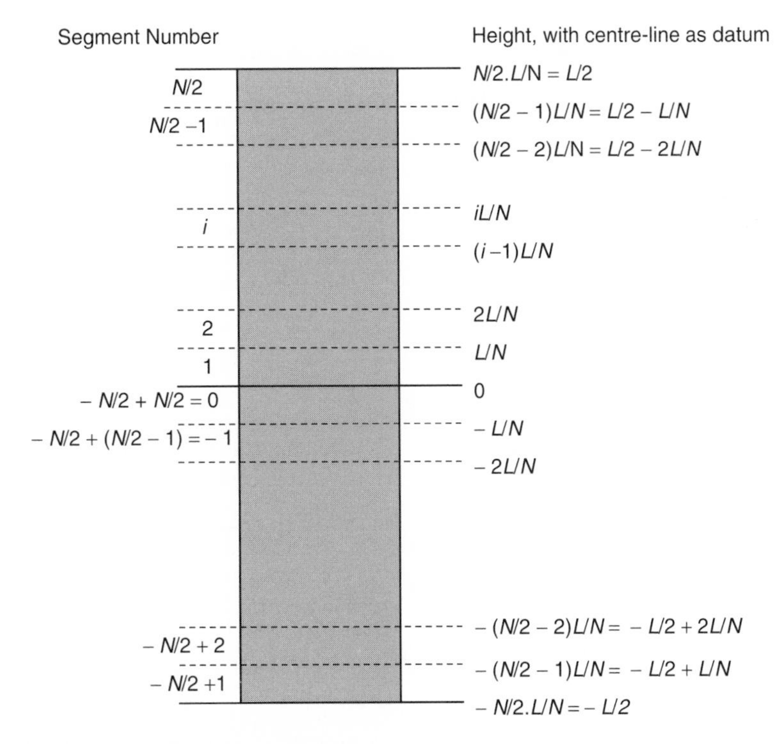

Figure 21.4 Dividing the reactor core into N axial segments.

Given the core-average neutron density, n, that results from integrating the neutronics equations (21.24) and (21.28), we may calculate the radial-average neutron density at any height from equation (21.45). At a height i segments up from the centre-line datum, namely $z = iL/N$, the radial-average neutron density will be:

$$\overline{n}_z\left(i\frac{L}{N}\right) = n\frac{\left(\frac{\pi}{2\varepsilon}\right)}{\sin\frac{\pi}{2\varepsilon}}\cos\frac{\pi i}{\varepsilon N} \qquad (21.46)$$

It is possible to estimate the average neutron density within the ith segment simply as half the sum of the radial-average neutron densities at heights iL/N and $(i-1)L/N$:

$$n_i = \frac{n}{2}\frac{\left(\frac{\pi}{2\varepsilon}\right)}{\sin\frac{\pi}{2\varepsilon}}\left(\cos\frac{\pi i}{\varepsilon N} + \cos\frac{\pi(i-1)}{\varepsilon N}\right)$$

$$\text{for } -\frac{N}{2}+1 \le i \le \frac{N}{2} \qquad (21.47)$$

The assumption inherent in equation (21.47) is that the variation in neutron density over the segment is linear. In fact, we know that its shape within the segment will actually be determined by the cosine function given by equation (21.45). We may use this fact to improve the estimate of the segment's average neutron density, which should follow an averaging equation similar to equation (21.43):

$$\begin{aligned} n_i\frac{L}{N} &= \int_{z=(i-1)L/N}^{z=i(L/N)} \overline{n}_0\cos\frac{\pi z}{\varepsilon L}\,dz \\ &= \frac{\varepsilon L}{\pi}\overline{n}_0\left[\sin\frac{\pi z}{\varepsilon L}\right]_{(i-1)L/N}^{i\,L/N} \qquad (21.48) \\ &= \frac{\varepsilon L}{\pi}\overline{n}_0\left(\sin\frac{\pi i}{\varepsilon N} - \sin\frac{\pi(i-1)}{\varepsilon N}\right) \end{aligned}$$

Substituting from equation (21.44) into (21.48) and rearranging gives the average neutron density over segment i as:

$$n_i = \frac{n}{2}\frac{N}{\sin\frac{\pi}{2\varepsilon}}\left(\sin\frac{\pi i}{\varepsilon N} - \sin\frac{\pi(i-1)}{\varepsilon N}\right)$$

$$\text{for } -\frac{N}{2}+1 \le i \le \frac{N}{2} \qquad (21.49)$$

This expression will give a slightly more accurate value that the arithmetic average of the neutron fluxes at either side of the segment.

[Equations (21.47) and (21.49) are asymptotically equivalent as the number of axial segments, N, grows very large, as may be shown by dividing the right-hand side of equation (21.49) by the right-hand side of equation (21.47). The ratio emerges as:

$$\begin{aligned} &\frac{2\varepsilon N}{\pi}\frac{\sin\frac{\pi i}{\varepsilon N} - \sin\frac{\pi(i-1)}{\varepsilon N}}{\cos\frac{\pi i}{\varepsilon N} + \cos\frac{\pi(i-1)}{\varepsilon N}} \\ &= \frac{2\varepsilon N}{\pi}\frac{2\cos\frac{1}{2}\left(\frac{\pi}{\varepsilon N}(2i-1)\right)\sin\frac{\pi}{2\varepsilon N}}{2\cos\frac{1}{2}\left(\frac{\pi}{\varepsilon N}(2i-1)\right)\cos\frac{\pi}{2\varepsilon N}} \qquad (21.50) \\ &= \frac{2\varepsilon N}{\pi}\tan\frac{\pi}{2\varepsilon N} \end{aligned}$$

where use has been made of the two trigonometric identities:

$$\sin C - \sin D = 2\sin\tfrac{1}{2}(C+D)\sin\tfrac{1}{2}(C-D)$$

$$\cos C + \cos D = 2\sin\tfrac{1}{2}(C+D)\cos\tfrac{1}{2}(C-D) \qquad (21.51)$$

As N grows large, so the argument of the tangent, $\pi/(2\varepsilon N)$, becomes small, allowing the tangent to be replaced by its angle in radians. Hence

$$\frac{2\varepsilon N}{\pi}\tan\frac{\pi}{2\varepsilon N} \rightarrow \frac{2\varepsilon N}{\pi}\frac{\pi}{2\varepsilon N} = 1 \qquad (21.52)$$

The difference in the estimates will be very small when $N = 10$, as is shown in Table 21.3.]

Dividing the reactor into 10 axial segments, the ratio of the segmental neutron density to the neutron

Table 21.3 Ratio of neutron density in segment to average neutron density over reactor core

Segment Number	*−4*	*−3*	*−2*	*−1*	*0*	*1*	*2*	*3*	*4*	*5*
$\frac{n_i}{n}$ (eqn. (21.49))	0.517	0.823	1.072	1.248	1.340	1.340	1.248	1.072	0.823	0.517
$\frac{n_i}{n}$ (eqn. (21.47))	0514	0.818	1.066	1.241	1.332	1.332	1.241	1.066	0.818	0.514

density averaged over the whole core will be as given in Table 21.3.

The power density, ξ, is proportional to the neutron density, n (equations (21.35) and (21.36)), with the constant of proportionality, α, depending on the enrichment of the fuel. The fuel enrichment will normally be invariant with axial height at fuel loading, but there will be a tendency for the fissile nuclei at the centre of the reactor to be used up preferentially during a reactor run. The calculation shown in Section 21.11 indicates the sort of change in the neutron-to-power coefficient, α, that is possible over the reactor as a whole.

The power density in segment i, ξ_i, will obey the equation:

$$\xi_i = \alpha_i n_i \tag{21.53}$$

where the segmental neutron-to-power coefficient, α_i, is subject to only a very slow rate of change, and may be regarded as constant during any transient. The segmental coefficient, α_i, may assumed to be the same as the average neutron-to-power coefficient across the core at the start of reactor life, i.e. $\alpha_i = \alpha$. This equation is sometimes assumed to be approximately valid further in the reactor run, although separately calculated coefficients will give a better representation.

21.14 Calculating the temperature of the fuel in each of the axial segments

As indicated in Section 21.12, there will be a variation in power deposition along any radius of the core. This will lead to differences in temperature in different pins at the same height in the core. In addition, despite the fact that the pins are normally narrow (typically of the order of 10 millimetres in diameter), there will be a very significant difference between the fuel temperature at the centre of the pin and the fuel temperature at the outer extremity of the pin: 500°C is not uncommon. But in keeping with the relatively simple model being developed in this chapter, we will work with the average temperature of the fuel in the axial segment. This will be taken to be the same as the radial-average fuel temperature of the notional average fuel pin between the same heights. The reader may wish to consult Appendix 8, which explores the heat transfer process from the fuel pin to the coolant, derives an overall heat transfer coefficient and provides calculations of the temperatures at various locations within a typical nuclear fuel pin.

The energy equation for an axial segment requires that the rate of change of energy is equal to the nuclear heating *minus* the heat transferred to the coolant. The heat capacity of the fuel will be much greater than that of the cladding, and so we will make the simplification that the cladding temperature is the same as the average fuel temperature in segment i, $T_{fuel,i}$. Hence we may write

$$(m_{fi} c_{pfi} + m_{cladi} c_{pcladi}) \frac{dT_{fuel,i}}{dt} = \Phi_{nuc,i} - \Phi_{trans,i} \tag{21.54}$$

where
- m_{fi} is the mass of the fuel in axial segment i of the core (kg),
- m_{cladi} is the mass of the cladding in axial segment i of the core (kg),
- c_{pfi} is the specific heat of the fuel in axial segment i of the core J/(kgK),
- c_{pcladi} is the specific heat of the cladding in axial segment i of the core J/(kgK),
- $\Phi_{nuc,i}$ is the nuclear heat deposited in axial segment i of the core (W), and
- $\Phi_{trans,i}$ is the heat transfer from axial segment i of the core to the coolant (W).

These gross, segmental variables may be found from the corresponding variables for the average fuel pin simply by multiplying the latter by the number of fuel pins in the core, N_{pins}.

The nuclear heat deposited may be calculated as the product of the neutron density in segment i, ξ_i, and the volume of the fuel:

$$\Phi_{nuc,i} = N_{pins} \pi a^2 \Delta L \xi_i \tag{21.55}$$

where a is the radius of the fuel pellets (m) and $\Delta L = L/N$ is the length of the axial segment (m).

The heat transferred from the fuel to the coolant is given by the heat transfer equation:

$$\Phi_{trans,i} = K_{over} A_{tot,i} (T_{fuel,i} - T_{cool,i}) \tag{21.56}$$

where
- $T_{cool,i}$ is the temperature of the coolant in segment i (K),
- $A_{tot,i}$ is the total heat transfer area between pins and coolant in segment i (m^2),
- K_{over} is the overall heat transfer coefficient between the fuel at average temperature and the coolant (W/(m^{2}K)).

The area, $A_{tot,i}$, is given by:

$$A_{tot,i} = 2\pi d \Delta L N_{pins} \tag{21.57}$$

where d is the outer radius of the fuel pin cladding (see Figure A8.1 of Appendix 8). The overall heat transfer coefficient, K_{over}, is given by equation (A8.32) in Appendix 8. It may be written directly as:

$$K_{over} = \frac{1}{\dfrac{d}{3k_f} + \dfrac{2d}{(a+b)K_{fc}} + \dfrac{d}{k_{clad}} \ln \dfrac{d}{b} + \dfrac{1}{K_{cc}}} \tag{21.58}$$

where:

b is the inner radius of the fuel-pin cladding (m) (see Figure A8.1 of Appendix 8),

k_f is the thermal conductivity of the fuel (W/(mK)),

K_{fc} is the heat-transfer coefficient between the fuel pellet and the clad (W/(m^{2}K)),

k_{clad} is the thermal conductivity of the fuel-pin cladding (W/(mK)),

K_{cc} is the heat-transfer coefficient between the cladding and the coolant (W/(m^{2}K)).

21.15 Calculating the coolant temperature

The coolant may be regarded as flowing in a duct subject to heat exchange, the equations for which were developed in Chapter 20, Sections 20.3, 20.4 and 20.5. Allowing for the fact that the flow through the reactor is likely to be vertical, the appropriate equation for a liquid coolant is

$$\frac{\partial T_{cool}}{\partial t} + c\frac{\partial T_{cool}}{\partial x} = \frac{v_{cool}}{A_{cool}c_{pcool}}\phi - \frac{cg}{c_{pcool}}\sin\theta \tag{21.59}$$

while the equation for a gas coolant is:

$$\frac{\partial T_{cool}}{\partial t} + \gamma c\frac{\partial T_{cool}}{\partial x} = \frac{v_{cool}}{A_{cool}c_{vcool}}\phi - \frac{cg}{c_{vcool}}\sin\theta \tag{21.60}$$

Here

A_{cool} is the flow area available to the coolant (m^2),

v_{cool} is the specific volume of the coolant (m^3/kg),

c_{pcool} is the specific heat of the liquid coolant (J/(kgK)),

c_{vcool} is the specific heat at constant volume of the gas coolant (J/(kgK)),

ϕ is the heat flux per unit length (W/m).

These equations are expressed in terms of heat flux per unit length, ϕ, rather than total heat flux, Φ. However, we may note that the fuel-generated heat flux per unit length in section i, $\phi_{trans,i}$, is given by:

$$\begin{aligned}\phi_{trans,i} &= \frac{\Phi_{trans,i}}{\Delta L} \\ &= K_{over}\frac{A_{tot,i}}{\Delta L}(T_{fuel,i} - T_{cool,i}) \\ &= 2\pi\, dN_{pins}K_{over}(T_{fuel,i} - T_{cool,i}) \\ &= N_{pins}K_{over}\pi D_{pins}(T_{fuel,i} - T_{cool,i})\end{aligned} \tag{21.61}$$

which is essentially the same form as the heat flux per unit length for the shell-side of the shell-and-tube heat exchanger (equation (20.17)). It is thus clear that the methods of calculating the shell-side temperatures in the heat exchanger may be applied to calculate the temperature of the coolant of the nuclear reactor. The reader is referred to Sections 20.7, 20.8 and 20.9 of Chapter 20.

21.16 Calculating the reactivity

At any given time into a reactor run, the reactivity, ρ, will depend both on the position of the control rods and on the temperature of the core and the coolant:

$$\rho = \rho(x_1, x_2 \ldots x_m, T_{fuel,1}, T_{fuel,2} \ldots T_{fuel,N}, T_{cool,1}, T_{cool,2} \ldots T_{cool,N}) \tag{21.62}$$

where x_i are the depths of insertion of the control rods, assumed to be m in number. The control rod worth will vary according to its location in the reactor. Its effect on reactivity will also depend on its current depth of insertion. The modeller will need to consult the reactor designers and operators for numerical data on control rod worth and control rod calibrations.

The temperature feedback mechanisms provide a link between the reactor's neutronics and its coolant systems independent of any action of the control system. The size and relative importance of the temperature effects will vary from reactor to reactor, but designers work hard to ensure that there is an overall negative coefficient of reactivity with temperature, which provides for automatic limiting or mitigation of temperature excursions. Some important temperature feedback mechanisms are listed below:

Doppler broadening, whereby the absorption of neutrons in the resonance region is increased with an increase in temperature. The increase in non-fissioning absorption by uranium-238 means that this effect reduces the reactivity of thermal reactors at higher temperatures;

coolant/moderator expansion, with increased temperature leading to increased neutron leakage, poorer moderation and hence a reduction in reactivity with temperature;

fuel expansion with temperature, reducing the density of fissile nuclei and hence the reactivity;

fuel support expansion, leading to reduced density of fissile nuclei and hence lower reactivity;

fuel pin bowing, which increases with temperature, leading to reduced density of fissile nuclei and hence lower reactivity;

control rod support structure expansion, leading to the control rods being inserted further into the core, reducing reactivity.

As was the case with control rods, the modeller will need to consult the reactor designers and operators for numerical data on the important temperature feedback mechanisms applicable to the reactor being modelled.

21.17 Bibliography

Fremlin, J.H. (1987). *Power Production: What are the Risks?*, Oxford University Press.

Glasstone, S. and Sesonske, A. (1981). *Nuclear Reactor Engineering*, 3rd edition, Van Nostrand Reinhold, New York.

Lewins, J. (1978). *Nuclear Reactor Kinetics and Control*, Pergamon Press, Oxford.

Lewis, C. (ed.) (1996). 'World nuclear status report '95', *Nuclear Engineering Worldscan*, **xvi**, No. 7–8, July–August.

22 Process controllers and control valve dynamics

22.1 Introduction

Despite its small size, the controller plays a crucial role in determining the behaviour of the process to which it is attached. It follows that the modeller needs to pay special attention to simulating controllers accurately. The most common types of controller found on process plant are proportional controllers, proportional plus integral (P + I) controllers and proportional plus integral plus derivative (PID) controllers. The characteristics of each will be discussed in turn from a simulation viewpoint. Particular attention will be paid to the most common forms of integral desaturation and the problems of simulating them using a simulation language.

The controller on a process plant is usually connected to a control valve, the opening of which is governed by the controller's output signal. Methods of calculating the flow through the valve have been described previously in Chapters 7, 8, 9 and 10. This chapter is concerned with modelling the dynamic movement of the valve stem, which determines the extent of valve opening. Valve-stem movement is subject to two important nonlinearities: the velocity limit and static friction. A simple way of allowing for the velocity limit will be given, and two broadly equivalent methods will be presented for modelling the effect of valve static friction.

22.2 The proportional controller

Figure 22.1 shows the action of the proportional controller in the feedback loop controlling the plant.

The controller is intended to keep the plant variable, θ_p, at or near to the setpoint value, θ_s. The plant variable passes through a measurement system, which produces the value θ_m, which is fed back in a negative sense to give the error, e:

$$e = \theta_s - \theta_m \tag{22.1}$$

The error signal is passed to the controller, which multiplies it by the gain, k, and transmits the resultant control signal, c, to the plant. Note, however, that the signal sent to the plant will be limited between a minimum value, $c_{\min}$, and a maximum value, $c_{\max}$, so that it may be calculated using a limiting function 'lim' (normally provided as standard within a simulation language):

$$c = \lim(c_{\min}, c_{\max}, ke) \tag{22.2}$$

The value and units of the controller's range will depend on the controller's construction and the manufacturer's preference. For example, a pneumatic controller will normally produce an output in the range 3 to 15 psi, whereas an electronic controller may produce a current output in the range 4 to 20 mA, or a voltage output in the range 0 to 10 volts. Alternatively a fully digital set of signals may be produced by the controller. It is usually convenient for the modeller to modify the gain in his simulation so as to normalize the output to fall in the range 0 to 1. Obviously this convention needs to be carried through consistently to the modelling of the final control element, the input of which will be the output from the controller.

Process plant custom sometimes expresses the controller gain in units of proportional band (PB). The proportional band is defined as the percentage of the

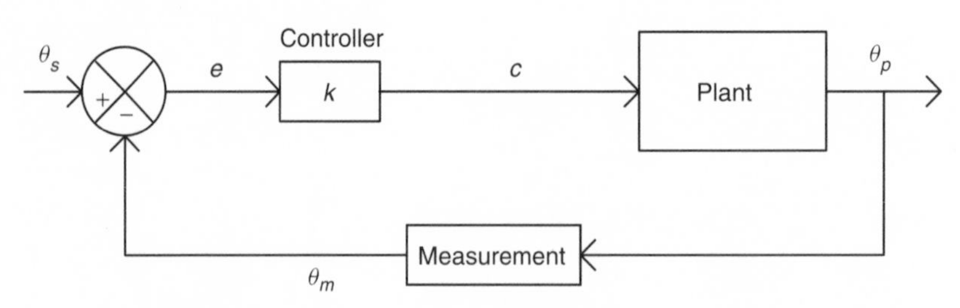

Figure 22.1 Proportional controller.

controller input (= error) range needed to cause a full-scale change in the controller output. The gain, k, is then given as:

$$k = \frac{\text{output range in output units}}{\frac{\text{PB}}{100} \times \text{input range in input units}} \tag{22.3}$$

For example, suppose that the measured variable is the level in a tank, that the measurement transducer has an effective range of 0.5 m and that the output from the controller is a current signal in the range 4–20 mA. If the proportional band is set at 250%, then the gain is given by

$$k = \frac{20 - 4}{\frac{250}{100} \times 0.5} = 12.8\,\text{mA/m} \tag{22.4}$$

Simulating this directly, the equation for the controller output may be calculated as

$$c = \lim(4, 20, 12.8e + 4)\ \text{mA} \tag{22.5}$$

Here the error, e, is metres and the controller output is in mA. The simulation of the actuator will now need to account for an input signal in mA, in close correspondence to the situation on the plant.

If the modeller chooses instead to work with a controller output normalized between 0 and 1, he will calculate the gain by setting the range in equation (22.3) as (1.0 – 0.0):

$$k = \frac{1.0 - 0.0}{\frac{250}{100} \times 0.5} = 0.8\,\text{m}^{-1} \tag{22.6}$$

so that the controller output is now calculated in dimensionless units as:

$$c = \lim(0, 1, 0.8e) \tag{22.7}$$

The simulation of the control actuator will now need to be adapted to accept this dimensionless signal from the controller.

Referring to the box marked 'Measurement' in Figure 22.1, the measurement signal, θ_m, will follow the true plant signal, θ_p, with a small lag, depending on the physics of the measurement system. It will normally be sufficient to model this as a simple exponential lag:

$$\frac{d\theta_m}{dt} = \frac{\theta_p - \theta_m}{\tau} \tag{22.8}$$

where τ is the time constant for the measurement system in question.

22.3 The basic operation of the proportional plus integral controller

The well-known deficiency of the proportional controller is that an error is necessary for the controller to sustain a non-zero output. Accordingly the plant variable, θ_p, will always be offset by a certain amount from the setpoint, θ_s. The proportional plus integral (P + I) controller avoids this problem by adding in an integral term that will build up as long as an error remains. Figure 22.2 shows the arrangement.

The controller output will now be composed of the two elements:

$$c = \lim(c_{\min}, c_{\max}, \text{P} + \text{I}) \tag{22.9}$$

where the proportional term remains as before:

$$P = ke \tag{22.10}$$

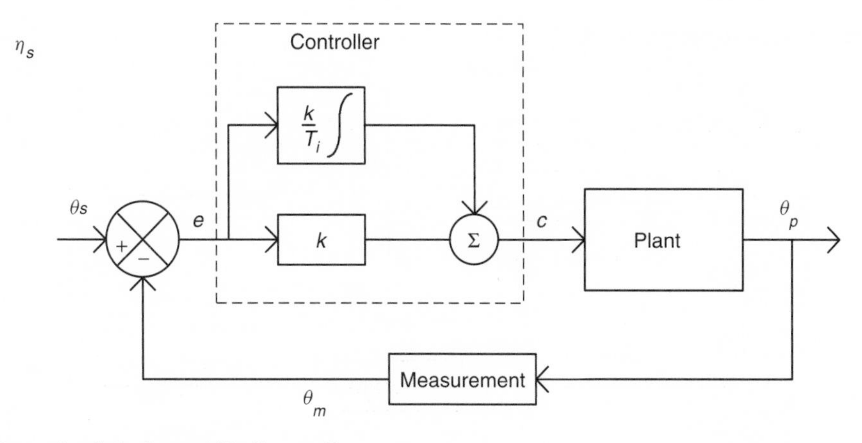

Figure 22.2 Proportional plus integral (P + I) controller.

while the integral term at the current time, t_n, is given by:

$$I(t_n) = I(t_0) + \int_{t_0}^{t_n} \frac{ke}{T_i}\,dt \qquad (22.11)$$

T_i is the integral action time, also known as the reset time. It is the time that would be taken for the change in the integral term to build up and equal the proportional term, given a sustained, constant error.

An important point to note is that the integral term defined by equation (22.11) is not bounded, and this clashes with the bounded nature of the controller output. Even if the controller output were not limited, an output signal growing without restriction would very soon encounter the saturation limits of the final control element: a valve can neither open more than wide open nor close more than tight shut. It is quite possible for an unlimited integral term to grow to a very large value if the error persists for a long time, for example under conditions of major plant upset. It would then take a comparably long time for the unlimited integral term to 'unwind' sufficiently to allow the controller output to come back within its normal working range. During all this time the control actuator would remain stuck at the extreme end of its range, providing no control action whatsoever, which is clearly an undesirable situation. It is to counter this that manufacturers normally provide the option of 'integral desaturation', which limits the integral term and allows speedier return to normal operation. A number of methods of providing integral desaturation are possible, but their simulation is not a trivial matter. We will return to this problem in Section 22.5.

22.4 The proportional plus integral plus derivative (PID) controller

The layout of a PID controller is shown in Figure 22.3.

A degree of anticipatory action has been added on to the P + I controller by including an additional term proportional to the derivative of the error, and this produces the most general form of the basic process controller used most extensively on process plants. As such, the tuning of the PID controller has been subject to a great deal of study. In point of fact, the additional complication of the PID controller over the P + I controller means that it is used very much less often on industrial process plant. Nevertheless, the control engineer needs to understand it and must be able to model it accurately.

The defining equation for the controller output is now:

$$c = \lim(c_{\max}, c_{\min}, P + I + D) \qquad (22.12)$$

where the proportional and integral terms are as defined in the previous section, and the derivative term is given by

$$D = kT_d \frac{de}{dt} \qquad (22.13)$$

Here T_d is known as the 'derivative action time'. Assuming that the error is increasing at a constant rate, the derivative action time is the time taken for

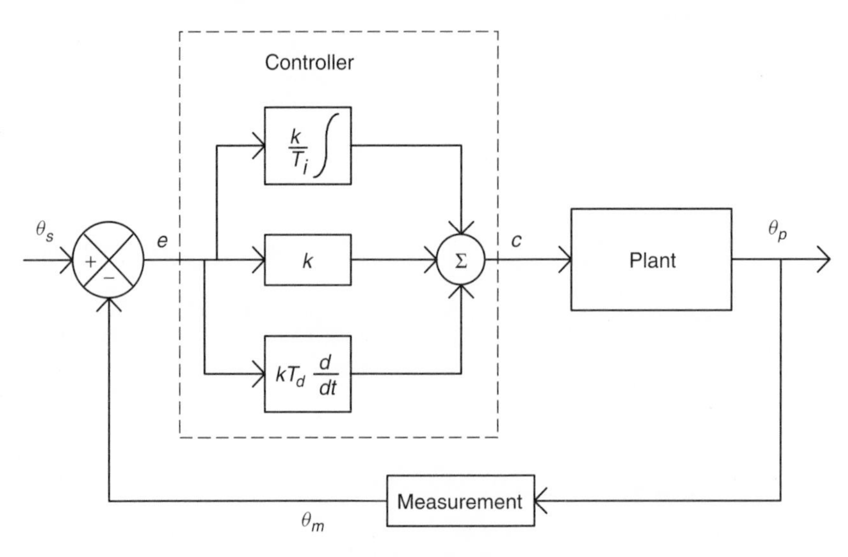

Figure 22.3 Proportional plus integral plus derivative (PID) controller.

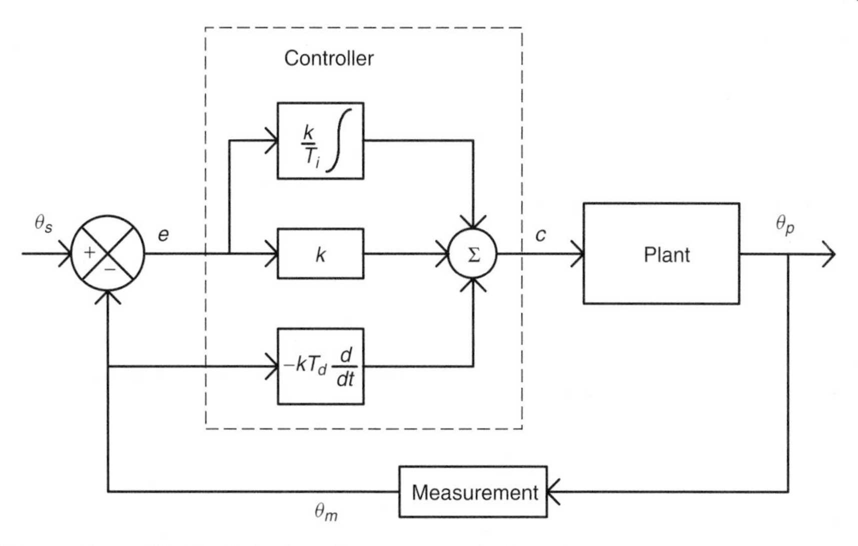

Figure 22.4 PID controller modified for derivative action on measured value only.

the proportional term to change by an amount equal to the derivative term.

A potential disadvantage of the arrangement shown in Figure 22.3 can be seen by expanding equation (22.13) to

$$D = kT_d\frac{d\theta_s}{dt} - kT_d\frac{d\theta_m}{dt} \qquad (22.14)$$

It can be seen from equation (22.14) that a rapid change, such as a step, in the setpoint, θ_s, will be transmitted into the plant as a large disturbance, which is likely to be counterproductive. An alternative arrangement that avoids this problem while maintaining the derivative action on plant signals is shown in Figure 22.4.

In the case of Figure 22.4, the derivative term is simply

$$D = -kT_d\frac{d\theta_m}{dt} \qquad (22.15)$$

Equation (22.14) is the most general equation for modelling the derivative term, which it splits into the derivative of the input and the derivative of the measurement. Clearly we may set $d\theta_s/dt = 0$ if the derivative acts on the measured value only. If the controller acts on derivative of the full error, then we need to calculate the derivative of the input also. This may be calculated to any desired degree of accuracy offline; to be sure, a step input will need to be replaced by a sharp ramp in order to produce a finite value, but the former will be more realistic physically.

We need the derivative of the measured value for both forms of controller, and here we are fortunate, since we have $d\theta_m/dt$ available already and given in equation (22.8). If the measurement time lag is very small it may be desirable to neglect it in order to reduce the stiffness of the simulation. In such a case we are assuming that $\theta_m = \theta_p$, which implies that $d\theta_m/dt = d\theta_p/dt$. Normally the plant variable will be either a state variable, with an equation defining its time derivative, or a simple function of such a state variable (for example, the state variable may be mass of liquid in a tank, while the controller may act on level, which is a simple function of mass. See Chapter 2, Section 2.2). Hence $d\theta_p/dt$ should be available directly, or following a simple calculation, within the main body of the process simulation. Some simulation programs include a numerical differentiator as a standard macro. The author's strong advice is not to use such a device: its use is highly inelegant, and often leads to numerical breakdown.

22.5 Integral desaturation

The essential features of integral desaturation can be explained by reference to a P + I controller. Three types will be considered and their describing equations given. Types 2 and 3 are implemented using digital techniques and so it has been found helpful to provide a subscripted label for time, t_n, which signifies a point in time n controller timesteps after the simulation start time, t_0. (The controller timestep need not be the same as the integration timestep of the simulation.)

22.5.1 Integral desaturation – Type 1

The form of integral desaturation which we shall call 'Type 1' specifies that the desaturated integral term should be held constant as soon as the controller output reached either its lower limit, c_{min}, or its upper limit, c_{max}. It can be shown that a Type 1 integral desaturation system leads to a controller that will come back out of saturation into its normal range when both of the following conditions are satisfied (see Appendix 9):

1. the error must have returned to the value present at entry to saturation.
2. the rate of change of error must satisfy

$$\frac{de}{dt} < -\frac{e}{T_i} \quad \text{for returning from upper limit saturation}$$

$$\frac{de}{dt} > -\frac{e}{T_i} \quad \text{for returning from lower limit saturation}$$

We may model Type 1 integral desaturation by replacing the pure integral term, I, in our simulation by the desaturated integral term, I_D, given at a general time, t, by:

$$I_D(t) = I_D(t_0) + \int_{t_0}^{t} B\,dt \tag{22.16}$$

where the integrand, B, is given by

$$\begin{aligned} B(t) &= 0 && \text{for } c_1(t) < c_{min} \\ &= \frac{ke(t)}{T_i} && \text{for } c_{min} \le c_1(t) \le c_{max} \\ &= 0 && \text{for } c_1(t) > c_{max} \end{aligned} \tag{22.17}$$

Here c_1 is the partially desaturated controller output, given by

$$c_1(t) = P(t) + I_D(t) \tag{22.18}$$

The (fully) desaturated controller output is then given by

$$c = \lim(c_{min}, c_{max}, c_1) \tag{22.19}$$

The equations presented above are sufficient to describe the action of a controller with Type 1 integral desaturation. They are also in a suitable form for direct implementation within a continuous system simulation language.

22.5.2 Integral desaturation – Type 2

A more sophisticated approach leads to a controller that returns from saturation more quickly. The Type 2 form of integral desaturation removes the first condition for the controller returning from saturation. It is now sufficient that the rate of change of error satisfies (see Appendix 9)

$$\frac{de}{dt} < -\frac{e}{T_i} \quad \text{for returning from upper limit saturation}$$

$$\frac{de}{dt} > -\frac{e}{T_i} \quad \text{for returning from lower limit saturation}$$

The Type 2 integral desaturation system checks the controller output continually. If its value is found to lie outside the output limits c_{min} and c_{max}, then the integral term is adjusted to ensure that the controller output lies on the nearest output limit.

The controller calculates the desaturated integral term, I_D, at a general time instant, t_n, from

$$I_D(t_n) = I_A(t_{n-1}) + \int_{t_{n-1}}^{t_n} \frac{ke}{T_i}\,dt \tag{22.20}$$

where I_A is the adjusted value of the integral term, given by

$$\begin{aligned} I_A(t_n) &= c_{min} - P(t_n) && \text{for } c_1(t_n) < c_{min} \\ &= I_D(t_n) && \text{for } c_{min} \le c_1(t_n) \le c_{max} \\ &= c_{max} - P(t_n) && \text{for } c_1(t_n) > c_{max} \end{aligned} \tag{22.21}$$

In equation set (22.21), c_1 is the partially desaturated output, given by

$$c_1(t_n) = P(t_n) + I_D(t_n) \tag{22.22}$$

The fully desaturated controller output, c, is then the sum of the proportional term and the adjusted integral term:

$$c(t_n) = P(t_n) + I_A(t_n) \tag{22.23}$$

Inspection of equations (22.21) to (22.23) shows that the fully desaturated controller output, c, is equal to the partially desaturated controller output, c_1, whenever the latter falls within the allowable controller output range, c_{min} to c_{max}. But if c_1 moves outside this range, then the controller output will remain at the closest limit, c_{min} or c_{max}.

The above equations constitute a full and sufficient description of the Type 2 integral desaturation system. However, equation (22.20) shows that the controller resets the integrator state, and this is not an option usually included within a continuous simulation language. Accordingly we need to find a different but equivalent formulation if we are to model Type 2 integral desaturation successfully. Let us reconsider equation (22.20), but move on in time to see the value of the desaturated integral term at time t_{n+1}:

$$I_D(t_{n+1}) = I_A(t_n) + \int_{t_n}^{t_{n+1}} \frac{ke}{T_i}\,dt \qquad (22.24)$$

This may be compared with the output of a completely unlimited integrator starting from the same starting condition $I(t_0)$ at time t_0:

$$I(t_{n+1}) = I(t_0) + \int_{t_0}^{t_{n+1}} \frac{ke}{T_i}\,dt \qquad (22.25)$$

or

$$I(t_{n+1}) = I(t_n) + \int_{t_n}^{t_{n+1}} \frac{ke}{T_i}\,dt \qquad (22.26)$$

Let us subtract equation (22.26) from equation (22.24):

$$I_D(t_{n+1}) - I(t_{n+1}) = I_A(t_n) - I(t_n) \qquad (22.27)$$

For convenience, we will define the right-hand side of equation (22.27) as the remainder, $R(t_n)$:

$$R(t_n) = I_A(t_n) - I(t_n) \qquad (22.28)$$

Rearranging equation (22.27) gives the desaturated integrator output at time, t_{n+1}, in terms of the unlimited integrator output at this time (easily derived by simple integration in the simulation) and the remainder term at time, t_n:

$$I_D(t_{n+1}) = I(t_{n+1}) + R(t_n) \qquad (22.29)$$

Thus we will be able to evaluate the desaturated integrator output provided we are able to evaluate $R(t_n)$. We shall now show how this may be done. First consider the behaviour of R in the three possible different circumstances for the partially desaturated output, c_1, namely:

Condition 1: $c_1(t_n) < c_{\min}$

Condition 2: $c_{\min} \le c_1(t_n) \le c_{\max}$

Condition 3: $c_1(t_n) > c_{\max}$

Condition 1. The partially desaturated output falls below the lower controller limit, i.e. $c_1(t_n) < c_{\min}$

In this case, from equation set (22.21), the adjusted integral term is given by

$$I_A(t_n) = c_{\min} - P(t_n) \qquad (22.30)$$

so that the remainder term is, by equation (22.28)

$$R(t_n) = c_{\min} - P(t_n) - I(t_n) \qquad (22.31)$$

The right-hand side of equation (22.31) may be calculated easily in the simulation at time, t_n.

Condition 2. The partially desaturated output lies between the controller limits, i.e. $c_{\min} \le c_1(t_n) \le c_{\max}$

Now, from equation set (22.21), the adjusted value of the integral term is equal to the desaturated integral term, i.e.

$$I_A(t_n) = I_D(t_n) \qquad (22.32)$$

The remainder term follows from equation (22.28) as

$$R(t_n) = I_D(t_n) - I(t_n) \qquad (22.33)$$

We may compare the right-hand side of equation (22.33) with equation (22.27). Shifting (22.27) back one instant, from time, t_{n+1}, to time, t_n, gives

$$I_D(t_n) - I(t_n) = I_A(t_{n-1}) - I(t_{n-1}) = R(t_{n-1}) \qquad (22.34)$$

where the definition of the remainder term given in equation (22.28) has also been used. Comparing (22.33) and (22.34) gives the result:

$$R(t_n) = R(t_{n-1}) \qquad (22.35)$$

Equation (22.35) may be implemented in a simulation by a delay function, incorporating a pure time delay equal to the controller timestep, τ_c:

$$\tau_c = t_n - t_{n-1} \qquad (22.36)$$

Condition 3. The partially desaturated output lies above the upper controller limit, i.e. $c_1(t_n) > c_{\max}$

Now, from equation set (22.21), the adjusted integral term is given by

$$I_A(t_n) = c_{\max} - P(t_n) \qquad (22.37)$$

so that the remainder term is, by equation (22.28)

$$R(t_n) = c_{\max} - P(t_n) - I(t_n) \qquad (22.38)$$

Again, this equation can be implemented without difficulty within a process simulation.

Summarizing for the three possible conditions, the behaviour of the remainder, R, is defined by:

$$\begin{aligned} R(t_n) &= c_{\min} - P(t_n) - I(t_n) && \text{for } c_1 < c_{\min} \\ &= R(t_{n-1}) && \text{for } c_{\min} \le c_1 \le c_{\max} \\ &= c_{\max} - P(t_n) - I(t_n) && \text{for } c_1 > c_{\max} \end{aligned} \qquad (22.39)$$

Having evaluated the remainder, R, it is now possible to combine equations (22.23) and (22.28) so as to produce the defining equation for the fully desaturated controller output:

$$c(t_n) = P(t_n) + I(t_n) + R(t_n) \qquad (22.40)$$

As noted above, the relationship (22.35) may be implemented in a continuous system simulation language with a delay function, using a small time delay, nominally the same as the controller timestep, although in practice some latitude may be allowed to avoid problems of stiffness. However, some languages have difficulty in implementing delay functions accurately and reliably, especially if they are using a variable-step-length, stiff-equation-solver. In such a case, the time-delay relation of (22.35) may be approximated by a first-order exponential lag:

$$\frac{dR_1}{dt}(t_n) = \frac{R(t_n) - R_1(t_n)}{\tau_1} \qquad (22.41)$$

where τ_1 is a short time constant. The equation set defining the remainder term, R, is then modified to:

$$\begin{aligned} R(t_n) &= c_{\min} - P(t_n) - I(t_n) \quad && \text{for } c_1 < c_{\min} \\ &= R_1(t_n) && \text{for } c_{\min} \le c_1 \le c_{\max} \\ &= c_{\max} - P(t_n) - I(t_n) && \text{for } c_1 > c_{\max} \end{aligned} \qquad (22.42)$$

22.5.3 Integral desaturation – Type 3

Type 3 integral desaturation is a variant of Type 2. The arrangement is shown in Figure 22.5.

This control structure places greater emphasis on the integral term of the controller through limiting the error used in calculating the proportional term. Such a procedure has the additional effect that once the controller has gone into saturation, it will not come out again until the error has returned to within the error limits set on the error being fed to the proportional term. The conditions for the controller returning from saturation are now (see Appendix 9):

1. the error must have returned to within the range $e_{\min} < e < e_{\max}$, where $e_{\min}$ and $e_{\max}$ are the lower and upper limits set on the error fed to the proportional term;
2. the rate of change of error must satisfy

$$\frac{de}{dt} < -\frac{e}{T_i} \quad \text{for returning from upper limit saturation}$$

$$\frac{de}{dt} > -\frac{e}{T_i} \quad \text{for returning from lower limit saturation}$$

The error fed to the proportional term is now modified to e_m, where

$$e_m = \lim(e_{\min}, e_{\max}, e) \qquad (22.43)$$

so that the modified proportional term, P_m, is now

$$P_m(t_n) = ke_m(t_n) \qquad (22.44)$$

The modified proportional term now replaces the normal proportional term in equation (22.22), so that the partially desaturated controller output, c_1, is now given by

$$c_1(t_n) = P_m(t_n) + I_D(t_n) \qquad (22.45)$$

The desaturated integral term remains as set down in equation (22.20), repeated below:

$$I_D(t_n) = I_A(t_{n-1}) + \int_{t_{n-1}}^{t_n} \frac{ke}{T_i}\, dt \qquad (22.20)$$

but the adjusted integral term must be changed to reflect the fact that the proportional term has now been modified:

$$\begin{aligned} I_A(t_n) &= c_{\min} - P_m(t_n) \quad && \text{for } c_1(t_n) < c_{\min} \\ &= I_D(t_n) && \text{for } c_{\min} \le c_1(t_n) \le c_{\max} \\ &= c_{\max} - P_m(t_n) && \text{for } c_1(t_n) > c_{\max} \end{aligned} \qquad (22.46)$$

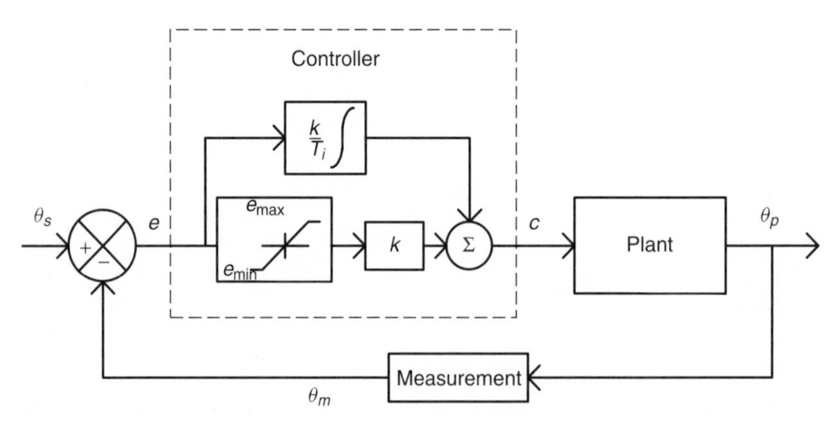

Figure 22.5 P + I controller with Type 3 integral desaturation.

The fully desaturated controller output is now given by the sum of the modified proportional term and the new form of the adjusted integral term:

$$c(t_n) = P_m(t_n) + I_A(t_n) \tag{22.47}$$

To implement this form of integral desaturation within a standard continuous simulation language, we introduce once again the concept of the remainder, R, just as we did in the case of Type 2 integral desaturation. The analysis then mirrors that of Section 22.5.2, and this gives the following equation set for R:

$$\begin{aligned} R(t_n) &= c_{\min} - P_m(t_n) - I(t_n) && \text{for } c_1 < c_{\min} \\ &= R(t_{n-1}) && \text{for } c_{\min} \le c_1 \le c_{\max} \\ &= c_{\max} - P_m(t_n) - I(t_n) && \text{for } c_1 > c_{\max} \end{aligned} \tag{22.48}$$

The defining equation for the fully desaturated controller output (cf. equation (22.40)) is then:

$$c(t_n) = P_m(t_n) + I(t_n) + R(t_n) \tag{22.49}$$

As before, it is possible to model the behaviour of R by a first-order exponential lag if necessary:

$$\frac{dR_1}{dt}(t_n) = \frac{R(t_n) - R_1(t_n)}{\tau_1} \tag{22.50}$$

where τ_1 is a short time constant, in which case the equation set (22.48) defining the remainder term, R, is modified to:

$$\begin{aligned} R(t_n) &= c_{\min} - P_m(t_n) - I(t_n) && \text{for } c_1 < c_{\min} \\ &= R_1(t_n) && \text{for } c_{\min} \le c_1 \le c_{\max} \\ &= c_{\max} - P_m(t_n) - I(t_n) && \text{for } c_1 > c_{\max} \end{aligned} \tag{22.51}$$

22.6 The dynamics of control valve travel

The final control element in a process plant is usually a control valve, the opening and closing of which is governed normally by a piston actuator. The actuator drives the valve stem to a position dictated by the signal, c, that it receives from the controller. As explained in Chapter 7, Sections 7.2 and 7.7, we use the term 'valve travel' to characterize the position of the piston actuator. The valve travel, x, is taken to be a normalized variable, with zero indicating that the valve opening is at its minimum and unity indicating that the valve is fully open. Figure 22.6 expands the section of the plant shown in Figures 22.1 to 22.5, linking the controller output, c, to the plant variable, θ_p.

The valve travel is lagged on the controller output, while the valve opening is a function, linear or nonlinear, of the valve travel. The change in valve opening effects a change in flow (which may be calculated using the methods of Chapters 7, 8, 9 and 10), and the balance of plant responds, causing a change in the plant variable, θ_p.

The dynamics of the valve-positioning system may be described in terms of its small-signal and large-signal characteristics. The small-signal characteristics may be represented to reasonable accuracy by a first-order exponential lag, which will obey the canonical equation:

$$\tau_v \frac{dx}{dt} + x = c \tag{22.52}$$

where τ_v is the valve's small-signal time constant (seconds). Equation (22.52) may be rearranged to give

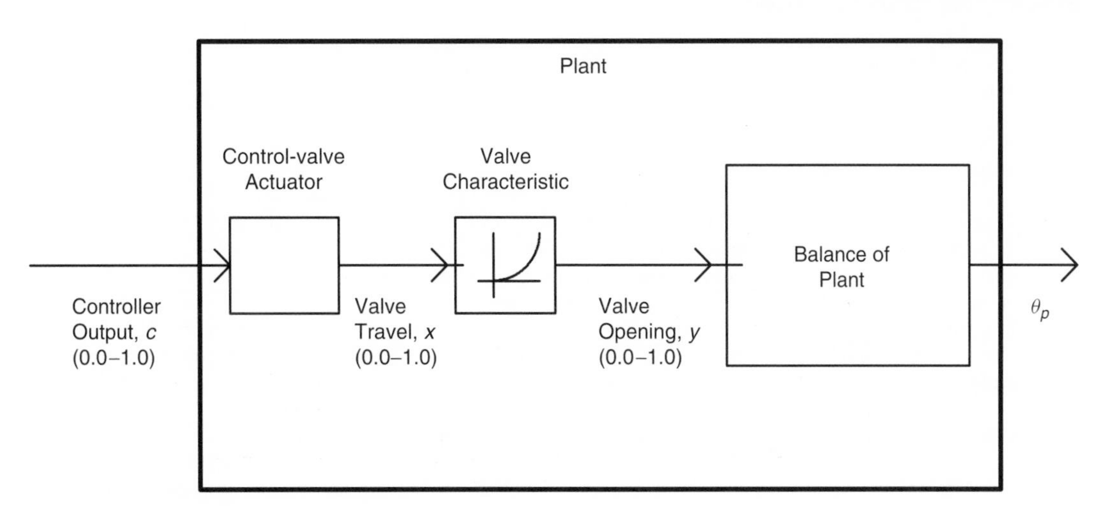

Figure 22.6 Block diagram showing the interface of the controller to the control-valve actuator.

the derivative of valve position explicitly as:

$$\frac{dx}{dt} = \frac{c - x}{\tau_v} \tag{22.53}$$

The large-signal characteristics of the valve-positioning system are dominated by the maximum speed/minimum stroke time of the valve. We may calculate the maximum speed by dividing the range of valve travel from fully closed to fully open by the minimum stroke time. Giving the minimum stroke time the symbol τ_{stroke}, and using the convention of normalized valve travel, the maximum velocity of the valve actuator will thus be

$$\left|\frac{dx}{dt}\right|_{max} = \frac{1}{\tau_{stroke}} \tag{22.54}$$

The transition from small-signal to large-signal behaviour occurs when the velocity predicted by equation (22.53) exceeds the limiting value given by equation (22.54). We may thus simulate the behaviour of valve travel easily using the 'lim' function introduced in Section 22.2, equation (22.2):

$$\frac{dx}{dt} = \lim\left(-\frac{1}{\tau_{stroke}}, \frac{1}{\tau_{stroke}}, \frac{c - x}{\tau_v}\right) \tag{22.55}$$

Valve manufacturers may usually be prevailed upon to provide the stroke time, τ_{stroke}, which is easily measured. The small-signal time constant, τ_v, being a little bit more difficult to measure, is less readily available. In the absence of other data, it may usually be taken as between 10 and 20% of the stroke time.

22.7 Modelling static friction: the velocity deadband method

Section 22.6 has accounted for one important nonlinearity, namely the velocity limit of the valve. Another nonlinearity that can cause control problems is static friction, sometimes referred to as 'stiction', although more properly known as Coulomb friction. Static friction acts to prevent or impede relative motion by opposing the force applied. Its effect on valve movement may be measured as the difference between the valve's demanded travel (equal to the normalized controller output) and the actual valve travel seen on the plant. This difference will lie normally in the normalized range 0.001 to 0.005 (0.1 to 0.5% of total valve travel) for valves fitted with a valve positioning system, although the author has had experience of an important control valve with a value measured at 0.0065. The situation is generally significantly worse for pneumatic valves not fitted with a positioning system, where the likely range is 0.005 to 0.05.

Experiments have shown that a force applied to two initially static objects in touching contact will encounter a countervailing frictional force that will prevent relative motion until a limiting value is reached. Motion will occur once this value has been exceeded, but it will still be opposed by a frictional force. This frictional force will be independent of velocity and will be only slightly less than the limiting value.

To model the effect of static friction on valve stem movement, we begin by noting that the driving function for valve motion is given by equation (22.53) as the difference between the controller output and current valve travel. This equation, which takes no account of static friction, may be plotted as the straight-line graph of Figure 22.7.

Static friction comes into play by requiring that the difference, $c - x$, exceed a certain value, b, before the valve stem can move. The characteristics of friction after motion has begun lead us to conclude that the driving function will be reduced by approximately the same amount, b, after movement has begun. This implies that Figure 22.7 should be modified to the form shown in Figure 22.8.

The equations describing Figure 22.8 are:

$$\begin{aligned}\frac{dx}{dt} &= \frac{c - x - b}{\tau_v} && \text{for } b < c - x;\ \text{or } x < c - b \\ &= 0 && \text{for } -b \le c - x \le b; \\ & && \text{or } c - b \le x \le c + b \\ &= \frac{c - x + b}{\tau_v} && \text{for } c - x < -b;\ \text{or } x > c + b\end{aligned} \tag{22.56}$$

The conditions in equation (22.56) are presented in terms of both the difference term, $c - x$, and the current valve travel, x, with the latter being often a more convenient form for application.

The derivative calculated using equation (22.56) is subject to the limitation on maximum velocity of equation (22.54), which may be implemented using a 'lim' function as before:

$$\frac{dx}{dt} = \lim\left(-\frac{1}{\tau_{stroke}}, \frac{1}{\tau_{stroke}}, \left.\frac{dx}{dt}\right|_{Eq.22.56}\right) \tag{22.57}$$

Simulation languages normally include a number of input/output nonlinear blocks as part of the package. One such is the deadband function, and it is immediately apparent from inspection of Figure 22.8 that the relationship between valve-stem velocity and demanded/actual position difference, $c - x$, could be programmed using a deadband block as an alternative to programming equation (22.56). This is the basis for naming this method of describing static friction the 'velocity deadband method'.

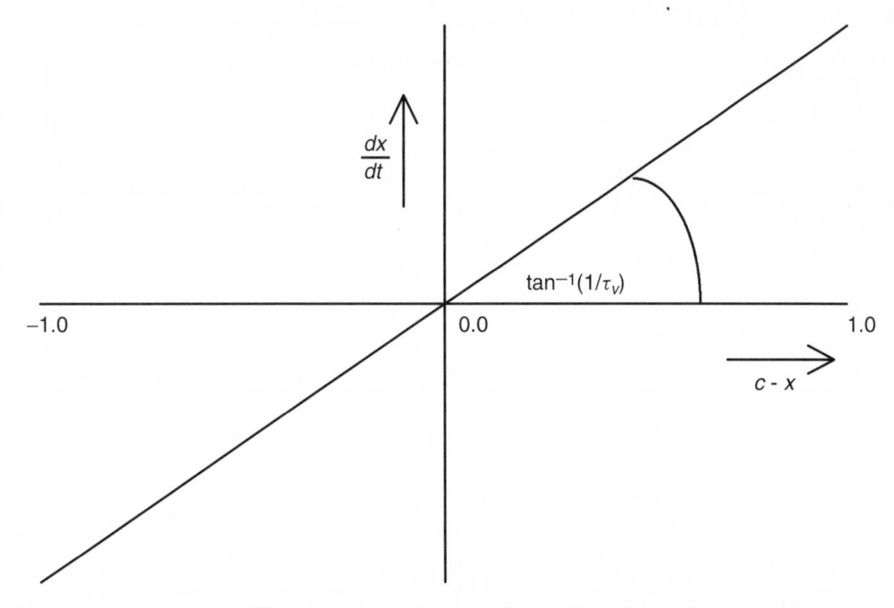

Figure 22.7 Valve velocity versus the difference between the controller output and current valve travel, when static friction is ignored.

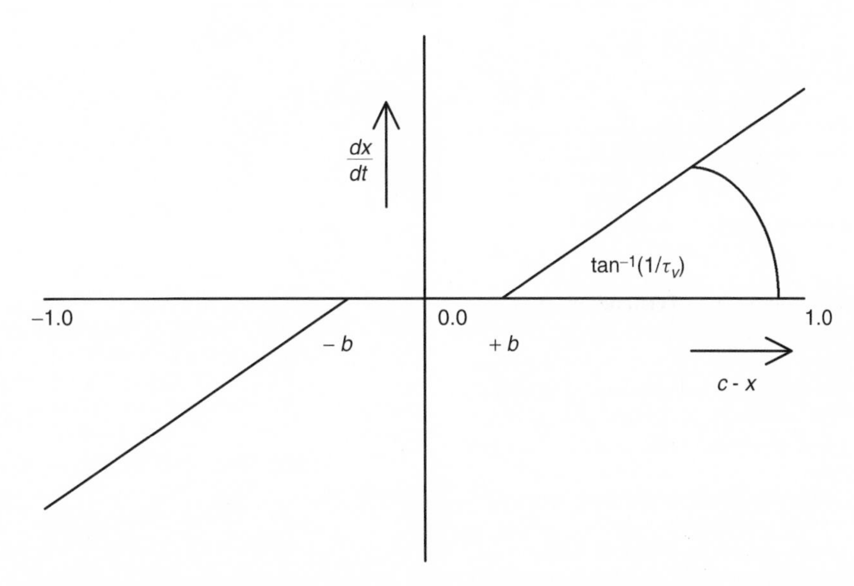

Figure 22.8 Valve velocity versus the difference between the controller output and current valve travel, allowing for static friction.

22.8 Using nonlinearity blocks: the backlash description of valve static friction

Nonlinear blocks can be very useful aids to programming controller functions, and their use will be discussed in this section. The backlash function will be discussed in some detail because it offers an alternative way of programming the effect of static friction.

The nonlinear block is often inserted in the simulation between the output of the controller and the input to the control-valve actuator, as shown in Figure 22.9. The nonlinear block produces an output, z, in response

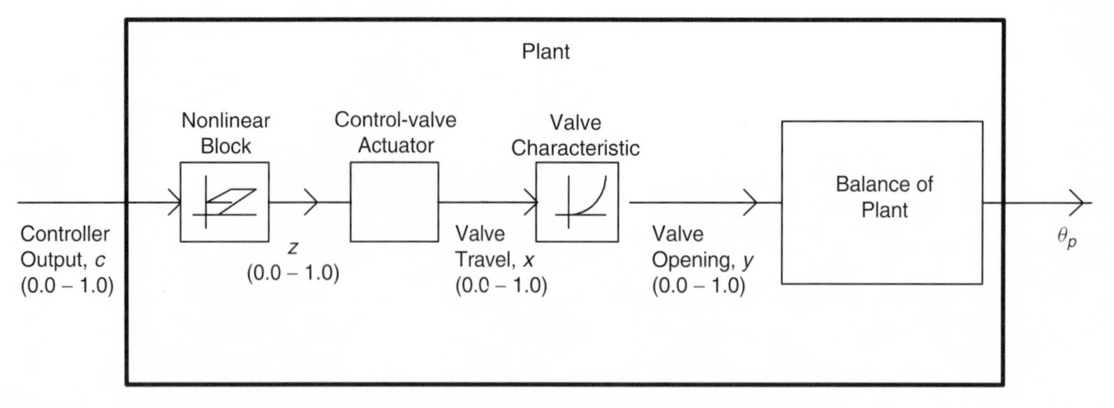

Figure 22.9 Introducing a nonlinear block between the controller and the control-valve actuator.

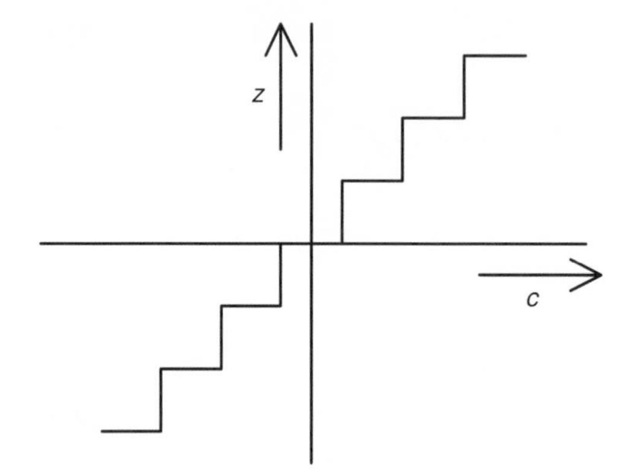

Figure 22.10 Quantizer nonlinear block.

to the controller signal fed to it, and the new signal, z, now governs the action of the control-valve actuator.

A simple nonlinear block is that of quantization, which would be useful in simulating the effect of a stepper motor. Figure 22.10 shows such a characteristic.

The valve travel, x, will follow the quantizer output, z, after an exponential time lag.

The backlash nonlinearity is shown in Figure 22.11. Such a characteristic is typically found in gear trains and mechanical linkages. As noted above it offers also an alternative way of modelling the effects of static friction on a valve positioning system. Any change in the controller output, c, will produce a proportional change in the backlash output, z, as long as the controller output continues to move in the same direction. If the controller output should reverse, however, there will be no change in the output of the backlash element until the backlash, of width $2b$, has been traversed. The graph of backlash is more complicated than that of quantization because it is multivalued, and there is an implicit dependence on past history and hence time. The backlash output depends not only on the signal from the controller but also on the value of the backlash output just past.

The graph illustrated in Figure 22.11 may be described by the following time-dependent equations:

$$\begin{aligned} z(t) &= c(t) - b \quad \text{if } z(t^-) < c(t) - b \\ &= z(t^-) \quad \text{if } c(t) - b \le z(t^-) \le c(t) + b \\ &= c(t) + b \quad \text{if } z(t^-) > c(t) + b \end{aligned} \tag{22.58}$$

Here t^- signifies the time immediately before time t. Figure 22.11 was constructed by applying the rules of equation (22.58) and tracing the behaviour of the backlash output from its initial value of zero as the controller output was increased from zero to unity and then cycled at intermediate levels. The half-width of the backlash, b, was set to the high value

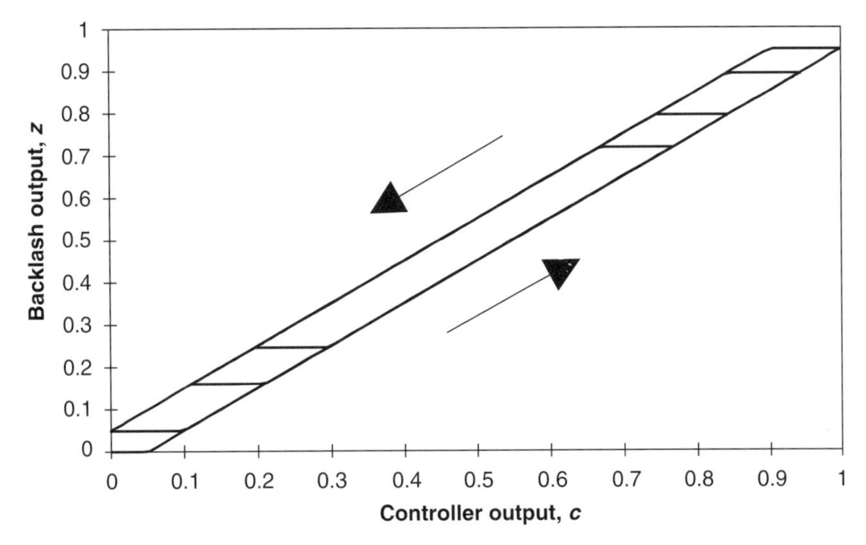

Figure 22.11 Backlash nonlinearity, with backlash width, 2*b*, = 0.1.

of 0.05 in order to make the structure of backlash clear. [Note the initial behaviour of z, shown in the extreme lower left-hand corner of the graph. Here backlash output, z, remains at zero until controller output, c, reaches 0.05, and then climbs linearly to 0.05 as c rises to 0.1. This portion of the graph can be traversed only once, since it is impossible thereafter to reduce z below the value of 0.05 given that the lowest value permitted for controller output, c, is 0.0.]

The control valve actuator will now respond to the backlash output. Assuming that the valve's velocity limit is not transgressed, the equation governing valve travel in a system with backlash is found by replacing c with z in equation (22.53):

$$\frac{dx}{dt} = \frac{z - x}{\tau_v} \tag{22.59}$$

We may substitute from equation (22.58) for z in equation (22.59) to find the following conditions for the value of dx/dt:

$$\begin{aligned} \frac{dx}{dt} &= \frac{c(t) - b - x(t)}{\tau_v} && \text{if } z(t^-) < c(t) - b \\ &= \frac{z(t^-) - x(t)}{\tau_v} && \text{if } c(t) - b \le z(t^-) \le c(t) + b \\ &= \frac{c(t) + b - x(t)}{\tau_v} && \text{if } z(t^-) > c(t) + b \end{aligned} \tag{22.60}$$

Now the valve travel follows the output of the backlash element, subject to an exponential time delay. Providing the valve time constant, τ_v, is small when judged against the frequency content of the signals driving the valve, the valve travel will represent an excellent approximation to the value the backlash output immediately past, i.e.:

$$x(t) \approx z(t^-) \tag{22.61}$$

Substituting from equation (22.61) into equation (22.60) gives:

$$\begin{aligned} \frac{dx}{dt} &= \frac{c(t) - b - x(t)}{\tau_v} && \text{if } x(t) < c(t) - b \\ &= 0 && \text{if } c(t) - b \le x(t) \le c(t) + b \\ &= \frac{c(t) + b - x(t)}{\tau_v} && \text{if } x(t) > c(t) + b \end{aligned} \tag{22.62}$$

which is, of course, mathematically identical to equation (22.56), the 'velocity deadband' description of valve static friction. This demonstrates that the combination of a backlash nonlinear block followed by an exponential lag provides an alternative way of modelling static friction in a valve, one that is broadly equivalent to the velocity deadband method.

It is interesting to compare the backlash description of static friction with the velocity deadband model. Equation (22.56) was programmed for the velocity deadband method, while equations (22.58) and (22.59) were programmed for the backlash description. The half-deadband/half-backlash width, b, was chosen as the largest likely value of 0.05. The two models were subjected to a controller output consisting of a damped oscillatory output of period 150 seconds. The exciting disturbance was assumed to be so slow that the valve's velocity limit would not be invoked, and no velocity limit was programmed for the valve in consequence.

Figure 22.12 shows the controller output and the valve travel calculated by both methods when the valve small-signal time constant was chosen as 3 seconds. Each model predicts a flat-topped response for valve travel, and the two predictions are indistinguishable from eachother on the scale of the graph, which is in line with expectation when the valve time constant is small.

Increasing the valve time constant to 25 seconds, one sixth of the period of the driving oscillation produced the results shown in Figure 22.13. The requirement of equation (22.61) is not met so well. Although the two models give generally similar results, there is now a visible difference between them, with the backlash model carrying out less 'peak-lopping.'

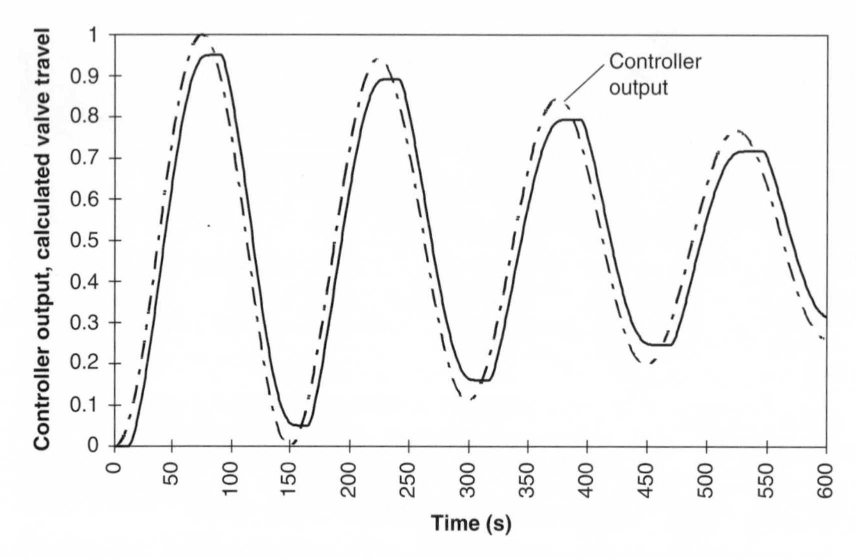

Figure 22.12 Simulated valve travel: backlash versus velocity deadband models. Valve small-signal time constant = 3 seconds.

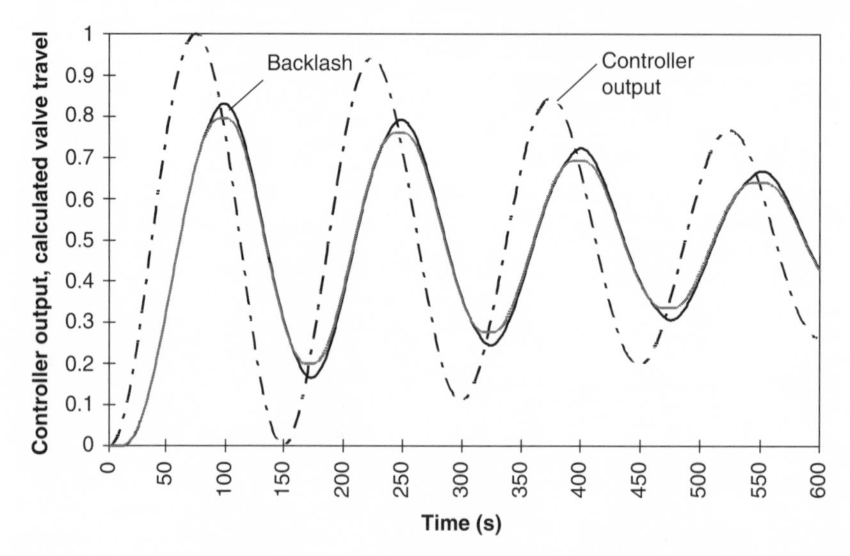

Figure 22.13 Simulated valve travel: backlash versus velocity deadband models. Valve small-signal time constant = 25 seconds.

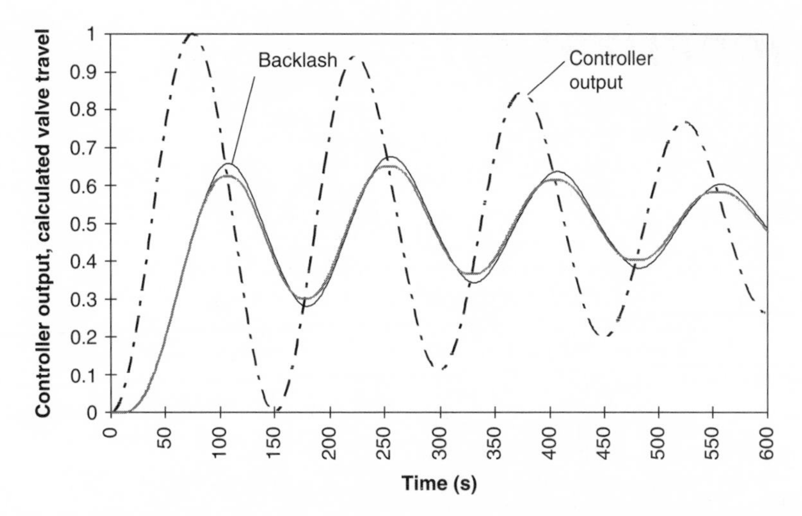

Figure 22.14 Simulated valve travel: backlash versus velocity deadband models. Valve small-signal time constant = 50 seconds.

Interestingly, however, the difference between the two models does not grow as the valve time constant is increased further. Figure 22.14 shows the case with the valve time constant set to 50 seconds. The fact that the response of the valve is now very sluggish indeed allows the condition of equation (22.61) to be met reasonably well once more.

22.9 Bibliography

Martin, E.N. (1975). Control valve movement limitations – an essay in practical control engineering, *Transactions Paper 8.75, Measurement and Control*, **8**, 363–370 (T67–T74)

While, C. and Hutchinson, S.J. (1979). Instrument models for process simulation, *Trans. Inst. MC*, **1**, 187–194.

23 Linearization

23.1 Introduction

The engineer engaged in dynamic modelling will very often find it useful to produce a simplified, analytically linearized model of an important part of the plant in addition to his main detailed model. Such a linearized model serves two important purposes:

(i) it affords a verifying check on the main model. The computer coding of the linearized model will of necessity be significantly different from that of the main model, bringing the benefits of 'N-version programming', a recognized technique for producing diverse software, recommended for high-integrity applications such as microprocessor-based protection systems.

(ii) the linearized model can be used for frequency-domain analysis and design of an important plant control system. (Techniques for calculating frequency responses for linear systems are not covered in this book, but are widely available in control engineering texts.)

A number of the more sophisticated dynamic simulation packages offer numerical linearization and some will produce a frequency response from the main model. While this is a very useful feature, the fact that it relies on the coding of the main model means that it cannot afford the verifying correctness-check that a substantially independent model can give. Moreover, the modeller may lose the additional 'feel' that analytical linearization can sometimes afford.

While analytic linearization is not recommended as appropriate in every case, it is a useful additional tool for critical situations. Its use on the difficult problems likely to be met in practice is being made easier now as a result of the availability of standard computer packages for symbolic, algebraic computation.

23.2 Principles of linearization

Analytic linearization relies on the application of a Taylor series, truncated after the first term. Let us consider the general case of a vector function, $\mathbf{g}$, of several variables, $\mathbf{x}$. When $\mathbf{x}$ is displaced by a small deviation $\tilde{\mathbf{x}}$ from a constant, central value $\bar{\mathbf{x}}$,

$$\mathbf{x} = \bar{\mathbf{x}} + \tilde{\mathbf{x}} \tag{23.1}$$

the Taylor series gives the following expansion:

$$\begin{aligned} \mathbf{g}(\mathbf{x}) &= \mathbf{g}(\bar{\mathbf{x}} + \tilde{\mathbf{x}}) \\ &= \bar{\mathbf{g}} + \tilde{\mathbf{g}} \\ &\approx \mathbf{g}(\bar{\mathbf{x}}) + \frac{\partial \mathbf{g}}{\partial \mathbf{x}}\tilde{\mathbf{x}} \end{aligned} \tag{23.2}$$

where the vector function has been decomposed into a constant part and a variable part:

$$\begin{aligned} \bar{\mathbf{g}} &= \mathbf{g}(\bar{\mathbf{x}}) = \text{constant} \\ \tilde{\mathbf{g}} &= \frac{\partial \mathbf{g}}{\partial \mathbf{x}}\tilde{\mathbf{x}} \end{aligned} \tag{23.3}$$

and where

$$\frac{\partial \mathbf{g}}{\partial \mathbf{x}} = \begin{bmatrix} \frac{\partial g_1}{\partial x_1} & \frac{\partial g_1}{\partial x_2} & \cdots & \frac{\partial g_1}{\partial x_n} \\ \frac{\partial g_2}{\partial x_1} & \frac{\partial g_2}{\partial x_2} & \cdots & \frac{\partial g_2}{\partial x_n} \\ \vdots & \vdots & \cdots & \vdots \\ \frac{\partial g_q}{\partial x_1} & \frac{\partial g_q}{\partial x_2} & \cdots & \frac{\partial g_q}{\partial x_n} \end{bmatrix} \tag{23.4}$$

for the case of the vector function, $\mathbf{g}$, having q components, g_i, $i = 1, 2, \ldots, q$, dependent on an $n \times 1$ vector, $\mathbf{x}$. The differentiation $\partial\mathbf{g}/\partial\mathbf{x}$ takes place at $\mathbf{x} = \bar{\mathbf{x}}$: $\partial\mathbf{g}/\partial\mathbf{x} = (\partial\mathbf{g}/\partial\mathbf{x})(\bar{\mathbf{x}})$.

When the vector, $\mathbf{x}$, is differentiated , we have from (23.1):

$$\begin{aligned} d\mathbf{x} &= d\bar{\mathbf{x}} + d\tilde{\mathbf{x}} \\ &= d\tilde{\mathbf{x}} \end{aligned} \tag{23.5}$$

since $\bar{\mathbf{x}}$ is made up of constant values.

Now the basic form of the time-invariant simulation problem (see equation (2.22)) is given by:

$$\frac{d\mathbf{x}}{dt} = \mathbf{f}(\mathbf{x}, \mathbf{u}) \tag{23.6}$$

Applying the principles above, this may be linearized to:

$$\frac{d\mathbf{x}}{dt} = \frac{d\tilde{\mathbf{x}}}{dt} = \mathbf{f}(\bar{\mathbf{x}}, \bar{\mathbf{u}}) + \frac{\partial \mathbf{f}}{\partial \mathbf{x}}\tilde{\mathbf{x}} + \frac{\partial \mathbf{f}}{\partial \mathbf{u}}\tilde{\mathbf{u}} \tag{23.7}$$

If the chosen values, $(\bar{\mathbf{x}}, \bar{\mathbf{u}})$ represent a steady state, then

$$\mathbf{f}(\bar{\mathbf{x}}, \bar{\mathbf{u}}) = \mathbf{0} \tag{23.8}$$

and so (23.7) becomes

$$\frac{d\tilde{\mathbf{x}}}{dt} = \frac{\partial \mathbf{f}}{\partial \mathbf{x}}\tilde{\mathbf{x}} + \frac{\partial \mathbf{f}}{\partial \mathbf{u}}\tilde{\mathbf{u}} \tag{23.9}$$

where $\partial\mathbf{f}/\partial\mathbf{x}$ is an $n \times n$ matrix usually named the $\mathbf{A}$ matrix in control engineering usage:

$$\frac{\partial \mathbf{f}}{\partial \mathbf{x}} = \mathbf{A} = \begin{bmatrix} \frac{\partial f_1}{\partial x_1} & \frac{\partial f_1}{\partial x_2} & \cdots & \frac{\partial f_1}{\partial x_n} \\ \frac{\partial f_2}{\partial x_1} & \frac{\partial f_2}{\partial x_2} & \cdots & \frac{\partial f_2}{\partial x_n} \\ \vdots & \vdots & \cdots & \vdots \\ \frac{\partial f_n}{\partial x_1} & \frac{\partial f_n}{\partial x_2} & \cdots & \frac{\partial f_n}{\partial x_n} \end{bmatrix} \tag{23.10}$$

Similarly $\partial\mathbf{f}/\partial\mathbf{u}$ is an $n \times l$ matrix usually named the $\mathbf{B}$ matrix in control engineering usage:

$$\frac{\partial \mathbf{f}}{\partial \mathbf{u}} = \mathbf{B} = \begin{bmatrix} \frac{\partial f_1}{\partial u_1} & \frac{\partial f_1}{\partial u_2} & \cdots & \frac{\partial f_1}{\partial u_l} \\ \frac{\partial f_2}{\partial u_1} & \frac{\partial f_2}{\partial u_2} & \cdots & \frac{\partial f_2}{\partial u_l} \\ \vdots & \vdots & \cdots & \vdots \\ \frac{\partial f_n}{\partial u_1} & \frac{\partial f_n}{\partial u_2} & \cdots & \frac{\partial f_n}{\partial u_l} \end{bmatrix} \tag{23.11}$$

While the above principles are easily laid out, applying analytical linearization in practice is not a trivial matter. It is for this reason that an extended case study will now be given involving the essential features of a real problem. It is hoped that this will convey a flavour of the detailed approach needed for the relatively complex systems likely to be met on a full-scale industrial plant.

23.3 Example of analytic linearization: the response of liquid flow to valve opening in a pumped liquid system

Figure 23.1 is a schematic showing the features of the system for linearization.

Liquid is pumped from the supply vessel into the discharge vessel via a liquid control valve, which might, for example, be acting to control level in the discharge vessel. In this linearization exercise, however, we shall not concern ourselves with the outer feedback loop, but only with the effect of changes in valve opening on liquid flow.

It will be seen from the figure that a differential pressure control system is present, which regulates the steam supply to a turbine that is directly geared to the pump. The behaviour of this control system will be as follows. An increase in the valve opening will tend to decrease the differential pressure across the feedvalve, and to counteract this, the differential pressure controller will increase the steam flow to the turbine. Allowing more steam into the turbine will increase its speed and hence cause an increase in the speed of the directly coupled pump. A higher pump speed will result in an increased pump discharge pressure, so that the tendency of differential pressure

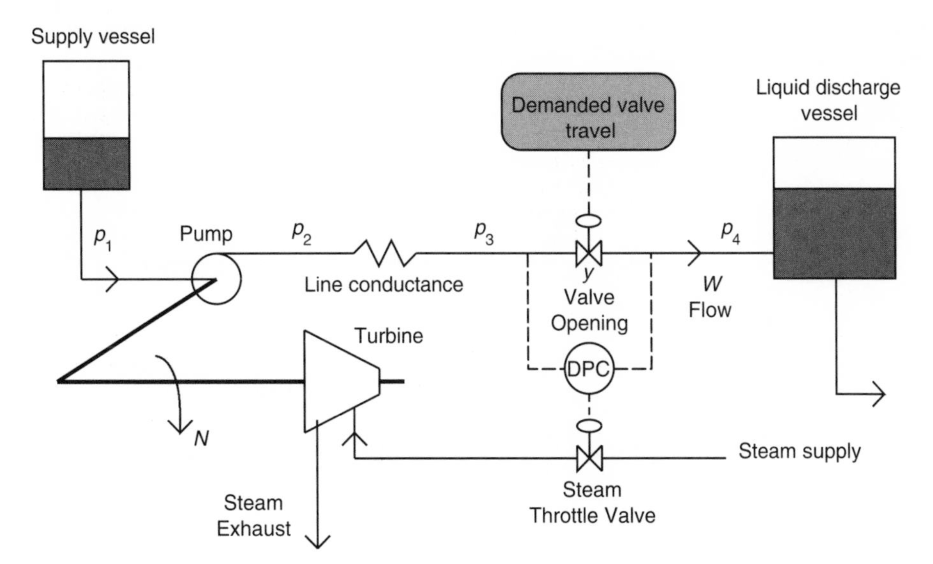

Figure 23.1 Pumped liquid system schematic.

to fall will be halted. Conversely, decreasing valve opening will lead to a temporary increase in differential pressure until the differential pressure controller has had time to reduce the speed of the pump by restricting the steam flow to the turbine.

If the differential pressure control loop were extremely fast, we could assume that the differential pressure across the valve remained constant, and the linearization exercise would be simplified enormously. But we shall examine the case where the differential pressure control loop is slow, and needs to be allowed for in calculating the response of flow to valve opening.

23.4 Response of flow to valve opening when the differential pressure controller is switched out

Let us consider the flow through the liquid control valve, W, which will be given by

$$W = C_V y \sqrt{\frac{p_3 - p_4}{v}} \tag{23.12}$$

where C_V is the valve conductance at fully open (m^2), y is the fractional valve opening, p_3 is the pressure just before the liquid control valve (Pa), p_4 is the pressure in the discharge vessel (Pa), and v is the specific volume of the liquid being pumped (m^3/kg). Assuming that specific volume of the liquid remains constant, and putting

$$\Delta p = p_3 - p_4 \tag{23.13}$$

we see that

$$W = W(y, \Delta p) \tag{23.14}$$

Hence we may expand using a Taylor series to give:

$$W(y, \Delta p) = \overline{W} + \tilde{W} = W(\overline{y}, \Delta\overline{p}) + \frac{\partial W}{\partial y}\tilde{y} + \frac{\partial W}{\partial \Delta p}\Delta\tilde{p} \tag{23.15}$$

so that the variable part of the flow is given by:

$$\tilde{W} = \frac{\partial W}{\partial y}\tilde{y} + \frac{\partial W}{\partial \Delta p}\Delta\tilde{p} \tag{23.16}$$

Assuming for simplicity that the pressure, p_4, of the discharge vessel is unaffected by small changes in feedflow, the differential pressure across the valve will be affected by just two factors: (i) the valve opening, and (ii) the action of the differential pressure control loop acting through the turbine and pump. The linearization principle is to regard these two factors as separate but additive.

Let us consider first the effect of valve opening when the differential pressure control loop is switched out. We may differentiate equation (23.16) with respect to $\tilde{y}$ and obtain the necessary total derivative:

$$\frac{d\tilde{W}}{d\tilde{y}} = \frac{\partial W}{\partial y} + \frac{\partial W}{\partial \Delta p}\frac{d\Delta\tilde{p}}{d\tilde{y}} \tag{23.17}$$

or, in view of equation (23.5),

$$\frac{d\tilde{W}}{d\tilde{y}} = \frac{dW}{dy} = \frac{\partial W}{\partial y} + \frac{\partial W}{\partial \Delta p}\frac{d\Delta p}{dy} \tag{23.18}$$

This form allows us to draw the block diagram for flow to valve opening as shown in Figure 23.2. [This diagram has been designed to incorporate features to take account later of the action of the differential controller when it is switched in. When this controller is functioning, any detected change in differential pressure will cause the controller eventually to produce a compensating addition, $\Delta\tilde{p}_2$, to the differential pressure. The total change in differential pressure, $\Delta\tilde{p}$,

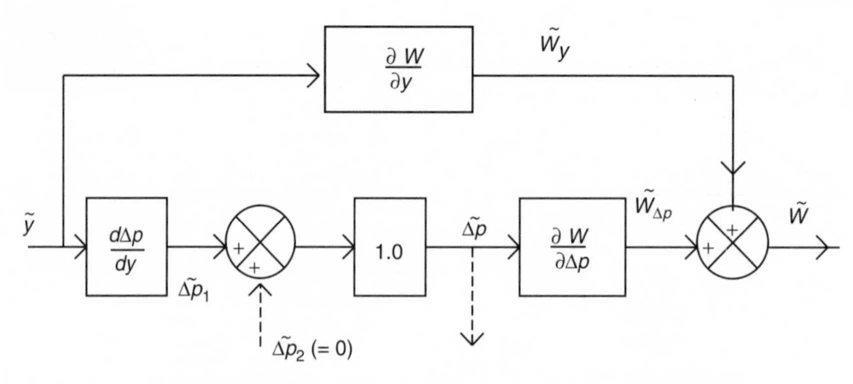

Figure 23.2 Block diagram of liquid flow to valve opening with the pressure controller switched out.

will then be the sum of the change caused directly by the change in valve opening, $\Delta\tilde{p}_1$, and this new term, $\Delta\tilde{p}_2$:

$$\Delta\tilde{p} = \Delta\tilde{p}_1 + \Delta\tilde{p}_2 \tag{23.19}$$

It is this equation that is embodied by the summing circle in the block diagram. In addition, a block with unity gain has been added for emphasis before the stub representing the pick-off leading to the differential pressure controller.]

The change in liquid control valve opening will cause an increase in liquid flow due to the increased flow conductance of the valve, but the increased valve opening will also lead to a decrease in differential pressure, which will tend to reduce the flow. The two contributions have been labelled $\tilde{W}_y$ and $\tilde{W}_{\Delta p}$ to indicate their origin, and the total flow change is the sum of the two:

$$\tilde{W} = \tilde{W}_y + \tilde{W}_{\Delta p} \tag{23.20}$$

It is clear from equation (23.18) that expressions for the partial derivatives, $\partial W/\partial y$ and $\partial W/\partial\Delta p$, and the total derivative $d\Delta p/dy$ need to be derived. It will be seen that the partial derivatives are found easily, but that it is a non-trivial exercise to find the total derivative $d\Delta p/dy$.

23.4.1 To find $\partial W/\partial y$

Differentiation of equation (23.12) with (23.13) with respect to valve opening, y, gives:

$$\begin{aligned}\frac{\partial W}{\partial y} &= C_V\sqrt{\frac{\Delta p}{v}}\\ &= \frac{\overline{W}}{\overline{y}}\end{aligned} \tag{23.21}$$

Thus $\partial W/\partial y$ can be found immediately from the steady-state values of flow and valve opening.

23.4.2 To find $\partial W/\partial\Delta p$

Differentiation of equation (23.12) (combined with equation (23.13)) with respect to Δp gives:

$$\begin{aligned}\frac{\partial W}{\partial\Delta p} &= \frac{1}{2\sqrt{v\Delta p}}C_V y\\ &= \frac{1}{2}\frac{\overline{W}}{\overline{\Delta p}}\end{aligned} \tag{23.22}$$

Hence $\partial W/\partial\Delta p$ can be found immediately from the steady-state values of flow and differential pressure.

23.4.3 To find $d\Delta p/dy$

We begin by rearranging equation (23.17) to give

$$\frac{d\Delta p}{dy} = \frac{\dfrac{dW}{dy} - \dfrac{\partial W}{\partial y}}{\dfrac{\partial W}{\partial\Delta p}} \tag{23.23}$$

Since we have just derived the two partial derivatives, $\partial W/\partial y$ and $\partial W/\partial\Delta p$, we have in equation (23.23) recast the problem in terms of the total derivative of flow to valve opening, dW/dy, when the differential pressure controller is switched out. This last condition implies that the steam throttle valve is held constant, and so, to a good approximation, that the turbine power remains constant. (The power may not stay quite constant because the turbine efficiency depends partly on the ratio of blade speed to steam speed, as described in Chapter 15, and the speed of the turbine and pump combination will undergo a degree of change as the liquid control valve is moved; the effect will, however, be second-order.)

Now from the sections on pumps in Chapter 17, and equations (17.41) and (17.42) in particular, the pumping power, P_P (watts), is related to the power supplied, P_S(watts), by:

$$P_P = Q(p_2 - p_1) = \eta_P P_S \tag{23.24}$$

where the pump efficiency, η_P, is a function of the ratio of volume flow rate to normalized speed, $U = Q(N_0/N)$:

$$\eta_P = f_{P3}(U) \tag{23.25}$$

The pump outlet pressure is equal to the pressure in the discharge vessel plus the differential pressure across the valve, plus the pressure drop through the pipeline:

$$p_2 = p_4 + \Delta p + \Delta p_L \tag{23.26}$$

where the pressure drop across the valve is given from equation (23.12) as:

$$\Delta p = \frac{v}{C_V^2 y^2}W^2 = \frac{1}{vC_V^2 y^2}Q^2 = a_V\frac{Q^2}{y^2} \tag{23.27}$$

where

$$a_V = \frac{1}{vC_V^2} \tag{23.28}$$

Similarly, applying the model of Chapter 4, Section 4.11, with height differences neglected, the line pressure drop is related to flow by:

$$\begin{aligned}\Delta p_L = p_2 - p_3 &= \frac{v}{C_L^2}W^2\\ &= \frac{1}{vC_L^2}Q^2 = a_L Q^2\end{aligned} \tag{23.29}$$

where

$$a_L = \frac{1}{vC_L^2} \tag{23.30}$$

Substituting into equation (23.24) from (23.25), (23.26), (23.27) and (23.29) gives the equation between power dissipated and power supplied as:

$$(p_4 - p_1)Q + a_L Q^3 + a_V \frac{Q^3}{y^2} = f_{P3}(U)P_S \tag{23.31}$$

We now differentiate equation (23.31) with respect to valve opening, y, with the following held constant:

$$\begin{aligned} p_1 &= \text{constant} \\ p_4 &= \text{constant} \\ P_S &= \text{constant} \end{aligned} \tag{23.32}$$

to give:

$$(p_4 - p_1)\frac{dQ}{dy} + 3a_L Q^2 \frac{dQ}{dy} + 3a_V \frac{Q^2}{y^2}\frac{dQ}{dy} - 2a_V \frac{Q^3}{y^3} = P_S \frac{df_{P3}}{dU}\frac{dU}{dy} \tag{23.33}$$

Using the definition of $U = Q(N_0/N)$, we may differentiate with respect to y and obtain

$$\begin{aligned} \frac{dU}{dy} &= \frac{N_0}{N}\frac{dQ}{dy} - Q\frac{N_0}{N^2}\frac{dN}{dQ}\frac{dQ}{dy} \\ &= \left(\frac{N_0}{N} - \frac{U}{N}\frac{dN}{dQ}\right)\frac{dQ}{dy} \end{aligned} \tag{23.34}$$

Accordingly, substituting into (23.33) and rearranging gives the total derivative of volume flow to valve opening as:

$$\frac{dQ}{dy} = \frac{2a_V \dfrac{Q^3}{y^3}}{\left(p_4 - p_1 + 3\left(a_L + \dfrac{a_V}{y^2}\right)Q^2 + P_S \dfrac{df_{P3}}{dU}\left(\dfrac{U}{N}\dfrac{dN}{dQ} - \dfrac{N_0}{N}\right)\right)} \tag{23.35}$$

The pump efficiency function, $f_{P3}(U)$, is frequently available as a low-order polynomial in U, in which case the differentiation df_{P3}/dU is a simple matter. Alternatively, the differentiation may be carried out graphically by finding the tangent to the $f_{P3}(U)$ vs. U curve. However, before we can solve equation (23.35), we still need to find the total derivative of pump speed to pump volume flow, dN/dQ, when the power supplied is kept constant. To do this we proceed as follows.

The pump head, H, at speed, N, is given by equation (17.28), re-expressed below as a function of the parameter $U = Q(N_0/N)$:

$$H = \frac{N^2}{N_0^2} f_{P1}(U) \tag{23.36}$$

The pressure rise across the pump is given by multiplying the head by the acceleration due to gravity, g, and dividing by the specific volume, v, of the liquid being passed. Hence

$$p_2 - p_1 = \frac{g}{v}\frac{N^2}{N_0^2} f_{P1}(U) \tag{23.37}$$

Substituting for $(p_2 - p_1)$ back into (23.24) from (23.37) and using also (23.25) gives:

$$P_S f_{P3}(U) = \frac{g}{v} Q \frac{N^2}{N_0^2} f_{P1}(U) \tag{23.38}$$

We now differentiate equation (23.38) with respect to volume flow, Q, under the condition that the power supplied by the turbine, P_S, is held constant:

$$P_S \frac{df_{P3}}{dU}\frac{dU}{dQ} = \frac{g}{v}\frac{N^2}{N_0^2} f_{P1} + \frac{g}{v} Q \frac{N^2}{N_0^2}\frac{df_{P1}}{dU}\frac{dU}{dQ} + 2\frac{g}{v} Q \frac{N}{N_0^2} f_{P1}\frac{dN}{dQ} \tag{23.39}$$

or, using equation (23.38):

$$P_S \frac{df_{P3}}{dU}\frac{dU}{dQ} = P_S \frac{f_{P3}}{Q} + P_S \frac{f_{P3}}{f_{P1}}\frac{df_{P1}}{dU}\frac{dU}{dQ} + 2P_S \frac{f_{P3}}{N}\frac{dN}{dQ} \tag{23.40}$$

so that

$$\left(\frac{df_{P3}}{dU} - \frac{f_{P3}}{f_{P1}}\frac{df_{P1}}{dU}\right)\frac{dU}{dQ} - \frac{f_{P3}}{Q} = 2\frac{f_{P3}}{N}\frac{dN}{dQ} \tag{23.41}$$

Since $U = Q(N_0/N)$, it follows that

$$\begin{aligned} \frac{dU}{dQ} &= \frac{N_0}{N} - \frac{QN_0}{N^2}\frac{dN}{dQ} \\ &= \frac{N_0}{N} - \frac{U}{N}\frac{dN}{dQ} \end{aligned} \tag{23.42}$$

This relationship may be substituted into equation (23.41) and, after rearrangement, the required total derivative, dN/dQ, emerges as:

$$\frac{dN}{dQ} = \frac{N}{Q}\frac{\left(\dfrac{U}{f_{P3}}\dfrac{df_{P3}}{dU} - \dfrac{U}{f_{P1}}\dfrac{df_{P1}}{dU} - 1\right)}{\left(\dfrac{U}{f_{P3}}\dfrac{df_{P3}}{dU} - \dfrac{U}{f_{P1}}\dfrac{df_{P1}}{dU} + 2\right)} \tag{23.43}$$

We may now use equation (23.35) to show that:

$$\frac{U}{N}\frac{dN}{dQ}-\frac{N_0}{N}$$

$$=\frac{N_0}{N}\left(\frac{\left(\frac{U}{f_{P3}}\frac{df_{P3}}{dU}-\frac{U}{f_{P1}}\frac{df_{P1}}{dU}-1\right)}{\left(\frac{U}{f_{P3}}\frac{df_{P3}}{dU}-\frac{U}{f_{P1}}\frac{df_{P1}}{dU}+2\right)}-1\right)$$

$$=\frac{N_0}{N}\frac{\frac{U}{f_{P3}}\frac{df_{P3}}{dU}-\frac{U}{f_{P1}}\frac{df_{P1}}{dU}-1-\frac{U}{f_{P3}}\frac{df_{P3}}{dU}+\frac{U}{f_{P1}}\frac{df_{P1}}{dU}-2}{\left(\frac{U}{f_{P3}}\frac{df_{P3}}{dU}-\frac{U}{f_{P1}}\frac{df_{P1}}{dU}+2\right)}$$

$$=\frac{-3\frac{N_0}{N}}{\left(\frac{U}{f_{P3}}\frac{df_{P3}}{dU}-\frac{U}{f_{P1}}\frac{df_{P1}}{dU}+2\right)} \qquad (23.44)$$

Substituting from equation (23.44) into equation (23.35) gives the total derivative of volume flow to valve opening:

$$\frac{dQ}{dy}=\frac{2a_V\frac{Q^3}{y^3}}{\left(p_4-p_1+3\left[\left(a_L+\frac{a_V}{y^2}\right)Q^2-\frac{N_0}{N}P_S\times\frac{\frac{df_{P3}}{dU}}{\left(\frac{U}{f_{P3}}\frac{df_{P3}}{dU}-\frac{U}{f_{P1}}\frac{df_{P1}}{dU}+2\right)}\right]\right)} \qquad (23.45)$$

Hence the total derivative of mass flow to fractional valve opening is given by:

$$\frac{dW}{dy}=\frac{\frac{2}{v^2C_V^2}\frac{\overline{Q}^3}{\overline{y}^3}}{\left(p_4-p_1+3\left[\left(\frac{1}{C_L^2}+\frac{1}{C_v^2\overline{y}^2}\right)\frac{\overline{Q}^2}{v}-\frac{N_0}{N}\overline{P}_S\times\frac{\frac{df_{P3}}{dU}}{\left(\frac{\overline{U}}{\overline{f}_{P3}}\frac{df_{P3}}{dU}-\frac{\overline{U}}{\overline{f}_{P1}}\frac{df_{P1}}{dU}+2\right)}\right]\right)} \qquad (23.46)$$

where equations (23.38) and (23.30) have been used to eliminate a_v and a_L and where the superimposed bar implies the value at the steady-state condition. The derivatives df_{P1}/dU and df_{P3}/dU are obviously taken at this condition also.

Equation (23.46) represents the change in flow to valve opening when no action is taken to control the differential pressure across the valve – a useful result in itself. Further, now that dw/dy has been found, substituting in equation (23.23) allows the total derivative $d\Delta p/dy$ to be found. Thus we are able to evaluate fully the response of the system with the differential pressure controller switched out, as shown in Figure 23.2.

23.5 Including the effect of the differential pressure controller

The fact that we have chosen a block diagram approach means that we may now proceed to evaluate also the response of flow to valve opening with the differential pressure controller switched in. Figure 23.2 includes stubs to allow for the change in differential pressure brought about by the action of the differential pressure controller. We will now connect blocks to those stubs and thus include the differential pressure control loop. The new blocks are shown in Figure 23.3.

The first block represents the time lag in measuring differential pressure. The measured value is then compared with the setpoint to form an error signal, which is supplied to the differential pressure controller. The output from the controller is a demand for steam throttle travel. The next block represents the response of turbine power to steam throttle demanded valve travel. This produces a signal of power change that is converted by the block following into a change in pump speed. Finally, the change in pump speed will cause the change in differential pressure due to the differential pressure control loop. This last process is assumed instantaneous because of the short time constants associated with flow establishment.

We shall deal with each of the elements in turn.

23.5.1 Δp measurement

The measurement of differential pressure will be an inherently linear operation, subject to an exponential lag of typically about 0.5 seconds. The mathematical description is thus:

$$T_m\frac{d\Delta\tilde{p}_m}{dt}+\Delta\tilde{p}_m=\Delta\tilde{p} \qquad (23.47)$$

It is possible to take the Laplace transform to give the transfer function in terms of the Laplace

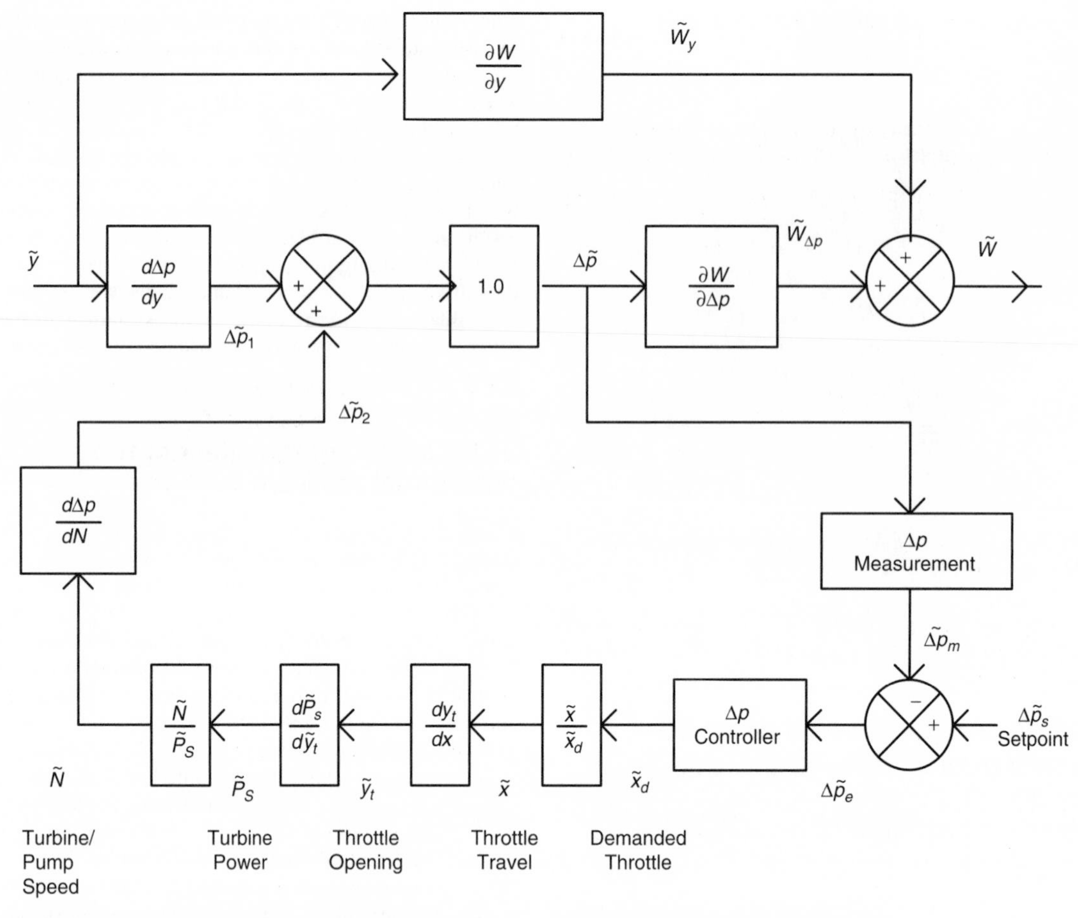

Figure 23.3 Block diagram of liquid flow to valve opening, with differential pressure controller included.

variable s as:

$$\frac{\Delta\tilde{p}_m}{\Delta\tilde{p}}(s) = \frac{1}{1+sT_m} \tag{23.48}$$

23.5.2 Δp comparator

In the linearization we can assume that no change in differential pressure setpoint occurs, so that

$$\Delta\tilde{p}_s = 0 \tag{23.49}$$

Accordingly the effect of the comparator is merely to multiply the incoming signal by a gain of -1:

$$\Delta p_e = -\Delta p_m \tag{23.50}$$

23.5.3 Δp controller

This again is an inherently linear process for small deviations. A $P+I$ controller will obey the mathematical description:

$$x_d = k\Delta p_e + \int_{t_0}^{t} \frac{k}{T_i}\Delta p_e\, dt + I(t_0) \tag{23.51}$$

where x is the demanded valve travel, k is the controller gain, T_i is the integral action time, and $I(t_0)$ is the integrator initial condition. To linearize this equation, we note that $\Delta p_e = 0$ in the steady state, so that

$$\bar{x}_d = I(t_o) \tag{23.52}$$

and the equation for deviations from the steady conditions is simply

$$\tilde{x}_d = k\Delta\tilde{p}_e + \int_{t_0}^{t} \frac{k}{T_i}\Delta\tilde{p}_e\, dt \tag{23.53}$$

Taking the Laplace transform gives the transfer function of the controller as:

$$\frac{\tilde{x}_d}{\Delta \tilde{p}_e} = \frac{k}{T_i}\left(\frac{1 + sT_i}{s}\right) \tag{23.54}$$

23.5.4 The response of steam throttle travel to demanded throttle travel, $\tilde{x}/\tilde{x}_d$

A valve may usually be modelled as an exponential lag, subject to rate limits. For small-signal, linear analysis, the rate limits will not be breached, so the model is simply:

$$T_v \frac{d\tilde{x}}{dt} + \tilde{x} = \tilde{x}_d \tag{23.55}$$

which has the Laplace transform:

$$\frac{\tilde{x}}{\tilde{x}_d} = \frac{1}{1 + sT_v} \tag{23.56}$$

23.5.5 Steam throttle opening to steam throttle travel, dy_t/dx

This derivative may be found from the valve characteristic. For example, for a linear valve, $dy_t/dx = 1$, while for a square-law valve, $dy_t/dx = 2\overline{x}$.

23.5.6 Response of turbine power to steam throttle opening, $d\tilde{P}_S/d\tilde{y}_t$

The fast dynamics of the turbine allow us to assume that the change in power is instantaneous. The power of the turbine may be written as:

$$P_S = W_{in} \sum_{i=1}^{n} \frac{W_i}{W_{in}} \eta_{s,i} D_i \tag{23.57}$$

where W_{in} is the steam flow through the steam throttle, W_i is the steam flow through the ith stage of the n stages making up the steam turbine, $\eta_{s,i}$ is the stage efficiency and D_i is the isentropic enthalpy drop across the stage. To a reasonable approximation, the stage flow ratios, W_i/W_{in}, the stage efficiencies, $\eta_{s,i}$, and the stage enthalpy drops, D_i, will be constant over a large range of conditions for a Rateau turbine. Assuming that the turbine is of this type, we may differentiate (23.57) with respect to W_{in} to give:

$$\frac{dP_S}{dW_{in}} \approx \frac{\overline{P}_S}{\overline{W}_{in}} \tag{23.58}$$

The inlet steam flow is given by the equation

$$W_{in} = y_t f(p_{\text{up}}, v_{\text{up}}, p_{\text{down}}) \tag{23.59}$$

where the subscript 'up' refers to steam conditions upstream of the throttle valve, while 'down' refers to conditions downstream. If we assume that these conditions do not change (a good assumption given the pressure stability of a boiler), then we may differentiate to give:

$$\frac{dW_{in}}{dy} = \frac{\overline{W}_{in}}{\overline{y}_t} \tag{23.60}$$

Combining (23.58) and (23.60) gives the required derivative:

$$\frac{dP_S}{dy_t} = \frac{\overline{P}_S}{\overline{W}_{in}} \frac{\overline{W}_{in}}{\overline{y}_t} = \frac{\overline{P}_S}{\overline{y}_t} \tag{23.61}$$

23.5.7 The response of turbine/pump speed to turbine power, $\tilde{N}/\tilde{P}_S$

The equation describing the dynamics of pump speed follows from equations (17.44) and (17.41) in Chapter 17:

$$4\pi^2 JN \frac{dN}{dt} = \eta_P P_S - P_P \tag{23.62}$$

where J is the moment of inertia of the pump and turbine combined. Substituting for the pumping power, P_P, from equation (17.32):

$$P_P = \frac{g}{v} Q \frac{N^2}{N_0^2} f_{P1}(U) \tag{23.63}$$

and for the pump efficiency, $\eta_P = f_{P3}(U)$, from equation (17.33) gives:

$$4\pi^2 JN \frac{dN}{dt} = f_{P3}(U) P_S - \frac{g}{v} Q \frac{N^2}{N_0^2} f_{P1}(U) \tag{23.64}$$

where $U = Q(N_0/N)$. This may be rearranged to give:

$$\frac{dN}{dt} = \frac{1}{4\pi^2 J}\left(\frac{f_{P3}(U) P_S}{N} - \frac{g}{v} Q \frac{N}{N_0^2} f_{P1}(U)\right) \tag{23.65}$$

This is a scalar form of equation (23.6), with

$$\begin{aligned} \mathbf{x} &= N \\ \mathbf{u} &= P_S \\ \mathbf{f}(\mathbf{x}, \mathbf{u}) &= f(N, P_S) = \frac{1}{4\pi^2 J} \\ &\quad \times \left(\frac{f_{P3}(U) P_S}{N} - \frac{g}{v} Q \frac{N}{N_0^2} f_{P1}(U)\right) \end{aligned} \tag{23.66}$$

Hence, by equation (23.9), the linearized version of this equation will be:

$$\frac{d\tilde{N}}{dt} = \frac{\partial f}{\partial N} \tilde{N} + \frac{\partial f}{\partial P_S} \tilde{P}_S \tag{23.67}$$

To find the partial derivatives, let us first define the expression, e, as

$$e = 4\pi^2 J f(N, P_s) = \frac{f_{P3}(U)P_S}{N} - \frac{g}{v}Q\frac{N}{N_0^2}f_{P1}(U) \qquad (23.68)$$

Taking the partial derivative of e with respect to P_S yields immediately

$$\frac{\partial e}{\partial P_S} = \frac{f_{P3}(\overline{U})}{\overline{N}} = \frac{\overline{P}_P}{\overline{N}\,\overline{P}_S} \qquad (23.69)$$

since in the steady state (see equation (17.42))

$$f_{P3}P_S = P_P \qquad (23.70)$$

Taking the partial derivative of e with respect to N gives:

$$\begin{aligned}\frac{\partial e}{\partial N} &= -\frac{1}{N^2}f_{P3}P_S + \frac{P_S}{N}\frac{df_{P3}}{dU}\frac{\partial U}{\partial N} - \frac{g}{v}Q\frac{1}{N_0^2}f_{P1} \\ &\quad - \frac{g}{v}Q\frac{N}{N_0^2}\frac{df_{P1}}{dU}\frac{\partial U}{\partial N} - \frac{g}{v}\frac{N}{N_0^2}f_{P1}\frac{\partial Q}{\partial N} \\ &= -\frac{1}{N^2}\left(f_{P3}P_S + \frac{g}{v}Q\frac{N^2}{N_0^2}f_{P1}\right) + \frac{1}{N}\frac{\partial U}{\partial N} \\ &\quad \times \left(P_S\frac{df_{P3}}{dU} - \frac{g}{v}Q\frac{N^2}{N_0^2}\frac{df_{P1}}{dU}\right) - \frac{g}{v}\frac{N}{N_0^2}f_{P1}\frac{\partial Q}{\partial N} \\ &= -\frac{1}{N^2}(P_P + P_P) + \frac{1}{N}\frac{\partial U}{\partial N} \\ &\quad \times \left(\frac{P_P}{f_{P3}}\frac{df_{P3}}{dU} - \frac{P_P}{f_{P1}}\frac{df_{P1}}{dU}\right) - \frac{P_P}{QN}\frac{\partial Q}{\partial N}\end{aligned} \qquad (23.71)$$

where extensive use has been made of equations (23.63) and (23.70) in the last line.

This expression may be simplified further by noting that differentiating U with respect to N gives:

$$\frac{\partial U}{\partial N} = -Q\frac{N_0}{N^2} = -\frac{U}{N} \qquad (23.72)$$

Accordingly, we may rewrite equation (23.71) as:

$$\begin{aligned}\frac{\partial e}{\partial N} &= -\frac{2P_P}{N^2} - \frac{P_P}{N^2} \\ &\quad \times \left(\frac{U}{f_{P3}}\frac{df_{P3}}{dU} - \frac{U}{f_{P1}}\frac{df_{P1}}{dU}\right) - \frac{P_P}{QN}\frac{\partial Q}{\partial N} \\ &= -\frac{P_P}{N^2}\left(\frac{U}{f_{P3}}\frac{df_{P3}}{dU} - \frac{U}{f_{P1}}\frac{df_{P1}}{dU}\right. \\ &\quad \left. + 2 + \frac{N}{Q}\frac{\partial Q}{\partial N}\right)\end{aligned} \qquad (23.73)$$

To proceed further, we need to evaluate $\partial Q/\partial N$ as follows. As noted in equation (23.26), the pump outlet pressure is equal to the pressure in the discharge vessel plus the differential pressure across the valve, plus the pressure drop through the pipeline:

$$p_2 = p_4 + \Delta p + \Delta p_L \qquad (23.26)$$

Subtracting the pump inlet pressure gives the pressure rise across the pump as

$$p_2 - p_1 = p_4 - p_1 + \Delta p + \Delta p_L \qquad (23.74)$$

But the pump pressure rise is given also in terms of the pump characteristic by equation (23.37), repeated below:

$$p_2 - p_1 = \frac{g}{v}\frac{N^2}{N_0^2}f_{P1}(U) \qquad (23.37)$$

Combining this with the pressure drop equation gives:

$$\frac{g}{v}\frac{N^2}{N_0^2}f_{P1}(U) = p_4 - p_1 + \Delta p + \Delta p_L \qquad (23.75)$$

Expressing the valve and line pressure drops in terms of flow using equations (23.27) to (23.30) gives:

$$\frac{g}{v}\frac{N^2}{N_0^2}f_{P1}(U) = p_4 - p_1 + \left(a_L + \frac{a_V}{y^2}\right)Q^2 \qquad (23.76)$$

We take the partial derivative with respect to N, noting that the pump inlet and discharge vessel pressures are constant, and that for the differential pressure control loop, the valve opening, y, is also considered constant. Hence:

$$2\frac{g}{v}\frac{N}{N_0^2}f_{P1} + \frac{g}{v}\frac{N^2}{N_0^2}\frac{df_{P1}}{dU}\frac{\partial U}{\partial N} = 2\left(a_L + \frac{a_V}{y^2}\right)Q\frac{\partial Q}{\partial N} \qquad (23.77)$$

Now $\partial U/\partial N = -U/N$ by equation (23.72), so that, after utilizing equation (23.37) and rearranging, equation (23.77) gives:

$$\frac{\partial Q}{\partial N} = \frac{(p_2 - p_1)\left(2 - \dfrac{U}{f_{P1}}\dfrac{df_{P1}}{dU}\right)}{2QN\left(a_L + \dfrac{a_V}{y^2}\right)} \qquad (23.78)$$

It follows that

$$\frac{N}{Q}\frac{\partial Q}{\partial N} = \frac{1}{2}\frac{(p_2 - p_1)\left(2 - \dfrac{U}{f_{P1}}\dfrac{df_{P1}}{dU}\right)}{Q^2\left(a_L + \dfrac{a_V}{y^2}\right)} \qquad (23.79)$$

or, using the pressure drop relations of equations (23.26) to (23.30):

$$\frac{N}{Q}\frac{\partial Q}{\partial N} = \frac{p_2 - p_1}{p_2 - p_4}\left(1 - \frac{1}{2}\frac{U}{f_{P1}}\frac{df_{P1}}{dU}\right) \qquad (23.80)$$

It follows that:

$$-\frac{U}{f_{P1}}\frac{df_{P1}}{dU}+2+\frac{N}{Q}\frac{\partial Q}{\partial N}$$
$$=-\frac{U}{f_{P1}}\frac{df_{P1}}{dU}+2+\frac{p_2-p_1}{p_2-p_4}\left(1-\frac{1}{2}\frac{U}{f_{P1}}\frac{df_{P1}}{dU}\right)$$
$$=\frac{1}{p_2-p_4}\left(-\frac{1}{2}\frac{U}{f_{P1}}\frac{df_{P1}}{dU}(2p_2-2p_4)+2p_2\right.$$
$$\left.-2p_4+p_2-p_1-\frac{1}{2}\frac{U}{f_{P1}}\frac{df_{P1}}{dU}(p_2-p_1)\right)$$
$$=\frac{1}{p_2-p_4}\left(3p_2-2p_4-p_1\right.$$
$$\left.-\frac{1}{2}\frac{U}{f_{P1}}\frac{df_{P1}}{dU}(3p_2-p_4-p_1)\right) \quad (23.81)$$

or

$$-\frac{U}{f_{P1}}\frac{df_{P1}}{dU}+2+\frac{N}{Q}\frac{\partial Q}{\partial N}=\frac{3p_2-2p_4-p_1}{p_2-p_4}\left(1-\frac{1}{2}\frac{U}{f_{P1}}\frac{df_{P1}}{dU}\right) \quad (23.82)$$

Substituting from equation (23.82) into equation (23.73) gives:

$$\frac{\partial e}{\partial N}=-\frac{P_P}{N^2}\left(\frac{U}{f_{P3}}\frac{df_{P3}}{dU}+\frac{3p_2-2p_4-p_1}{p_2-p_4}\times\left(1-\frac{1}{2}\frac{U}{f_{P1}}\frac{df_{P1}}{dU}\right)\right) \quad (23.83)$$

Having evaluated $\partial e/\partial P_S$ and $\partial e/\partial N$, we may now use the definition of the expression, e, equation (23.68), to write down the formulae for $\partial f/\partial P_S$ and $\partial f/\partial N$ as:

$$\frac{\partial f}{\partial P_s}=\frac{1}{4\pi^2 J\overline{N}}\frac{\overline{P}_P}{\overline{P}_S}$$
$$\frac{\partial f}{\partial N}=-\frac{\overline{P}_P}{4\pi^2 J\overline{N}^2}\left(\frac{\overline{U}}{\overline{f}_{P3}}\frac{df_{P3}}{dU}+\frac{3\overline{p}_2-2\overline{p}_4-\overline{p}_1}{\overline{p}_2-\overline{p}_4}\times\left(1-\frac{1}{2}\frac{\overline{U}}{\overline{f}_{P1}}\frac{df_{P1}}{dU}\right)\right) \quad (23.84)$$

Hence the linearized differential equation defining turbine/pump speed is, from equation (23.67):

$$\frac{d\tilde{N}}{dt}=\frac{1}{4\pi^2 J\overline{N}}\frac{\overline{P}_P}{\overline{P}_S}\tilde{P}_S-\frac{\overline{P}_P}{4\pi^2 J\overline{N}^2}\times\left(\frac{\overline{U}}{\overline{f}_{P3}}\frac{df_{P3}}{dU}+\frac{3\overline{p}_2-2\overline{p}_4-\overline{p}_1}{\overline{p}_2-\overline{p}_4}\times\left(1-\frac{1}{2}\frac{\overline{U}}{\overline{f}_{P1}}\frac{df_{P1}}{dU}\right)\right)\tilde{N} \quad (23.85)$$

This equation may be rearranged to give the alternative form:

$$\frac{4\pi^2 J\overline{N}^2}{\overline{P}_P\left(\frac{\overline{U}}{\overline{f}_{P3}}\frac{df_{P3}}{dU}+\frac{3\overline{p}_2-2\overline{p}_4-\overline{p}_1}{\overline{p}_2-\overline{p}_4}\times\left(1-\frac{1}{2}\frac{\overline{U}}{\overline{f}_{P1}}\frac{df_{P1}}{dU}\right)\right)}\frac{d\tilde{N}}{dt}+\tilde{N}$$
$$=\frac{1}{\left(\frac{\overline{U}}{\overline{f}_{P3}}\frac{df_{P3}}{dU}+\frac{3\overline{p}_2-2\overline{p}_4-\overline{p}_1}{\overline{p}_2-\overline{p}_4}\times\left(1-\frac{1}{2}\frac{\overline{U}}{\overline{f}_{P1}}\frac{df_{P1}}{dU}\right)\right)}\frac{\overline{N}}{\overline{P}_S}\tilde{P}_S \quad (23.86)$$

which is of the form

$$T_P\frac{d\tilde{N}}{dt}+\tilde{N}=k_P\tilde{P}_s \quad (23.87)$$

where the pump time constant is given by

$$T_P=\frac{4\pi^2 J\overline{N}^2}{\overline{P}_P\left(\frac{\overline{U}}{\overline{f}_{P3}}\frac{df_{P3}}{dU}+\frac{3\overline{p}_2-2\overline{p}_4-\overline{p}_1}{\overline{p}_2-\overline{p}_4}\times\left(1-\frac{1}{2}\frac{\overline{U}}{\overline{f}_{P1}}\frac{df_{P1}}{dU}\right)\right)} \quad (23.88)$$

and the pump gain is given by

$$k_P=\frac{1}{\left(\frac{\overline{U}}{\overline{f}_{P3}}\frac{df_{P3}}{dU}+\frac{3\overline{p}_2-2\overline{p}_4-\overline{p}_1}{\overline{p}_2-\overline{p}_4}\times\left(1-\frac{1}{2}\frac{\overline{U}}{\overline{f}_{P1}}\frac{df_{P1}}{dU}\right)\right)}\frac{\overline{N}}{\overline{P}_s} \quad (23.89)$$

Equation (23.87) has the Laplace transform:

$$\frac{\tilde{N}}{\tilde{P}_s}(s)=\frac{k_p}{1+sT_P} \quad (23.90)$$

23.5.8 The change in differential pressure with speed, $d\Delta p/dN$

The response is assumed instantaneous because of the very fast establishment of liquid flow. It is a total derivative from the perspective of the differential pressure control loop, where the liquid control valve position is assumed constant.

We begin with the pressure drop equation, (23.75) repeated below

$$\frac{g}{v}\frac{N^2}{N_0^2}f_{P1}(U) = p_4 - p_1 + \Delta p + \Delta p_L \quad (23.75)$$

We may re-express the line pressure drop in terms of the valve pressure drop by using the results of equations (23.27) and (23.29):

$$\Delta p = \frac{a_V}{y^2}Q^2$$
$$\Delta p_L = a_L Q^2 \quad (23.91)$$

so that

$$\Delta p_L = \frac{a_L}{a_V}y^2\Delta p \quad (23.92)$$

Accordingly we may rewrite equation (23.75) as:

$$\frac{g}{v}\frac{N^2}{N_0^2}f_{P1}(U) = p_4 - p_1 + \left(1 + \frac{a_L}{a_V}y^2\right)\Delta p \quad (23.93)$$

Differentiating with respect to N under the condition that p_1, p_4 and y are all constant yields:

$$2\frac{g}{v}\frac{N}{N_0^2}f_{P1} + \frac{g}{v}\frac{N^2}{N_0^2}\frac{df_{P1}}{dU}\frac{dU}{dN} = \left(1 + \frac{a_L}{a_V}y^2\right)\frac{d\Delta p}{dN} \quad (23.94)$$

or, using the pump pressure rise equation (23.37):

$$p_2 - p_1 = \frac{g}{v}\frac{N^2}{N_0^2}f_{P1}(U) \quad (23.37)$$

to re-express the equation:

$$\frac{2}{N}(p_2 - p_1) + \frac{(p_2 - p_1)}{f_{P1}}\frac{df_{P1}}{dU}\frac{dU}{dN} = \left(1 + \frac{a_L}{a_V}y^2\right)\frac{d\Delta p}{dN} \quad (23.95)$$

Now the total derivative

$$\frac{dU}{dN} = \frac{d}{dN}\left(Q\frac{N_0}{N}\right)$$

is given by:

$$\frac{dU}{dN} = \frac{\partial U}{\partial N} + \frac{\partial U}{\partial Q}\frac{dQ}{dN} = -\frac{U}{N} + \frac{N_0}{N}\frac{dQ}{dN} \quad (23.96)$$

But from equation (23.27),

$$Q^2 = \frac{y^2}{a_V}\Delta p \quad (23.97)$$

which may be differentiated to give:

$$\frac{dQ}{dN} = \frac{1}{2Q}\frac{y^2}{a_V}\frac{d\Delta p}{dN} \quad (23.98)$$

It follows that

$$\frac{N_0}{N}\frac{dQ}{dN} = \frac{N_0}{2QN}\frac{y^2}{a_V}\frac{d\Delta p}{dN} = \frac{1}{2}Q\frac{N_0}{N}\frac{y^2}{a_V Q^2}\frac{d\Delta p}{dN} = \frac{1}{2}\frac{U}{\Delta p}\frac{d\Delta p}{dN} \quad (23.99)$$

Combining equations (23.96) and (23.99) yields the total differential of U to N as:

$$\frac{dU}{dN} = -\frac{U}{N} + \frac{1}{2}\frac{U}{\Delta p}\frac{d\Delta p}{dN} \quad (23.100)$$

We may substitute this into equation (23.95):

$$\frac{2}{N}(p_2 - p_1) + \frac{(p_2 - p_1)}{f_{P1}}\frac{df_{P1}}{dU} \times \left(-\frac{U}{N} + \frac{1}{2}\frac{U}{\Delta p}\frac{d\Delta p}{dN}\right) = \left(1 + \frac{a_L}{a_V}y^2\right)\frac{d\Delta p}{dN} \quad (23.101)$$

or

$$\frac{(p_2 - p_1)}{N}\left(2 - \frac{U}{f_{P1}}\frac{df_{P1}}{dU}\right) = \frac{d\Delta p}{dN}\left(1 + \frac{a_L}{a_V}y^2 - \frac{1}{2}\frac{(p_2 - p_1)}{\Delta p}\frac{U}{f_{P1}}\frac{df_{P1}}{dU}\right) \quad (23.102)$$

Multiplying throughout by Δp,

$$\frac{\Delta p}{N}(p_2 - p_1)\left(2 - \frac{U}{f_{P1}}\frac{df_{P1}}{dU}\right) = \frac{d\Delta p}{dN}\left(\Delta p + \frac{a_L}{a_V}y^2\Delta p - \frac{1}{2}(p_2 - p_1)\frac{U}{f_{P1}}\frac{df_{P1}}{dU}\right) = \frac{d\Delta p}{dN}\left(p_2 - p_4 - \frac{1}{2}(p_2 - p_1)\frac{U}{f_{P1}}\frac{df_{P1}}{dU}\right) \quad (23.103)$$

in which equation (23.91) has been used in the expansion:

$$\Delta p + \frac{a_L}{a_V}y^2\Delta p = p_3 - p_4 + p_2 - p_3 = p_2 - p_4 \quad (23.104)$$

Rearranging equation (23.103) the required relationship emerges as:

$$\frac{d\Delta p}{dN} = \frac{\Delta \overline{p}}{\overline{N}} \frac{2 - \dfrac{\overline{U}}{\overline{f}_{P1}} \dfrac{df_{P1}}{dU}}{\left(\left(\dfrac{\overline{p}_2 - \overline{p}_4}{\overline{p}_2 - \overline{p}_1} \right) - \dfrac{1}{2} \dfrac{\overline{U}}{\overline{f}_{P1}} \dfrac{df_{P1}}{dU} \right)} \tag{23.105}$$

At this stage the operators in each of the blocks in the block diagram of Figure 23.3 have been found.

23.6 Using the linear block diagram

A linear, time-domain simulation can easily be written once the mathematical description of each block in the block diagram is available. A very important use of such a linear simulation is to act as a check against the main, fully nonlinear, simulation program. While the derivation of the linear model is based on similar principles to that of the main nonlinear model, it is clear from the foregoing analysis that its programming as a simulation must necessarily be completely diverse. Accordingly, it can act as a very good, diverse check on the correctness of the programming of the main model. Tests can be made on both the linear and nonlinear models for the same small changes in forcing functions, and any significant discrepancy will need to be investigated. The use of the block diagram approach to linearization means that responses of each of the subsystems represented in the blocks can be tested against the main model, and so the causes of any discrepancies can be found quickly.

A further use for the control engineer is that the linear analysis allows him to use frequency response methods for control system design. The fact that the system has been broken down into subsystems also means that any nonlinearities such as deadzone that are known to exist may be introduced easily into the block diagram layout and represented by describing functions.

It will be clear from the detailed analysis of a realistic process system that analytic linearization is unlikely to be a trivial exercise. The amount of detailed mathematical analysis often needed means that it cannot be recommended for every case. And certainly it should never be regarded as an alternative to full, nonlinear simulation. However, if the control engineer does carry out such a linear, block-diagram analysis and conducts successfully the subsequent linear/nonlinear model comparisons, he will be rewarded not only by a greater confidence in his main simulation model but also by a significantly increased quantitative feel for the physical interactions taking place in his process.

The reader interested in more details of the industrial problem on which the worked example was based is referred to this chapter's references concerning the control of a boiler feedwater system.

23.7 Bibliography

Harrison, T.A. and Thomas, P.J. (1986). Dynamic interactions between recirculation boilers connected in parallel, *Boiler Dynamics and Control in Nuclear Power Stations 3, Proceedings of the Third International Conference*, Harrogate, 21–25 October. 1985, Paper 18, British Nuclear Energy Society.

Holywell, P.D. (1986). Multivariable control system analysis of a drum level control system, *Nuclear Energy*, **25**, 163–168.

Munro, N. (ed.) (1997). Special Feature on Symbolic Computation, *Computer & Control Engineering Journal*, **8**, No 2, 50–76.

Owens, D.H. (1981). *Multivariable and Optimal Systems*, Academic Press.

Thomas, P.J. and Evans, N.J. (1991). Introduction to design and implementation issues of microprocessor based systems, Chapter 2 of *Microprocessor Based Protection Systems*, edited by A. Churchley, Elsevier Applied Science, London.

Thomas, P.J., Harrison, T.A. and Holywell, P.D. (1986). Analysis of limit cycling on a boiler feedwater control system, *Boiler Dynamics and Control in Nuclear Power Stations 3, Proceedings of the third international conference* held in Harrogate, 21–25 October 1985, Paper 10, British Nuclear Energy Society.

Thomas, P.J., Harrison, T.A. and Holywell, P.D. (1987). Using advanced control techniques to improve the stability and performance of nuclear power station boiler, *IFAC 10th World Congress*, Munich, Germany, July.

Thomas, P.J., Harrison, T.A. and Holywell, P.D. (1988). Stabilizing the controls of a boiler feedwater system, *Trans. Inst. Measurement and Control*, **10**, 15–20.

24 Model validation

24.1 Introduction

The earliest models devised by control engineers were simplified, analytic linearizations of system behaviour about an operating point, and were used more or less exclusively for the selection of control parameters. Not too much was expected from the dynamic model, and so the requirement for rigorous model validation, as opposed to intuitive feel, was small. But the advent of first analogue and then digital computers extended very considerably the range of simulation studies possible, with the result that the expectations of today's dynamic simulations are much greater. The control engineer's model will be expected to cope with highly nonlinear phenomena and may be expected to give predictions of plant behaviour in fault conditions and following a trip, thus moving into the area of safety assessment. Accordingly there is an increasing demand to demonstrate that the model is valid, and there should be, ideally, a quantitative measure of the model's validity.

This chapter will explore the philosophy of model validation, and goes on to describe the Model Distortion Method for quantitative model validation.

24.2 The philosophy of model validation

In considering what we mean by model validation, it is instructive to review the procedure by which we set up a computer model of a general physical process, for example a boiler. We begin by writing equations to describe the dynamics of the system, and in so doing set down a detailed theory of how the system will behave. We code the mathematical equations embodying this theory into a computer simulation language to form a program: let us call this the 'generic model'. To apply it to a particular plant, we supply the program with parameters specific to the plant in question to form the 'specific model'. Then we add initial conditions and forcing functions particular to the transient in question to form the 'particular model' (cf. the analogous mathematical term 'particular integral'). This process of adding detail to build up first the specific model and then the particular model is shown diagrammatically in Figure 24.1, where it can be seen that the particular model includes the specific model and that the specific model includes the generic model.

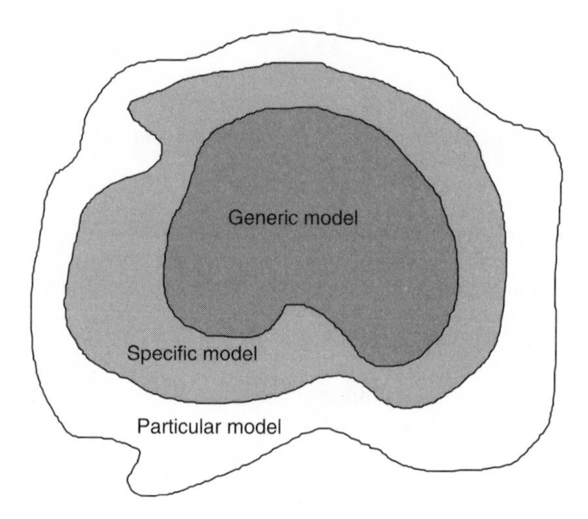

Figure 24.1 Relationships between generic, specific and particular models.

Suppose we run the particular model to obtain model predictions. Let us now enquire: how much logical justification can we have for regarding these predictions as correct?

Clearly the predictions will be correct if the particular model is correct. But, by reference to Figure 24.1, the particular model will be correct only if the specific model is correct, which will in turn be correct only if the generic model is correct. This leads us to the fundamental philosophical question: can we prove that our generic model is correct?

The basis for an answer to this question was provided by the philosopher Sir Karl Popper in 1934, who was interested in the characteristics of a scientific theory. (We may note here that a dynamic model embodied in a computer program is, of course, a scientific theory – a very well documented theory.) Popper's conclusions were that:

(1) to be deemed scientific, a theory must be 'falsifiable'. That is to say, it must be empirically testable, in principle at least, and there must be some test that we can set for the theory in which an unfavourable outcome will prove the theory wrong; and
(2) there can never be a rigorous logical justification for any scientific theory. The best we can do is to set empirical tests for the theory – fair

> tests, but the more severe the better – and continue to use the theory only so long as the theory passes all the tests.

A theory can never be proved by any of its successes, since a new test, perhaps as yet not thought of, may come along that it will fail. Failure in any fair test, on the other hand, indicates a fault in the theory: 'falsifies' it. At this stage, effort will need to be spent in improving the theory or devising a completely new theory.

As already noted, the generic model mentioned above represents a well-documented, scientific theory about how the physical process will behave. Following Popper's conclusions, we see that we will never be able to prove beyond doubt that any generic model for any physical process is correct. We should, however, test it rigorously against as many empirical data as possible. Each successful test will give confidence that we may continue to use the generic model to predict process behaviour. Failure of any fair test, on the other hand, will indicate that the model is in some way defective, and needs correcting. It is not the successes but the failures that provide us with the spur and the additional information needed to improve our generic models.

Model validation is a more fundamental and far-reaching process than that of model verification, where the object is merely to ensure that the model as programmed reflects accurately the intentions of the modeller and for which similar predictions from, for example, a linearized model based on the same physical theory can prove an adequate check, as discussed in the previous chapter. Model validation needs to test whether the model's physical basis is correct, and for this data from the real world are required – either from the plant or from an experimental rig. It is desirable to test the generic model against data from as many different plants and experimental rigs as possible, subject to the constraints of economics: all the information gained is 'grist to the mill'. A distinct specific model will need to be set up for each new plant, of course, and it will be necessary to supply the model with the initial conditions and forcing functions that particularize the transient in question. However, if a fair test is failed, and, moreover, no fault is found in the specific or the particular features of the model (that is to say, the plant-specific parameters, the initial conditions and the forcing functions are all found to be programmed correctly), then we have no option but to conclude that the fault lies with the generic model. The error in the generic model must then be found and corrected, and all users of the generic model informed of the correction.

We have talked of tests for simulation models, but so far we have not discussed what constitutes a fair test for a dynamic computer model. Clearly a comparison with empirical data from rig or plant is needed, but the assessment process should ideally not rely solely on a visual comparison between the recorded transients and the corresponding model predictions, since this leaves too much room for subjective opinion and dispute. An objective, quantitative assessment is clearly preferable, and this is provided by the Model Distortion Method, which will be discussed below.

24.3 The concept of model distortion

A mathematical model of suitably high order can be made to follow *any* set of recorded transients exactly by varying appropriately the nominally constant parameters as functions of time – 'distorting the model'. Clearly, the better the model the less will be the required distortion. More particularly, if the required distortion comes within acceptable limits (in view of approximations made in the model), then the model may be deemed capable of explaining the recorded transients. This notion accommodates instrument error and signal noise in a pessimistic way, since both make reconciliation between the model and the record more difficult.

The concept of time-dependent model distortion becomes clearer through considering a typical modelling approximation. For example, steam flow, W, through a valve might be modelled as proportional to the upstream pressure, p, and valve opening, y:

$$W = kyp \tag{24.1}$$

where k is the constant of proportionality. The flow/pressure characteristics are in reality likely to be rather more complex, certainly for unchoked flow, and using recorded values of W, y and p, we may back-calculate the time-varying gain, $k(t)$, by:

$$k(t) = \frac{W(t)}{y(t)p(t)} \tag{24.2}$$

where, in view of the rather naive model used, we could expect $k(t)$ to vary significantly. In attempting to improve the model we might introduce suitable power laws and perhaps other variables. By doing so we would produce a model in which less parameter variation was necessary – less model distortion.

Another way of expressing the idea of model distortion is to regard the recorded plant transients as continuous, indirect measurements of the model's constant parameters. If these new, inferential measurements of the model parameters fall outside the ranges set by previous, more direct measurements then there is an inconsistency. We should reject the model in its current state and seek to improve it. By this formulation, it becomes clear that the model distortion approach consists of mapping the complex dynamic behaviour of recorded plant variables onto the simple (i.e. unchanging) behaviour of constants before performing our tests.

To translate these ideas into mathematics, let us suppose that we have produced a model of the plant defined by

$$\frac{d\mathbf{x}}{dt} = \mathbf{f}(\mathbf{x}, \mathbf{u}, \mathbf{a}, t) \tag{24.3}$$

where
- $\mathbf{x}$ is the n-dimensional vector of state variables,
- $\mathbf{u}$ is the l-dimensional vector of forcing variables or inputs,
- $\mathbf{a}$ is the p-dimensional vector of constant parameters, and
- $\mathbf{f}$ is an n-dimensional, nonlinear, vector function.

We shall call this the 'nominal model'. Comparing equation (24.3) with the standard simulation equation (2.22), it will be seen that the dependence of the vector function, $\mathbf{f}$, on its constant parameters, $\mathbf{a}$, has been made explicit in equation (24.3).

For simplicity, let us suppose that some of the n states, $\mathbf{x}$, correspond to available plant measurements. Assuming we have k measured plant variables, we find the corresponding k model variables as a subset of the states, $\mathbf{x}$. Let $\mathbf{y}$ be a k-dimensional vector of corresponding model variables, given by

$$\mathbf{y} = \mathbf{C}\mathbf{x} \tag{24.4}$$

where $\mathbf{C}$ is a $k \times n$ matrix with one unity element per row and the rest zeros.

If we define the measured variables, recorded from a plant transient, to be elements of the k-dimensional vector, $\mathbf{z}(t)$, then we may define a k-dimensional, error vector, $\mathbf{e}(t)$, by

$$\mathbf{e}(t) = \mathbf{z}(t) - \mathbf{y}(\mathbf{a}, t) \tag{24.5}$$

The vector $\mathbf{y}$ has been shown deliberately as a function of the constants, $\mathbf{a}$, because the size of the components of this vector will depend on the selection of those parameters. A procedure often adopted is to minimize some norm of the error, $\mathbf{e}$, by a best choice of the parameters, $\mathbf{a}$, for example time-domain least-squares minimization. Application of this process in itself provides a check on the validity of the model, since the optimal choice of parameters, $\mathbf{a}_{opt}$, should lie within ranges expected on physical grounds. The corresponding transient states will be $\mathbf{y}(\mathbf{a}_{opt}, t)$, and thus we may define the minimum error as

$$\mathbf{e}_{\min}(t) = \mathbf{z}(t) - \mathbf{y}(\mathbf{a}_{opt}, t) \tag{24.6}$$

where the subscripts *min* and *opt* refer to the specific error norm that has been minimized. Note, however, that the error described by equation (24.6) is not generally zero, even for the best choice of constant parameters.

Following the model-distortion idea described above, let us vary dynamically those parameters hitherto deemed constant. The values of the state variables will now be different from those of the nominal model, with the difference dependent on the extent of the variation in the parameters. Letting the new states be denoted by the n-dimensional vector, $\mathbf{X}$, they will follow the equation

$$\frac{d\mathbf{X}}{dt} = \mathbf{f}(\mathbf{X}, \mathbf{u}, \mathbf{a} + \boldsymbol{\alpha}, t) \tag{24.7}$$

where $\boldsymbol{\alpha}$ is a p-dimensional vector of variations, while $\mathbf{f}$ is the same nonlinear vector function as used in equation (24.3), and $\mathbf{u}$ is the same input vector.

Let us denote by $\mathbf{Y}$ the corresponding k-dimensional output vector, which will be given by

$$\mathbf{Y} = \mathbf{C}\mathbf{X} \tag{24.8}$$

In most cases it will be possible to drive the system given by equations (24.7) and (24.8) with parameter variations such that the k plant states, $\mathbf{z}$, are matched by the outputs, $\mathbf{Y}$, of the distorted model, i.e.:

$$\mathbf{Y} = \mathbf{z} \tag{24.9}$$

The general problem is now to solve for the vector of parameter variations, $\boldsymbol{\alpha}$, given by the equation set:

$$\frac{d\mathbf{X}}{dt} = \mathbf{f}(\mathbf{X}, \mathbf{u}, \mathbf{a} + \boldsymbol{\alpha}, t) \tag{24.7}$$

$$\mathbf{z} = \mathbf{C}\mathbf{X} \tag{24.10}$$

Once the vector, $\boldsymbol{\alpha}$, has been found, the size of its elements may be investigated and a judgment made on whether the model using constant parameters is reasonable or not.

Before considering methods of finding $\boldsymbol{\alpha}$, let us review how the concept of model distortion has allowed us to move from comparison between model and plant to a comparison between two models, in one of which we allow the nominally constant parameters to be varied. Suppose we drive the system to follow the recorded transients exactly. Using the superscript '0' to label the vectors associated with exact matching, the parameter variations will be given by $\boldsymbol{\alpha}^0(t)$, the states by $\mathbf{X}^0(t)$ and the outputs by $\mathbf{Y}^0(t)$. Exact matching implies:

$$\mathbf{Y}^0 = \mathbf{z} \tag{24.11}$$

where the matched outputs are related to the states of the distorted model by:

$$\mathbf{Y}^0 = \mathbf{C}\mathbf{X}^0 \tag{24.12}$$

We may substitute for $\mathbf{z}$ from equation (24.11) into equation (24.6) and thus allow the minimum error vector, $\mathbf{e}_{\min}$, to be rewritten:

$$\mathbf{e}_{\min}(t) = \mathbf{Y}^0(t) - \mathbf{y}(\mathbf{a}_{opt}, t) \tag{24.13}$$

or, in terms of the state variables:

$$\mathbf{e}_{\min}(t) = \mathbf{CX}^0(t) - \mathbf{Cx}(\mathbf{a}_{opt}, t) \qquad (24.14)$$

The significance of equation (24.14) is that consideration of the difference between plant and model has been transposed into a consideration of the difference between two models, in one of which the parameters are fixed, while in the other the parameters are varied. The advantage of this transposition is that whereas the mathematical structure of the plant can be known only approximately, the structure of the two models is known precisely. This will prove useful in studying the variation in the parameters (the distortion in the model) necessary to follow the recorded transients.

It is possible to derive $\boldsymbol{\alpha}(t)$ directly from equations (24.7) and (24.10), step by step. Alternatively, an estimate may be made from the results of the exercise to establish the parameters yielding the minimum squared error. We shall deal with this approximate method, which is based on transfer-function notions, in the next section. The mathematical derivation is rather complex, but the method finally derived is straightforward to apply.

24.4 Transfer-function-based technique for model distortion

24.4.1 Defining the companion model

Let us linearize equation (24.7). Provided that $\mathbf{X}$ does not differ greatly from $\mathbf{x}$ and $\boldsymbol{\alpha}$ is small, we may approximate the right-hand side by a first-order Taylor series about a point defined by $\mathbf{a} = \mathbf{a}_{opt}$:

$$\frac{d\mathbf{X}}{dt} = \mathbf{f}(\mathbf{x}, \mathbf{u}, \mathbf{a}_{opt}, t) + \mathbf{J}_\mathbf{x}(\mathbf{X} - \mathbf{x}) + \mathbf{J}_\mathbf{a}\boldsymbol{\alpha} \qquad (24.15)$$

where $\mathbf{J}_\mathbf{x}$ is the Jacobian matrix:

$$\mathbf{J}_\mathbf{x} = \begin{bmatrix} \frac{\partial f_1}{\partial x_1} & \frac{\partial f_1}{\partial x_2} & \cdots & \frac{\partial f_1}{\partial x_n} \\ \frac{\partial f_2}{\partial x_1} & \frac{\partial f_2}{\partial x_2} & \cdots & \frac{\partial f_2}{\partial x_n} \\ \vdots & \vdots & & \vdots \\ \frac{\partial f_n}{\partial x_1} & \frac{\partial f_n}{\partial x_2} & \cdots & \frac{\partial f_n}{\partial x_n} \end{bmatrix} \qquad (24.16)$$

and $\mathbf{J}_\mathbf{a}$ is the Jacobian matrix:

$$\mathbf{J}_\mathbf{a} = \begin{bmatrix} \frac{\partial f_1}{\partial a_1} & \frac{\partial f_1}{\partial a_2} & \cdots & \frac{\partial f_1}{\partial a_p} \\ \frac{\partial f_2}{\partial a_1} & \frac{\partial f_2}{\partial a_2} & \cdots & \frac{\partial f_2}{\partial a_p} \\ \vdots & \vdots & & \vdots \\ \frac{\partial f_n}{\partial a_1} & \frac{\partial f_n}{\partial a_2} & \cdots & \frac{\partial f_n}{\partial a_p} \end{bmatrix} \qquad (24.17)$$

Defining $\boldsymbol{\phi}$ as the state difference vector:

$$\boldsymbol{\phi} = \mathbf{X} - \mathbf{x} \qquad (24.18)$$

yields on differentiation:

$$\frac{d\boldsymbol{\phi}}{dt} = \frac{d\mathbf{X}}{dt} - \frac{d\mathbf{x}}{dt} \qquad (24.19)$$

Substituting into equation (24.15) gives

$$\frac{d\mathbf{x}}{dt} + \frac{d\boldsymbol{\phi}}{dt} = \mathbf{f}(\mathbf{x}, \mathbf{u}, \mathbf{a}_{opt}, t) + \mathbf{J}_\mathbf{x}\boldsymbol{\phi} + \mathbf{J}_\mathbf{a}\boldsymbol{\alpha} \qquad (24.20)$$

But

$$\frac{d\mathbf{x}}{dt} = \mathbf{f}(\mathbf{x}, \mathbf{u}, \mathbf{a}_{opt}, t) \qquad (24.21)$$

is the equation of the nominal model with best-estimate constant parameters. Subtracting equation (24.21) from equation (24.20) yields

$$\frac{d\boldsymbol{\phi}}{dt} = \mathbf{J}_\mathbf{x}\boldsymbol{\phi} + \mathbf{J}_\mathbf{a}\boldsymbol{\alpha} \qquad (24.22)$$

It is natural to define the outputs, $\boldsymbol{\psi}$, of the linearized system to correspond to those of the original system:

$$\boldsymbol{\psi} = \mathbf{Y}(t) - \mathbf{y}(\mathbf{a}_{opt}, t) = \mathbf{CX} - \mathbf{Cx}(\mathbf{a}_{opt}, t) \qquad (24.23)$$

Hence

$$\boldsymbol{\psi} = \mathbf{C}\boldsymbol{\phi} \qquad (24.24)$$

Equations (24.22) and (24.24) define a linear system by which the state differences are driven by parameter variations about their optimal value, and we name this the 'companion model'. The properties of this companion model will clearly depend on $\mathbf{J}_\mathbf{x}$ and $\mathbf{J}_\mathbf{a}$. A controllability test will show whether all the elements of the $\boldsymbol{\phi}$ vector can be influenced by variations in the $\boldsymbol{\alpha}$ vector. Provided that the system is controllable, then it should be possible to drive $\boldsymbol{\phi}$ to any transient desired, in the absence of limits on the amplitude or frequency of variation of $\boldsymbol{\alpha}$.

The outputs, $\boldsymbol{\psi}$, of the companion model may be related to the error between the distorted model and the plant by the following steps:

$$\begin{aligned} \mathbf{e}(t) &= \mathbf{z}(t) - \mathbf{y}(\mathbf{a}_{opt} + \boldsymbol{\alpha}, t) \\ &= \mathbf{z}(t) - \mathbf{y}(\mathbf{a}_{opt}, t) - (\mathbf{y}(\mathbf{a}_{opt} + \boldsymbol{\alpha}, t) - \mathbf{y}(\mathbf{a}_{opt}, t)) \\ &= \mathbf{z}(t) - \mathbf{y}(\mathbf{a}_{opt}, t) - (\mathbf{Y}(t) - \mathbf{y}(\mathbf{a}_{opt}, t)) \\ &= \mathbf{e}_{\min}(t) - \boldsymbol{\psi}(t) \end{aligned} \qquad (24.25)$$

Thus when $\boldsymbol{\alpha} = \mathbf{0}$, then $\boldsymbol{\psi} = \mathbf{0}$ and hence $\mathbf{e}(t) = \mathbf{e}_{\min}(t)$. However driving the companion model with model-matching parameter variations, $\boldsymbol{\alpha}^0(t)$, will cause $\mathbf{Y}(t) = \mathbf{z}(t)$, and hence $\mathbf{e}(t) = \mathbf{z}(t) - \mathbf{Y}(t) = \mathbf{0}$. Accordingly, from equation (24.25), model matching through parameter variations implies

$$\boldsymbol{\psi}(t) = \mathbf{e}_{\min}(t) \tag{24.26}$$

Known results from control theory give the transfer function matrix for the companion model in terms of the model's matrices, $\mathbf{J_a}$, $\mathbf{J_x}$ and $\mathbf{C}$, as:

$$\mathbf{G}(s) = \mathbf{C}(s\mathbf{I}_n - \mathbf{J_x})^{-1}\mathbf{J_a} \tag{24.27}$$

where s is the Laplace variable. Hence the Laplace transform of the output vector of the companion model is given by

$$\boldsymbol{\psi}(s) = \mathbf{G}(s)\alpha(s) \tag{24.28}$$

Thus we may express the influence on the ith component of the error vector, $\boldsymbol{\psi}$, coming from the jth component of the parameter vector, $\boldsymbol{\alpha}$, by

$$\psi_i(s) = g_{ij}(s)\alpha_j(s) \tag{24.29}$$

where $g_{ij}(s)$ is the element of the transfer-function matrix, $\mathbf{G}(s)$ at the intersection of row i and column j. Based on the premise that the systems found on a process plant can usually be represented well by second-order transfer functions, we may expect this transfer function to have the form

$$g_{ij}(s) = \frac{c_{ij}}{\dfrac{s^2}{\omega_{ij}^2} + 2\dfrac{\varsigma_{ij}}{\omega_{ij}}s + 1} \tag{24.30}$$

with appropriately chosen gain, c_{ij}, undamped natural frequency, ω_{ij}, and damping ratio, ς_{ij}.

24.4.2 Estimation of parameter variance needed for model matching

Let us now investigate the parameter variation needed to match model to plant, assuming that the matching is brought about by each model parameter in turn, acting alone. We will begin by choosing to eliminate the mismatch for just one measured output. Considering only one measured variable means that we need minimize only one mean-squared error. Suppose this corresponds to the ith output variable, then we choose the values of $\mathbf{a}_{opt}$ so as to minimize the mean squared error, $\overline{e_i^2}$, given by:

$$\overline{e_i^2} = \frac{1}{t_1}\int_0^{t_1} e_i^2\,dt \tag{24.31}$$

where t_1 is the duration of the transient. Having found the optimal set of constant parameters, we may apply a range of different values of the jth constant parameter near its optimal value, $a_{opt,j}$, while maintaining the other parameters at their optimal values. The resulting values of $\overline{e_i^2}$ may then be plotted against a_j, and the procedure repeated for all p constant parameters. Such a plot will look somewhat like Figure 24.2.

Inserting a parameter value of $a_j \neq a_{opt,j}$ into the nominal model will produce the same output as applying a steady value $a_j = a_{opt,j} + \alpha_{js}$ to the distorted model. Further, the procedure is equivalent to applying a steady value α_{js} to the companion model. From the form of equation (24.30), it is clear that if a constant value of parameter variation $\alpha_j(t) = \alpha_{js}$ is applied to the companion model, a constant output $\psi_i(t) = \psi_{is}$

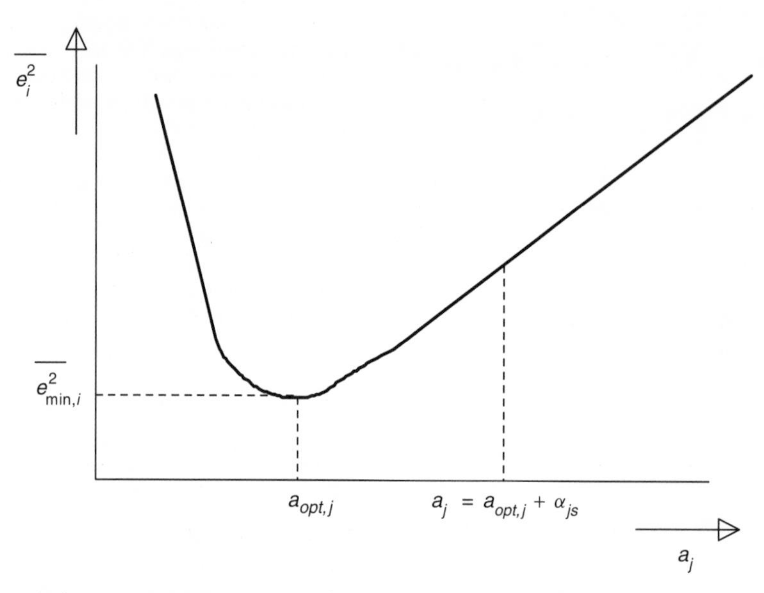

Figure 24.2 Mean squared error versus parameter value for the nominal model and also for the distorted model when driven with a constant, non-optimal parameter value.

will be brought about after an initial settling period, where

$$\Psi_{is} = c_{ij}\alpha_{js} \tag{24.32}$$

Using equation (24.25), the error associated with the ith output of the distorted model, which will be the same as the corresponding error for the nominal model, is

$$\begin{aligned} e_i &= e_{\min,i} - \Psi_{is} \\ &= e_{\min,i} - c_{ij}\alpha_{js} \end{aligned} \tag{24.33}$$

The mean squared error for this constant parameter excursion is found using equation (24.31):

$$\begin{aligned} \overline{e_i^2} &= \frac{1}{t_1}\int_0^{t_1} (e_{\min,i} - c_{ij}\alpha_{js})^2\, dt \\ &= \frac{1}{t_1}\int_0^{t_1} e_{\min,i}^2\, dt - \frac{2c_{ij}\alpha_{js}}{t_1}\int_0^{t_1} e_{\min,i}\, dt \\ &\quad + \frac{c_{ij}^2\alpha_{js}^2}{t_1}\int_0^{t_1} dt \\ &= \overline{e_{\min,i}^2} + c_{ij}^2\alpha_{js}^2 \end{aligned} \tag{24.34}$$

where we have used the result that

$$\bar{e}_i = \frac{1}{t_1}\int_0^{t_1} e_i\, dt = 0 \tag{24.35}$$

because minimization of the mean squared error implies that the mean error is zero.

Using equation (24.34) along with the minimization plot of Figure 24.2 allows us to evaluate the gain, c_{ij}, from

$$c_{ij} = \frac{\sqrt{\overline{e_i^2} - \overline{e_{\min,i}^2}}}{\alpha_{js}} \tag{24.36}$$

Equation (24.36) may be applied to any point $(a_{opt,j} + \alpha_{js}, \overline{e_i^2})$ on the curve of Figure 24.2 to obtain a value for c_{ij}. Bearing in mind the fact that we are invoking linear concepts to model a nonlinear response, experience has shown that a reasonable range of system behaviour is captured if we choose $\alpha_{js} = \alpha_{jd}$ where $\overline{e_i^2}$ is double $\overline{e_{\min,i}^2}$. In this case equation (24.36) simplifies to

$$c_{ij} = \frac{\sqrt{\overline{e_{\min,i}^2}}}{\alpha_{jd}} \tag{24.37}$$

While we have now derived from the minimization graph a good estimate of the steady-state gain from parameter variation to measured variable, our ultimate aim is to relate the variance of the parameter to the variance of the output of the distorted model. Recalling the first equality in equation (24.23), repeated below:

$$\boldsymbol{\psi} = \mathbf{Y}(t) - \mathbf{y}(\mathbf{a}_{opt}, t) \tag{24.23}$$

it is clear that the variance of the output of the companion model $\boldsymbol{\psi}$ must be identical to the variance of the output of the distorted model, $\mathbf{Y}$, since $\mathbf{y}$ is independent of parameter variations, $\boldsymbol{\alpha}$. We may further our analysis of the companion model by making informed assumptions about the dynamics of the system.

Let us make the usual assumption that the range of frequencies generated by the variable α_j in controlling the $\mathbf{Y}$ vector to match the $\mathbf{z}$ vector may be represented by filtered white noise. By this condition we are assuming implicitly that the parameter varies randomly about a central value $(\alpha_{opt,j})$ and that the parameter variation has a limited bandwidth. We envisage the companion model being excited as in Figure 24.3.

White noise of intensity Φ is filtered to produce a signal α_j of variance $\sigma_{\alpha,j}^2$. This signal is used to drive the companion model to produce the signal ψ_i, which has variance $\sigma_{\psi,i}^2$.

It is a standard result (Douce, 1963) that we may calculate the mean squared value or variance of α_j in terms of the white noise intensity Φ by integrating from one extreme of the imaginary axis to the other:

$$\sigma_{\alpha,j}^2 = \frac{\Phi}{2\pi i}\int_{s=-i\infty}^{+i\infty} \left| \frac{c_{a,j}}{1 + \dfrac{s}{\omega_{a,j}}} \right|^2 ds \tag{24.38}$$

Using standard integrals, we find that

$$\sigma_{\alpha,j}^2 = \frac{c_{a,j}^2\omega_{a,j}}{2}\Phi \tag{24.39}$$

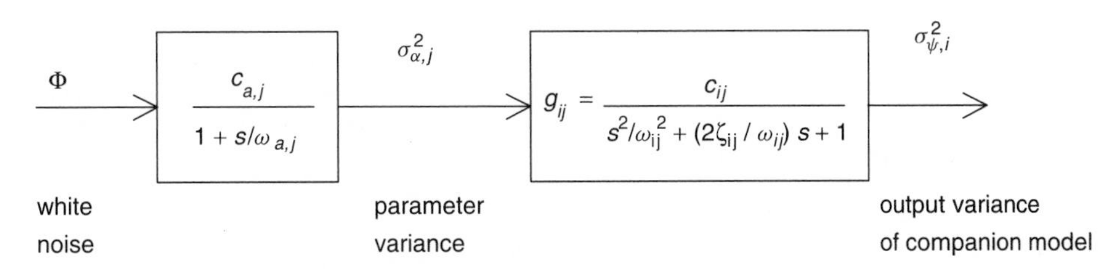

Figure 24.3 Companion model being excited by filtered white noise.

We may also relate the mean squared value or variance of ψ_i to Φ:

$$\sigma^2_{\psi,i} = \frac{\Phi}{2\pi i}\int_{s=-i\infty}^{+i\infty}\left|\frac{c_{a,j}}{1+\dfrac{s}{\omega_{a,j}}}\times\frac{c_{ij}}{\dfrac{s^2}{\omega_{ij}^2}+\dfrac{2\varsigma_{ij}}{\omega_{ij}}s+1}\right|^2 ds$$

$$= \frac{\Phi}{2\pi i}\times\int_{s=-i\infty}^{+i\infty}\left|\frac{c_{a,j}c_{ij}}{\dfrac{s^3}{\omega_{a,j}\omega_{ij}^2}+\dfrac{2\varsigma_{ij}\omega_{ij}+\omega_{a,j}}{\omega_{a,j}\omega_{ij}^2}s^2+\dfrac{2\varsigma_{ij}\omega_{a,j}\omega_{ij}+\omega_{ij}^2}{\omega_{a,j}\omega_{ij}^2}s+1}\right|^2 ds \quad (24.40)$$

Once again standard integrals may be used, and they give the solution as

$$\sigma^2_{\psi,i} = \frac{c^2_{a,j}c^2_{ij}(2\varsigma_{ij}\omega_{ij}+\omega_{a,j})}{4\varsigma_{ij}\left(2\varsigma_{ij}+\dfrac{\omega_{ij}}{\omega_{a,j}}+\dfrac{\omega_{a,j}}{\omega_{ij}}\right)}\Phi \quad (24.41)$$

Hence dividing equation (24.39) by (24.41) gives the relationship between $\sigma^2_{\alpha,j}$ and $\sigma^2_{\psi,i}$ as:

$$\frac{\sigma^2_{\alpha,j}}{\sigma^2_{\psi,i}} = \frac{2\varsigma_{ij}\left(2\varsigma_{ij}+\dfrac{\omega_{ij}}{\omega_{a,j}}+\dfrac{\omega_{a,j}}{\omega_{ij}}\right)}{c^2_{ij}\left(2\varsigma_{ij}\dfrac{\omega_{ij}}{\omega_{a,j}}+1\right)} \quad (24.42)$$

Making the assumption that the frequency contents of the driving signal, α_j, and the output signal, ψ_j, are roughly the same allows us to put $\omega_{a,j}\approx\omega_{ij}$, so that equation (24.41) may be simplified to:

$$\frac{\sigma^2_{\alpha,j}}{\sigma^2_{\psi,i}} = \frac{4\varsigma_{ij}(2\varsigma_{ij}+1)}{c^2_{ij}(2\varsigma_{ij}+1)} \quad (24.43)$$

As a matter of experience, most process systems are sluggish and exhibit little or no overshoot. We may use this knowledge to estimate a range of values for ς_{ij}. From servomechanism theory, $\varsigma_{ij} = 1/2$ corresponds to about 16% overshoot and substituting this value into equation (24.43) gives

$$\frac{\sigma^2_{\alpha,j}}{\sigma^2_{\psi,i}} = \frac{1.5}{c^2_{ij}}$$

A value $\varsigma_{ij} = 1/\sqrt{2}$ corresponds to about 4% overshoot and yields

$$\frac{\sigma^2_{\alpha,j}}{\sigma^2_{\psi,i}} = \frac{2}{c^2_{ij}}$$

while $\varsigma_{ij} = 1$ corresponds to no overshoot, when

$$\frac{\sigma^2_{\alpha,j}}{\sigma^2_{\psi,i}} = \frac{2.7}{c^2_{ij}}$$

Choosing the last value on the basis of conservatism, the relationship between the variances $\sigma^2_{\alpha,j}$ and $\sigma^2_{\psi,i}$ may be written as:

$$\sigma^2_{\alpha,j} = \frac{2.7}{c^2_{ij}}\sigma^2_{\psi,i} \quad (24.44)$$

We may eliminate c_{ij} from equation (24.44) by using equation (24.37):

$$\sigma^2_{\alpha,j} = \frac{2.7}{\overline{e^2_{\min,i}}}\alpha^2_{jd}\sigma^2_{\psi,i} \quad (24.45)$$

This is a general relationship illustrating how the variance of the output of the companion system is related to the variance of one of the nominally constant parameters. But we are interested in the situation where the parameter is varied so as to cause exact matching. In this case we have shown that equation (24.26) applies:

$$\boldsymbol{\psi}(t) = \mathbf{e}_{\min}(t) \quad (24.26)$$

which implies immediately that

$$\sigma^2_{\psi,i} = {\sigma^0_{\psi,i}}^2 = \overline{e^2_{\min,i}} \quad (24.46)$$

where we have applied the superscript '0' to denote the variance in ψ_i when model-matching has been achieved. The corresponding terms may therefore be cancelled in equation (24.45) to give

$$\sigma^2_{\alpha,j} = 2.7\alpha^2_{jd} = (\Delta a^{(i)}_j)^2 \quad (24.47)$$

where we have assigned the symbol $(\Delta a^{(i)}_j)^2$ to the variance of parameter a_j needed to cause the model to match the ith measured variable on the plant exactly. This gives the standard deviation, $\Delta a^{(i)}_j$, as:

$$\Delta a^{(i)}_j = \sqrt{2.7}\alpha_{jd} \quad (24.48)$$

In fact we know that the real system is unlikely to be symmetrical about the mean of the mean-squared-error graph, and so it is better to take an average of the steady variations needed to double the mean squared error on each side of the minimum. Therefore we put

$$\alpha_{jd} = \frac{\alpha_{jd,1}+\alpha_{jd,2}}{2} \quad (24.49)$$

where $\alpha_{jd,1}$ is the lower deviation that causes doubling of the minimum mean squared error, while $\alpha_{jd,2}$ is the

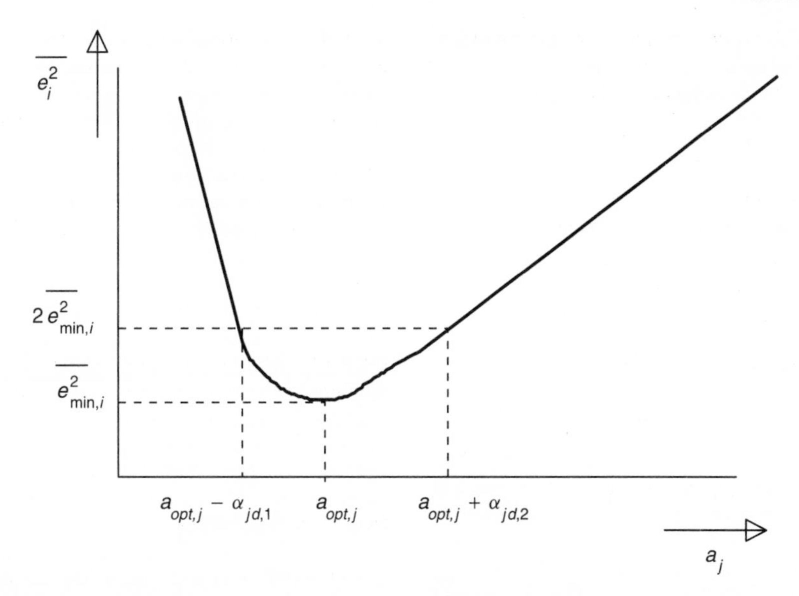

Figure 24.4 Parameter optimization curve illustrating non-symmetrical error doubling.

upper deviation that causes doubling of the minimum mean squared error (see Figure 24.4).

Substituting back into equation (24.48) gives the final estimate of the standard deviation in the parameter a_j needed for the model to match the ith measured variable on the plant exactly as:

$$\Delta a_j^{(i)} = \frac{\sqrt{2.7}}{2}(\alpha_{jd,1} + \alpha_{jd,2}) = 0.82(\alpha_{jd,1} + \alpha_{jd,2}) \tag{24.50}$$

Clearly a number of approximations have been used in deriving this estimate, which cannot be claimed as exact. However, a degree of corroboration has been given by Cameron's alternative statistical treatment based on applying an F-distribution to a linear system sampled at discrete time intervals (Cameron *et al.*, 1998). Putting

$$k_\alpha = \frac{\Delta a_j^{(i)}}{\alpha_{jd,1} + \alpha_{jd,2}} \tag{24.51}$$

information based on Cameron *et al.*'s Table 1 is given in Table 24.1.

The values of k_α contained in Table 24.1 are broadly in-line with the derived value of $k_\alpha = 0.82$ implicit in equation (24.50).

24.4.3 Model acceptance for transfer-function-based technique: explainability

Let us consider first 'explainability': whether or not the model is able to explain the observed plant transient.

Table 24.1 Estimates of $k_\alpha = \Delta a_j^{(i)}/(\alpha_{jd,1} + \alpha_{jd,2})$ for different numbers of nominally constant parameters and different numbers of data points, taken from Cameron *et al.* (1998).

Number of nominally constant parameters, p	*Number of sampled data points in transient*	*Significance level for F-test*	$k_\alpha = \dfrac{\Delta a_j^{(i)}}{\alpha_{jd,1} + \alpha_{jd,2}}$
1	10	5%	0.74
3	20	5%	0.74
5	40	5%	0.59
10	60	5%	0.63
15	60	5%	0.78
20	60	5%	0.93
3	20	1%	0.93
5	40	1%	0.71
10	60	1%	0.73
15	60	1%	0.89
20	60	1%	1.05

The first test is whether the optimal, constant value of each parameter lies within a reasonable range of the value expected in advance, $\hat{a}_j$:

$$\hat{a}_j - n\tau_j \leq a_{opt,j} \leq \hat{a}_j + n\tau_j \qquad \text{for } j = 1 \text{ to } p \tag{24.52}$$

where τ_j is the standard deviation expected in advance (the symbol τ has been chosen to connote 'tolerance') and n may be chosen as, for instance, 2 or 3.

The second test makes uses of the analysis of the previous section. It is a standard result that the mean squared value of the output of the companion system described by equation (24.29) will be related to the mean squared value or variance of the input by:

$$\sigma_{\psi,i}^2 = |g_{ij}(i\omega)|^2 \sigma_{\alpha,j}^2 \quad (24.53)$$

When the input variance, $\sigma_{\alpha,j}^2$ is made equal to $(\Delta a_j^{(i)})^2$, the variance of the jth parameter, a_j, needed for the model to match the ith measured variable on the plant, then the output variance will be the model-matching output variance:

$$\sigma_{\psi,i}^{0\ 2} = |g_{ij}|^2(\Delta a_j^{(i)})^2 \qquad \text{for } j = 1 \text{ to } p \quad (24.54)$$

However, it is reasonable to assume that variations may take place in more than just one parameter at a time. Making the assumption, usually reasonable, that the model's nominally constant parameters are independent of eachother, we may add weighted contributions:

$$\sigma_{\psi,i}^{0\ 2} = \sum_{j=1}^{p} |g_{ij}|^2 d_j (\Delta a_j^{(i)})^2 \quad (24.55)$$

where d_j is the fraction of the jth parameter variance needed on its own to eliminate the mismatch. Substituting from equation (24.54) into (24.55) gives:

$$\sigma_{\psi,i}^{0\ 2} = \sum_{j=1}^{p} d_j \sigma_{\psi,i}^{0\ 2} = \sigma_{\psi,i}^{0\ 2} \sum_{j=1}^{p} d_j \quad (24.56)$$

indicating that the sum of the weighting coefficients, d_j, is unity:

$$\sum_{j=1}^{p} d_j = 1 \quad (24.57)$$

The contributory variance required for model matching should be less than or equal to this expected tolerance, implying the p conditions:

$$d_j(\Delta a_j^{(i)})^2 \leq \tau_j^2 \qquad \text{for } j = 1 \text{ to } p \quad (24.58)$$

or

$$\frac{\tau_j^2}{(\Delta a_j^{(i)})^{22}} \geq d_j \qquad \text{for } j = 1 \text{ to } p \quad (24.59)$$

We may add these p inequalities together:

$$\sum_{j=1}^{p} \frac{\tau_j^2}{(\Delta a_j^{(i)})^2} \geq \sum_{j=1}^{p} d_j \quad (24.60)$$

Using equation (24.57), this simplifies to

$$\sum_{j=1}^{p} \frac{\tau_j^2}{(\Delta a_j^{(i)})^2} \geq 1 \quad (24.61)$$

This last equation represents the second test for explainability. The allowable fractional contribution in variance from the parameters should combine to more than unity if the mismatch between model and plant measurement i is to be explainable in terms of known model approximations.

In many simulation studies there is one plant variable that is overridingly important and which is available from a test recording, in which case applying the procedure indicated above just once will provide the necessary test of explainability. In some cases, however, there will be several recorded plant variables of interest, when it becomes necessary to repeat the procedure. Noting that a new set of optimal parameters is needed to minimize the mismatch for each recorded variable, it is necessary to carry out the parameter-optimization procedure k times, once for each measured output. We may then carry out the first test for explainability k times:

$$\hat{a}_j - n\tau_j \leq a_{opt,j}^{(i)} \leq \hat{a}_j + n\tau_j$$
$$\text{for } j = 1 \text{ to } p, i = 1 \text{ to } k \quad (24.62)$$

The upper and lower doubling distortions may be found for each parameter at the end of each parameter-optimization exercise, and we may then use equation (24.50) to determine $\Delta a_j^{(i)}$ for $j = 1, \ldots, p$; $i = 1, \ldots, k$. This allows us to apply the second test of explainability k times, where k is the number of recorded variables:

$$\sum_{j=1}^{p} \frac{\tau_j^2}{(\Delta a_j^{(i)})^2} \geq 1 \qquad \text{for } i = 1 \text{ to } k \quad (24.63)$$

24.4.4 Model acceptance for transfer-function-based technique: predictability

Explainability is only the first criterion for model acceptance. To be useful, the model must be able to predict the behaviour one or more key variables to a reasonable level of accuracy. Supposing the model is to be used to predict the behaviour of a certain important variable, θ, (e.g. the peak value of reactor exit temperature during a transient) then clearly the uncertainty in the nominally constant parameters will affect the tolerance to be placed on the calculated value of θ. Providing equation (24.61) or the more general equation (24.63) is satisfied, then a conservative estimate of the variance, σ_θ^2, to be associated with the variable, θ, is

$$\sigma_\theta^2 = \sum_{j=1}^{p} \left(\frac{\partial \theta}{\partial a_j} \right)^2 \tau_j^2 \quad (24.64)$$

On the other hand, we may give full weight to knowledge derived from the results of the transient by replacing τ_j^2 in equation (24.64) by the weighted contribution $d_j(\Delta a_j^{(i)})^2$:

$$\sigma_\theta^2 = \sum_{j=1}^{p} \left(\frac{\partial\theta}{\partial a_j}\right)^2 d_j(\Delta a_j^{(i)})^2 \tag{24.65}$$

where the variances $(\Delta a_j^{(i)})^2$ in the multiple-recorded-transient case would need to be chosen as a matter of judgement as those most likely to be relevant. For example, if θ represented peak reactor exit temperature for a new postulated transient and reactor exit temperature had been one of the recorded transients in a previous plant test, then it would make sense to use the estimates of parameter variances derived from matching the model to the previous record of reactor exit temperature. Unfortunately the analysis does not provide us with firm values for the weighting coefficients, d_j. An average value is

$$d_j = \frac{1}{p} \qquad \text{for all } j \tag{24.66}$$

while a highly pessimistic estimate is

$$d_j = 1 \qquad \text{for all } j \tag{24.67}$$

Naturally, for us to have high confidence in the model's predictions, the plant transient against which the model has been tested has to be sufficiently wide-ranging to traverse all the regions of the model upon which later predictions will be based.

24.5 Time-domain technique for the solution of the model distortion equations

24.5.1 Finding the parameter variations needed to match the behaviour of all recorded variables

The time-domain method presented here tackles head-on the problem of finding the parameter variations necessary to cause model matching for all k recorded variables. In some cases, the derivative of the measured state will itself be a state in the model, in which case the behaviour of two state variables may be found from a single record using the simple procedure of differentiation, which may be carried out to very good accuracy off-line. For example, suppose both position, x, and velocity, $v = dx/dt$, are state variables

$$\mathbf{x} = \begin{bmatrix} x \\ v \end{bmatrix} \tag{24.68}$$

in a model defined by:

$$\frac{d\mathbf{x}}{dt} = \begin{bmatrix} \dfrac{dx}{dt} \\ \dfrac{dv}{dt} \end{bmatrix} = \mathbf{f}(\mathbf{x}, \mathbf{u}, \mathbf{a}, t) \tag{24.69}$$

Given a measurement record of distance versus time, it will be possible to differentiate this record to obtain $v = dx/dt$, the time history of the second state, v. We shall give the name 'inferred-state history' to the behaviour of a state variable deduced from a measurement record either directly or by differentiation.

Using the procedure outlined above, we may generate a total of $k_1 \geq k$ 'inferred state histories', $\mathbf{z}_1$ from the k original instrument outputs, $\mathbf{z}$. We now sort the differential equations (DEs) defining the nominal model (equation (24.3)) into sets on the basis of the state variables they contain:

- the first set should contain the DEs for the k_1 states corresponding to the inferred-state histories;
- the second set should contain the DEs for the k_2 states that appear as 'inputs' to the first set of DEs;
- the third set should contain the DEs for the k_3 states that appear as 'inputs' to the first set of DEs;

and so on. The sorting process should continue until all the DEs have been sorted into sets, with a total number of N sets, with $N \leq n$, where n is the total number of model DEs in the nominal model. The total number of equations will be unaltered by this allocation process:

$$\sum_{m=1}^{N} k_m = n$$

The first set of DEs will have the form

$$\frac{d\mathbf{x}^{(1)}}{dt} = \mathbf{f}^{(1)}(\mathbf{x}^{(1)}, \mathbf{x}^{(2)}, \mathbf{u}, \mathbf{a}^{(1)}, t) \tag{24.70}$$

where the vector $\mathbf{x}^{(1)}$ contains the k_1 inferred states:

$$\mathbf{x}^{(1)} = \begin{bmatrix} x_1 \\ x_2 \\ \vdots \\ x_{k_1} \end{bmatrix} \tag{24.71}$$

while the vector $\mathbf{x}^{(2)}$ contains the k_2 unmeasured states having a direct effect on the differential $d\mathbf{x}^{(1)}/dt$:

$$\mathbf{x}^{(2)} = \begin{bmatrix} x_{k_1+1} \\ x_{k_1+2} \\ \vdots \\ x_{k_1+k_2} \end{bmatrix} \tag{24.72}$$

and the vector $\mathbf{a}^{(1)}$ contains the p_1 parameters that appear explicitly in equation (24.70):

$$\mathbf{a}^{(1)} = \begin{bmatrix} a_1 \\ a_2 \\ \vdots \\ a_{p_1} \end{bmatrix} \quad (24.73)$$

Similarly, the second set of DEs will have the form:

$$\frac{d\mathbf{x}^{(2)}}{dt} = \mathbf{f}^{(2)}(\mathbf{x}^{(1)}, \mathbf{x}^{(2)}, \mathbf{x}^{(3)}, \mathbf{u}, \mathbf{a}^{(1)}, \mathbf{a}^{(2)}, t) \quad (24.74)$$

and so on.

By this ordering process we replace the original set of DEs of equation (24.3) by the N subsystems:

$$\frac{d\mathbf{x}^{(1)}}{dt} = \mathbf{f}^{(1)}(\mathbf{x}^{(1)}, \mathbf{x}^{(2)}, \mathbf{u}, \mathbf{a}^{(1)}, t)$$

$$\frac{d\mathbf{x}^{(2)}}{dt} = \mathbf{f}^{(2)}(\mathbf{x}^{(1)}, \mathbf{x}^{(2)}, \mathbf{x}^{(3)}, \mathbf{u}, \mathbf{a}^{(1)}, \mathbf{a}^{(2)}, t)$$

$$\vdots$$

$$\frac{d\mathbf{x}^{(i)}}{dt} = \mathbf{f}^{(i)}(\mathbf{x}^{(1)}, \mathbf{x}^{(2)}, \ldots \mathbf{x}^{(i)}, \mathbf{x}^{(i+1)}, \mathbf{u}, \mathbf{a}^{(1)}, \mathbf{a}^{(2)} \ldots \mathbf{a}^{(i)}, t)$$

$$\vdots$$

$$\frac{d\mathbf{x}^{(N)}}{dt} = \mathbf{f}^{(i)}(\mathbf{x}^{(1)}, \mathbf{x}^{(2)}, \ldots \mathbf{x}^{(N)}, \mathbf{x}^{(N+1)}, \mathbf{u}, \mathbf{a}^{(1)}, \mathbf{a}^{(2)} \ldots \mathbf{a}^{(N)}, t) \quad (24.75)$$

The vector of all the state variables, $\mathbf{x}$, is related to the subsystems' state vectors by:

$$\mathbf{x} = \begin{bmatrix} \mathbf{x}^{(1)} \\ \mathbf{x}^{(2)} \\ \vdots \\ \mathbf{x}^{(N)} \end{bmatrix} \quad (24.76)$$

while the vector, $\mathbf{a}$, of nominally constant parameters is related to the subvectors, $\mathbf{a}^{(i)}$, $i = 1, 2, \ldots N$, by:

$$\mathbf{a} = \begin{bmatrix} \mathbf{a}^{(1)} \\ \mathbf{a}^{(2)} \\ \vdots \\ \mathbf{a}^{(N)} \end{bmatrix} \quad (24.77)$$

The total number of nominally constant parameters remains unchanged of course:

$$\sum_{i=1}^{N} p_i = p$$

We may now deal with each of the N subsystems in sequence, starting with subsystem 1. We require that the states of this subsystem will match at all times the k_1 inferred-state histories, i.e.

$$\mathbf{x}^{(1)}(t) = \mathbf{z}_1(t) \quad (24.78)$$

It follows directly by differentiation that

$$\frac{d\mathbf{x}^{(1)}}{dt}(t) = \frac{d\mathbf{z}_1}{dt}(t) \quad (24.79)$$

But, as was noted earlier, off-line differentiation can be an accurate process, and so the fact that we have the inferred-state histories, $\mathbf{z}_1$, means that we may deduce their derivatives, $d\mathbf{z}_1/dt$. We may use this fact to restate our problem as the need to satisfy the equation:

$$\frac{d\mathbf{z}_1}{dt} = \mathbf{f}^{(1)}(\mathbf{x}^{(1)}, \mathbf{x}^{(2)}, \mathbf{u}, \mathbf{a}^{(1)}, t) \quad (24.80)$$

given $d\mathbf{z}_1/dt$. We may vary the nominally constant parameters, $\mathbf{a}^{(1)}$, but we should note also that, in the absence of direct measurements, we have no reason to suppose that our central estimates for $\mathbf{x}^{(2)}$ are precisely correct. Accordingly we may, in addition, impose variations from the central values on the state subvector, $\mathbf{x}^{(2)}$. Hence the model-matching problem is to find, at each time instant, t, parameter variations, $\boldsymbol{\alpha}^{(1)}$, and state deviations, $\chi^{(2)}$, that will satisfy the condition:

$$\mathbf{f}^{(1)}(\mathbf{z}_1, \hat{\mathbf{x}}^{(2)} + \chi^{(2)}, \mathbf{u}, \hat{\mathbf{a}}^{(1)} + \boldsymbol{\alpha}^{(1)}, t) - \frac{d\mathbf{z}_1}{dt} = \mathbf{0} \quad (24.81)$$

Here $\hat{\mathbf{x}}^{(2)}$ represents the central trajectories of the k_2 states in $\mathbf{x}^{(2)}$ as found from the undistorted model and $\hat{\mathbf{a}}^{(1)}$ represents the central estimates of the p_1 parameters in $\mathbf{a}^{(1)}$.

Equation (24.81) is a system of k_1 nonlinear, simultaneous equations in $(k_2 + p_1)$ unknowns. If there are more equations than unknowns, i.e. $k_1 > (k_2 + p_1)$, then the possibility of inconsistency arises, and a solution may not be possible. If the number of equations equals the number of unknowns, i.e. $k_1 = (k_2 + p_1)$, then any solution will be unique. If there are more unknowns than equations, i.e. $k_1 < (k_2 + p_1)$, then the best that can be done is to define the k_1 unknowns in terms of the remaining $(k_2 + p_1) - k_1$ unknowns. The last possibility is almost certainly the one we will meet, and although the solution appears weak at first glance, nevertheless it will prove sufficient for our purposes.

The essence of our problem is that we want to match the plant transients, but to maintain our parameter variations within as narrow bands as possible. Certainly we will wish to keep the mean-squared value of the parameter variations below the expected parameter variance, i.e.:

$$\overline{(\alpha_j^{(1)})^2} \leq \tau_j^2 \qquad \text{for } j = 1 \text{ to } p_1 \tag{24.82}$$

Furthermore, the deviations, $\boldsymbol{\chi}^{(2)}$, of the unmeasured state variables should be small, since any large deviations would imply that large variations would be needed in the parameters, $\mathbf{a}^{(2)}$, of the second subsystem of equation (24.75). Accordingly we shall need to keep the mean-squared values of $\boldsymbol{\chi}^{(2)}$ small:

$$\overline{(\chi_j^{(2)})^2} \leq \tau_{x,j}^2 \qquad \text{for } j = 1 \text{ to } k_2 \tag{24.83}$$

where $\tau_{x,j}^2$ is the variance we may assign in advance to the deviation from its original value of the jth state variable in the second set.

Combining our requirement to minimize parameter variations and unmeasured-state deviations with the requirement for model-plant matching, we pose the problem mathematically as: at each time instant, minimize the scaled norm $h(\boldsymbol{\alpha}^{(1)}, \boldsymbol{\chi}^{(2)})$ of parameter variations and state deviations where

$$h(\boldsymbol{\alpha}^{(1)}, \boldsymbol{\chi}^{(2)}) = \sum_{j=1}^{p_1} (w_j^{(1)})^2 (\alpha_j^{(1)})^2 + \sum_{j=1}^{k_2} (w_{x,j}^{(2)})^2 (\chi_j^{(2)})^2 \tag{24.84}$$

subject to the k_1 constraints implied by equation (24.81):

$$\boldsymbol{\Phi}^{(1)}(\boldsymbol{\alpha}^{(1)}, \boldsymbol{\chi}^{(2)}) = \mathbf{f}^{(1)}(\mathbf{z}_1, \hat{\mathbf{x}}^{(2)} + \boldsymbol{\chi}^{(2)}, \mathbf{u}, \hat{\mathbf{a}}^{(1)} + \boldsymbol{\alpha}^{(1)}, t) - \frac{d\mathbf{z}_1}{dt} = \mathbf{0} \tag{24.85}$$

The weighting factors in equation (24.84) could be chosen to scale the variations in terms of the expected variances, i.e.:

$$w_j^{(1)} = \frac{1}{\tau_j^{(1)}} \qquad w_{x,j}^{(2)} = \frac{1}{\tau_{x,j}^{(2)}} \tag{24.86}$$

or else in terms of the central values:

$$w_j^{(1)} = \frac{1}{\hat{a}_j^{(1)}} \qquad w_{x,j}^{(2)} = \frac{1}{\hat{x}_j^{(2)}} \tag{24.87}$$

The problem may be seen as that of minimizing an objective function, subject to a set of nonlinear constraints. It may be solved by Lagrange's method as follows. Form the function:

$$H = h + \sum_{i=1}^{k_1} \lambda_i^{(1)} \phi_i^{(1)} \tag{24.88}$$

and determine the k_1 parameters, $\lambda_i^{(1)}$, the p_1 parameters, $\alpha_i^{(1)}$, and the k_2 parameters, $\chi_i^{(2)}$ from the p_1 equations:

$$\frac{\partial H}{\partial \alpha_j^{(1)}} = 0 \qquad \text{for } j = 1 \text{ to } p_1 \tag{24.89}$$

the k_2 equations

$$\frac{\partial H}{\partial \chi_j^{(2)}} = 0 \qquad \text{for } j = 1 \text{ to } k_2 \tag{24.90}$$

and the k_1 constraint equations

$$\phi_j^{(1)} = 0 \qquad \text{for } j = 1 \text{ to } k_1 \tag{24.91}$$

These $p_1 + k_2 + k_1$ nonlinear simultaneous equations in $p_1 + k_2 + k_1$ unknowns may be solved at each timestep using a nonlinear equation solver of the type discussed in Chapter 2, Section 2.10 or by the Method of Referred Derivatives, discussed in Section 2.11 and in Chapter 18, Sections 18.7 to 18.9. The application of the latter method will be discussed further in the next section.

The solution will allow a picture to be assembled of the time behaviour of each of the nominally constant parameters, now given by:

$$\mathbf{a}^{(1)}(t) = \hat{\mathbf{a}}^{(1)} + \boldsymbol{\alpha}^{(1)}(t) \tag{24.92}$$

Statistical measures such as mean-squared value may then be evaluated and used in later tests. In addition, the modified time-histories of the state variables in $\mathbf{x}^{(2)}$ in will have been calculated:

$$\mathbf{x}^{(2)}(t) = \hat{\mathbf{x}}^{(2)}(t) + \boldsymbol{\chi}^{(2)}(t) \tag{24.93}$$

The derivatives $d\mathbf{x}^{(2)}/dt$ may be found by numerical differentiation:

$$\frac{d\mathbf{x}^{(2)}}{dt}(t) = \frac{d\hat{\mathbf{x}}^{(2)}}{dt}(t) + \frac{d\boldsymbol{\chi}^{(2)}}{dt}(t) \tag{24.94}$$

Then the values of $\mathbf{x}^{(2)}(t)$ and $d\mathbf{x}^{(2)}/dt$ may be supplied to the second subsystem of equation (24.75), where they take on roles analogous to those played by $\mathbf{z}_1(t)$ and $(d\mathbf{z}_1(t)/dt)$ in the first subsystem. The second subsystem will also be supplied with the time-histories of the first set of nominally constant parameters, $\mathbf{a}^{(1)}(t)$.

A similar procedure is now applied to the second set of DEs in order to find $\mathbf{a}^{(2)}(t)$ and $\mathbf{x}^{(3)}(t)$. We seek to minimize the scaled norm given by:

$$h(\boldsymbol{\alpha}^{(2)}, \boldsymbol{\chi}^{(3)}) = \sum_{j=1}^{p_2} (w_j^{(2)})^2 (\alpha_j^{(2)})^2 + \sum_{j=1}^{k_3} (w_{x,j}^{(3)})^2 (\chi_j^{(3)})^2 \quad (24.95)$$

subject to the k_2 constraints:

$$\boldsymbol{\Phi}^{(2)}(\boldsymbol{\alpha}^{(2)}, \boldsymbol{\chi}^{(3)}) = \mathbf{f}^{(2)}(\mathbf{z}_1, \mathbf{x}^{(2)}, \hat{\mathbf{x}}^{(3)}, +\boldsymbol{\chi}^{(3)}, \mathbf{u}, \mathbf{a}^{(1)}, \hat{\mathbf{a}}^{(2)}, +\boldsymbol{\alpha}^{(2)}, t) - \frac{d\mathbf{x}^{(2)}}{dt} = \mathbf{0} \quad (24.96)$$

Hence we form the Lagrange function:

$$H = h + \sum_{i=1}^{k_2} \lambda_i^{(2)} \phi_i^{(2)} \quad (24.97)$$

and find the k_2 parameters, $\lambda_i^{(2)}$, the p_2 parameters, $\alpha_i^{(2)}$, and the k_3 parameters, $\chi_i^{(3)}$ from the $p_2 + k_3 + k_2$ equations:

$$\frac{\partial H}{\partial \alpha_j^{(2)}} = 0 \qquad \text{for } j = 1 \text{ to } p_2 \quad (24.98)$$

$$\frac{\partial H}{\partial \chi_j^{(3)}} = 0 \qquad \text{for } j = 1 \text{ to } k_3 \quad (24.99)$$

$$\phi_j^{(2)} = 0 \qquad \text{for } j = 1 \text{ to } k_2 \quad (24.100)$$

Thus $\mathbf{a}^{(2)}(t)$, $\mathbf{x}^{(3)}(t)$ and, by numerical differentiation, also $d\mathbf{x}^{(3)}/dt$, may now be calculated, and supplied to the third set of DEs.

The process may then be applied repeatedly until in the final subsystem, the Nth, the only free variables will be contained in the vector, $\mathbf{a}^{(N)}$. The process will come to an end when the behaviour of these variables has been calculated. At this stage the behaviour of the recorded transients will have been mapped completely onto the behaviour of the system's nominally constant parameters.

A check on the accuracy of the mapping process may be achieved by re-running the model with the calculated parameter distortions, and comparing the calculated states with the measurements. Ideally they should be identical, but in any case statistical measures of mismatch may be calculated and used to judge how closely the distorted model fits the plant records.

24.5.2 Calculating the parameter variations using the Method of Referred Derivatives

We will introduce the following notation to simplify the analysis. Let the p_1-dimensional vector, $\mathbf{g}_1$, be given by:

$$\mathbf{g}_1 = \frac{\partial H}{\partial \boldsymbol{\alpha}^{(1)}} = \begin{bmatrix} \dfrac{\partial H}{\partial \alpha_1^{(1)}} \\ \dfrac{\partial H}{\partial \alpha_2^{(1)}} \\ \vdots \\ \dfrac{\partial H}{\partial \alpha_{p1}^{(1)}} \end{bmatrix} \quad (24.101)$$

Let the k_2-dimensional vector, $\mathbf{g}_2$, be given by:

$$\mathbf{g}_2 = \frac{\partial H}{\partial \boldsymbol{\chi}^{(2)}} = \begin{bmatrix} \dfrac{\partial H}{\partial \chi_1^{(2)}} \\ \dfrac{\partial H}{\partial \chi_2^{(2)}} \\ \vdots \\ \dfrac{\partial H}{\partial \chi_{k_2}^{(2)}} \end{bmatrix} \quad (24.102)$$

Let the k_1-dimensional vector, $\mathbf{g}_3$, be given by:

$$\mathbf{g}_3 = \boldsymbol{\phi}^{(1)} = \begin{bmatrix} \phi_1^{(1)} \\ \phi_2^{(1)} \\ \vdots \\ \phi_{k_1}^{(1)} \end{bmatrix} \quad (24.103)$$

Let these vectors now be combined to form the $(p_1 + k_2 + k_1)$-dimensional vector, $\mathbf{G}$:

$$\mathbf{G} = \begin{bmatrix} \mathbf{g}_1 \\ \mathbf{g}_2 \\ \mathbf{g}_3 \end{bmatrix} \quad (24.104)$$

Using this notation, equations (24.89) to (24.91) may be expressed in the form

$$\mathbf{G}(\boldsymbol{\alpha}^{(1)}(t), \boldsymbol{\chi}^{(2)}(t), \boldsymbol{\lambda}^{(1)}(t), \mathbf{z}_1(t), \dot{\mathbf{z}}_1(t), \hat{\mathbf{x}}^{(2)}(t), \mathbf{u}(t), \hat{\mathbf{a}}^{(1)}, t) = \mathbf{0} \quad (24.105)$$

where the dependence on time has been shown for the parameter variations, $\boldsymbol{\alpha}^{(1)}$, state deviations, $\boldsymbol{\chi}^{(2)}$, and Lagrange multipliers, $\boldsymbol{\lambda}^{(1)}$.

Applying the Method of Referred Derivatives, we argue that since $\mathbf{G}$ is identically zero at all times, it follows that its derivative will be zero also:

$$\frac{d\mathbf{G}}{dt} = \frac{\partial \mathbf{G}}{\partial \boldsymbol{\alpha}^{(1)}} \frac{d\boldsymbol{\alpha}^{(1)}}{dt} + \frac{\partial \mathbf{G}}{\partial \boldsymbol{\chi}^{(2)}} \frac{d\boldsymbol{\chi}^{(2)}}{dt} + \frac{\partial \mathbf{G}}{\partial \boldsymbol{\lambda}^{(1)}} \frac{d\boldsymbol{\lambda}^{(1)}}{dt} + \frac{\partial \mathbf{G}}{\partial \mathbf{z}_1} \frac{d\mathbf{z}_1}{dt} + \frac{\partial \mathbf{G}}{\partial \dot{\mathbf{z}}_1} \frac{d^2\mathbf{z}_1}{dt^2} + \frac{\partial \mathbf{G}}{\partial \hat{\mathbf{x}}^{(2)}} \frac{d\hat{\mathbf{x}}^{(2)}}{dt} + \frac{\partial \mathbf{G}}{\partial \mathbf{u}} \frac{d\mathbf{u}}{dt} + \frac{\partial \mathbf{G}}{\partial t} = \mathbf{0} \quad (24.106)$$

In the usual case where the model differential equations have no explicit dependence on time, we may simplify by putting $\partial \mathbf{G}/\partial t = 0$. Hence equation (24.106) may be rearranged to:

$$\begin{bmatrix} \dfrac{\partial \mathbf{G}}{\partial \boldsymbol{\alpha}^{(1)}} & \dfrac{\partial \mathbf{G}}{\partial \boldsymbol{\chi}^{(2)}} & \dfrac{\partial \mathbf{G}}{\partial \boldsymbol{\lambda}^{(1)}} \end{bmatrix} \begin{bmatrix} \dfrac{d\boldsymbol{\alpha}^{(1)}}{dt} \\ \dfrac{d\boldsymbol{\chi}^{(2)}}{dt} \\ \dfrac{d\boldsymbol{\lambda}^{(1)}}{dt} \end{bmatrix} = -\frac{\partial \mathbf{G}}{\partial \mathbf{z}_1}\frac{d\mathbf{z}_1}{dt} - \frac{\partial \mathbf{G}}{\partial \dot{\mathbf{z}}_1}\frac{d^2\mathbf{z}_1}{dt^2} - \frac{\partial \mathbf{G}}{\partial \hat{\mathbf{x}}^{(2)}}\frac{d\hat{\mathbf{x}}^{(2)}}{dt} - \frac{\partial \mathbf{G}}{\partial \mathbf{u}}\frac{d\mathbf{u}}{dt} \tag{24.107}$$

This represents a set of $p_1 + k_2 + k_1$ *linear*, simultaneous equations in the $p_1 + k_2 + k_1$ unknown derivatives, $d\boldsymbol{\alpha}^{(1)}/dt$, $d\boldsymbol{\chi}^{(2)}/dt$ and $d\boldsymbol{\lambda}^{(1)}/dt$. The derivatives on the right-hand side of equation (24.107) are already known or may be found by numerical differentiation. Equation (24.107) is in the canonical form of equation (2.98), and is thus amenable to a similar method of solution. The derivatives, $d\boldsymbol{\alpha}^{(1)}/dt$, $d\boldsymbol{\chi}^{(2)}/dt$ and $d\boldsymbol{\lambda}^{(1)}/dt$ may be found at each timestep and, from an initial solution for $\boldsymbol{\alpha}^{(1)}$, $\boldsymbol{\chi}^{(2)}$, and $\boldsymbol{\lambda}^{(1)}$ at time t_0, these derivatives may be integrated numerically with respect to time in the normal way. For further details see also Chapter 18, Section 18.7. As a result we may deduce the time-varying behaviour of the first set of parameters, $\mathbf{a}^{(1)}(t)$, needed for model matching.

The second set of DEs given in equation (24.75) may now be supplied with the generated values of the first set of constants, $\mathbf{a}^{(1)}(t)$, and the second set of state variables $\mathbf{x}^{(2)}(t)$, together with their derivatives, formed by:

$$\frac{d\mathbf{a}^{(1)}}{dt}(t) = \frac{d\boldsymbol{\alpha}^{(1)}}{dt}(t) \tag{24.108}$$

$$\frac{d\mathbf{x}^{(2)}}{dt}(t) = \frac{d\hat{\mathbf{x}}^{(2)}}{dt}(t) + \frac{d\boldsymbol{\chi}^{(2)}}{dt}(t) \tag{24.109}$$

The form of the Lagrange function, H, for the second set of DEs is now given by equations (24.95) to (24.97). Hence our $(p_2 + k_3 + k_2)$-dimensional $\mathbf{G}$ vector is given by:

$$\underset{\sim}{\mathbf{G}} = \begin{bmatrix} \dfrac{\partial H}{\partial \boldsymbol{\alpha}^{(2)}} \\ \dfrac{\partial H}{\partial \boldsymbol{\chi}^{(3)}} \\ \boldsymbol{\phi}^{(2)} \end{bmatrix} \tag{24.110}$$

Thus equations (24.98) to (24.100) may be written:

$$\mathbf{G}(\boldsymbol{\alpha}^{(2)}(t), \boldsymbol{\chi}^{(3)}(t), \boldsymbol{\lambda}^{(2)}(t), \mathbf{z}_1(t), \mathbf{x}^{(2)}(t), \dot{\mathbf{x}}^{(2)}(t), \hat{\mathbf{x}}^{(3)}(t), \mathbf{u}(t), \mathbf{a}^{(1)}(t), \hat{\mathbf{a}}^{(2)}, t) = \mathbf{0} \tag{24.111}$$

Applying the Method of Referred Derivatives to the case where the model has no explicit dependence on time (i.e. $\partial \mathbf{G}/\partial t = \mathbf{0}$) produces the linear equation

$$\begin{bmatrix} \dfrac{\partial \mathbf{G}}{\partial \boldsymbol{\alpha}^{(2)}} & \dfrac{\partial \mathbf{G}}{\partial \boldsymbol{\chi}^{(3)}} & \dfrac{\partial \mathbf{G}}{\partial \boldsymbol{\lambda}^{(2)}} \end{bmatrix} \begin{bmatrix} \dfrac{d\boldsymbol{\alpha}^{(2)}}{dt} \\ \dfrac{d\boldsymbol{\chi}^{(3)}}{dt} \\ \dfrac{d\boldsymbol{\lambda}^{(2)}}{dt} \end{bmatrix} = -\frac{\partial \mathbf{G}}{\partial \mathbf{z}_1}\frac{d\mathbf{z}_1}{dt} - \frac{\partial \mathbf{G}}{\partial \mathbf{z}_2}\frac{d\mathbf{x}^{(2)}}{dt} - \frac{\partial \mathbf{G}}{\partial \dot{\mathbf{z}}_2}\frac{d^2\mathbf{x}^{(2)}}{dt^2} - \frac{\partial \mathbf{G}}{\partial \hat{\mathbf{x}}^{(3)}}\frac{d\hat{\mathbf{x}}^{(3)}}{dt} - \frac{\partial \mathbf{G}}{\partial \mathbf{a}^{(1)}}\frac{d\mathbf{a}^{(1)}}{dt} - \frac{\partial \mathbf{G}}{\partial \mathbf{u}}\frac{d\mathbf{u}}{dt} \tag{24.112}$$

This equation may be solved for the derivatives $d\boldsymbol{\alpha}^{(2)}/dt$, $d\boldsymbol{\chi}^{(3)}/dt$ and $d\boldsymbol{\lambda}^{(2)}/dt$, which may then be integrated numerically as discussed above. By this method we shall have deduced the behaviour of the second set of nominally constant parameters, $\mathbf{a}^{(2)}(t) = \hat{\mathbf{a}}^{(2)} + \boldsymbol{\alpha}^{(2)}(t)$. In addition, we shall have determined the modified behaviour of the second set of state variables: $\mathbf{x}^{(3)}(t) = \hat{\mathbf{x}}^{(3)}(t) + \boldsymbol{\chi}^{(3)}(t)$.

We may pass on to the third set of DEs the time-histories: $\mathbf{a}^{(2)}(t)$, $\mathbf{x}^{(3)}(t)$ and their derivatives, found according to:

$$\frac{d\mathbf{a}^{(2)}}{dt}(t) = \frac{d\boldsymbol{\alpha}^{(2)}}{dt}(t) \tag{24.113}$$

$$\frac{d\mathbf{x}^{(3)}}{dt}(t) = \frac{d\hat{\mathbf{x}}^{(3)}}{dt}(t) + \frac{d\boldsymbol{\chi}^{(3)}}{dt}(t) \tag{24.114}$$

A similar procedure to the one described may be followed for this third set of DEs; and so on until the final mapping to give $\mathbf{a}^{(N)}(t)$ after considering the Nth set of DEs.

24.5.3 Model acceptance criteria for the time-domain technique: explainability

Unlike the transfer-function-based technique, the time-domain technique does not require the central values of the nominally constant parameters to be determined from a minimization exercise. Nevertheless, we will expect the modeller to use sensible estimates, which may be expressed as a condition similar to inequality (24.52):

$$a_j^0 - n\tau_j \le \hat{a}_j \le a_j^0 + n\tau_j \qquad \text{for } j = 1 \text{ to } p \tag{24.115}$$

where $\hat{a}_j$ is now taken as the value of the jth parameter used in the nominal model, while a_j^0 is the expected value of the jth parameter, and n is a low number, say 2 or 3. Usually, of course, the modeller will make

$$\hat{a}_j = a_j^0 \qquad \text{for } j = 1 \text{ to } p \tag{24.116}$$

Using the time-domain method described above, it is possible to calculate the time-behaviour of the model-matching parameter variations, and hence it is a simple matter to determine the contributory variance $(\delta a_j)^2$ of each parameter that is necessary to match model to all the transient measurements recorded from the plant. To satisfy the explainability criterion, the variance of each parameter should be within the variance anticipated in advance, i.e.:

$$(\delta a_j)^2 \leq \tau_j^2 \qquad \text{for } j = 1 \text{ to } p \tag{24.117}$$

Provided that this condition is satisfied for each of the p parameters, the model is acceptable in the first instance in the sense that the deviations of the model outputs from the plant records are explainable in terms of known approximations.

24.5.4 Model acceptance criteria for the time-domain technique: predictability

As outlined in Section 24.4.4, we are likely to want to use the model to predict the behaviour of a certain important variable, θ (e.g. the peak value of reactor exit temperature during a transient). Providing equation (24.117) is satisfied, then a conservative estimate of the variance, σ_θ^2, to be associated with the variable, θ, is given by equation (24.64), repeated below:

$$\sigma_\theta^2 = \sum_{j=1}^{p} \left(\frac{\partial \theta}{\partial a_j} \right)^2 \tau_j^2 \tag{24.64}$$

An estimate for σ_θ^2 giving full weight to knowledge derived from the results of the transients used as the basis for the model-distortion exercise would be

$$\sigma_\theta^2 = \sum_{j=1}^{p} \left(\frac{\partial \theta}{\partial a_j} \right)^2 (\delta a_j)^2 \tag{24.118}$$

The same caveat as uttered in Section 24.4.4 applies when the model is to be used for subsequent prediction purposes, namely that the recorded plant transients against which the model has been tested should be sufficiently wide-ranging to traverse all the regions of the model utilized by the later model calculations if we are to have a high confidence in the model's predictions.

24.6 Applications

The techniques outlined above have been applied to solve real problems on a number of plants. Some examples are detailed in Butterfield and Thomas (1986) and Li (1986).

24.7 Bibliography

Butterfield, M.H. and Thomas, P.J. (1986). Methods of quantitative validation for dynamic simulation models – Part 1: Theory, *Trans. Inst. Measurement and Control*, **8**, 182–200.

Butterfield, M.H. and Thomas, P.J. (1986). Methods of quantitative validation for dynamic simulation models – Part 2: Examples, *Trans. Inst. Measurement and Control*, **8**, 201–219.

Cameron, R.G. (1992). Model validation by the distortion method: linear state space systems, *IEE Proceedings-D*, **139**, 296–300.

Cameron, R.G., Marcos, R.L. and de Prada, C. (1998). Model validation of discrete transfer functions using the distortion method, *Mathematical and Computer Modelling of Dynamical Systems*, **4**, 58–72.

Douce, J.L. (1963). *An Introduction to the Mathematics of Servomechanisms*, The English Universities Press, London.

Li, C.L.R. (1986). Application of distortion technique for model validation to nuclear power plant – AGR scatter plug, *Trans. Inst. Measurement and Control*, **8**, 220–232.

Appendix 1 Comparative size of energy terms

A1.1 Introduction

The energy equation (3.34) below was derived in Chapter 3, Section 3.5:

$$\frac{dE}{dt} = \Phi - P + W_1\left(h_1 + \frac{1}{2}c_1^2 + gz_1\right) - W_2\left(h + \frac{1}{2}c_2^2 + gz\right) \qquad (3.34)$$

where the energy, E, contained in the bounded volume consisted of internal energy, kinetic energy and potential energy:

$$E = m\left(u + \tfrac{1}{2}c^2 + gz\right) \qquad (3.35)$$

Here m is the mass of the fluid in the bounded volume, c is its bulk velocity, and z is the height of its centre of gravity above the datum. Many textbooks make the assumption that the internal energy dominates, but often they do not provide the justification. This Appendix will offer justifications appropriate to the situations normally found on a process plant, where we are usually concerned with the contents of vessels or tanks fixed in space.

A1.2 Bulk kinetic energy

The contents of a process vessel will be subject to some degree of turbulence or mixing, and so we may normally regard their bulk velocity, gas or liquid, as zero. Thus

$$\tfrac{1}{2}c^2 \approx 0 \qquad (A1.1)$$

Hence equation (3.35) is immediately reduced to:

$$E = m(u + gz) \qquad (A1.2)$$

A1.3 The relative size of the potential energy term

Possible sizes of the potential energy term, gz, may be compared with typical values of specific internal energy, u. A large process vessel might conceivably be 20 m high, and have the centre of gravity of its contents located 10 m above its base, which we will take as the datum. The specific potential energy of the vessel contents, liquid or gas, would then be given by $gz = 9.8 \times 10 \approx 100$ J/kg. This figure is likely to be dwarfed by the specific internal energy.

For the first comparison, let us take the example of the specific internal energy of a low-temperature liquid: say water at 20°C and 1 bar. From the steam tables, this has a specific internal energy of 83 900 J/kg. This is nearly three orders of magnitude higher that the potential energy term, justifying amply the neglect of the latter.

The specific internal energy will rise with temperature for both liquids and gases. Thus the comparison will be more unequal if we move on to consider a gas at a higher temperature. Compressed air at 10 bar and 350 K has a specific internal energy of 248 200 J/kg, more than three orders of magnitude higher than the potential energy term quoted above. For superheated steam at 350°C and 100 bar, the specific internal energy is 2 702 000 J/kg, which is four orders of magnitude larger than the potential energy term.

Thus $u \gg gz$ in the cases considered, and on the basis that they are typical or even conservative process-plant examples, we may disregard the term gz in equation (A1.2) with negligible loss of accuracy.

However, it might be pointed out, with fairness, that equation (3.34) deals with the derivative of the total energy term, rather than the energy term itself. Is it possible that the derivative will allow the potential energy term more influence? This can be the case, but nevertheless disregarding the potential energy term will still introduce little error into the final result because the simulation will require the derivative to be integrated to give the final value. To illustrate this, let us differentiate formally equation (A1.2):

$$\frac{dE}{dt} = (u + gz)\frac{dm}{dt} + m\left(\frac{du}{dt} + g\frac{dz}{dt}\right) \qquad (A1.3)$$

Let us make the most general assumption that the specific internal energy is a function of both temperature and pressure (as might be the case for a non-ideal gas):

$$\frac{dE}{dt} = (u + gz)\frac{dm}{dt} + m \times \left(\frac{\partial u}{\partial T}\frac{dT}{dt} + \frac{\partial u}{\partial p}\frac{dp}{dt} + g\frac{dz}{dt}\right) \qquad (A.14)$$

Now assume that the mass, temperature and pressure of the fluid in the vessel stay constant, but that the height of its centre of gravity changes. This is a possible situation for the cover-gas over a liquid. In this instance equation (A1.4) reduces to simply:

$$\frac{dE}{dt} = mg\frac{dz}{dt} \tag{A1.5}$$

Now it is clear that the *only* energy change experienced by the gas in the bounded volume is the change in its potential energy due to the repositioning of the centre of gravity. The temperature does not change at all, and so the internal energy cannot change at all. If we persist in making the assumption that $E = mu$, it follows that, for constant mass,

$$\frac{dE}{dt} = m\frac{du}{dt} = mg\frac{dz}{dt} \tag{A1.6}$$

and hence that

$$\frac{du}{dt} = g\frac{dz}{dt} \tag{A1.7}$$

Integrating equation with respect to time, we find that

$$\begin{aligned} u(t) &= u(0) + \int_0^t g\frac{dz}{dt}\,dt \\ &= u(0) + g(z(t) - z(0)) \end{aligned} \tag{A1.8}$$

Equation (A1.8) must be incorrect, since it implies $u = u(z)$ instead of the true situation, $u = u(p, T)$. Nevertheless, while equation (A1.8) will introduce an error, that error cannot be large since the height differences must be bounded to the small values afforded by the limits of the vessel, and $u \gg gz$ as demonstrated above.

This bounding argument provides a general justification for making the assumption

$$E = mu \tag{A1.9}$$

and hence

$$\frac{dE}{dt} = m\frac{du}{dt} + u\frac{dm}{dt} \tag{A1.10}$$

for normal process plant situations.

However, it is useful to examine typical process configurations in more detail. Doing so reveals that the situation is often even more clear-cut in practice. Three representative cases will be considered:

(i) a liquid or a gas completely filling a vessel;
(ii) a liquid partially filling a vessel;
(iii) a gas partially filling a vessel and contained above a movable interface, such as a gas blanket over a liquid, undergoing a near adiabatic expansion or compression.

A1.4 Vessel filled with liquid or gas

The term $g(dz/dt)$ in equation (A1.3) is disposed of very quickly for a vessel that is always completely filled with either liquid or gas: assuming a homogeneous mixture, the centre of gravity will stay in the same position, and so $g(dz/dt) = 0$. Hence equation (A1.3) reduces to

$$\begin{aligned} \frac{dE}{dt} &= (u + gz)\frac{dm}{dt} + m\frac{du}{dt} \\ &\approx u\frac{dm}{dt} + m\frac{du}{dt} \end{aligned} \tag{A1.11}$$

since $u \gg gz$ as shown in Section A1.3.

A1.5 Liquid partially filling a vessel

When the liquid in a vessel has a free boundary, the centre of gravity of the liquid will move as the level changes. There will normally be a level controller to ensure that this movement is small, but in fact the term $g(dz/dt)$ can be expected to be insignificant, even

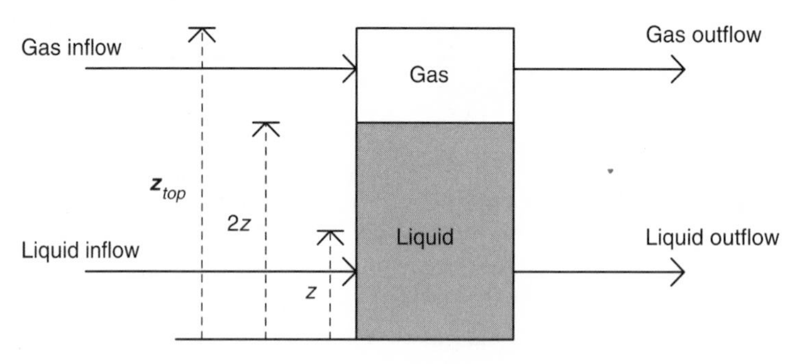

Figure A1.1 Liquid partially filling a vessel.

without level control. The height relative to the base of the vessel of the centre of gravity of the liquid will be

$$z = \frac{vm}{2A} \tag{A1.12}$$

for a vessel of uniform cross-sectional area, A, containing a mass of liquid, m, of specific volume, v (Figure A1.1).

A formal differentiation of (A1.12) with respect to time gives:

$$\frac{dz}{dt} = \frac{v}{2A}\frac{dm}{dt} + \frac{m}{2A}\frac{dv}{dt} \tag{A1.13}$$

Now for a liquid, both the specific volume, v, and the specific energy, u, are overwhelmingly dependent on temperature, with only a weak dependence on pressure. Hence we may use the approximations:

$$\frac{dv}{dt} \approx \frac{dv}{dT}\frac{dT}{dt}$$

$$\frac{du}{dt} \approx \frac{du}{dT}\frac{dT}{dt} \tag{A1.14}$$

Using equations (A1.12) and (A1.14), we may re-express equation (A1.13) as:

$$\frac{dz}{dt} = \frac{z}{m}\frac{dm}{dt} + \frac{z}{v}\frac{dv}{dT}\frac{dT}{dt} \tag{A1.15}$$

Substituting from equations (A1.14) and (A1.15) into equation (A1.3), we have:

$$\frac{dE}{dt} = (u + 2gz)\frac{dm}{dt} + m\left(\frac{du}{dT} + \frac{gz}{v}\frac{dv}{dT}\right)\frac{dT}{dt} \tag{A1.16}$$

It has already been shown that $u \gg gz$ for normal process-plant systems, so that the first term on the right-hand side of (A1.16) may be simplified to merely $u(dm/dt)$. We now need to examine the relative sizes of the terms in the second bracket, with the aim of showing that

$$\frac{du}{dT} \gg \frac{gz}{v}\frac{dv}{dT} \tag{A1.17}$$

Let us take as an example the case of water at 20°C at 1 bar as above. Assuming the water level, $2z$, is 10 m, the centre of gravity of the water will have a height $z = 5$ m and so we have the following values:

$$u = 83,900\,\text{J/kg}, \quad \frac{du}{dT} = 4185\,\text{J/(kg K)}$$

$$v = 0.001002\,\text{m}^3/\text{kg}, \quad \frac{dv}{dT} = 2 \times 10^{-7}\,\text{m}^3/(\text{kg K})$$

$$gz = 9.81 \times 5 = 49.05\,\text{m}^2/\text{s}^2$$

$$\frac{gz}{v}\frac{dv}{dT} = 9.8 \times 10^{-3}\,\text{J/(kg K)}$$

Hence $(gz/v)(dv/dT)$ is five orders of magnitude less than du/dT, and may safely be ignored. Hence equation (A1.16) reduces to equation (A1.10) once more.

A1.6 Gas partially filling a vessel, contained above a movable interface, e.g. a liquid surface, undergoing a near-adiabatic expansion or compression

This may be regarded as the complementary situation to that considered in Section A1.5: as the liquid surface rises in a partially filled vessel, the gas-space above it is reduced, and vice versa when the liquid surface falls. (Figure A1.2).

The argument concerning $g(dz/dt)$ for the gas volume is more complicated than in Section A1.5 because the specific volume of a gas will depend on both temperature and pressure.

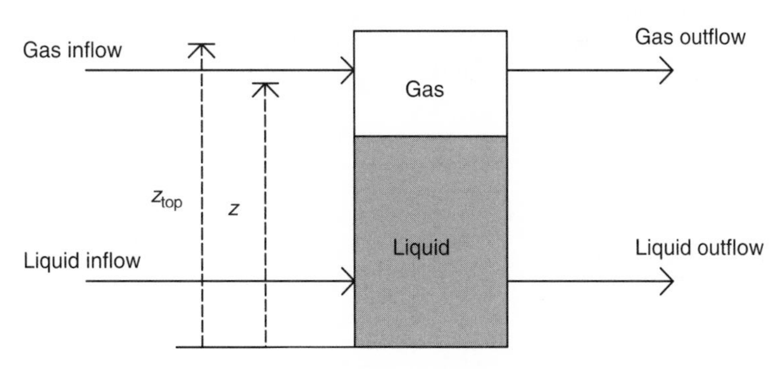

Figure A1.2 Gas partially filling a vessel, contained above a liquid surface.

The centre of gravity of the gas will be at a height, z, above the base of a vessel taken as the datum, given by

$$z = z_{top} - \frac{vm}{2A} \qquad \text{(A1.18)}$$

where m is the mass of the gas, v is its specific volume, A is the cross-sectional area of the vessel and z_{top} is the height of the top of the vessel. Hence

$$\begin{aligned}\frac{dz}{dt} &= -\left(\frac{v}{2A}\frac{dm}{dt} + \frac{m}{2A}\frac{dv}{dt}\right)\\ &= \frac{(z - z_{top})}{m}\frac{dm}{dt} + \frac{(z - z_{top})}{v}\frac{dv}{dt}\end{aligned} \qquad \text{(A1.19)}$$

The gas will conform to the characteristic equation of state:

$$pv = ZR_wT \qquad \text{(3.2)}$$

and during the expansion or compression will obey the general polytropic equation

$$pv^n = \text{constant} \qquad \text{(A1.20)}$$

where n is the polytropic exponent, the value of which will depend on the conditions of the expansion/compression. Differentiating equation (3.2) with respect to time gives

$$p\frac{dv}{dt} + v\frac{dp}{dt} = ZR_w\frac{dT}{dt} \qquad \text{(A1.21)}$$

while differentiating (A1.20) gives

$$npv^{n-1}\frac{dv}{dt} + v^n\frac{dp}{dt} = 0 \qquad \text{(A1.22)}$$

so that

$$v\frac{dp}{dt} = -np\frac{dv}{dt} \qquad \text{(A1.23)}$$

Substituting from equation (A1.23) into equation (A1.21) gives the differential for specific volume in terms of the temperature differential:

$$\frac{dv}{dt} = \frac{ZR_w}{(1-n)p}\frac{dT}{dt} = \frac{v}{(1-n)T}\frac{dT}{dt} \qquad \text{(A1.24)}$$

Then substituting into equation (A1.19) gives the differential of height as:

$$\frac{dz}{dt} = \frac{(z - z_{top})}{m}\frac{dm}{dt} + \frac{(z_{top} - z)}{(n-1)T}\frac{dT}{dt} \qquad \text{(A1.25)}$$

Recalling equation (A1.3):

$$\frac{dE}{dt} = (u + gz)\frac{dm}{dt} + m\left(\frac{du}{dT}\frac{dT}{dt} + g\frac{dz}{dt}\right) \qquad \text{(A1.3a)}$$

and substituting for the height differential from (A1.25) gives:

$$\begin{aligned}\frac{dE}{dt} &= (u + g(2z - z_{top}))\frac{dm}{dt}\\ &\quad + m\left(\frac{du}{dT} + g\frac{(z_{top} - z)}{(n-1)T}\right)\frac{dT}{dt}\end{aligned} \qquad \text{(A1.26)}$$

The term $(2z - z_{top})$ is the height of the lower gas interface and the magnitude of the term $g(2z - z_{top})$ will be very much smaller than the specific internal energy by the arguments above. Hence the first term on the right-hand side of equation (A1.26) may be simplified to $u(dm/dt)$.

We would also like to neglect the term $g(z - z_{top})/(n-1)T$ after showing that it is small in comparison with du/dT. For an near-ideal gas, the differential du/dT is

$$\frac{du}{dT} = \frac{N_F}{2}ZR_w \qquad \text{(A1.27)}$$

(by equations (3.13) and (3.21) in Chapter 3), in which the number of degrees of freedom, N_F, depends on the number of atoms in the molecule:

$$\begin{aligned}N_F &= 3 \text{ for a monatomic gas}\\ &= 5 \text{ for a diatomic gas}\\ &= 6 \text{ for a polyatomic gas}\end{aligned} \qquad \text{(3.22)}$$

Thus for an ideal gas the ratio, r_{PI}, of the potential energy term to the internal energy term is given by

$$r_{PEIE} = \frac{1}{\frac{du}{dT}}g\frac{(z_{top} - z)}{(n-1)T} = \frac{2g(z_{top} - z)}{N_F(n-1)ZR_wT} \qquad \text{(A1.28)}$$

It is clear from equation (A1.28) that the ratio, r_{PEIE}, will decrease as the initial temperature of the gas rises and as the expansion or compression nears adiabatic conditions.

Let us take the case of a gas blanket with its centre of gravity 5 m below the top of the vessel. We will assume the gas to be nitrogen at 20°C, and treat it as an ideal, diatomic gas, with a compressibility factor of unity. If the compression or expansion is adiabatic, then $n = 1.4$. As a result the ratio, r_{PEIE}, is given by

$$\begin{aligned}r_{PEIE} &= \frac{2g(z_{top} - z)}{N_F(n-1)ZR_wT}\\ &= \frac{2 \times 9.81 \times 5}{5 \times (1.4 - 1) \times 1 \times \frac{8314}{28} \times 293}\\ &= 5.64 \times 10^{-4}\end{aligned}$$

Taking an example of a non-ideal gas, let us assume that the nitrogen is replaced by superheated steam at 250°C and 10 bar. We then have

$$g\frac{(z_{top} - z)}{(n-1)T} = \frac{9.81 \times 5}{(1.3 - 1) \times 523} = 0.31 \text{ J/(kg K)}$$

By comparison, data from steam tables gives $du/dT =$ 1700 J/(kg K), so that $r_{PEIE} = 1.82 \times 10^{-4}$.

Clearly it will normally be quite safe to assume that $g(z - z_{top})/(1 - n)T \ll (du/dT)$ for all cases of adiabatic or near-adiabatic expansions or compressions, so that we may use the simplified form of equation (A1.3), namely:

$$\frac{dE}{dt} = u\frac{dm}{dt} + m\frac{du}{dt} \tag{A1.10}$$

Appendix 2 Explicit calculation of compressible flow using approximating functions

A2.1 Introduction

The method given in Sections 6.2 to 6.4 of Chapter 6 allows an iterative calculation of steady-state, compressible flow. However, the control engineer is likely to need to calculate the flow of a gas or vapour under a changing set of conditions, implying iteration at each timestep. An explicit method is obviously to be preferred, and this Appendix shows how it is possible to use the established basis of the exact but implicit method to construct an approximate but explicit method of calculation which preserves most of the accuracy.

A2.2 Applying dimensional analysis to compressible flow

Inspection of the full equations defining compressible flow summarized in Section 6.4 of Chapter 6 shows that the process may be characterized using the following seven quantities involving the three dimensions, mass, length and time, M, L, T:

$$W,\ p_1,\ v_1,\ A,\ \frac{p_4}{p_1},\ K_T,\ \gamma$$

where $K_T = 4fL_{eff}/D$ is the number of velocity heads. The Buckingham Pi theorem tells us that when a process involves seven quantities and three dimensions, that process may be described by an equation relating $7 - 3 = 4$ dimensionless groups, Π_i, formed from the seven quantities. Since three of the quantities, $p_4/p_1, K_T, \gamma$ are already dimensionless, we may seek the fourth dimensionless group by combining the other four quantities:

$$\Pi_4 = [p_1]^a[v_1]^b[A]^c[W]^1 \tag{A2.1}$$

Applying MLT dimensions gives:

$$M^0L^0T^0 = [M^1L^{-1}T^{-2}]^a[M^{-1}L^3]^b[L^2]^c[M^1T^{-1}]^1 \tag{A2.2}$$

Equating indices gives the equations:

$$\begin{aligned} 0 &= a - b + 1 \\ 0 &= -a + 3b + 2c \\ 0 &= -2a - 1 \end{aligned} \tag{A2.3}$$

which has the solution: $a = -1/2$, $b = 1/2$, $c = -1$. Hence the fourth dimensionless group is the dimensionless flow function:

$$\Pi_4 = \frac{W}{A\sqrt{\dfrac{p_1}{v_1}}} \tag{A2.4}$$

As a result, compressible flow may be described by an equation of the form:

$$\phi_0\left(\frac{W}{A\sqrt{\dfrac{p_1}{v_1}}},\ \frac{p_4}{p_1},\ K_T,\ \gamma\right) = 0 \tag{A2.5}$$

or, equivalently,

$$\frac{W}{A\sqrt{\dfrac{p_1}{v_1}}} = \phi_1\left(\frac{p_4}{p_1},\ K_T,\ \gamma\right) \tag{A2.6}$$

The analysis of Chapter 6, Sections 6.2 to 6.4, has provided us a way of calculating the flow, W, for any combination of parameters $p_1, v_1, A, p_4/p_1, K_T, \gamma$. Hence we may evaluate the dimensionless flow function at each point simply by dividing the mass flow calculated by the factor $A\sqrt{p_1/v_1}$. For brevity, we will give this dimensionless flow function the symbol, f_{pipe}:

$$f_{pipe} = \frac{W}{A\sqrt{\dfrac{p_1}{v_1}}} \tag{A2.7}$$

A2.3 The shape of the dimensionless flow function, f_{pipe}

Given the specific-heat ratio, γ, the pipe-flow function, f_{pipe}, may be calculated for flow through pipes with a wide range of total frictional losses, K_T, and over the full range of pressure ratios. Tables A2.1 to A2.4 give the results of calculations for a monatomic gas, a diatomic gas, superheated steam/polyatomic gas

Table A2.1 Dimensionless flow function, f_{pipe}, for a monatomic gas: $\gamma = 1.67$

K_T	0.03125	0.0625	0.125	0.25	0.5	1	2	4	8
p_3/p_1 crit	0.476683	0.468498	0.454716	0.432886	0.400879	0.358108	0.306649	0.251188	0.19742
p_4/p_1									
0.01	0.711751	0.699531	0.678952	0.646358	0.598566	0.534703	0.457868	0.375057	0.294775
0.05	0.711751	0.699531	0.678952	0.646358	0.598566	0.534703	0.457868	0.375057	0.294775
0.1	0.711751	0.699531	0.678952	0.646358	0.598566	0.534703	0.457868	0.375057	0.294775
0.15	0.711751	0.699531	0.678952	0.646358	0.598566	0.534703	0.457868	0.375057	0.294775
0.2	0.711751	0.699531	0.678952	0.646358	0.598566	0.534703	0.457868	0.375057	0.294774
0.25	0.711751	0.699531	0.678952	0.646358	0.598566	0.534703	0.457868	0.375057	0.294201
0.3	0.711751	0.699531	0.678952	0.646358	0.598566	0.534703	0.457868	0.374394	0.29255
0.35	0.711751	0.699531	0.678952	0.646358	0.598566	0.534703	0.457178	0.372271	0.289766
0.4	0.711751	0.699531	0.678952	0.646358	0.598566	0.533872	0.454573	0.368584	0.285793
0.45	0.711751	0.699531	0.678952	0.646148	0.597114	0.530552	0.449856	0.363218	0.280565
0.5	0.711153	0.698544	0.677155	0.642967	0.592386	0.52442	0.44282	0.356038	0.273996
0.55	0.705464	0.692519	0.670533	0.635489	0.583926	0.515134	0.43322	0.346878	0.265978
0.6	0.693103	0.680296	0.658305	0.623064	0.571205	0.502283	0.420753	0.335525	0.256362
0.65	0.673432	0.661126	0.639683	0.604951	0.553583	0.48535	0.405026	0.321696	0.244951
0.7	0.645738	0.634188	0.61377	0.58025	0.530245	0.463657	0.385515	0.305007	0.231468
0.75	0.608974	0.598374	0.579386	0.547773	0.500085	0.436264	0.361479	0.284906	0.215514
0.8	0.561446	0.551966	0.534786	0.505784	0.461483	0.401773	0.331804	0.260553	0.196474
0.85	0.500209	0.492028	0.477054	0.451441	0.411798	0.357898	0.29465	0.230548	0.17332
0.9	0.419434	0.412795	0.400537	0.37931	0.346012	0.30029	0.246497	0.192203	0.144067
0.95	0.304112	0.299456	0.29079	0.275613	0.251484	0.217996	0.178453	0.138679	0.103652
0.98	0.19512	0.192193	0.186719	0.177069	0.161608	0.140006	0.11443	0.088751	0.066224
1	0	0	0	0	0	0	0	0	0

K_T	8	16	32	64	128	256	512	1024	2048
p_3/p_1 crit	0.19742	0.149914	0.110938	0.080651	0.057968	0.041374	0.02941	0.020857	0.014772
p_4/p_1									
0.01	0.294775	0.223842	0.165646	0.120423	0.086553	0.061777	0.043913	0.031142	0.022056
0.05	0.294775	0.223842	0.165646	0.120423	0.086553	0.061774	0.043902	0.031125	0.022039
0.1	0.294775	0.223842	0.165646	0.120393	0.086454	0.061641	0.043776	0.031023	0.021961
0.15	0.294775	0.223842	0.165478	0.120042	0.08608	0.061319	0.043524	0.030836	0.021826
0.2	0.294774	0.223461	0.164767	0.119298	0.08544	0.060819	0.043152	0.030566	0.021632
0.25	0.294201	0.222298	0.163493	0.118169	0.08454	0.060141	0.042658	0.030211	0.021379
0.3	0.29255	0.220333	0.161656	0.116656	0.08338	0.059285	0.042039	0.029768	0.021064
0.35	0.289766	0.217545	0.159248	0.114754	0.081952	0.058244	0.041291	0.029235	0.020686
0.4	0.285793	0.213904	0.15625	0.112446	0.080245	0.057009	0.040408	0.028607	0.02024
0.45	0.280565	0.209372	0.152636	0.109716	0.078244	0.055569	0.03938	0.027876	0.019722
0.5	0.273996	0.203893	0.148365	0.106531	0.075927	0.053906	0.038195	0.027036	0.019127
0.55	0.265978	0.197391	0.143384	0.102848	0.073264	0.052	0.036839	0.026074	0.018446
0.6	0.256362	0.189763	0.137618	0.09862	0.070216	0.049824	0.035293	0.024978	0.01767
0.65	0.244951	0.180868	0.130965	0.093771	0.066732	0.04734	0.03353	0.023729	0.016785
0.7	0.231468	0.170508	0.123283	0.088199	0.062739	0.044498	0.031513	0.0223	0.015774
0.75	0.215514	0.158395	0.114367	0.081757	0.058134	0.041223	0.02919	0.020655	0.014611
0.8	0.196474	0.144087	0.1039	0.07422	0.052755	0.037401	0.026481	0.018738	0.013254
0.85	0.17332	0.126842	0.091351	0.065211	0.046334	0.032843	0.023252	0.016452	0.011637
0.9	0.144067	0.105222	0.07569	0.053997	0.038353	0.027181	0.019242	0.013614	0.009629
0.95	0.103652	0.075558	0.05429	0.038706	0.027483	0.019475	0.013785	0.009753	0.006898
0.98	0.066224	0.04822	0.034625	0.024677	0.017519	0.012412	0.008786	0.006216	0.004396
1	0	0	0	0	0	0	0	0	0

Table A2.2 Dimensionless flow function, f_{pipe}, for a diatomic gas: $\gamma = 1.4$

K_T	0.03125	0.0625	0.125	0.25	0.5	1	2	4	8
p_3/p_1 crit	0.51893	0.511098	0.497693	0.476043	0.443584	0.399164	0.344442	0.284176	0.224681
p_4/p_1									
0.01	0.67261	0.662458	0.645084	0.617022	0.57495	0.517376	0.446448	0.368334	0.29122
0.05	0.67261	0.662458	0.645084	0.617022	0.57495	0.517376	0.446448	0.368334	0.29122
0.1	0.67261	0.662458	0.645084	0.617022	0.57495	0.517376	0.446448	0.368334	0.29122
0.15	0.67261	0.662458	0.645084	0.617022	0.57495	0.517376	0.446448	0.368334	0.29122
0.2	0.67261	0.662458	0.645084	0.617022	0.57495	0.517376	0.446448	0.368334	0.29122
0.25	0.67261	0.662458	0.645084	0.617022	0.57495	0.517376	0.446448	0.368334	0.291071
0.3	0.67261	0.662458	0.645084	0.617022	0.57495	0.517376	0.446448	0.368256	0.289889
0.35	0.67261	0.662458	0.645084	0.617022	0.57495	0.517376	0.446435	0.366966	0.287502
0.4	0.67261	0.662458	0.645084	0.617022	0.57495	0.517375	0.445166	0.36402	0.28387
0.45	0.67261	0.662458	0.645084	0.617022	0.574924	0.515996	0.441691	0.359318	0.278939
0.5	0.67261	0.662458	0.645079	0.616568	0.572807	0.511742	0.435813	0.352737	0.272632
0.55	0.671465	0.660821	0.642446	0.612443	0.566954	0.504264	0.427292	0.344118	0.264846
0.6	0.664291	0.653329	0.634345	0.603412	0.556802	0.493143	0.415826	0.333255	0.255437
0.65	0.649704	0.638851	0.619832	0.588652	0.541674	0.477851	0.401025	0.31987	0.244211
0.7	0.626836	0.616406	0.597881	0.567186	0.520718	0.457697	0.382364	0.30358	0.230892
0.75	0.594556	0.58479	0.567224	0.537758	0.492793	0.431727	0.3591	0.283836	0.215084
0.8	0.551111	0.542216	0.526042	0.498575	0.45624	0.398527	0.330115	0.259798	0.196172
0.85	0.493488	0.485681	0.471354	0.446737	0.408381	0.355792	0.293561	0.230065	0.173127
0.9	0.415773	0.409334	0.397425	0.376741	0.344146	0.299145	0.245909	0.191943	0.143964
0.95	0.302818	0.29823	0.289689	0.274701	0.250823	0.217592	0.178246	0.138588	0.103615
0.98	0.194793	0.191881	0.186439	0.176838	0.161441	0.139904	0.114378	0.088728	0.066215
1	0	0	0	0	0	0	0	0	0

K_T	8	16	32	64	128	256	512	1024	2048
p_3/p_1 crit	0.224681	0.171375	0.127206	0.092657	0.066674	0.047621	0.033864	0.02402	0.017014
p_4/p_1									
0.01	0.29122	0.222128	0.164878	0.120097	0.086419	0.061724	0.043893	0.031134	0.022053
0.05	0.29122	0.222128	0.164878	0.120097	0.086419	0.061724	0.043884	0.031119	0.022037
0.1	0.29122	0.222128	0.164878	0.120092	0.08635	0.061606	0.043764	0.031019	0.02196
0.15	0.29122	0.222128	0.164814	0.119809	0.085999	0.061291	0.043515	0.030833	0.021824
0.2	0.29122	0.221988	0.164229	0.11911	0.085374	0.060795	0.043144	0.030563	0.021631
0.25	0.291071	0.221072	0.163048	0.118012	0.084485	0.060122	0.042651	0.030208	0.021378
0.3	0.289889	0.219303	0.161283	0.116524	0.083333	0.059269	0.042033	0.029766	0.021064
0.35	0.287502	0.216675	0.158932	0.114641	0.081912	0.05823	0.041286	0.029234	0.020685
0.4	0.28387	0.213169	0.155983	0.112351	0.080212	0.056997	0.040403	0.028605	0.02024
0.45	0.278939	0.208753	0.152411	0.109634	0.078216	0.055558	0.039376	0.027875	0.019722
0.5	0.272632	0.203375	0.148177	0.106461	0.075903	0.053897	0.038192	0.027034	0.019126
0.55	0.264846	0.196962	0.143228	0.102792	0.073244	0.051993	0.036837	0.026073	0.018446
0.6	0.255437	0.189413	0.13749	0.098574	0.0702	0.049818	0.035291	0.024977	0.01767
0.65	0.244211	0.180588	0.130863	0.093733	0.066719	0.047336	0.033528	0.023728	0.016785
0.7	0.230892	0.17029	0.123204	0.088169	0.062729	0.044494	0.031511	0.022299	0.015774
0.75	0.215084	0.158233	0.114308	0.081735	0.058126	0.04122	0.029189	0.020655	0.014611
0.8	0.196172	0.143973	0.103859	0.074205	0.052749	0.037399	0.026481	0.018737	0.013254
0.85	0.173127	0.126769	0.091324	0.065201	0.046331	0.032842	0.023252	0.016452	0.011637
0.9	0.143964	0.105183	0.075676	0.053991	0.038351	0.027181	0.019242	0.013614	0.009629
0.95	0.103615	0.075544	0.054285	0.038704	0.027483	0.019474	0.013785	0.009753	0.006898
0.98	0.066215	0.048217	0.034623	0.024676	0.017518	0.012412	0.008786	0.006216	0.004396
1	0	0	0	0	0	0	0	0	0

Table A2.3 Dimensionless flow function, f_{pipe}, for superheated steam: $\gamma = 1.3$

K_T	0.03125	0.0625	0.125	0.25	0.5	1	2	4	8
p_3/p_1 crit	0.53666	0.529	0.5158	0.49432	0.46181	0.41687	0.36094	0.29874	0.23684
p_4/p_1									
0.01	0.656174	0.646808	0.630674	0.604408	0.56466	0.509713	0.44132	0.365273	0.289582
0.05	0.656174	0.646808	0.630674	0.604408	0.56466	0.509713	0.44132	0.365273	0.289582
0.1	0.656174	0.646808	0.630674	0.604408	0.56466	0.509713	0.44132	0.365273	0.289582
0.15	0.656174	0.646808	0.630674	0.604408	0.56466	0.509713	0.44132	0.365273	0.289582
0.2	0.656174	0.646808	0.630674	0.604408	0.56466	0.509713	0.44132	0.365273	0.289582
0.25	0.656174	0.646808	0.630674	0.604408	0.56466	0.509713	0.44132	0.365273	0.289539
0.3	0.656174	0.646808	0.630674	0.604408	0.56466	0.509713	0.44132	0.365272	0.288598
0.35	0.656174	0.646808	0.630674	0.604408	0.56466	0.509713	0.44132	0.364402	0.28641
0.4	0.656174	0.646808	0.630674	0.604408	0.56466	0.509713	0.440658	0.361824	0.282946
0.45	0.656174	0.646808	0.630674	0.604408	0.56466	0.509104	0.437788	0.357447	0.27816
0.5	0.656174	0.646808	0.630674	0.604381	0.563648	0.505738	0.432469	0.351156	0.27198
0.55	0.655962	0.646326	0.629528	0.601748	0.558994	0.499114	0.424465	0.342799	0.264305
0.6	0.651033	0.640895	0.623251	0.59425	0.55003	0.48881	0.413478	0.332171	0.254995
0.65	0.638721	0.628519	0.610591	0.581023	0.536062	0.474294	0.39912	0.318999	0.243858
0.7	0.618042	0.608119	0.590454	0.561051	0.51622	0.454868	0.380864	0.3029	0.230618
0.75	0.58782	0.578433	0.561518	0.533043	0.489344	0.429572	0.357968	0.283327	0.21488
0.8	0.546263	0.537637	0.521927	0.495172	0.453756	0.396984	0.32931	0.259439	0.196027
0.85	0.490323	0.482689	0.468663	0.444511	0.406759	0.354791	0.293042	0.229835	0.173036
0.9	0.414043	0.407698	0.395952	0.375521	0.34326	0.2986	0.245629	0.191819	0.143915
0.95	0.302204	0.297649	0.289164	0.274268	0.250508	0.217399	0.178148	0.138545	0.103598
0.98	0.194637	0.191734	0.186305	0.176728	0.161361	0.139855	0.114353	0.088717	0.06621
1	0	0	0	0	0	0	0	0	0

K_T	8	16	32	64	128	256	512	1024	2048
p_3/p_1 crit	0.23684	0.18102	0.13455	0.0981	0.07063	0.05046	0.03589	0.02546	0.01803
p_4/p_1									
0.01	0.289582	0.221331	0.164519	0.119944	0.086357	0.061699	0.043883	0.03113	0.022051
0.05	0.289582	0.221331	0.164519	0.119944	0.086357	0.061699	0.043876	0.031117	0.022036
0.1	0.289582	0.221331	0.164519	0.119944	0.0863	0.061589	0.043758	0.031017	0.021959
0.15	0.289582	0.221331	0.164488	0.119697	0.085961	0.061278	0.04351	0.030831	0.021824
0.2	0.289582	0.221266	0.163968	0.119019	0.085343	0.060785	0.04314	0.030561	0.02163
0.25	0.289539	0.220478	0.162834	0.117937	0.084459	0.060113	0.042648	0.030207	0.021377
0.3	0.288598	0.218807	0.161104	0.116461	0.083311	0.059261	0.04203	0.029765	0.021063
0.35	0.28641	0.216257	0.158781	0.114588	0.081893	0.058224	0.041284	0.029233	0.020685
0.4	0.282946	0.212817	0.155856	0.112306	0.080196	0.056992	0.040401	0.028605	0.020239
0.45	0.27816	0.208457	0.152304	0.109596	0.078202	0.055554	0.039374	0.027874	0.019722
0.5	0.27198	0.203127	0.148087	0.106429	0.075891	0.053893	0.03819	0.027034	0.019126
0.55	0.264305	0.196757	0.143154	0.102765	0.073234	0.05199	0.036836	0.026073	0.018445
0.6	0.254995	0.189246	0.13743	0.098552	0.070192	0.049815	0.03529	0.024977	0.017669
0.65	0.243858	0.180455	0.130814	0.093716	0.066713	0.047334	0.033527	0.023728	0.016785
0.7	0.230618	0.170188	0.123166	0.088156	0.062724	0.044493	0.031511	0.022299	0.015774
0.75	0.21488	0.158156	0.114281	0.081725	0.058123	0.041219	0.029189	0.020655	0.01461
0.8	0.196027	0.143919	0.103839	0.074198	0.052747	0.037398	0.026481	0.018737	0.013254
0.85	0.173036	0.126735	0.091312	0.065197	0.04633	0.032842	0.023252	0.016452	0.011637
0.9	0.143915	0.105165	0.075669	0.053989	0.038351	0.02718	0.019242	0.013614	0.009629
0.95	0.103598	0.075538	0.054283	0.038703	0.027483	0.019474	0.013785	0.009753	0.006898
0.98	0.06621	0.048215	0.034623	0.024676	0.017518	0.012412	0.008786	0.006216	0.004396
1	0	0	0	0	0	0	0	0	0

Table A2.4 Dimensionless flow function, f_{pipe}, for dry, saturated steam: $\gamma = 1.135$

K_T	*0.03125*	*0.0625*	*0.125*	*0.25*	*0.5*	*1*	*2*	*4*	*8*
p_3/p_1 *crit*	*0.5689*	*0.56158*	*0.54884*	*0.5278*	*0.49542*	*0.44982*	*0.39195*	*0.32641*	*0.26014*
p_4/p_1									
0.01	0.626203	0.618153	0.604127	0.580968	0.545329	0.49513	0.431432	0.359295	0.28635
0.05	0.626203	0.618153	0.604127	0.580968	0.545329	0.49513	0.431432	0.359295	0.28635
0.1	0.626203	0.618153	0.604127	0.580968	0.545329	0.49513	0.431432	0.359295	0.28635
0.15	0.626203	0.618153	0.604127	0.580968	0.545329	0.49513	0.431432	0.359295	0.28635
0.2	0.626203	0.618153	0.604127	0.580968	0.545329	0.49513	0.431432	0.359295	0.28635
0.25	0.626203	0.618153	0.604127	0.580968	0.545329	0.49513	0.431432	0.359295	0.28635
0.3	0.626203	0.618153	0.604127	0.580968	0.545329	0.49513	0.431432	0.359295	0.285916
0.35	0.626203	0.618153	0.604127	0.580968	0.545329	0.49513	0.431432	0.359091	0.284156
0.4	0.626203	0.618153	0.604127	0.580968	0.545329	0.49513	0.431401	0.357298	0.281049
0.45	0.626203	0.618153	0.604127	0.580968	0.545329	0.49513	0.429797	0.353606	0.276564
0.5	0.626203	0.618153	0.604127	0.580968	0.545313	0.49357	0.425636	0.34792	0.270647
0.55	0.626203	0.618153	0.604126	0.580525	0.543021	0.488676	0.418696	0.340102	0.263202
0.6	0.62494	0.616376	0.601283	0.575984	0.536408	0.480023	0.408691	0.329959	0.254096
0.65	0.616961	0.608014	0.592192	0.565752	0.524749	0.467075	0.395236	0.317222	0.243138
0.7	0.600536	0.591593	0.575601	0.54873	0.507135	0.449123	0.377808	0.301514	0.230059
0.75	0.57435	0.565705	0.550066	0.523545	0.482366	0.425194	0.355661	0.282289	0.214463
0.8	0.536529	0.528433	0.513641	0.488302	0.448724	0.393849	0.327673	0.258706	0.195734
0.85	0.483945	0.476657	0.463229	0.440008	0.403469	0.352753	0.291986	0.229365	0.172848
0.9	0.410544	0.404388	0.392968	0.37305	0.34146	0.297492	0.245058	0.191567	0.143814
0.95	0.300958	0.29647	0.288101	0.273388	0.249869	0.217008	0.177948	0.138457	0.103563
0.98	0.194321	0.191436	0.186035	0.176505	0.161199	0.139756	0.114303	0.088695	0.066201
1	0	0	0	0	0	0	0	0	0

K_T	*8*	*16*	*32*	*64*	*128*	*256*	*512*	*1024*	*2048*
p_3/p_1 *crit*	*0.26014*	*0.19964*	*0.14881*	*0.10869*	*0.07834*	*0.05601*	*0.03985*	*0.02827*	*0.02003*
p_4/p_1									
0.01	0.28635	0.219748	0.163801	0.119637	0.086231	0.061649	0.043863	0.031123	0.022048
0.05	0.28635	0.219748	0.163801	0.119637	0.086231	0.061649	0.043859	0.031111	0.022034
0.1	0.28635	0.219748	0.163801	0.119637	0.086196	0.061554	0.043746	0.031013	0.021958
0.15	0.28635	0.219748	0.163801	0.119464	0.085882	0.061251	0.043501	0.030828	0.021823
0.2	0.28635	0.219748	0.163428	0.118834	0.085279	0.060763	0.043132	0.030559	0.021629
0.25	0.28635	0.219243	0.162394	0.117785	0.084406	0.060094	0.042641	0.030205	0.021377
0.3	0.285916	0.217784	0.160737	0.116332	0.083266	0.059245	0.042025	0.029763	0.021063
0.35	0.284156	0.215402	0.158474	0.114479	0.081855	0.05821	0.041279	0.029231	0.020684
0.4	0.281049	0.212098	0.155597	0.112214	0.080163	0.05698	0.040397	0.028603	0.020239
0.45	0.276564	0.207854	0.152086	0.109519	0.078175	0.055544	0.039371	0.027873	0.019721
0.5	0.270647	0.202623	0.147905	0.106364	0.075868	0.053885	0.038188	0.027033	0.019126
0.55	0.263202	0.196341	0.143002	0.102712	0.073215	0.051983	0.036833	0.026072	0.018445
0.6	0.254096	0.188906	0.137306	0.098508	0.070176	0.04981	0.035288	0.024976	0.017669
0.65	0.243138	0.180185	0.130716	0.093681	0.0667	0.047329	0.033525	0.023727	0.016785
0.7	0.230059	0.169977	0.12309	0.088128	0.062715	0.044489	0.03151	0.022299	0.015774
0.75	0.214463	0.157999	0.114223	0.081705	0.058115	0.041216	0.029188	0.020654	0.01461
0.8	0.195734	0.143809	0.103799	0.074184	0.052742	0.037397	0.02648	0.018737	0.013254
0.85	0.172848	0.126664	0.091286	0.065187	0.046326	0.03284	0.023251	0.016451	0.011637
0.9	0.143814	0.105127	0.075656	0.053984	0.038349	0.02718	0.019241	0.013614	0.009629
0.95	0.103563	0.075525	0.054278	0.038701	0.027482	0.019474	0.013785	0.009753	0.006898
0.98	0.066201	0.048212	0.034621	0.024675	0.017518	0.012412	0.008786	0.006216	0.004396
1	0	0	0	0	0	0	0	0	0

and dry, saturated steam ($\gamma = 1.67$, 1.4, 1.3 and 1.135, respectively). These calculations were made at intervals over the joint range

$$0.03125 \le K_T \le 2048$$
$$0 \le \frac{p_4}{p_1} \le 1 \tag{A2.8}$$

(The frictional range is wide, corresponding to an effective length range for standard industrial pipe.of $\sim 3.5D \le L_{eff} \le \sim 114\,000D$.) Figure A2.1 below illustrates the shape of the f_{pipe} function for a diatomic gas, with $\gamma = 1.4$. Also plotted is the corresponding flow function for a nozzle undergoing a pure, isentropic expansion, found from equations (5.27) and (5.59) as:

$$f_{noz}\left(\frac{p_4}{p_1}\right) = \sqrt{\frac{2\gamma}{\gamma-1}\left(\left(\frac{p_4}{p_1}\right)^{2/\gamma} - \left(\frac{p_4}{p_1}\right)^{(\gamma+1)/\gamma}\right)} \quad \text{for } \frac{p_4}{p_1} > \left(\frac{2}{\gamma+1}\right)^{\gamma/(\gamma-1)}$$
$$= \sqrt{\gamma\left(\frac{2}{\gamma+1}\right)^{(\gamma+1)/(\gamma-1)}} \quad \text{for } \frac{p_4}{p_1} \le \left(\frac{2}{\gamma+1}\right)^{\gamma/(\gamma-1)} \tag{A2.9}$$

It is clear from the figure that the pipe function, f_{pipe}, approaches the limiting case of the pure isentropic nozzle function, f_{noz}, at extremely low values of frictional loss, as one would expect. It is possible to use f_{noz} as the basis for approximating f_{pipe} at very small values of frictional loss, K_T, but the figure indicates that this basis is likely to be problematic for the more usual case of moderate to large frictional losses ($K_T > 1$). Fornately it is possible to develop a long-pipe approximating function, instead, which may be used in conjunction with a relatively simple set of correction factors to give an explicit calculation of the flow of gas through a pipe. The accuracy of this approximate method compared with the full, implicit method is better than 2%.

A2.4 Developing a long-pipe approximation to the full compressible flow equations

To derive an approximation to the compressible flow equations for a long pipe, we may take as our starting point the incompressible flow equation (4.81), revised so that the outlet is taken as the point just inside the pipe outlet in Figure 6.1:

$$W = C_L\sqrt{\frac{p_1 - p_3 - \dfrac{g(z_3 - z_1)}{v_z}}{v_1}} \tag{A2.10}$$

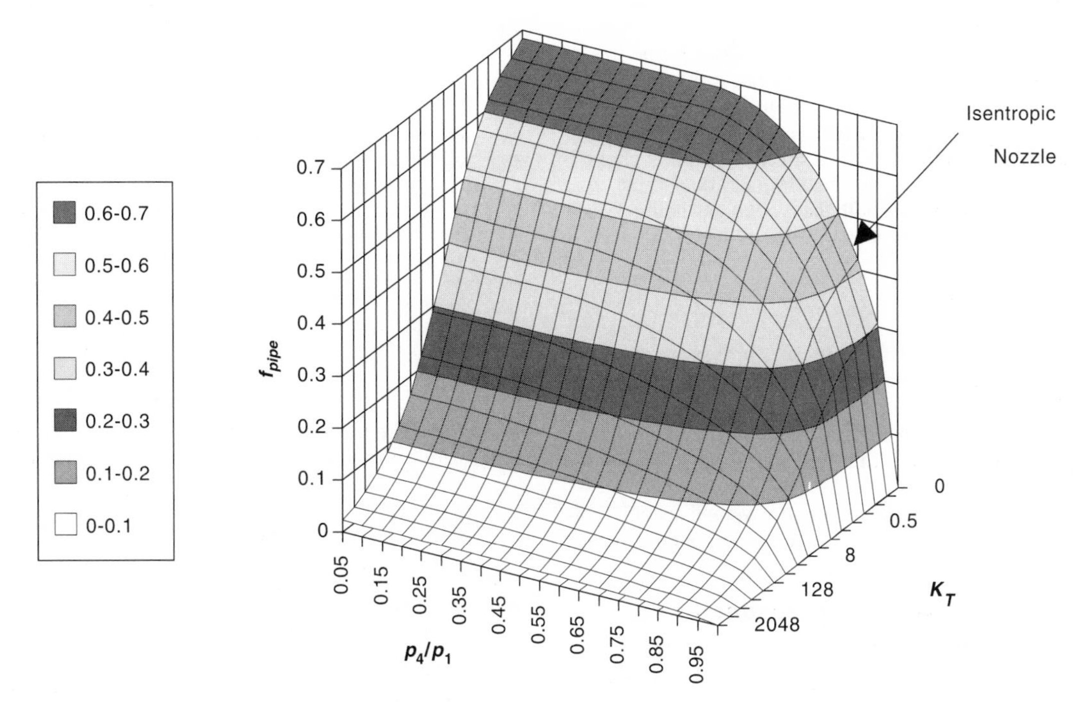

Figure A2.1 Flow function, f_{pipe}, versus overall pressure ratio, p_4/p_1, and total frictional loss in velocity heads, $K_T = 4fL_{eff}/D$.

where C_L is the pipe conductance. To apply this to compressible flow, let us first replace both the inlet specific volume and the specific volume in the rising section of pipe by an average specific volume: $v_1 = v_z = v_{ave}$. Next let us include a factor $b_o = b_o(p_4/p_1, K_T, \gamma)$ to account for the compressibility of the gas. Thus the mass flow will be given by

$$W = b_0 C_L \sqrt{\frac{p_1 - p_3 - \dfrac{g(z_3 - z_1)}{v_{ave}}}{v_{ave}}} \tag{A2.11}$$

By equation (4.43), the conductance, C_L, is related to the total velocity head drop, K_T, by:

$$C_L = A\sqrt{\frac{2}{4f\dfrac{L_{eff}}{D}}} = A\sqrt{\frac{2}{K_T}} \tag{A2.12}$$

where the effective pipe length is as defined in equation (6.59). As a result, we may rewrite equation (A2.11) as:

$$W = b_0 A \sqrt{\frac{2}{K_T}\left(\frac{p_1 - p_3 - \dfrac{g(z_3 - z_1)}{v_{ave}}}{v_{ave}}\right)} \tag{A2.13}$$

Note that this flow model assumes that the thermodynamic properties of the fluid just inside the inlet are equal to those in the upstream vessel, so that no distinction need be made between fluid properties at station '1' and station '2'. However, it is not possible to assume that the pressure just inside the outlet of the pipe, p_3, will be equal to the pressure of the receiving vessel, p_4, at all times. Whenever the pipe is choked, the pressure, p_3, will be equal to its critical value, p_{3c}, which will depend on both the type of gas being passed and on the frictional loss in velocity heads. Hence:

$$\frac{p_3}{p_1} = \max\left(\frac{p_4}{p_1}, \frac{p_{3c}}{p_1}(\gamma, K_T)\right) \tag{A2.14}$$

Fortunately the relationship p_{3c}/p_1 versus K_T may be calculated in advance and values are included in Tables A2.1 to A2.4 for four important specific-heat ratios, γ. Furthermore, it is a relatively simple matter to approximate the resulting function using polynomials, as is shown in Section A2.6.

But it remains to determine the average specific volume in the pipe, v_{ave}, and the correction parameter, b_0.

A2.4.1 Deriving an approximate expression for v_{ave}

For liquid flow, which is very close to incompressible, we may for all practical purposes put $v_{ave} = v_1$, the upstream specific volume. But for a gas, the specific volume will vary continuously as the pressure falls, and the change from pipe inlet to pipe outlet may be large. To estimate the value of v_{ave} suitable for gas flow, we begin by simplifying equation (A2.13) by assuming height differences can be neglected. We then recall equations (4.19) and (4.21), repeated below:

$$\frac{dp}{dx} = -\frac{g}{v}\frac{dz}{dx} - \frac{1}{v}\frac{dF}{dx} \tag{4.19}$$

$$\frac{dF}{dx} = \frac{2f}{D}\frac{W^2}{A^2}v^2 \tag{4.21}$$

Combining these under the assumption that $dz = 0$ gives:

$$dp = -\frac{2f}{D}G^2 v\,dx \tag{A2.15}$$

where G is the mass velocity, $G = W/A$.

If the change of pressure with distance, dp/dx, is gradual, then little energy will be lost to friction and the expansion will be close to isentropic. We will take this as a reasonable starting assumption, applicable to at least some pipe expansions, and correct any deficiencies later through the choice of the b_0 function. Hence we may write the specific volume at a downstream point as:

$$v = v_1 p_1^{1/\gamma}\frac{1}{p^{1/\gamma}} \tag{A2.16}$$

Substituting into (A2.15) yields the equation:

$$p^{1/\gamma}dp = -p_1^{1/\gamma}v_1\frac{2f}{D}G^2 dx \tag{A2.17}$$

We may integrate equation (A2.17) from the start of the pipe to a distance x along the effective length:

$$\int_{p_1}^{p} p^{1/\gamma}dp = -p_1^{1/\gamma}v_1\frac{2f}{D}G^2\int_0^x dx \tag{A2.18}$$

to give:

$$\frac{\gamma}{\gamma+1}\left(p^{1+(1/\gamma)} - p_1^{1+(1/\gamma)}\right) = -p_1^{1/\gamma}v_1\frac{2f}{D}G^2 x \tag{A2.19}$$

Rearranging equation (A2.19) yields the following expression for pressure at a distance x downstream:

$$p = p_1\left(1 - \frac{\gamma+1}{\gamma}\frac{v_1}{p_1}\frac{2f}{D}G^2 x\right)^{\gamma/(\gamma+1)} \tag{A2.20}$$

Having determined the pressure distribution in the pipe, we may now proceed to evaluate the average specific volume, given formally by:

$$v_{ave} = \frac{1}{L_{eff}}\int_0^{L_{eff}} v\,dx \tag{A2.21}$$

Substituting from (A2.16) and (A2.20) into (A2.21), we achieve:

$$v_{ave} = \frac{1}{L_{eff}} \int_0^{L_{eff}} \frac{p_1^{1/\gamma} v_1}{p_1^{1/\gamma} \left(1 - \frac{\gamma+1}{\gamma} \frac{v_1}{p_1} \frac{2f}{D} G^2 x\right)^{1/(\gamma+1)}} dx \qquad \text{(A2.22)}$$

or

$$v_{ave} = \frac{v_1}{L_{eff}} \int_0^{L_{eff}} \frac{1}{\left(1 - \frac{\gamma+1}{\gamma} \frac{v_1}{p_1} \frac{2f}{D} G^2 x\right)^{1/(\gamma+1)}} dx \qquad \text{(A2.23)}$$

This may be integrated using the standard form:

$$\int \frac{dx}{(a+bx)^{1/n}} = \frac{n}{(n-1)b} (a+bx)^{(n-1)/n} \qquad \text{(A2.24)}$$

to give:

$$v_{ave} = \frac{v_1}{L_{eff}} \frac{p_1}{v_1} \frac{1}{\frac{2f}{D} G^2} \times \left[\left(1 - \frac{\gamma+1}{\gamma} \frac{v_1}{p_1} \frac{2f}{D} G^2 x\right)^{\gamma/(\gamma+1)} \right]_0^{L_{eff}} \qquad \text{(A2.25)}$$

or

$$v_{ave} = \frac{p_1}{\frac{2fL_{eff}}{D} G^2} \times \left[1 - \left(1 - \frac{\gamma+1}{\gamma} \frac{v_1}{p_1} \frac{2fL_{eff}}{D} G^2\right)^{\gamma/(\gamma+1)} \right] \qquad \text{(A2.26)}$$

Now $p = p_3$ when $x = L_{eff}$, so that from equation (A2.20),

$$\frac{p_3}{p_1} = \left(1 - \frac{\gamma+1}{\gamma} \frac{v_1}{p_1} \frac{2f}{D} G^2 L_{eff}\right)^{\gamma/(\gamma-1)} \qquad \text{(A2.27)}$$

Putting $x = L_{eff}$ and $p = p_3$ into equation (A2.19) gives:

$$\frac{2fL_{eff}}{D} G^2 = \frac{\gamma}{\gamma+1} \frac{p_1}{v_1} \left(1 - \left(\frac{p_3}{p_1}\right)^{(\gamma+1)/\gamma}\right) \qquad \text{(A2.28)}$$

Substituting from (A2.27) and (A2.28) into equation (A2.26) gives the required expression for the average specific volume in terms of the inlet specific volume and the upstream and downstream pressures:

$$v_{ave} = v_1 \frac{\gamma+1}{\gamma} \frac{\left(1 - \frac{p_3}{p_1}\right)}{\left(1 - \left(\frac{p_3}{p_1}\right)^{(\gamma+1)/\gamma}\right)} \qquad \text{(A2.29)}$$

This value of v_{ave} may be used in equation (A2.13).

A2.4.2 The correction factor b_0

To enable us to compare the long-pipe approximation with the exact solution valid for a horizontal pipe, we substitute equation (A2.29) into (A2.13) with the additional condition that $z_1 = z_3$, i.e.:

$$W = b_0 A \sqrt{\frac{2}{K_T} \frac{(p_1 - p_3)}{v_1} \frac{\gamma}{\gamma+1} \frac{1 - \left(\frac{p_3}{p_1}\right)^{(\gamma+1)/\gamma}}{\left(1 - \frac{p_3}{p_1}\right)}} \qquad \text{(A2.30)}$$

which simplifies to:

$$W = b_0 A \sqrt{\frac{p_1}{v_1}} \sqrt{\frac{1}{K_T} \frac{2\gamma}{\gamma+1} \left(1 - \left(\frac{p_3}{p_1}\right)^{(\gamma+1)/\gamma}\right)} \qquad \text{(A2.31)}$$

Defining the long-pipe approximation function, f_{lpa}, as the dimensionless function

$$f_{lpa} = \sqrt{\frac{1}{K_T} \frac{2\gamma}{\gamma+1} \left(1 - \left(\frac{p_3}{p_1}\right)^{(\gamma+1)/\gamma}\right)} \qquad \text{(A2.32)}$$

it is clear that equation (A2.31) may be written in the dimensionless form

$$\frac{W}{A\sqrt{\frac{p_1}{v_1}}} = b_0 f_{lpa} \qquad \text{(A2.33)}$$

Comparing equations (A2.33) and (A2.7), it is clear that the correction factor, b_0, is simply the ratio of the exact pipe-flow function, f_{pipe}, to its approximate equivalent just defined:

$$b_0\left(\frac{p_4}{p_1}, K_T, \gamma\right) = \frac{f_{pipe}\left(\frac{p_4}{p_1}, K_T, \gamma\right)}{f_{lpa}\left(\frac{p_4}{p_1}, K_T, \gamma\right)} \qquad \text{(A2.34)}$$

(Inspection of equation (A2.32) seems to indicate that f_{lpa} is a function of $p_3/p_1, K_T, \gamma$ rather than $p_4/p_1, K_T, \gamma$, but equation (A2.14) shows that p_3/p_1 depends on p_4/p_1.)

A2.5 Calculation of b_0

Equation (A2.31) can be made to reproduce precisely the mass flow calculated by the full, compressible flow equations if b_0 is made to vary according to equation (A2.34). This may be done by taking the results of the exact calculation of f_{pipe} contained in Tables A2.1 to A2.4 and dividing by the corresponding value of f_{lpa}, thus giving b_0 over a broad range of pressure ratios and frictional losses. Figure A2.2 shows the behaviour of b_0 over a range of pressure ratio, p_4/p_1, of 0.05 to 0.95, and with frictional loss, K_T, varied from 1 to 2048 velocity heads.

As we would expect from the assumptions made in the derivation of the long-pipe approximation, b_0 is close to unity when the change in pressure with distance, dp/dx, is relatively low, namely when the pipe is long and/or the pressure drop small. The contours show how b_0 increases towards unity as $p_4/p_1 \to 1.0$ and $K_T \to \infty$. In fact at $K_T = 128$ and $p_4/P_1 = 0.95$, b_0 is already greater than 0.99. But the deviations from unity grow when dp/dx is large, namely when the pipe is short, or the pressure ratio is small.

A2.6 Using polynomial functions to characterize the b_0 surface

Inspection of the behaviour of the surface shown in Figure A2.2 indicates that the dependence of b_0 on K_T follows a roughly log-linear pattern. Accordingly it is natural to choose the descriptive polynomials to be functions of $\ln K_T$ rather than simply K_T.

It is found that for a given value of specific-heat ratio, γ, the relationship between p_{3c}/p_1 and K_T may be modelled adequately by a fourth-order polynomial in $\ln K_T$:

$$\frac{p_{3c}}{p_1} = \pi_0 + \pi_1(\ln K_T) + \pi_2(\ln K_T)^2 + \pi_3(\ln K_T)^3 + \pi_4(\ln K_T)^4 \quad \text{(A2.35)}$$

The relationship between b_0 and p_4/p_1 is shown in Figure A2.3 below for various values of K_T for a diatomic gas, where $\gamma = 1.4$. The nearly linear nature of the relationship for pressure ratios above critical means that it can be approximated by linear deviations from the value of b_0 at $p_4/p_1 = 0.95$, with the slope of the line depending on the frictional loss, K_T.

The model is thus:

$$b_0\left(K_T, \frac{p_4}{p_1}\right) = b_0(K_T, 0.95) + m_b(K_T)\left(\frac{p_3}{p_1} - 0.95\right) \quad \text{(A2.36)}$$

Here $b_0(K_T, 0.95)$ may be calculated for a given value of γ from a a fourth-order polynomial in $\ln K_T$:

$$b_o(K_T, 0.95) = \beta_0 + \beta_1(\ln K_T) + \beta_2(\ln K_T)^2 + \beta_3(\ln K_T)^3 + \beta_4(\ln K_T)^4 \quad \text{(A2.37)}$$

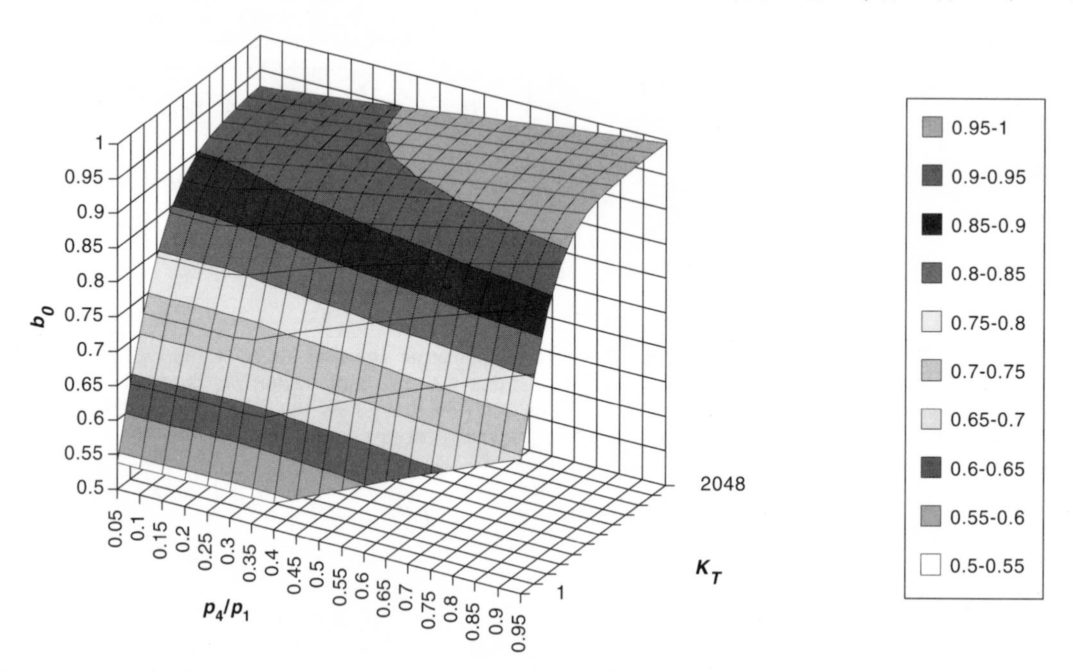

Figure A2.2 Plot of long-pipe approximation parameter, b_0, against frictional pressure loss in velocity heads, K_T, and pressure ratio, p_4/p_1.

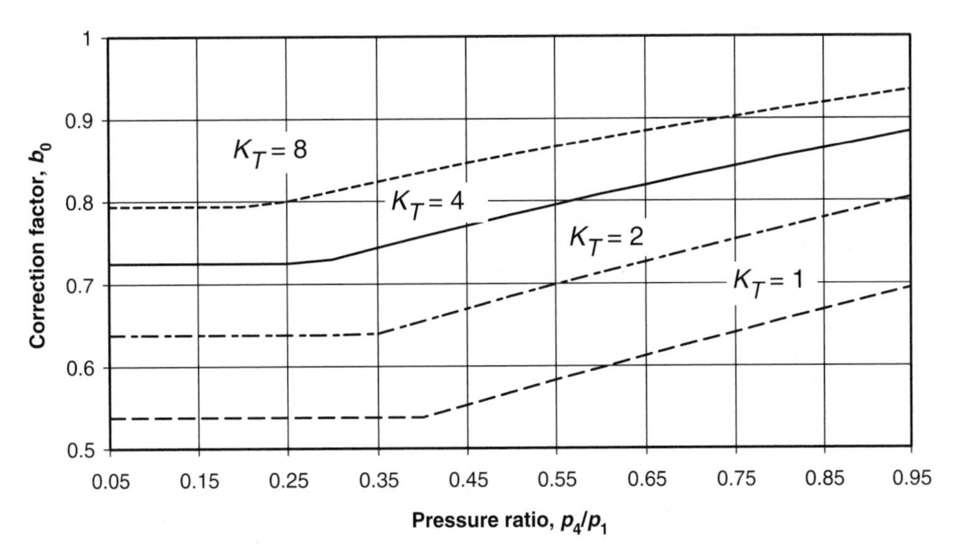

Figure A2.3 The relationship between correction factor, b_0, and pressure ratio at various values of frictional loss, K_T, for a diatomic gas, with $\gamma = 1.4$.

while $m_b(K_T)$ may be calculated for a given value of γ from a fifth-order polynomial in $\ln K_T$

$$m_b(K_T) = \mu_0 + \mu_1(\ln K_T) + \mu_2(\ln K_T)^2 + \mu_3(\ln K_T)^3 + \mu_4(\ln K_T)^4 + \mu_5(\ln K_T)^5 \quad \text{(A2.38)}$$

A list of polynomial coefficients is given in Tables A2.5 to A2.8.

A2.7 Size of errors using approximating functions

The percentage error introduced by calculating the correction factor, b_0, using polynomials will be

$$\frac{\hat{b}_0 f_{lpa} - f_{pipe}}{f_{pipe}} \times 100$$

where $\hat{b}_0$ is the estimate of b_0. This error is found to be less than 2% over the ranges of p_4/p_1, K_T and γ

Table A2.5 Polynomial coefficients for a monatomic gas: $\gamma = 1.67$

Table A2.5(a) $K_T \leq 1$

Polynomial coefficients	Power, *i*, of $\ln K_T$: 0	1	2	3	4	5
π_i	0.357339	−0.07459	−0.01597	−0.00118	2.73E-05	0.0
β_i	0.69463	0.17503	−0.0227	−0.0136	−0.0015	0.0
μ_i	0.274056	0.017077	−0.04335	−0.01357	−0.00126	0.0

Table A2.5(b) $1 \leq K_T \leq 2048$

Polynomial coefficients	Power, *i*, of $\ln K_T$: 0	1	2	3	4	5
π_i	0.35889	−0.07394	−0.00636	0.002641	−0.00017	0.0
β_i	0.69379	0.19261	−0.0472	0.00518	−0.0002	0.0
μ_i	0.27786	0.005452	−0.02986	0.007003	−5.68E-04	1.32E-05

Table A2.6 Polynomial coefficients for a diatomic gas: $\gamma = 1.4$

Table A2.6(a) $K_T \leq 1$

Polynomial coefficients	Power, i, of $\ln K_T$					
	0	*1*	*2*	*3*	*4*	*5*
π_i	0.398919	−0.07546	−0.01595	−0.00118	7.98E-06	0.0
β_i	0.69463	0.18078	−0.0152	−0.0103	−0.001	0.0
μ_i	0.283636	0.002084	−0.05475	−0.01675	−0.0016	0.0

Table A2.6(b) $1 \leq K_T \leq 2048$

Polynomial coefficients	Power, i, of $\ln K_T$					
	0	*1*	*2*	*3*	*4*	*5*
π_i	0.399854	−0.0767	−0.0093	0.003245	−0.00021	0.0
β_i	0.69348	0.19344	−0.0473	0.00517	−0.0002	0.0
μ_i	0.289121	−0.01076	−0.02967	0.007741	−6.92E-04	1.95E-05

Table A2.7 Polynomial coefficients for superheated steam: $\gamma = 1.3$

Table A2.7(a) $K_T \leq 1$

Polynomial coefficients	Power, i, of $\ln K_T$					
	0	*1*	*2*	*3*	*4*	*5*
π_i	0.416819	−0.07565	−0.01595	−0.00118	1.94E-06	0.0
β_i	0.69446	0.18094	−0.0151	−0.0103	−0.001	0.0
μ_i	0.28872	0.001523	−0.0514	−0.01455	−0.00126	0.0

Table A2.7(b) $1 \leq K_T \leq 2048$

Polynomial coefficients	Power, i, of $\ln K_T$					
	0	*1*	*2*	*3*	*4*	*5*
π_i	0.417795	−0.0788	−0.00981	0.00333	−0.00021	0.0
β_i	0.69351	0.19344	−0.0471	0.00513	−0.0002	0.0
μ_i	0.291405	−0.01117	−0.03245	0.008318	−7.18E-04	1.87E-05

considered. Figure A2.4 shows the error surface for a diatomic gas, with $\gamma = 1.4$.

A2.8 Simplified approximation using a constant value of b_0

It is evident from the 5% contour bands for b_0 shown in Figure A2.2 that it may often be possible to choose a constant value of b_0 to give a simpler calculation of flow. This may be sufficiently accurate for the modeller's purposes provided the variations in pressure ratio and in pipe effective length is small. Denoting this constant value $\overline{b}_0$, we may calculate it from a set of reference conditions using equations (A2.11), and (A2.29), modified to refer to the steady-state operating conditions, denoted by an overscript bar:

$$\overline{b}_0 = \frac{\overline{W}}{\overline{C}_L}\sqrt{\frac{\overline{v}_{ave}}{\overline{p}_1 - \overline{p}_3 - \dfrac{g(\overline{z}_3 - \overline{z}_1)}{\overline{v}_{ave}}}} \tag{A2.39}$$

Table A2.8 Polynomial coefficients for dry, saturated steam: $\gamma = 1.135$

Table A2.8(a) $K_T \leq 1$

Polynomial coefficients	Power, i, of $\ln K_T$					
	0	*1*	*2*	*3*	*4*	*5*
π_i	0.449784	−0.07695	−0.01662	−0.00115	2.72E-05	0.0
β_i	0.69413	0.18133	−0.0151	−0.0103	−0.001	0.0
μ_i	0.298632	−0.00896	−0.0573	−0.01546	−0.0013	0.0

Table A2.8(b) $1 \leq K_T \leq 2048$

Polynomial coefficients	Power, i, of $\ln K_T$					
	0	*1*	*2*	*3*	*4*	*5*
π_i	0.450736	−0.0799	−0.01244	0.003823	−0.00024	0.0
β_i	0.69304	0.19504	−0.0477	0.00523	−0.0002	0.0
μ_i	0.302007	−0.02798	−0.02925	0.007534	−5.68E-04	8.66E-06

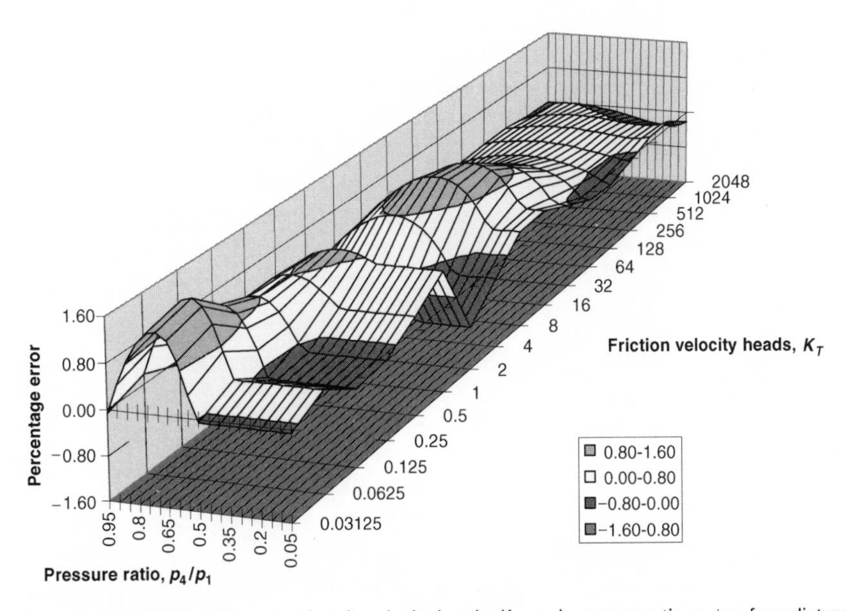

Figure A2.4 Flow error versus frictional pressure loss in velocity heads, K_T, and pressure ratio, p_4/p_1, for a diatomic gas, $\gamma = 1.4$.

where

$$\overline{v}_{ave} = \overline{v}_1 \frac{\gamma + 1}{\gamma} \frac{\left(1 - \dfrac{\overline{p}_3}{\overline{p}_1}\right)}{\left(1 - \left(\dfrac{\overline{p}_3}{\overline{p}_1}\right)^{(\gamma+1)/\gamma}\right)} \tag{A2.40}$$

and, from equation (A2.12)

$$\overline{C}_L = A\sqrt{\frac{2}{\overline{K}_T}} \tag{A2.41}$$

Here the pressure ratio, $\overline{p}_3/\overline{p}_1$, follows from equation (A2.14):

$$\frac{\overline{p}_3}{\overline{p}_1} = \max\left[\frac{\overline{p}_4}{\overline{p}_1}, \frac{\overline{p}_{3c}}{\overline{p}_1}\right] \tag{A2.42}$$

and $\overline{p}_1$, $\overline{p}_3$, $\overline{p}_{3c}$, $\overline{p}_4$, $\overline{v}_1$, $\overline{v}_{ave}$, $\overline{C}_L$, $\overline{K}_T$ and $\overline{W}$ are the values in the reference conditions. $\overline{p}_{3c}/\overline{p}_1$ may be derived from equation (A2.35), with the reference value $K_T = \overline{K}_T$.

For a horizontal pipe, equation (A2.43) below may be used in place of equations (A2.39) and (A2.40)

$$\overline{b}_0 = \frac{f_{pipe}\left(\dfrac{\overline{p}_4}{\overline{p}_1}\right)}{f_{lpa}\left(\overline{K}_T, \dfrac{\overline{p}_3}{\overline{p}_1}\right)} = \frac{1}{A}\frac{\overline{W}}{f_{lpa}\left(\overline{K}_T, \dfrac{\overline{p}_3}{\overline{p}_1}\right)}\sqrt{\frac{\overline{v}_1}{\overline{p}_1}} \qquad \text{(A2.43)}$$

A2.9 Bibliography

Streeter, V.L., and Wylie, E.B. (1983). *Fluid Mechanics, First SI Metric Edition*, McGraw-Hill, New York, Chapter 4, Dimensional analysis and dynamic similitude.

Appendix 3 Equations for control valve flow in SI units

A3.1 Introduction

The historical dominance of US valve manufacturers has led to valve characteristics often being given in standard US units (US gallons, standard cubic feet per hour, etc.). A particularly important influence because of its extensive research into and testing of control valves was the Fisher Controls Company of Marshalltown, Iowa (now part of Fisher-Rosemount Ltd.) This appendix converts the basic liquid-flow equation from US to SI units. The appendix then goes on to derive Fisher's 'Universal Gas-Sizing Equation' ('FUGSE') from the basic liquid-flow equation and shows how this may be converted into SI units.

Some of the important variables in the text to follow will be superscripted with an asterisk when given in US units in order to differentiate them from the equivalent or related variables given in standard SI units.

A3.2 Liquid flow through the valve

Liquid flow through the valve in US units is given by the following equation :

$$Q^* = C_v^* \sqrt{\frac{\Delta p^*}{G}} \tag{A3.1}$$

where

Q^* is the volume flow in US gallons per minute,

Δp^* is the pressure difference across the valve in psi,

$G = (v_w(520))/v_1$ is the specific gravity of the fluid relative to water at 60° Fahrenheit = 520° Rankine = 15.56°C,

C_v^* is the valve capacity for water at 60°F at a given valve travel/opening, in US gall/min/(psi)$^{1/2}$.

On the basis that

$$1 \text{ US gall} = 3.78543 \times 10^{-3}\, \text{m}^3$$

$$1 \text{ psi} = 6895\, \text{Pa}$$

$$v_w(520) = 1.016 \times 10^{-3}\, \text{m}^3/\text{kg}$$

we may convert equation (A3.1) to give a flow, Q, in m^3/s:

$$Q = \frac{3.78543 \times 10^{-3}}{60 \times \sqrt{6895}\sqrt{v_w(520)}} C_v^* \sqrt{v_1 \Delta p} \tag{A3.2}$$

where v_1 is in m^3/kg, and $\Delta p = p_1 - p_2$ is in Pa. Equation (A3.2) may be re-expressed as:

$$Q = C_v \sqrt{v_1 \Delta p} \tag{A3.3}$$

where C_v is the valve conductance given by:

$$\begin{aligned} C_v &= \frac{3.78543 \times 10^{-3}}{60 \times \sqrt{6895}\sqrt{v_w(520)}} C_v^* \\ &= \frac{3.78543 \times 10^{-3}}{60 \times \sqrt{6895}\sqrt{1.016 \times 10^{-3}}} C_v^* \\ &= 2.3837 \times 10^{-5} C_v^* \end{aligned} \tag{A3.4}$$

Note that the valve conductance, C_v, has units of m^2, and is thus different in dimension as well as units from C_v^*. Finally, we may convert to the desired SI form in mass flow (kg/s) by dividing equation (A3.3) by specific volume, v_1:

$$W = C_v \sqrt{\frac{\Delta p}{v_1}} \tag{A3.5}$$

A3.3 Gas flow at small pressure drops in US units

Manufacturers found that equation (A3.1) served well for gas flow at high-pressure ratios as well as for liquid flow. The equation is often quoted with the flow units of standard cubic feet per hour, with the specific gravity standard changed from water to air, which requires the somewhat involved conversion route set down below. The specific gravity with respect to water is related to the specific gravity with respect to air by:

$$\begin{aligned} G &= \frac{v_w(520)}{v_a(p_1^*, T_1^*)} \frac{v_a(p_1^*, T_1^*)}{v_1^*(p_1^*, T_1^*)} \\ &= \frac{v_w(520)}{v_a(p_1^*, T_1^*)} G_a \end{aligned} \tag{A3.6}$$

where

$v_1(p_1^*, T_1^*)$ is the specific volume of the gas at inlet pressure p_1^* psia and inlet temperature T_1^* °R,

$v_a(p_1^*, T_1^*)$ is the specific volume of air at the temperature and pressure of the gas entering the valve,

$v_w(520)$ is the specific volume of water at 60°F = 520°R (assumed independent of pressure), and

$G_a = (v_a(p_1^*, T_1^*))/(v_1(p_1^*, T_1^*))$ is the specific gravity with respect to air.

Assuming air can be treated as a perfect gas, the equation of state gives (in US units):

$$\frac{p_1^* v_a(p_1^*, T_1^*)}{T_1^*} = \frac{14.7 v_a(14.7, 520)}{520} \quad \text{(A3.7)}$$

or:

$$v_a(p_1^*, T_1^*) = v_a(14.7, 520)\frac{14.7}{p_1^*}\frac{T_1^*}{520} \quad \text{(A3.8)}$$

Substituting into (A3.6) gives:

$$G = \frac{v_w(520)}{v_a(14.7, 520)}\frac{p_1^*}{14.7}\frac{520}{T_1^*}G_a \quad \text{(A3.9)}$$

Taking $v_a(14.7, 520) = 13.09\,\text{ft}^3/\text{lb}$ and $v_w(520) = 0.0161\,\text{ft}^3/\text{lb}$, equation (A3.9) gives:

$$G = 1.2299 \times 10^{-3}\frac{p_1^*}{14.7}\frac{520}{T_1^*}G_a \quad \text{(A3.10)}$$

Using the conversion factor 1 US gallon per minute = 8.0208 cubic feet per hour together with (A3.10) in equation (A3.1) gives the flow in cubic feet per hour as:

$$Q_{cfh} = 8.0208C_v^*\sqrt{\frac{1}{1.2299 \times 10^{-3}}\frac{\Delta p^*}{G_a}\frac{14.7}{p_1^*}\frac{T_1^*}{520}}$$

$$= 228.7C_v^*\sqrt{\frac{\Delta p^*}{G_a}\frac{14.7}{p_1^*}\frac{T_1^*}{520}} \quad \text{(A3.11)}$$

We transform this to units of standard cubic feet per hour (scfh) by noting that a mass that has volume V at (p_1^*, T_1^*) will have volume V_{scf} at $p^* = 14.7$ psia, $T^* = 520$°R, where:

$$V_{scf} = V\frac{520}{T_1^*}\frac{p_1^*}{14.7} \quad \text{(A3.12)}$$

so that volume flow will obey the relationship

$$Q_{scfh} = Q_{cfh}\frac{520}{T_1^*}\frac{p_1^*}{14.7} \quad \text{(A3.13)}$$

Applying this to the volume flow rate of equation (A3.11) yields:

$$Q_{scfh} = 228.7C_v^*\sqrt{\frac{\Delta p^*}{G_a}\frac{14.7}{p_1^*}\frac{T_1^*}{520}}\frac{520}{T_1^*}\frac{p_1^*}{14.7} \quad \text{(A3.14)}$$

or, after rearrangement, the desired final equation:

$$Q_{scfh} = 59.64C_v^* p_1\sqrt{\frac{\Delta p^*}{p_1^*}}\sqrt{\frac{520}{G_a T_1^*}} \quad \text{(A3.15)}$$

where Q_{scfh} is the flow rate in standard cubic feet per hour (14.7 psia, 60°F).

A3.4 Gas flow at very large pressure drops

Equation (A3.15) possesses the limitation that it does not predict the choking effect that will occur when the flow becomes sonic in the valve throat. For such conditions, US manufacturers predict choked flow, Q_{crit}, in terms of the measured 'gas sizing coefficient', C_g^*:

$$Q_{crit} = C_g^* p_1^*\sqrt{\frac{520}{G_a T_1^*}} \quad \text{(A3.16)}$$

where

Q_{crit} is the critical flow rate in standard cubic feet per hour,

C_g^* is the gas sizing coefficient (scf/hr/psia) for a given valve opening.

A3.5 Gas flow at intermediate pressure drops: the Fisher Universal Gas Sizing Equation (FUGSE)

The FUGSE links the two equations for low and high-pressure drops in a simple way by exploiting the characteristics of the sine function that $\sin(\pi/2) = 1$ and $\sin\theta \cong \theta$ in radians for low values of θ:

$$Q_{scfh} = \sqrt{\frac{520}{G_a T_1^*}}C_g^* p_1^* \sin\theta \quad \text{(A3.17)}$$

where

$$\theta = \min\left(\frac{\pi}{2}, \frac{59.64}{C_1^*}\sqrt{\frac{\Delta p^*}{p_1^*}}\right) \quad \text{(A3.18)}$$

in which $C_1^* = (C_g^*/C_v^*)$ in units (scf/US gall).(min/h)/psia$^{1/2}$. It will be seen that the FUGSE reproduces the flow rates of equations (A3.15) and (A3.16) at high and low values of the valve pressure ratio. There is no strong basis for expecting the FUGSE to perform particularly well at intermediate pressure ratios, and, indeed, its performance here is less good.

A3.6 Converting the Fisher Universal Gas Sizing Equation to SI units

To convert the equation pair (A3.17) and (A3.18) into SI units, we begin by converting back from scf/h to cubic feet per hour using the inverse of equation (A3.13), namely:

$$Q_{cfh} = Q_{scfh} \frac{14.7}{p_1^*} \frac{T_1^*}{520} \tag{A3.19}$$

so that the volume flow in ft^3/hr is given by

$$\begin{aligned} Q_{cfh} &= \frac{14.7}{p_1^*} \frac{T_1^*}{520} \sqrt{\frac{520}{G_a T_1^*}} C_g^* p_1^* \sin\theta \\ &= 14.7 \sqrt{\frac{T_1^*}{520 G_a}} C_g^* \sin\theta \end{aligned} \tag{A3.20}$$

The specific gravity may be expanded as follows:

$$\begin{aligned} G_a &= \frac{v_a(p_1^*, T_1^*)}{v_1^*(p_1^*, T_1^*)} \\ &= \frac{v_a(14.7, 520)}{v_1^*(p_1^*, T_1^*)} \frac{14.7}{p_1^*} \frac{T_1^*}{520} \\ &= \frac{13.09}{v_1^*(p_1^*, T_1^*)} \frac{14.7}{p_1^*} \frac{T_1^*}{520} \end{aligned} \tag{A3.21}$$

so substituting back into (A3.20) gives:

$$Q_{cfh} = 1.06 C_g^* \sqrt{p_1^* v_1^*} \sin\theta \tag{A3.22}$$

Dividing by the specific volume in ft^3/lb gives the mass flow in lb/h:

$$W^* = 1.06 C_g^* \sqrt{\frac{p_1^*}{v_1^*}} \sin\theta \tag{A3.23}$$

We may convert this equation to SI units by noting the following conversion factors:

$$p_1^*(\text{psia}) = 1.45308 \times 10^{-4} p_1(\text{Pa})$$

$$v_1^*(\text{ft}^3/\text{lb}) = 16.0179 v_1(\text{m}^3/\text{kg})$$

$$1 \text{ lb/h} = \frac{1}{2.205} \times \frac{1}{3600} \text{ kg/s}$$

so that flow in kg/s is given by:

$$W = C_g \sqrt{\frac{p_1}{v_1}} \sin\theta \tag{A3.24}$$

where the inlet pressure, p_1, is in Pa, the inlet specific volume, v_1, is in m^3/kg, while the new SI gas coefficient, C_g, is given by

$$C_g = 4.02195 \times 10^{-7} C_g^* \tag{A3.25}$$

and has dimensions of m^2. Note that C_g has a different dimension from C_g^*, as well as different units. It will be shown in Appendix 4 thatC_g (and hence also C_g^*) will depend on the nature of the gas being passed. The test gas is normally air, which is composed overwhelmingly of the diatomic gases nitrogen and oxygen, and hence has a specific-heat ratio, $\gamma = 1.4$. Thus the quoted C_g^*-value will normally be C_g^* (1.4).

The term θ in equation (A3.18) is dimensionless, and so it has not been necessary to change it in the conversion process. It may, however, be rearranged in terms of the SI conductance ratio, $C_1 = C_g/C_v$, which has the advantage over C_1^* that it is dimensionless. C_1 may be related to C_1^* through equations (A3.25) and (A3.4):

$$\begin{aligned} C_1 &= \frac{4.02195 \times 10^{-7} C_g^*}{2.3837 \times 10^{-5} C_v^*} \\ &= 1.6873 \times 10^{-2} C_1^* \end{aligned} \tag{A3.26}$$

Since C_g^* depends on specific-heat ratio, γ, but C_v^* does not, it follows that C_1^* and hence C_1 are dependent on γ in the same way as C_g^* and C_g . Substituting from equation (A3.26) into equation (A3.18) gives an alternative formula for θ:

$$\theta = \min\left(\frac{\pi}{2}, \frac{1.0063}{C_1} \sqrt{\frac{\Delta p}{p_1}}\right) \tag{A3.27}$$

A.3.7 Summary of conversions between SI and US valve coefficients

The SI valve conductance, C_v (m^2), is related to the US valve flow coefficient, C_v^* (USgall/min/(psi)$^{1/2}$) by

$$C_v = 2.3837 \times 10^{-5} C_v^* \tag{A3.4}$$

The SI limiting gas conductance, C_g (m^2), is related to the US gas sizing coefficient, C_g^* (scf/h/psia) by:

$$C_g = 4.02195 \times 10^{-7} C_g^* \tag{A3.25}$$

The SI conductance ratio, C_1 (dimensionless) is related to the US liquid-gas-coefficient ratio, C_g^* ((scf/US gall).(min/h)/psia$^{1/2}$) by

$$C_1 = 1.6873 \times 10^{-2} C_1^* \tag{A3.26}$$

As discussed above, both C_g and C_1 are functions of the ratio of specific heats, γ. The quoted values of C_g^* and C_1^* can be expected to be quoted for air, with. $\gamma = 1.4$.

Appendix 4 Comparison of Fisher Universal Gas Sizing Equation, FUGSE, with the nozzle-based model for control valve gas flow

A4.1 Introduction

The FUGSE blends the liquid-flow algorithm for high-pressure ratios with a choking equation valid at low valve pressure ratios by using a sine function, the argument of which is chosen to be proportional to the square root of the fractional pressure drop, $\sqrt{\Delta p/p_1} = \sqrt{1 - p_2/p_1}$. The resulting equations for flow in the US units of standard cubic feet per hour are:

$$Q_{scfh} = \sqrt{\frac{520}{G_a T_1^*}} C_g^* p_1^* \sin\theta \quad \text{(A4.1)}$$

$$\theta = \min\left(\frac{\pi}{2}, \frac{59.64}{C_1^*}\sqrt{\frac{\Delta p^*}{p_1^*}}\right) \quad \text{(A4.2)}$$

where the variables marked with asterisks are in US units. Appendix 3 shows how these equations may be converted to SI units, giving the mass flow, W, in kg/s as:

$$W = C_g \sqrt{\frac{p_1}{v_1}} \sin\theta \quad \text{(A4.3)}$$

$$\theta = \min\left(\frac{\pi}{2}, \frac{1.0063}{C_1}\sqrt{1 - \frac{p_2}{p_1}}\right) \quad \text{(A4.4)}$$

where C_1 is the dimensionless ratio of the limiting gas conductance to the valve conductance:

$$C_1 = \frac{C_g}{C_v} \quad \text{(A4.5)}$$

The angle, θ, will be small when the valve pressure ratio, p_2/p_1, is high, allowing application of the approximation $\sin\theta \approx \theta$, which tranforms equations (A4.3) and (A4.4) essentially into the liquid flow equation (9.1). At very high-pressure drops, on the other hand, the limit of equation (A4.4) is invoked, so that $\theta = \pi/2$ and $\sin\theta = 1$, which results in equation (A4.3) reducing to the choked gas equation (9.2). But while the FUGSE will be valid at either end of the range of pressure ratios, the fact that the sine function joins the extremes in a conveniently smooth manner does not guarantee that the formula will be accurate at intermediate values. This appendix examines how closely the FUGSE matches direct data and the nozzle-based method for calculating control valve gas flow.

A4.2 Comparison of the Fisher Universal Gas Sizing Equation, FUGSE, with direct data

Direct data are presented in the Fisher Control Valve Handbook for the low valve pressure ratios associated with the onset of choking. The Fisher data are in the form of a graph of $\Delta p/p_1 (= 1 - p_2/p_1)$ versus C_1^* ($= C_g^*/C_v^*$, see Appendix 3) at the onset of choking. Under the assumption that the tests were based on the passage of air, which is a diatomic gas ($\gamma = 1.4$), we may convert the data first into a relationship between $p_2/p_1|_{choke}$ and C_1, and then, using

$$C_1 = C_{1t}(1.4) C_{fgh} \quad \text{(A4.6)}$$

(see equation (9.50)), where $C_{1t}(1.4) = 0.4842$ (Table 9.1), into a relationship between $p_2/p_1|_{choke}$ and C_{fgh}.

Meanwhile the FUGSE predicts the onset of choked flow in the valve when the two terms on the right-hand side of equation (A4.4) are equal:

$$\frac{1.0063}{C_1}\sqrt{\frac{p_1 - p_2}{p_1}} = \frac{\pi}{2} \quad \text{(A4.7)}$$

Hence the valve pressure ratio at choking may be calculated from the quadratic curve

$$\left.\frac{p_2}{p_1}\right|_{choke} = 1 - 2.4366 C_1^2 \quad \text{(A4.8)}$$

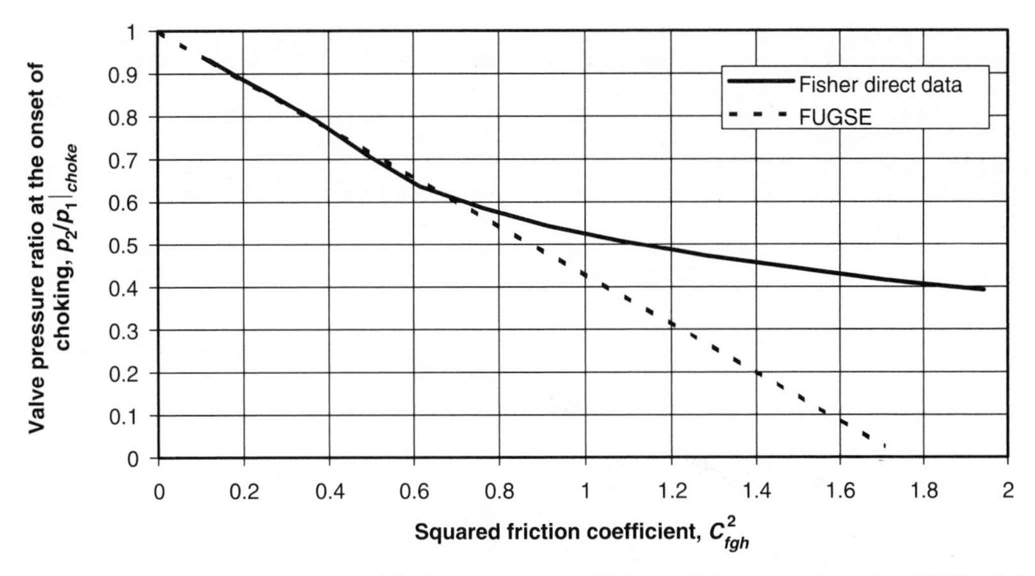

Figure A4.1 Valve pressure ratio at the onset of choking versus squared friction coefficient: comparison of the FUGSE with Fisher direct data.

Using equation (A4.6) for flowing air gives the quadratic relationship:

$$\left.\frac{p_2}{p_1}\right|_{choke} = 1 - 0.5712C^2_{fgh} \tag{A4.9}$$

Equation (A4.9) is compared with the direct data in Figure A4.1, where it can be seen that the match is good only up to $C^2_f \approx 0.7$, the lower edge of the range of regulating duty for control valves.

An increasing divergence is evident after this point, and it is noticeable that the FUGSE does not reproduce the important direct data point (1.0, 0.5283).

Further examination of equation (A4.9) shows that negative values of the valve pressure ratio at choking are predicted for $C_f > 1.323$, suggesting that choked flow is impossible at high friction coefficients. This does not reflect physical reality, since, inter alia, it would preclude choked flow occurring in large numbers of globe valves, however low the valve pressure ratio. The discrepancy arises as a result of the long tail generated by the sine function of equation (A4.3) at low pressure ratios and high friction coefficients.

A4.3 Comparison of the FUGSE with the nozzle-based model for control valve gas flow

We may compare the flow-predictions of the FUGSE with the nozzle-based model by taking out the common factor $C_g\sqrt{p_1/v_1}$ from equation (A4.3) and comparing the result with the equivalent nozzle function, f_{NV}, given by equation set (9.61), repeated below:

$$f_{NV} = 2.065\sqrt{\frac{\gamma}{\gamma-1}\left(\left(\frac{p_t}{p_1}\right)^{2/\gamma} - \left(\frac{p_t}{p_1}\right)^{(\gamma+1)/(\gamma)}\right)} \quad \text{for } \frac{p_t}{p_1} > \frac{p_{tc}}{p_1}$$

$$= 1.0 \quad \text{for } \frac{p_t}{p_1} \leq \frac{p_{tc}}{p_1} \tag{9.61}$$

where the throat pressure ratio, p_t/p_1, in equation (9.61) is found as a function of C_{fgh} and p_2/p_1 by the procedure given in Chapter 9, Section 9.9.

The corresponding FUGSE function, f_{FUGSE}, is simply:

$$f_{FUGSE} = \sin\theta \tag{A4.10}$$

where the angle θ is given in terms of valve pressure ratio, p_2/p_1, and friction coefficient, C_{fgh}, by substituting from equation (A4.6) into equation (A4.4):

$$\theta = \min\left(\frac{\pi}{2}, \frac{1.0063}{C_{1t}C_{fgh}}\sqrt{1-\frac{p_2}{p_1}}\right) \tag{A4.11}$$

Using the FUGSE function, f_{FUGSE}, and the nozzle-valve function, f_{NV}, allows a general comparison to be made for an arbitrary set of upstream conditions and limiting gas conductances. Assuming that the gas being passed is diatomic, so that $\gamma = 1.4$, Figure A4.2 shows the behaviour of the two functions as the valve pressure ratio is decreased from unity for two, low values of the friction coefficient, namely $C_{fgh} = 0.7$ and $C_{fgh} = 1.0$. The flows predicted by both calculational routes will be very close at $C_{fgh} = 0.7$, typical of a rotary valve at or near fully open, but a disparity begins to emerge at lower valve pressure ratios when the friction coefficient is at its isobaric value, $C_{fgh} = 1.0$, typical of a rotary valve at low valve travel.

Figure A4.3 shows f_{FUGSE} and f_{NV} against valve pressure ratio, p_2/p_1, when the friction coefficient is $C_{fgh} = 1.25$, typical of a globe valve. The FUGSE function predicts flows significantly smaller than the nozzle-valve function as the valve pressure ratio falls below about 0.9. The FUGSE function does not predict choked flow until p_2/p_1 has fallen to about 0.15, approximately a third of the valve pressure ratio,

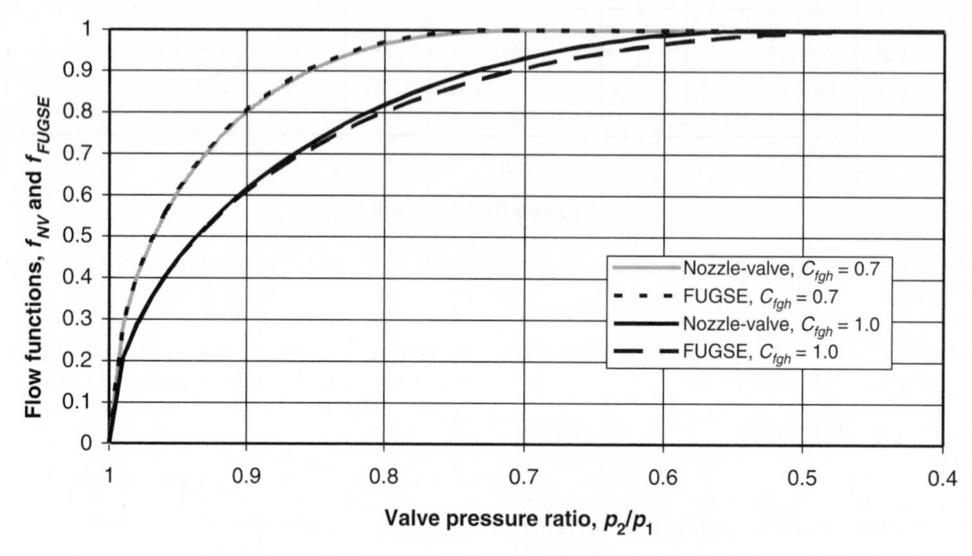

Figure A4.2 Comparison of nozzle-valve and FUGSE flow functions for friction coefficients, C_f, of 0.7 and 1.0.

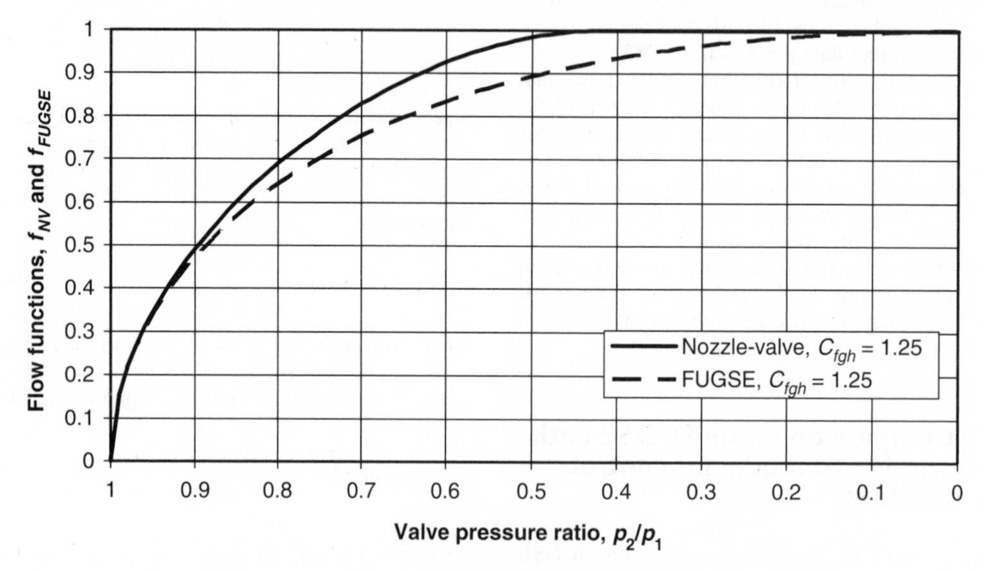

Figure A4.3 Comparison of nozzle-valve and FUGSE functions when the friction coefficient, C_f, is 1.25.

~0.45, at which the direct data (see Figure A4.1) have indicated choking to occur.

The FUGSE has the merit of mathematical simplicity (especially in the SI form of equations (A4.3) and (A4.4)). Further, provided allowance is made for the change in limiting gas conductance, C_g, when non-diatomic gases are being passed (equation (9.56)), it can be expected to produce reasonably accurate predictions of flow for the low friction coefficients associated with rotary valves, particularly at high valve openings. But it will perform less well for the large friction coefficients typical of globe valves, especially as the valve pressure ratio approaches its choked value.

Appendix 5 Measurement of the internal energy of reaction and the enthalpy of reaction using calorimeters

A5.1 Introduction

There is sometimes confusion concerning the difference between the internal energy of reaction, ΔU_j, and the enthalpy of reaction, ΔH_j. This appendix has been included to give a physical interpretation of the difference between the two concepts. The principles of two pieces of laboratory equipment are explained: the 'bomb' calorimeter, which measures internal energy change, and the Boys' open-system calorimeter, which measures enthalpy change. Measurements taken from the two systems allow the two parameters to be distinguished clearly.

A5.2 Measuring the internal energy of reaction using the bomb calorimeter

The bomb calorimeter is so called because there is no provision for an inflow or outflow, and because the reaction is often exothermic. Unlike a true bomb, however, the walls are designed to have sufficient strength to withstand the increased pressure due to rapid heating! The apparatus is shown schematically in Figure A5.1.

Measured quantities of the reactants are introduced into the bomb calorimeter, which is sometimes divided into separate compartments to prevent reaction during filling. Thermal equilibrium is established between the bomb and the water bath. Then the reaction is initiated, sometimes by allowing mixing, sometimes by introducing a spark, and it runs its course inside the calorimeter. The heat flow, Φ, is marked in the figure as going from the water bath into the bomb, in accordance with the convention of equation (13.34) of Chapter 13, Section 13.7, but the physical heat flow will be in the opposite direction for the exothermic reactions for which the device is most commonly used. Because of the high thermal inertia of the water bath, the final temperature of the contents of the bath

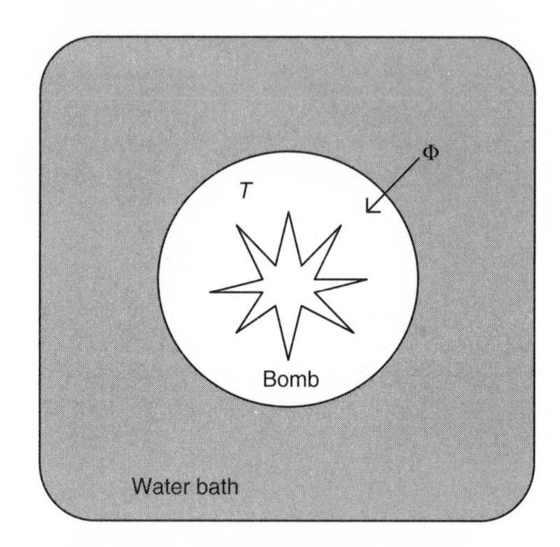

Figure A5.1 Schematic of bomb calorimeter.

and bomb will be almost the same as the starting temperature. The integrated heat flow is found by measuring the small change in the temperature of the water in the water bath.

We may apply equation (13.43) to the system, with the conditions that there is no inflow or outflow, and no mechanical power abstracted. Since specific internal energy is a function of temperature only, we may use equation (13.40) also to give:

$$\sum_{i=1}^{N} M_i \frac{d\hat{u}_i}{dT}\frac{dT}{dt} = \Phi - V\sum_{j=1}^{M} r_j \Delta U_j \qquad \text{(A5.1)}$$

Integrating this equation between the start and finish times of the measurement, we have:

$$\sum_{i=1}^{N} M_i \frac{d\hat{u}_i}{dT}\int_{T_1}^{T_2} dT = \sum_{i=1}^{N} M_i \frac{d\hat{u}_i}{dT}(T_2 - T_1) = \int_{t_1}^{t_2} \Phi\, dt - V\sum_{j=1}^{M} \Delta U_j \int_{t_1}^{t_2} r_j\, dt \qquad \text{(A5.2)}$$

Since the start and finish temperatures are approximately equal, $T_2 - T_1 \approx 0$, so that:

$$\int_{t_1}^{t_2} \Phi\, dt - V \sum_{j=1}^{M} \Delta U_j \int_{t_1}^{t_2} r_j\, dt \approx 0 \qquad \text{(A5.3)}$$

For the case where there is only one reaction, $M = 1$, and we may omit the j subscript. Using the definition of the reaction rate density, r, from equations (13.27) and (13.28), the second term in equation (A5.3) becomes:

$$V\Delta U \int_{t_1}^{t_2} \frac{1}{a_i V} \frac{dM_i}{dt}\bigg|_r dt = \frac{1}{a_i} \Delta U \int_{M_i(t_1)}^{M_i(t_2)} dM_i|_r$$
$$= \Delta U \frac{M_i(t_2) - M_i(t_1)}{a_i} \quad \text{for all } i \qquad \text{(A5.4)}$$

Meanwhile the first term is found from the small change in water temperature:

$$\int_{t_1}^{t_2} \Phi\, dt = -m_w c_{pw}(T_w(t_2) - T_w(t_1)) \qquad \text{(A5.5)}$$

where m_w is the mass of water (kg), c_{pw} is the specific heat of the water (J/(kgK)), and T_w is the temperature of the water (K). Due allowance has been made for the sign convention for heat flow. Hence the internal energy of reaction is:

$$\Delta U = -a_i m_w c_{pw} \frac{T_w(t_2) - T_w(t_1)}{M_i(t_2) - M_i(t_1)} \quad \text{for all } i \qquad \text{(A5.6)}$$

Data associated with any of the chemical species, i, involved in the reaction may be used in equation (A5.6).

A5.3 Measuring the enthalpy of reaction using an open-system calorimeter

The open-system calorimeter allows a constant flow to pass through a reaction tube surrounded by a water bath (Figure A5.2). The reaction is initiated and the apparatus is run until a steady state is reached, with the temperature in the centre of the reaction tube held equal to the inlet temperature by a control system (not shown). The temperature drop across the reaction tube will thus be rendered negligible. The heat flow marked on the diagram is from the water bath to the reaction tube, in accordance once more with the sign convention of equation (13.34), although an exothermic reaction will cause the heat to flow out of the reaction tube and into the water.

We may apply the energy equation (13.35) to the system. Since no mechanical power is abstracted from the device, we may set $P = 0$. Further, since the temperature in the reaction tube is the same as the inlet and the outlet temperatures, it follows that the inlet enthalpy of component i will equal the outlet enthalpy and the enthalpy inside the reaction tube:

$$h_{1,i} = h_{2,i} = h_i \qquad \text{(A5.7)}$$

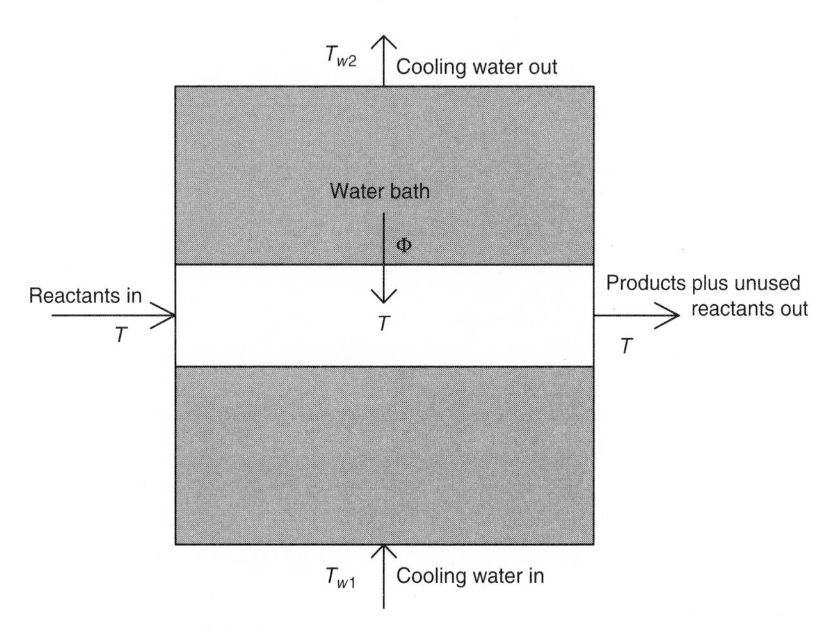

Figure A5.2 Open system calorimeter.

Also, in the steady state, $dE/dt = 0$. Accordingly, the energy equation yields

$$\Phi + \sum_{i=1}^{N} \hat{h}_i (F_{1,i} - F_{2,i}) = 0 \qquad \text{(A5.8)}$$

But the equation for the conservation of mass (13.33) gives for the steady-state condition, when $dM_i/dt = 0$,

$$F_{1,i} - F_{2,i} = -V \sum_{j=1}^{M} a_{ij} r_j \qquad \text{for } i = 1 \text{ to } N \text{ (A5.9)}$$

Combining equations (A5.8) and (A5.9) gives

$$\Phi - V \sum_{j=1}^{M} r_j \sum_{i=1}^{N} a_{ij} \hat{h}_i = 0 \qquad \text{(A5.10)}$$

or, using equation (13.46)

$$\Phi - V \sum_{j=1}^{M} r_j \Delta H_j = 0 \qquad \text{(A5.11)}$$

The heat transferred may be measured from the change in water temperature across the water bath:

$$\Phi = -W_w c_{pw} (T_{w2} - T_{w1}) \qquad \text{(A5.12)}$$

where W_w is the flow rate of water (kg/s). Hence we may evaluate the composite term for several simultaneous reactions:

$$\sum_{j=1}^{M} r_j \Delta H_j = -\frac{W_w c_{pw} (T_{w2} - T_{w1})}{V} \qquad \text{(A5.13)}$$

When there is a single reaction only, then equation (A5.9) simplifies to

$$F_{1,i} - F_{2,i} = -V a_i r \qquad \text{for } i = 1 \text{ to } N \quad \text{(A5.14)}$$

and equation (A5.13) then becomes:

$$r \Delta H = -\frac{W_w c_{pw} (T_{w2} - T_{w1})}{V} \qquad \text{(A5.15)}$$

Hence we may combine equations (A5.14) and (A5.15) to give the enthalpy of reaction, ΔH (J/kmol rxn), as:

$$\Delta H = -a_i W_w c_{pw} \frac{T_{w2} - T_{w1}}{F_{2,i} - F_{1,i}} \qquad \text{(A5.16)}$$

As was the case with equation (A5.6), it does not matter which chemical, i, is chosen for use in equation (A5.16), reactant or product.

Having found the enthalpy of reaction, ΔH, equation (13.48) may, of course, be used to find the internal energy of reaction, ΔU for a gas.

Appendix 6 Comparison of efficiency formulae with experimental data for convergent-only and convergent–divergent nozzles

A6.1 Experimental results

In his paper 'Reaction tests of turbine nozzles for supersonic velocities', Keenan (1949) presented the results of experimental tests on five turbine nozzles: one is convergent-only and four are convergent–divergent. He measured the outlet velocity using a force balance and then calculated the velocity coefficient as the ratio of measured velocity downstream of the nozzle, c_1, to the velocity, c_{1s} that would occur following an isentropic expansion from the stagnation state at nozzle inlet to the exhaust chamber pressure, p_1:

$$C_{vel} = \frac{c_1}{c_{1s}} \tag{A6.1}$$

Keenan plotted the velocity coefficient against isentropic outlet velocity, but we shall find it more convenient to display his results by replotting velocity coefficient against pressure ratio. This transformation has been carried out by relating isentropic velocity to pressure ratio using

$$c_{1s} = \sqrt{2\frac{\gamma}{\gamma - 1} p_{0T} v_{0T} \left(1 - \left(\frac{p_1}{p_{0T}}\right)^{(\gamma-1)/\gamma}\right)} \tag{A6.2}$$

(from equation (14.45), with γ replacing m because the whole process is isentropic).

The inlet stagnation state for all the tests was 1.38 MPa and 305.5°C. Most of the expansion takes place in the superheated region and so the value $\gamma = 1.3$ was used in equation (A6.2), which assumes implicitly that the vapour remains supersaturated below the saturation line.

Figure A6.1 shows Keenan's results replotted against pressure ratio.

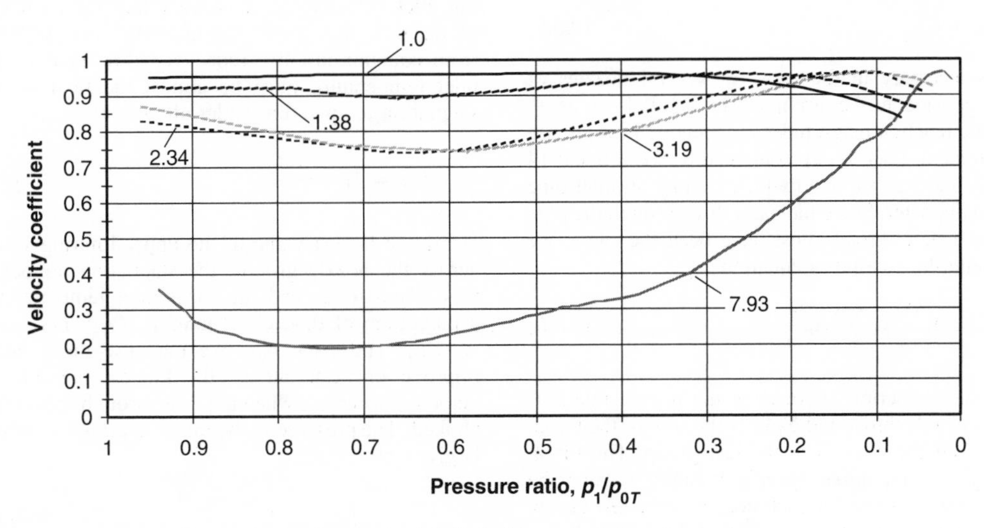

Figure A6.1 Velocity coefficients for the five nozzles, labelled with nominal divergence ratios.

Keenan classified his nozzles according to the nominal divergence ratio, the ratio of a nominal exit area to the throat area. The curves in Figure A6.1 are labelled with these ratios, and we shall continue to use these labels as identifiers.

The nozzle efficiency is given by equation (14.5) in terms of the inlet velocity, c_0, and the outlet velocity, c_1:

$$\eta_N = \frac{\frac{1}{2}c_1^2 - \frac{1}{2}c_0^2}{\frac{1}{2}c_{1s}^2 - \frac{1}{2}c_0^2} = \frac{\left(\frac{c_1}{c_{1s}}\right)^2 - \left(\frac{c_0}{c_{1s}}\right)^2}{1 - \left(\frac{c_0}{c_{1s}}\right)^2} = \frac{C_{vel}^2 - \left(\frac{c_0}{c_{1s}}\right)^2}{1 - \left(\frac{c_0}{c_{1s}}\right)^2} \tag{A6.3}$$

When c_0/c_{1s} is small, we may expand using the binomial theorem to give the nozzle efficiency as:

$$\eta_N \approx \left(C_{vel}^2 - \left(\frac{c_0}{c_{1s}}\right)^2\right)\left(1 + \left(\frac{c_0}{c_{1s}}\right)^2\right) \approx C_{vel}^2 - \left(1 - C_{vel}^2\right)\left(\frac{c_0}{c_{1s}}\right)^2 \tag{A6.4}$$

where terms in c_0/c_{1s} to the fourth power and above have been neglected because of their small magnitude. Provided either the inlet velocity, c_0 , is small, or the velocity coefficient, C_{vel}, is large, or if both conditions exist, then the nozzle efficiency is given to good accuracy by the square of the velocity coefficient:

$$\eta_N \approx C_{vel}^2 \tag{A6.5}$$

It was necessary in the tests for several nozzles of the same size to be arranged in parallel so as to make the reaction magnitudes large enough for precise measurement. Each group contained between 6 and 14 nozzles, and was supplied with steam by an inlet pipe of large diameter. Since the mass flow through the pipe and the nozzle throats must have been the same, we may write the continuity equation:

$$W = A_0\frac{c_0}{v_0} = \sum_i A_{ti}\frac{c_t}{v_t} \tag{A6.6}$$

where the subscript '0' denotes the nozzle inlet, 't' denotes nozzle throat and $\Sigma_i A_{ti}$ is the sum of the throat areas of all the nozzles in the test group. The throat velocity, c_t, and throat specific volume, v_t, may be taken to be the same for each nozzle in the test group. It follows from equation (A6.6) that the inlet velocity will obey the following equation and inequalities:

$$c_0 = \frac{\sum_i A_{ti}}{A_0}\frac{v_0}{v_t}c_t < \frac{\sum_i A_{ti}}{A_0}c_t < \frac{\sum_i A_{ti}}{A_0}c_{tc} \tag{A6.7}$$

The inequalities derive from the facts that (i) the specific volume after expansion to the throat, v_t, will always be greater than the specific volume at the nozzle inlet, v_0, and (ii) the velocity at the throat, c_t, will reach its maximum, sonic value, c_{tc}, in critical conditions. Now the supply velocity, c_0, will attain its greatest value when the nozzle flow is choked, i.e. when the throat velocity has reached critical. The critical throat velocity, c_{tc}, is given by equation (14.56) as:

$$c_{tc} = \sqrt{2\frac{\gamma}{\gamma+1}p_{0T}v_{0T}} = \sqrt{\frac{2 \times 1.3}{2.3} \times 1.38 \times 10^6 \times 0.1872} = 540.3 \text{ m/s} \tag{A6.8}$$

The inlet pipe has a diameter of 8 inches, and an area $A_0 = 50.27\,\text{in}^2$, which is between 33 and 66 times larger than the combined throat areas of the nozzle groups. On using equation (A6.8) and the inequality of (A6.7), it is found that the upper limiting value of the inlet velocity, c_0, must very low for each group of nozzles: less than $540/33 = 16.3$ m/s. It follows from equation (14.40), repeated below that the local inlet temperature must be almost identical with the stagnation temperature under all conditions:

$$T_{0T} = T_0 + \frac{\frac{1}{2}c_0^2}{c_p} \tag{A6.9}$$

Using the largest value of the upper limiting velocity across the nozzle groups, the stagnation temperature was found to exceed the local inlet temperature by a maximum of 0.06°C . Therefore $T_0 \approx T_{0T}$, so that $v_0 \approx v_{0T}$ and $p_0 \approx p_{0T}$. We may use these facts to improve our estimate of the highest possible inlet velocity, occurring when the flow through the nozzle is choked. The pressure in the throat at critical conditions is given by

$$\frac{p_{tc}}{p_{0T}} = \left(\frac{2}{\gamma+1}\right)^{m_{0c}/(m_{0c}-1)} \tag{14.54}$$

while, from (14.23), the specific volume at the throat at critical is found from

$$\frac{v_{0T}}{v_{tc}} = \left(\frac{p_{tc}}{p_{0T}}\right)^{1/m_{0c}} \tag{A6.10}$$

Combining equations (14.54) and (A6.10) gives the ratio of specific volumes as:

$$\frac{v_{0T}}{v_{tc}} = \left(\frac{2}{\gamma+1}\right)^{1/(m_{0c}-1)} = \left(\frac{2}{1.3+1.0}\right)^{1/(1.27-1.0)} = 0.6 \tag{A6.11}$$

(Here we have taken the efficiency of the convergent section of the nozzle as 0.92.) Returning to equation (A6.7), the maximum possible inlet velocity is:

$$c_{0\,\max} = \frac{\sum_i A_{ti}}{A_0}\frac{v_0}{v_t}c_{tc} \approx \frac{\sum_i A_{ti}}{A_0}\frac{v_{0T}}{v_t}c_{tc} \approx \frac{\sum_i A_{ti}}{A_0} \times 0.6 \times 540.3\ \text{m/s} \tag{A6.12}$$

Table A6.1 lists the total area of each nozzle group, the upper limiting value of inlet velocity, c_0, deduced from equation (A6.7) and the best estimate of the highest possible inlet velocity.

It is clear from the very low values of even the highest possible inlet velocities that the calculation of efficiency as the square of the velocity coefficient (Equation A6.5) will introduce very little error. Thus we may derive simply the measured nozzle efficiencies and plot them as functions of pressure ratio (Figure A6.2).

Table A6.1 Maximum possible nozzle inlet velocities

Nominal divergence ratio	*1.0*	*1.38*	*2.34*	*3.19*	*7.93*
$\sum_i A_{ti}$ in^2	1.381	1.370	1.518	1.214	0.754
Upper limit of c_0 (m/s) (equation A6.7)	14.86	14.75	16.32	13.02	8.10
Best estimate of $c_{0\,\max}$ (m/s) (equation A6.12)	8.88	8.81	9.78	7.80	4.85

A6.2 Theory versus experiment for the convergent-only nozzle

Inspection of Figure A6.2 reveals that the measured efficiency for the convergent-only nozzle increases very slightly (from about 0.91 to about 0.92) as the pressure ratio is reduced to critical, and it continues to rise as the pressure ratio drops, reaching a maximum of 0.925 at $p_1/p_{0T} = 0.45$. The efficiency then declines gradually, being 0.9 at $p_1/p_{0T} = 0.29$ and 0.75 at

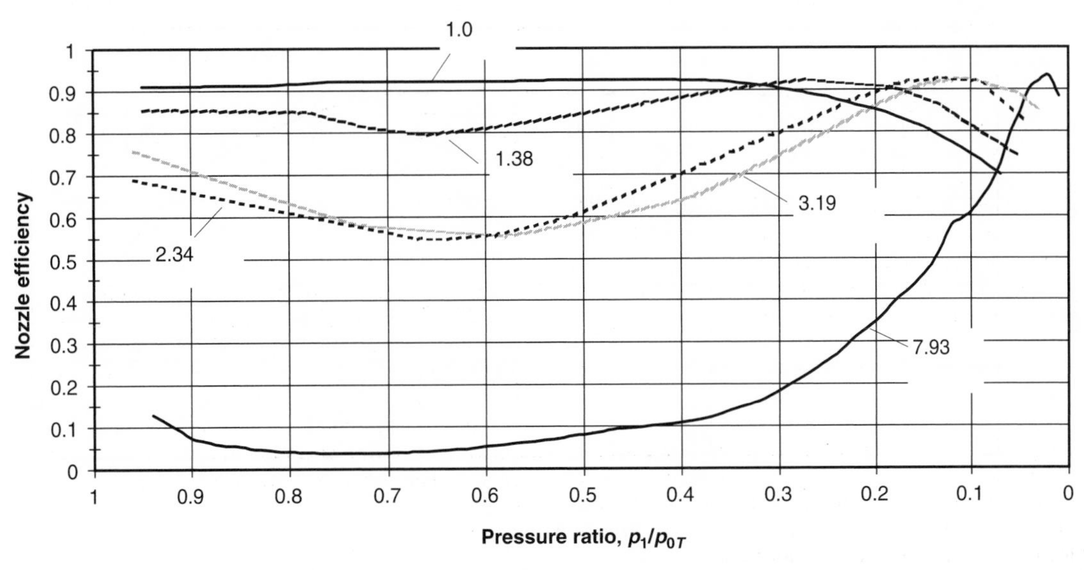

Figure A6.2 Nozzle efficiencies for the five nozzles, labelled with nominal divergence ratios.

$p_1/p_{0T} = 0.1$. This indicates that the measured steam velocity continues to increase past the speed of sound and well into the supersonic region.

This behaviour may be compared with the theoretical model of efficiency, given by equation set (14.71), repeated below:

$$\eta_{N0} = \eta_{N0D} \quad \text{for } \frac{p_1}{p_{0T}} \geq r_{\lim} \left(\frac{2}{\gamma+1}\right)^{m_{0D}/(m_{0D}-1)}$$

$$= \frac{1 - \dfrac{T_{0T}}{T_0} \dfrac{2}{\gamma+1} r_{\lim}^{\eta_{N0D}(\gamma-1)/\gamma}}{1 - \left(\dfrac{p_1}{p_0}\right)^{(\gamma-1)/\gamma}}$$

$$\text{for } \frac{p_1}{p_{0T}} < r_{\lim} \left(\frac{2}{\gamma+1}\right)^{m_{0D}/(m_{0D}-1)} \tag{14.71}$$

The nozzle efficiency is assumed constant down to a pressure ratio

$$r_{\lim} \left(\frac{2}{\gamma+1}\right)^{m_{0D}/(m_{0D}-1)}$$

If we allow for a further expansion at the nozzle exit, i.e. choose $r_{\lim} < 1$, the theoretical model predicts a continuously increasing, supersonic velocity for pressure ratios bounded by

$$\left(\frac{2}{\gamma+1}\right)^{m_{0D}/(m_{0D}-1)} \leq \frac{p_1}{p_{0T}} \leq r_{\lim} \left(\frac{2}{\gamma+1}\right)^{m_{0D}/(m_{0D}-1)}$$

A constant velocity is then predicted at lower pressure ratios.

To compare the theoretical model with the experiment, the 'design' velocity coefficient, C_{vel}, was taken as 0.96, which was approximately the mean measured level down to a pressure ratio of 0.35. This figure implies a design efficiency (= efficiency at critical) for the convergent-only nozzle given by $\eta_{NOD} = 0.922 = \eta_{N0c}$. The value of $m_{0D} = m_{0c}$ is then:

$$m_{0D} = \frac{\gamma}{\gamma - \eta_{0D}(\gamma - 1)}$$

$$= \frac{1.3}{1.3 - 0.922 \times 0.3} = 1.27$$

(see equation (14.50)), while the critical pressure ratio for the frictionally resisted expansion is $(2/(\gamma + 1))^{m_{0D}/(m_{0D}-1)} = 0.5183$. Since the measured nozzle efficiency remained substantially high down to a pressure ratio of 0.26, a value of $r_{\lim} = 0.26/0.5183 \approx 0.5$ was chosen as representative. Figure A6.3 shows the resultant theoretical curve for efficiency against the measured curve.

It is clear that the theoretical curve captures the main features. However, it is striking that, while the experimental curve begins to fall significantly at a pressure ratio of about 0.35, its fall is relatively gentle. In contrast, the theoretical efficiency is held constant until the pressure ratio is 0.26, but then it falls away rapidly as a result of the velocity being fixed thereafter. Comparing the two curves, the fact that the measured efficiency declines more slowly than the theoretical efficiency implies that the measured speed must continue to increase down to a pressure ratio of 0.1 and

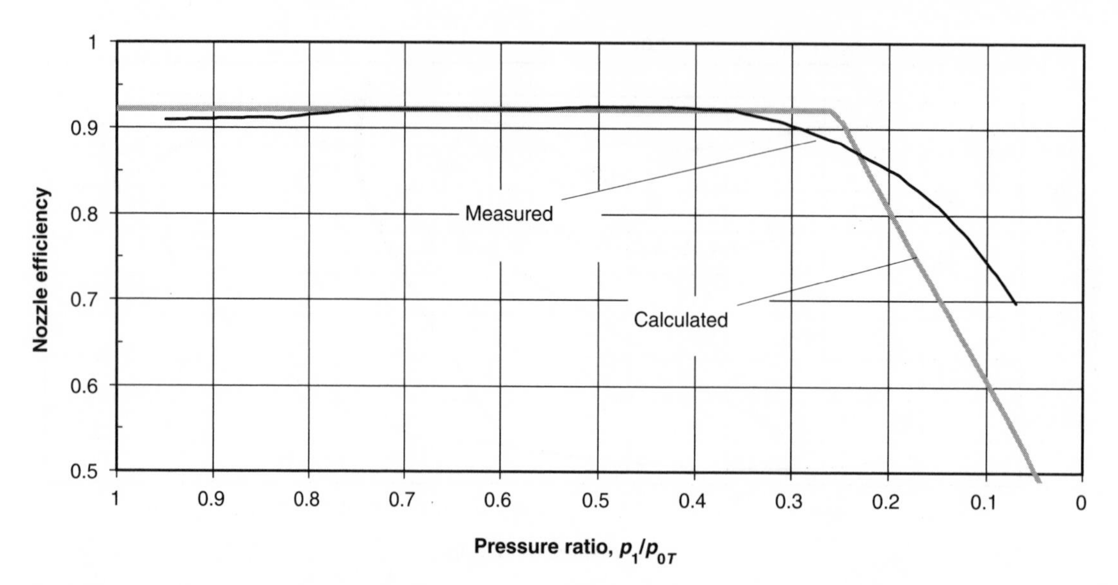

Figure A6.3 Efficiency of convergent-only nozzle.

below. Apparently the large exhaust chamber is allowing the steam to expand without any well-defined limit as the exhaust pressure is lowered. The usual penalty of chaotic flow (and hence flow interference) at very low pressure ratios seems to have been mitigated by the presence of a complicated network of coarse screen discs at the nozzle exit. These were designed to prevent the dynamic action of discharged steam on the nozzle arm forming part of the force balance, and appear to have had the secondary effect of promoting a well-ordered further expansion downstream of the nozzle exit.

It is not normal for a nozzle to discharge into a large volume, since, in a turbine, the nozzle discharges almost directly into the blade, and there is little space for a further expansion. An untypically low value of r_{lim} was chosen, namely 0.5, in an attempt to account for the unusual exit conditions present in the experiment. As is clear from Figure A6.3, this does not produce perfect matching, but this is not considered a major disadvantage because the conditions of the experiment are unlikely to be repeated in a real turbine.

A6.3 Divergence ratio for the convergent–divergent nozzles

A6.3.1 Keenan's method of estimating divergence ratio

Keenan classified the nozzles according to their 'nominal divergence ratio', a figure described as arbitrary and defined as the ratio of a nominal exit area to the throat area. For all except the nozzle with the largest divergence ratio, the nominal area was obtained by projecting the nozzle passage to a plane normal to the passage axis at the intersection of the axis with the exit plane. The concept is illustrated in Figure A6.4 below.

For the three nozzles of smaller divergence, the nominal exit area was taken at the section C–C, so that the nominal divergence ratio was:

$$\text{Nominal divergence ratio} = \frac{\text{Area at C–C}}{\text{Area at A–A}}$$

The nozzle with a nominal divergence ratio of 1.38 was rectangular in cross-section at the throat and had divergence in a single plane. The nozzles with divergence ratios of 2.34 and 3.19 were also rectangular in cross-section at the throat, but had divergence in two planes, so that the section B–B was a quadrilateral prism. The nozzle with nominal divergence ratio of 7.93 was circular in cross-section. Interestingly, Keenan took the nominal divergence ratio for this nozzle as:

$$\text{Nominal divergence ratio} = \frac{\text{Area at B–B}}{\text{Area at A–A}}$$

Now taking the exit area at the section B–B rather than C–C seems physically more intuitive for all the nozzles, since B–B represents the limit of physical constraint on the flow. Therefore we shall define a new term, the 'physical divergence ratio', as:

$$\text{Physical divergence ratio} = \frac{\text{Area at B–B}}{\text{Area at A–A}}$$

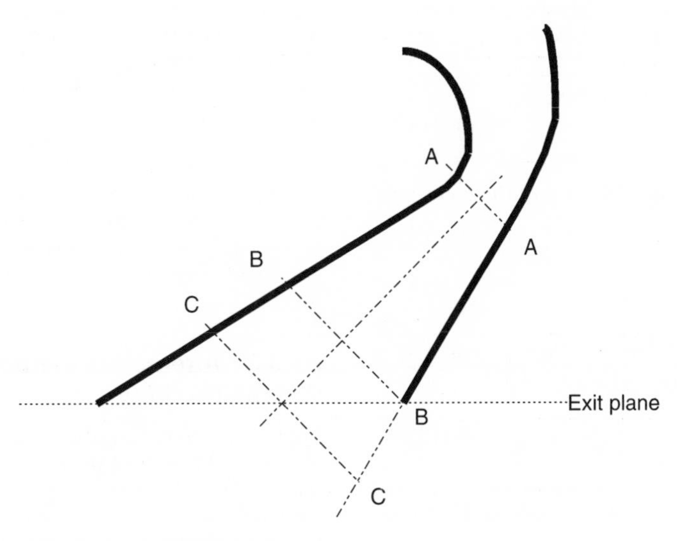

Figure A6.4 Schematic of convergent–divergent nozzle.

Calculations for all four convergent–divergent nozzles based on the engineering drawings contained in Keenan's paper allowed Table A6.2 to be drawn up.

Table A6.2 Nominal and physical divergence ratios

Nominal divergence ratio	*Physical divergence ratio*
1.38	1.16
2.34	1.67
3.19	2.04
7.93	7.93

In attempting to explain his measurements, Keenan eschewed the nominal divergence ratio, but chose instead the 'assumed divergence ratio', determined from the pressure ratio at which the nozzle efficiency reached its maximum value. Keenan did not detail the equations he used, but we will outline the type of approach possible.

The divergence ratio, A_1/A_t, may be deduced using equation (14.80):

$$\left(\frac{p_1}{p_{0T}}\bigg|_D\right)^{2/m_D} - \left(\frac{p_1}{p_{0T}}\bigg|_D\right)^{(m_D+1)/m_D} = \left(\frac{A_t}{A_1}\right)^2 \frac{\gamma-1}{2}\left(\frac{2}{\gamma+1}\right)^{(m_{0c}+1)/(m_{0c}-1)} \quad (14.80)$$

where, following Keenan, we take $(p_1/p_{0T})|_D = (p_{1\,crit\,2})/p_{0T}$ as the pressure ratio at maximum efficiency. The index, m_D, may be calculated from the maximum measured nozzle efficiency, η_{ND}, according to

$$m_D = \frac{\gamma}{\gamma - \eta_{ND}(1-\gamma)} \quad (14.81)$$

while m_{0c} depends on the efficiency of the convergent section at choking:

$$m_{0c} = \frac{\gamma}{\gamma - \eta_{N0c}(1-\gamma)} \quad (14.82)$$

η_{N0c} may be estimated from the experimental data at low pressure ratios using equation (14.88):

$$\eta_{Ncrit\,1} \approx \eta_{N0c}^2 \quad (14.88a)$$

As a result of calculations presumably similar to those above, Keenan derives a set of assumed divergence ratios (Table A6.3).

Table A6.3 Nominal and Keenan's assumed divergence ratio

Nominal divergence ratio	*Keenan's assumed divergence ratio*
1.38	1.3
2.34	1.86
3.19	2.8
7.93	6.0

A problem with this method of estimating divergence ratio is that it gives most weight to the low-pressure-ratio end of the experimental results, an area influenced significantly by the unusually large exhaust chamber, as discussed in Section A6.2 for the convergent-only nozzle. The best (i.e. highest) efficiency for that nozzle occurred at a pressure ratio of 0.45, which is significantly below the critical ratio of 0.5183. On the face of it, this would suggest, erroneously, that the nozzle was of convergent–divergent type. If we apply equations (14.80), (14.81), (14.82) and (14.88a), we calculate a divergence ratio of 1.045, which is, of course, higher than the true value of 1.0. This gives rise to the suspicion that Keenan's method will tend to overestimate the effective ('assumed') divergence ratio.

Keenan sketches out a method for estimating nozzle efficiency at pressure ratios above $(p_1/p_{0T})|_D$, although the full details are not presented. Using his assumed divergence ratio, he presents his calculated efficiency curve for the nozzle with nominal divergence of 2.34. While the method captures the general trend of the efficiency dip for pressure ratios in the region $p_{1\,crit\,1}/p_{0T} < p_1/p_{0T} < p_{1\,crit\,2}/p_{0T}$, the quantitative match is poor. In particular, Keenan's analysis predicts a minimum efficiency that is about half the measured value: 0.29 predicted, 0.55 measured. Only for the largest divergence ratio is there a reasonably good match between Keenan's predicted minimum efficiency value and that measured. The disparities found between predicted and measured efficiency curves tend to confirm the view that calculating the divergence ratio on the basis of the point of maximum efficiency produces overestimates.

A6.3.2 Alternative method of estimating divergence ratio

We may observe, first of all, that no additional expansion will take place at the nozzle outlet if the pressure ratio is above the nozzle's lower critical pressure ratio, $(p_{1\,crit\,2})/p_{0T}$. The large size of the exhaust chamber will have no effect on the measured efficiency when the pressure in the exhaust chamber is above this point. This suggests that we should estimate the divergence ratio from experimental data at pressure ratios some way above $(p_{1\,crit\,2})/p_{0T}$. It will be shown that we can accomplish this through calculating the divergence ratio that causes the minimum predicted efficiency to match the minimum measured value.

Chapter 14 gives the efficiency in the region $(p_{1\,crit\,2}/p_{0T}) < (p_1/p_{0T}) < (p_{1\,crit\,1}/p_{0T})$ as

$$\eta_N = \frac{1 - \dfrac{T_{0T}}{T_0}\dfrac{T_1}{T_{0T}}}{1 - \left(\dfrac{p_1}{p_0}\right)^{(\gamma-1)/\gamma}}$$

$$\text{for } \frac{p_{1\,crit\,2}}{p_{0T}} < \frac{p_1}{p_{0T}} < \frac{p_{1\,crit\,1}}{p_{0T}} \tag{14.106}$$

where putting $T = T_1$, $p = p_1$ and $A = A_1$ in equation (14.104) produces:

$$\frac{T_1}{T_{0T}} = \frac{1}{\gamma-1}\left(\frac{2}{\gamma+1}\right)^{(m_{0c}+1)/(1-m_{0c})}\left(\frac{A_1}{A_t}\right)^2\left(\frac{p_1}{p_{0T}}\right)^2 \times\left[\sqrt{1+2(\gamma-1)\left(\frac{2}{\gamma+1}\right)^{(m_{0c}+1)/(m_{0c}-1)}\times\left(\frac{A_t}{A_1}\right)^2\left(\frac{p_{0T}}{p_1}\right)^2}-1\right] \tag{A6.13}$$

We may estimate η_{N0c} for the purposes of these equations from the measured efficiency at low pressure ratios, and hence find m_{0C}. As discussed at the end of Section A6.1, the local and stagnation temperatures will be almost identical, so that $T_{0T}/T_0 \approx 1$.

We know from calculus that the minimum value of efficiency occurs when

$$\frac{d\eta_N}{d\left(\dfrac{p_1}{p_{0T}}\right)} = 0 \tag{A6.14}$$

Carrying out the necessary differentiation of equations (14.106) and (A6.13) gives the condition:

$$2\left(1-\left(\frac{p_1}{p_{0T}}\right)^{(\gamma-1)/\gamma}\right)\left[Y\left(\frac{A_1}{A_t}\right)^2\frac{p_1}{p_{0T}}\{1-X\}+\frac{1}{\dfrac{p_1}{p_{0T}}X}\right] + \frac{\gamma-1}{\gamma}\left(\frac{p_1}{p_{0T}}\right)^{(-1)/\gamma}\left[Y\left(\frac{A_1}{A_t}\right)^2\left(\frac{p_1}{p_{0T}}\right)^2\{1-X\}+1\right]=0 \tag{A6.15}$$

where

$$Y = \frac{1}{\gamma-1}\left(\frac{2}{\gamma+1}\right)^{(m_{0c}+1)/(1-m_{0c})} \tag{A6.16}$$

and

$$X = \sqrt{1+\frac{2}{Y\left(\dfrac{A_1}{A_t}\right)^2\left(\dfrac{p_1}{p_{0T}}\right)^2}} \tag{A6.17}$$

Equation (A6.15) relates the divergence ratio, A_1/A_t, to the pressure ratio at minimum nozzle efficiency.

In addition, the minimum calculated efficiency must match the minimum measured value. Using equations (A6.13) and (14.106), this condition may be written:

$$\frac{1-Y\left(\dfrac{A_1}{A_t}\right)^2\left(\dfrac{p_1}{p_{0T}}\right)^2(X-1)}{1-\left(\dfrac{p_1}{p_{0T}}\right)^{(\gamma-1)/\gamma}} - \eta_{meas\ \min} = 0 \tag{A6.18}$$

Equations (A6.15) and (A6.18), taken together with the auxiliary equations (A6.16) and (A6.17) represent two nonlinear simultaneous equations in the two unknowns, A_1/A_t and p_1/p_{0T}, the latter at the point of minimum calculated efficiency. The fact that a solution is possible, albeit iterative, demonstrates that the necessary procedure for calculating the divergence ratio has been developed. The minimum calculated nozzle efficiency will now match the minimum measured efficiency.

A6.4 Interpreting the experimental results for convergent–divergent nozzles

Nozzle with nominal divergence = 1.38

The curve of velocity coefficient, C_{vel}, for this nozzle shows a maximum value in the supersonic region of 0.96 at a pressure ratio of 0.27, implying $\eta_{Ncrit\,2} = \eta_{ND} = 0.96^2 = 0.922$. The velocity coefficient falls with increasing pressure ratio, then shows a sharp rise to a value of 0.92 at a pressure ratio of 0.78; then it rises very gradually to 0.925 as the pressure ratio falls to 0.95. We will assume that the upper critical pressure ratio occurs at the end of the sharp rise, i.e. at a value of 0.78. Its associated efficiency value is $\eta_{Ncrit\,1} = 0.92^2 = 0.846$.

We may estimate the efficiency of the convergent section of the nozzle at choking using equation (14.88):

$$\eta_{Ncrit\,1} = \min(\eta_{N0c}^2, \eta_{ND}) \tag{14.88}$$

Since $\eta_{N0c}^2 < \eta_{ND}$, we may take $\eta_{N0c} = \sqrt{\eta_{Ncrit\,1}} = \sqrt{0.846} = 0.92$. This figure is very similar to the subsonic efficiency, 0.922, of the convergent-only nozzle. The fact that the throat of the convergent-only nozzle had a comparable area and was also rectangular in cross-section gives confirmation that 0.92 is a reasonable figure to use for η_{N0c}.

The value of the polytropic index at choking for the convergent-only nozzle may be found from equation (14.50) as

$$m_{0c} = \frac{\gamma}{\gamma - \eta_{N0c}(1-\gamma)} = \frac{1.3}{1.3 - 0.92(1.0-1.3)} = 1.27$$

We may now apply equations (A6.15) to (A6.18) and solve iteratively for the two unknowns, divergence ratio, A_1/A_t, and pressure ratio at minimum efficiency, p_1/p_{0T}, resulting in the values:

$$\frac{A_1}{A_t} = 1.098$$

$$\left.\frac{p_1}{p_{0T}}\right|_{\min \eta_N} = 0.566$$

Once the divergence ratio has been found, an iterative solution of equation (14.75) for the conditions at the nozzle end gives the upper and lower critical pressure ratios:

$$\left(\frac{p_1}{p_{0T}}\right)^{2/m} - \left(\frac{p_1}{p_{0T}}\right)^{(m+1)/m} = \left(\frac{A_t}{A_1}\right)^2 \frac{\gamma-1}{2}\left(\frac{2}{\gamma+1}\right)^{(m_{0c}+1)/(m_{0c}-1)} \quad \text{(A6.19)}$$

where the index, m, is taken as $m(\eta_{Ncrit\,1})$ for the upper critical pressure ratio, $p_{1\,crit\,1}/p_{0T}$ while it is taken as $m(\eta_{Ncrit\,2})$ for the lower critical pressure ratio, $p_{1\,crit\,2}/p_{0T}$. The two solutions were found to be:

$$\frac{p_{1\,crit\,1}}{p_{0T}} = 0.692$$

$$\frac{p_{1\,crit\,2}}{p_{0T}} = 0.350$$

Finally, we choose $r_{\lim} = 0.5$, recognizing a similar exhaust effect as was found with the convergent-only nozzle. Table A6.4 summarizes these figures, and compares them with the measured values.

Nozzles with nominal divergence ratios of 2.34, 3.19 and 7.93

The same methods may be used to find the principal features for the nozzles of larger divergence ratio, except that the data in the low pressure range are insufficient to allow identification of the upper critical pressure ratio and its associated efficiency. But since the efficiencies of all the convergent–divergent nozzles at the lower critical pressure ratio are similar, we will make the further assumption that their efficiencies will be similar at the upper critical pressure ratio. Hence we will use the figure derived for the nozzle of 1.38 nominal divergence, namely $\eta_{Ncrit\,1} = 0.846$, to analyse the performance of the nozzles with nominal divergence ratios of 2.34, 3.19 and 7.93. Performing the necessary calculations allows us to construct Tables A6.5, A6.6 and A6.7.

It is now possible to calculate the nozzle efficiency throughout the pressure range by applying a variant of equation (14.107), and then equations (A6.13) and (14.114):

Table A6.4 Principal features for convergent–divergent nozzle with nominal divergence ratio of 1.38. Bold type indicates the values taken from the experimental data

	$\frac{A_1}{A_t}$	η_{N0c}	$\eta_{Ncrit\,1}$	$\eta_{N\min}$	$\eta_{Ncrit\,2}$	$\frac{p_{1\,crit\,1}}{p_{0T}}$	$\left.\frac{p_1}{p_{0T}}\right\|_{\min \eta_N}$	$\frac{p_{1\,crit\,2}}{p_{0T}}$	$r_{\lim}$
Calculated	1.098	0.92	**0.846**	**0.79**	**0.922**	0.692	0.566	0.350	0.5
Measured	1.16	–	0.846	0.79	0.922	0.78	0.66	0.27	–

Table A6.5 Principal features for convergent–divergent nozzle with nominal divergence ratio of 2.34

	$\frac{A_1}{A_t}$	η_{N0c}	$\eta_{Ncrit\,1}$	$\eta_{N\min}$	$\eta_{Ncrit\,2}$	$\frac{p_{1\,crit\,1}}{p_{0T}}$	$\left.\frac{p_1}{p_{0T}}\right\|_{\min \eta_N}$	$\frac{p_{1\,crit\,2}}{p_{0T}}$	$r_{\lim}$
Calculated	1.365	0.92	**0.846**	**0.546**	**0.925**	0.844	0.587	0.214	0.5
Measured	1.67	–	–	0.546	0.925	–	0.65	0.13	–

Table A6.6 Principal features for convergent–divergent nozzle with nominal divergence ratio of 3.19

	$\frac{A_1}{A_t}$	η_{N0c}	$\eta_{Ncrit\,1}$	$\eta_{N\,min}$	$\eta_{Ncrit\,2}$	$\frac{p_{1\,crit\,1}}{p_{0T}}$	$\left.\frac{p_1}{p_{0T}}\right\vert_{min\,\eta_N}$	$\frac{p_{1\,crit\,2}}{p_{0T}}$	r_{lim}
Calculated	1.359	0.92	**0.846**	**0.55**	**0.924**	0.842	0.586	0.216	0.5
Measured	2.04	–	–	0.55	0.924	–	0.58	0.11	–

Table A6.7 Principal features for convergent–divergent nozzle with nominal divergence ratio of 7.93

	$\frac{A_1}{A_t}$	η_{N0c}	$\eta_{Ncrit\,1}$	$\eta_{N\,min}$	$\eta_{Ncrit\,2}$	$\frac{p_{1\,crit\,1}}{p_{0T}}$	$\left.\frac{p_1}{p_{0T}}\right\vert_{min\,\eta_N}$	$\frac{p_{1\,crit\,2}}{p_{0T}}$	r_{lim}
Calculated	5.57	0.92	**0.846**	**0.037**	**0.931**	0.992	0.621	0.023	0.5
Measured	7.93	–	–	0.037	0.931	–	0.62	0.024	–

$$\eta_N = \eta_{N0c}^2 \quad \text{for } \frac{p_1}{p_{0T}} \geq \frac{p_{1\,crit\,1}}{p_{0T}}$$

$$= \frac{1 - \dfrac{T_{0T}}{T_0}\dfrac{T_1}{T_{0T}}}{1 - \left(\dfrac{p_1}{p_0}\right)^{(\gamma-1)/(\gamma)}} \quad \text{for } \frac{p_{1\,crit\,2}}{p_{0T}} < \frac{p_1}{p_{0T}} < \frac{p_{1\,crit\,1}}{p_{0T}} \qquad \text{(A6.20)}$$

$$= \eta_{ND} \quad \text{for } r_{lim}\frac{p_{1\,crit\,2}}{p_{0T}} \leq \frac{p_1}{p_{0T}} \leq \frac{p_{1\,crit\,2}}{p_{0T}}$$

where

$$\frac{T_1}{T_{0T}} = \frac{1}{\gamma-1}\left(\frac{2}{\gamma+1}\right)^{(m_{0c}+1)/(1-m_{0c})}\left(\frac{A_1}{A_t}\right)^2\left(\frac{p_1}{p_{0T}}\right)^2 \times \left[\sqrt{\begin{matrix} 1+2(\gamma-1)\left(\dfrac{2}{\gamma+1}\right)^{(m_{0c}+1)/(m_{0c}-1)} \\ \times\left(\dfrac{A_t}{A_1}\right)^2\left(\dfrac{p_{0T}}{p_1}\right)^2 \end{matrix}} - 1\right] \qquad \text{(A6.13)}$$

$$\eta_N = \frac{1 - \dfrac{T_{0T}}{T_0} r_{lim}^{\eta_D(\gamma-1)/\gamma}\left(\dfrac{p_{1\,crit\,2}}{p_{0T}}\right)^{\eta_D(\gamma-1)/\gamma}}{1 - \left(\dfrac{p_1}{p_0}\right)^{(\gamma-1)/\gamma}} \quad \text{for } \frac{p_1}{p_{0T}} < r_{lim}\frac{p_{1\,crit\,2}}{p_{0T}} \qquad \text{(14.114)}$$

Applying these equations to the four convergent–divergent nozzles enables us to plot Figures A6.5 to A6.8.

A6.5 Comparing calculated efficiency curves with measured efficiency curves

There is clearly good general agreement, with the efficiency dip between the upper and lower critical pressure ratios being brought out well.

The lower critical pressure ratio tends to be overestimated for the nozzles with nominal divergence ratios of 1.38, 2.34 and 3.19, with the calculated value being 0.08, 0.084 and 0.106 larger than the measured value. However, these discrepancies are of the same order and in the same direction as the difference between the theoretical, calculated pressure ratio at critical, 0.52, for the convergent-only nozzle, and the pressure ratio of maximum measured efficiency, namely 0.45. The difference may well be a consequence of the very large exhaust chamber. The effect is absent in the nozzle with nominal divergence of 7.93 perhaps as a result of its already large divergence ratio, or else due to the circular geometry of this test nozzle.

The pressure ratio at minimum efficiency is calculated to be somewhat below the measured value for the nozzles with nominal divergence ratios of 1.38 and 2.34 (with discrepancies of 0.094 and 0.063, respectively). But the measured behaviour of the nozzles with nominal divergences of 3.19 and 7.93 is captured well in this regard.

The nozzle efficiency at pressure ratios above the calculated upper critical point is overestimated for nozzles with 2.34 and 3.19 nominal divergence, but is reproduced reasonably well for the nozzle of 1.38 nominal divergence. However, the calculated upper critical pressure ratio at 0.69 is 0.09 lower than the measured value for this last nozzle. Data does not exist for the upper critical pressure for the nozzle of 7.93 nominal divergence ratio, but the match between

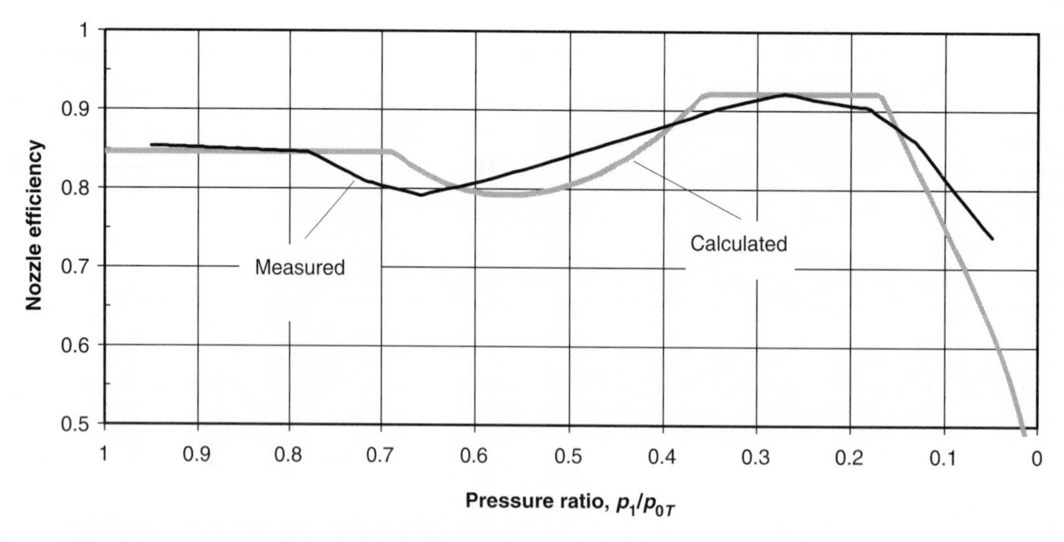

Figure A6.5 Measured versus calculated efficiency for the convergent–divergent nozzle of 1.38 nominal divergence ratio.

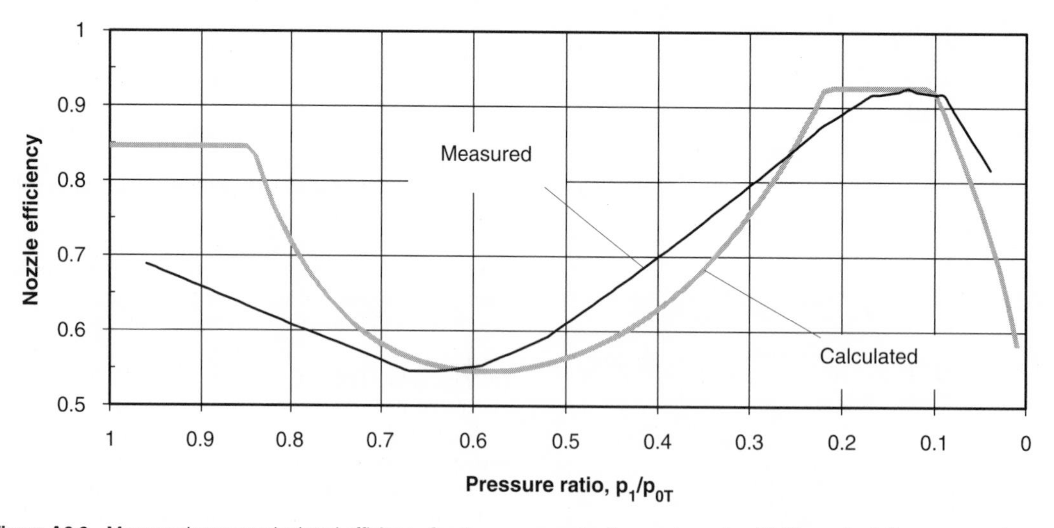

Figure A6.6 Measured versus calculated efficiency for the convergent–divergent nozzle of 2.34 nominal divergence ratio.

measured and calculated curves is excellent between pressure ratios of 0.65 and 0.95.

The behaviour of efficiency at very low pressure ratios shows the same trend as exhibited with the convergent-only nozzle: the measured efficiency drops away more slowly than the calculated value, indicating that the outlet velocity is continuing to increase well below the lower critical pressure ratio.

It should be noted that the calculated behaviour for efficiency depends heavily on the value of divergence ratio calculated from the iterative solution of equations (A6.15) to (A6.18). It is noticeable that the calculated divergence ratios come out below Keenan's assumed divergence ratios, and are generally nearer the physical divergence ratios. Intuitively we would expect the calculated ratio to be strongly related to the physical divergence ratio. Figure A6.9 compares the two.

While it would be wrong to claim too much from a graph with only four points, there does appear to be a reasonable correlation between the calculated and the physical divergence ratio (correlation coefficient = 0.998). The plotted line has the equation

$$\left.\frac{A_1}{A_t}\right|_{calc} = 0.676\left.\frac{A_1}{A_t}\right|_{phys} + 0.185 \tag{A6.21}$$

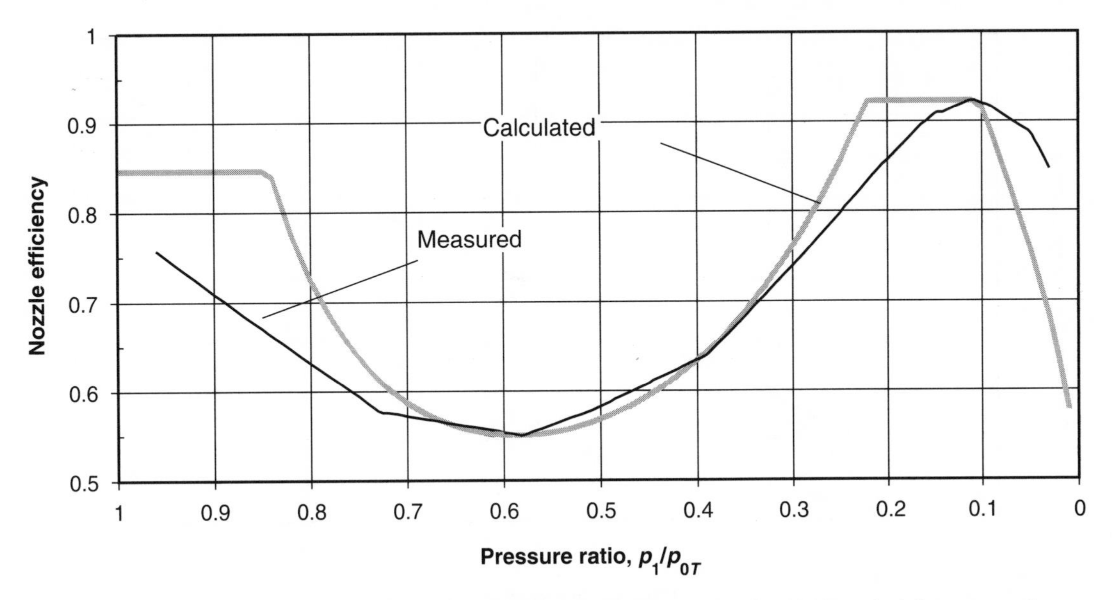

Figure A6.7 Measured versus calculated efficiency for the convergent–divergent nozzle of 3.19 nominal divergence ratio.

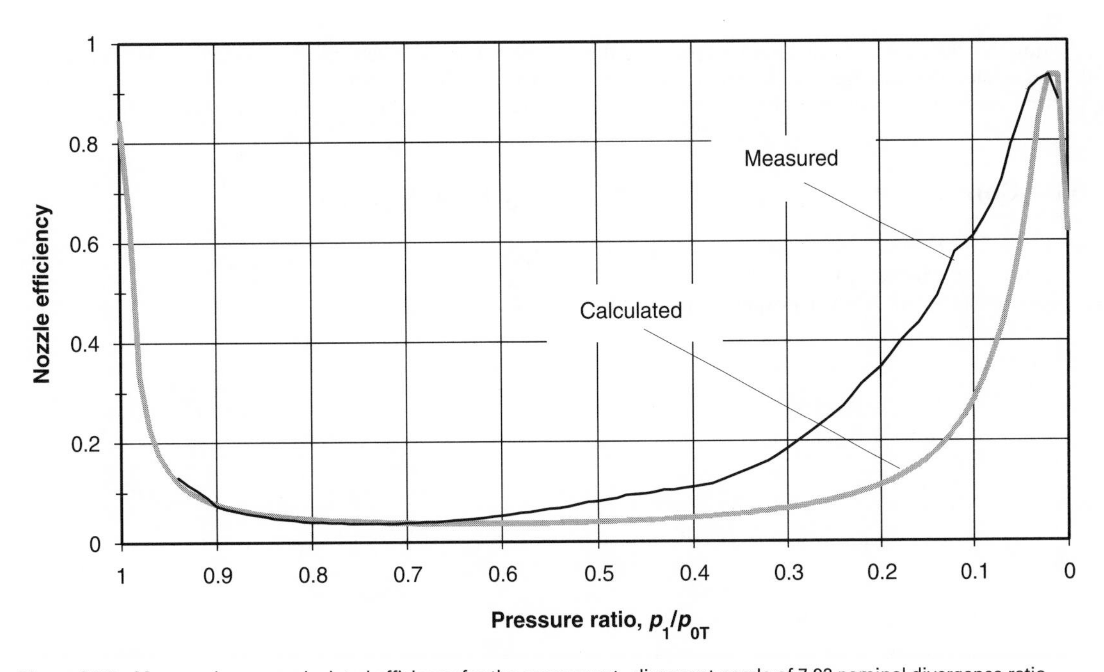

Figure A6.8 Measured versus calculated efficiency for the convergent–divergent nozzle of 7.93 nominal divergence ratio.

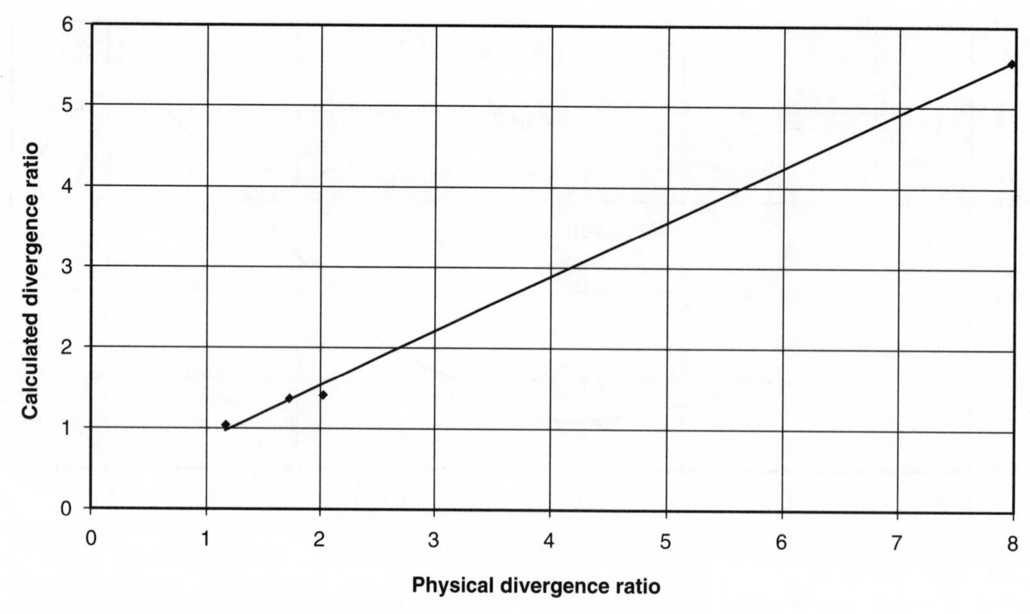

Figure A6.9 Calculated divergence ratio versus physical divergence ratio.

indicating that the steam appears not to experience the full expansion in the divergent section that one would expect on the basis of geometry alone.

They may be expected to provide the control engineer with useful predictions of the efficiency of turbine nozzles at conditions well away from the design point.

A6.6 Conclusions

The theoretical calculations of efficiency based on the methods of Chapter 14 have shown a good, if not perfect, match with Keenan's experimental results.

A6.7 Reference

Keenan, J. H. (1949). Reaction tests of turbine nozzles for supersonic velocities, *Trans. Amer. Soc. Mech. Eng.*, **71**, 781–787.

Appendix 7 Approximations used in modelling turbine reaction stages in off-design conditions

A7.1 Axial velocity over the fixed blades at off-design conditions for a 50% reaction stage

The turbine designer will normally ensure that, at the design point, the axial velocity at the inlet to the fixed blades will be equal both to the axial velocity at the outlet of the fixed blades and to the axial velocity at the outlet of moving blades:

$$c_0 \sin \alpha_0 = c_1 \sin \alpha_1 = c_2 \sin \alpha_2 \qquad \text{(A7.1)}$$

where it has been shown in Chapter 15 that the design conditions for a 50% reaction stage are $\alpha_0 = \alpha_2 = \beta_1$, while $\beta_1 = 90°$ ensures optimum efficiency. Achieving this condition throughout the turbine requires the design engineer to iterate on his design, matching the velocity diagrams for all the stages through the appropriate selection of stage parameters. A set of calculations of similar complexity would be needed to find the true relationship between the axial velocities at the inlet and outlet of the fixed blades at each off-design condition, but it would be impractical to attempt to carry out all these calculations at each timestep of a simulation. Fortunately we may use an approximate analysis to reach a simple result.

Let us consider a 50% reaction stage that is preceded by a notional upstream stage that is representative of a general turbine stage. We need take no account of the quantity of work done by this notional stage; we will assume only that the expansion over this stage is polytropic, characterized by polytropic index, n. Let us assume that the stagnation pressure at the inlet to the notional upstream stage is p_{up} and the corresponding stagnation specific volume is v_{up}. Figure A7.1 gives a diagram of the two stages.

We will assume for simplicity that the expansion through the notional upstream stage is followed by a further expansion with the same polytropic index, n, through to the outlet of the fixed blades of the reaction stage. Using equation (5.22), the velocity at the fixed blades' inlet, c_0, will be:

$$c_0 = \sqrt{2\frac{n}{n-1} p_{up} v_{up} \left(1 - \left(\frac{p_0}{p_{up}}\right)^{(n-1)/n}\right)} \qquad \text{(A7.2)}$$

while the velocity at the outlet of the fixed blades, c_1, will be:

$$c_1 = \sqrt{2\frac{n}{n-1} p_{up} v_{up} \left(1 - \left(\frac{p_1}{p_{up}}\right)^{(n-1)/n}\right)} \qquad \text{(A7.3)}$$

We know that equation (A7.1) will hold at the design point, and so we may substitute the velocity expressions from equations (A7.2) and (A7.3) into equation (A7.1). This gives the following relationship between

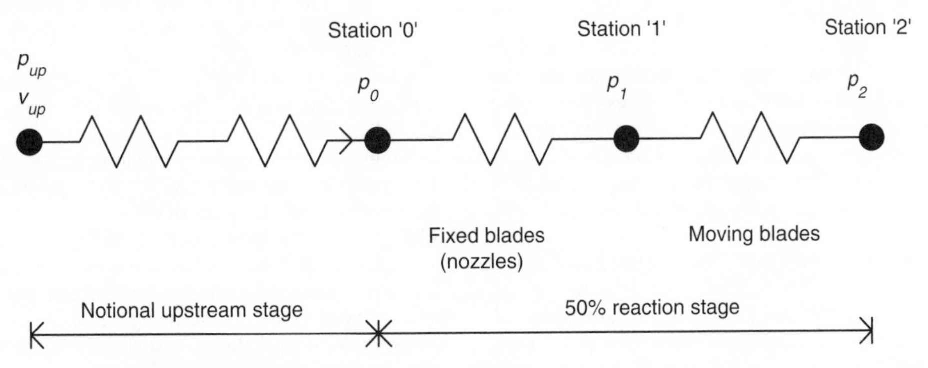

Figure A7.1 50% reaction stage preceded by notional upstream stage.

the ratios of the pressures at stations '0' and '1' to the upstream pressure at the design point, which we characterize with the subscript 'D':

$$\left.\frac{p_0}{p_{up}}\right|_D = \left[1 - \frac{\sin^2\alpha_1}{\sin^2\alpha_0}\left(1 - \left(\left.\frac{p_1}{p_{up}}\right|_D\right)^{(n-1)/n}\right)\right]^{n/(n-1)} \tag{A7.4}$$

Continuity requires that the mass flow through the notional upstream stage is equal to the mass flow through the fixed blades, so that, using equation (5.25):

$$A_0\sqrt{2\frac{n}{n-1}\frac{p_{up}}{v_{up}}\left(\left(\frac{p_0}{p_{up}}\right)^{2/n} - \left(\frac{p_0}{p_{up}}\right)^{(n+1)/n}\right)} = A_1\sqrt{2\frac{n}{n-1}\frac{p_{up}}{v_{up}}\left(\left(\frac{p_1}{p_{up}}\right)^{2/n} - \left(\frac{p_1}{p_{up}}\right)^{(n+1)/n}\right)} \tag{A7.5}$$

where A_0 is the area normal to the flow at the inlet to the fixed blades and A_1 is the area normal to the flow at the outlet of the fixed blades. Hence the continuity equation becomes:

$$\left(\left(\frac{p_0}{p_{up}}\right)^{2/n} - \left(\frac{p_0}{p_{up}}\right)^{(n+1)/n}\right) - \frac{A_1^2}{A_0^2}\left(\left(\frac{p_1}{p_{up}}\right)^{2/n} - \left(\frac{p_1}{p_{up}}\right)^{(n+1)/n}\right) = 0 \tag{A7.6}$$

Since equation (A7.6) holds generally, it must be valid at the design point in particular. Hence we may use it in conjunction with equation (A7.4) to define the area ratio as soon as we know either $(p_0/p_{up})|_D$ or $p_1/p_{up}|_D$:

$$\left(\frac{A_1}{A_0}\right)^2 = \frac{\left(\left.\frac{p_0}{p_{up}}\right|_D\right)^{2/n} - \left(\left.\frac{p_0}{p_{up}}\right|_D\right)^{(n+1)/n}}{\left(\left.\frac{p_1}{p_{up}}\right|_D\right)^{2/n} - \left(\left.\frac{p_1}{p_{up}}\right|_D\right)^{(n+1)/n}} \tag{A7.7}$$

(Alternatively, if we knew the area ratio and the ratio of fixed blade inlet and outlet pressures at the design point: $(p_0/p_1)|_D = (p_0/p_{up})|_D \div (p_1/p_{up})|_D$, then this last equation forms a simultaneous pair with equation (A7.6) in the ratios of mid-stage and inlet pressures to upstream pressure at the design point, $(p_1/p_{up})|_D$ and $(p_0/p_{up})|_D$.)

Given the fixed area ratio, the continuity equation (A7.6) may be solved for the ratio of stage inlet pressure to upstream pressure, (p_0/p_{up}), at any value of mid-stage to upstream pressure ratio, (p_1/p_{up}). Then we may find the axial velocity ratio at any condition, design or off-design, through dividing equation (A7.2) by equation (A7.3), and multiplying both sides by $(\sin\alpha_0/\sin\alpha_1)$ to give:

$$\frac{c_0\sin\alpha_0}{c_1\sin\alpha_1} = \frac{\sin\alpha_0}{\sin\alpha_1}\sqrt{\frac{1-\left(\frac{p_0}{p_{up}}\right)^{(n-1)/n}}{1-\left(\frac{p_1}{p_{up}}\right)^{(n-1)/n}}} \tag{A7.8}$$

We may use equation (A7.8) in conjunction with equation (A7.6) to estimate the effect on axial velocity ratio of changes in upstream and mid-stage pressure, as transmitted by the mid-stage to upstream pressure ratio, (p_1/p_{up}).

Let us take as an example a 50% reaction stage with a stage inlet angle, $\alpha_0 = 90°$, a fixed blade outlet angle, $\alpha_1 = 20°$, and a ratio of mid-stage to upstream pressure at the design point, $(p_1/p_{up})|_D = 0.875$. We will assume superheated steam is being passed isentropically through reaction and upstream stages, so that $n = 1.3$.

Using equation (A7.4), the ratio of stage inlet to upstream pressure at the design point emerges as $(p_0/p_{up})|_D = 0.985$, which, taken in conjunction with $(p_1/p_{up})|_D = 0.875$, implies a mid-stage to stage inlet pressure ratio at the design point of

$$\left.\frac{p_1}{p_0}\right|_D = \left.\frac{p_1}{p_{up}}\right|_D \div \left.\frac{p_0}{p_{up}}\right|_D = \frac{0.875}{0.985} = 0.889$$

Using equation (15.21), with the degree of reaction set at $\rho = 0.5$ and with the exponents: $m = m_S = n = 1.3$ gives a ratio of stage outlet to stage inlet pressure at the design point of

$$\left.\frac{p_2}{p_0}\right|_D = \left(\frac{\left(\left.\frac{p_1}{p_0}\right|_D\right)^{(m-1)/m} - \rho}{1-\rho}\right)^{(m_S/m_S-1)} = 0.787$$

This implies a fractional pressure drop $(p_0 - p_2)/p_0 = 1 - 0.787 = 0.213$ across the reaction stage, which is about half the critical value for superheated steam $(1 - 0.546 = 0.454)$.

Substituting $(p_1/p_{up})|_D = 0.875$ and $(p_0/p_{up})|_D = 0.985$ in equation (A7.7) yields the area ratio as $(A_1/A_0) = 0.375$.

Having established the constant data, we may solve the continuity equation (A7.6) for (p_0/p_{up}) for a sequence of off-design values of (p_1/p_{up}) between 0.75 and 1.0, the former corresponding to a low pressure ratio over the stage as a whole: $(p_2/p_0) \approx 0.5$. The axial velocity ratio over the fixed blades may be calculated in each case using equation (A7.8), and plotted against the ratio of mid-stage to upstream pressure, (p_1/p_{up}) (see Figure A7.2).

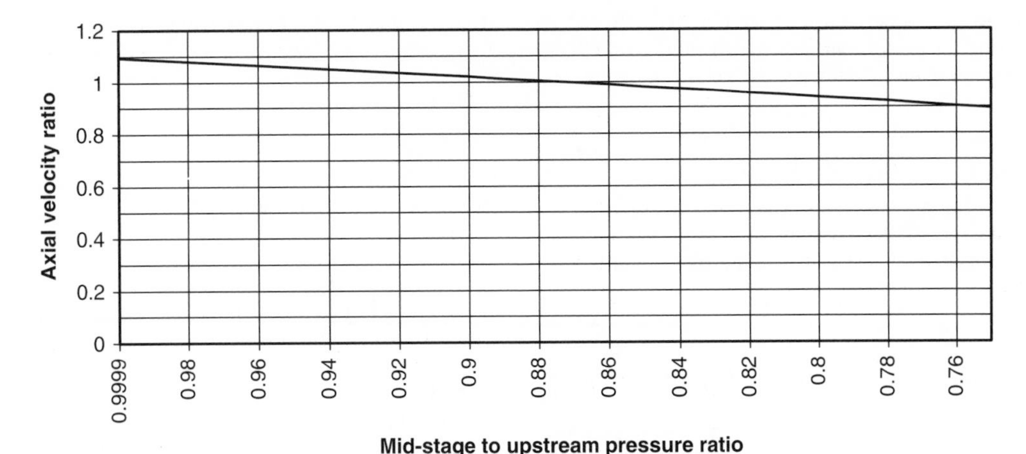

Figure A7.2 Axial velocity ratio over the fixed blades versus mid-stage to upstream pressure ratio.

It is clear from Figure A7.2 that the axial velocity ratio stays within +/ − 10% of unity for a very wide range of mid-stage to upstream pressure ratios, corresponding to a wide variation in conditions above and below the reaction stage being considered. Thus we may assume that the axial velocity at the inlet to the fixed blades will always be approximately equal to the axial velocity at the outlet from the fixed blades:

$$c_0 \sin \alpha_0 \approx c_1 \sin \alpha_1 \tag{A7.9}$$

A7.2 Degree of reaction at off-design conditions for a 50% reaction stage

The performance of an operational turbine stage will be governed by the total stage pressure ratio, (p_2/p_0), and to the blade speed to nozzle-outlet gas speed ratio, $R_B = c_B/c_1$. Changes in the conditions away from the design point will be characterized by alterations to either or both these parameters. We will now consider how the degree of reaction is affected by such changes.

Applying equation (5.20) with $z_1 = z_2$ to the expansion from the fixed-blade inlet to the fixed-blade outlet gives the velocity of the gas leaving the fixed blades, c_1, as:

$$c_1^2 = 2\frac{n}{n-1} p_0 v_0 \left(1 - \left(\frac{p_1}{p_0}\right)^{(n-1)/n}\right) + c_0^2 \tag{A7.10}$$

where a polytropic expansion with exponent n has been assumed. Applying the result of the previous section embodied in equation (A7.9), we may rewrite equation (A7.10) as:

$$c_1^2 = 2S\frac{n}{n-1} p_0 v_0 \left(1 - \left(\frac{p_1}{p_0}\right)^{(n-1)/n}\right) \tag{A7.11}$$

where

$$S = \frac{\sin^2 \alpha_0}{\sin^2 \alpha_0 - \sin^2 \alpha_1} \tag{A7.12}$$

The mass flow at the exit of the fixed blades is given by:

$$W_1^2 = \frac{A_1^2 c_1^2}{v_1^2} = \frac{A_1^2 c_1^2}{v_0^2}\left(\frac{p_1}{p_0}\right)^{2/n} \tag{A7.13}$$

since the specific volume at the fixed-blade outlet is related to that at the inlet by the polytropic condition:

$$v_1 = v_0 \left(\frac{p_0}{p_1}\right)^{1/n} \tag{A7.14}$$

The area A_1 is the area normal to the flow at the outlet of the fixed blades.

A similar set of equations may be written for the moving blades, although there is no condition equivalent to equation (A7.9). Thus the velocity relative to the moving blade at its outlet is given by:

$$c_{ro}^2 = 2\frac{n}{n-1} p_1 v_1 \left(1 - \left(\frac{p_2}{p_1}\right)^{(n-1)/n}\right) + c_{ri}^2 \tag{A7.15}$$

where the relative velocity at blade inlet is found from the velocity diagram, Chapter 15, Figure 15.7, as

$$\begin{aligned} c_{ri}^2 &= c_1^2 - 2c_1 c_B \cos \alpha_1 + c_B^2 \\ &= c_1^2(1 - 2R_B \cos \alpha_1 + R_B^2) \end{aligned} \tag{A7.16}$$

(This is the same equation as (15.47), derived in analysing the performance of the impulse blade.)

The mass flow calculated at the exit from the moving blades will be:

$$W_2^2 = \frac{A_2^2 c_{ro}^2}{v_2^2} = \frac{A_2^2 c_{ro}^2}{v_0^2}\left(\frac{p_2}{p_0}\right)^{2/n} \qquad \text{(A7.17)}$$

since, assuming the same polytropic index over the complete stage, the specific volume at the moving-blade outlet is related to that at the stage inlet by:

$$v_2 = v_0\left(\frac{p_0}{p_2}\right)^{1/n} \qquad \text{(A7.18)}$$

The area A_2 is the cross-sectional area normal to the flow at the outlet of the moving blades.

Continuity demands that the mass flow rate calculated at the fixed-blade outlet is the same as the mass flow at the moving-blade outlet, so that from equations (A7.13) and (A7.18) we have

$$\frac{A_1^2 c_1^2}{v_0^2}\left(\frac{p_1}{p_0}\right)^{2/n} = \frac{A_2^2 c_{ro}^2}{v_0^2}\left(\frac{p_2}{p_0}\right)^{2/n} \qquad \text{(A7.19)}$$

or

$$\frac{A_1^2}{A_2^2} c_1^2 = \left(\frac{p_2}{p_1}\right)^{2/n} c_{ro}^2 \qquad \text{(A7.20)}$$

Substituting for c_1^2 from equation (A7.11) and for c_{ro}^2 from equation (A7.16) gives, after rearrangement:

$$S\left(1-\left(\frac{p_1}{p_0}\right)^{(n-1)/n}\right)\left(\frac{A_1^2}{A_2^2}-\left(\frac{p_2}{p_1}\right)^{2/n}\times(1-2R_B\cos\alpha_1+R_B^2)\right) = \frac{p_1 v_1}{p_0 v_0}\left(\left(\frac{p_2}{p_1}\right)^{2/n}-\left(\frac{p_2}{p_1}\right)^{(n+1)/n}\right) \qquad \text{(A7.21)}$$

We may recast this equation in terms of the temperature ratio, (T_1/T_0), by noting that:

$$\frac{p_1 v_1}{p_0 v_0} = \frac{T_1}{T_0} \qquad \text{(A7.22)}$$

from the gas characteristic equation, while the combination of the gas characteristic equation and the condition for polytropic expansion gives:

$$\frac{p_k}{p_i} = \left(\frac{T_k}{T_i}\right)^{n/(n-1)} \quad \text{for } k > i \qquad \text{(A7.23)}$$

Applying the reaction equation (15.20) relates the stage temperature ratios to the degree of reaction:

$$\frac{T_2}{T_0} = \frac{\dfrac{T_1}{T_0}-\rho}{1-\rho} \qquad \text{(A7.24)}$$

But $(T_2/T_1) = (T_2/T_0)/(T_1/T_0)$, so that

$$\frac{T_2}{T_1} = \frac{\dfrac{T_1}{T_0}-\rho}{\dfrac{T_1}{T_0}(1-\rho)} \qquad \text{(A7.25)}$$

It follows from equation (A7.23) that

$$\frac{p_2}{p_1} = \left(\frac{\dfrac{T_1}{T_0}-\rho}{\dfrac{T_1}{T_0}(1-\rho)}\right)^{n/(n-1)} \qquad \text{(A7.26)}$$

Putting

$$\tau = \frac{T_1}{T_0} \qquad \text{(A7.27)}$$

and substituting for the various terms above into equation (A7.21) gives:

$$S(1-\tau)\left(\frac{A_1^2}{A_2^2}-\left(\frac{\tau-\rho}{\tau(1-\rho)}\right)^{2/(n-1)}\times(1-2R_B\cos\alpha_1+R_B^2)\right)-\tau\left(\left(\frac{\tau-\rho}{\tau(1-\rho)}\right)^{2/(n-1)}-\left(\frac{\tau-\rho}{\tau(1-\rho)}\right)^{(n+1/n-1)}\right)=0 \qquad \text{(A7.28)}$$

Equation (A7.28) is an implicit equation in the three variables: blade to gas speed ratio, R_B, degree of reaction, ρ, and temperature ratio, τ. The temperature ratio, τ, itself may be regarded as a function of degree of reaction, ρ, and stage pressure ratio (p_2/p_0) since, from (A7.23) and (A7.24),

$$\frac{p_2}{p_0} = \left(\frac{\tau-\rho}{1-\rho}\right)^{n/(n-1)} \qquad \text{(A7.29)}$$

or

$$\tau = (1-\rho)\left(\frac{p_2}{p_0}\right)^{(n-1)/n} + \rho \qquad \text{(A7.30)}$$

We may therefore find consistent values of (p_2/p_0), ρ and R_B, by specifying both the pressure ratio, (p_2/p_0), and the blade to gas speed ratio, R_B, and then adjust the reaction ratio, ρ, iteratively so that equations (A7.30) and (A7.28) are each satisfied.

Let us take a typical example where superheated steam is passing through a 50% reaction stage with a fixed-blade outlet angle, $\alpha_1 = 20°$, and a total stage pressure ratio at the design point of $(p_2/p_0) = 0.875$.

Taking polytropic index for superheated steam as $n = 1.3$, and applying equation (15.21) with $\rho = 0.5$

and $m = m_S = n = 1.3$ gives the design ratio of mid-stage to inlet pressure as $(p_1/p_0) = 0.9359$. At the design point for a 50% reaction stage, $c_1 = c_{ro}$, so that from equation (A7.20), the area ratio is given by $(A_1/A_2) = (p_2/p_1)^{1/n} = 0.9496$. Assuming a stage inlet angle $\alpha_0 = 90°$, we may use equation (A7.12) to calculate the value of $S = 1.1325$.

With the values of α_1, (A_1/A_2) and S now available, we may calculate the degree of reaction as a function of stage pressure ratio and blade-to-gas speed ratio. Setting the stage pressure ratio at the design value of $(p_2/p_0) = 0.875$, we vary the blade-to-gas speed ratio, R_B, between 0.1 and 1.8 and solve equations (A7.30) and (A7.28) iteratively to find the corresponding degrees of reaction, ρ. Then we alter the stage pressure ratio first to $(p_2/p_0) = 0.9375$ and then to $(p_2/p_0) = 0.75$ so that the normalized stage pressure drop, $(p_0 - p_2)/p_0$, is first halved and then doubled. In both cases R_B is varied from 0.1 to 1.8 at each value of stage pressure ratio, and the associated degree of reaction, ρ, is found. The results are plotted in Figure A7.3.

It is clear from the figure that the graph of reaction against blade to gas speed ratio is hill-shaped at each stage pressure ratio, with a relatively flat hump lying symmetrically about the design value $R_B = \cos\alpha_1$. The degree of reaction stays close to the design value of 0.5 even in the face of substantial deviations from the design blade-to-gas speed ratio. For instance, the degree of reaction stays within the range $0.45 \le \rho \le 0.5$ for $+/-40\%$ deviations in speed ratio at the design pressure ratio. The curve is shifted downwards with increasing stage pressure ratio, and upwards with decreasing pressure ratio, but the shift is not large even for the significant changes in stage pressure ratio shown.

We may conclude from the above that a typical 50% reaction stage will maintain its degree of reaction at roughly 0.5 for fairly substantial deviations from the design conditions.

The main turbine of a power station is an important special case, since it will be running at essentially constant, synchronous speed as a result of its directly coupled alternator being connected to an electricity supply grid. Hence the blade speed will be constant at c_B^*:

$$c_B = c_B^* \tag{A7.31}$$

As a result, the blade speed to gas speed ratio will depend solely on the gas speed leaving the fixed blades:

$$R_B = \frac{c_B^*}{c_1} \tag{A7.32}$$

The gas speed, c_1, is given by equation (A7.11), which may be re-expressed in terms of inlet temperature, T_0, and temperature ratio (T_1/T_0) using the gas characteristic equation (3.2) and equation (A7.23):

$$c_1^2 = 2S\frac{n}{n-1}ZR_wT_0\left(1 - \frac{T_1}{T_0}\right) \tag{A7.33}$$

Using the definition of τ from equation (A7.27), we see that R_B emerges as a function of τ and T_0:

$$R_B = \frac{c_B^*}{\sqrt{2S\frac{n}{n-1}ZR_wT_0(1-\tau)}} \tag{A7.34}$$

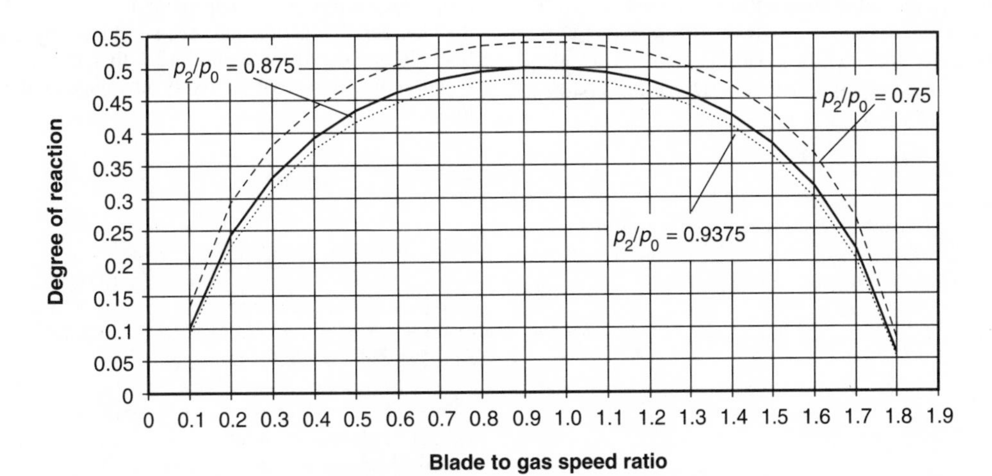

Figure A7.3 Degree of reaction versus blade to gas speed ratio, with stage pressure ratio as parameter, for a typical 50% reaction stage.

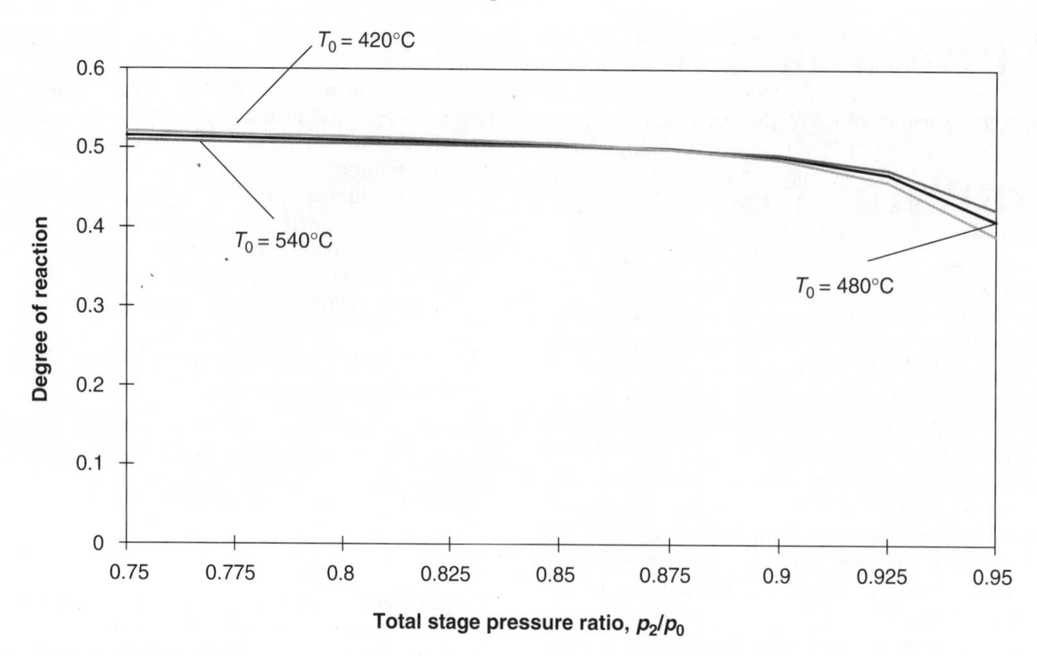

Figure A7.4 Degree of reaction versus total stage pressure ratio when the blade speed is kept constant.

It is now possible to see equations (A7.28), (A7.30) and (A7.34) as a defining set for degree of reaction, ρ, in terms of stage pressure ratio, (p_2/p_0), and inlet temperature, T_0.

We will take the example of a 50% reaction stage just considered, and assume that at the design condition it is being fed with steam at 480°C.

Taking the inlet pressure as 40 bar, and using the steam-table value for specific volume, we have $ZR_wT_0 = p_0v_0 = 40 \times 10^5 \times 0.08381 = 3.3524 \times 10^5$ J/kg. If the inlet temperature is raised to 540°C, then $R_wT_0 = p_0v_0 = 40 \times 10^5 \times 0.09137 = 3.6548 \times 10^5$ J/kg, while lowering it to 420°C gives $R_wT_0 = p_0v_0 = 40 \times 10^5 \times 0.07602 = 3.0408 \times 10^5$ J/kg.

We now vary the stage pressure ratio from 0.75 to 0.95 at each value of ZR_wT_0, and calculate the associated degree of reaction through the iterative solution of equations (A7.34), (A7.30) and (A7.28). The results are shown in Figure A7.4.

It is clear that the degree of reaction stays very close to 0.5 even when the total stage pressure ratio is reduced substantially. It falls away faster when the stage pressure ratio is increased above its design value of 0.875, but even here the normalized stage pressure drop, $(p_0 - p_2)/p_0$ is approximately halved before the degree of reaction has been reduced by 10% to 0.45. It is clear from Figure A7.4 that the degree of reaction is insensitive to likely changes in the inlet temperature, T_0.

Appendix 8 Fuel pin average temperature and effective heat transfer coefficient

A8.1 Introduction

Consider a single fuel pin, shown in cross-section in Figure A8.1.

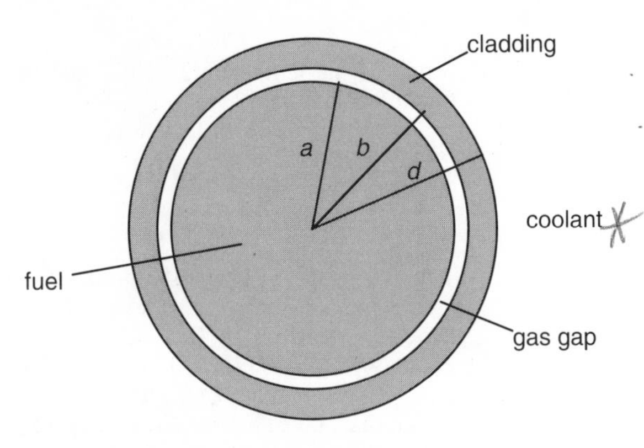

Figure A8.1 Cross-section of fuel pin.

The nuclear heating will be distributed evenly throughout the fuel, and to a first approximation, the heat will be conducted away from the fuel pellet solely in the radial direction and not in the axial direction. Since the heat transfer out of the fuel pellet occurs only at its outer surface, it follows that the pellet will be very much hotter at its centre than at its edge. Typically the temperature at the centre of the fuel pellet will be 500°C higher than the temperature at the pellet boundary. For simplicity, we wish to use a single value to characterize the temperature of the fuel in each axial segment, and it is important to use a well-selected average value. The difference between this average temperature and the temperature of the coolant will then constitute the driving force for heat transfer. We will need to associate an effective heat transfer coefficient with this temperature difference in order to calculate the instantaneous heat transfer from the fuel to the coolant.

A8.2 Applying Fourier's law of heat conduction to the fuel

Fourier's law states that the heat flow, Φ (W), is proportional to the area of flow, A (m^2), and the temperature gradient, dT/dx (K/m). In one dimension, this may be written:

$$\Phi = -kA\frac{dT}{dx} \tag{A8.1}$$

where the constant of proportionality, k, is called the thermal conductivity of the material, with units W/(mK). There will be no build-up of heat at any point along the path of heat flow in the steady state, so that the heat flow through successive points on the path must be the same. Hence the heat flow from point '1' on the path to point '2' may be found from integrating equation (A8.1) with Φ, k and A constant:

$$\Phi = -kA\frac{T_2 - T_1}{x_2 - x_1} \tag{A8.2}$$

It may be shown (Rogers and Mayhew, 1992) that applying Fourier's law to heat conduction in a three-dimensional solid subject to a distributed internal heat source produces the following equation characterizing the steady state:

$$\frac{\partial^2 T}{\partial x^2} + \frac{\partial^2 T}{\partial y^2} + \frac{\partial^2 T}{\partial z^2} = -\frac{\xi}{k} \tag{A8.3}$$

where ξ is the internal heat source density (W/m^3), which may vary with position in the general case. Equation (A8.3) expresses in Cartesian coordinates the general equation for an arbitrary set of coordinates:

$$\nabla^2 T = -\frac{\xi}{k} \tag{A8.4}$$

Its cylindrical nature means that cylindrical coordinates are the most convenient for analysing heat conduction in the fuel pin. A formal statement of equation (A8.4) in cylindrical coordinates is:

$$\frac{1}{r}\frac{\partial}{\partial r}\left(r\frac{\partial T}{\partial r}\right)+\frac{1}{r^2}\frac{\partial^2 T}{\partial^2\theta}+\frac{\partial^2 T}{\partial^2 z}=-\frac{\xi}{k} \qquad \text{(A8.5)}$$

Considering the complete fuel pin, we may allow for the axial variation in nuclear heating by dividing the core into a number of axial segments (see Chapter 21, Section 21.13). This done, we choose to neglect the relatively small degree of heat conduction in the axial, z, direction, and thus reduce the dimension of the system we are examining from three to two. The heat deposition in each axial segment will be essentially uniform across any cross-section of the fuel pin, so that the power density in each segment of fuel pin, ξ_i, will be independent of radial position. This fact brings a second simplification in its train, namely that the temperature at the same distance along any radius in the the cross-sectional plane will be the same, implying $\partial^2 T_i/\partial\theta^2 = 0$, where T_i is the temperature of the fuel in the ith axial pin-segment. Thus the segment of fuel pin is reduced to a one-dimensional system along any radial direction. Hence applying equation (A8.5) gives:

$$\frac{1}{r}\frac{d}{dr}\left(r\frac{dT_i}{dr}\right)=-\frac{\xi_i}{k_f} \qquad \text{(A8.6)}$$

where partial derivatives have been replaced by total derivatives along the radial distance, r, and where k_f is the thermal conductivity of the fuel (W/(mK)). This will vary with temperature, and hence will vary along the radial dimension. However, the predominantly linear mode of variation of k_f with temperature can be handled adequately by taking an average of fuel temperature. The total axial variation in temperature at any height in the fuel pin will tend to be less than the temperature difference between centre and edge. However, the effects of axial variations could be accommodated easily by choosing a different value of k_f for each axial segment.

Integrating equation (A8.6) with respect to radial distance, r, gives:

$$r\frac{dT_i}{dr}=-\frac{\xi_i}{k_f}\frac{r^2}{2}+C \qquad \text{(A8.7)}$$

Physically, we know that the maximum temperature will occur at the centre of the pin, i.e. at $r = 0$, implying $(dT_i/dr)(0) = 0$. Substituting these conditions into equation (A8.7) shows that the constant of integration must be zero. Hence

$$\frac{dT_i}{dr}=-\frac{\xi_i}{k_f}\frac{r}{2} \qquad \text{(A8.8)}$$

Integrating a second time gives

$$T_i=-\frac{\xi_i}{k_f}\frac{r^2}{4}+C \qquad \text{(A8.9)}$$

At the edge of the fuel pin, $r = a$ (see Figure A8.1), and we will name the associated temperature $T_{a,i}$. Hence

$$C=T_{a,i}+\frac{\xi_i}{k_f}\frac{a^2}{4} \qquad \text{(A8.10)}$$

It follows that the difference between the temperature, T_i, at any radial location, r, in the fuel and the temperature at the edge of the fuel segment, $T_{a,i}$, is given by:

$$T_i-T_{a,i}=\frac{\xi_i}{4k_f}(a^2-r^2) \qquad \text{(A8.11)}$$

The temperature at the centre of the fuel pin may be found by putting $r = 0$ into equation (A8.11):

$$T_{0,i}=T_{a,i}+\frac{\xi_i}{4k_f}a^2 \qquad \text{(A8.12)}$$

The average value of the temperature of the fuel in segment i is found from:

$$\begin{aligned}T_{ave,i}&=\frac{1}{a}\int_{r=0}^{a}\left(T_{a,i}+\frac{\xi_i}{4k_f}(a^2-r^2)\right)dr\\&=\frac{1}{a}\left[rT_{a,i}+r\frac{\xi_i}{4k_f}a^2\right]_0^a-\frac{1}{a}\frac{\xi_i}{4k_f}\int_{r=0}^{a}r^2\,dr\\&=T_{a,i}+\frac{\xi_i}{4k_f}a^2-\frac{1}{a}\frac{\xi_i}{12k_f}[r^3]_0^a\\&=T_{a,i}+\frac{\xi_i}{6k_f}a^2\end{aligned} \qquad \text{(A8.13)}$$

The amout of heat deposited in the fuel in segment i will be the product of the volume of the segment and the internal heat-source density:

$$\Phi_i=\pi a^2\Delta L\xi_i \qquad \text{(A8.14)}$$

where ΔL is the length of the segment. In the steady state, the heat flow out of the segment must be the same as the heat deposition. Using equation (A8.14), equation (A8.13) may be re-expressed in terms of the heat flow, Φ_i, as

$$T_{ave,i}-T_{a,i}=\frac{1}{6\pi k_f\Delta L}\Phi_i \qquad \text{(A8.15)}$$

It is instructive to ascertain the position of the average temperature, which may be found by substituting $T_i = T_{ave,i}$ from equation (A8.13) into equation (A8.11):

$$r^2=\tfrac{1}{3}a^2 \qquad \text{(A8.16)}$$

so that

$$r=\frac{\sqrt{3}}{3}a=0.577a \qquad \text{(A8.17)}$$

A8.3 Heat transfer across the gas gap

The gas gap is normally filled with helium gas, which has good heat transfer properties. The gas composition changes somewhat during a reactor run due to the addition of fission product gases. However, heat transfer predictions are affected more by the nuclear-induced swelling and cracking of the fuel pellets, leading to partial and non-uniform contact between pellet and cladding. Modelling the heat transfer process must inevitably involve a certain degree of uncertainty as a result. The model used is based on Newton's law of convective cooling, which may be expressed as

$$\Phi = KA\Delta T \tag{A8.18}$$

where K is the heat transfer coefficient (W/(m^{2}K)), A is the area at the heat-exchange surface, while ΔT is the temperature difference. In the steady state, the amount of heat passing through the gas gap will be the same as the total heat deposited in the fuel.

Assuming a very narrow gas gap (as is the normal case), the radii a and b in Figure A8.1 are very similar, and we may take the area A to be located at the centre of the disc of radius $(a+b)/2$. Thus the area will be the circumference of a disc of this radius multiplied by the segment length:

$$A = 2\pi\frac{a+b}{2}\Delta L = \pi(a+b)\Delta L \tag{A8.19}$$

Substituting from equation (A8.19) into equation (A8.18) gives the following relation for the temperature difference between fuel and clad in segment i:

$$T_{a,i} - T_{clad\,1,i} = \frac{1}{\pi(a+b)K_{fc}\Delta L}\Phi_i \tag{A8.20}$$

where $T_{clad\,1,i}$ is the temperature of the cladding inner wall in segment i, while K_{fc} is the heat transfer coefficient between fuel pellet and clad, likely to take approximately the same value for all axial segments.

A8.4 Heat transfer through the cladding

There is usually a fairly significant temperature difference across the cladding. The heat transfer area, A, at an arbitrary distance, r, along a radius will be given by

$$A = 2\pi r\Delta L \tag{A8.21}$$

Noting that in the steady state the heat passing through the cladding will the same as the heat passing through the gas gap and the heat generated in the fuel, we may apply equation (A8.1) along a radius within the cladding in segment i to give

$$\Phi_i = -k_{clad}2\pi r\Delta L\frac{dT}{dr} \tag{A8.22}$$

where k_{clad} is the thermal conductivity of the cladding material (W/(mK)). This equation may be rearranged to:

$$\frac{\Phi_i}{k_{clad}2\pi\Delta L}\frac{dr}{r} = -dT \tag{A8.23}$$

Since the heat passing, Φ_i, will be constant in the steady state, it is a simple matter to integrate equation (A8.22) from one side of the cladding to the other, i.e. from $r = b$ to $r = d$ and thus produce:

$$\begin{aligned} T_{clad\,1,i} - T_{clad\,2,i} &= \frac{\Phi_i}{2\pi k_{clad}\Delta L}[\ln r]_b^d \\ &= \left(\frac{1}{2\pi k_{clad}\Delta L}\ln\frac{d}{b}\right)\Phi_i \end{aligned} \tag{A8.24}$$

where $T_{clad\,2,i}$ is the temperature of the cladding outer wall in segment i.

A8.5 Heat transfer from the cladding to the coolant

This convective heat transfer process follows the treatment of the gas gap, in that the descriptive equation is:

$$\Phi = K_{cc}A\Delta T \tag{A8.25}$$

where K_{cc} is the heat transfer coefficient between cladding and coolant. The heat-transfer area within segment i is given by:

$$A = 2\pi d\Delta L \tag{A8.26}$$

while the heat being passed in the steady state is the same as the heat being passed through the gas gap and through the cladding, Φ_i. Hence we may substitute from (A8.26) into (A8.25) to give:

$$T_{clad\,2,i} - T_{cool,i} = \frac{1}{2\pi dK_{cc}\Delta L}\Phi_i \tag{A8.27}$$

A8.6 The overall heat transfer coefficient

We may add equations (A8.15), (A8.20), (A8.24) and (A8.27) to produce

$$T_{ave,i} - T_{cool,i} = \frac{\Phi_i}{\pi\Delta L}\left[\frac{1}{6k_f} + \frac{1}{(a+b)K_{fc}} + \frac{1}{2k_{clad}}\ln\frac{d}{b} + \frac{1}{2dK_{cc}}\right] \tag{A8.28}$$

where in the steady state, the heat being transferred is related to the nuclear heat deposition density by equation (A8.14), repeated below:

$$\Phi_i = \pi a^2 \Delta L \xi_i \tag{A8.14}$$

Equation (A8.28) may be rearranged into the form:

$$\Phi_i = \frac{\pi \Delta L}{\left[\frac{1}{6k_f} + \frac{1}{(a+b)K_{fc}} + \frac{1}{2k_{clad}} \ln \frac{d}{b} + \frac{1}{2dK_{cc}}\right]} \times (T_{ave,i} - T_{cool,i}) \tag{A8.29}$$

Multiplying top and bottom of the right-hand side of equation (A8.29) by the outside diameter of the fuel pin cladding, $2d$, gives:

$$\Phi_i = \frac{2\pi d \Delta L}{2d\left[\frac{1}{6k_f} + \frac{1}{(a+b)K_{fc}} + \frac{1}{2k_{clad}} \ln \frac{d}{b} + \frac{1}{2dK_{cc}}\right]} \times (T_{ave,i} - T_{cool,i})$$

$$= \frac{A_i}{\left[\frac{d}{3k_f} + \frac{2d}{(a+b)K_{fc}} + \frac{d}{k_{clad}} \ln \frac{d}{b} + \frac{1}{K_{cc}}\right]} \times (T_{ave,i} - T_{cool,i}) \tag{A8.30}$$

where $A_i = 2\pi d \Delta L$ (m^2) is the heat transfer area for segment i, being the area of cladding in contact with the coolant.

Equation (A8.30) may be compared with the canonical form of the heat transfer equation, namely:

$$\Phi_i = K_{over} A_i (T_{ave,i} - T_{cool,i}) \tag{A8.31}$$

where K_{over} is the overall heat transfer coefficient from the average fuel pin temperature to the coolant, referred to the outside of the fuel cladding. Comparing equations (A8.30) and (A8.31), it is clear that:

$$\frac{1}{K_{over}} = \frac{d}{3k_f} + \frac{2d}{(a+b)K_{fc}} + \frac{d}{k_{clad}} \ln \frac{d}{b} + \frac{1}{K_{cc}} \tag{A8.32}$$

The heat transfer coefficient between pin cladding and coolant, K_{cc}, will vary with mass flow raised to the power of 0.8 as explained in Chapter 20, Section 20.6.

A8.7 Example of calculating average fuel temperatures in a PWR

It is often easiest for the control engineer to produce a starting condition for the simulation by estimating first the segment temperatures on the coolant side and then using equation (A8.28) to calculate the corresponding average fuel temperature in segment i. Let us calculate the average fuel temperature at the start of life in 10 axial segments of the average fuel pin in the PWR core described in the example of Chapter 21, Section 21.11. Assume that the zircaloy fuel pins have the dimensions laid out in Table A8.1, while the thermal parameters are as given in Table A8.2.

Table A8.1 Fuel pin dimensions (all mm)

Outside diameter	*Cladding thickness*	*Pellet diameter*	*Fuel pin length*
9.5	0.57	8.19	4270

Assume the core-average neutron density, n, is 1.6×10^{14} neutrons/m^3.

From the dimensions table, $a = 8.19 \times 10^{-3}/2 = 4.095 \times 10^{-3}$ m, $b = (9.5 - 2 \times 0.57) \times 10^{-3}/2 = 4.18 \times 10^{-3}$ m and $d = 9.5 \times 10^{-3}/2 = 4.25 \times 10^{-3}$ m, while $\Delta L = 4.27/10 = 0.427$ m. Thus the volume of fuel in one fuel pin segment is $\pi a^2 \Delta L = \pi \times 4.095^2 \times 10^{-6} \times 0.427 = 22.5 \times 10^{-6}\, m^3$.

The relationship between neutron density and power density at the start of life has been derived in equation (21.39), namely

$$\phi = 2.074 \times 10^{-6} n \tag{21.39}$$

Hence for a typical core-average neutron density, $n = 1.6 \times 10^{14}$, the core-average power density is $\xi =$

Table A8.2 Thermal parameters of fuel pin

Fuel thermal conductivity, k_f W/(mK)	*Effective heat transfer coefficient between fuel and cladding, K_{fc} W/(m^2K)*	*Cladding thermal conductivity, k_{clad} W/(mK)*	*Heat transfer coefficient between cladding and coolant, K_{cc} W/(m^2K)*
2.8	6000	13	34 000

Table A8.3 Nuclear heating and temperatures in fuel pin segments

Segment number	*Power density ratio* $\frac{\xi_i}{\xi}$	*Power density in segment* ξ_i *MW/m*^{3}	*Heat deposited in/transferred from segment of pin,* Φ_i*, W*	*Coolant temp* $^\circ C$	*Average fuel temp* $^\circ C$	*Centre fuel temp* $^\circ C$
5	0.517	172	3860	330	582	668
4	0.823	273	6141	326	727	864
3	1.072	356	8003	322	845	1023
2	1.248	414	9319	318	927	1134
1	1.340	445	10001	314	968	1190
0	1.340	445	10001	310	964	1186
−1	1.248	414	9319	306	915	1122
−2	1.072	356	8003	302	825	1003
−3	0.823	273	6141	298	699	836
−4	0.517	172	3860	294	546	632

$2.074 \times 10^{-6} \times 1.6 \times 10^{14} = 331.8 \times 10^6$ W/m^3. The power densities in each of the segments will follow the ratios of neutron densities given in Table 21.3 of Chapter 21. For example the power densities in the outer segments '-4' and '5' will be $\xi_{-4} = \xi_5 = 0.517 \times 331.8 \times 10^6 = 171.5 \times 10^6$ W/m^3, while the power densities in segments '-1' and '2' will be $\xi_{-1} = \xi_2 = 1.248 \times 331.8 \times 10^6 = 414.1 \times 10^6$ W/m^3, and so on. The total heat deposited in each segment is given by the heat density times the fuel pin segment volume: $\Phi_i = 22.5 \times 10^{-6} \xi_i$ W.

The coolant temperature entering the reactor will have a temperature of 290°C, while the coolant outlet temperature will be 330°C, and we will make the simplifying assumption that its temperature rises linearly through the core. We will assume that the coolant in each segment may be characterized by its outlet temperature. Hence the coolant temperature in segment '-4' will be 294°C, that of segment '-3' will be 298°C etc. Thus we may use equation (A8.28) to calculate the average fuel temperature in each segment.

We may also calculate the fuel temperature at the centre of each segment through replacing $T_{ave,i}$ by $T_{0,i}$ and the term $1/6k_f$ by $1/4k_f$ in equation (A8.28):

$$T_{0,i} - T_{cool,i} = \frac{\Phi_i}{\pi \Delta L}\left[\frac{1}{4k_f} + \frac{1}{(a+b)K_{fc}} + \frac{1}{2k_{clad}} \ln\frac{d}{b} + \frac{1}{2dK_{cc}}\right] \quad \text{(A8.33)}$$

[This transformation arises from using equation (A8.12) instead of equation (A8.13) as the reference basis for heat flow. Equation (A8.15) is then replaced by:

$$T_{0,i} - T_{a,i} = \frac{1}{4\pi k_f \Delta L}\Phi_i \quad \text{(A8.34)}$$

The rest of the steps in the analysis are common.]

Results are shown in Table A8.3.

A8.8 Bibliography

Glasstone, S. and Sesonske, A. (1981). *Nuclear Reactor Engineering*, 3rd edition, Van Nostrand Reinhold, New York.

Rogers, G.F.C. and Mayhew, Y.R. (1992). *Engineering Thermodynamics, Work and Heat Transfer*, 4th edition, Longman Scientific & Technical, London.

Appendix 9 Conditions for emergence from saturation for P + I controllers with integral desaturation

A9.1 Introduction

Chapter 22 considered three types of integral desaturation algorithms for P + I controllers, giving them the names Type 1, Type 2 and Type 3, in order of increasing complexity. Each algorithm leads to a slightly different controller response. In particular, the controller will emerge from saturation under different conditions, depending on which algorithm is employed. This appendix analyses when each of the controllers with integral desaturation will come out of saturation.

A9.2 Type 1 integral desaturation

A9.2.1 Size of the error at emergence from controller saturation

With Type 1 integral desaturation, the integral term is limited when the partially desaturated output, $c_1(t)$, takes a value outside the controller's output limits (see equations (22.16) and (22.17)).

Suppose that the error is positive and non-decreasing so that the partially desaturated output, c_1, will be increasing. Eventually, at time, t_m, say, the value of c_1 will reach the upper output limit of the controller, c_{max}. At this point using equation (22.18):

$$c_1(t_m) = P(t_m) + I_D(t_m) = c_{max} \qquad \text{(A9.1)}$$

The integral term, I_D, will now be held, and will retain thereafter the value:

$$I_D(t) = I_D(t_m) = c_{max} - P(t_m) \quad \text{for } t \geq t_m \qquad \text{(A9.2)}$$

unless and until c_1 falls back below c_{max}.

Now suppose that after a further time, the error begins to fall, and that at a later time, t_n

$$c_1(t_n) = c_{max} \qquad \text{(A9.3)}$$

Using equation (22.18) once more, equation (A9.3) implies that:

$$P(t_n) + I_D(t_n) = c_{max} \qquad \text{(A9.4)}$$

But the integrator term will have been held fixed during this time of saturation, i.e. $I_D(t_n) = I_D(t_m)$. Substituting into equation (A9.4) gives

$$P(t_n) + I_D(t_m) = c_{max} \qquad \text{(A9.5)}$$

Comparing equation (A9.5) with equation (A9.1), it is clear that the proportional term at the time of the controller's emergence from saturation, t_n, must be the same as the proportional term at the time of entering saturation, t_m:

$$P(t_n) = P(t_m) \qquad \text{(A9.6)}$$

Bearing in mind that $P(t) = ke(t)$, it follows immediately that:

$$e(t_n) = e(t_m) \qquad \text{(A9.7)}$$

This means that the integrator will not begin to come out of saturation until the error has returned to the value it had when the controller went into saturation.

An essentially identical analysis will hold for emergence from saturation at the lower limit, c_{min}. We now suppose that the error is negative and non-increasing and hence the partially desaturated output, c_1, will be decreasing. Eventually, at say time, t_m, the value of c_1 will reach the lower output limit of the controller, c_{min}. At this point using equation (22.18), we have:

$$c_1(t_m) = P(t_m) + I_D(t_m) = c_{min} \qquad \text{(A9.8)}$$

The integral term, I_D, will now be held at:

$$I_D(t) = I_D(t_m) = c_{min} - P(t_m) \quad \text{for } t \geq t_m \qquad \text{(A9.9)}$$

until c_1 rises back above c_{min}.

Supposing that the error later begins to rise so that at time, t_n, the partially desaturated output is given by:

$$c_1(t_n) = c_{min} \qquad \text{(A9.10)}$$

Using equation (22.18), equation (A9.10) implies that:

$$P(t_n) + I_D(t_n) = c_{min} \qquad \text{(A9.11)}$$

But the integrator term will not have altered during this time of (low) saturation. Hence $I_D(t_n) = I_D(t_m)$, and substituting into equation (A9.11) gives

$$P(t_n) + I_D(t_m) = c_{min} \qquad \text{(A9.12)}$$

Comparing equations (A9.8) and (A9.12), it is clear that the proportional term at the time of emerging from saturation, t_n, must be the same as the proportional term at the time of entering saturation, t_m:

$$P(t_n) = P(t_m) \quad \text{(A9.13)}$$

Since $P(t) = ke(t)$, it follows again that:

$$e(t_n) = e(t_m) \quad \text{(A9.14)}$$

This condition is identical to that of equation (A9.7). Thus it is clear that if the controller is to emerge from either high or low saturation, the error must have returned to the value it held when the controller went into saturation.

A.9.2.2 Conditions on the rate of change of error at emergence from controller saturation

The condition of equation (A9.7) is a necessary but not sufficient condition for the emergence from saturation for a controller with a Type 1 integral desaturation algorithm.

Considering the case of the controller emerging from high saturation, when the partially desaturated output, $c_1(t)$, has come back down to the controller output saturation level, c_{max}. It is clear that the controller can come out of saturation now, but only if $c_1(t)$ is falling, i.e. if:

$$\frac{dc_1}{dt}(t_n) < 0 \quad \text{(A9.15)}$$

Substituting from equation (22.18) into equation (A9.15) gives the condition:

$$\frac{dP}{dt}(t_n) + \frac{dI_D}{dt}(t_n) < 0 \quad \text{(A9.16)}$$

Since $P(t) = ke(t)$, it follows that

$$\frac{dP}{dt}(t_n) = k\frac{de}{dt}(t_n) \quad \text{(A9.17)}$$

Further, from equation (22.16),

$$\frac{dI_D}{dt}(t_n) = B(t_n) \quad \text{(A9.18)}$$

where, from equation (22.17)

$$B(t_n) = k\frac{e(t_n)}{T_i} \quad \text{(A9.19)}$$

for the emergent condition, $c_1(t_n) = c_{max}$. Hence substituting from (A9.17), (A9.18) and (A9.19) into equation (A9.16) gives:

$$\frac{de(t_n)}{dt} < -\frac{e(t_n)}{T_i} \quad \text{(A9.20)}$$

This inequality expresses the notion that the error must be falling fast enough for the decreasing proportional term to overcome the effect of a still-increasing integral term.

Turning to the case of the controller emerging from low saturation, a very similar analysis pertains. Suppose now that the partially desaturated output, $c_1(t)$, has come back up to the controller output saturation level, c_{min}. It is clear that the controller can come out of saturation now only if $c_1(t)$ is rising, i.e. if:

$$\frac{dc_1}{dt}(t_n) > 0 \quad \text{(A9.21)}$$

Substituting from equation (22.18) into equation (A9.21) gives the condition:

$$\frac{dP}{dt}(t_n) + \frac{dI_D}{dt}(t_n) > 0 \quad \text{(A9.22)}$$

Equations (A9.17) and (A9.18) are still valid for $(dP/dt)(t_n)$ and $(dI_D/dt)(t_n)$. Further, equation (A9.19) still holds for $B(t_n)$ for the lower emergent condition, $c_1(t_n) = c_{min}$. Hence substituting into equation (A9.22) gives:

$$\frac{de(t_n)}{dt} > -\frac{e(t_n)}{T_i} \quad \text{(A.23)}$$

A9.2.3 Summary of conditions for a controller with Type 1 integral desaturation emerging from controller saturation

To summarize, the conditions for the controller returning to its normal range are:

(1) the error must have returned to the value current at entry to saturation;
(2) the rate of change of error must satisfy:

$$\frac{de}{dt} < -\frac{e}{T_i}$$ for returning from upper-limit saturation

$$\frac{de}{dt} > -\frac{e}{T_i}$$ for returning from lower-limit saturation

A9.3 Type 2 integral desaturation

With the Type 2 algorithm for integral desaturation the controller output, $c(t)$, is made up of the proportional term, P, and the adjusted integral term, I_A:

$$c(t_n) = P(t_n) + I_A(t_n) \quad \text{(22.23)}$$

where the integral term I_A is adjusted to ensure that the controller output, $c(t)$, remains at the boundary of the normal range (at either c_{max} or c_{min}) whenever the partially desaturated output, $c_1(t)$, falls outside the normal range. This allows the controller to return from saturation faster than the Type 1 algorithm.

Suppose that the controller goes into high saturation at time, t_m, and that it is still in saturation at a later time, t_n, so that the controller output is at its maximum value:

$$c(t_n) = c_{max} \tag{A9.24}$$

The partially desaturated output, $c_1(t)$, will be little different from c_{max} because of the resetting of the desaturated integrator specified by equations (22.20) and (22.21), repeated below:

$$I_D(t_n) = I_A(t_{n-1}) + \int_{t_{n-1}}^{t_n} \frac{ke}{T_i}\,dt \tag{22.20}$$

where

$$\begin{aligned} I_A(t_n) &= c_{min} - P(t_n) \quad \text{for } c_1(t_n) < c_{min} \\ &= I_D(t_n) \quad \text{for } c_{min} \le c_1(t_n) \le c_{max} \\ &= c_{max} - P(t_n) \quad \text{for } c_1(t_n) > c_{max} \end{aligned} \tag{22.21}$$

Hence, during controller saturation between times t_m and t_n, the output of the desaturated integrator, I_D, will be given by:

$$I_D(t_n) = c_{max} - P(t_{n-1}) + \int_{t_{n-1}}^{t_n} \frac{ke}{T_i}\,dt \tag{A9.25}$$

and the partially desaturated controller output, c_1, follows from equation (22.22) as:

$$\begin{aligned} c_1(t_n) &= P(t_n) + I_D(t_n) \\ &= c_{max} + P(t_n) - P(t_{n-1}) + \int_{t_{n-1}}^{t_n} \frac{ke}{T_i}\,dt \end{aligned} \tag{A9.26}$$

Expanding the proportional terms using $P(t) = ke(t)$ and the integral term using the Euler integration algorithm valid for small time intervals, $\Delta t = t_n - t_{n-1}$:

$$\begin{aligned} c_1(t_n) &= c_{max} + ke(t_n) - ke(t_{n-1}) + ke(t_{n-1})\frac{\Delta t}{T_i} \\ &= c_{max} + k\left(e(t_n) - e(t_n - \Delta t)\left(1 - \frac{\Delta t}{T_i}\right)\right) \\ &= c_{max} + \varepsilon \end{aligned} \tag{A9.27}$$

where

$$\varepsilon = k\left(e(t_n) - e(t_n - \Delta t)\left(1 - \frac{\Delta t}{T_i}\right)\right) \tag{A9.28}$$

In the limit as $\Delta t \to 0$, $\varepsilon \to 0$. For small but nevertheless finite Δt, ε will clearly be small. The quantity, ε, will be positive during high saturation as long as the error is constant or increasing.

The condition for ε becoming negative is:

$$e(t_n) - e(t_n - \Delta t)\left(1 - \frac{\Delta t}{T_i}\right) < 0 \tag{A9.29}$$

Dividing by Δt and rearranging gives:

$$\frac{e(t_n) - e(t_n - \Delta t)}{\Delta t} < -\frac{e(t_n - \Delta t)}{T_i} \tag{A9.30}$$

In the limit as $\Delta t \to 0$, condition (A9.30) becomes:

$$\frac{de}{dt}(t_n) < -\frac{e(t_n)}{T_i} \tag{A9.31}$$

Thus a rate of decrease of error of the magnitude stipulated by condition (A9.31) will cause the quantity, ε, to become negative, which will cause the term $c_1(t_n)$ to fall just below c_{max}:

$$c_1(t_n) < c_{max} \tag{A9.32}$$

Suppose for the moment that the error begins to decrease, but not as rapidly as stipulated by condition (A9.31). Now $c_1(t_n)$ will still be just above c_{max}, i.e. $c_1(t_n) > c_{max}$. Hence, from equation (22.21),

$$I_A(t_n) = c_{max} - P(t_n) \tag{A9.33}$$

Substituting this value of the adjusted integral term into equation (22.23), it is clear that the controller output, $c(t_n)$, will be unaltered from its high limit:

$$\begin{aligned} c(t_n) &= P(t_n) + I_A(t_n) \\ &= P(t_n) + c_{max} - P(t_n) = c_{max} \end{aligned} \tag{A9.34}$$

But if the error is falling at the rate given in condition (A9.31) or faster, then $c_1(t_n) < c_{max}$, and, by equation (22.21), this will cause the adjusted integrator, I_A, to switch to the value of the desaturated integrator, I_D:

$$I_A(t_n) = I_D(t_n) \tag{A9.35}$$

As a result, from equation (22.23), the controller output will be given by:

$$c(t_n) = P(t_n) + I_D(t_n) = c_1(t_n) \tag{A9.36}$$

where the second equality follows from equation (22.22). A (sustained) fall in the controller output, $c(t)$, implies

$$\frac{dc}{dt}(t_n) < 0 \tag{A9.37}$$

or, using equation (A9.29):

$$\frac{dP}{dt}(t_n) + \frac{dI_D}{dt}(t_n) < 0 \tag{A9.38}$$

Using $P = ke$ and equation (22.20), this translates to:

$$k\frac{de}{dt}(t_n) + k\frac{e(t_n)}{T_i} < 0 \quad \text{(A9.39)}$$

or:

$$\frac{de}{dt}(t_n) < -\frac{e(t_n)}{T_i} \quad \text{(A9.40)}$$

which is the same condition as (A9.31).

A similar analysis applied to emergence from low saturation shows that recovery there requires:

$$\frac{de}{dt}(t_n) > -\frac{e(t_n)}{T_i} \quad \text{(A9.41)}$$

A9.4 Type 3 integral desaturation

This algorithm is the same as the Type 2 algorithm, except that the proportional term is replaced by the error-limited term, P_m:

$$P_m(t_n) = ke_m(t_n) \quad \text{(22.44)}$$

where e_m is the modified error term, given by:

$$e_m = \lim(e_{\min}, e_{\max}, e) \quad \text{(22.43)}$$

It follows that the analysis of Section A9.3 for Type 2 emergence from high saturation transfers across directly to Type 3, except that P_m now replaces P. Equation (A9.38) then becomes:

$$\frac{dP_m}{dt}(t_n) + \frac{dI_D}{dt}(t_n) < 0 \quad \text{(A9.42)}$$

which may be simplified to:

$$\frac{de_m}{dt}(t_n) < -\frac{e(t_n)}{T_i} \quad \text{(A9.43)}$$

But e_m will be constant at the high or low limit outside the range $e_{\min} \le e \le e_{\max}$, implying

$$\frac{de_m}{dt}(t_n) = 0 \quad \text{for } e(t_n) \ge e_{\max} \text{ or } e(t_n) \le e_{\min} \quad \text{(A9.44)}$$

so that the condition of inequality (A9.43) cannot possibly be satisfied until the error has fallen back to within the range $e_{\min} \le e \le e_{\max}$. When the error is back within range, then $e_m = e$. Hence inequality (A9.43) becomes the familiar condition:

$$\frac{de}{dt}(t_n) < -\frac{e(t_n)}{T_i} \quad \text{(A9.45)}$$

A similar analysis reveals that the controller with Type 3 integral desaturation cannot come out of low saturation until

(1) the error has returned to the normal range, $e_{\min} \le e \le e_{\max}$;
(2) the rate of increase of error satisfies

$$\frac{de}{dt}(t_n) > -\frac{e(t_n)}{T_i} \quad \text{(A9.46)}$$

Index